Handbook of Forensic Statistics

Chapman & Hall/CRC
Handbooks of Modern Statistical Methods

Series Editor

Garrett Fitzmaurice, *Department of Biostatistics, Harvard School of Public Health, Boston, MA, U.S.A.*

The objective of the series is to provide high-quality volumes covering the state-of-the-art in the theory and applications of statistical methodology. The books in the series are thoroughly edited and present comprehensive, coherent, and unified summaries of specific methodological topics from statistics. The chapters are written by the leading researchers in the field and present a good balance of theory and application through a synthesis of the key methodological developments and examples and case studies using real data.

Published Titles

Handbook of Design and Analysis of Experiments
Angela Dean, Max Morris, John Stufken, and Derek Bingham
Handbook of Cluster Analysis
Christian Hennig, Marina Meila, Fionn Murtagh, and Roberto Rocci
Handbook of Discrete-Valued Time Series
Richard A. Davis, Scott H. Holan, Robert Lund, and Nalini Ravishanker
Handbook of Big Data
Peter Bühlmann, Petros Drineas, Michael Kane, and Mark van der Laan
Handbook of Spatial Epidemiology
Andrew B. Lawson, Sudipto Banerjee, Robert P. Haining, and María Dolores Ugarte
Handbook of Neuroimaging Data Analysis
Hernando Ombao, Martin Lindquist, Wesley Thompson, and John Aston
Handbook of Statistical Methods and Analyses in Sports
Jim Albert, Mark E. Glickman, Tim B. Swartz, and Ruud H. Koning
Handbook of Methods for Designing, Monitoring, and Analyzing Dose-Finding Trials
John O'Quigley, Alexia Iasonos, and Björn Bornkamp
Handbook of Quantile Regression
Roger Koenker, Victor Chernozhukov, Xuming He, and Limin Peng
Handbook of Statistical Methods for Case-Control Studies
Ørnulf Borgan, Norman Breslow, Nilanjan Chatterjee, Mitchell H. Gail, Alastair Scott, and Chris J. Wild
Handbook of Environmental and Ecological Statistics
Alan E. Gelfand, Montserrat Fuentes, Jennifer A. Hoeting, and Richard L. Smith
Handbook of Approximate Bayesian Computation
Scott A. Sisson, Yanan Fan, and Mark Beaumont
Handbook of Graphical Models
Marloes Maathuis, Mathias Drton, Steffen Lauritzen, and Martin Wainwright
Handbook of Mixture Analysis
Sylvia Frühwirth-Schnatter, Gilles Celeux, and Christian P. Robert
Handbook of Infectious Disease Data Analysis
Leonhard Held, Niel Hens, Philip O'Neill, and Jacco Walllinga
Handbook of Forensic Statistics
David Banks, Karen Kafadar, David H. Kaye, and Maria Tackett

For more information about this series, please visit: https://www.crcpress.com/Chapman--HallCRC-Handbooks-of-Modern-Statistical-Methods/book-series/CHHANMODSTA

Handbook of Forensic Statistics

Edited by
David Banks
Karen Kafadar
David H. Kaye
Maria Tackett

CRC Press
Taylor & Francis Group
Boca Raton London New York

CRC Press is an imprint of the
Taylor & Francis Group, an **informa** business
A CHAPMAN & HALL BOOK

First edition published 2021
by CRC Press
6000 Broken Sound Parkway NW, Suite 300, Boca Raton, FL 33487-2742

and by CRC Press
2 Park Square, Milton Park, Abingdon, Oxon, OX14 4RN

First issued in paperback 2022

**Visit the Taylor & Francis Web site at
http://www.taylorandfrancis.com**

**and the CRC Press Web site at
http://www.crcpress.com**

Library of Congress Cataloging-in-Publication Data
Library of Congress Control Number: 2020934402

ISBN: 978-0-367-52772-3 (pbk)
ISBN: 978-1-138-29540-7 (hbk)
ISBN: 978-0-367-52770-9 (ebk)

DOI: 10.1201/9780367527709

Typeset in Palatino
by Nova Techset Private Limited, Bengaluru & Chennai, India

In memory of Stephen E. Fienberg (1942–2016), a pioneer in the field of forensic statistics and the use of statistics in the courtroom, and his beloved wife, Joyce Fienberg (1943–2018), who tragically perished in the Tree of Life mass shooting in Pittsburgh in 2018.

Contents

Foreword..ix
Preface...xi
Editors..xiii
Contributors...xv

Section I   Perspectives on Forensic Statistics

1. **The History of Forensic Inference and Statistics: A Thematic Perspective** 3
   Colin Aitken and Franco Taroni

Section II   General Concepts and Methods

2. **Frequentist Methods for Statistical Inference** 39
   David H. Kaye

3. **Bayesian Methods and Forensic Inference** 73
   David Banks and Maria Tackett

4. **Comparing Philosophies of Statistical Inference** 91
   Hal S. Stern

5. **Decision Theory** 103
   Franco Taroni, Silvia Bozza, and Alex Biedermann

6. **Association Does Not Imply Discrimination: Clarifying When Matches Are (and Are Not) Meaningful** 131
   Maria Cuellar, Lucas Mentch, and Cliff Spiegelman

7. **Validation of Forensic Automatic Likelihood Ratio Methods** 143
   Daniel Ramos, Didier Meuwly, Rudolf Haraksim, and Charles E.H. Berger

8. **Bayesian Networks in Forensic Science** 165
   A. Philip Dawid and Julia Mortera

Section III   Legal and Psychological Dimensions

9. **How Well Do Lay People Comprehend Statistical Statements from Forensic Scientists?** 201
   Kristy A. Martire and Gary Edmond

10. Forensic Statistics in the Courtroom 225
    David H. Kaye

Section IV   Applications of Statistics to Particular Fields in Forensic Science

11. DNA Frequencies and Probabilities 251
    Bruce S. Weir

12. Kinship 265
    Bruce S. Weir

13. Statistical Support for Conclusions in Fingerprint Examinations 277
    Cedric Neumann, Jessie Hendricks, and Madeline Ausdemore

14. Probabilistic Considerations When Interpreting Database Search
    and Selection Effects 325
    M.J. Sjerps

15. Comparing Handwriting in Questioned Documents 341
    Alan Julian Izenman

16. An Introduction to Firearms Examination for Researchers in Statistics 365
    Susan VanderPlas, Alicia Carriquiry, Heike Hofmann, James Hamby, and Xiao Hui Tai

17. Shoeprints: The Path from Practice to Science 391
    Sarena Wiesner, Naomi Kaplan-Damary, Benjamin Eltzner, and Stephan Huckemann

18. Forensic Glass Evidence 411
    Karen Pan, Junqi Chen, and Karen Kafadar

19. Estimation of Insect Age for Assessing Minimum Post-Mortem Interval
    in Forensic Entomology Casework 443
    Davide Pigoli, Martin J.R. Hall, and John A.D. Aston

20. Statistical Models in Forensic Voice Comparison 451
    *Geoffrey Stewart Morrison, Ewald Enzinger, Daniel Ramos,
    Joaquín González-Rodríguez, and Alicia Lozano-Díez*

21. Bringing New Statistical Approaches to Eyewitness Evidence 499
    Alice J. Liu, Karen Kafadar, Brandon L. Garrett, and Joanne Yaffe

Index..541

Foreword

The subject of this book is forensic statistics—the application of statistical and probabilistic reasoning to the discovery and proof of facts in legal settings. It focuses on those aspects of forensic science that are used primarily in criminal cases.

The book begins with thematic and methodological chapters. The first chapter is essential reading for a broad understanding of current developments. It sketches the recent history of forensic inference and statistics, identifies today's issues and interpretive methods, and points to where they are heading.

Chapters 2–8 (Section II) discuss general concepts, methods, and philosophies of statistical inference, including basic introductions (intended for readers with limited knowledge of statistics) to such topics as hypothesis testing, confidence intervals and credible regions, likelihood ratios, and simulation methods. Additional chapters in this section discuss decision theory, Bayesian networks, and validation of likelihood ratios. Within these chapters, a reader can find the underlying statistical theory used by later chapters on applications of statistics to various domains of forensic science.

Chapters 9 and 10 (Section III) are on statistics in the courtroom and the psychological dimensions of statistical testimony. They complete the background for Chapters 11–21 (Section IV) on the statistical aspects of methods employed in specific areas of forensic science. The latter chapters include the analysis of DNA samples, friction ridge skin patterns, shoe prints, glass fragments, firearms and ammunition, and much more. In this section, the authors lay out the current state of statistical knowledge and practice in many different areas, but forensic science is so broad that this handbook by no means exhausts the field. Nonetheless, Section IV addresses domain-specific applications that are commonly encountered in criminal investigations and trials, and the statistical issues that arise in one subfield frequently transfer to others.

The chapters were commissioned, compiled, reviewed, and edited with a broad readership in mind. We believe they will be of value to students, statisticians, forensic scientists, and lawyers. Previous familiarity with statistical thinking and methods will be a strong asset in benefiting from the book as a whole, but every chapter can be read on its own. Some chapters are relatively mathematical and technical; others have no formulas, or just a few. Some chapters are wide ranging; others focus on particular types of forensic-science evidence. Each chapter describes, to varying extents, past literature on completed studies, while some present new research. We hope the handbook will be useful to broad audiences with diverse interests.

David Banks
Duke University

Karen Kafadar
University of Virginia

David H. Kaye
Pennsylvania State University

Maria Tackett
Duke University

Preface

During the 2015–2016 academic year, the Statistical and Applied Mathematical Sciences Institute (SAMSI) held a research program called Statistics and Applied Mathematics in Forensic Science. That program had many positive scientific outcomes, and the editors hope that this book will count among them.

The SAMSI research effort was part of a broader movement triggered, to a large degree, by the 2009 report from a committee of the National Academies entitled *Strengthening Forensic Science in the United States: A Path Forward*. Among other issues, that report called attention to gaps in the scientific and statistical research into common forensic science methods for analyzing various forms of trace evidence, to problems with the presentation of findings and opinions in the courtroom, and to the need for education and training to enable criminalists and other practicing forensic scientists to understand probability and the limits of decision making under uncertainty.

These issues continue to be central to the transformation of forensic-science disciplines from skilled crafts into fully scientific endeavors. For example, the limited validation of pattern-matching evidence (a category that includes tool marks on firearms and other items, shoe prints, hair microscopy, and handwriting) remains a matter of concern to a growing number of forensic scientists and courts. In many of these fields, examiners are taught to follow broadly defined steps known as "Analysis, Comparison, Evaluation, and Verification", or ACE-V. The steps constitute a high-level description of the process of identifying the features of a pair of items, assessing the degree of similarity between two items, and coming to a decision (or not) about their origins. In latent fingerprinting, for example, the analysis step produces a determination of whether a latent print has enough clear detail to merit further study. The comparison step is a side-by-side examination of the features of the latent print and an exemplar print from a known finger. The evaluation step yields a final determination as to whether the prints come from the same finger, whether they do not, or whether the similarities and dissimilarities in the patterns are inconclusive. Finally, the verification step requires a second practitioner (typically unblinded to the information in the case and to the first examiner's work and conclusions) to evaluate the prints.

Of course, "the devil is in the details". Merely dividing the process into these phases does not make it a reproducible and valid method—particularly when the interpretive process is highly personal and judgmental. Thus, the FBI had to repudiate its thrice-verified conclusion that a latent fingerprint on a bag containing detonators and explosives found in the aftermath of the 2004 train bombing in Madrid belonged to Brandon Mayfield, an Oregon lawyer, who claimed he had not left the United States for ten years and did not own a passport. The retraction came after Spanish National Police matched the latent print to Ouhnane Daoud, an Algerian national living in Spain.

To be sure, there is experimental evidence that supports the ability of examiners following the general ACE-V steps for friction-ridge matching to reach correct conclusions, but only with accuracies below the level that many members of the public would expect. And, the "black box studies", as they are sometimes called, do not reveal the accuracy and reliability of examiners engaged in real casework, all the way from the beginning (evidence collection) to the end (final assessment) of the process. This limitation is especially significant when one considers the wide range of settings in which friction ridge analysis is performed—crime laboratories, consultants, police "identification units", and non-accredited facilities. And, as the 2009 National Academies report also noted, training may be done formally or through informal mentoring. It can consist of just a short course, and there is no unified curriculum.

As a result, little is known about error rates in practice. Although at least one laboratory has experimented with introducing a small number of blind test prints into the flow of casework as a quality control measure, and although analysts in accredited laboratories (who know when they are being tested) are periodically given proficiency tests, well-powered double-blind experiments using realistically smudged or partial prints to determine a practitioner's error rate are rarely used.

In addition to the issues of validation and estimation of error rates for forensic-science methods and examiners, how to accurately and comprehensibly communicate the probative value of forensic-science findings has moved to center stage. For more than thirty years, the paradigm of presenting firm (or even overtly probabilistic) conclusions on source hypotheses has been questioned. The 2009 National Academies report noted literature advocating having the expert describe the weight of evidence in favor of one hypothesis relative to another instead of deciding between the two. That approach has been adopted in several countries. Yet, issues of validation pertain in all approaches (e.g., frequentist, likelihood, or Bayesian), and unresolved questions remain regarding what type of presentation on weight of evidence is best understood by judges and jurors.

This book contains discussions of all these topics—and more. Nevertheless, as a short handbook rather than an encyclopedia, it does not purport to review every statistical aspect of every method within the sprawling field of forensic science. As editors, we aimed for a collection that would provide cross-cutting statistical and historical background for all readers as well as reviews of the state of the statistical science in certain fields. Although every chapter was peer-reviewed by other experts and revised in response to their comments, inclusion does not necessarily imply that the editors agree with all the views and positions expressed in each chapter. Rather than impose our own opinions, we allowed the authors to speak in their own voices. We hope that the final result is a convenient resource for students and practitioners of statistics, forensic science, and law—and one that will improve the standard of practice and the communication of uncertainty in the courtroom.

The editors thank the referees who volunteered their expertise and time to this task. Besides the four editors, the referees were Colin Aitken, Alicia Carriquiry, Maria Cuellar, Ernest Fokoué, Christopher Glynn, Alice Liu, Lucas Mentch, Roi Naveiro Flores, Cedric Neumann, Karen Pan, and William Thompson.

David Banks
Duke University

Karen Kafadar
University of Virginia

David H. Kaye
Pennsylvania State University

Maria Tackett
Duke University

Editors

David Banks is a professor in the Department of Statistical Science at Duke University. He is a former coordinating editor of the *Journal of the American Statistical Association*, director of the Statistical and Applied Mathematical Sciences Institute, and a Fellow of the American Statistical Association and the Institute of Mathematical Statistics.

Karen Kafadar is a Commonwealth Professor and the chair of the Department of Statistics at the University of Virginia. She is a former president of the ASA; a Fellow of the International Statistics Institute, the ASA, and the AAAS; and a former member of the Forensic Science Standards Board (FSSB) of the Organization of Scientific Area Committees for Forensic Science (OSAC).

David H. Kaye is Distinguished Professor of Law Emeritus at Pennsylvania State University and Regents' Professor of Law and Life Sciences Emeritus at Arizona State University. He is a former editor of *Jurimetrics Journal*; a member of the FSSB; and the 2020 recipient of the Association of American Law Schools' Wigmore Lifetime Achievement Award for contributions to the understanding of the proof process and the rules of evidence.

Maria Tackett is an assistant professor of the practice in the Department of Statistical Science at Duke University.

Contributors

Colin Aitken
University of Edinburgh
Edinburgh, Scotland

John A.D. Aston
University of Cambridge
Cambridge, England

Madeline Ausdemore
South Dakota University
Brookings, South Dakota

David Banks
Duke University
Durham, North Carolina

Charles E.H. Berger
Netherlands Forensic Institute
Leiden University
Leiden, the Netherlands

Alex Biedermann
University of Lausanne
Lausanne, Switzerland

Silvia Bozza
Ca' Foscari University of Venice
Venice, Italy

Alicia Carriquiry
Iowa State University
Ames, Iowa

Junqi Chen
University of Virginia
Charlottesville, Virginia

Maria Cuellar
University of Pennsylvania
Philadelphia, Pennsylvania

A. Philip Dawid
University of Cambridge
Cambridge, England

Gary Edmond
University of New South Wales
Sydney, Australia

Benjamin Eltzner
University of Göttingen
Göttingen, Germany

Ewald Enzinger
Aston University
Birmingham, United Kingdom
and
Eduworks
Corvallis, Oregon

Brandon L. Garrett
Duke University
Durham, North Carolina

Joaquín González-Rodríguez
Higher Polytechnic School
Autonomous University of Madrid
Madrid, Spain

Martin J.R. Hall
Museum of Natural History
London, England

James Hamby
International Forensic Science Laboratory &
Training Centre
Indianapolis, Indiana

Rudolf Haraksim
European Commission
Ispra, Italy

Jessie Hendricks
South Dakota University
Brookings, South Dakota

Heike Hofmann
Iowa State University
Ames, Iowa

Stephan Huckemann
University of Göttingen
Göttingen, Germany

Alan Julian Izenman
Temple University
Philadelphia, Pennsylvania

Karen Kafadar
University of Virginia
Charlottesville, Virginia

Naomi Kaplan-Damary
University of California Irvine
Irvine, California

David H. Kaye
Pennsylvania State University
University Park, Pennsylvania

and

Arizona State University
Phoenix, Arizona

Alice J. Liu
University of Virginia
Charlottesville, Virginia

Alicia Lozano-Díez
Higher Polytechnic School
Autonomous University of Madrid
Madrid, Spain

Kristy A. Martire
University of New South Wales
Sydney, Australia

Lucas Mentch
University of Pittsburgh
Pittsburgh, Pennsylvania

Didier Meuwly
Netherlands Forensic Institute
and
University of Twente
Enschede, the Netherlands

Geoffrey Stewart Morrison
Aston University
and
Forensic Evaluation Ltd.
Birmingham, United Kingdom

Julia Mortera
Roma Tre University
Rome, Italy

Cedric Neumann
South Dakota State University
Brookings, South Dakota

Karen Pan
University of Virginia
Charlottesville, Virginia

Davide Pigoli
King's College
London, England

Daniel Ramos
Higher Polytechnic School
Autonomous University of Madrid
Madrid, Spain

M.J. (Marjan) Sjerps
Netherlands Forensic Institute
and
University of Amsterdam
Amsterdam, the Netherlands

Cliff Spiegelman
Texas A&M University
College Station, Texas

Hal S. Stern
University of California Irvine
Irvine, California

Maria Tackett
Duke University
Durham, North Carolina

Xiao Hui Tai
Carnegie Mellon University
Pittsburgh, Pennsylvania

Franco Taroni
University of Lausanne
Lausanne, Switzerland

Susan VanderPlas
Iowa State University
Ames, Iowa

Bruce S. Weir
University of Washington
Seattle, Washington

Sarena Wiesner
Israeli Police Force
Jerusalem, Israel

Joanne Yaffe
University of Utah
Salt Lake City, Utah

Section I

Perspectives on Forensic Statistics

1

The History of Forensic Inference and Statistics: A Thematic Perspective

Colin Aitken and Franco Taroni

CONTENTS

1.1   Introduction 4
1.2   Forensic Science and the Evaluation of Evidence 5
1.3   The Need for an Interpretative Model 6
1.4   Support of Judicial Disciplines for a Scientific Presentation
      of the Value of Evidence 8
1.5   Probability of Proposition Given Evidence and of Evidence Given Proposition . . . 11
1.6   Quantification of the Value of Evidence Using Alternative Numerical
      Summaries 12
1.7   Change from Two-Stage Approach to Continuous Approach 13
1.8   Presentation of Evidence: New Challenges to Solve 15
      1.8.1   The Island Problem and Results of a Database Selection 15
      1.8.2   Profile Probability vs Conditional Profile Probability 16
      1.8.3   Evaluation by Taking Errors into Account 17
1.9   A Minimum Value for the Profile Probability 18
1.10 Propositions and Pre-Assessment 19
      1.10.1 The Choice of Propositions 19
      1.10.2 The Pre-Assessment 20
1.11 Translation of a Numerical Value into a Verbal Equivalent 20
1.12 Assessment of Performance 22
1.13 Role for Likelihood Ratio as a Measure for Investigation as Well as
      for Evaluation 25
1.14 Probabilistic Graphical Models 26
      1.14.1 Bayesian Networks 26
      1.14.2 Bayesian Networks to Manage 'Masses' of Evidence 26
      1.14.3 Bayesian Networks in Judicial Contexts 27
      1.14.4 Bayesian Networks in Forensic Science: Particular Case Modeling 27
      1.14.5 Bayesian Networks in Forensic Science: Generic Patterns of Inference 28
1.15 Not Only Inference: The Way to Make a Decision 28
      1.15.1 The Objectives and Ingredients of Decision Theory 29
      1.15.2 Graphical Models 29
1.16 The Existence or Otherwise of a True Value of the Evidence 30
Acknowledgments 31
References 31

1.1 Introduction

The historical development of forensic inference and statistics is presented through fifteen important themes. The themes have been chosen as the ones that created, and in some cases are still creating, important debates. The choice of themes is a personal choice of the authors and some readers may not agree. It is hoped this form of presentation will help clarify thinking around current problems and suggest ways in which the subject may develop further.

It is only the role of statistics in the evaluation of evidence in criminal cases that is discussed. No reference is made to civil law, such as examples of jury selection and employment discrimination. Preference is given instead to the development since Dennis Lindley's seminal paper in *Biometrika* in 1977 (Lindley, 1977).

Fifteen themes are identified as important in the development of the ideas for probabilistic and statistical reasoning in forensic science. The first theme (Section 1.2) is the recognition in the early 20th century of the need for the interpretation of scientific findings in the administration of criminal justice. The next two themes (Sections 1.3 and 1.4) concern ideas for the integration of scientific information with other relevant information from a particular criminal case and the increasing support of judicial disciplines for the scientific presentation of evidence. These ideas led to recognition of the importance of the separation of evidence from propositions and the correct conditioning of one on the other depending on the role of the person making the judgement (Section 1.5). Various attempts to quantify the value of the evidence before the general acceptance of the likelihood ratio as the best way to do this are described in Section 1.6, followed by the description of the discrediting of the idea of a match (Section 1.7). The advent of DNA profiling in the mid-1980s led to consideration of many new factors in the evaluation of evidence which are outlined in Section 1.8. One factor in particular, that of the possibility of extremely small probabilities for a DNA profile and correspondingly large values of the likelihood ratio merits a section on its own (Section 1.9). The concept of propositions was developed further in the late 1990s with the introduction of differing levels of propositions (Section 1.10). The general use of the likelihood ratio and the difficulty jurists had with its interpretation led to attempts to summarise its numerical value verbally (Section 1.11). Though the role of the likelihood ratio was generally accepted, there could be several different values in a particular case arising from the use of different assumptions and statistical models. Recognition of these differences led to consideration of methods for the assessment of the performance of different models (Section 1.12). The role for the forensic scientist in the investigation of a crime (the *investigative role*) before they are asked to evaluate evidence in a trial (the *evaluative role*) was recognised in a separate development in the 1990s, a role that is described in Section 1.13. Statistical research in the late 20th century led to probabilistic graphical networks for complicated problems of inference. These networks had an intuitively satisfying application in forensic science, in particular for the management of many different pieces of evidence, and this application is described in Section 1.14. Ultimately a decision has to be reached by the jurist (jury or judge) concerning the outcome of a criminal trial. Scientists also have decisions to make, earlier in the process, for example concerning sample size or choice of analysis. The role of decision theory for the scientific process is described in Section 1.15. Finally, the early years of the 21st century have seen questioning of the presentation of a single value for evidential value with the likelihood ratio. The alternative suggestion is that the single value of the likelihood ratio should be replaced by an interval for, or a lower bound on, its value. A comment on this debate is given in Section 1.16.

1.2 Forensic Science and the Evaluation of Evidence

In the early 1960s, the forensic science community started to take a more explicit position with respect to the problems of interpretation and evaluation of scientific data. In a now widely known quote, Kirk and Kingston (1964) from the University of California, Berkeley, note:

> When we claim that criminalistics is a science, we must be embarrassed, for no science is without some mathematical background, however meagre. This lack must be a matter of primary concern to the educator [...]. Most, if not all, of the amateurish efforts of all of us to justify our own evidence interpretations have been deficient in mathematical exactness and philosophical understanding. (at pp. 435–436)

Today, interpretation and data evaluation are still a neglected area, mainly in fields that involve so-called *physical* evidence. This neglect continues to exist despite an important paper by Stoney (1984) that gave the relevant questions a scientist should ask in their analyses. This neglect is now acknowledged, for instance, by reports such as that of the National Research Council of the US (National Research Council, 2009) and of the President's Council of Advisors on Science and Technology (PCAST, 2016). In its report the NRC Council notes that '[t]here is a critical need in most fields of forensic science to raise standards for reporting and testifying about the results of investigations' (at p. 185). In many contexts this perception is reinforced by the fact that scientists' assessments of evidential value consist of a largely subjective component with a connotation of arbitrariness. As mentioned by Kirk and Kingston (1964), it was indeed rare at their time of writing that a scientist's opinion was based on a quantitative study. With some notable exceptions, for example transferred material such as DNA, fibres or glass fragments, this is still the situation today even though quantification with the use of probabilities was suggested by earlier forensic scientists, for example Bertillon (1893, 1898) and Locard (1920, 1940) in areas such as questioned documents and anthropology. See, for the sake of illustration, a quote from Bertillon (1898) on the need for a quantification:

> This writing, characterized by the set of unique features we have enumerated, can only be encountered in one individual among a hundred, among a thousand, among ten thousand or among a million individuals. (at p. 20)

Perhaps the first probabilistic approach for evidence evaluation was the approach used in the Howland Will case between 1865 and 1868 (Meier and Zabell, 1980). A general probabilistic summary of the evidence was given by scientists in the early 20th century (Darboux et al., 1908) in the Dreyfus case. This approach was supported later by Kingston (1965a,b) and echoed by Saks and Koehler (2005) who concluded that 'Although obstacles exist both inside and outside forensic science, the time is ripe for the traditional forensic sciences to replace antiquated assumptions of uniqueness and perfection with a more defensible empirical and probabilistic foundation' (at p. 895). However, this approach is still viewed as controversial in some quarters.

The final word in this section comes from an article some 25 years before those of Kingston. Locard (1940) proposed some inspired guidelines for the interpretation of scientific evidence. These guidelines remain pertinent to scientists and lawyers even today.

The physical certainty provided by scientific evidence rests upon evidential values of different orders. These are measurable and can be expressed numerically. Hence the expert knows and argues that he knows the truth, but only within the limits of the risks of error inherent to the technique. This numbering of adverse probabilities should be explicitly indicated by the expert. The expert is not the judge: he should not be influenced by facts of a moral sort. His duty is to ignore the trial. It is the judge's duty to evaluate whether or not a single negative evidence, against a sextillion of probabilities, can prevent him from acting. And finally, it is the duty of the judge to decide if the evidence is in that case, proof of guilt. (at pp. 286–287)

1.3 The Need for an Interpretative Model

Data are fundamental for the reduction of uncertainty about propositions of interest for a court. Uncertainty should be expressed by the concept of probability. Uncertainty is inevitable because the role of a court is to reconstruct what has happened in the past (i.e., the commission of the crime) based on incomplete knowledge. Such a reconstruction inevitably results in an uncertain representation of the true state of affairs. Information may be gained by enquiry, analysis and experimentation. As a consequence of this, a method is required to adjust existing beliefs in the light of newly acquired evidence. Inferences, if they are to be taken seriously, must be approached within a probabilistic framework. The revision of beliefs should be made according to Bayesian procedures. This is not controversial; it is a logical consequence of the basic rules of probability. Reference will be made often to Bayes's theorem; it is a very important result that helps one understand how beliefs should be adjusted in the light of new evidence. Although the theorem has a history of about 250 years, the associated approach to inference has gained more widespread use only since the end of the 20th century. This is the case even though, historically, practical applications of patterns of reasoning corresponding to a Bayesian approach can be found, for example, as early as the beginning of the 20th century. For example, at the Dreyfus's military trial held in 1908, Henri Poincaré invoked Bayes's theorem as the only way in which the court ought to revise its opinion about the issue of forgery (Darboux et al., 1908). He described its applications as follows:

> An effect may be the product of either cause A or cause B. The effect has already been observed; one wants to know the probability that it is the result of cause A; this is the *a posteriori* probability. But, I'm not able to calculate this if an accepted convention does not permit me to calculate in advance the *a priori* probability for the cause producing the effect; I want to speak of the probability of this eventuality, for one who has never before observed the result.

Given the difficulty - for a scientist - in dealing with the *a priori* probability, Poincaré and his colleagues supported the use of the likelihood ratio, the expression that is the connection between prior and posterior probabilities:

> Since it is absolutely impossible for us [the experts] to know the *a priori* probability, we cannot say: this coincidence proves that the ratio of the forgery's probability to the inverse probability is a real value. We can only say: following the observation of this coincidence, this ratio becomes X times greater than before the observation. (at p. 504)

This quotation is a statement of the odds form of Bayes's theorem, namely

$$\frac{Pr(H_p \mid E, I)}{Pr(H_d \mid E, I)} = \frac{Pr(E \mid H_p, I)}{Pr(E \mid H_d, I)} \times \frac{Pr(H_p \mid I)}{Pr(H_d \mid I)}. \tag{1.1}$$

Statements about the probability of the evidence E if the suspect is not the forger H_d, with background information I, are part of the likelihood ratio $Pr(E \mid H_p, I)/Pr(E \mid H_d, I)$ (the X of the statement of Poincaré's and his colleagues). The proposition that the suspect is the forger is denoted by H_p, the proposition that the suspect is not the forger by H_d. '[T]he ratio of the forgery's probability to the inverse probability' is a statement of the odds in favour of H_p. '[F]ollowing the observation of this coincidence, this ratio becomes X times greater than before the observation.' The probability the suspect is the forger given the evidence is the numerator of the posterior odds $Pr(H_p \mid E, I)/Pr(H_d \mid E, I)$. As H_p and H_d are complementary (a situation that does not always hold in the evaluation of evidence), it is possible to determine the probability $Pr(H_p \mid E, I)$ from the posterior odds. The likelihood ratio X and the posterior odds are related by the prior odds $Pr(H_p \mid I)/Pr(H_d \mid I)$.

Whereas Poincaré and his colleagues refused to evaluate prior probabilities, an anonymous commentator of the Dreyfus trial (Darboux et al., 1907) opined that reasoning remained valid and sound if probabilities could be put on past acts. The commentator developed the inferential reasoning adopting a shoe print example:

> Burglary is committed in a house surrounded by a park. A suspect is apprehended because of his appearance and his criminal history record. These elements alone are not sufficient to allow conviction. However, shoe prints are recovered at the scene and they correspond to the soles of the suspect's shoes. This is sufficient proof. The juror's opinion is established and the conviction is delivered. But nothing proves with certainty that two different shoes could not produce an identical shoe print, and that the shoe prints from the scene could not come from shoes worn by someone other than the suspect. The juror's logic is sound only if the probability of other explanations is extremely small. The juror's reasoning is as follows: the perpetrators of the burglary are either the suspect or a person or persons unknown. The *a priori* probabilities of these possibilities are fixed only by moral criteria. This possibility cannot be expressed accurately by a single number, but it can be said to fall within certain limits or a given range. However, after the verification of the shoe prints, everything changes. If the prints are those of the accused, the probability of observing such evidence is very high, the functional equivalent of certainty. Conversely, the probability of finding these prints made by someone else cannot be precisely determined, but is is extremely low. In sum, the *a priori* probabilities have been modified, permitting the 'a beyond reasonable doubt' conviction required by the law. (at pp. 19–20)

Despite the problem encountered by Bertillon during the Dreyfus case (e.g., an early example of what is now known as 'the prosecutor's fallacy', Section 1.6), Bertillon can be nominated as the first Bayesian forensic scientist. In fact, after expressing the need of a quantification of the observed features (Bertillon, 1898), he completed his reasoning by affirming that the only way to accept an expert's categorical conclusions was to consider not only the statistical evidence provided by the examination of the document, but also other information pertaining to the inquiry. He described how the number of people who could be the author of the questioned document size is reduced by the inquiry (i.e., the testimonies and circumstances of the case). This description introduces the general idea of a relevant population, a concept expanded and discussed by Lempert (1991), Robertson and Vignaux (1993a), Champod et al. (2004) and Kaye (2004). An important contribution to the role of scientific evidence is that of Fienberg et al. (1996). He and his co-authors note that (a) what is treated as a relevant population may only be a conveniently available population and (b) the event that evidence associated with the crime came from the defendant is not necessarily the same as the event that the defendant committed the crime.

Therefore, the evidentiary value of the scientific observations, even if not totally confirmatory of guilt, could supply sufficient information to allow a conviction when the case is considered as a whole. Other examples of similar reasoning were published by Balthazard (1911) and Souder (1934) in the fields of fingermarks and typewriting machines, respectively. A complete historical summary of the relationship of forensic scientists to Bayesian ideas is presented in Taroni et al. (1998).

The Bayesian interpretative model has attracted many supporters in the area of interpretation and evaluation of evidence in forensic science. The model contains all the ingredients necessary for the required inferences.

> [...] the only argument we can adduce is to ask the reader to pursue it and see where it leads - the proof of the pudding is in the eating. (Lindley, 1985, p. 101)

1.4 Support of Judicial Disciplines for a Scientific Presentation of the Value of Evidence

A crucial factor in the progressive acceptance of a probabilistic presentation of the value of evidence has been the support of the judiciary. This support was not gathered by statisticians or forensic scientists but by jurists.

The support dates back to 1897. Mr Justice Holmes, then of the Supreme Judicial Court of Massachusetts and latterly of the Supreme Court of the United States (Holmes, 1897) wrote

> For the rational study of the law the black-letter man may be the man of the present, but the man of the future is the man of statistics and the master of economics. (at p. 469)

He was echoed almost a century later by Twining who affirmed that

> The lawyer of today needs to be a master of elementary statistics. (Twining, 1994, p. 209)

Other supporters amongst jurists for a probabilistic presentation of the value of the evidence include Sir Richard Eggleston. Through a series of examples he underlined the fundamental role played by probabilities in legal settings. He wrote in the introduction to his book:

> It is plain from this example that probabilities must play a very large part in the decision of cases in the courts. Even in criminal cases, where the jury must be satisfied beyond reasonable doubt, probability theory is often of the highest importance. The acceptance of fingerprint evidence, for example, depends essentially on the 'multiplication rule', to be discussed hereafter. Yet the legal profession as a whole has been notably suspicious of the learning of mathematicians and actuaries, and ignorant of the work of philosophers in this field. (Eggleston, 1983, p. 2)

A justification for the use of probabilities can also be seen through the definition of 'relevant evidence' as defined by The U.S. Federal Rule of Evidence 401. The Rule states:

> Relevant evidence means evidence having any tendency to make the existence of any fact that is of consequence to the determination of the action more probable or less probable than it would be without the evidence.

The use of a probabilistic line of reasoning is supported by Lempert who wrote:

> One of the main areas of interest of the so-called *new evidence* scholarship is the application of probability theory to arguments about facts in legal cases. As a preliminary to making a decision, courts have to 'find facts' which require them to reason under uncertainty. In some cases it may be the reasoning process itself which is examined in an appeal. The result may be a statement by the court about how facts ought to be thought about. Alternatively the way facts are thought about in a particular case may be seized upon as a precedent for future cases. Should there be rules about how facts are to be thought about? And, if so, does probability theory offer a prescription for those rules? (Lempert, 1986, p. 457)

There is a need for a clarification here. Lempert's statement supports the use of probabilistic reasoning by noting it is about the structure of reasoning and not particularly about numbers. This perspective has been reaffirmed by Robertson and Vignaux (1993a). Numbers are not important in themselves: what really matters is that numbers allow us to use powerful rules of reasoning which can be implemented by computer programs. What is important is not whether the numbers are 'precise', whatever the meaning of 'precision' may be in reference to subjective degrees of belief based upon personal knowledge. What is important is that we are able to use sound rules of reasoning to check the logical consequences of our propositions, and consider the consequences with respect to the degree of belief in one proposition of assuming a certain degree of belief in another proposition.

The legal system is concerned with making decisions, and decisions must often be made in a situation of uncertainty, either as to what has happened in the past, or as to what is going to happen in the future. Eggleston emphasised that

> We are not concerned here with uncertainty as to the legal rule, though this is frequently a matter of anxiety, especially to those who have to make the decision whether or not to commence proceedings. Our interest is in uncertainty as to the facts to which the law must be applied.[...] [Therefore] students and practitioners, on the one hand, and non-lawyers, on the other, need to understand the judicial approach to probability. (Eggleston, 1983, pp. 3–4, 10)

It seems therefore that jurists support some of the concerns and desiderata expressed by forensic scientists such as Kingston and Kirk (1964) in the 1960s. They wrote:

- It can be fairly stated that there is no form of evidence whose interpretation is so definite that statistical treatment is not needed or desirable.
- The statistical analysis provides the criminalist with a basis for his opinion, and an evaluation of the likelihood that his testimony reflect the truth, rather than his personal belief or bias.

- This is not proposed in the belief that such accepted evaluations will be changed, but more in the hope that firmer lines of reasoning might replace the arbitrary justifications upon which many such evaluations now rest.

The first point is restated by Robertson and Vignaux (1995b, p. 12) in the following terms:

> An ideal piece of evidence would be something that always occurs when what we are trying to prove is true and never occurs otherwise. If we are trying to demonstrate the truth of a hypothesis or assertion we would like to find as evidence something which always occurs when the hypothesis is true and never occurs when the hypothesis is not true. In real life, evidence this good is almost impossible to find.

The necessity of probabilistic reasoning in law has been discussed. The next step is discussion of its use in the presentation of the value for a piece of evidence. The following quote from 1977 is an early suggestion for lawyers of a form of words for the presentation of a numerical value of the likelihood ratio*:

> [...] the defendant's thumb print was found on the gun the killer used. [...] assume that the fact-finder believes that the presence of this evidence is 500 times more likely if the defendant is guilty than if he is not guilty. [...] Now suppose that the prosecution wished to introduce evidence proving that a print matching the defendant's index finger was found on the murder weapon. If this were the only fingerprint evidence in the case, it would lead the fact-finder to increase his estimated odds on the defendant's guilt to the same degree that the proof of the thumb print did. Yet, it is intuitively obvious that another five hundredfold increase is not justified when evidence of the thumb print has already been admitted. (Lempert, 1977, p. 1043)

This idea is novel for a lawyer. However, two forensic scientists had already used such a probabilistic metric for the value of the evidence. A practical example is offered by Kingston and Kirk (1964, p. 514):

> Now consider a problem of evaluating the significance of the coincidence of several properties in two pieces of glass. Suppose that the probability of two fragments from different sources having this coincidence of properties is 0.005, and that the probability of such a coincidence when they are from the same source is 0.999. What do these figures mean? They are simply guides for making a decision about the origin of the fragments [...]

Not all jurists appreciated the statistical approach to the evaluation of evidence. In the early 1970s a counter-argument was well-aired and debated; see Finkelstein and Fairley (1970), Tribe (1971) and Finkelstein and Fairley (1971). Notwithstanding this debate some jurists did appreciate the approach. For example, the likelihood ratio and the role background information plays in the assignment of probabilities was described by Richard Friedman in 1996:

> Suppose, for example, you are sitting in a restaurant when you hear a voice that you do not recognize yell, 'There's been an accident outside!' You know nothing about the declarant and her relationship to what either did or did not happen outside, apart from what you know in general about the world and what you can infer from her voice. But

---

* A form of words that is not to be confused with a verbal summary of the likelihood ratio, a form of words that is discussed in Section 1.11.

you know enough to make a preliminary assessment of how likely she would make the statement if it were true, and how likely she would do so if it were not - that is, you have enough information to make an assessment, albeit very tentative, of the likelihood ratio. If you turn to look and find that she is obviously drunk or obviously joking or, on the other hand, shaken and bloodied, you rapidly and radically may reassess the likelihoods in light of this further information. (Friedman, 1996, p. 1817)

In conclusion, it is affirmed that a probabilistic approach and particularly one based on Bayesian modelling should be considered as a valuable tool for reasoning about evidence (Redmayne, 1996); this is a consequence, as suggested by Robertson and Vignaux (1991), of Bayes's theorem as a formalisation of logic and common sense.

Examples of the applications of Bayesian analyses to legal matters include Cullison (1969), Fairley (1973), Finkelstein and Fairley (1970), Finkelstein and Fairley (1971), Fienberg and Kadane (1983), Kaye (1986) and Anderson and Twining (1998).

1.5 Probability of Proposition Given Evidence and of Evidence Given Proposition

One of the most important topics that created debate in the development of forensic statistics in the 1980s is the difference between the probability of a proposition given evidence and the probability of evidence given a proposition. To a statistician, the difference is clear. To a statistical layman the difference does not seem to be so clear. The interpretation of the probability of evidence given a proposition as the probability of the proposition given the evidence has been termed the prosecutor's fallacy (Thompson and Schumann, 1987) or inversion fallacy (Kaye, 1993) or transposed conditional (Evett, 1995). This confusion can be expressed more clearly as the interpretation of a small probability of finding the evidence on a person who is innocent of a crime as a small probability that a person on whom the evidence is found is innocent of the crime.

The error is easily exposed through the use of the odds form of Bayes's theorem. Consider a suspect and evidence E and mutually exclusive propositions:

- H_p: the suspect is guilty;
- H_d: the suspect is innocent.

These propositions are also exhaustive. In general it is not necessary for the evaluation of evidence for the propositions to be exhaustive but it is helpful here to expose the fallacy. For the evaluation of evidence there will always be a framework of circumstances or background information I to bear in mind for the evaluation.

The odds form of Bayes's theorem is given in (1.1). Statements about the probability of the evidence if the suspect is innocent are part of the likelihood ratio $Pr(E \mid H_p, I)/Pr(E \mid H_d, I)$. Statements about the probability of innocence of the suspect given the evidence are part of the posterior odds $Pr(H_p \mid E, I)/Pr(H_d \mid E, I)$. The likelihood ratio and the posterior odds are related by the prior odds $Pr(H_p \mid I)/Pr(H_d \mid I)$.

There is a very good medical analogy. The propositions of guilt and innocence may be replaced with diagnoses of presence or absence of disease. The evidence in the criminal case is replaced with medical symptoms or medical test results. In medicine there are often data on the incidence of symptoms in the presence of the disease, a so-called *incidence rate* and on the incidence of symptoms in the absence of the disease. However, given these data it is not possible to estimate the probability a patient has the disease without knowledge of the so-called *base rate* of the disease in some background population from which the prior odds can be assigned.

Similarly in forensic science, knowledge of the incidence of a certain characteristic amongst a relevant population as well as in the criminal is not sufficient for a determination of guilt in a possessor of the profile. It is also necessary to have an assignment of the prior probability of guilt, the equivalent of a base rate in the medical analogy.

The propositions of guilt and innocence are what are known as *offence-level* propositions (see Section 1.10). It is a large step, for example, to move from an inference about the DNA profile of a suspect to the guilt of the suspect. It may be that the only inference possible is that the DNA of the suspect was present at the crime scene. Propositions about the source of the DNA are known as *source-level* propositions (see Section 1.10).

There are many variations of the prosecutor's fallacy and these are discussed as other errors of logic in Koehler (1993).

Thompson and Schumann (1987) introduced also a *defence attorney's fallacy* to balance the prosecutor's fallacy. Consider a crime in which a relevant population to which the criminal is deemed to belong is of size 100,000. For example, this could be a rape in which it is thought there are 100,000 males who could have committed the crime. A degraded DNA profile of the criminal obtained from semen found on the victim has an occurrence of 1 in 2,000. A suspect is identified in a manner independent of the profile and found to have a profile which is indistinguishable from that of the one known to have come from the criminal. The prosecutor's fallacy interprets the value of 1 in 2,000 as a probability of 1 in 2,000 that the suspect is innocent. The defence argue that there are 100,000 people who could have been the criminal. The occurrence of the profile is 1 in 2,000, thus there are fifty people in the relevant population who could have this profile. The suspect is one of fifty people so the probability of innocence is forty nine out of fifty. So far, the argument is correct. The argument becomes fallacious when it is extended to argue that the evidence is thus irrelevant. Before consideration of the evidence, the suspect was one of 100,000 men, after consideration of the evidence the suspect is one of fifty men. Such a consequence is very relevant.

1.6 Quantification of the Value of Evidence Using Alternative Numerical Summaries

The use of the likelihood ratio and functions of it, such as the logarithm (Peirce, 1878; Good, 1950), for the evaluation of evidence increased following the seminal paper of Lindley (1977). Before 1977, several attempts had been made to summarise the value numerically.

Consider categorical evidence, such that the evidence manifests itself as one, and only one, of a set K of exhaustive and mutually exclusive categories. The probability a particular evidential item has category k is p_k, $k = 1, \ldots, K$: $\sum_{k=1}^{K} p_k = 1$. Various suggestions might be offered for the value of evidence of category k found at a crime scene.

- The probability that two people have the same category k is p_k^2;

- The probability that two people have the same category, without specifying the category is

$$p_1^2 + \cdots + p_K^2.$$

- Given that one person, the *control* person, is of category k, the probability another person, chosen at random from a relevant population is also of category k is p_k.

The first two suggestions are of limited importance for the evaluation of evidence. One piece of evidence will have a known source. That evidence is known as *control* evidence. Another piece of evidence which is to be compared with the control evidence will have an unknown source, this evidence is known as *recovered* evidence. It is the third suggestion that is of importance. The numerator of a likelihood ratio is the probability the recovered evidence is of the same category k as the control evidence assuming, for example, that the recovered evidence and the control evidence come from the same source and in an idealised scenario this probability is 1. The denominator of a likelihood ratio is the probability the recovered evidence is of the same category as the control evidence assuming the recovered evidence and the control evidence come different sources; this probability is p_k. Thus p_k^{-1} is the value of the evidence.

A general assessment of the evidential value of a method is *discriminating power* (DP). It is related to the second probability with $DP = 1 - (p_1^2 + \cdots + p_K^2)$. DP is the probability that two people chosen at random will belong to different categories. For example, if everybody is of the same category, say category 1 without loss of generality, then $p_1 = 1$ and $p_2 + \cdots + p_K = 0$ and $DP = 0$; no-one can be discriminated. Conversely, if all categories are equally likely, $p_1 = \cdots = p_K = 1/K$ and $p_1^2 + \cdots + p_K^2 = 1/K$ and it can be shown that DP is maximised.

Discriminating power is a measure of the general worth of an evidential type. A high value for *DP* is indicative of evidence of a good discriminatory type. A low value for *DP* is indicative of evidence of a poor discriminatory type. Discriminating power is not a measure of evidential value in a particular case. An example of the use of discriminating power for hair examinations is given in Gaudette and Keeping (1974) with a critical discussion in Aitken and Robertson (1987).

DNA evidence introduced new challenges, notably that of a DNA mixture. An approach related to discriminating power, known as *random man not excluded* was proposed, discussed and criticised. A debate on this topic is given in Buckleton et al. (2016b).

1.7 Change from Two-Stage Approach to Continuous Approach

Procedures for the evaluation of evidence in forensic science were changed dramatically by a paper by Dennis Lindley in 1997 (Lindley, 1977). Previous to the publication of that paper a common procedure was a two-stage approach.

- Similarity: In a comparison of characteristics of evidence found at a crime scene and in the environment of a suspect, are the characteristics similar or dissimilar?

- Rarity: If the characteristics are dissimilar then the evidence is not considered any further, the evidence associated with the suspect is deemed not to be associated with the crime. If the characteristics are similar then the evidence associated with the suspect is deemed to be associated with the crime. The strength of the evidence under source level propositions (Section 1.10) is measured by the rarity of the characteristic; the more rare the characteristic, the stronger the association.

This description of the two-stage approach begs the questions of what is meant by similarity and what is meant by rarity.

Often, similarity was defined in relation to the result of a significance test. The characteristic takes the form of a continuous measurement such as that of the refractive index of a fragment of glass. The comparison of characteristics was made with a significance test of a null hypothesis of common source for the measurements of the crime scene characteristic and measurements of the characteristic (e.g., refractive index of fragments of glass) found in association with a suspect (e.g., upon their clothing). If the result of the test were such that the null hypothesis was rejected at some pre-specified level (e.g., 5%) then the two pieces of evidence would be deemed dissimilar and the alternative hypothesis of different sources would be accepted, in the sense that a decision would be made not to consider this evidence further. There are two problems with this procedure. The first problem is that the null hypothesis is one of common source. The null hypothesis is conventionally taken as the status quo and it will only be rejected if there is sufficient evidence against it (e.g., at a significance level of 5%). It is normally the prosecution that proposes a hypothesis of common source and it is the prosecution that wishes to show the suspect is guilty. A proposition that the suspect is guilty until there is sufficient evidence to show them innocent is contrary to the presumption of innocence. The second problem is the effect that has been called that of falling off a cliff (Robertson and Vignaux, 1995b; Robertson et al., 2016). Assume a pre-specified level of 5% for rejection of the hypothesis of common source. Evidence for which the comparison gives a result which is significant at the 5.1% level will be deemed to have a common source. Evidence for which the comparison gives a result which is significant at the 4.9% level will be deemed to have different sources. This is unsatisfactory.

Assessment of rarity in such a procedure is difficult. One procedure proposed in the late 1970s is that of a *coincidence probability*; see Evett (1977). This probability was defined as the probability that the characteristics taken from evidence selected at random from some item selected in turn at random from some relevant population of items would be found to be similar, in some sense, to a control item with a particular value of the characteristic. An example is that of the refractive index of glass fragments from a window. The coincidence probability would be the probability that the refractive indexes of a number of fragments selected at random from a window selected in turn at random from some relevant population of windows would be found to be similar, in some sense, to a control window, at a crime scene say, with a set of refractive indexes from a particular sample of fragments from the control window.

This approach can be compared with that of discriminating power. For discriminating power, the recovered and control fragments are both taken to be random samples from some underlying population. The probability is that of two random samples having similar characteristics. For a coincidence probability the data from the control window are taken to be fixed. The concern is with the assignment of the probability that one sample, the recovered sample, is found to be similar to the sample from the control window. Any variability in the values of the data from the control window is ignored.

The problems associated with the two-stage approach were overcome by Lindley (1977). A procedure was developed which accounted for the similarity and the rarity, with associated variation, in one statistic based on the likelihood ratio. The likelihood ratio developed provided a continuous measure of the value of evidence. Consider evidence E which is a set of continuous measurements, E_c and E_r, on control and recovered material, i.e., material for which the source is known and material for which the source is unknown, respectively, with $E = \{E_c, E_r\}$. The propositions for comparison are

H_p: E_c and E_r have the same source;
H_d: E_c and E_r have different sources.

The likelihood ratio is then

$$\frac{f(E \mid H_p)}{f(E \mid H_d)},$$

where the probability Pr of (1.1) is replaced by probability density functions in recognition of the continuous nature of the evidence. Further details are given in Lindley (1977).

1.8 Presentation of Evidence: New Challenges to Solve

In the early 1990s, interest in the probabilistic evaluation of DNA profiling results grew considerably. Topics such as the effect of database searches to select persons of interest, the role of sub-populations in the assignment of conditional probabilities and the consideration of error for false inclusions, were responsible for an increase in papers focusing on forensic inference.

1.8.1 The Island Problem and Results of a Database Selection

One important debate was one that became known as the *island problem* to which various solutions have been proposed (Eggleston, 1983; Yellin, 1979; Lindley, 1987). The problem relates to the determination of the probability of guilt. The problem has often been approached by consideration of a finite population such as may be found on an island, of population size $(N + 1)$ say. A crime is committed and evidence of a characteristic (e.g., a blood stain of DNA profile Γ, with occurrence γ amongst some larger population) is found at the scene of the crime. A person is found who possesses this characteristic and the probability of their guilt is of interest.

Determination of the probability is related to the manner in which this person has been selected (become a person of interest - POI). There are different ways a POI comes into consideration in a criminal investigation. One of them is through selection from a database. The compilation of DNA databases could enable police forces to collect samples taken during investigations of unsolved criminal cases, as well as samples from convicted felons, in order that such stored information could be used to select a person of interest in a way similar to the collection and storage of fingerprint records.

A debate appeared around the question 'should the fact that the person of interest was selected through a database search affect the value of the evidence?' Confusion surrounding the evaluation of the outcome of such a search can arise because the probability of a match increases as the database gets larger. Robertson and Vignaux (1995a) explain this confusion by stating that '[i]t is commonly claimed that the evidential value of a match, when a POI is selected through a search in a database, is affected by the number of comparisons one has made. Certainly, the larger the database the more likely we are to find a match' (at p. 122). This leads to the erroneous conclusion that the larger the database the weaker the evidence. This erroneous approach was proposed in a report of the National Research Council of the United States (National Research Council, 1996) which published the recommendation:

> When the suspect is found by a search of DNA databases, the random match probability should be multiplied by N, the number of persons in the database. (Recommendation 5.1 at p. 40)

It has been shown that the application of such a recommendation produces illogical results with a drastic dilution of the strength of the evidence. Balding and Donnelly (1996) and Evett and Weir (1998) showed that the likelihood ratio is higher following a search than in a case where the size of the potential criminal population is known and no sequential searches have been performed. Each person who does not match the DNA profile of the recovered trace is excluded. The exclusion of these individuals from the pool of the potential culprits increases the probability of involvement of the individual who matches. The argument was developed further in Dawid and Mortera (1996) and potential solutions were given in Dawid (2001), Balding (2002) and Kaye (2009).

1.8.2 Profile Probability vs Conditional Profile Probability

Imagine the likelihood ratio calculation in a common situation involving a single biological stain where there is the DNA profile E_r (r for 'recovered' as the origin of the sample is not known) of the crime sample and the profile E_c (c for 'control' as the origin of the sample is known) of a POI. Let I represent the background information and let the propositions be

H_p: the POI is the source of the stain,

H_d: another person, unrelated to the POI, is its source; *i.e.*, the POI is not the source of the stain.

Both profiles are of genotype A, say. The likelihood ratio can then be expressed as

$$\frac{Pr(E_r = A \mid E_c = A, H_p, I)}{Pr(E_r = A \mid E_c = A, H_d, I).}$$

Assume that the DNA typing system is sufficiently reliable that two samples from the same person will be found to match when the POI is the donor of the stain (proposition H_p), and that there are no false negatives. If it is known that the POI is of type A and if H_p is assumed true then it follows that the recovered sample is of type A and $Pr(E_c = A \mid E_p = A, H_p, I) = 1$.

It is widely assumed that the DNA profiles from two different people (the POI and the donor of the stain when proposition H_d is true) are independent. Then $Pr(E_c = A \mid E_p = A, H_d, I) = Pr(E_c = A \mid I)$. In such a case only the so-called *profile probability*, γ_A, with which an unknown person would have the profile A is needed. This is a widely accepted over-simplification. In reality, the evidential value of a match between the profile of the recovered sample and that of the POI needs to take into account the fact that there is a person (the POI) who has already been seen to have that profile (type A). So, the probability of interest is $Pr(E_r = A \mid E_c = A, H_d, I)$ and this can be different from $Pr(E_r = A \mid I)$. In fact, observing one genotype in the population increases the chance of observing another of the same type. Hence, within a population, DNA profiles with matching allele types are more common than suggested by the independence assumption, even when two individuals are not directly related. The conditional probability (also called *conditional profile probability* or *conditional match probability*) incorporates the effect of population structure and other dependencies between individuals. The more common source of dependency is a result of a membership in the same population and having similar evolutionary histories. Populations are finite in size thus two people taken at random from a population have a non-zero chance of having relatively recent common ancestors. Disregarding this correlation of alleles in the calculation of the value of the evidence results in an exaggeration of the strength of the evidence against the compared person. This aspect was presented by Balding and Nichols (1994, 1995) and mainly supported by Weir (1996), Curran et al. (2003), Buckleton and Triggs (2005) and Curran and Buckleton (2007).

1.8.3 Evaluation by Taking Errors into Account

The evaluation of scientific evidence has to consider the role of error. Error has been mentioned by many scholars in scientific and legal literature (i.e., Koehler, 1996, 1997 and reiterated Koehler, 2018), including in the PCAST report to the US President on 'Forensic science in criminal courts; ensuring scientific validity of feature-comparison methods' (President's Council of Advisors on Science and Technology PCAST, 2016), that states:

> Without appropriate estimate of accuracy, an examiner's statement that two samples are similar - or even indistinguishable - is scientifically meaningless: it has no probative value, and considerable potential for prejudicial impact. Nothing - not training, personal experience nor professional practices - can substitute for adequate empirical demonstration of accuracy. (at p. 46)

Therefore, when evaluating the strength of DNA evidence for supporting that two samples have a common source, one must consider two factors. One factor is the conditional profile probability. A coincidental match occurs when two different people share the same DNA profile. The second factor is the probability of a false positive. A false positive occurs when a laboratory erroneously reports a DNA correspondence between two samples that actually have different profiles. A false positive may occur due to error in the collection or handling of samples, misinterpretation of test results, or incorrect reporting of test results (Thompson, 1995). Either a coincidental match or a false positive could cause a forensic scientist to report a DNA match between samples from different individuals. Thus, the conditional profile probability and the false positive probability should both be considered in order to make a fair evaluation of the evidence (Koehler et al., 1995). Proficiency testing performances do not necessarily equate to the false positive probability in a particular case. This aspect represents a first practical difficulty. A second practical difficulty is the presentation of a logical framework which takes account of both probabilities. Various suggestions have been made (Robertson and Vignaux, 1995b; Balding and Donnelly, 1995; Balding, 2000). A likelihood ratio framework for considering the role that error may play in determining the value of forensic DNA evidence in a particular case is presented in Thompson et al. (2003) and Buckleton et al. (2005). Even a small false positive probability can, in some circumstances, be highly influential so serious consideration has to be given to its estimation. Recognition is needed that accurate assignments for false positive probabilities can be crucial for the assessment of the value of DNA evidence.

1.9 A Minimum Value for the Profile Probability

At the end of the 1990s, the increasing use of DNA evidence put forward new questions. One of those that is still of importance was 'What figure should be presented in court if the aim is to provide the judge and jury with the best numerical measure of the actual probability that the defendant's DNA profile match is accidental?' This is purely a quantitative question.

It was (and still is) common that experts quoted astronomical figures such as 1 in ten billion or trillion and insisted that such figures were both accurate and reliable. It can be agreed that anyone who makes such a statement in Court is either numerically naïve or is deliberately trying to deceive a (possibly numerically unsophisticated) judge and jury because such figures are well beyond the accuracy of human science and technology for several reasons. Scientific and legal literature critically discussed this point giving the practical range of values for a profile probability when a variety of specified alternatives (possible sources of the stain other than the POI), corresponding to individuals who exhibit different degrees of relatedness to the POI, presented full matching profiles. Simulations have been used to generate such values (Foreman et al., 1997; Foreman and Evett, 2001). At that time, using the most common DNA profiling system (10-locus Short Tandem Repeat, or STR, profiling system), the authors recommended general conditional match probability values for use when reporting full profile matches. They supported the use of a value of 1 in a billion (10^{-9}) when an alternative individual (unrelated and coming from the same sub-population of the real donor of the recovered stain) is considered. A general discussion of this topic was clearly presented in Evett et al. (2000a). The same minimum value was also given by Hopwood et al. (2012) using 15-plex STR profiling systems. The use of values more extreme than 10^{-9} for a profile probability or conditional match probability is widely and constantly criticized (see, for example, Kaye, 1993, Lambert et al., 1995, Saks and Koehler, 2008 and Curran, 2010). The main reason for such a criticism is expressed by Hopwood et al. (2012) who noticed that

> Such values [values lesser than 10^{-9}] invoke independence assumptions to a scale of robustness that we cannot demonstrate empirically, given the size of available databases. [...] In addition to the empirical evidence for the reliability of DNA evidence interpretation, we recognise also that such numbers are difficult to conceptualise and require unreasonable real life comparisons. (p. 188)

A related practical problem with which forensic scientists are faced is the estimate of genotype proportions. A method was outlined by Balding (1995) taking advantage of Bayesian statistical methods that was able to deal with situations where the genotype may not appear in the sample at a given locus. A development by taking account of sampling error using what is called the size-bias correction was proposed by Curran et al. (2002). Work is still in progress at the time of writing to deal with other forms of DNA evidence such as mt-DNA or Y-Chromosome (Buckleton et al., 2016a).

1.10 Propositions and Pre-Assessment

In the late 1990s increasing recognition of different aspects, or levels, of forensic examinations and reports of their outcomes, led to series of important papers that introduced a framework known as 'Case Assessment and Interpretation'.

1.10.1 The Choice of Propositions

For an inferential process to be balanced, or in the words of some authors, impartial (Jackson, 2000), attention cannot be restricted to only one side of an argument. Evett (1996) noted, for instance, that

> a scientist cannot speculate about the truth of a proposition without considering at least one alternative proposition. Indeed, an interpretation is without meaning unless the scientist clearly states the alternatives he has considered. (at p. 122)

The requirement for the consideration of alternative propositions is a general one that applies equally in many instances of daily life, but in legal contexts its role is fundamental. This criterion requires a scientist to consider at least two competing propositions. The exact phrasing of propositions is important, an importance that underlies a concept known in the context as the level of propositions or *hierarchy of propositions* (Cook et al., 1998a). The reasons for this are twofold. Firstly, a proposition's content crucially affects the degree to which that proposition is helpful for the courts. For example, the pair of propositions 'the suspect (some other person) is the source of the crime stain' (known in this context as a *source-level* proposition) addresses a potential link between an item of evidence and an individual (that is, a suspect) on a rather general level. Generally, *activity-level* (e.g., 'the suspect (some other person) attacked the victim') or *offence-level* (e.g., 'the suspect (some other person) is the offender') propositions tend to meet a court's need more closely. Secondly, the level of proposition defines the extent of circumstantial information that is needed to address a proposition meaningfully. For example, when reasoning from a source- to an activity-level proposition, phenomena such as transfer of material during the alleged or alternative actions should be taken into account. Under an offence-level pair of propositions, consideration needs to be given to the relevance of a crime stain, in other words, whether or not it has been left by the offender (Stoney, 1991a, 1994). The concept of relevance is not necessarily needed when attention is confined to a source-level or an activity-level proposition.

The scientist should insist on a well-defined framework of propositions. This is in sharp contrast to the opinion that data should be allowed to 'speak for themselves', a suggestion that evidential value represents some sort of attribute that is intrinsic to data and independent of circumstances. Such an opinion is viewed cautiously in legal reasoning, where the following position has been reached:

> In court, as elsewhere, the data cannot 'speak for itself'. It has to be interpreted in the light of the competing hypotheses put forward and against a background of knowledge and experience about the world. (Robertson and Vignaux, 1993a, p. 470)

As may be seen, the concept of levels for propositions is important because it is closely tied to the notion of evidential value. The latter is a subjective function of the former, in the sense that the value assigned to evidence by a particular individual depends on (a) the propositions amongst which the individual seeks to discriminate and (b) contextual information that is available to the individual. Evidential value is to be seen neither as an abstract property of the external world nor as one that can be elicited in a uniquely defined way (Evett et al., 2000b).

1.10.2 The Pre-Assessment

An evaluation process starts when the scientist first meets the case. It is at this stage that the scientist thinks about the questions that are to be addressed and the outcomes that may be expected. The scientist should attempt to frame propositions of interest and think about potential outcomes and the value of evidence that is expected (Evett et al., 2000c). There is a wide tendency to consider evaluation of evidence as one of the last steps of a casework examination, notably at the time of preparing the formal report. This is so even if an earlier interest in the process would enable the scientist to make better decisions about the allocation of resources. A first approach to decision-making in an operational forensic science problem has been proposed by Cook et al. (1998b). It is based on a model embodying the principle of likelihood ratio as a measure of the value of evidence. In that spirit, Cook et al. (1998b) proposed a model for enhancing the cost-effectiveness of a casework activity from initial contact with the customer. The aim is to enable the customer to make better decisions. In routine work, an assignment of the expected likelihood ratio is often requested by forensic laboratories, before, for example, the performance of any analytical tests. Such an assignment will help the scientist to support a better decision for the customer. Imagine, for the sake of illustration, a situation in paternity testing where the alleged father is unavailable but a cousin of the alleged father could be considered and tested. In such a case, the two propositions of interest may be of the form of H_p: the tested person is a cousin of the true father, and H_d: the tested person is unrelated to the child. Two questions are of interest: (1) can we obtain a value supporting the hypothesis H_p or H_d in this scenario?, and (2) how can the laboratory or the customer take a rational decision on the necessity to perform tests after an assignment of possible values of likelihood ratio? The first question refers to the pre-assessment process, the second to decision-making (Section 1.15). Answers to the first question are proposed in Cook et al. (1998b, 1999) and Evett et al. (2000c). A review of case assessment is given in Jackson et al. (2015).

1.11 Translation of a Numerical Value into a Verbal Equivalent

A verbal scale for a numerical ratio of probabilities in the context of hypothesis testing was discussed by Jeffreys (1983) (first edition in 1939). The ratio, denoted as K by Jeffreys, is that of the probability of the null hypothesis given evidence and background information to the probability of the alternative hypothesis given evidence and background information. Jeffreys takes the prior probabilities of the two hypotheses to be equal so that the posterior odds equals the likelihood ratio. The verbal summary is then phrased in the form of support provided by the evidence against the null hypothesis. Jeffreys comments that K need not be known with much accuracy. If $K > 1$ the null hypothesis is supported. Jeffreys further comments that interest is with the values of $K < 1$ when the null hypothesis may be rejected. A logarithmic scale with so-called grades and the associated verbal descriptors proposed by Jeffreys is given in Table 1.1.

TABLE 1.1

Verbal Scale of Support K for a Null Hypothesis Proposed by Jeffreys (1983) Where K is the Ratio of the Probability of the Null Hypothesis Given Evidence and Background Information to the Probability of the Alternative Hypothesis Given Evidence and Background Information

Grade	Value of K	Verbal Descriptor
0	$K > 1$	Null hypothesis *NH* supported
1	$1 > K > 10^{-1/2}$	Evidence against *NH* but not worth more than a bare mention
2	$10^{-1/2} > K > 10^{-1}$	Evidence against *NH* substantial
3	$10^{-1} > K > 10^{-3/2}$	Evidence against *NH* strong
4	$10^{-3/2} > K > 10^{-2}$	Evidence against *NH* very strong
5	$10^{-2} > K$	Evidence against *NH* decisive

TABLE 1.2

Verbal Scale of Support for a Likelihood Ratio Proposed by Nordgaard et al. (2012) and Nordgaard and Rasmusson (2012)

Scale	Interval of Likelihood Ratio LR	Degree of Support
+4	$10^6 \leq LR$	extremely strong support
+3	$6000 \leq LR < 10^6$	strong support
+2	$100 \leq LR < 6000$	support
+1	$6 \leq LR < 100$	support to some extent
0	$1/6 \leq LR < 6$	support neither … nor
−1	$1/100 \leq LR < 1/6$	support to some extent
−2	$1/6000 < LR \leq 1/100$	support
−3	$1/10^6 < LR \leq 1/6000$	strong support
−4	$LR \leq 1/10^6$	extremely strong support

Another approach is to use an ordinal scale for the likelihood ratio. In forensic science such a procedure was introduced by Evett (1991) for ease of communication. A similar scale was proposed in Nordgaard et al. (2012) and Nordgaard and Rasmusson (2012) with a response for the likelihood ratio close to 1 and then a 4-point scale for the likelihood ratio > 1 with its reciprocal for the likelihood ratio < 0 and is given in Table 1.2. The choice of a verbal scale is initially subjective but not arbitrary. However, if the scale is to have credibility and be acceptable to the courts then a particular choice has to be agreed amongst the scientists and ideally published in a peer-reviewed journal.

A 6-point verbal scale for values of the likelihood ratio greater than 1, reciprocated for values of the likelihood ratio less than 1 is given as an illustration in ENFSI (2015) with six adjectives for support of weak, moderate, moderately strong, strong, very strong and extremely strong and corresponding numerical ranges for the logarithm of the likelihood ratio of $\{0 - 1, 1 - 2, 2 - 3, 3 - 4, 4 - 6, > 6\}$.

Note that while it is permissible to interpret a numerical value verbally, it is not meaningful to interpret a verbal scale numerically. Also, there are still several aspects of the use of verbal scales to be considered. First, there is the nature of the assistance that a verbal scale might offer to the fact-finder (judge or jury). The second is whether the numerical value of a likelihood ratio is a sufficient summary of the value of the evidence. Thirdly, the limitations of the use of verbal scales need to be recognised. A discussion of the disadvantages of verbal scales is to be found in Marquis et al. (2016). At the time of writing the topic is still under discussion (Berger and Stoel, 2018).

1.12 Assessment of Performance

The development of models for the evaluation of evidence led inevitably to the development of measures for the assessment of their performance. An assessment of their performance enabled the scientist to justify their choice of model and hence their value for the evidence. Measures of performance are generic. They are applied to the general performance of the model with measures obtained from datasets where the correct answer is known. It is not possible to assess a performance in a particular case since the correct answer is not known. The quality of a model is defined as the ability of the model to support the correct result.

The choice of a model and the assessment of the performance of a model require the existence of at least one and preferably two datasets in which the correct answers are known. The first dataset is used as a training set to determine the best model for the data and for estimation of parameters, if any, in the chosen model. The second dataset is used as a validation set to assess the performance of the chosen model and parameter values. In the absence of two datasets then the training set can be used as the validation set with due care to allow for any bias in the results that may arise from this double use.

Consider source propositions with two sets of trace evidence, one set nominally from a known source and one set nominally from an unknown source. The prosecution proposition H_p is that the unknown source is the known source. The defence proposition H_d is that the unknown source is a different source to the known source. An example of such trace evidence is that of window glass evidence, with measurements of refractive index and elemental compositions from fragments within the same windows and with many windows that provide the opportunity to compare measurements from fragments from different windows. Comparisons may then be made of measurements taken from the validation dataset on two sets of different fragments from within the same window (same-source comparisons) and on two sets of fragments from between different windows (the fragments then of necessity being different and the comparisons being of different-source). For each comparison one set may be chosen in the model as the one of known source and the other may be chosen in the model as the one of unknown source (even though its source is actually known). As the source (window) of each set is known, the correct proposition in any comparison, H_p, same-source, or H_d, different-source, is known. For each comparison, a likelihood ratio using a model determined from the training set, is calculated. A value of the *LR* greater than 1 is said to support the prosecution proposition H_p. A value of the *LR* less than 1 is said to support the defence proposition H_d. However, unlike a court case where it is not known which proposition is correct, the correct answer is known. Performance of the chosen model can then be assessed with a comparison of the results (supports) with the type of comparison (same-source or different-source) which had been made.

Several measures of performance have been developed.

- *Rates of misleading evidence:* these are
  (a) for comparisons, there are two rates, first, the number of same-source comparisons with $LR < 1$ divided by the total number of same-source comparisons and, second, the number of different-source comparisons with $LR > 1$ divided by the total number of different-source comparisons; if a conclusion that the two items of evidence being compared come from the same source is considered a positive result and if a conclusion that the two items of evidence being compared come from different sources is considered a negative result then a same-source comparison with $LR < 1$ may be thought of as a *false negatives* and a different-source comparison with $LR > 1$ may be thought of as a *false positive*;
  (b) for discrimination, the number of members of the validation set that are allocated to the wrong group, divided by the total number of allocations. For discrimination between two groups, A and B say, two rates can be determined, first, the number of members of A allocated to B divided by the total number of members of A and, second, the number of members of B allocated to A divided by the total number of members of B. If there are more than two groups, various possible combinations of rates may be calculated.

  Histograms of the values of $\log(LR)$ from all possible same-source comparisons from the validation dataset and, separately, of the values of $\log(LR)$ from all possible different-source comparisons from the validation dataset can be drawn on the same axes. The quality of a method at a particular value of $\log(LR)$ is the amount of overlap of the histograms at that value. Ideally, all values of $\log(LR)$ for same-source comparisons will be greater than 0, all values of $\log(LR)$ for different-source comparisons will be less than 0 and there will be no overlap. An analogous use of histograms may be made for discrimination.

- *Tippett plots:* these are generalisations of rates of misleading evidence for comparisons (Evett and Buckleton, 1996; Tippett et al., 1968). They are the complement of empirical cumulative distribution functions for same-source and different-source comparisons. The plots come in pairs, one for same-source comparisons and one for different-source comparisons. The $\log(LR)$ is plotted on the x-axis and, for a particular value x_0 of the $\log(LR)$, the y-axis is the relative frequency of the number of comparisons greater than x_0. For same-source comparisons, it is to be hoped that all $\log(LR)$ values are greater than 0. Thus for $x < 0$, it is hoped the corresponding value on the y-axis will be 1 (or 100%). Similarly, for different-source comparisons, it is to be hoped that all $\log(LR)$ values are less than 0. Thus for $x > 0$, it is hoped the corresponding value on the y-axis will be 0 (or 0%).

  The distance from the intersection of the same-source plot with the line $\log(LR) = 0$ and the line $y = 1(100\%)$ is the rate of misleading evidence for same-source comparisons, the proportion of same-source comparisons that have a value of $\log(LR) < 0$ ($LR < 1$). The distance from the intersection of the different-source plot with the line $\log(LR) = 0$ and the line $y = 0(0\%)$ is the rate of misleading evidence for different-source comparisons, the proportion of different-source comparisons that have a value of $\log(LR) > 0$ ($LR > 1$).

- *Empirical cross-entropy:* The measure used here is known as a *score* and the definition is such that low scores are good. In a comparison of the performance of two models, the model with the lower score is deemed to be the better model. A quadratic scoring rule was used by Lindley (1991) to justify the use of probability as the only measure of uncertainty. The scoring rule used for evaluation of evidence is the logarithmic rule (Good, 1952).

  In the context of forensic science, consider the prosecution and defence propositions H_p and H_d, respectively, and in this context, assume $\Pr(H_p) = 1 - \Pr(H_d)$. For evidence evaluation, the logarithmic rule with base 2 is used for reasons associated with information theory where the common unit of information is the bit. Given a particular model, let p be the posterior probability obtained for H_p given evidence E and background information I. Then $(1 - p)$ is the posterior probability for H_d given evidence E and background information I. The logarithmic scoring rule states that

  (a) If H_p is true, score $-\log_2 p = -\log_2 \Pr(H_p \mid E, I)$;
  (b) If H_d is true, score $-\log_2(1 - p) = -\log_2 \Pr(H_d \mid E, I)$.

  If p is high and H_p is true then the score is low. If p is high and H_d is true then the score is high. For example, consider $p = 0.9$ and H_p true; the score is $-\log_2(0.9) = +0.15$. If H_d true; the score is $-\log_2(1 - 0.9) = +3.32$*

  The measure of performance for evidence evaluation is then a weighted average value of the logarithmic scoring rule, and is known as the *empirical cross-entropy* (*ECE*) empirical cross-entropy:

$$ECE = -\frac{\Pr(\theta_p \mid I)}{N_p} \sum_{i \in \text{true } \theta_p} \log_2 \Pr(\theta_p \mid E_i, I)$$

$$-\frac{\Pr(\theta_d \mid I)}{N_d} \sum_{j \in \text{true } \theta_d} \log_2 \Pr(\theta_d \mid E_j, I)$$

$$= \frac{\Pr(\theta_p \mid I)}{N_p} \sum_{i \in \text{true } \theta_p} \log_2 \left(1 + \frac{1}{LR_i \times O(\theta_p)}\right)$$

$$+ \frac{\Pr(\theta_d \mid I)}{N_d} \sum_{j \in \text{true } \theta_d} \log_2 \left(1 + LR_j \times O(\theta_p)\right),$$

where E_i and E_j denote the evidence and LR_i and LR_j denote the corresponding LR values in the training set (or validation set if one exists) with N_p members when θ_p is true and N_d members when θ_d is true and $O(\theta_p)$ denotes the prior odds $Pr(H_p \mid I) = Pr(H_d \mid I)$. (Meuwly et al., 2017) and (Ramos and Gonzalez-Rodriguez, 2013).

This measure indicates better performance when the likelihood ratio leads to the correct decision. The numerical value will be lower as the performance increases. The value of the ECE can be plotted showing its value for a certain range of priors.

---

* Note for calculation purposes, $\log_2 x = \log_{10}(x)/\log_{10}(2)$.

1.13 Role for Likelihood Ratio as a Measure for Investigation as Well as for Evaluation

The description of the role of the likelihood ratio for the evaluation of evidence in the form of continuous measurements by Lindley (1977) led to an upsurge of interest in the area amongst statisticians and forensic scientists. Another role for the likelihood ratio in forensic science was introduced in the late 1990s by Cook et al. (1998b). This role was to provide assistance to the police in the investigation of a crime. When the scientist was presenting evidence in court, their role was that of evidence evaluation and the mode of operation was said to be *evaluative*. When the scientist was assisting police in an investigation, their role was said to be *investigative* (Jackson et al., 2015).

The investigative role of the scientist is part of the procedure known as *case assessment and interpretation* (CAI), much of which is outlined in Section 1.10. The CAI procedure has clarified considerably the information that is required by investigators to aid the provision of the characteristics of balance, logic, transparency and robustness (Association of Forensic Science Providers, 2009).

The evaluative role of the scientist provides a measure of support, either numerically or verbally (see Section 1.11), which is in the form of a likelihood ratio. The numerator and denominator of the likelihood ratio are the probabilities of the evidence given the prosecution and defence propositions, respectively. The scientist offers no opinion on the probabilities of these propositions; opinion on these is the role of the fact-finder be they judge or jury.

In contrast, the investigative role of the scientist is to aid in the investigation of a crime. The aid is provided with the use of abductive reasoning to generate explanations for what is being discovered during the course of the investigation. Abduction as defined by Jackson et al. (2015) is the intellectual and imaginative process of generating possible explanations to account for an expert's actual or anticipated scientific observations. Explanations generated by abduction can then be tested against observed data. Note that the word 'explanation' is used during an investigation to describe particular hypotheses for what may have happened to produce what is discovered during the investigation. The word 'proposition' is used during the evaluative stage of the process, for example for the interpretation in court. The propositions are the hypotheses put forward by the prosecution and defence.

During the investigative phase it is permissible for a scientist to provide what may be called posterior probabilities for explanations. A likelihood ratio for evaluation is best used with two and only two mutually exclusive propositions. With more than two propositions it is necessary to include prior probabilities for the propositions to obtain an evaluation. For investigations, the scientist can use prior probabilities for explanations, even if there are only two. Only relative values for the posterior probabilities will be obtained. It is unlikely a scientist can be certain that the choice of explanations is not only mutually exclusive but also exhaustive. Overall, the purpose of the investigative use of the likelihood ratio is to reduce uncertainty about events material to the investigation and thus to help direct the investigation.

Sometimes it may be possible to assess the potential evidential value or investigative impact of a test with the use of experimental data in which the test has been used with known results. If the value or impact is high then the investigators may deem it worthwhile to conduct the test. An example is described in Jackson et al. (2015) based on a German case reported by Oesterhelweg et al. (2008). The case involved the use of a cadaver dog. Data were available from experiments with cadaver dogs from which it was possible to provide estimates of the probabilities of a positive (+) signal from the dog if a cadaver scent were or were not present (H_p or H_d) and the probabilities of a negative (−) signal from the dog if a cadaver scent were or were not present (H_p or H_d), using the appropriate proportions in the resulting table. The corresponding likelihood ratios, $Pr(+ \mid H_p)/Pr(+ \mid H_d)$ and $Pr(− \mid H_p)/Pr(− \mid H_d)$ can be calculated, for the value of a positive or negative result. The investigator can then decide whether to use the test or not, given the relative values of the likelihood ratios.

The investigative role of the likelihood ratio has also been used for the examination of questioned documents, notably when a person of interest is not available. The evidence is described by trace characteristics only. The forensic scientist can assign conditional probabilities under the assumptions of pairs of explanations. For example, these might be that the writer was male or was female, or that the writer was left-handed or was right-handed. The likelihood ratios may then be calculated for each of these pairs of explanations and the results used to inform the investigation (Taroni et al., 2012).

1.14 Probabilistic Graphical Models

Starting from 1989, a series of papers showed that intricate frameworks of circumstances - situations involving many variables - require a logical assistance and should be approached in a formal way, pointing out the utility of graphical methods that deal with an analysis of rational thinking.

1.14.1 Bayesian Networks

Methods of formal reasoning to assist forensic scientists to understand better the various dependencies which may exist among different aspects of evidence have been developed. Probabilistic graphical models, in particular Bayesian networks (or, Bayes nets for short), emerged from such research as an important approach, capable of providing a valuable aid for representing relationships among target characteristics and propositions (hypotheses) in situations of uncertainty.

They can assist the user not only in describing challenging practical problems, and communicating information about their structural properties, but also in actually computing the effect of knowing the truth or otherwise of one proposition, or one item of evidence, on the probability of other propositions - avoiding possibly difficult algebraic case-by-case calculations. In addition, Bayesian networks have the potential of clarifying the rationale behind particular probabilistic solutions.

Based on the elements of graph and probability theory, Bayesian networks can roughly be defined as a pictorial representation of the dependencies and influences (represented by arcs) among variables (represented by nodes) deemed to be relevant for a particular probabilistic inference problem.

1.14.2 Bayesian Networks to Manage 'Masses' of Evidence

The advances in formalization and computational support for rational thinking are highly valuable because they contribute to the coherent use of forensic information in the legal process. However, at their current level of development, probabilistic approaches still focus essentially on single items of evidence. Difficulties with the combination of evidence have been discussed under the concept of conjunction; see Cohen (1977, 1988) and Dawid (1987). The inability to evaluate more than a few items of evidence is currently felt as a major limitation, as already noticed some time ago by Schum (1994):

> What is clear is that no probabilist had ever given attention to the task of weighing entire masses of evidence given at trial. As Wigmore noted (1937 at p. 9) 'The logicians have furnished us in plenty with canons of reasoning for specific single inferences; but for a total mass of contentious evidence in judicial trials, they have offered no system'. (at p. 61)

Lindley (2004) reiterated this point when he wrote that a

> [...] problem that arises in courtroom, affecting both lawyers, witnesses and jurors, is that several pieces of evidence have to be put together before a reasoned judgement can be reached. [...] probability is designed to effect such combinations but the accumulation of simple rules can produce complicated procedures. Methods of handling sets of evidence have been developed: for example, Bayes nets. (at p. xxiv)

'Bayesian networks provide a solution.'

1.14.3 Bayesian Networks in Judicial Contexts

The study of representational schemes for assisting reasoning about evidence in legal settings has a remarkably long history. In this context the charting method developed by Wigmore (1913) is a frequently referenced (though essentially deterministic) predecessor of modern network approaches to inference and decision analyses that can be traced back to the beginning of the 20th century. It is only about three decades ago that researchers have begun to show an interest in graphical approaches with the valid incorporation of probability theory. Examples include decision trees and a modified, more compact version of these, called 'route diagrams' (Friedman, 1986a,b). Since the early 1990s, however, it is Bayesian networks that have advanced to a preferred formalism among researchers and practitioners engaged in the joint study of probability and evidence in judicial contexts, notably because of the efficient representational capacity of computer systems to handle multiple pieces of information. Thus, researchers in law – compared to those in other domains of applications - were among the pioneers who realized the practical potential of Bayesian networks. In judicial contexts, two different ways in which Bayesian networks are used as a modelling technique can be distinguished. Legal scholars focus on Bayesian networks as a means for structuring cases as a whole whereas forensic scientists concentrate primarily on the evaluation of selected issues that pertain to scientific evidence (Robertson and Vignaux, 1993b). Many studies with an emphasis on legal applications thus rely on Bayesian networks as a method for the retrospective analysis of complex and historically important *causes celèbres*, such as the Collins case (Edwards, 1991), the Sacco and Vanzetti case (Kadane and Schum, 1996), the Omar Raddad case (Levitt and Blackmond Laskey, 2001) or the O.J. Simpson case (Thagart, 2003).

1.14.4 Bayesian Networks in Forensic Science: Particular Case Modeling

The first study in print on Bayesian networks applied to forensic case settings was published in the late 1980s (Aitken and Gammerman, 1989). It focused on a hypothetical murder scenario where the authors showed how a network approach might be applied to cases involving several, possibly complicated, interrelated issues. They provided a detailed discussion on how (i) relevant propositions can be extracted from a scenario, (ii) relationships between propositions are represented qualitatively in terms of a directed graph, and (iii) subjective beliefs are incorporated as probabilities and used for inference. The authors noticed that

> [a] probabilistic reasoning system has been developed and implemented [...] The ideas behind the current development of the system have an obvious application in the assessment of evidence which is illustrated with an artificial example in the criminal legal field [...].

This paper (and a second one focused on specific case analysis (offender profiling) Aitken et al., 1996) have opened up new horizons in law and forensic science.

1.14.5 Bayesian Networks in Forensic Science: Generic Patterns of Inference

Instead of focusing on a particular scenario, as outlined in the previous section, it is also possible to pursue a modelling approach that aims at a standard analysis of recurrent patterns of inference concerning scientific evidence. This perspective concentrates on some more generic and fundamental issues that forensic scientists should account for if they seek to evaluate their evidence in the light of propositions that are of judicial interest. The modelling concentrates on aspects of case settings that determine the general pattern of inference (e.g., the relevance of evidence (Garbolino and Taroni, 2002)), irrespective of details about a particular situation.

Many inferential problems in the analysis of DNA profiling were solved by Dawid et al. (1999, 2002). Within the branch of DNA evidence, an extensive body of knowledge (accepted biological theory) is available and on which one can rely during network construction. For example, consideration of Mendelian laws of inheritance allows one to obtain clear indications on how nodes in a network ought to be combined. In this way, basic sub-models have been proposed and repeatedly used for logically structuring larger networks. An extension to deal with an important category that covers studies focusing on small quantities of DNA has been discussed by Evett et al. (2002). Finally, a hierarchical approach, notably where analyses lead to large network topologies (e.g., when information pertaining to different genetic markers needs to be combined), has been proposed in Dawid (2003) and Mortera et al. (2003).

Forensic applications of Bayesian networks range from offender profiling to single and complex configurations of different kinds of trace evidence. A detailed collection of models is available in (Taroni et al., 2014). Bayesian networks allow their users to engage in probabilistic analyses of much higher complexity than what would be possible through traditional approaches that mostly rely on rather rigid, purely arithmetic developments. The graphical nature of Bayesian networks facilitates the formal discussion and clarification of probabilistic arguments.

1.15 Not Only Inference: The Way to Make a Decision

Courts typically seek to reduce the uncertainty of their knowledge about a defendant's true connection with a criminal act. Often, part of this search is based on the evidence offered by forensic scientists. According to this view, inference provides contributory information to judicial decision-making (e.g., the decision as to whether a defendant should be found guilty of the offence of which they have been charged). Assessment of this contributory information reflects the intention to promote accurate decision-making. This aspect to judicial decision-making was recognised in the legal literature in a seminal paper published by Kaplan (1968). Later, Kaye (1979), Fienberg and Schervish (1986) and Kaye (1988) described the decision-making process, a process that plays a key role in everyday routine life. This process consists of the rational choice, given personal objectives, between two or more possible outcomes when the consequences of the choice are uncertain. Decision analysis helps individuals better to understand the problem they are faced with and to make clearer and more consistent decisions. The approach has also been applied in forensic science to deal with situations where decisions are required (Taroni et al., 2005; Biedermann et al., 2008). These situations include the identification process, earlier discussed by Stoney (1991b), later by Champod (2000), then further explored by Cole (2009) and supported by Champod et al. (2016).

1.15.1 The Objectives and Ingredients of Decision Theory

Given a set of beliefs about states of the world which cannot be known, the general objective is to identify an available course of action that is logically consistent with a person's personal preferences for consequences. This is an expression of a view according to which one decides on the basis of essentially two ingredients: one's beliefs about past, present or future happenings and, secondly, one's valuation of the consequences. The former ingredient will be expressed by probability and the latter ingredient will be captured by invoking an additional concept, known as *utility*. Both concepts can operate within a general theory of decision that involves the practical rule which says that one should select that decision which has the highest expected utility (or, alternatively, which minimizes expected loss). When the class of such operations is based on beliefs that have been informed with Bayesian updating (statistical inference), then this process is called Bayesian decision analysis.

A decision-based approach can help (i) to clarify the fundamental differences between the value of evidence as reported by an expert and the final decision that is to be reached by a jurist, and (ii) to provide a means to show a way ahead as of how these two distinct roles (evaluation and decision) can be conceptualized to interface neatly with each other. These are topics that are unfortunately viewed differently rather than in a unified manner as already suggested by De Finetti (1968):

> Probabilities are chiefly tools for inference (induction) and for decisions (under conditions of uncertainty) [...] The subjectivistic approach is simple and natural (it seems common-sense): every new information leads to the inference of a new distribution of probabilities from the old one (according to Bayes' theorem) and so we have at any moment a probability distribution which gives the basis (i.e., the weights for the expected utility to be maximized) for every decision including that of deferring the final decision in order to collect previously, in any specific way, useful additional information. Inference and decisions are thus (as they must be) logically independent problems, only related by the output of the first serving as input for the latter. (at p. 48)

1.15.2 Graphical Models

An extension of Bayesian Networks provides the scientists with an aid to support decision-making as illustrated in Taroni et al. (2014) and Gittelson (2013). The addition of an explicit representation of the decisions under consideration and the value (utility) of the resulting outcomes (the states that may result from a decision) leads to Bayesian decision networks, also called influence diagrams. These networks combine probabilistic reasoning with the utility theory to make decisions using the criteria of maximizing the expected utility. Therefore, an influence diagram consists of three types of nodes: (1) the chance nodes which represent random variables (as in Bayesian networks); (2) the decision nodes with the states of a decision node being the different actions that are the outcomes of the various decisions amongst which the decision-maker must choose; and (3) the utility nodes which represent the decision-maker's utility function. They are characterized by utility tables specified for every outcome.

1.16 The Existence or Otherwise of a True Value of the Evidence

Recent (late 2010s) debate has concerned the inclusion, or otherwise, of a measure of uncertainty for a statement about the value of evidence. The best measure of the value of evidence is the likelihood ratio. The determination of a value for the likelihood ratio requires choices by the analyst of (a) the model to be used and, if a parametric model, (b) the the parameters of the model, and (c) the dataset for training purposes for the model and the parameters. A Bayesian approach to evaluation will also involve the choice of a prior distribution and associated hyperparameters. The argument in favour of the provision of a measure of uncertainty is that all these choices suggest possible variability in the value ultimately calculated. This variability should then be represented in a summary of the value of the evidence. For example, such a summary might be a point estimate and a lower confidence bound (e.g., 95%) on this estimate in order to favour the defence. A series of papers in *Law, Probability and Risk* (Nordgaard, 2016; Sjerps et al., 2016; Taroni et al., 2016a,b) and in *Science and Justice* (Morrison, 2016) debate the issue.

There are inferential difficulties with the incorporation of a measure of uncertainty in the summary of the value of the evidence. Determination of the value of the evidence already incorporates the uncertainty in model choice, parameter choice and choice of training data. The addition of a another layer of uncertainty is akin to asking for a probability on a probability: a person is uncertain about the occurrence of an event and then uncertain about their uncertainty.

> It is nonsense for you to have a belief about your belief if only because to do so leads to an infinite regress of beliefs about beliefs about beliefs . . . (Lindley, 2006, p. 115)

Secondly, the example of the provision of a lower confidence bound requires justification for the strength of the confidence. Thirdly, it is also difficult to incorporate a value based on a lower confidence bound with the values obtained from other evidence. Even provision of an estimate of variability separate from a point estimate means difficulty for a fact-finder wishing to combine these two estimates. Finally, it is not possible to know the true value of the evidence in a particular case. All that can be done is determination of the best estimate of the value. This is done with the likelihood ratio, as justified by Good (1952).

Acknowledgments

Colin Aitken gratefully acknowledges the support of the Leverhulme Trust with an Emeritus Research Fellowship, grant number EM2016-027. Franco Taroni gratefully acknowledges the support of the Swiss National Science Foundation through grant No. IZSEZ0-19114.

References

Aitken, C.G.G., Connolly, T., Gammerman, A., Zhang, G., Bailey, D., Gordon, R., and Oldfield, R. Statistical modelling in specific case analysis. *Science & Justice*, **36**, 245–255, 1996.

Aitken, C.G.G. and Gammerman, A. Probabilistic reasoning in evidential assessment. *Journal of the Forensic Science Society*, **29**, 303–316, 1989.

Aitken, C.G.G. and Robertson, J. A contribution to the discussion of probabilities and human hair comparisons. *Journal of Forensic Sciences*, **32**, 684–689, 1987.

Anderson, T. and Twining, W. *Analysis of Evidence: How to do Things with Facts based on Wigmore's Science of Judicial Proof*. Northwestern University Press, Evanston, IL, 1998.

Association of Forensic Science Providers. Standards for the evaluation of evaluative forensic science expert opinion. *Science & Justice*, **49**, 161–164, 2009.

Balding, D.J. Estimating products in forensic identification using DNA profiles. *Journal of the American Statistical Association*, **90**, 839–844, 1995.

Balding, D.J. Interpreting DNA evidence: can probability theory help? In J.L. Gastwirth, editor, *Statistical Science in the Courtroom*, pages 51–70. Springer-Verlag, New York, 2000.

Balding, D.J. The DNA database search controversy. *Biometrics*, **58**, 241–244, 2002.

Balding, D.J. and Donnelly, P. Inferring identity from DNA profile evidence. *Proceedings of the National Academy of Sciences USA*, **92**, 11741–11745, 1995.

Balding, D.J. and Donnelly, P. Evaluating DNA profile evidence when the suspect is identified through a database search. *Journal of Forensic Sciences*, **41**, 603–607, 1996.

Balding, D.J. and Nichols, R.A. DNA profile match probability calculation: how to allow for population stratification, relatedness, database selection and single bands. *Forensic Science International*, **64**, 125–140, 1994.

Balding, D.J. and Nichols, R.A. A method for quantifying differentiation between populations at multiallelic loci and its implications for investigating identity and paternity. In B.S. Weir, editor, *Human Identification: The Use of DNA Markers*, pages 3–12. Kluwer Academic, 1995.

Balthazard, V. De l'identification par les empreintes digitales. *Comptes rendus des séances de l'Académie des sciences 1864*, **25**, 683–684, 1911.

Berger, C.E.H. and Stoel, R.D. Letter to the Editor – Response to 'A study of the perception of verbal expression of the strength of evidence'. *Science & Justice*, **58**, 76–77, 2018.

Bertillon, A. Instructions Signalétiques. *Imprimerie Administrative*, Melun, 1893.

Bertillon, A. La comparaison des écritures et l'identification graphique. In *Revue Scientifique, Dec. 18, 1897–Jan. 1, 1898*. Typographie Chamerot et Renouard, Paris, 1898.

Biedermann, A., Bozza, S., and Taroni, F. Decision-theoretic properties of forensic identification: underlying logic and argumentative implications. *Forensic Science International*, **177**, 120–132, 2008.

Buckleton, J.S., Taylor, D., Bright, J.-A., and Curran, J.M. Sampling effect. In J.S. Buckleton, J.-A. Bright, and D. Taylor, editors, *Forensic DNA Evidence Interpretation*, pages 181–202. CRC Press, Boca Raton, 2016a.

Buckleton, J.S., Taylor, D., Gill, P., Curran, J.M., and Bright, J-A. Complex profiles. In J.S. Buckleton, J.-A. Bright, and D. Taylor, editors, *Forensic DNA Evidence Interpretation*, pages 229–276. CRC Press, Boca Raton, FL, 2016b.

Buckleton, J.S. and Triggs, C.M. Relatedness and DNA: are we taking it seriously enough? *Forensic Science International*, **152**, 115–119, 2005.

Buckleton, J.S., Triggs, C.M., and Walsh, S.J. *Forensic DNA Evidence Interpretation*. CRC Press, Boca Raton, FL, 2005.

Champod, C. Identification/individualization. In J.A. Siegel, P.J. Saukko, and G.C. Knupfer, editors, *Encyclopaedia of Forensic Sciences*, pages 1077–1084. Academic Press, San Diego, USA, 2000.

Champod, C., Evett, I.W., and Jackson, G. Establishing the most appropriate databases for addressing source level propositions. *Science & Justice*, **44**, 153–164, 2004.

Champod, C., Lennard, C., Margot, P., and Stoilovic, M., editors. *Fingerprints and Other Ridge Skin Impressions* (2nd ed.). CRC Press, Boca Raton, FL, 2016.

Cohen, L.J. *The Probable and the Provable*. Clarendon Press, Oxford, UK, 1977.

Cohen, L.J. The difficulty about conjunction in forensic proof. *The Statistician*, **37**, 415–416, 1988.

Cole, S.A. Forensics without uniqueness, conclusions without individualization: the new epistemology of forensic identification. *Law, Probability and Risk*, **8**, 233–255, 2009.

Cook, R., Evett, I.W., Jackson, G., Jones, P.J., and Lambert, J.A. A hierarchy of propositions: deciding which level to address in casework. *Science & Justice*, **38**, 231–239, 1998a.

Cook, R., Evett, I.W., Jackson, G., Jones, P.J., and Lambert, J.A. A model for case assessment and interpretation. *Science & Justice*, **38**, 151–156, 1998b.

Cook, R., Evett, I.W., Jackson, G., Jones, P.J., and Lambert, J.A. Case pre-assessment and review in a two-way transfer case. *Science & Justice*, **39**, 103–111, 1999.

Cullison, A.D. Probability analysis of judicial fact-finding: a preliminary outline of the subjective approach. *University of Toledo Law Review*, **1**, 538–598, 1969.

Curran, J.M. Are DNA profiles as rare as we think? Or can we trust DNA statistics? *Significance*, **6**, 62–66, 2010.

Curran, J.M. and Buckleton, J.S. The appropriate use of subpopulation corrections for differences in endogamous communities. *Forensic Science International*, **168**, 106–111, 2007.

Curran, J.M., Buckleton, J.S., and Triggs, C.M. What is the magnitude of the subpopulation effect? *Forensic Science International*, **135**, 1–8, 2003.

Curran, J.M., Buckleton, J.S., Triggs, C.M., and Weir, B.S. Assessing uncertainty in DNA evidence caused by sampling effects. *Science & Justice*, **42**, 29–37, 2002.

Darboux, J.G., Appell, P.E., and Poincaré, J.H. *Le rapport de MM. Darboux, Appell et Poincaré*. L'Action Française, Paris, 1907.

Darboux, J.G., Appell, P.E., and Poincaré, J.H. Examen critique des divers systèmes ou études graphologiques auxquels a donné lieu le bordereau. In *L'affaire Dreyfus - La révision du procès de Rennes - Enquête de la chambre criminelle de la Cour de Cassation*. Ligue française des droits de l'homme et du citoyen, Paris, 1908.

Dawid, A.P. The difficulty about conjunction. *The Statistician*, **36**, 91–97, 1987.

Dawid, A.P. Comment on Stockmarr's 'Likelihood ratios for evaluating DNA evidence, when the suspect is found through a database search'. *Biometrics*, **57**, 976–978, 2001.

Dawid, A.P. An object-oriented Bayesian network for estimating mutation rates. Technical Report 226, Department of Statistical Science, University College London, 2003.

Dawid, A.P. and Mortera, J. Coherent analysis of forensic identification evidence. *Journal of the Royal Statistical Society*, Series B, **58**, 425–443, 1996.

Dawid, A.P., Mortera, J., Pascali, V.L., and van Boxel, D. Probabilistic expert systems for forensic inference from genetic markers. *Scandinavian Journal of Statistics*, **29**, 577–595, 2002.

Dawid, A.P., van Boxel, D.W., Mortera, J., and Pascali, V.L. Inference about disputed paternity from an incomplete pedigree using a probabilistic expert system. *Bulletin of the International Statistical Institute*, **58**(Book 1), 241–242, 1999.

De Finetti, B. Probability: the subjectivistic approach. In R. Klibansky, editor, *La Philosophie Contemporaine (Tome 2) – Philosophie des Sciences*, pages 45–53. La Nuova Italia Editrice, Firenze, Italy, 1968.

Edwards, W. Influence diagrams, Bayesian imperialism, and the Collins case: an appeal to reason. *Cardozo Law Review*, **13**, 1025–1074, 1991.

Eggleston, R. Evidence, *Proof and Probability* (2nd ed.). Weidenfeld and Nicolson, 1983.

ENFSI. ENFSI guideline for evaluative reporting in forensic science, 2015. http://enfsi.eu/wp-content/uploads/2016/09/m1_guideline.pdf.

Evett, I.W. The interpretation of refractive index measurements. *Forensic Science International*, **9**, 209–217, 1977.

Evett, I.W. Interpretation: a personal odyssey. In C.G.G. Aitken and D.A. Stoney, editors, *The Use of Statistics in Forensic Science*, pages 9–22. Ellis Horwood, Chichester, 1991.

Evett, I.W. Avoiding the transposed conditional. *Science & Justice*, **35**, 127–131, 1995.

Evett, I.W. Expert evidence and forensic misconceptions of the nature of exact science. *Science & Justice*, **36**, 118–122, 1996.

Evett, I.W. and Buckleton, J.S. Statistical analysis of STR data. In A. Carracedo, B. Brinkmann, and W. Bär, editors, *Advances in Forensic Haemogenetics* 6, pages 79–86. Springer Verlag, Berlin, 1996.

Evett, I.W., Foreman, L.A., Jackson, G., and Lambert, J.A. DNA profiling: A discussion of issues relating to the reporting of very small match probabilities. *The Criminal Law Review*, pages 341–355, 2000a.

Evett, I.W., Gill, P., Jackson, G., Whitaker, J., and Champod, C. Interpreting small quantities of DNA: the hierarchy of propositions and the use of Bayesian networks. *Journal of Forensic Sciences*, **47**, 520–530, 2002.

Evett, I.W., Jackson, G., and Lambert, J.A. More on the hierarchy of propositions: exploring the distinction between explanations and propositions. *Science & Justice*, **40**, 3–10, 2000b.

Evett, I.W., Jackson, G., Lambert, J.A., and McCrossan, S. The impact of the principles of evidence interpretation and the structure and content of statements. *Science & Justice*, **40**, 233–239, 2000c.

Evett, I.W. and Weir, B.S. *Interpreting DNA Evidence*. Sinauer Associates Inc., Sunderland, 1998.

Fairley, W.B. Probabilistic analysis of identification evidence. *Journal of Legal Studies*, II, 493–513, 1973.

Fienberg, S.E. and Kadane, J.B. The presentation of Bayesian statistical analyses in legal proceedings. *The Statistician*, **32**, 88–98, 1983.

Fienberg, S.E., Krislov, S.H., and Straf, M.L. Understanding and evaluating statistical evidence in litigation. *Jurimetrics Journal*, **36**, 1–32, 1996.

Fienberg, S.E. and Schervish, M.J. The relevance of Bayesian inference for the presentation of statistical evidence and for legal decision making. *Boston University Law Review*, **66**, 771–798, 1986.

Finkelstein, M.O. and Fairley, W.B. A Bayesian approach to identification evidence. *Harvard Law Review*, **83**, 489–517, 1970.

Finkelstein, M.O. and Fairley, W.B. A comment on 'Trial by mathematics'. *Harvard Law Review*, **84**, 1801–1809, 1971.

Foreman, L.A. and Evett, I.W. Statistical analyses to support forensic interpretation for a new ten-locus STR profiling system. *International Journal of Legal Medicine*, **114**, 147–155, 2001.

Foreman, L.A., Smith, A.F.M., and Evett, I.W. A Bayesian approach to validating STR multiplex databases for use in forensic casework. *International Journal of Legal Medicine*, **110**, 244–250, 1997.

Friedman, R.D. A close look at probative value. *Boston University Law Review*, **66**, 733–759, 1986a.

Friedman, R.D. A diagrammatic approach to evidence. *Boston University Law Review*, **66**, 571–622, 1986b.

Friedman, R.D. Assessing evidence. *Michigan Law Review*, **94**, 1810–1838, 1996.

Garbolino, P. and Taroni, F. Evaluation of scientific evidence using Bayesian networks. *Forensic Science International*, **125**, 149–155, 2002.

Gaudette, B.D. and Keeping, E.S. An attempt at determining probabilities in human scalp hair. *Journal of Forensic Sciences*, **19**, 599–605, 1974.

Gittelson, S. *Evolving from inferences to decisions in forensic science*. PhD thesis, The University of Lausanne, School of Criminal Justice, Lausanne, 2013.

Good, I.J. Rational decisions. *Journal of the Royal Statistical Society, Series B*, **14**, 107–114, 1952.

Good, I.J. *Probability and the Weighing of Evidence*. Griffin, London, 1950.

Holmes, O.W. Path of the law. *Harvard Law Review*, **10**, 457–478, 1897.

Hopwood, A.J., Puch-Solis, R., Tucker, V.C., Curran, J.M., Skerrett, J., Pope, S., and Tully, G. Consideration of the probative value of single donor 15-plex STR profiles in UK populations and its presentation in UK courts. *Science & Justice*, **52**, 185–190, 2012.

Jackson, G. The scientist and the scales of justice. *Science & Justice*, **40**, 81–85, 2000.

Jackson, G., Aitken, C.G.G., and Roberts, P. *Case Assessment and Interpretation of Expert Evidence*. Royal Statistical Society, 2015. www.rss.org.uk/statsandlaw; www.maths.ed.ac.uk/~cgga.

Jeffreys, H. *Theory of Probability* (3rd ed.). Clarendon Press, Oxford, 1983.

Kadane, J.B. and Schum, D.A. *A Probabilistic Analysis of the Sacco and Vanzetti Evidence*. John Wiley & Sons, New York, 1996.

Kaplan, J. Decision theory and the factfinding process. *Stanford Law Review*, **20**, 1065–1092, 1968.

Kaye, D.H. Probability theory meets res ipsa loquitur. *Michigan Law Review*, **77**, 1456–1484, 1979.

Kaye, D.H. Quantifying probative value. *Boston University Law Review*, **66**, 761–766, 1986.

Kaye, D.H. What is Bayesianism? In P. Tillers and E.D. Green, editors, *Probability and Inference in the Law of Evidence, The Uses and Limits of Bayesianism* (Boston Studies in the Philosophy of Science), pages 1–19. Springer, Dordrecht, 1988.

Kaye, D.H. DNA evidence: probability, population genetics, and the courts. *Harvard Journal of Law & Technology*, **7**, 101–172, 1993.

Kaye, D.H. Logical relevance: problems with the reference population and DNA mixtures in people v. pizarro. *Law, Probability and Risk*, **3**, 211–220, 2004.

Kaye, D.H. Rounding up the usual suspects: a logical and legal analysis of DNA trawling cases. *North Carolina Law Review*, **87**, 425–503, 2009.

Kingston, C.R. Application of probability theory in criminalistics. *Journal of the American Statistical Association*, **60**, 70–80, 1965a.

Kingston, C.R. Application of probability theory in criminalistics – II. *Journal of the American Statistical Association*, **60**, 1028–1034, 1965b.

Kingston, C.R. and Kirk, P.L. The use of statistics in criminalistics. *The Journal of Criminal Law, Criminology and Police Science*, **55**, 514–521, 1964.

Kirk, P.L. and Kingston, C.R. Evidence evaluation and problems in general criminalistics. *Journal of Forensic Sciences*, **9**, 434–444, 1964.

Koehler, J.J. Error and exaggeration in the presentation of DNA evidence at trial. *Jurimetrics*, **34**, 21–39, 1993.

Koehler, J.J. On conveying the probative value of DNA evidence: frequencies, likelihood ratios, and error rates. *University of Colorado Law Journal*, **67**, 859–886, 1996.

Koehler, J.J. Why DNA likelihood ratios should account for error (even when a National Research Council report says they should not). *Jurimetrics Journal*, **37**, 425–437, 1997.

Koehler, J.J. Forensics or Fauxrensics? ascertaining accuracy in the forensic sciences. *Arizona State Law Journal*, **49**, 1369–1416, 2018.

Koehler, J.J., Chia, A., and Lindsey, S. The random match probability in DNA evidence: irrelevant and prejudicial? *Jurimetrics Journal*, **35**, 201–219, 1995.

Lambert, J.A., Scranage, J.K. and Evett, I.W. Large scale database experiments to assess the significance of matching DNA profiles. *International Journal of Legal Medicine*, **108**, 8–13, 1995.

Lempert, R.O. Modelling relevance. *Michigan Law Review*, **75**, 1021–1057, 1977.

Lempert, R.O. The new evidence scholarship: analyzing the process of proof. *Boston University Law Review*, **66**, 439–477, 1986.

Lempert, R.O. Some caveats concerning DNA as criminal identification evidence: with thanks to the Reverend Bayes. *Cardozo Law Review*, **13**, 303–341, 1991.

Levitt, T.S. and Blackmond Laskey, K. Computational inference for evidential reasoning in support of judicial proof. *Cardozo Law Review*, **22**, 1691–1731, 2001.

Lindley, D.V. A problem in forensic science. *Biometrika*, **64**, 207–213, 1977.

Lindley, D.V. *Making Decisions* (2nd ed.). John Wiley & Sons, Chichester, 1985.

Lindley, D.V. The probability approach to the treatment of uncertainty in artificial intelligence and expert systems. *Statistical Science*, **2**, 17–24, 1987.

Lindley, D.V. Probability. In C.G.G. Aitken and D.A. Stoney, editors, *The Use of Statistics in Forensic Science*, pages 27–50. Ellis Horwood, Chichester, 1991.

Lindley, D.V. Foreword. In Aitken, C.G.G. and Taroni, F. *Statistics and the Evaluation of Evidence for Forensic Scientists*. John Wiley & Sons, New York, 2nd edition, 2004.

Lindley, D.V. *Understanding Uncertainty*. John Wiley & Sons, Hoboken, 2006.

Locard, E. *L'enquête criminelle et les méthodes scientifiques*. Flammarion, Paris, 1920.

Locard, E. *L'enquête criminelle. Traité de criminalistique*. Desvigne, Lyon, 1940. Tome septième, Livre VIII.

Marquis, R., Biedermann, A., Cadola, L., Champod, C., Gueissaz, L., Massonnet, G., Mazzella, W.D., Taroni, F., and Hicks, T. Discussion on how to implement a verbal scale in a forensic laboratory: Benefits, pitfalls and suggestions to avoid misunderstandings. *Science & Justice*, **56**, 364–370, 2016.

Meier, P. and Zabell, S. Benjamin Peirce and the Howland will. *Journal of the American Statistical Association*, **75**, 497–506, 1980.

Meuwly, D., Ramos, D., and Haraksim, R. A guideline for the validation of likelihood ratio methods used for forensic evidence evaluation. *Forensic Science International*, **276**, 142–153, 2017.

Morrison, G.S. Special issue on measuring and reporting the precision of forensic likelihood ratios: introduction to the debate. *Science and Justice*, **56**, 371–373, 2016.

Mortera, J., Dawid, A.P., and Lauritzen, S.L. Probabilistic expert systems for DNA mixture profiling. *Theoretical Population Biology*, **63**, 191–205, 2003.

National Research Council. *NRC II – The Evaluation of Forensic DNA Evidence*. National Academy Press, Washington, DC, 1996.

National Research Council. *Strengthening Forensic Science in the United States: A Path Forward*. National Academy Press, Washington, DC, 2009.

Nordgaard, A. Comment on 'Dismissal of the illusion of uncertainty in the assessment of a likelihood ratio'. *Law, Probability and Risk*, **15**, 17–22, 2016.

Nordgaard, A., Ansell, R., Drotz, W., and Jaeger, L. Scale of conclusions for the value of evidence. *Law, Probability and Risk*, **11**, 1–24, 2012.

Nordgaard, A. and Rasmusson, B. The likelihood ratio as value of evidence – more than a question of numbers. *Law, Probability and Risk*, **11**, 303–315, 2012.

Oesterhelweg, L., Kröber, S., Rottman, K., Willhöft, J., Bfraun, C., Thies, N., Püschel, K., Silkenath, J., and Gehl, A. Cadaver dogs – a study on detection of contaminated carpet squares. *Forensic Science International*, **174**, 35–39, 2008.

Peirce, C.S. The probability of induction. In J.R. Newman, editor, *The World of Mathematics, 1956*, volume 2, Simon Schuster, New York, 1878.

President's Council of Advisors on Science and Technology (PCAST). *Forensic Science in Criminal Courts: Ensuring Scientific Validity of Feature-Comparison Methods*. Washington, DC, 2016. https://www.nitrd.gov/pubs/PCAST-NITRD-report-2015.pdf.

Ramos, D. and Gonzalez-Rodriguez, J. Reliable support: measuring calibration of likelihood ratios. *Forensic Science International*, **230**, 156–169, 2013.

Redmayne, M. Science, evidence and logic. *The Modern Law Review*, **59**, 747–760, 1996.

Robertson, B.W. and Vignaux, G.A. Extending the conversation about Bayes. *Cardozo Law Review*, **13**, 629–645, 1991.

Robertson, B.W. and Vignaux, G.A. Probability – the logic of the law. *Oxford Journal of Legal Studies*, **13**, 457–478, 1993a.

Robertson, B.W. and Vignaux, G.A. Taking fact analysis seriously. *Michigan Law Review*, **91**, 1442–1464, 1993b.

Robertson, B.W. and Vignaux, G.A. DNA evidence: wrong answers or wrong questions? In B.S. Weir, editor, *Human Identification: The Use of DNA Markers*, pages 145–152. Kluwer Academic, Dordrecht, 1995a.

Robertson, B.W. and Vignaux, G.A. *Interpreting Evidence. Evaluating Forensic Science in the Courtroom*. John Wiley & Sons, Chichester, 1995b.

Robertson, B.W., Vignaux, G.A., and Berger, C.E.H. *Interpreting Evidence. Evaluating Forensic Science in the Courtroom* (2nd ed.). John Wiley & Sons, Chichester, 2016.

Saks, M.J. and Koehler, J.J. The coming paradigm shift in forensic identification science. *Science*, **309**, 892–895, 2005.

Saks, M.J. and Koehler, J.J. The individualization fallacy in forensic science evidence. *Vanderbilt Law Review*, **61**, 199–219, 2008.

Schum, D.A. *Evidential Foundations of Probabilistic Reasoning*. John Wiley & Sons, Inc., New York, 1994.

Sjerps, M.J., Alberink, I., Bolck, A., Stoel, R.D., Vergeer, P., and van Zanten, J.H. Uncertainty and LR: to integrate or not to integrate: that's the question. *Law, Probability and Risk*, **15**, 23–29, 2016.

Souder, W. The merits of scientific evidence. *Journal of the American Institute of Criminal Law and Criminology*, **25**, 683–684, 1934–1935.

Stoney, D.A. Evaluation of associative evidence: choosing the relevant question. *Journal of the Forensic Science Society*, **24**, 473–482, 1984.

Stoney, D.A. Transfer evidence. In C.G.G. Aitken and D.A. Stoney, editors, *The Use of Statistics in Forensic Science*, pages 107–138. Ellis Horwood, New York, 1991a.

Stoney, D.A. What made us ever think we could individualize using statistics? *Journal of the Forensic Science Society*, **31**, 197–199, 1991b.

Stoney, D.A. Relaxation of the assumption of relevance and an application to one-trace and two-trace problems. *Journal of the Forensic Science Society*, **34**, 17–21, 1994.

Taroni, F., Biedermann, A., Bozza, S., Garbolino, P., and Aitken, C.G.G. *Bayesian Networks for Probabilistic Inference and Decision Analysis in Forensic Science*. Statistics in Practice. John Wiley & Sons, Chichester, 2nd edition, 2014.

Taroni, F., Bozza, S., and Aitken, C.G.G. Decision analysis in forensic science. *Journal of Forensic Sciences*, **50**, 894–905, 2005.

Taroni, F., Bozza, S., Biedermann, A., and Aitken, C.G.G. Dismissal of the illusion of uncertainty in the assessment of a likelihood ratio. *Law, Probability and Risk*, **15**, 1–16, 2016a.

Taroni, F., Bozza, S., Biedermann, A., and Aitken, C.G.G. Rejoinder. *Law, Probability and Risk*, **15**, 31–34, 2016b.

Taroni, F., Champod, C., and Margot, P. Forerunners of Bayesianism in early forensic science. *Jurimetrics Journal*, **38**, 183–200, 1998.

Taroni, F., Marquis, R., Schmittbuhl, M., Biedermann, A., Thiéry, A., and Bozza, S. The use of the likelihood ratio for evaluative and investigative purposes in comparative handwriting examinations. *Forensic Science International*, **214**, 189–194, 2012.

Thagart, P. Why wasn't O.J. convicted? Emotional coherence and legal inference. *Cognition and Emotion*, **17**, 361–383, 2003.

Thompson, W.C. Subjective interpretation, laboratory error and the value of DNA evidence: three case studies. *Genetica*, **96**, 153–168, 1995.

Thompson, W.C. and Schumann, E.L. Interpretation of statistical evidence in criminal trials: the prosecutor's fallacy and the defense attorney's fallacy. *Law and Human Behaviour*, **11**, 167–187, 1987.

Thompson, W.C., Taroni, F., and Aitken, C.G.G. How the probability of a false positive affects the value of DNA evidence. *Journal of Forensic Sciences*, **48**, 47–54, 2003.

Tippett, C.F., Emerson, V.J., Fereday, M.J., Lawton, F., and Lampert, S.M. The evidential value of the comparison of paint flakes from sources other than vehicles. *Journal of the Forensic Science Society*, **8**, 61–65, 1968.

Tribe, L. Trial by mathematics: precision and ritual in the legal process. *Harvard Law Review*, **84**, 1329–1393, 1971.

Twining, W. *Rethinking Evidence*. Northwestern University Press, Evanston, Ill. USA, 1994.

Weir, B.S. *Genetic Data Analysis II*. Sinauer Associates Inc., Sunderland, MA, 1996.

Wigmore, J.H. The problem of proof. *Illinois Law Review*, **8**, 77–103, 1913.

Yellin, J. Book review of *Evidence, Proof and Probability*, 1st edition, Eggleston, R. (1978). Weidenfeld and Nicolson. *Journal of Economic Literature*, **583**, 583–584, 1979.

Section II

General Concepts and Methods

2

Frequentist Methods for Statistical Inference

David H. Kaye

CONTENTS

2.1 Introduction 39
2.2 Definitions and Notation 40
    2.2.1 Data and Evidence 40
2.3 Random Variables and Probability Distributions 40
    2.3.1 Sampling from a Distribution or Population 42
2.4 Estimation 44
    2.4.1 Properties of Point Estimators 44
    2.4.2 Estimating Allele Proportions 44
    2.4.3 Estimating a False Positive Probability Through an Experiment 46
    2.4.4 Interpreting Confidence Intervals 49
2.5 p-Values 50
    2.5.1 p-Values in a Comparison of Glass Fragments 50
    2.5.2 Interpreting p-Values 52
2.6 Hypothesis Tests 53
    2.6.1 Classical Hypothesis Tests for Refractive Index Matching 54
    2.6.2 Hypothesis Testing with p-Values 59
    2.6.3 Hypothesis Testing with Confidence Intervals 59
2.7 Issues in Interpreting the Results of Hypothesis Tests, p-Values, and Confidence Coefficients 60
    2.7.1 Transposition 60
    2.7.2 Multiple Tests: Proof of the Null Hypothesis and Adjusted p-Values 61
    2.7.3 Arbitrary Lines 63
    2.7.4 Alternatives and Likelihoods 64
2.8 Resampling Methods 65
    2.8.1 Bootstrap Estimates 66
    2.8.2 Permutation Tests 68
Acknowledgments 70
References 70

2.1 Introduction

Statistics textbooks promise that methods for statistical inference can assist in "making valid generalizations from samples" (Freedman et al., 1998, p. xvi); in "draw[ing] conclusions about a population or process based on sample data" (Moore and McCabe, 1993, p. 427); or in answering the question, "[g]iven the outcomes, what can we say about the process that generated the data?" (Wasserman, 2004, p. ix). This chapter outlines the logic of "classical" or "frequentist" methods for such inference. These methods do not purport to remove all threats to the validity of inferences, but they do provide a clear way to think about and respond to statistical error—the ubiquitous variability in outcomes that results from random influences.

This chapter describes three commonly used concepts for assessing statistical error—confidence intervals, p-values, and hypothesis tests. Our exposition emphasizes the basics of the reasoning behind these devices rather than the procedures for computing quantities of interest. We also outline the logic underlying resampling methods. We identify common misinterpretations of computed quantities, and we discuss some of the comparative advantages and disadvantages of using confidence intervals, p-values, classical hypothesis tests, and likelihood ratios for various purposes in forensic science. Along with idealized, simple examples of probabilistic processes, we use two principal examples from forensic science to illustrate the classical methods in this context. The first involves an experiment to ascertain the validity and false positive probability of identifications made by latent fingerprint examiners. The second involves measurements of the refractive index of glass fragments.

Before examining particular statistical assessments in the forensic-science field, we give some basic definitions in Section 2.2. Readers familiar with statistical inference can skip over this section or refer back to it if the terminology or notation in later sections is not immediately clear.

2.2 Definitions and Notation

2.2.1 Data and Evidence

Forensic scientists or examiners collect and interpret *data*—measurements (including observations) of properties of items of interest, such as fingerprints, hairs, glass fragments, bodily fluids, soil, pollen, toolmarks, suspected drugs, breath samples from drivers, and so on. The word *evidence* has several possible meanings. To a forensic examiner, "evidence" may refer to the items themselves, as in the phrase "trace evidence"; to a lawyer, it may refer to testimony or reports that are part of a judicial record giving the data and the interpretations of the items the examiner has studied; to a statistician, it may refer to the data. Usually, the meaning should be clear from the context.

2.3 Random Variables and Probability Distributions

Because the measured properties tend to vary from one evidentiary item to another, we use the word *variable* to refer to numerical values for the properties. A *random variable*, denoted by some capital letter, is a variable whose possible values occur according to some probability mechanism. For example, imagine a coin with the number 0 on one side and 1 on the other. The number that comes up on each flip of the coin is a random variable W that can take the values $w = 0$ and $w = 1$. The fancy way to say this is that W is a Bernouilli random variable.

If the coin is fair (evenly balanced), the probability of each possible value w (0 and 1) is $f(w) = {}^1/_2$, and the probability of all other values is $f(w) = 0$. This assignment of probabilities to values of W is a *probability distribution*. The distribution gives probabilities for possible values or ranges of values of the random variable. It is a *discrete* distribution because W has only a finite number of possible values (indeed, only two, making it a binary variable). When a random variable pertains to a *continuous* quantity, say the dimensions of bones in human remains,* the probability distribution is a *probability density function.*† The area under the graph of the function for an interval (say, 15–20 cm for the length of a bone) is the probability that the variable will have a value within this interval.

Of course, nature does not demand that the coin be fair. It could have been manufactured so it is more likely for one side to turn up than the other. Let the probability of a 1 be denoted by θ. Then the probability of 0 is $1 - \theta$. Because θ could have many different values (depending on how the coin was made or how it has been worn down), there is a family of distributions that depend on θ. If we were to graph any particular one of these distributions with w along the horizontal axis and the probability of w on the vertical axis, the graph $f(w)$ would have height $1 - \theta$ above 0, θ above 1, and 0 everywhere else, as shown in Figure 2.1.

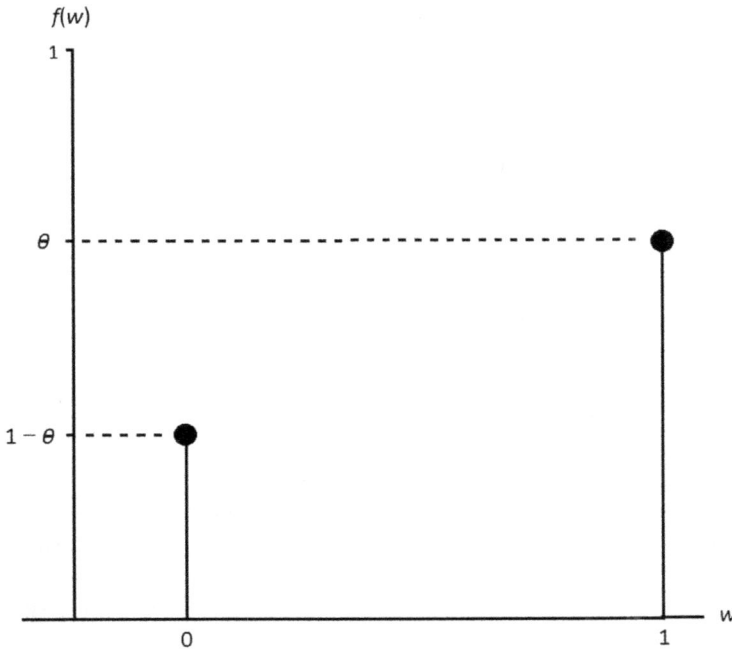

FIGURE 2.1
A Bernoulli random variable takes on the values 1 with probability θ and 0 with probability $1 - \theta$.

---

* For a statistical analysis of this variable intended to resolve the most famous mystery in aviation history, see Jantz, 2018 (discussed in Kaye, 2018).
† Some authors reserve the term "distribution" for a cumulative probability up to a given value. This cumulative distribution is a non-decreasing sum for a discrete variable and a non-decreasing area for a continuous one (as the variable goes from its lowest to its highest value). It indicates percentiles for the values.

The number θ is called a *parameter* for the distributions, and the set of all its possible values is the *parameter space*. The distributions for the variables in our later forensic-science examples will have different shapes than the very simple Bernoulli distribution, but they too will have parameters (which will be written as Greek letters). To emphasize that the Bernouilli distribution depends on the parameter θ, we can write it as $f(w;\theta)$ or $f_\theta(w)$. Equations for probability distributions like $f(w;\theta)$ are also known as *probability models*.

The *expected value* of a random variable is a weighted average of all the possible values, where the weights are the probabilities. Denoting the possible values of the discrete random variable W with the index j, the expected value of W is

$$E(W) = \sum_{j=0}^{1} w_j f(w_j;\theta) = 0(1-\theta) + 1(\theta) = \theta.$$

When plotted on the horizontal axis in the graph of $f(w;\theta)$, θ would lie between 0 and 1. It is a measure of the center of the distribution. This particular measure of central tendency is the mean of the probability distribution, often denoted as μ.

Another expected value is the *variance*. The variance is a measure of how spread out the distribution is. It is defined as the probability-weighted squares of the distances between every possible value of the random variable and its mean. For the Bernoulli distribution, the variance is

$$Var(w) = E[(w-\theta)^2] = (0-\theta)^2 f(0) + (1-\theta)^2 f(1) = \theta(1-\theta).$$

The *standard deviation* is the square root of the variance. Denoting the standard deviation by σ, we can write $Var(w) = \sigma^2$.

2.3.1 Sampling from a Distribution or Population

An observation from a Bernoulli distribution can be thought of as a random draw from an infinite *population* of 1s and 0s in which the ratio of 1s to 0s is θ to $1-\theta$. We can imagine flipping the coin ad infinitum and recording the outcomes in a list. After some number n of tosses, we will have a *random sample* of size n drawn from the Bernoulli distribution. The sample consists of n values of W_i (denoted $w_1, w_2, \ldots, w_n$), where the subscript is the number of the toss. For example, a sample of size $n = 8$ might consist of the list 0,0,0,0,1,0,1,1. As $n \to \infty$, the proportion p of 1s in the sample approaches θ. So we can refer to θ as either the parameter of a probability distribution or the parameter of a population.

Statistical inference takes us from the data to the population. That is, we want to gain insight into the true value of an unknown population parameter θ from the sample values of the random variable. Sample statistics are usually used in this process. A *statistic* is a function of the data. For instance, adding all the w_i together and dividing by n gives the proportion of 1s in the sample. We can designate this quantity $p = x/n$, where x is the number of 1s in the sample of size n. This sample proportion and the count of the 1s are also random variables (denoted P and X, respectively).

The key idea in frequentist inference is to imagine drawing not one, but infinitely many samples of size n. A *statistical model* is a mathematical statement of how the data come from the population. Here, the model is that P is the *binomial* random variable* X divided by n:

$$f(p; \theta, n) = \binom{x}{n} \theta^x (1 - \theta)^{n-x}, \quad \text{for } p = 0, 1/n, 2/n, \ldots, 1.$$

Figure 2.2 shows the distribution of the proportion P for a sample of size $n = 8$ drawn from a Bernoulli distribution with parameter $\theta = {}^1/_2$.

The distribution in this figure is the *sampling distribution* of the sample proportion. If we were to draw more and more samples of size 8 from the Bernoulli distribution, we could plot the resulting sample proportions in a histogram. As the number of samples grows, the histogram would eventually converge to the distribution in Figure 2.2. This mathematical result is one reason for using the sample proportion as an estimator of the population proportion θ. The next section discusses estimation of some important population proportions in forensic science.

Because any given sample is unlikely to be a perfect microcosm of the whole population, the sample statistic $p = x/n$ could differ from the population value θ. That difference is *sampling error*. The tools discussed below enable us to account for this sampling error—at

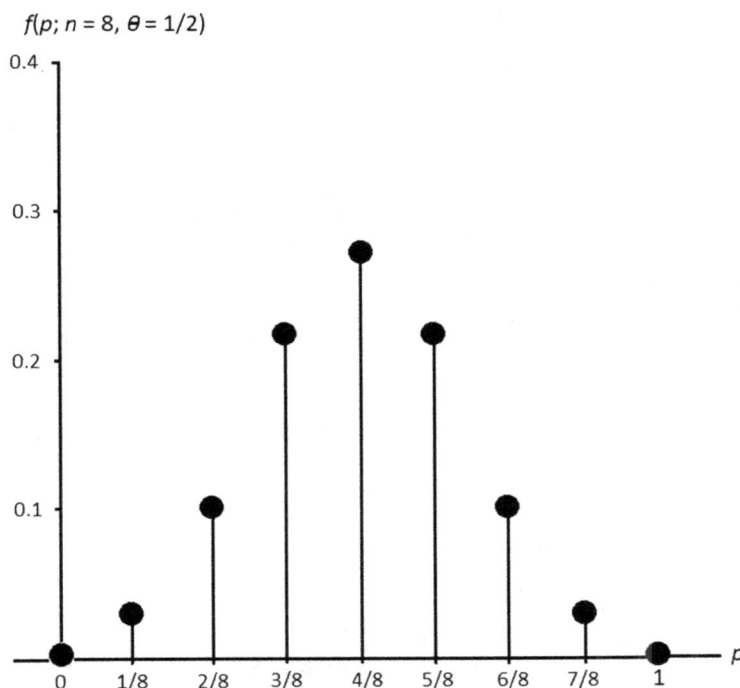

$f(p; n = 8, \theta = 1/2)$

FIGURE 2.2
The distribution of the binomial proportion P for samples of size $n = 8$ when the binomial probability $\theta = {}^1/_2$. The heights are the probability that P will attain the discrete values on the p-axis.

---

* This sum X of the n Bernoulli variables is $X = W_1 + W_2 + \ldots W_n$. It can take on values from 0 to n, and the probability for these values follows the binomial distribution $f(x; n, \theta) = \binom{x}{n} \theta^x (1 - \theta)^{n-x}$. The mean $\mu = n\theta$ and variance $\sigma^2 = n\theta(1 - \theta)$ determine the location and shape of the distribution of X.

least to the extent that the general form of our probabilistic model of the data-generating process tracks the real-world phenomenon that we have modeled.

To summarize the main points, a *statistic* is a function of the data. The data, and the statistic from the data, come from a *sample*. The sample comes from a *population* with numerical characteristics called *parameters* that correspond to the parameters of the probability distribution of a *random variable*. A *statistical model* uses random variables to describe the probabilistic process that gives rise to the sample statistics. Statistical inference uses a sample statistic to shed light on the unknown value of a parameter in the statistical model. As we shall see, frequentist methods are based strictly on their long-run properties—how they would behave for repeated sample data. They are a way to recognize, and, hopefully, to avoid being tricked or misled by the random fluctuations in statistics that would occur in repeated studies.

2.4 Estimation

2.4.1 Properties of Point Estimators

The objective of frequentist estimation is to provide an estimate of the value of the parameter based solely on the sample data. The first step is to choose an estimator. An *estimator t* is a statistic that can be derived from each possible sample. Its value for the sample at hand is a *point estimate*.

Estimators may be selected for their long-run properties. For example, an *unbiased estimator t* of θ will give estimates whose errors eventually should average out to zero. *Error* is simply the difference between the estimate and the true value. For an unbiased estimator, the expected value of the errors is $E(t - \theta) = 0$. A *maximum likelihood* estimator selects the parameter value that maximizes the probability of the data given the proposed parameter value. Estimators also may be chosen to minimize expected losses associated with errors. For example, if losses vary as the square of the error, one might seek a "minimum variance estimator" to keep the mean squared error $E[(t - \theta)^2]$ small. Still other criteria exist. It is not always possible to satisfy all of them with a single estimator. For the simple examples in this section, though, we can use an estimator that has all these properties.

2.4.2 Estimating Allele Proportions

2.4.2.1 A Point Estimate

Forensic DNA evidence requires estimation. Since its earliest days, forensic geneticists reported a "random match probability" to explain how improbable it would be to find a "match" between to forensic DNA profiles from two unrelated individuals (Kaye, 2010). Because every profile is extremely rare, it is not feasible to estimate their relative frequencies by counting how often each profile occurs in a sample from the general population. But it is feasible to estimate the population proportion θ_j for each more common allele A_j that is part of the full profile. The full profile probability is then a function of the θ_j. The details of the function need not concern us here. Our immediate problem is merely to estimate the population parameters θ_j.

Population geneticists do so by computing the proportions p_j in DNA collected in modestly sized samples from blood banks or other convenient sources. To the extent that the people whose DNA is represented in these samples are selected independently of their forensic DNA profiles, the samples can be treated as if they are random. Suppose we have only a single sample of size $n = 100$ and that one particular allele, A_1, is seen in $X_1 = 10$ of them. Then the sample proportion is $p_1 = 10/100$. Since we are focusing on only one allele, we can drop the subscripts.

As an estimator of the proportion θ in the population from which a simple random sample has been drawn, the sample proportion p is an unbiased, minimum variance, maximum likelihood estimator, so we will use it for a point estimator. The value $p = 1/10$ is our point estimate for the population proportion θ. But how good is this estimate? Interval estimates can be constructed to help assess the possible random sampling error. The standard error is the standard deviation of the sampling distribution. Because the allele proportion $p = X/n$, and X is binomial with mean $n\theta$ and variance $n\theta(1 - \theta)$, the standard error for p is $\sigma_p = \sqrt{\theta(1 - \theta)/n}$. Unfortunately, we do not know the value of θ, but if we plug in our point estimate $p = x/n$, we can estimate it as $s_p = \sqrt{p(1 - p)/n} = 0.03$. This small number for the standard error indicates that in repeated experiments, sampling error usually would not cause p to stray dramatically from the sample value, giving us some "confidence" in the single sample value of p (namely, 0.10) that came from the only sample we have.

2.4.2.2 Constructing a Confidence Interval

Confidence intervals make this notion of "confidence" more precise, and they are one way to define how far from the estimate the population parameter plausibly might be. For large n, p is approximately normally distributed with the mean θ and standard error $\sigma_p = \sqrt{\theta(1 - \theta)/n}$. Because 95% of the area under the normal curve lies within ± 1.96 standard deviations of its mean, the sample proportion in repeated sampling will fall within $\theta \pm 1.96\sigma_p$ in about 95% of the samples. Again, we do not know the value of θ—we are trying to estimate it—but we can consider what would happen if we were to collect an infinite series of samples and form an interval around the proportion p from each sample according to the same recipe. Suppose we use the following recipe: For each p, the interval will be $p \pm 1.96s_p = p \pm 1.96\sqrt{p(1 - p)/n}$. These intervals will vary from sample to sample, but in the very long run, 95% will cover the unknown, fixed value θ. The interval for the allele proportion $p = 0.10$, by this recipe for the population proportion, is about 0.10 ± 0.06, which can also be written as the interval [0.04, 0.16].

The normal approximation for a binomial proportion is a standard, textbook procedure, but it does not always work well (and it is not recommended for small samples). Better formulas for confidence intervals for binomial proportions are available (see Brown et al., 2001; Zhou et al., 2008). For our example, the improved methods give a slightly wider interval.

2.4.2.3 Choosing a Confidence Coefficient

A 95% confidence coefficient is common in the scientific literature, but other coefficients can be used. Table 2.1 gives the intervals (using the Wilson (1927) method) for three different confidence coefficients.

Figure 2.3 shows that the width of the interval increases as we demand more confidence. Wider intervals are more accurate—they are more likely to cover the true value of the parameter—but they are less precise. By being sufficiently vague, one can be highly accurate. One should decide what degree of confidence and what degree of precision are

TABLE 2.1

Wilson Confidence Intervals for Different Choices of
Confidence γ for Estimating the Allele Proportion When
$p = 1/10$ and $n = 100$

Confidence Coefficient γ	CI Lower Limit	CI Upper Limit
0.90	0.0607	0.1604
0.95	0.0552	0.1744
0.99	0.0460	0.2038

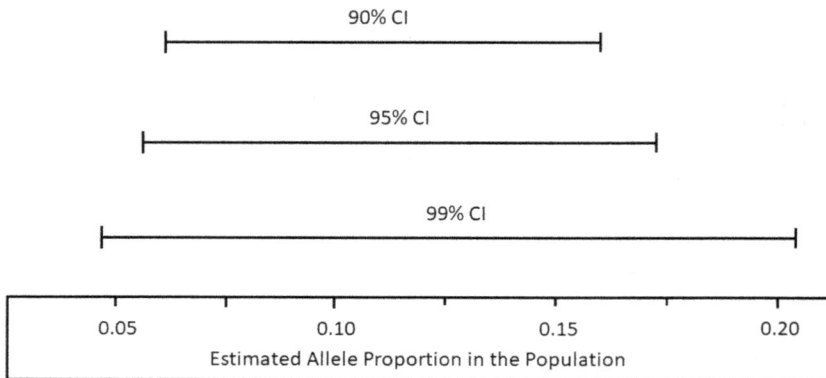

FIGURE 2.3
Confidence intervals from Table 2.1. The interval estimate becomes less precise (wider intervals) as we demand more confidence.

appropriate for an application. Ideally, the interval will have a high confidence coefficient *and* be narrow enough to pin down the estimate to a useful range. The ranges here are not negligible, but they all convey a sense of the uncertainty in the point estimate. When all the intervals are very wide, we can see that the data are not adequate for making precise statements.

In this case, the relevant uncertainty for the full profile may be less than the intervals for the separate allele proportions suggest. That is because the function that estimates the probability for the full profile frequency involves a product of the estimated allele proportions. The estimates for some allele proportions will be too large; others will be too small. Across many loci, these errors will tend to counteract one another. A method for finding confidence limits for the full profile frequency can be found in NRC Committee (1996, pp. 146–147).

2.4.3 Estimating a False Positive Probability Through an Experiment

2.4.3.1 The Design of Experiments to Test Categorical Source Attributions

For a second illustration of the estimation of a parameter in forensic science, we turn to a report of the President's Council of Advisors on Science and Technology (PCAST, 2016). The report recommended more experiments to examine the validity of largely subjective comparisons of features of "known" and "questioned" items (such as a fingerprint taken from a suspect under known conditions and a latent print found at a crime scene). By presenting examiners with pairs of prints resembling those that occur in actual cases, blinding

TABLE 2.2

Examiners" Conclusions (− or +) by Source Condition S (0 or 1). The Table Indicates That the Examiners Made b False-Positive Associations, d True Positives, a True Negatives, and c False Negatives

Examiner's Finding	True State of Affairs	
	$S = 0$	$S = 1$
−	a	c
+	b	d

both them and the person presenting the pairs to the examiners to the origins of each pair, and having the examiners reach conclusions as to whether the prints come from the same finger or from fingers from different individuals, the experimenters can gauge how well the tested examiners can differentiate between true (same-source) pairs and false (different-source) pairs. The data from such an experiment (disregarding cases in which an examiner refuses to classify the pairs by source because he or she finds them unsuitable for examination or inconclusive) can be arranged in a two-by-two table for the reported classifications. Table 2.2 is a generic version of a classification table, with + denoting a finding of a positive association, i.e., a report of the same source, and − denoting a negative association, i.e., a report of a different source. S refers to the true situation regarding the source of the pair: $S = 1$ denotes same-source pairs, and $S = 0$ designates a different-source pair presented to the examiner.

Several probabilities are commonly used to indicate the accuracy achieved by the examiners (Kafadar, 2015, pp. 114–116):

- Sensitivity $\Pr(+|S) = 1$. If the two items come from the same source ($S = 1$), what is the probability that the examiner will report "same source"? The opposite of sensitivity is the false-negative probability: $\Pr(-|S = 1) = 1 - \Pr(+|S = 1)$.
- Specificity $\Pr(-|S = 0)$. If the two items come from different sources ($S = 0$), what is the probability that the examiner will report "different source"? The opposite of specificity is the false-positive probability: $\Pr(+|S = 0) = 1 - \Pr(-|S = 0)$.

The PCAST report focused almost exclusively on estimating the false-positive probability (FPP). Within the sample of examiners and pairs of prints studied, the examiners had $n = a + b$ opportunities to classify a different-source pair, and they did so incorrectly b times. Hence, the proportion of false positives out of all n findings when $S = 0$ is the number $p = b/n$.

This proportion p would vary from one experiment to another (even if the repeated experiments are as identical as possible) if only because of differences within and across the examiners performing the same task in ways that the researchers cannot control. Once again, if we model the variation in the sample statistic p as in Section 2.1, then p is a binomial random variable B (with parameters θ and n) divided by n.

The PCAST Report suggests forming a 95% one-sided confidence interval estimate for θ and emphasizes informing a judge or jury of the upper confidence limit. If the purpose of the interval is to convey to a judge interested in knowing what the true FPP might be—regardless of whether it is above or below the sample value p—then a two-sided interval is appropriate, and we will use that in the discussion here.

2.4.3.2 An Experiment to Test Categorical Judgments of Latent Print Examiners

In 2011, the FBI and the Noblis Corporation published the first large-scale experiment to assess the accuracy of latent fingerprint examiners (Ulery et al., 2011). The researchers presented 169 latent print examiners with pairs of prints—one latent print and one exemplar. Some of the pairs (520) came from the same finger. The rest (224) came from different fingers. The latent prints were intended to be representative of case work, and the different-source pairs were designed to be difficult comparisons. The test subjects were volunteers attending meetings of the International Association for Identification (IAI). They were relatively experienced and proficient latent print examiners. The nature of the samples and the examiners affects the generalizability of the data to all examiners and all fingerprints, but not the statistical error in going from the sample to the population (of all examiners who attend IAI meetings and would be volunteers for experiments like this one using the same pairs of prints).

The IAI examiners worked through a total of 17,121 presentations of 744 image pairs (100 pairs per examiner). Only 10,052 (59%) were deemed "of value for individualization."[*] For these presentations, $a = 3622$, $b = 6$, $c = 450$, and $d = 3663$.[†] Table 2.3 gives the resulting sample proportions that can be used to estimate the probabilities for the pairs deemed of value for individualization in which the examiners reached a definite conclusion.[‡]

TABLE 2.3

Sample Proportions for Estimating Specificity, Sensitivity, and the Corresponding Error Probabilities

	True State of Affairs	
Examiner's Finding	$S = 0$	$S = 1$
−	$a/(a+b) = 99.83\%$	$c/(c+d) = 10.9\%$[a]
+	$b/(a+b) = 0.17\%$[b]	$d/(c+d) = 89.1\%$

[a] Adding inconclusives to the denominator lowers the false negative rate to $450/5969 = 7.5\%$.

[b] Adding inconclusives lowers the false positive proportion to $6/10052 = 0.1\%$. In no case did two examiners make the same false positive error. The errors occurred on image pairs where a large majority of examiners made correct exclusions; one occurred on a pair where the majority of examiners judged the comparisons to be inconclusive. Thus, the six erroneous identifications probably would have been detected if independent, blind verification were performed as part of the operational examination process.

---

[*] In addition, there were 3,122 comparisons based on latent fingerprints deemed of value only for exclusion. Because standard operating procedures typically include only value-for-individualization comparisons, these other outcomes are not considered here.

[†] Five examiners ($5/169 = 3\%$) made false identifications. One of the five examiners made two false identifications.

[‡] There were also 455 "inconclusives" in the different-source condition and 1,856 in the same-source condition.

TABLE 2.4

Wilson Confidence Intervals for Different Choices
of Confidence for Estimating the FPP When
$p = 6/3628 = 0.17\%$

Confidence Coefficient	CI Lower Limit	CI Upper Limit
0.90	0.09%	0.32%
0.95	0.08%	0.36%
0.99	0.06%	0.45%

2.4.3.3 Constructing Confidence Intervals

The numbers in Table 2.3 are our point estimates. We can proceed as in the allele-frequency example.[§] The false-positive proportion is $p = b/n$, and B is binomial with mean $n\theta$ and variance $n\theta(1 - \theta)$. The plug-in estimated standard error for p is 0.07%. For the asymptotically normal distribution, the 95% CI is again $p \pm 1.96\sqrt{p(1 - p)/n}$. The interval for the Noblis-FBI experiment, by this recipe for the true FPP, is 0.17% ±0.13%, or [0.04%, 0.30%]. Other interval estimates with better procedures (calculated using EpiTools from Ausvet, 2019) produce only minor differences. Results from the Wilson method, for three different confidence coefficients, are given in Table 2.4. Even at the 99% level, the estimated false-positive probability is less than half a percent.

2.4.4 Interpreting Confidence Intervals

Confidence intervals are useful for presenting an estimate of the parameter value and for conveying a sense of the statistical error that is involved in making that estimate from the sample data. Nonetheless, they are prone to misinterpretation. First, because they have sharp boundaries, people may think that there is an important difference between a value at the edge of the interval and one just beyond it. Second, people may think that values near the center of the interval are more likely to be closer to the true value than those at the extremes. Yet the statement of confidence applies equally to all the values in the interval. Third, people may think that a C% interval describes what would happen some C% of the time if a new sample were drawn and a new value for the statistic were computed. In fact, it has been shown that the confidence coefficient for a normal variable tends to overstate the probability that the next sample statistic will lie within the interval. For example, Cumming and Maillardet (2006) found that "[o]n average, a 95% CI will include just 83.4% of future replication means." Finally, and perhaps most pervasively, people misconstrue "confidence" as the probability that the true value is within the interval based on the one sample statistic. Statements to this effect appear in court opinions (Kaye et al., 2011; Kaye and Freedman, 2011) and in the forensic science literature (Weir, 1992; Kaye, 2016, 2019). Indeed, it has been suggested that the word "confidence" be replaced with less suggestive terms (Amrhein et al., 2019, p. 266; Kaye, 1986, p. 1349). The interpretative difficulty is that in the frequentist paradigm, the parameter is not a random variable, but rather is a fixed, unknown value. Because the uncertainty about the parameter is epistemic rather than aleatory, one must step outside the frequentist perspective on probability as a limiting relative frequency to speak

---

[§] But see Chapter 13 (Neumann et al., 2021), questioning whether the independence assumptions apply to the Noblis-FBI experiment.

of the probability that a parameter has a particular value (see Chapters 3 (Banks and Tackett, 2021) and 4 (Stern, 2021)). As a result, the frequentist confidence interval cannot generally be interpreted as the probability that the parameter falls into the interval computed from the single sample that was drawn (e.g., DeGroot and Schervish, 2002, p. 412; Hubert and Wainer, 2013, p. 202; Wasserman, 2004, p. 92). The particular interval either contains the value of the unknown parameter, or it does not. Other intervals (from other samples) would vary in their locations and widths, and it would be incoherent to say that the parameter has the same probability of being in two intervals that are not the same.

Despite these caveats, the CI is valuable as a way to see the quality of a point estimate. Using the CI to decide that the parameter value falls inside or outside of the interval is also a way to control the probability of errors, and that is the framework in which it was developed. In the long run, if we declare that every $C\%$ CI covers the true value, we will be right about $C\%$ of the time. We pursue this decision-making perspective on inference in Section 2.6.

2.5 p-Values

Estimating with confidence intervals gives regions that, it is hoped, include the true value of the parameter being estimated. They take us from a sample statistic to plausible values for the parameter. In contrast, p-values require us to start from a claim—a hypothesis that we can call H_0—about the parameter. We assume for the sake of the analysis that H_0 is true and we ask how compatible the data are with that assumption. To put it another way, the p-value answers the question of how surprising it would be to encounter such data if H_0 is correct. If the data are pretty much the same as would be expected under H_0, then there is little reason to question the hypothesis. But if they are very different from what would be expected, they count as evidence against that hypothesis. The p-value is a probability that indicates (or is supposed to indicate) how strongly the data disconfirm the hypothesis. Everything being equal, smaller p-values reflect stronger evidence against H_0. That is the rough idea. Let's look at how a p-value is computed to reveal some of its nuances and limitations.

2.5.1 p-Values in a Comparison of Glass Fragments

Suppose a broken window is part of a crime scene, and shards of glass are recovered from a suspect's clothing. Measurements of the refractive index are taken for 10 small fragments of the window and for 5 fragments found on the suspect—one measurement per fragment. These measurements are shown in Table 2.5 (adapted from Evett, 1977, p. 215).

Given the small variability in the measurements from the glass recovered from the suspect, it is reasonable to assume that the glass fragments on the suspect came from the same source. The question, then, is whether they came from the broken window. As a preliminary step in answering this question, it is traditional to ask whether the two sets of measurements are so different that they are strong evidence against the statistical hypothesis that the suspect glass comes from glass that has the same refractive index as the window (either because the source is the window or is, by an unlucky coincidence, another source with the same refractive index as the window). For simplicity, we will assume that

TABLE 2.5

Refractive Index Measurements k_i for 10 Glass Fragments from a Broken Window and q_i for 5 Fragments from a Suspect's Clothing

i	Window k_i	Suspect q_i
1	1.51844	1.51848
2	1.51848	1.51850
3	1.51844	1.51848
4	1.51850	1.51844
5	1.51840	1.51846
6	1.51848	
7	1.51846	
8	1.51846	
9	1.51844	
10	1.51848	
Mean	1.518458	1.518472

the broken window has the same index of refraction in every part of it and that the source of the suspect glass also is homogenous. All the variability then is the result of measurement error. Following Evett, we will conveniently assume that the measurement error ε_i is independent and normally distributed with mean $\mu = 0$ and known standard error $\sigma = 4 \times 10^{-5}$ for each measurement i. This assumption can be written more compactly as $\varepsilon_i = \varepsilon \sim N(\mu, \sigma^2) = N(0, (4 \times 10^{-5})^2)$.

Let K_i be a random variable whose values k_i are independent measurements of the refractive index in the known window. Let Q_i be the measured value q_i for each glass particle associated with the suspect. The statistical model is

$$k_i = \theta_k + \varepsilon \quad \text{and} \quad q_i = \theta_q + \varepsilon$$

where ε was defined above, θ_k is the true refractive index of the known window glass, and θ_q is the true refractive index of the unknown source of the glass on the suspect's clothing.

Given this model, how compatible are the data with hypothesis H_0 that $\theta_k = \theta_q$? We can use the difference in the two sample means $d = \bar{k} - \bar{q}$ to find a p-value. The expected value of the random variable D if H_0 is true is $E(D) = E(\bar{K} - \bar{Q}) = \theta_k - \theta_q = 0$. The variance σ_d^2 is

$$\sigma_d^2 = Var(\bar{K} - \bar{Q}) = \frac{\sigma^2}{10} + \frac{\sigma^2}{5} = \frac{3\sigma^2}{10},$$

and the standard error is $\sigma_d = 2.19 \times 10^{-5}$.

The observed difference is $d = 1.518458 - 1.518472 = -1.4 \times 10^{-5}$. Consequently, the difference in the means is $z = d/\sigma_d = -0.639$ standard errors. For a normal random variable like D, the probability of a difference of this magnitude or greater (either above or below) the expected value of 0 is 0.52. Such a p-value is often called a tail-end probability because it comes from the probability mass in the tails of the distribution of the random variable. Figure 2.4 shows the two tails for the p-value in this case.*

---

* When the variances are unknown and estimated from the sample data, the statistic d/s_p follows the Student's t-distribution with the variance computed as in a two-sample t-test.

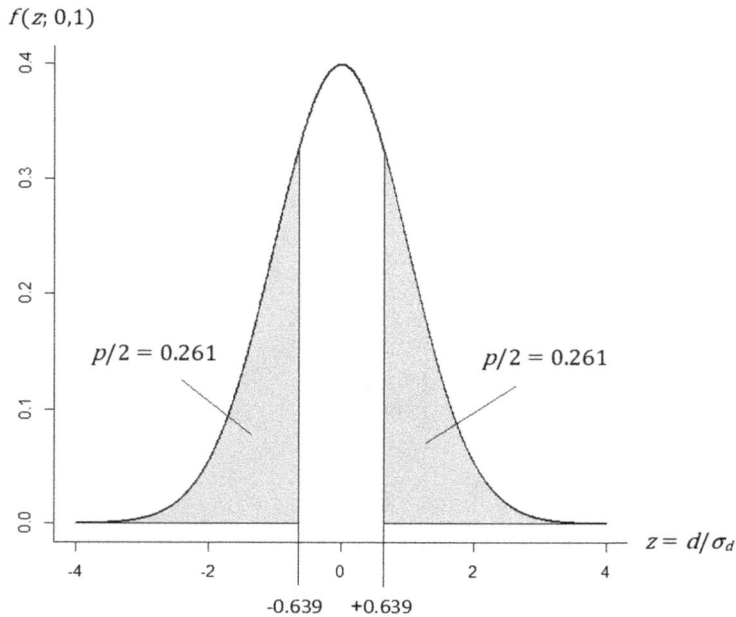

FIGURE 2.4
The sampling distribution for the difference D between the sample means expressed as the number Z of standard errors. The p-value 0.52 comes from the shaded areas under the tails of the standard normal curve $f(z; 0, 1)$.

The observed difference in sample means is not a surprising event under the null hypothesis H_0 that the two sets of glass fragments came from sources with the same refractive index. Differences at least as large as this one would occur, in the long run, about half of the time when making 10 and 5 measurements, respectively, for known and questioned glass fragments from such sources. This large p-value tells us that the data do not come close to excluding the broken window or another one like it as the source of the glass on the suspect.

2.5.2 Interpreting p-Values

We have seen that small p-values argue against the hypothesis used in computing it. But a large p-value should not necessarily be taken as indicating strong evidence that the hypothesis is true. The data may be too weak to say much. If we flip a coin thought to be fair only $n = 4$ times and observe $X = 3$ heads, the p-value for the hypothesis that the coin is fair ($\theta = 1/2$) is the probability mass in the large tails of the binomial distribution $f(x; \theta = 1/2, n = 4)$:

$$\Pr[(X \geq 3) \cup (X \leq 1) \mid \theta = {}^1/_2] = 1 - \Pr(X = 2 \mid \theta = {}^1/_2) = {}^5/_8.$$

As before, the p-value is greater than one-half, but with only four tosses, even the most extreme data ($X = 4$ or $X = 0$) would not be surprising for a fair coin. The p-value for each of those outcomes is $1/8$, so we should not be particularly impressed with one of those outcomes either. The experiment is incapable of producing strong evidence against the hypothesis that the coin is fair, and it would be a mistake to interpret a large p-value as indicative of strong evidence that the coin is fair.

It also would be a mistake to think that very small p-values necessarily signal "significant" or "highly significant" differences—in the sense of either the magnitude or the practical importance of the differences. As for the magnitude of a difference, small p-values can occur for large and small differences alike, making the quantity a poor measure of effect size. As for practical importance, the statistical hypothesis may not really matter. For example, a p-value for the hypothesis of equal refractive index parameters could be small, signaling strong evidence of a slightly different value in the questioned and known samples, but if the index varied by more than that amount in the region of the window that was broken, the "highly significant" p-value would not signify anything very useful. A test for a particular protein in a blood sample might produce a measurement with a small p-value with regard to the hypothesis that the protein is not in the blood, but size of the p-value does not speak to probative value of the data as an identification tool. After all, if 99.9% of the population has the protein, the fact that the p-value is small merely leads to discovery of a fact that has almost no evidentiary value. Similarly, even the tiniest p-value in the original glass example does not demonstrate that the suspect came into contact with the broken window. That inference depends on how many other glass sources in the relevant population of glass have similar refractive indices (e.g., Evett, 1977; Kaye, 2015).

Finally, we should note that, as with the confidence coefficient, transposition of the p-value is common. It occurs in the forensic science literature (Kaye, 1994) and appears in many court opinions and much expert testimony that transforms a p-value for trace evidence into a source probability (Kaye et al., 2011; Kaye and Freedman, 2011). Indeed, it has been said that "[b]ecause people have a hard time understanding the difference between $\Pr(A|B)$ and $\Pr(B|A)$, there are as many incorrect interpretations of p-values in books and scientific literature as there are correct ones" (Westfall and Henning, 2013, p. 396). A p-value is the probability of extreme data given the hypothesis, not the probability of the hypothesis given the extreme data. "Because p is calculated by assuming that the null hypothesis is correct, p does not give the chance that the null is true. The p-value merely gives the chance of getting evidence against the null hypothesis as strong as or stronger than the evidence at hand. Chance affects the data, not the hypothesis. According to the frequency theory of statistics, there is no meaningful way to assign a numerical probability to the null hypothesis." (Kaye and Freedman, 2011, p. 250). As the American Statistical Association (2016) cautioned, "P-values do not measure the probability that the studied hypothesis is true, or the probability that the data were produced by random chance alone." Although "[r]esearchers often wish to turn a p-value into a statement about the truth of a null hypothesis, or about the probability that random chance produced the observed data, [t]he p-value is neither. It is a statement about data in relation to a specified hypothetical explanation, and is not a statement about the explanation itself" (ibid).

2.6 Hypothesis Tests

The third major frequentist statistical tool for inference is hypothesis testing. Rather than estimate the parameter or gauge the strength of the evidence against a hypothesized value (or set of values) for the parameter, frequentist hypothesis testing prescribes epistemic decisions with regard to these values. It gives us decision rules that are shaped by the probabilities of erroneous decisions. A report of a committee of the National Academy of Sciences (National Research Council Committee, 2004) summarized the basic idea for formulating a statistical decision rule for matching bullet lead fragments on the basis of measurements of the concentrations of various trace elements as follows (Kafadar and Spiegelman, 2004, pp. 170–71):

> The classical approach to deciding between the two hypotheses was developed in the 1930s. The standard hypothesis-testing procedure consists of these steps:
>
> 1. Set up the two hypotheses. The "assumed" state of affairs is generally the *null hypothesis*, for example, "drug is no better than placebo." In the compositional analysis of bullet lead (CABL) context, the null hypothesis is "bullets do not match" or "mean concentrations of materials from which these two bullets were produced are not the same" (assume "not guilty"). The converse is called the *alternative hypothesis*, for example, "drug is effective" or in the CABL context, "bullets match" or "mean concentrations are the same."
>
> 2. Determine an *acceptable level of risk* posed by rejecting the null hypothesis when it is actually true. The level is set according to the circumstances. Conventional values in many fields are 0.05 and 0.01; that is, in one of 20 or in one of 100 cases when this test is conducted, the test will erroneously decide on the alternative hypothesis ("bullets match") when the null hypothesis actually was correct ("bullets do not match"). The preset level is considered inviolate; a procedure will not be considered if its "risk" exceeds it....
>
> 3. Calculate a quantity based on the data (for example, involving the sample mean concentrations of the seven elements in the two bullets), known as a *test statistic*. The value of the test statistic will be used to test the null hypothesis versus the alternative hypothesis.
>
> 4. The preset level of risk and the test statistic together define two regions, corresponding to the two hypotheses. If the test statistic falls in one region, the decision is to fail to reject the null hypothesis; if it falls in the other region (called the *critical region*), the decision is to reject the null hypothesis and conclude the alternative hypothesis.
>
> The critical region has the following property: Over the many times that this protocol is followed, the probability of falsely rejecting the null hypothesis does not exceed the preset level of risk. [Ideally, a] test procedure ... has a further property: if the alternative hypothesis holds, the procedure will have the greatest chance of correctly rejecting the null hypothesis.

We will show how this procedure works using the glass example of Section 2.5 and how it relates to the p-value described there as well as to confidence intervals.

2.6.1 Classical Hypothesis Tests for Refractive Index Matching

2.6.1.1 Type I Errors and the Size of a Test

As in Section 2.5, the null hypothesis is H_0: $\theta_k = \theta_q$, which is equivalent to saying that δ, the difference in the parameter for the two groups, is $\delta = \theta_k - \theta_q = 0$. Now we explicitly add the two-sided alternative hypothesis H_1: $\theta_k \neq \theta_q$, which is to say that their difference is not zero ($\delta \neq 0$). In deciding between the two hypotheses, we can make two types of error, as indicated in Table 2.6.

If we conclude that the two sets of glass do not have the same refractive index (a nonmatch) when they do (a false rejection), we falsely exclude the window as the source of the glass on the suspect. That falsely exonerates the suspect. If we conclude that the two sets

TABLE 2.6

Two Types of Error and Two Types of Correct Decisions for a
Hypothesis Test of the Equality of the Refractive Index

	H_0 is true $(\delta = 0)$	H_1 is true $(\delta \neq 0)$
Do not reject H_0	Correct decision	Type II error
	True acceptance	False acceptance
	True negative	False negative
	True match (inclusion)	False match (inclusion)
Reject H_0	Type I error	Correct decision
	False rejection	True rejection
	False positive	True positive
	False nonmatch (exclusion)	True nonmatch (exclusion)

have the same refractive index (a match) when they do not, we falsely include the window as a possible source of the questioned glass. That false match incriminates the suspect, since it implies (albeit incorrectly) that the glass on the suspect either came from this window or another one like it.

The usual nomenclature for statistical hypothesis testing labels a false rejection as a Type I error, a false positive, or a false alarm, and the researcher tries to avoid it. To do that here, we must choose a small number for the tolerable risk of a false rejection. Suppose we select $\alpha = 0.05$. This false rejection probability is the *size* or *significance level* of the test, and data that lead to rejection at this level will be called *statistically significant*.

Our test statistic is D, the difference in the two sample means. Under H_0, D is normal with mean $\mu_d = \delta = 0$ and standard error $\sigma_d = 2.19 \times 10^{-5}$ (Section 2.5.1). Because 5% of the area under the standard normal curve lies in the region $|z| \geq 1.96$, the critical region for D with $\alpha = 0.05$ is $|d| \geq 1.96\sigma_d = 4.29 \times 10^{-5}$. If d falls into this critical region (also called the *rejection region R*), we will reject H_0 and conclude that H_1 is true—the broken window will be excluded as a possible source of the questioned glass on the suspect. This outcome will occur (in the long run) in no more than 5% of the cases in which a window is the source. These false rejections are known as Type I errors, and defining R in this fashion protects us from a larger expected rate of these errors.

If we wanted still more protection from Type I errors, we could pick a smaller value for α. Suppose we choose $\alpha = 0.01$. The rejection region then is $|d| \geq 2.58\sigma_d = 5.65 \times 10^{-5}$. The data must be more extreme (in terms of what H_0 predicts) to permit us to reject H_0. We are more attached to the null hypothesis because we are more averse to prematurely exonerating guilty suspects.

With respect to the test size α, writers sometimes draw an analogy between a classical hypothesis test and a criminal trial in which the defendant is not guilty until proven innocent (e.g., Chihara and Hesterberg, 2011, p. 221; Wasserman, 2004; Saks and Neufeld, 2012, p. 150). The analogy is questionable (and has proved misleading) because the significance level relates to $\Pr(\text{data}|H_0)$, whereas proof beyond a reasonable doubt pertains to $\Pr(H_0|\text{data})$ (Kaye, 1987). In any event, the analogy does not apply to the hypothesis H_0 that $\delta = 0$ in this forensic context. If we demand extreme data (a small rejection region so as to keep α small), it will be harder to dislodge a "presumption" of guilt, or more accurately, an assumption that a defendant is associated with the crime (Zadora et al., 2013, p. 23). This is the opposite of the situation with examiners decisions on "a match" for a fingerprint. There, a false positive meant a false match—a false inclusion—whereas here it means a false exclusion. The low specificity compared to the sensitivity of the examiners in the FBI-Noblis experiment, if extrapolated to case work, would protect defendants from false convictions (at the cost of increasing the rate of false acquittals).

This difference in the implications of Type I and Type II error in the two situations does not imply that it is wrong to use $\delta = 0$ for the null hypothesis and to demand a low risk of false rejection for that hypothesis (Kaye, 2017). The selection of $\delta = 0$ for H_0 might be understood better by regarding the refractive index comparison as a screening test for more elaborate chemical tests (of the elemental composition of glass) or further investigation of the suspect (Kaye, 2015). We would not want such a screening test to be too demanding, so we elect to keep α small.

For the data in Table 2.5, the value of D is $d = -1.4 \times 10^{-5}$. As shown in Figure 2.5, it lies solidly in the acceptance region. The glass evidence seems to have incriminated the suspect.

2.6.1.2 Type II Errors and the Power of a Test

The flip side of Type I error is Type II error—failing to reject H_0 when H_1 is true. Letting A stand for the acceptance region and T be the test statistic, the probability of making a Type II error (false acceptance) is $\beta = \Pr(t \text{ in } A|H_1) = 1 - \Pr(t \text{ in } R|H_1)$. The probability of rejecting H_0 when H_1 is true is called the *power*: Power $= \Pr(t \text{ in } R|H_1) = 1 - \beta$.

It is relatively simple to compute the probability for a statistic when the hypothesis states a specific value for the parameter, as does the null hypothesis that the true difference

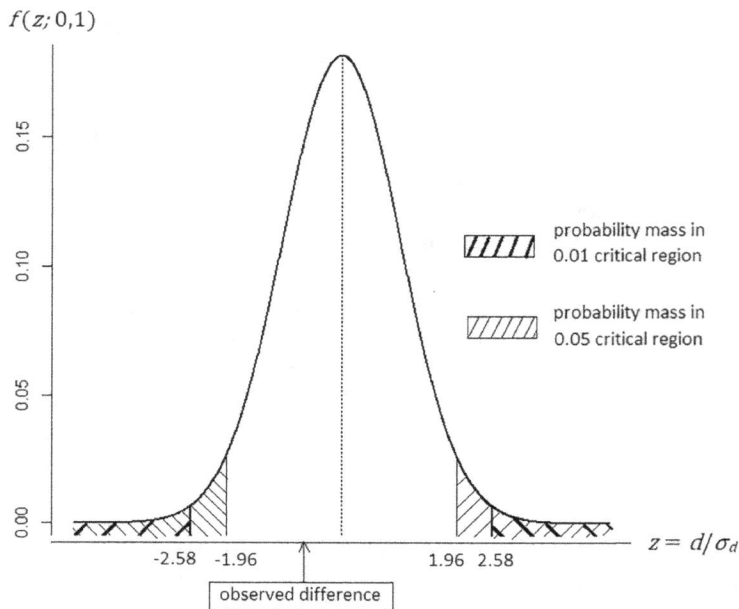

FIGURE 2.5
Rejection regions for a test of size 0.05 and one of size 0.01 for the glass data. The observed value of the test statistic falls into the acceptance region, so the window cannot be ruled out as the source of the glass on the suspect's clothing.

is zero. It is not so straightforward when the alternative H_1 is a set of many possible values. H_1: $\theta_k \neq \theta_q$ covers the entire parameter space for δ except for the single point $\delta = \theta_k - \theta_q = 0$. Being composed of multiple possibilities for δ, it is called a *composite hypothesis*. In contrast, the hypothesis of equality (that $\theta_k \neq \theta_q$) or no difference ($\delta = \theta_k - \theta_q = 0$) is a *simple hypothesis* (also called a point hypothesis or an exact hypothesis). There is no single error probability for a composite hypothesis. Instead, one has to compute the power Pr(t in $A|\delta$) as a function of the unknown difference $\delta = \theta_k - \theta_q$. The bigger this difference, the greater the probability that the test will detect it. A powerful test has a good probability of detecting even a small true difference. A weak test does not. Data that produce a rejection under a test that lacks such power are not much evidence that the no-difference null hypothesis is true.

Size and power are prospective concepts. The level α is set before the test is conducted, and the best region for maximizing power that is consistent with that choice becomes the rejection region (if possible).[*] How powerful are the tests with our rejection regions $|d| \geq 1.96\sigma_d = 4.29 \times 10^{-5}$ (for $\alpha = 0.05$ and $|d| \geq 2.58\sigma_d = 5.65 \times 10^{-5}$ (for $\alpha = 0.01$)? Because power is a function that varies across all the values of $\delta \neq 0$, there is no single answer. But we will compute the power for one of these values to give a limited answer. Suppose the true difference is some number Δ of standard errors above the value of zero proposed by H_0. That is, $\Delta = z(\delta_1) - z(\delta_0)$, where δ_1 is a point value for D in the zone demarcated by H_1. For our example, we will use $\delta_1 = \sigma_d = 2.19 \times 10^{-5}$. Because the true difference δ is the mean μ_d of the sampling distribution of D, this choice for δ shifts the standard normal curve for $Z = D/\sigma_d$ one unit to the right ($\Delta = 1$). The critical region is the same because it comes from H_0. Figure 2.6 sketches the general picture.

The two shaded areas in the lower curve in Figure 2.6 represent the probability for obtaining measured differences d in the rejection region. If D falls into R, we accept the alternative hypothesis, which is the correct decision when $\delta \neq 0$. So these areas are the power—the probability of a correct rejection when the true difference δ has the particular alternative value δ_1.[†]

Table 2.7 lists this power and the complementary false-acceptance probability β for $\delta_1 = \sigma_d = 2.19 \times 10^{-5}$ and for three rejection regions: $|d| \geq 1.96\sigma_d = 4.29 \times 10^{-5}$ (for $\alpha = 0.05$), $|d| \geq 2.58\sigma_d = 5.65 \times 10^{-5}$ (for $\alpha = 0.01$), and $|d| \geq 3.29\sigma_d = 7.21 \times 10^{-5}$ (for $\alpha = 0.001$). Although the choice of this alternative parameter value is arbitrary, computations across a range of values would show the kind of differences that the statistical test has a reasonable chance of detecting. The observed difference here falls into the region for which power is small. This suggests that the failure to exclude the window as the source of the questioned fragments provides only weak evidence that the true refractive indices for the known and questioned fragments are very different.[‡]

---

[*] There could be several regions R for which Pr(t in $A|H_0$) $= \alpha$. In that case, when testing two simple hypotheses, we choose the region that has the largest power. In other words, we pick a rejection region that minimizes β for the fixed α. When the alternative hypothesis is composite, we select the region R for which the test is more powerful than any other rejection region *for every value of the parameter as given in* H_1. Although such uniformly most powerful tests do not always exist, they do for the examples in this chapter, and the test statistics we have used result in regions that minimize β (maximize power) for the stipulated . The proof relies on likelihood ratios.

[†] By symmetry, the power is the same for an equal but negative δ_1. When $\Delta = -1$, for example, the normal curve for the null distribution is shifted one standard unit to the left to obtain the alternative distribution.

[‡] One might be tempted to compute only the power or false-acceptance probability for the simple alternative hypothesis that the true difference δ is the observed difference $d = -1.4 \times 10^{-5}$ ($Z = -0.639$ standard errors). Simulation studies (e.g., Yuan and Maxwell, 2005) have shown that this post hoc power is not a reliable indicator of the true power of a study. It is not a recommended procedure (Gelman, 2019).

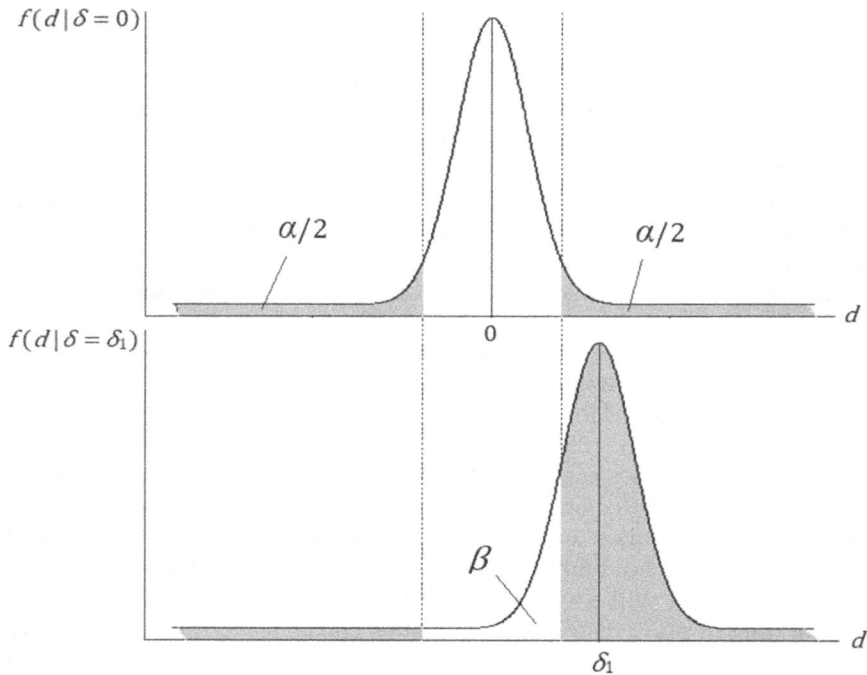

FIGURE 2.6
The probability distribution of the measured difference $Z = D/\sigma_d$ of the means for the two sets of glass fragments when the true difference is not $\delta = 0$, but an alternative value δ_1.

TABLE 2.7

Power and False Acceptance Probability (β) of the Test for $\delta = 0$ Against the Alternative $\delta = -1$ Standard Error When the False Rejection Probabilities Are $\alpha = 0.05, 0.01$, and 0.001

		Alternative Parameter Value $\delta_1 = \sigma_d$	
Significance Level α	Rejection Region	Power for $\Delta = 1$	β for $\Delta = 1$
0.05	$\pm 1.96\sigma_d$	0.17	0.83
0.01	$\pm 2.58\sigma_d$	0.058	0.94
0.001	$\pm 3.29\sigma_d$	0.011	0.99

Notice that as the significance level α decreases (making it less likely to have a false positive), the power also decreases (making it more likely to have a false negative). This tradeoff is a general phenomenon (for a given sample size). Looking back at Figure 2.6, as we stretch out the acceptance region for the null distribution to guard against false rejections, there is more area under the alternative probability density curve for the region:

$$\text{Smaller } \alpha \rightarrow \text{bigger rejection region} \rightarrow \text{less power} \rightarrow \text{larger } \beta$$

Conversely, if we are willing to tolerate a greater risk of false positives (here, false exclusions), we will have fewer false negatives (here, false matches) and thus greater power.

2.6.2 Hypothesis Testing with p-Values

A p-value can be used to perform a hypothesis test. Instead of explicitly defining the critical region, we compute the p-value and compare it to the pre-set threshold α for rejection.[*] If $p \leq \alpha$, we reject H_0 and declare the data to be "statistically significant" at the α level; if $p > \alpha$, we do not reject H_0. For the glass data, we saw that the p-value was $p = 0.52$. The difference obviously is "not statistically significant" at the 0.05 level (or any of the levels that are normally used).

A p-value sometimes is called an "attained significance probability" because it corresponds to a test whose size α is the p-value for the observed data. For $\alpha = 0.52$, the observed difference $d = 1.4 \times 10^{-4}$ would have been (barely) significant. This idea of an attained significance level is misleading. The error probabilities α and β are only applicable when the rejection region is known in advance of the data. The p-values are better understood as indicating how strongly the data seem to refute the null hypothesis. Nevertheless, there is nothing wrong with performing a hypothesis test with a preset α via a p-value.

2.6.3 Hypothesis Testing with Confidence Intervals

Confidence intervals also are intimately related to hypothesis tests. Let the null hypothesis H_0 be that a parameter θ, such as the proportion of a given DNA allele in the population, has the value θ_0. We want to test H_0 at a level α. We form a confidence interval using a confidence coefficient γ of $1 - \alpha$ and reject H_0 if and only if the resulting interval covers θ_0. That does the trick. For example, Section 2.4.3.3 reported that the 95% CI for the false positive probability in the fingerprint study was [0.0008, 0.0036]. At a significance level of 0.05, we could not reject the null hypothesis that the long-run error rate is 0.36%. On the other hand, if we were willing to tolerate a greater risk of a false rejection, namely 0.10, we could reject this hypothesis; the 90% CI is [0.09%, 0.32%], which does not cover the proposed parameter value of 0.36%. Similarly, the 95% CI for estimating the true difference δ between the two sample means for the glass data (Table 2.5) is the observed difference d bracketed by nearly two standard errors: $d \pm 1.96\sigma_d = [-5.69, 2.89]$.[†] Because the interval includes zero, we cannot reject the null hypothesis that $\delta = 0$ at the 0.05 level.

Figure 2.7 displays the fact that that "a coefficient γ confidence set ... can be thought of as a set of null hypotheses that would be accepted at significance level $1 - \gamma$" (DeGroot and Schervish, 2002, p. 457). All the sample intervals in the figure do not cover θ_0. Rejecting H_0 for all such intervals, but not for the intervals (excluded from the picture) that cover θ_0, creates a gap that is the acceptance region A for the test of H_0 whose size is $\alpha = 1 - \gamma$. The midpoints of all the sample CIs that lead to rejection thus fill the rejection region R for the test.

Another way to say it is that, in the long run, if many CIs are formed for an estimator in repeated samples, an expected fraction γ of them will cover θ_0 (when it is the true value for θ), so using them to say whether θ_0 is within them will lead us astray (when H_0 is

---

[*] This approach is a hybrid of the perspectives associated with Sir Ronald Fisher, on the one hand, and Jerzy Neyman and Egon Pearson, on the other. How completely the Fisherian conception of p-values can be reconciled with the Neyman-Pearson theory of decision-oriented hypothesis tests is open to question (e.g., Lehmann, 1993).

[†] In the glass example, the standard error is known with good precision from separate studies. In most applications, it has to be estimated from the one set of sample data at hand. As a result, the width of the confidence intervals in the hypothetical ensemble of repeated intervals will vary. See §2.4.2.2. That is why the hypothetical intervals in Figure 2.7 have different widths.

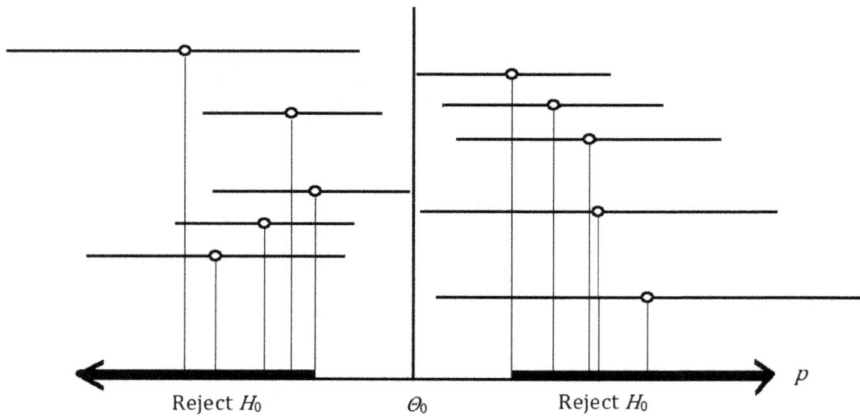

FIGURE 2.7
Using confidence intervals for a sample proportion to test the hypothesis that the population proportion is the value θ_0 proposed by the null hypothesis. The confidence (coverage probability γ) that corresponds to a significance level α is $1 - \alpha$.

true) only in the remaining fraction $1 - \gamma = \alpha$ of the samples. Hence, the decision procedure falsely rejects the claim that $\theta = \theta_0$ about 100α % of the time. The confidence-interval procedure keeps the long run error rate at or below the desired level.

2.7 Issues in Interpreting the Results of Hypothesis Tests, p-Values, and Confidence Coefficients

We identified a number of issues and potential sources of confusion in working with p-values (Section 2.5.2) and confidence intervals (Section 2.4.4). Hypothesis tests are subject to the same misunderstandings as well as to additional pitfalls.

2.7.1 Transposition

As with p-values and confidence coefficients, in legal and other settings there is a tendency to transpose a conditional probability. With hypothesis tests, that probability is the significance level α, which is often misconstrued as the probability that a parameter has a given value rather than as a statement of the probability of a Type I error conditional on the statistical model and the parameter value or values assumed in the null hypothesis. That a finding is "statistically significant" at the 0.05 level does *not* mean there is only a 5% chance that the ... ratio is the result of chance alone" (In re Zoloft (Sertraline Hydrochloride) Products Liability Litigation, 26 F.Supp.3d 449 (E.D. Pa. 2014)). "Scientists"—at least those who appreciate the logic of significance testing—do not "consider two standard deviations in *either* direction to be ... a level at which there is a 95% probability the discrepancy cannot be due to chance" (State v. Lilly, 930 N.W.2d 293, 304 (Iowa 2019)). The frequentist ideas outlined in this chapter do not include calculations of probabilities that chance (as captured in a statistical model) is the cause of an outcome. Yet, jurists tend to confuse the complement of the significance level with the different kind of probability implicit in a legal burden of persuasion.* As we have explained throughout this chapter, the probability in frequentist probability models pertains to the variability in the data that are modeled rather than to the models about the origin of the data and hypotheses embedded in them. Efforts to resolve the problem of "inverse probability"—to give the probability of a cause in light of the data on an effect have spawned the branch of statistical thought known as Bayesian inference summarized in the next two chapters (Banks and Tackett, 2021; Stern, 2021).

2.7.2 Multiple Tests: Proof of the Null Hypothesis and Adjusted p-Values

As discussed in Section 2.6.1.2, the failure of a hypothesis test to reject a null hypothesis is not necessarily an indication of the truth of the alternative hypothesis (let alone the truth of alternatives that the researcher has not considered). The test may lack power. Speaking more generally, if a theory does not pass a severe test, the testing does not warrant its acceptance (Mayo, 2018); to make the alternative hypothesis convincing, other evidence may be required. (Of course, even passing a single severe test may not be sufficient to justify acceptance of a claim. Scientists may ask for replication of the result via independent studies (PCAST, 2016; National Research Council Committee, 2019).[†])

When multiple studies are conducted, or when multiple tests are performed in the course of a single study, none might be significant. This pattern could be an indication that there is no significant difference to be found, but it could be that the collective power of the studies is such that, if they somehow could be combined into a single study, the p-value would have been much smaller than the isolated p-values. Methods for *meta-analysis* try to combine the results of individual studies to ascertain whether their totality is adequate to reject the null hypothesis.

Multiple tests also can show some small p-values along with many larger ones. Testing more than one hypothesis or computing more than one p-value changes the implications of the significance level α and the p-value p. Suppose we conduct a coin-tossing experiment with 10 tosses and α set at 0.01 to test whether a coin is fair. The only data that will allow us to reject H_0 ($\theta = 1/2$) versus the alternative H_1 ($\theta \neq 1/2$) at this level are one of the two most extreme results—0 or 10 heads. These values for X, the number of heads, comprise the critical region R. We flip the coin ten times and find all heads ($x = 10$). Consequently, we conclude the coin is not fair.

Of course, we can reach the same conclusion without mentioning a rejection region by computing the p-value for the data summarized by the statistic $x = 10$. The p-value is the probability of data this far or farther from the expected value $x = 5$—namely, $p = \Pr(x \leq 0) + \Pr(x \geq 10) = 2/10^{10} = 0.002$. Because $p < 0.01$, we reject H_0. The ten heads are explicable only as either a surprising event (if we maintain H_0 is true) or as the outcome for a

---

* The result is a statement such as "[m]ore-probable-than-not might be likened to P < 0.5, so that preponderance of the evidence is nearly ten times less significant (whatever that might mean) than the scientific standard" In re Ephedra Prods. Liab. Litig., 393 F.Supp.2d 181, 193 n. 9 (S.D.N.Y. 2005). On the relationship between a significance level and the legal burdens of persuasion (see Kaye 1983, 1987).

† "Replicability" can mean "obtaining consistent results across studies aimed at answering the same scientific question, each of which has obtained its own data." In this sense, it "involves new data collection and similar methods used by previous studies" (NRC Committee on Reproducibility and Replicability in Science, 2019). However, the word also can be used to refer to replicate measurements on a single unit (for example, a glass fragment or a hair) under the same or very similar conditions. Metrologists use the term "repeatability" to refer to the extent to such repeated measurements have similar values. In trace-evidence comparisons, "repeatability" may refer to intra-examiner reliability, and "reproducibility" to inter-examiner reliability in reaching conclusions on the same specimens (Ulery et al., 2012). A distinct concept, generalizability, "refers to the extent that results of a study apply in other contexts or populations that differ from the original one" (NRC Committee on Reproducibility and Replicability in Science, 2019).

coin for which $\theta \neq {}^1/_2$ (not so surprising if H_0 is false and θ is near 1). So the statistic is significant at the pre-data $\alpha = 0.01$ level.

But imagine that we had tossed the coin 100,000 times and searched the list of resulting heads and tails until—lo and behold—we found a string of ten heads. This $x = 10$ is not at all surprising—there were a great many opportunities for this rare event to occur. Now the evidentiary import of p is totally different. The value $p = 0.002$ grossly overstates the significance of $x = 10$. Likewise for $\alpha = 0.01$. We have not kept the long-run false-rejection rate at or below 1%. Repeated multiple testing of this sort with a fair coin will generate sample data for which $x = 10$ somewhere in the 100,000 tosses nearly every time.

To take a more realistic example, consider an investigation of correlations between a set of traits and 100,000 genes. This study would have a good chance of finding some correlations (at a significance level of $\alpha = 0.01$) for at least some of the genes even if absolutely none of them are truly correlated with any of the traits. If we were to read that these genes were found to be correlated at the 0.01 level and did not know of all the other negative (or, more precisely, non-positive) findings, we would have a false impression of what the data as a whole prove. We could be overly impressed with a report that the genes have been found to be correlated with a particular trait.

What should be done to correct for multiple testing? We defined the probability of making a Type I error (incorrectly rejecting H_0) for a single hypothesis test as its significance level α. But, as the coin flipping and multiple gene-trait testing examples show, when multiple tests are conducted, the probability of a Type I error in at least one of them can be much greater than α. This combined Type I error rate, the probability of making a Type I error in *at least* one test, is the *family-wise error rate*. For m statistically independent tests at the same level α, the family-wise error rate is

$$\alpha^* = 1 - (1 - \alpha)^m.^*$$

For example, if two hypothesis tests are conducted, each with a significance level of $\alpha = 0.05$, then the family-wise error rate is $\alpha^* = 1 - (1 - 0.05)^2 = 0.0975$. If five tests are conducted, then the family-wise error rate is $1 - (1 - 0.05)^5 = 0.2262$. A 22.62% chance of making at least one Type I error in the five tests is much larger than the 5% probability for a single hypothesis test.

The most straightforward way to control the family-wise error rate is to apply a Bonferroni correction. To keep α^* at or below 0.05, the probability of Type I error for each individual hypothesis test is set at α/m. To have a family-wise error rate of 0.05 for two hypothesis tests, the Type I error rate for each individual test should be $\alpha = 0.05/2 = 0.025$; for five hypothesis tests, it should be $0.05/5 = 0.01$.

Because significance levels underlie confidence intervals (Cox, 2006, p. 40), adjustments also should be made when using confidence intervals to test the plausibility of claims. For example, if there are five 95% confidence intervals calculated, then the overall confidence coefficient is approximately 77.3%—there is only a 77.3% chance that all five of the 95% confidence intervals contain the true parameter. To achieve coverage of 95%, the confidence coefficient for each individual interval should be $1 - \alpha/m = 1 - 0.05/5 = 99\%$.

When the tests are not independent, the Bonferroni correction will be conservative—it will overcorrect for the number of tests.[†] It also is conservative because it seeks to make even a single false rejection very unlikely. Sometimes it may be more reasonable to control

---

[*] The probability of a rejection when H_0 is true is α, and the probability of an acceptance is $1 - \alpha$. The probability of m acceptances in m independent tests is $(1 - \alpha)^m$. Therefore, the probability of no such false acceptances (which is to say at least one false rejection) is $1 - (1 - \alpha)^m$.

[†] It is based on a generalization of the inequality $\Pr(A \cup B) = \Pr(A) + \Pr(B) - \Pr(A \cap B) = \Pr(A) + \Pr(B)$.

the *false discovery rate* (FDR). The FDR is the mean of the number of false rejections divided by the number of rejections. Adjusting α to control the FDR is more complicated, but it usually allows more null hypotheses to be rejected than does the more stringent Bonferroni method (Benjamini and Hochberg, 1995; Efron, 2010, pp. 41–61; Reiner et al., 2003; Wasserman, 2004, 166–168). Still more corrective procedures beside the Bonferroni correction and selecting α to control the expected FDR are available (Efron, 2010).

For a time, the multiple testing issue proved controversial in connection with trawls of DNA databases. A National Research Council Committee (1996) proposed a Bonferroni adjustment to the estimated random match probability when the defendant emerged as a suspect because of a trawl through a database of the DNA profiles of previous offenders (a "cold hit"). The committee reasoned that under the null hypothesis that no one in the database is the source of a crime-scene sample, the probability of a match to at least one profile is bounded above by approximately Np when the database size N is a small fraction of the number of people in the population. It invoked the example of a selective search for an impressive string of heads when flipping coins and recommended presenting the inflated probability Np instead of p for a cold hit.[*] The intuition was that with $m = N$ independent opportunities for a false match to the crime-scene DNA profile, the random-match probability p understates the false rejection (Type I error) probability by as much as a factor of N. The formal claim is that the random-match probability p is a p-value for evidence ("the profiles match") as against the null hypothesis that a defendant is not the source[†] and that we should make a Bonferroni adjustment to the p-value because the database trawl tests the null hypothesis $m = N$ times.

There is, however, a crucial difference between the usual problem with a search for significance and the discovery of a suspect through a database trawl. Consider (1) the search for a string of ten heads when flipping a coin and (2) the trawl through a database of DNA profiles, each one of which has a probability $1/1024$ of a match to a crime-scene DNA sample. When we ignore all the other series of coin tosses that did not generate a string of ten heads and report only the probability of $1/1024$ for the one sequence we have cherry picked, we are suppressing evidence *supportive of the null hypothesis* that the coin is unbiased. We are not using that evidence in the calculation of $1/1024$ as a measure of surprise under the null hypothesis. When we ignore all the nonmatches in the database and report only the random-match probability of $1/1024$, we are suppressing evidence *supportive of the alternative hypothesis* that the defendant is the source. Think about it: Every exclusion of a member of the population as the source increases the probability that the source is within the nonexcluded portion of the population—a group that includes the defendant. Thus, many writers have concluded that if we do use the evidence excluding other possible sources, the total picture makes a *stronger* case against the null hypothesis for the defendant as the source. The situation is just the opposite of what occurs with the typical search for significance. See Kaye (2009) for references and Chapter 14 (Sjerps, 2021) for further analysis.

2.7.3 Arbitrary Lines

A particular feature of hypothesis testing that has prompted criticism is the sharp boundary it creates between values of a statistic that lie near the "cliff" between "significant" and

---

[*] The author of this chapter introduced this analogy into the report.

[†] To frame this hypothesis as a statement about a parameter, let S be a Bernoulli variable with possible values 0 (defendant matches) and 1 (defendant does not match). Let ω be a parameter for source hypotheses: $\omega = 0$ means that the defendant is not the source, and $\omega = 1$ means that the defendant is the source. For the one-suspect match case, the null distribution is $f(S = 0; \omega = 0) = 1$ and $f(S = 1; \omega = 0) = p$. The observed match ($S = 1$) is strong evidence against H_0 ($\omega = 0$) to the extent that p is very small.

"not significant." Obviously, there is no fundamental difference between a point estimate that barely makes its way out of an acceptance region of, say, $\mu_0 \pm 3\sigma$ and one that just fails to cross the $\pm 3\sigma$ lines for rejection of H_0. Nevertheless, hypothesis testing is useful for making decisions when one has to jump off the cliff or stay put. It gives us some assurance that the decision to jump will not be wrong too frequently when we should not be taking that leap.

But what is "too frequently"? In considering whether to be impressed with the results of a study, Sir Ronald Fisher, who introduced the p-value to statistics, proposed "a low standard of significance at the 5 per cent point" (Fisher, 1926, p. 504). But he did not propose this figure as if it were the only reasonable choice,* and "[e]ven in the 19th century, we find people such as Francis Edgeworth taking values "like" 5%—namely 1%, 3.25%, or 7%—as a criterion for how firm evidence should be before considering a matter seriously" (Stigler, 2008). Today, it seems doubtful that there can be a uniform value for α—a single criterion for "significant" versus "not significant"—for all sciences (and we would suggest, for all of forensic science).

As explained in Section 2.6.1, the choice of the critical region involves a tradeoff between Type I and Type II errors. The appropriate tradeoff plainly depends on the context. The numbers in the 5% or so range were proposed in the context of scientific theory formation and evaluation generally. Hypothesis tests in forensic analyses may have different consequences. It is one thing to decide to continue to study questioned glass fragments (by measuring their other potentially discriminating properties) to see if the known broken window can be excluded right off as its plausible source. It is quite another to inform judges or juries that the two samples match because they show no statistically significant difference in all the chemical tests to which they were ultimately subjected.

2.7.4 Alternatives and Likelihoods

Not only can any single value for the threshold for statistical significance be criticized as somewhat arbitrary, but in the trial context it has been argued that legal decision-makers should not receive a binary match/no-match decision from the forensic analyst in order to use the data in forming their own opinions. Approaches that characterize the strength of the forensic evidence without adding a categorical conclusion about a null hypothesis might be better suited to the legal factfinders' task.

Quoting a p-value is a step in this direction. For a given study, smaller p-values connote stronger evidence against H_0, but many articles and books on forensic inference complain that p-values do not adequately reveal how strongly the data point toward one hypothesis as opposed to another (e.g., Robertson et al., 2016; Taroni et al., 2016). Indeed, the data can produce a large p-value even when an alternative hypothesis is more congruent with the data than the null hypothesis. Consider once more the coin tossing experiment, this time with $n = 9$ tosses of one of two possible coins. One coin (the "null coin") is fair ($\theta = 0.5$). The "alternative coin" is biased towards heads ($\theta = 0.8$). The data are summarized by

---

* Fisher's longer explanation (Fisher, 1926, p. 504) was that

> [I]t is convenient to draw the line at about the level at which we can say: 'Either there is something in the treatment, or a coincidence has occurred such as does not occur more than once in twenty trials.' ... If one in twenty does not seem high enough odds, we may, if we prefer it, draw the line at one in fifty (the 2 per cent point), or one in a hundred (the 1 per cent point). Personally, the writer prefers to set a low standard of significance at the 5 per cent point, and ignore entirely all results which fail to reach that level. A scientific fact should be regarded as experimentally established only if a properly designed experiment rarely fails to give this level of significance.

$X = 7$ heads. The p-value is $\Pr(X \leq 2) + \Pr(X \geq 7) = 0.18$, which is not very impressive (and is far from significant at conventional values like 0.01 or 0.05). Yet, the actual result $X = 7$ is more probable when flipping the alternative coin than the null one: $f(x = 7; 9, 0.8) = 0.30$ compared to $f(x = 7; 9, 0.5) = 0.07$.

To formalize and generalize this example, we can define the likelihood function as an arbitrary constant c times the probability distribution function $f(x; \theta)$. That is,

$$l(\theta | x) = cf(x; \theta).$$

The order of X and the θ are reversed from the order in the probability distribution because we regard the data as fixed—they are what they are for the experiment—and we think of the parameter as varying—because we are asking how compatible different values of the parameter would be with the given data. The likelihood is *not* a probability. In frequentist statistics, parameter values do not have probabilities attached to them. Nonetheless, we can think of them as having different possible values for the limited purpose of assessing how compatible different values are with the data, and we can compare the likelihoods of different parameter values by forming a *likelihood ratio* for any pair of possible parameter values:

$$L_{10} = \frac{l(\theta_1 | x)}{l(\theta_0 | x)}.$$

In our coin-tossing experiment, the "not significant" finding ($p = 0.18$) that $X = 7$ has a likelihood ratio of

$$L_{10} = \frac{c \binom{9}{7} (0.8)^7 (1 - 0.8)^2}{c \binom{9}{7} (0.5)^7 (1 - 0.5)^2} = \frac{0.30}{0.07} = 4.3.$$

On average, we would expect the number of heads to take on the observed value of 7 more than four times as often for the biased coin than for the null coin. The data are that much more compatible with H_1 than H_0.

This relative-likelihood measure does not privilege the null hypothesis. It is simply an expression of how strongly the data support the hypothesis in the numerator as opposed to the one in the denominator. If one prefers to have H_0 appear in the numerator, the ratio would be the reciprocal L_{01}.

Likelihood ratios can be placed on a logarithmic scale to make them additive for conditionally independent items of evidence. Edwards (1992, p. 32) defines *support* for a hypothesis as $\ln(L_{10})$. Good (1991, pp. 98–99) argues for defining *weight of evidence* as the base-ten logarithm of the Bayes factor for simple hypotheses H_0 and H_1, namely, the posterior odds for $\theta_1 : \theta_0$ divided by the prior odds for $\theta_1 : \theta_0$. For the purpose of providing evaluations of trace evidence to legal factfinders, the likelihood-ratio-as-support or the Bayes-factor-as-weight approach has the imprimatur of many organizations and writers (see Chapter 10 (Kaye, 2021)). Studies on communicating the degree of support by quoting a likelihood ratio are discussed in Chapter 9 (Martire and Edmond, 2021).

2.8 Resampling Methods

In general, simulation studies are computer experiments that involve creating data by pseudo-random sampling from known probability distributions (Morris et al., 2019). In explaining the logic of statistical inference from the frequentist perspective, we relied on examples with well-defined parametric probabilistic and statistical models. But analytic solutions based on stated distribution functions are not always possible, and some statistics, such as the median, require other methods of analysis. With modern computer power, *resampling* methods need not make assumptions about the distributions of the sample statistics or their asymptotic properties. Two common simulation-based resampling methods are bootstrap estimates and permutation tests.

2.8.1 Bootstrap Estimates

The bootstrap is a method for estimating percentiles, central values, and measures of dispersion for a population distribution. It can be used to estimate standard errors and to compute confidence intervals. It relies on the hope that the sample is representative of the population. If that is the case, we can use simulation instead of theory to learn about characteristics of the population. Consider a simple random sample of size n drawn from a large population. Previous sections used formulas for estimating population parameters from sample statistics. The formulas are based on what happens eventually when drawing more and more random samples of this size. We do not actually draw repeated samples because resampling from the big population can be expensive or impractical.

With bootstrapping, we need not rely on the formulas and the assumptions that they demand. Instead, we resample (with replacement) from a "little population"—the sample itself. Let's call this original sample S and the resamples we obtain from it $S_1, S_2, S_3, \ldots$. Each S_j is known as a bootstrap sample. It has size n, and it comes from the relatively few members of the real population that appear in S.

Because a bootstrap sample is taken with replacement, some observations from the original data set may be in the bootstrap sample multiple times. For example, if S is $\{0, 1, 1, 2\}$, the 3^4 possible bootstrap samples are $\{0, 0, 0, 0\}$, $\{0, 0, 0, 1\}$, $\{0, 0, 1, 1\}$, $\ldots$, $\{2, 2, 2, 2\}$. Values (such as the numeral 1) that appear more frequently in S are more likely to occur more frequently in the bootstrap samples as well, so the bootstrap sampling distribution tends to reflect the frequency distribution of the values in the original sample. In turn, a statistic $T(S_j)$ for all the bootstrap samples will have its own distribution. With enough bootstrap samples, we can obtain a pretty good idea of the nature of this bootstrap distribution. For example, we can estimate its mean and variance, or draw boxplots or other pictures that reflect or represent the bootstrap distribution. In this way, the bootstrap samples can be used to calculate a bootstrap interval that is similar to the theory-based confidence interval.

The simplest steps for producing an interval estimate by bootstrapping are as follows:[*]

1. Draw a bootstrap sample S_1 (a sample of size n drawn with replacement from the original and only data we have).

2. Calculate the estimate $T(S_j)$ for the population feature of interest. For example, calculate the sample median if the objective is to estimate the population median or the sample variance if the goal is to estimate the population variance.

3. Repeat steps (1) and (2) thousands of times to approximate the bootstrap distribution.

4. Calculate the upper and lower limits for the $C\%$ bootstrap interval. For example, to calculate a 90% bootstrap interval, the lower limit of the interval is the value at the 5th percentile, and the upper limit is the value at the 95th percentile.

---

[*] There are refinements that generate better performing bootstrap confidence intervals.

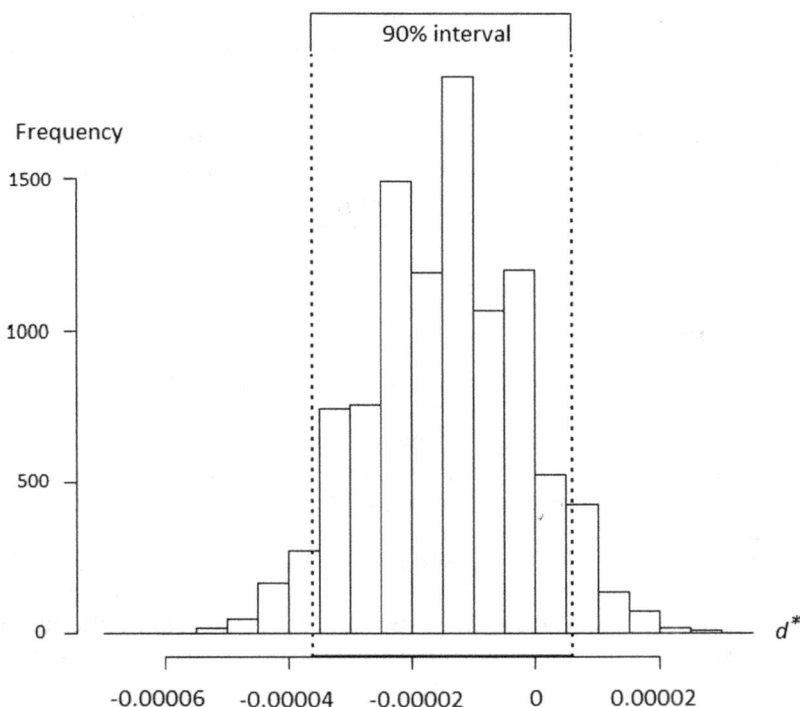

FIGURE 2.8
Histogram of the distribution of the difference d^* in the sample means for $N = 10,000$ pairs of bootstrap samples drawn from the glass data in Table 2.5. The value for the same-refractive-index hypothesis ($\delta = 0$) lies in the main body of this simulated distribution rather than in a small tail, as indicated by the fact that the 90% coverage interval includes 0.

Appling this procedure, we drew 10,000 bootstrap samples of the measurements for the glass from the window and suspect's clothing given in Table 2.5. For each pair of bootstrap samples, we computed the difference d^* in the sample means. As expected, the mean of the bootstrap distribution of d^* was the same as the difference in the one pair of real sample means (-1.4×10^{-5}). However, the bootstrap standard error (SE) was smaller than the SE computed from Evett's postulated figure for the standard error of measurement (1.28×10^{-5} compared to 2.19×10^{-5}). The 90% bootstrap CI is $[-3.6 \times 10^{-5}, 0.6 \times 10^{-5}]$, and the 95% CI is $[-3.8 \times 10^{-5}, 1.0 \times 10^{-5}]$. Because these intervals include 0, at the 0.10 and 0.05 levels for significance, we cannot reject the hypothesis that the two sets of measurements come from glass with the same refractive index. The coverage has to drop to 70% for the bootstrap CI to just graze $\delta = 0$. (That interval is $[-2.8 \times 10^{-5}, 0.00000]$.) Hence, the bootstrap estimate for the p-value is 0.30 instead of the 0.52 reported in Section 2.5.1 on the basis of Evett's assumptions regarding normality. Even so, the observed difference is still compatible with the null hypotheses (see Figure 2.8).

In short, the interpretations and uses for the bootstrap interval are similar to those for confidence intervals. But the results are dependable only if the original sample S used as a "little population" is reasonably large and thus likely to be representative of the big population (Freedman, 2005, p. 148). For the small number of measurements in Table 2.5, the bootstrap CIs and p-value are not terribly trustworthy. We have supplied them simply to illustrate how bootstrapping can help evaluate hypotheses about population parameters without making strong assumptions about the distribution of the test statistic in the population.

The distribution of the values in the sample S is sometimes called the "empirical distribution of the sample,"* but, as we have emphasized, it is not the distribution from which the real sample came. The empirical distribution is conditional on the data S. It is, in other words, an approximation to the distribution that gave rise to S. This approximation treats each of the n sample points in S as equally likely and puts probability $1/n$ on each of them. "Lacking other information, this is perhaps the best we can do" (Freedman, 2005, p. 151).

2.8.2 Permutation Tests

The permutation test is a nonparametric method for testing whether two distributions are the same. It dates back to the 1930s (David, 2008). Again, we start with a sample of size n. This sample could be very small, and just how we obtained it is not that important. It does not have to be a random sample. The null hypothesis does not have to be expressed in terms of a statement about a population parameter, and no assumptions involving properties such as normality or constant variance are necessary.

Suppose that $k_1, k_2, \ldots, k_m$ are m repeated, independent measurements of the refractive index of a known glass fragment and $q_1, q_2, \ldots, q_n$ are n repeated measurements of a questioned sample (compare Table 2.5 of Section 2.5.1). Let H_0 be the hypothesis that the two sample sets of measurements come from identical distributions, and let H_1 be that they come from different distributions. Let $T(K_1, K_2, \ldots, K_m, Q_1, Q_2, \ldots, Q_n)$ be some test statistic. For example, let T be the absolute value of the difference in the sample means: $T(K_1, \ldots, K_n, Q_1, \ldots, Q_m) = |\bar{K}_n - \bar{Q}_m|$. Finally, let t_{obs} denote the value of T for the observed sample of measurements.

Now "shuffle" the $N = n + m$ observations that comprise the sample $S = \{k_1, k_2, \ldots, k_m, q_1, q_2, \ldots, q_n\}$ to obtain a permuted sample S_1. For example S_1 might be $\{k_2, k_1, \ldots, k_3, q_2, q_1, \ldots, q_n\}$. Calculate the test statistic T for the permuted sample. Repeat this process for every possible permutation of the original sample S. This yields $N!$ test statistics $t_1, t_2, \ldots, t_{N!}$.

Under the null hypothesis, each t_j is equally probable, and the distribution that puts probability $1/N!$ on each t_j is called the *permutation distribution*. Therefore, to test H_0 against the alternative H_1, we ask whether t_{obs} is very different from all the other t values. That outcome would be surprising if H_0 is true. Thus, the fraction of the t_j's that are as extreme, or more extreme, than t_{obs} in the permutation distribution is the p-value.

For the data in Table 2.5, there are $15! = 1,307,674,368,000$ permuted samples to consider. Rather than find t for all of them, we can randomly sample a large number of them. When a sample of all possible permutations is used, the resulting test usually is referred to as a "randomization test." For one run of 9,999 permuted samples, we computed a p-value of 0.44 for the glass data in Table 2.5. This is less than the $p = 0.52$ computed with the normal error model and Evett's value for the standard error of measurement, but, as with the bootstrap samples (for which $p = 0.30$), it still is compatible with the null hypothesis—the

---

* Here, "distribution" refers to a cumulative distribution. Its value at a given point is equal to the proportion of observations from the sample that are less than or equal to that point. More formally, the empirical distribution associated with a vector of numbers $x = (x_1, \ldots, x_N)$ is the probability distribution with expectation operator $E_n\{F(X)\} = (1/N)\sum F(x_i)$. The function $F(X)$ is the distribution that arises in finite population sampling. Suppose we have a population of size N whose members have values $x_1, \ldots, x_n$ for a particular measurement. The value of that measurement for a randomly drawn member of this population has a probability distribution that is this empirical distribution.

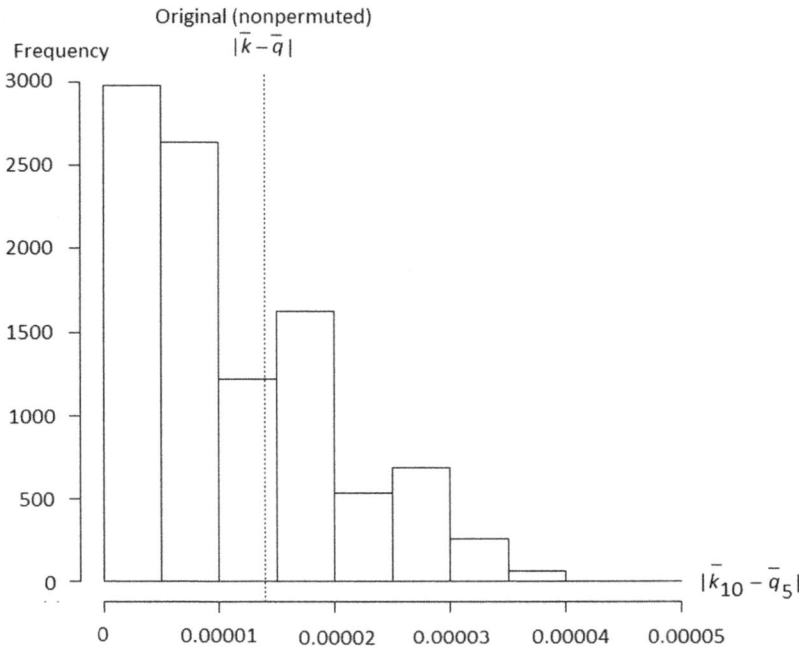

FIGURE 2.9

Location of the absolute value of the difference between the sample mean of the measurements on the window glass and the sample mean of the suspect glass within a histogram for a random sample ($N = 9{,}999$) of the permutation distribution. The p-value is the fraction of such differences as large or larger than the observed difference (approximately the area to the right of the dotted line).

observed difference in the refractive index measurements of the window glass and the suspect glass is not an outlier for the permutation distribution (see Figure 2.9).

The general algorithm for the permutation test of the null hypothesis that two samples have been drawn from the same probability distribution (same population) is the following (Wasserman, 2004, p. 163):

1. Compute the observed value of the test statistic: $t_{obs} = T(X_1, X_2, \ldots, X_m, Y_1, Y_2, \ldots, Y_n)$.
2. Randomly permute the data. Compare the statistic again using the permuted data.
3. Repeat the previous step B times and let $t_1, \ldots, t_B$ denote the resulting values.
4. The approximate p-value is

$$\frac{1}{B} \sum_{j=1}^{B} I(t_j > t_{obs}),$$

where $I(t_j > t_{obs})$ is 1 when its argument is true and 0 otherwise.

For this permutation test to work, the data must be "exchangeable." That is, shuffling the observed data points must keep the data set just as probable as the original.* When every S_j

---

* In addition, the procedure is not applicable to all testing problems. For example, if we have a single sample S and our test statistic is the same for every permuted sample (as with the sample mean or median), then every permuted S will have the same value as t_{obs}.

is as likely to arise as any other one, observed differences will tend to be small and random if H_0 is true. If t is unusual, as indicated by our p-value, then H_0 may well be false.

Acknowledgments

Maria Tackett made important contributions to Sections 10.2.4–10.2.8, and David Banks and Hal Stern supplied helpful comments. Sections 10.2.2–10.2.7 were influenced by the author's previous work with the late David A. Freeman. All mistakes are the author's.

References

American Statistical Association. ASA Statement on Statistical Significance and P-values, 2016, Feb. 5. doi:10.1080/00031305.2016.1154108

Amrhein, V., Trafimow, D., and Greenland, S. Inferential statistics as descriptive statistics: There is no replication crisis if we don't expect replication. *The American Statistician*, **73**(S1), 262–270, 2019, https://doi.org/10.1080/00031305.2019.1543137

Ausvet. 2019. Calculate confidence limits for a sample proportion. In EpiTools, http://epitools.ausvet.com.au/content.php?page=CIProportion.

Banks, D., and Tackett, M. Bayesian methods and forensic inference. In *Handbook of Forensic Statistics*, D. Banks, K. Kafadar, D.H. Kaye, and M. Tackett, editors. CRC Press, Boca Raton Florida, 2021.

Benjamini, Y. and Hochberg, Y. Controlling the false discovery rate: A practical and powerful approach to multiple testing. *Journal of the Royal Statistical Society, Series B, Methodological*, **57**, 289–300, 1995.

Brown, L.D., Cai, T.T., and DasGupta, A. Interval estimation for a binomial proportion. *Statistical Science*, **16**, 101–133, 2001.

Chihara, L., and Hesterberg, T. *Mathematical Statistics with Resampling and R*. John Wiley & Sons, Hoboken, 2011.

Cox, D.R. *Principles of Statistical Inference*. Cambridge University Press, Cambridge, U.K., 2006.

Cumming, G., and Maillardet, R. Confidence intervals and replication: Where will the next mean fall? *Psychological Methods*, **11**, 217–227, 2006.

David, H.A. The beginnings of randomization tests. *American Statistician*, **62**, 70–72, 2008.

DeGroot, M.H., and Schervish, M.J. *Probability and Statistics* (3rd ed.). Addison-Wesley, Boston, 2002.

Edwards, A.W.F. *Likelihood: Expanded Edition*. Cambridge University Press, Cambridge, U.K., 1992.

Efron, B. *Large-Scale Inference: Empirical Bayes Methods for Estimation, Testing, and Prediction*. Cambridge University Press, Cambridge U.K., 2010.

Evett, I.W. The interpretation of refractive index measurements. *Forensic Science International*, **9**, 209–217, 1977.

Fisher, R. The arrangement of field experiments. *Journal of the Ministry Agriculture of Great Britain*, **33**, 503–515, 1926.

Freedman, D. *Statistical Models: Theory and Practice*. Cambridge University Press, Cambridge, U.K., 2005.

Freedman, D., Pisani, R., and Purves, R. *Statistics* (3rd ed.). W.W. Norton & Co., New York, 1998.

Gelman, A. Don't calculate post-hoc power using observed estimate of effect size. *Annals of Surgery*, **269**(1), e9–e10, 2019. doi:10.1097/SLA.0000000000002908 (letter)

Good, I.J. Weight of evidence and the Bayesian likelihood ratio. In *The Use of Statistics in Forensic Science*, C.G.G. Aitken, and D.A. Stoney, editors, pages 55–106. Ellis Horwood, Chichester, U.K., 1991.

Hubert, L., and Wainer, H. *A Statistical Guide of the Ethically Perplexed*. CRC Press, Boca Raton, 2013.

Jantz, R.L. Amelia Earhart and the Nikumaroro bones: A 1941 analysis versus modern quantitative techniques. *Forensic Anthropology* **1**(2), 83–96, 2018, http://journals.upress.ufl.edu/fa/article/view/525/518.

Kafadar, K. Statistical issues in assessing forensic evidence. *International Statistical Review*, **83**(1), 111–134, 2015.

Kafadar, K. and Spiegelman, C. Statistical analysis of bullet lead data. Appendix K in the National Research Council Committee Report. *Forensic Analysis: Weighing Bullet Lead Evidence* (K. O. MacFadden, Chair). The National Academies Press, Washington, DC, 2004.

Kaye, D.H. Statistical significance and the burden of persuasion. *Law and Contemporary Problems*, **46**, 13–23, 1983.

Kaye, D.H. Is proof of statistical significance relevant? *Washington Law Review*, **61**(4), 1333–1365, 1986.

Kaye, D.H. Hypothesis testing in the courtroom. In *Contributions to the Theory and Application of Statistics*, A. Gelfand, editor, page 331–356. Academic Press, Orlando, 1987.

Kaye, D.H. Commentary on "proficiency of professional document examiners in writer identification". *Journal of Forensic Sciences*, **39**(6), 1344, 1994.

Kaye, D.H. Rounding up the usual suspects: A logical and legal analysis of DNA trawling cases. *North Carolina Law Review*, **87**(2), 2009. https://ssrn.com/abstract=1134205.

Kaye, D.H. *The Double Helix and the Law of Evidence*. Harvard University Press, Cambridge, MA, 2010.

Kaye, D.H. Reflections on glass standards: Statistical tests and legal hypotheses. *Statistica Applicata – Italian Journal of Applied Statistics*, **27**(2), 73–186, 2015.

Kaye, D.H. PCAST's sampling errors (part 1). *Forensic Science, Statistics and the Law*, 2016, Oct. 24. http://for-sci-law.blogspot.com/2016/10/pcasts-sampling-errors.html.

Kaye, D.H. Hypothesis testing in law and forensic science: A memorandum. *Harvard Law Review Forum*, **130**(5), 127–136, 2017.

Kaye, D.H. Panel remarks: Bringing statistics into the courtroom: The Nikumaroro bones: How can forensic science assist factfinders? *Virginia Journal of Criminal Law*, **6**, 101–118, 2018.

Kaye, D.H. Confidence intervals – if only it were that simple. *Forensic Science, Statistics and the Law*, 2019, Nov. 3, http://for-sci-law.blogspot.com/2016/11/the-false-positive-fallacy-in-first.html.

Kaye, D.H. The ultimate-issue rule and forensic science identification. *Jurimetrics: The Journal of Law, Science, and Technology*, **60**(2), 2021.

Kaye, D.H., Bernstein, D.E., and Mnookin, J.L. *The New Wigmore: A Treatise on Evidence: Expert Evidence* (2nd ed.). Aspen Pub., New York, 2011.

Kaye, D.H., and Freedman, D.A. Reference guide on statistics. In *Reference Manual on Scientific Evidence* (3rd ed.), pages 241–302. National Academy Press, Washington, DC, 2011.

Ke-Hai Yuan, K.H., and Maxwell, S. On the post hoc power in testing mean differences. *Journal of Educational and Behavioral Statistics*, **32**(2): 141–167, 2005, https://doi.org/10.3102/10769986030002141.

Lehmann, E.L. The Fisher, Neyman-Pearson theories of testing hypotheses: One theory or two? *Journal of the American Statistical Association*, **88**, 1242–1249, 1993.

Martire, K.A., and Edmond, G. How well do lay people comprehend statistical statements from forensic scientists? In *Handbook of Forensic Statistics*, D. Banks, K. Kafadar, D.H. Kaye, and M. Tackett, editors. CRC Press, Boca Raton Florida, 2021.

Mayo, D. *Statistical Inference as Severe Testing: How to Get Beyond the Statistics Wars*. Cambridge U. Press, Cambridge, U.K., 2018.

Moore, D.S., and McCabe, G.P. *Introduction to the Practice of Statistics* (2nd ed.). W.H. Freeman & Co., New York, 1993.

Morris, T.P., White, I.R., and Crowther, M.J. Using simulation studies to evaluate statistical methods. *Statistics in Medicine*, **38**(11), 2074–2102, 2019.

National Research Council Committee Report. *The Evaluation of Forensic DNA Evidence.* National Academy Press, Washington D.C., 1996.

National Research Council Committee Report. *Forensic Analysis: Weighing Bullet Lead Evidence.* National Academies Press, Washington, D.C., 2004.

National Research Council Committee Report. *Reproducibility and Replicability in Science.* National Academies Press, Washington, D.C., 2019.

PCAST (President's Council of Advisors on Science and Technology). *Report to the President: Forensic Science in Criminal Courts: Ensuring Scientific Validity of Feature-Comparison Methods.* Washington D.C., 2016.

Reiner, A., Yekutieli, D., and Benjamini, Y. Identifying differentially expressed genes using false discovery rate controlling procedures. *Bioinformatics*, **19**, 368–375, 2003.

Robertson, B., Vignaux, G.A., and Berger, C.E.H. *Interpreting Evidence: Evaluating Forensic Science in the Courtroom* (2nd ed.). John Wiley & Sons, Chichester, U.K, 2016.

Saks, M.J., and Neufeld, S. Parallels in law and statistics: Decision making under uncertainty. *Jurimetrics: The Journal of Law, Science and Technology*, **52**, 117–122, 2012.

Sjerps, M.J. Probabilistic considerations when interpreting database search and selection effects. In *Handbook of Forensic Statistics*, D. Banks, K. Kafadar, D.H. Kaye, and M. Tackett, editors. CRC Press, Boca Raton Florida, 2021.

Stern, H.S. Comparing philosophies of statistical inference. In *Handbook of Forensic Statistics*, D. Banks, K. Kafadar, D.H. Kaye, and M. Tackett, editors. CRC Press, Boca Raton, Florida, 2021.

Stigler, S. Fisher and the 5% Level. *Chance*, **21**(4), 12, 2008. doi:10.1007/s00144-008-0033-3.

Taroni, F., Biedermann, A., and Bozza, S. Statistical hypothesis testing and common misinterpretations: Should we abandon p-value in forensic science applications?, *Forensic Science International*, 259:e32–e36, 2016.

Ulery, B.T., Hicklin, R.A., Buscaglia, J., and Roberts, M.A. Accuracy and reliability of forensic latent fingerprint decisions. *Proceedings of the National Academy of Sciences (USA)*, **108**(19), 7733–7738, 2011, http://www.pnas.org/content/108/19/7733.full.pdf

Ulery, B.T., Hicklin, R.A., Buscaglia, J., and Roberts, M.A. Repeatability and reproducibility of decisions by latent fingerprint examiners. *PLoS ONE*, **7**(3):e32800, 2012. https://doi.org/10.1371/journal.pone.0032800.

Wasserman, L. *All of Statistics: A Concise Course in Statistical Inference.* Springer, New York, NY, 2004.

Weir, B.S. Population genetics in the forensic DNA debate. *Proceedings of the National Academy of Sciences (USA)*, **89**, 11654, 1992.

Westfall, P., and Henning, K.S. *Understanding Advanced Statistical methods.* CRC Press, Boca Raton, FL, 2013.

Wilson, E.B. Probable inference, the law of succession, and statistical inference. *Journal of the American Statistical Association*, **22**(158): 209–212, 1927.

Zadora, G., Martyna, A., Ramos, D., and Aitken, C. *Statistical Analysis in Forensic Science: Evidential Value of Multivariate Physicochemical Data.* John Wiley & Sons; New York, NY, 2013.

Zhou, X.H., Li, C.M., and Yang, Z. Improving interval estimation of binomial proportions. *Philosophical Transactions A Mathematics, Physics and Engineering Science*, **366**(1874), 2405–2418, 2008. doi: 10.1098/rsta.2008.0037, PMID: 18407898

3

Bayesian Methods and Forensic Inference

David Banks and Maria Tackett

CONTENTS

3.1 Introduction 73
3.2 The Basics 74
    3.2.1 A Beta-Binomial Mock Example 76
    3.2.2 A Gamma-Poisson Mock Example 79
3.3 Markov Chain Monte Carlo 85
3.4 Broad Applications 86
3.5 Summary 89
Acknowledgments 89
References 89

3.1 Introduction

Bayesian methods are well-established analytical tools. Along with frequentist methods (see Chapter 2), these comprise the two dominant approaches to statistical inference. (A third group of methods, called fiducial or likelihood inference has found less favor.) Reviews of all three perspectives can be found in Young et al. (2005).

The Bayesian approach differs from the frequentist approach in several important ways. First, a Bayesian expresses uncertainty as a probability, which a frequentist is not allowed to do. For example, when a frequentist sets a confidence interval, he may not say something such as "the probability that the interval contains the true mean of the population is 0.95." Instead, he must say that "95% of similarly constructed intervals will contain the true mean." This is because the interval either contains the true mean or it does not. The frequentist does not know which of these is correct, but he may not express his uncertainty as a probability.

In contrast, a Bayesian seeing the same data will construct a *credible interval*, a region which, to a Bayesian, has some pre-specified probability of containing the unknown parameter value. In some cases a credible interval with probability 0.95 of containing the parameter value is in exact numerical agreement with the 95% confidence interval. But the philosophy and interpretation of a credible interval is very different from that of a confidence interval.

A second difference is that the Bayesian may integrate over the parameter space. This is important in several ways, but one application is in hypothesis testing. A frequentist will test the null hypothesis against the alternative, and he will either reject or fail to reject the null hypothesis at some pre-specified alpha level (often 0.05 or 0.01). In contrast, the Bayesian will calculate her *posterior distribution*, a probability distribution that describes her belief about the value of the unknown parameter after observing the data. Then she can integrate her posterior distribution over the region corresponding to the null hypothesis, and thus calculate her probability that the null hypothesis is true.

The third difference is the most controversial. Bayesians often start with subjective beliefs about the probabilities of events. For example, based on life experience or just plain prejudice, a Bayesian juror might have an initial belief that the chance that the defendant is guilty is 0.9, which should then be modified through the use of a mathematical tool called "Bayes' Rule" as evidence is presented. This initial belief is difficult to square with the legal tenet that a defendant should be presumed innocent until guilt is proven, but it may be a more realistic representation of juror thinking than to assume each juror is a blank slate upon which the lawyers paint their narratives.

Savage (1954) proved that under a very general and reasonable definition of rationality, any rational person must act as a Bayesian. Specifically, if agents have beliefs that can be characterized through a probability function on some space, then rational agents should make decisions that maximize their expected utility function (i.e., decisions that, on average, provide the largest benefit or least loss where the benefit or loss need not be monetary). And this compels them to act as Bayesians. On the other hand, another theorem shows that a committee (such as a jury) cannot act collectively as a Bayesian unless they initially hold identical opinions (*prior beliefs* or simply *priors*) about the probability distributions for all relevant unknowns in the problem and also agree on the mechanism for generating the observations (cf. Kadane et al., 1993). The usual way of addressing this problem is to invoke another theorem, which states that under general conditions, two Bayesians who observe the same stream of data will converge in their beliefs. Thus, in the context of a jury trial, each will juror have separate prior beliefs, and the lawyers would attempt to present enough evidence that all members of the jury would converge towards unanimity (as required in many states), so that all would vote for acquittal or conviction. Some jurors might need little evidence, others a great deal, but all opinions should be changing in the same direction.

3.2 The Basics

Bayes' Theorem (or Bayes' Rule) is a completely standard and universally accepted consequence of elementary probability. Recall that the conditional probability of an event A given that event B is observed is defined as

$$\mathbb{P}[A|B] = \frac{\mathbb{P}[A \text{ and } B]}{\mathbb{P}[B]}$$

where $\mathbb{P}[A \text{ and } B]$ is the probability that both events A and B occur, and $\mathbb{P}[B]$ is the probability that event B occurs.

Let $A_1, \ldots, A_k$ be a finite partition of the set of possible outcomes, where a finite partition means that one of $A_1, \ldots, A_k$ must happen, but the events are mutually exclusive (disjoint), meaning that it is impossible for two or more of them to happen at the same time. In drawing a card from a standard deck, one possible finite partition is Clubs, Diamonds, Hearts and Spades; another finite partition is Red and Black; a third is aces, twos, $\ldots$, kings.

Bayes' Rule says that if $A_1, \ldots, A_k$ is a finite partition and one observes the event B, then the probability of A_i given that B is

$$\mathbb{P}[A_i|B] = \frac{P[B|A_i] \times \mathbb{P}[A_i]}{\sum_{j=1}^{k} P[B|A_j] \times \mathbb{P}[A_j]}. \tag{3.1}$$

One can show that the numerator is just $\mathbb{P}[A_i$ and $B]$ and, for a finite partition, the denominator is just $\mathbb{P}[B]$, so this is just a rewriting of the definition of conditional probability. Richard Price published this theorem in 1763, on behalf of the Reverend Thomas Bayes, who had discovered it, in the *Philosophical Transactions of the Royal Society* (Bayes and Price, 1763). To illustrate an application of Bayes's Theorem, consider the enzyme-linked immunosorbent assay (ELISA) test for HIV. Chou et al. (2005) of the U.S. Preventive Services Task Force studied the ELISA test carefully, and found that if one has HIV, ELISA signals with probability 0.997 (the sensitivity), and if one does not have HIV, then it does not signal with probability 0.985 (the specificity). And it is known that about 0.34% of the U.S. population is HIV positive (CDC, 2017).

Suppose a random person is given a blood test (e.g., the person wants to join the army, and the blood test is a routine part of the physical examination). And suppose the ELISA test signals that the person has HIV. What is the chance that the ELISA test is correct and the subject is truly HIV positive? Note that having HIV or not having HIV constitutes a finite partition with just two events: one of the two must be true, but it is not possible for both to be true. Using Bayes's Rule, one finds that the probability of being HIV positive given a positive ELISA test is

$$\mathbb{P}[\text{ HIV } | \text{ pos test }]$$

$$= \frac{\mathbb{P}[\text{ pos test } | \text{ HIV }] * \mathbb{P}[\text{ HIV }]}{\mathbb{P}[\text{ pos test } | \text{ HIV }] * \mathbb{P}[\text{ HIV }] + \mathbb{P}[\text{ pos test } | \text{ no HIV }] * \mathbb{P}[\text{ no HIV }]}$$

$$= (0.997) * (0.0034) / [(0.997) * (0.0034) + (1 - 0.985) * (1 - 0.0034)]$$

$$= 0.1848.$$

So even though the test signals HIV, because the prevalence of HIV in the population is so low, the actual chance of being HIV positive is fairly small.

This calculation requires that the person being tested is chosen at random (i.e., in a manner that is unrelated to HIV status). It would not apply to someone who came to the physician because he felt ill. Also note that this example can be generalized to breathalyzer tests, fingerprint matches, and other forensic-science tests.

In this example, the Bayesian analyst did not need to use subjective probability, because she knew the baseline prevalence of HIV in the U.S. population and the problem assumed that person being tested was a random draw from that population, allowing her to use an objective probability. However, if the person were being tested because he felt ill or had engaged in risky behavior, then a Bayesian physician would have a different, probably subjective, probability of HIV infection, and that probability would be larger than 0.0034.

Although Bayesian inference has compelling intellectual properties, in its mathematical form, it generally requires the evaluation of complex integrals. Specifically, the integral form of the Bayes' Rule, which is the analogue of the discrete form given in (3.1), is

$$\pi^*(\theta|x_1, \ldots, x_n) = \frac{f(x_1, \ldots, x_n|\theta)\pi(\theta)}{\int_{-\infty}^{\infty} f(x_1, \ldots, x_n|\theta)\pi(\theta)\, d\theta} \tag{3.2}$$

where $\pi(\theta)$ is the prior density for the unknown parameter θ, $\pi^*(\theta|x_1, \ldots, x_n)$ is the posterior density for θ after observing the data $x_1, \ldots, x_n$, and $f(x_1, \ldots, x_n|\theta)$ is the density for the distribution of the data given the value of θ. A few examples will clarify all this, but the main point is that the analyst had prior beliefs that were updated after seeing data to form posterior beliefs, and that this updating was done according to Bayes' Rule.

The integral in the denominator of (3.2) was a stumbling block to the use of Bayesian inference for several centuries. In most cases, the integral could not be evaluated, and thus the updating calculation could not be completed. Essentially, the only circumstances in which tractable calculation was possible were with conjugate families. In conjugate families, the prior belief on the value of the unknown parameter is assumed to be a member of a convenient family, and the distribution that generates the data, conditional on the value of the parameter, is also assumed to be a member of a convenient family. The convenience of these two families derives from the fact that when the prior belief about the parameter is updated by the data to obtain the new (posterior) belief about the parameter, it turns out that the new belief is in the same distributional family as the prior. The three most prominent of these conjugate families are the beta-binomial, the normal-normal, and the gamma-Poisson families.

Two hypothetical applications illustrate different facets of Bayesian reasoning. These examples will use the beta-binomial and gamma-Poisson families.

3.2.1 A Beta-Binomial Mock Example

The beta-binomial is used when making inferences about a proportion or a probability. For example, suppose a forensic accountant wants to estimate the proportion of fraudulent claims filed with an insurance company by an automobile repair chain. Formally, one would have to assume that an infinite number of claims have been filed, and that the forensic accountant is sampling this population of claims at random. In practice, it is sufficient to simply assume that a very large number of claims have been filed, but it is still essential that the sampling be random.

The binomial distribution describes the number of "successes" in a fixed number of trials, where the chance of success on each trial is independent with constant probability θ. For n trials, the probability of exactly k successes is

$$\mathbb{P}[k \text{ successes }] = \left(\begin{array}{c} n \\ k \end{array} \right) \theta^k (1 - \theta)^{n-k}$$

where

$$\left(\begin{array}{c} n \\ k \end{array} \right) = \frac{n!}{k!(n-k)!} = \frac{n(n-1) \cdots 1}{[k(k-1) \cdots 1][(n-k)(n-k-1) \cdots 1]}.$$

(Deeper mathematics shows that $0! = 1$, which is a case that can arise in practice; for example, if none of the trials is a success, $k = 0$.) The binomial distribution is iconically used to describe the number of Heads obtained in n tosses of a coin that has probability θ of coming up Heads.

In our example, the forensic accountant decided how many claims to sample; this is her n. She then draws a random sample of n claims and audits each. A success is finding a fraudulent claim. If she records a 0 for a failure and a 1 for a success, then, in the notation of (3.2), she observes $x_1, \ldots, x_n$, where each observation x_i is either a one or a zero. Based on this sample, she now wants to make an inference about θ, the proportion of fraudulent claims.

As a Bayesian, she has a prior opinion about the value of θ. If that opinion happens to be expressible as a beta distribution, then she is in a conjugate family and can calculate the awkward integral in the denominator of (3.2).

A beta random variable may take any value between 0 and 1, inclusive. Since the beta distribution describes a continuous random variable, the probability that it takes any specific value is zero, but if one integrates the density function of a beta distribution between, say, 0 and 0.5, one obtains the probability the value p is less than or equal to 0.5.

The beta distribution is indexed by two parameters, traditionally represented as α and β. Both α and β must be greater than zero. The density function of a random variable θ with Beta(α, β) distribution is

$$\pi(\theta) = \frac{\Gamma(\alpha + \beta)}{\Gamma(\alpha)\Gamma(\beta)} \theta^{\alpha-1} (1 - \theta)^{\beta-1} \tag{3.3}$$

where $\Gamma(z) = (z - 1)!$ for all positive z (the definition in the case of non-integer z values derives from an integral representation of this function).

The beta family is quite flexible, and can represent many possible beliefs about θ. Figure 3.1 shows the shape of the density function for three possible pairs of indices (α, β).

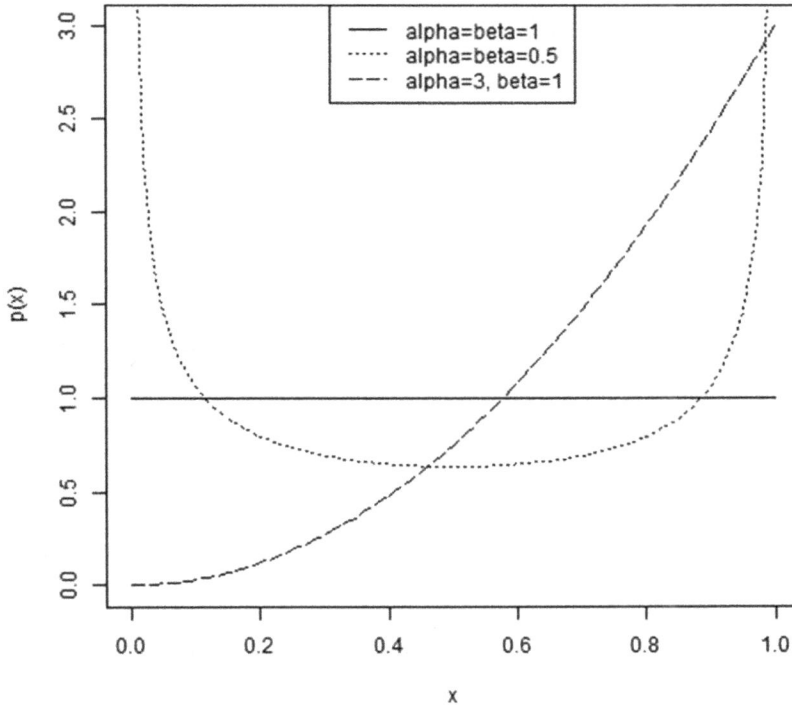

FIGURE 3.1
This figure shows three different densities for the beta distribution, which illustrate the wide range of prior beliefs that can be incorporated. The case $\alpha = \beta = 1$ is uninformative about the probability of success θ; the case $\alpha = \beta = 0.5$ represents the belief that θ is probably close to zero or one; and the case $\alpha = 3$, $\beta = 1$ puts more weight on larger values of θ.

Note that when $\alpha = \beta = 1$, the density is flat. The forensic accountant might use this density as her prior if she were completely agnostic about θ and thought that all values between 0 and 1 were equally likely. This would be a noninformative prior, since no value of θ is favored. (For a discussion of using a flat prior with data from a Supreme Court case, see Kaye, 1982, p. 779.) Alternatively, it might be reasonable to think that the automobile repair chain is either mostly honest or mostly dishonest, in which case she could use the informative beta prior whose density has $\alpha = \beta = 0.5$. Or, based on other evidence, she might believe that the chain is a criminal enterprise, and then she might pick the density corresponding to $\alpha = 3$ and $\beta = 1$.

At this point the Bayesian forensic accountant has collected the sample $x_1, \ldots, x_n$ and selected the α and β that specify her beta prior. She now wants to solve (3.2) in order to determine her new belief about the density of θ, the probability of a fraudulent claim given what she has learned from her sample. She can replace $x_1, \ldots, x_n$ by the number of successes, i.e., the number of fraudulent claims. (In statistics, there is often a *sufficient statistic*, one that summarizes all the relevant information in the data; in this case one can show that $k = \sum x_i$ is sufficient for inference on θ.) Then the posterior density is

$$\pi^*(\theta \mid k) = \frac{f(k \mid \theta)\pi(\theta)}{\int_{-\infty}^{\infty} f(k \mid \theta)\pi(\theta)\, d\theta}$$

$$= \frac{\left[\binom{n}{k}\theta^k(1-\theta)^{n-k}\right] * \left[\frac{\Gamma(\alpha+\beta)}{\Gamma(\alpha)\Gamma(\beta)}\theta^{\alpha-1}(1-\theta)^{\beta-1}\right]}{\int_0^1 \left[\binom{n}{k}\theta^k(1-\theta)^{n-k}\right] * \left[\frac{\Gamma(\alpha+\beta)}{\Gamma(\alpha)\Gamma(\beta)}\theta^{\alpha-1}(1-\theta)^{\beta-1}\right]\, d\theta}$$

$$= \frac{\Gamma(n+\alpha+\beta)}{\Gamma(k+\alpha)\Gamma(n-k+\beta)}\theta^{k+\alpha-1}(1-\theta)^{n-k+\beta-1}$$

which is the Beta$(\alpha + k, \beta + n - k)$ density.

The trick to making this conjugate family work is to recognize that the denominator has integrated out θ, so it is some constant. And the numerator is a member of the beta family, since its kernel (the terms in the integrand that depend upon θ) is the product of θ to a power times $1 - \theta$ to a power. So the constant in the denominator combined with the constants in the numerator have to be the value that forces the integral of the density to equal 1. From (3.3), it is clear that the constant must be $\Gamma(n+\alpha+\beta)/\Gamma(k+\alpha)\Gamma(n-k+\beta)$.

The posterior density is the complete expression of the forensic accountant's new belief. She can use it in several ways. For example, the district attorney may have an informal rule of thumb that if the proportion of invalid claims is less than 3%, then it is probably error rather than fraud, and not worth prosecutorial effort. In that case, the forensic accountant would integrate her posterior for θ over the region from 0 to 0.03, to calculate her probability that the district attorney should decline to prosecute.

But it is instructive to look at the mean and variance of her posterior distribution, to see how the new data have affected her previous opinion. One can show that the mean μ and variance σ^2 of a beta distribution with parameters α and β are

$$\mu = \frac{\alpha}{\alpha + \beta} \qquad \sigma^2 = \frac{\alpha\beta}{(\alpha + \beta)^2(\alpha + \beta + 1)}.$$

Looking first at the posterior mean μ^*, which is a beta distribution with parameters $\alpha^* = \alpha + k$ and $\beta^* = \beta + n - k$, we can write

$$\mu^* = \frac{\alpha + k}{\alpha + \beta + n} = c\frac{\alpha}{\alpha + \beta} + (1 - c)\frac{k}{n}$$

where $c = (\alpha + \beta)/(n + \alpha + \beta)$. This representation of μ^* is a weighted average of the mean of the prior and the mean of the data (k out of n successes implies the average of the data is k/n). As the sample size increases, more weight is placed upon upon the empirical average, and less upon the prior mean, which makes sense. The forensic accountant might have silly prior beliefs, but if she observes lots of data and follows Bayes' Rule, her opinions will shift toward what the data are revealing. This type of shift, to take more account of the data as the sample size grows, is standard in Bayesian analysis.

Looking at the posterior variance, one sees that

$$\sigma^{*2} = \frac{(\alpha + k)(\beta + n - k)}{(\alpha + \beta + n)^2(\alpha + \beta + n + 1)}.$$

This formula shows that as n increases, the posterior variance shrinks, so the forensic accountant's uncertainty becomes smaller and her probability concentrates upon a shrinking region within the interval $[0, 1]$ in which θ, the probability of success, must lie.

Before leaving this mock example, there are a few points to underscore regarding the use of subjective priors. First, Savage's results on the rationality of Bayesian decision makers requires the analyst to use her true subjective prior, rather than an noninformative (or objective) prior. Second, noninformative priors are not simple; they may not exist, they may not be unique, and they may be improper (meaning that their integral is infinite, rather than equal to 1). For the beta-binomial example, the Haldane prior, the Jeffreys prior, and the maximum entropy prior are all, in different technical senses, objective priors (cf. Kass and Wasserman, 1996). Third, as the sample size grows, all Bayesians who put non-zero probability on the entire domain of θ will converge to the same opinion, no matter which starting point prior each chose.

3.2.2 A Gamma-Poisson Mock Example

For another mock case study that illustrates several facets of Bayesian reasoning, suppose that a hospital review system flags a nurse who has had an unusually large number of his patients die during his shift. The police are consulted, and the district attorney wants to determine how improbable that observed number of deaths might be. The matter is complicated since well-known cases, such as that of Lucia de Berk (cf. Derksen and Meijsing, 2009), have highlighted the fact that, just by chance, an innocent nurse may have an unusual number of fatalities on his shifts. And if hospitals monitor many, many nurses, then it is certain that some innocent nurses will seem suspicious.

The analysis in this situation could entail the use of the gamma-Poisson conjugate family. The Poisson distribution describes the number of independent events that occur in a fixed amount of time, a fixed area, or a fixed volume. Assuming that an innocent nurse has independent deaths on his watch, is reasonable to model the number of deaths as a Poisson random variable with unknown mean θ. For expository simplicity, it is helpful to assume that the period of time in this application is one year.

A Poisson random variable can only take the values $0, 1, \ldots$, without any upper limit. (Obviously, in applications, there is generally some upper bound, so this is technically only an approximation, but it is widely used and broadly accurate.) The probability of observing exactly k deaths is

$$\mathbb{P}[X = k] = \frac{\theta^k}{k!} \exp(-\theta).$$

Here $\exp(-\theta)$ is Euler's constant, approximately 2.718, raised to the power $-\theta$. Using a Taylor series, one can show that the average number of events is equal to θ.

To form a conjugate family, the prior belief about θ must be represented by a gamma distribution. The investigator would have to form that prior based on personal experience, other data, or input from experts. For example, if the nurse in question worked with geriatric patients, then one would expect more natural deaths than if he worked in the pediatric unit. And hospital administrators could provide data on death rates in similar units at similar hospitals. The investigator would combine this information, with her own subjective beliefs, to specify her prior.

The gamma family is a group of distributions indexed by two parameters, α and β, both of which must be greater than zero. The gamma density function for a random variable θ is written as

$$f(\theta) = \begin{cases} \beta^\alpha \theta^{\alpha-1} \exp(-\beta\theta)/\Gamma(\alpha) & \text{for } \theta > 0 \\ 0 & \text{otherwise.} \end{cases}$$

Be warned that some textbooks write this density function with an alternate parameterization, so the mathematical expression, although equivalent, looks a little different. If one is working with a gamma density, be sure to check which parameterization is being used.

The gamma family includes the exponential distributions and the chi-squared distributions as special cases. Figure 3.2 illustrates some of the beliefs about θ that can be described through a gamma distribution.

The mean of a gamma distribution is α/β, and the variance is α/β^2. When specifying a subjective prior, it is often helpful to try to quantify one's uncertainty by thinking about the mean v of the random variable θ and one's uncertainty about its value (perhaps expressed as the range $[L, U]$ that one believes has probability 0.95 of including the true mean of the distribution). Then one can solve the system of equations

$$v = \alpha/\beta \text{ and } [(U - L)/4]^2 = \alpha/\beta^2 \tag{3.4}$$

to find values for α and β that reflect one's personal beliefs. This strategy for eliciting one's belief is simple and based upon confidence intervals for normal distributions. But there are alternatives—the literature upon elicitation of Bayesian subjective beliefs is large. A standard reference is O'Hagan et al. (2006).

If there are data on the number of deaths for nurses who are not under suspicion, say $x_1, \ldots, x_n$, then the analyst can update her subjective prior through Bayes' Rule to find her posterior belief about the distribution of θ for innocent nurses. Suppose the initial subjective prior has parameters α and β. Then it turns out that the posterior will be a gamma density with parameters $\alpha^* = \alpha + \sum x_i$ and $\beta^* = \beta + n$. This calculation can be a little tricky, since one must use data from other nurses who work with similar patients (e.g., geriatric or pediatric), and one must take care to adjust for the differing number of hours each nurse worked during the year.

Returning to the hypothetical case, suppose the investigator wanted to find the posterior distribution for the annual death rate θ of the suspect nurse. As will be seen, this is not the best thing to do, but it is a starting point for a naive Bayesian analysis.

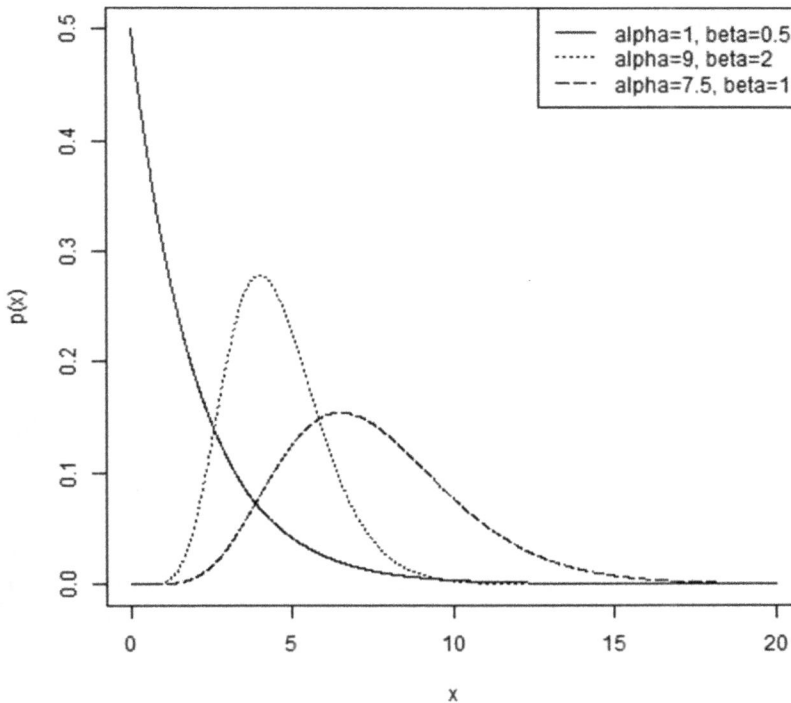

FIGURE 3.2
This figure shows three different densities for the gamma distribution, which illustrate the kinds of prior belief that may be represented. The case $\alpha = 1$, $\beta = 0.5$ corresponds to an exponential distribution; the other two cases show different means, variances, and skewness.

Her use of a prior implicitly accords with the fact that not all nurses have the same value of θ. A well-trained and experienced nurse might have a θ value slightly smaller than a less experienced nurse. This is related to random effects models that are discussed in the next section.

If x people have died on the suspect nurse's watch, then the conjugacy property shows that the posterior distribution for the suspect nurse's annual death rate θ has gamma distribution with parameters $\alpha^* = \alpha + x$ and $\beta^* = \beta + 1$. If this density has a mean that is very much larger than the prior mean α/β, then that could be interpreted as evidence against the nurse.

A frequentist might want to decide whether the nurse's annual death rate θ is less than or equal to θ_0, the average value for all similar nurses, or perhaps a value that is set by hospital policy as a proficiency standard. The frequentist would then test the null hypothesis that the suspect nurse's θ is less than or equal to θ_0 against the alternative hypothesis that the nurse's θ is greater than θ_0. This entails calculation of a test statistic and, from that, a significance probability. If the significance probability were very low, say less than 0.01, then the frequentist would reject the null hypothesis and conclude that the nurse's mean annual death rate was greater than θ_0 (see Chapter 2). Technically, the significance probability is the chance of observing a death count as high or higher than the suspect nurse's count when the null hypothesis is true.

From a legal standpoint, one must be careful about the conclusion drawn from rejection of the null hypothesis. The investigator decides that the suspect nurse's fatality rate is larger than the specified null value, but this could be due to incompetence rather than criminality, or a management policy that assigns the most senior nurse to the sickest patients. Nonetheless, such analyses are sometimes undertaken as part of an investigation (for cases and discussions, see Fienberg and Kaye, 1991; Loue, 2010).

Although it may not be the best approach, the Bayesian analogue to this hypothesis test would calculate the posterior probability that the nurse's θ was greater than θ_0. The Bayesian calculates the posterior distribution for the nurse's θ, and then integrates over the parameter space from θ_0 to infinity, to find

$$\mathbb{P}[\theta > \theta_0 | x] = \int_{\theta_0}^{\infty} \pi^*(\theta | x) \, d\theta,$$

which is the posterior probability that the nurse's θ exceeds the hospital's threshold of θ_0. Note that (1) the Bayesian analysis sidesteps the awkwardly expressed significance probability, (2) the solution was found by integrating over the parameter space, and (3) the uncertainty about θ is directly expressed as a probability, which frequentists may not do.

A different analysis might consider the probability of observing x or more deaths if the distribution for the nurse's death rate θ were a draw from the prior distribution. This is not actually a Bayesian analysis, since, among other things, Bayesians do not condition on unobserved events (i.e., Bayesians do not find the probability seeing more deaths than were counted). Instead, it is an attempt to harness a Bayesian tool to a frequentist purpose. However, note that in this situation, the frequentist logic forces the analyst to calculate the probability of x or more deaths, since the probability of seeing exactly x deaths can be misleadingly small; the probability of seeing exactly ten ants in a garden is low, but the chance of seeing ten or more could be large.

As before, this calculation requires integration over the parameter space. This is not permitted in frequentist analyses, but it allows the Bayesian investigator to calculate the probability of observing exactly x deaths when the uncertainty about the death rate is expressed as a distribution, which can then be repeated for $x + 1$, $x + 2$, and so forth. This hybrid Bayes-frequentist approach incorporates the fact that not all nurses have the same value θ—innocent nurses presumably have θ values that are close to some lower value than a guilty nurse.

Specifically, the Bayesian investigator solves

$$\mathbb{P}[x \text{ deaths }] = \int_0^{\infty} \mathbb{P}[x \text{ deaths } | \theta] \pi(\theta) \, d\theta$$

$$= \int_0^{\infty} \frac{\theta^x}{x!} \exp(-\theta) \beta^{\alpha} \theta^{\alpha-1} \exp(-\beta\theta) / \Gamma(\alpha) \, d\theta$$

$$= \frac{\beta^{\alpha}}{x! \Gamma(\alpha)} \int_0^{\infty} \theta^{\alpha+x-1} \exp[-(\beta+1)\theta] \, d\theta$$

$$= \frac{\beta^{\alpha}}{x! \Gamma(\alpha)} \frac{\Gamma(\alpha+x)}{(\beta+1)^{\alpha+x}}.$$

The last step follows by recognizing the kernel of the gamma density in the integral, and recalling that the gamma density, integrated from zero to infinity, equals 1. This calculation can be repeated for the values $x + 1, x + 2, \ldots$, and the sum of a small number of these probabilities quickly approximates $\mathbb{P}[x$ or more deaths].

If the probability of observing a nurse with x or more deaths in, say, a year is very small, then this could be considered evidence of homicide. A numerical example is helpful. Suppose the investigator assumes that nurses in this kind of ward will have an average of 4 patient deaths per year, with a standard deviation of 2. Then, from (3.4), the gamma prior has $\alpha = 1$ and $\beta = 0.5$. Suppose that the suspect nurse has had 12 patients die on during his shift this year. Then, for this prior, the probability of exactly that many deaths is

$$\frac{\beta^\alpha}{x!\Gamma(\alpha)} \frac{\Gamma(\alpha + x)}{(\beta + 1)^{\alpha+x}} = \frac{0.5^1}{12!\Gamma(1)} \frac{\Gamma(13)}{(1.5)^{13}} = 0.00257. \tag{3.5}$$

So the chance of a good nurse having exactly 12 deaths in one year is very small.

To find the probability of 12 or more deaths in a year, the investigator calculates

$$\mathbb{P}[x \text{ or more deaths }] = \sum_{x}^{\infty} \frac{\beta^\alpha}{x!\Gamma(\alpha)} \frac{\Gamma(\alpha + x)}{(\beta + 1)^{\alpha+x}}. \tag{3.6}$$

For our conveniently chosen values of α and β, (3.6) is a geometric series and can be summed exactly; it equals 0.00771. For less convenient values, the summation can be easily approximated, but that detail lies outside the ambit of this discussion.

However, recall that there are probably many, many nurses who work in hospitals that monitor staff for unusual fatality patterns. Just by chance, one unlucky nurse will surely accumulate a suspicious number of deaths. To correct for this, one must calculate the probability that among m monitored nurses who are innocent, at least one of them will have 12 or more fatalities during the year.

There are different strategies for making this correction. In principle, a well-calibrated Bayesian need not make any adjustment, because her prior for the guilt of the nurse takes account of the fact that there are many nurses being monitored and thus the chance that this specific nurse is criminal is automatically low. Alternatively, the *False Discovery Rate* of Benjamini and Hochberg (1995) admits a Bayesian interpretation (cf. Muller et al., 2006). Scott and Berger (2006) develop a Bayesian solution for multiple testing when each of the hypotheses is independent, in the sense that the guilt or innocence of one nurse is unaffected by the guilt or innocence of another, and Gelman et al. (2012) tackles the more complicated case when dependence is present. These approaches are generally somewhat technical.

For an easier path, consider a Bayesian perspective on adjustment for multiple testing (Westfall et al., 1997). Let p be the probability that an innocent nurse has k or more fatalities in a year. Then among m monitored nurses who are innocent of murder, the probability that all of them will have fewer than k fatalities is $(1 - p)^m$. This is simple probability, and only assumes that the nurses are independent. From this, it follows that the probability that one or more nurses has k or more fatalities is $1 - (1 - p)^m$.

Recall the mock example, in which the nurse's probability of having 12 or more deaths in one year is 0.00771. If 10 comparable and innocent nurses are being monitored, then the probability that at least one of them reaches this level of suspicion is 0.0745. If there are 20 comparable nurses, then the probability that one or more seems suspicious is 0.1434, and if there are 100, then the probability is 0.5388. In a large hospital system, it seems quite likely that many nurses are monitored, and so the evidence in this example is probably not determinative on its own.

Finally, before leaving this example, consider the classical Bayesian approach, which avoids the awkward hybrids and compromises discussed previously. It is directly comparable to the ELISA example. Either the nurse is innocent or guilty—that provides a finite partition. The investigator (or a juror) has a personal subjective prior probability for innocence or guilt, which is based upon all that they know about the circumstances before seeing the data (i.e., the number of fatalities during the year on that nurse's shift).

To find the posterior probability of guilt, a Bayesian must build a probability model for the number of deaths for an innocent nurse and the number of deaths for a guilty nurse. Different people would likely build that model in different ways. Earlier in this section, there was a model that found the probability of exactly 12 deaths for an innocent nurse as 0.00257. Similar reasoning could build a model for the probability of a guilty nurse having 12 fatalities in one year. It is likely that conjugate families would not seem reasonable to some of these Bayesians, in which case they would use Markov chain Monte Carlo methods, as discussed in the next section.

In any case, after eliciting their personal priors for innocence and guilt, and calculating the probability of exactly 12 deaths under the model for innocent and guilty nurses, all of the Bayesians would then find the posterior probability as

$\mathbb{P}[$ guilt $|12$ deaths$]$

$$= \frac{\mathbb{P}[\text{guilt}]\mathbb{P}[12 \text{ deaths} \mid \text{guilt}]}{\mathbb{P}[\text{guilt}]\mathbb{P}[12 \text{ deaths} \mid \text{guilt}] + \mathbb{P}[\text{innocence}]\mathbb{P}[12 \text{ deaths} \mid \text{innocence}]}.$$

Here $\mathbb{P}[$ guilt $]$ is the prior probability of guilt and $\mathbb{P}[$ innocence $] = 1 - \mathbb{P}[$ guilt $]$. Note that this traditional Bayesian analysis did not attempt to find the probability of observing 12 or more deaths. Instead, it simply used the observed count.

In Bayesian hypothesis testing, one decides which of two models (e.g., those for guilt and innocence) is more likely. The *Bayes factor* measures the degree to which the data support one or the other model, and it corresponds to a likelihood ratio (see Chapter 2). Formally, the Bayes factor $BF(H_0, H_A)$ is the ratio of the posterior odds of the null hypothesis H_0 and alternative hypothesis H_A to the prior odds of the null and alternative. Specifically,

$$BF(H_0, H_A) = \frac{\mathbb{P}[H_A|D]/\mathbb{P}[H_0|D]}{\mathbb{P}[H_A]/\mathbb{P}[H_D]}$$

where D indicates the data. It is not obvious from this expression, but it is nonetheless true, that when two models or two hypotheses explain the data equally well, the Bayes factor automatically penalizes the more complex model, and thus applies Occam's Razor (Kass and Raftery, 1995).

A Bayes factor of, say, 5 means that after seeing the data, the alternative hypothesis is five times more likely relative to the null hypothesis than it was *a priori*, before seeing the data. In scientific settings, a Bayes factor of 10 or more is generally considered to be strong evidence for the alternative hypothesis (Jeffreys, 1998 [1961], p. 432). But in forensic settings, the standard is generally much more stringent. The Department of Justice (2018, p. 4) proposed uniform language for legal testimony in which a Bayes factor between 2 and 100 is described as "limited support," a factor between 100 and 10,000 corresponds to moderate support, a factor between 10,000 and 1,000,000 is strong support, and a value greater than 1,000,000 is very strong support.

The logarithm of the Bayes factor is called the *weight of evidence*. Good (1985) has a slightly theoretical discussion of this concept, partly in the context of the legal system. Also, for more extensive discussions of the use of Bayesian reasoning in court cases, see Fienberg and Kadane (1983); Kaye et al. (2011).

3.3 Markov Chain Monte Carlo

Conjugacy is sometimes helpful, but most realistic applications are more complex. It is likely that distribution of the data and distribution of the honest Bayesian belief do not correspond to a conjugate family, and thus, in general, there is no solution that can be expressed as a finite number of standard mathematical terms—some numerical approximation is needed, and these can be computationally intractable or difficult to interpret. This limited the use of Bayesian methods for almost a century.

Gelfand and Smith (1990) found a way forward. Instead of trying to approximate the posterior density analytically, they developed a computer-intensive procedure that could sample repeatedly from the posterior distribution. Variants of this procedure are called Gibbs Sampling and Markov chain Monte Carlo (MCMC). The former is a special case of the latter, and these procedures have been further developed and improved by many researchers.

When one can draw large samples from the posterior density, it is possible to estimate the density with as much accuracy as the application requires. With a sample of size n, one can put weight $1/n$ on each observation to obtain an estimator that is a (possibly multivariate) histogram that converges to the true density function. If more accuracy is needed, one can simply increase n.

These procedures are not foolproof, and a number of technical problems that can arise (although these pathologies generally do not arise in reasonably posed problems—MCMC is robust).

One potential problem is that the Markov chain can take a long time to converge to its stationary distribution, the point at which the samples that are drawn are taken from the true posterior distribution. The usual solution is to let the chain run for a several thousand iterations, discard those draws, and then begin recording the sample values (this initial period is called the *burn-in*). To confirm that the chain has reached its stationary distribution, the statistician will typically monitor *trace plots*. These plots help diagnose when convergence has been reached, and other technical problems that can occur.

A second issue is that, at convergence, the sampled observations are draws from the posterior, but they are not an independent (random) sample. Consecutive draws are correlated, and so one needs to draw very large samples to ensure that the empirical distribution closely approximates the true posterior distribution.

A third potential issue arises when the support of the posterior density (i.e., the region where the density is greater than zero or, equivalently, the region where there it is possible to observe a data point) has a complex shape. Recall that the support of the beta density was the interval from 0 to 1, and the support of the gamma density was the positive numbers. These are simple supports. But in multivariate settings, certain geometries of the support lead to slow mixing (i.e., slow exploration of the support), so the analyst could be deceived into thinking that the histogram of the data accurately reflects the true distribution when it does not.

Multivariate parameter spaces arise when measuring the amounts of several trace elements in a glass fragment or bullet lead. For example, the FBI used to perform a test called Compositional Analysis of Bullet Lead (CABL). The test estimated the proportion of seven trace elements in a bullet: antimony, copper, arsenic, silver, tin, bismuth, and cadmium. This implies a seven-dimensional parameter space, and the goal is to estimate $(\theta_1, \ldots, \theta_7)$, the proportions of the trace elements in bullet. When these means were very similar to the estimates from another bullet, the FBI experts used to testify that the bullets were manufactured on the same day, from the same "melt" of lead.

CABL testimony is no longer given, following scientific criticism (National Research Council, 2004), but it is illustrative to note that in this example the support of the posterior density will seven-dimensional simplex, since $\theta_i \geq 0$ for all $i = 1, \ldots, 7$ and $\sum \theta_i \leq 1$. This simplicial region is not problematic for MCMC, but more exotic regions are.

Suppose, for example, that the support of the multivariate density looked like a dumbbell. Then the MCMC algorithm might easily wander around in one of the end weights for a very long time before finding the bar that connects to the other end weight. In that case the traceplot would look fine for a long time, and the analyst might wrongly conclude that convergence had been achieved, even though less than half of the support had been explored. (But, as is probably evident, this kind of pathology arises very rarely, and should not be problematic for routine Bayesian applications.)

The main message is that Bayesian statisticians are now able to analyze problems that are much more complex and realistic than used to be possible. It has been used in many cases to interpret complex DNA mixtures (cf. Chapter 11). MCMC, the computational algorithm for doing such analyses, is a mature and reliable technique, but it should be implemented by a trained statistician since there are certain kinds of pitfalls that require special expertise to avoid.

For additional discussion of MCMC, there are a number of possible books, at a range of technical detail. One that is both popular and fairly accessible is Gelman et al. (2013). Gamerman and Lopes (2006) is a bit older and more theoretical. McElreath (2018) is recent, fairly introductory, and has many examples.

3.4 Broad Applications

Modern Bayesian analysis can address essentially any question that frequentist methods do, but there are some areas where the Bayesian perspective is especially elegant. These situations are probably rare in legal contexts, but it may be useful to sketch a few examples.

First, consider *random effects* models. To return to the automobile repair example, suppose one suspects that a particular shop routinely overcharges for, say, an engine tune-up. A sample of data from different repair shops reveals what each has charged for tune-ups performed during the previous year. Some shops have done many repairs, but for others, there are only a few observations. The goal is to calculate each shop's average cost, borrowing strength from those with many observations to improve estimates for those with few.

Let Y_{ij} be the charge for the ith tune-up at the jth shop. Then the *fixed effects* model is

$$Y_{ij} = \mu + b_j + \epsilon_{ij}$$

where μ is the average of all shops, b_j is the effect of the jth shop, and $\epsilon_{ij} \sim N(0, \sigma^2)$. The fixed effects estimate of b_j is $\bar{Y}_j - \bar{Y}$.

In a fixed effects model, it is very difficult to borrow strength. But in a random effects model, it is easy. Fixed effects models are not Bayesian hierarchical models, but random effects models are. One can use the Durbin-Wu-Hausman test (Hausman, 1978) to decide whether to use a fixed effects or random effects model.

The random effects model is similar to the fixed effects model but allows the parameters to be random. The model is

$$Y_{ij} = \mu + b_j^* + \epsilon_{ij}$$

where the prior is that $b_j^* \sim N(0, \tau^2)$. In this case, the Best Linear Unbiased Predictor (BLUP) of b_j^* is

$$\hat{b}_j^* = \frac{\tau^2}{(\sigma^2/n + \tau^2)}(\bar{Y}_j - \bar{Y}).$$

So the estimate shrinks towards zero by an amount that depends on the magnitudes of τ^2 and σ^2. More completely, one obtains a posterior distribution for each shop's effect:

$$b_j^* | Y \sim N\left(\frac{\tau^2}{(\sigma^2/n + \tau^2)}(\bar{Y}_j - \bar{Y}), \frac{\tau^2 \sigma^2/n}{\tau^2 + (\sigma^2/n)}\right).$$

Shrinkage borrows strength by saying that the charges at the jth shop tend to be similar to those at the other shops. This could enable an attorney understand how much variation there is in average tune-up fees across different shops, and thus to better assess whether the suspect shop's charges are extreme. This methodology would be directly applicable to cases alleging wage discrimination, excessive pollution, or medical malpractice, where the observations would be salaries in similar jobs, daily emissions of nitrogen oxide, or survival times of patients following a procedure, respectively.

Besides fixed effects and random effects models, there are *mixed effects* models (cf. Sahai and Ageel, 2012, for a discussion of all three types of models). These arise when some of the variables are not random, but others are. In the repair shop example, where older vehicles would be expected to require more labor, there might be information on the age of the vehicle, A_i, which is a fixed effect. In that case, the model is

$$Y_{ij} = \mu + b_j^* + \beta A_i + \epsilon_{ij}$$

where β is a non-random unknown fixed effect and the b_j^* are random effects.

Let the vector of scores be Y, the vector of ages be a, and the vector of errors (individual effects) be ϵ (all are $n \times 1$). There are k shops, and let the $k \times 1$ vector of shop effects be b^*. Suppose $b^* \sim N(0, \tau^2 I)$, $\epsilon \sim N(0, \sigma^2 I)$, and that these two random vectors are independent.

With these assumptions (which can be made much more general), the point estimates are:

$$\begin{pmatrix} \hat{\beta} \\ \hat{b}^* \end{pmatrix} = \begin{bmatrix} a'a/\tau^2 I & a'Z/\tau^2 I \\ Z'a/\tau^2 I & Z'Z/\tau^2 I + \sigma^{-2} I \end{bmatrix}^{-1} \begin{bmatrix} a'Y/\tau^2 I \\ Z'Y/\tau^2 I \end{bmatrix}$$

where I is the identity matrix (i.e., a square matrix that has ones on the diagonal from top left to bottom right, and zeroes for all other entries) and Z is the $n \times k$ design matrix that assigns vehicles to shops. Specifically, the i, jth entry of Z is zero unless vehicle i is tuned-up at shop j, in which case it the entry is 1.

The important point is that one can improve estimates using Bayesian methods to borrow strength from similar cases, while still incorporating known covariates. Thus, an age discrimination suit could control for fixed effects, such as years of education, while still using random effects models to identify inequities. Or one could use fixed effects such as hours of operation to assess the amount of pollution from a factory, or the severity of illness when looking at patient survival times.

A second useful technique is hierarchical Bayesian models. In these, one places a prior over the parameters in the prior distribution, and one can extend this to higher levels by placing a third prior on the parameters in the second prior (forming a hierarchy). These models tend to be robust, in the sense that the posterior distribution is less sensitive to the prior specification. Additionally, there are many situations where this approach facilitates the construction of reasonable models.

The random and mixed effects repair-shop cases were two-level hierarchical models. There was a prior on the the distribution of the effect, which was input to the model for the data. In general, a three-level hierarchical model finds

$$\mathbb{P}(\theta, \phi | X) = \frac{\mathbb{P}[X | \theta, \phi] \mathbb{P}[\theta, \phi]}{\mathbb{P}[X]} = \frac{\mathbb{P}[x | \theta] \mathbb{P}[\theta | \phi] \mathbb{P}[\phi]}{\mathbb{P}[X]}.$$

There can be many more levels, but the mathematics is essentially the same—there are just more parameters.

A third area in which Bayesian methods are attractive is variable selection. A famous example in which this would apply is the case brought against Harris Trust & Savings Bank for wage discrimination against women (U.S. Department of the Treasury v. Harris Trust and Savings Bank, 1979). The bank denied discriminating and contended that wages were based only on seniority, experience, education, and age.

The case was complex and had many facets, but a statistical analysis by Harold Roberts presented during the trial used a multiple linear regression model in which the response variable was salary and the explanatory variables were seniority, experience, education, age, and gender. Roberts did a standard frequentist t-test to determine whether the gender coefficient was significantly different from zero. The test showed that the coefficient was very different from zero, which strongly suggested that gender was indeed a factor in determining wages. (Although experts continue to use such multiple linear regression in discrimination litigation, there are better tools.)

The Bayesian alternative (not used in the case), has a prior that puts probability p on the value of the zero for the gender coefficient, and probability $1 - p$ on the gender coefficient being a random draw from a very dispersed distribution, such a normal distribution with mean 0 and standard deviation 5,000. The Bayesian would use the salary data to get a posterior distribution on the coefficient, and presumably would find that the posterior probability p^* that the coefficient is zero is quite small, and that the posterior probability that the coefficient is negative (indicating discrimination against women) and large. The analyst would also calculate the posterior mean of the coefficient, which would also be useful in assessing damages.

The Bayesian world is big, and there are many other kinds of methods and applications that are relevant to legal questions. But Bayesian tools should not be applied without guidance from experts with advanced degrees in statistics. Nor, for that matter, should frequentist analyses be done by non-professionals.

3.5 Summary

Bayesian inference is a versatile and well-established method of modern statistical inference. People are generally more likely to express opinions in terms of probability, as a Bayesian does, than in the contorted language of frequentist exposition.

However, Bayesian methods are less common in legal cases than they are in published scientific articles. There are several reasons for this: courts have accepted the use of other methods in the past; realistic applications are computer-intensive; and using prior information in framing an analysis and defending its conclusions can create difficulties for lawyers.

A much richer discussion of the Bayesian perspective in the context of the law is given in the book *Statistics and the Law*, edited by De Groot et al. (1986). It is dated only in that it precedes the invention of MCMC techniques, but the fundamental issues and philosophy are addressed from many angles. Case law and evidentiary issues are discussed in Kaye et al. (2011).

Acknowledgments

This work was done as part of the research program on Forensic Statistics that was organized by the Statistical and Applied Mathematical Sciences Institute, with support from the National Science Foundation.

References

Bayes, T. and Price, R. An essay towards solving a problem in the doctrine of chance. By the Late Rev. Mr. Bayes, Communicated by Mr. Price, in a Letter to John Canton, A. M. F. R. S. *Philosophical Transactions of the Royal Society of London*, **53**, 370–418, 1763.

Benjamini, Y. and Hochberg, Y. Controlling the false discovery rate: A practical and powerful approach to multiple testing. *Journal of the Royal Statistical Society Series B*, **57**, 289–300, 1995.

CDC. Estimated HIV incidence and prevalence in the United States, 2010–2016. HIV Surveillance Supplemental Report 2019;24(No. 1), 2017.

Chou, R., Huffman, L.H., Fu, R., Smits, A.K., and Korthuis, P.T. Screening for HIV: A review of the evidence for the U.S. preventive services task force. *Annals of Internal Medicine*, **143**(1), 55–73, July 2005.

DeGroot, M.H., Fienberg, S.E., and Kadane, J.B. (Eds.). *Statistics and the Law*. New York: Wiley, 1986.

Department of Justice. Uniform language for testimony and reports using forensic autosomal DNA examinations using probabilistics genotyping systems. https://www.justice.gov/olp/page/file/1095961/download. Verified on January 12, 2020, 2018.

Derksen, T. and Meijsing, M. The fabrication of facts: the lure of the incredible coincidence. In *Legal Evidence and Proof Statistics, Stories, Logic*, pages 39–70, 2009.

Fienberg, S.E. and Kadane, J.B. The presentation of Bayesian statistical analyses in legal proceedings. *Journal of the Royal Statistical Society: Series D (The Statistician)*, **32**(1–2), 88–98, 1983.

Fienberg, S.E. and Kaye, D.H. Legal and statistical aspects of some mysterious clusters. *Journal of the Royal Statistical Society: Series A*, **154**, 61–74, 1991.

Gamerman, D. and Lopes, H.F. *Markov Chain Monte Carlo: Stochastic Simulation for Bayesian Inference*. Boca Raton, FL: Chapman and Hall/CRC, 2006.

Gelfand, A.E. and Smith, A.F. Sampling-based approaches to calculating marginal densities. *Journal of the American Atatistical Association*, **85**, 398–409, 1990.

Gelman, A., Carlin, J.B., Stern, H.S., Dunson, D.B., Vehtari, A., and Rubin, D.B. *Bayesian Data Analysis*. Boca Raton, FL: Chapman and Hall/CRC, 2013.

Gelman, A., Hill, J., and Yajima, M. Why we (usually) don't have to worry about multiple comparisons. *Journal of Research on Educational Effectiveness*, **5**, 189–211, 2012.

Good, I.J. "Weight of evidence: A brief survey" (with discussion). In *Bayesian Statistics 2*, J. Bernardo, M. DeGroot, D. Lindley, and A. Smith, editors, pages 249–269. Amsterdam: Elsevier Science Publishers, 1985.

Hausman, J. Specification tests in econometrics. *Econometrica*, **46**, 1251–1271, 1978.

Jeffreys, H. *The Theory of Probability* (3rd ed.). Oxford University Press, England, 1998 [1961].

Kadane, J.B., Girón, J., Peña, D., Fishburn, P., French, S., Lindley, D.V., Parmigiani, G., and Winkler, R.L. Several Bayesians: A review. *Test*, **2**, 1–32, 1993.

Kass, R.E. and Raftery, A.E. Bayes factors. *Journal of the American Statistical Association*, **90**, 773–795, 1995.

Kass, R.E. and Wasserman, L. Formal rules for selecting prior distributions: A review and annotated bibliography. *Journal of the American Statistical Association*, **91**, 1343–1370, 1996.

Kaye, D.H. The numbers game: Statistical inference in discrimination cases. *Michigan Law Review*, **80**, 833–856, 1982.

Kaye, D.H., Bernstein, D.E., and Mnookin, J.L. *The New Wigmore: A Treatise on Evidence: Expert Evidence* (2nd ed.). Aspen Pub., New York, 2011.

Loue, S. *Forensic Epidemiology: Integrating Public Health and Law Enforcement*. Jones & Bartlett Learning, 2010.

McElreath, R. *Statistical Rethinking: A Bayesian Course with Examples in R and Stan*. Chapman and Hall/CRC, 2018.

Muller, P., Parmigiani, G., and Rice, K. FDR and Bayesian multiple comparisons rules, Johns Hopkins University, Department of Biostatistics Working Paper 115, 2006.

National Research Council. *Forensic Analysis: Weighing Bullet Lead Evidence*. The National Academies Press, Washington, DC, 2004. https://doi.org/10.17226/10924.

O'Hagan, A., Buck, C.E., Daneshkhah, A., Eiser, J.R., Garthwaite, P.H., Jenkinson, D.J., Oakley, J., and Rakow, T. *Uncertain Judgements: Eliciting Experts' Probabilities*. John Wiley & Sons, 2006.

Sahai, H. and Ageel, M.I. *The Analysis of Variance: Fixed, Random and Mixed Models*. Springer Science and Business Media, 2012.

Savage, L.J. *The Foundations of Statistics*. Wiley, New York, NY, 1954.

Scott, J.G. and Berger, J.O. An exploration of aspects of Bayesian multiple testing. *Journal of Statistical Planning and Inference*, **136**, 2144–2162, 2006.

U.S. Department of the Treasury v. Harris Trust and Savings Bank. U.S. Department of Labor Administrative Proceeding, Case No. 78-OFCCP-2 (Burrow, ALJ), 1979.

Westfall, P., Johnson, W., and Utts, J. A Bayesian perspective on Bonferroni adjustment. *Biometrika*, **84**, 419–427, 1997.

Young, G.A., Severini, T.A., and Smith, R.L. *Essentials of Statistical Inference*, Vol. **16**. Cambridge University Press, 2005.

4

Comparing Philosophies of Statistical Inference

Hal S. Stern

CONTENTS

4.1 Inferential Philosophies 91
    4.1.1 Frequentist Inference 92
    4.1.2 Bayesian Inference 92
    4.1.3 Other Approaches to Inference 93
4.2 Comparing the Approaches 94
    4.2.1 Planning Studies Using Frequentist Inference 95
    4.2.2 Challenges for Frequentist Inference 95
    4.2.3 Flexible Inference with Bayesian Methods 96
    4.2.4 Model Modifications and Adjustments 96
    4.2.5 The Prior Distribution and the Definition of Probability 97
4.3 Relevance to Forensic Statistics 97
    4.3.1 Likelihood Ratios and Bayes Factors 98
    4.3.2 Two-Stage Procedures in Forensic Science 99
    4.3.3 Forensic Evidence as Expert Opinion and Error Rates 100
4.4 Summary 100
References 100

4.1 Inferential Philosophies

Previous chapters in the volume (Banks and Tackett, 2020; Kaye, 2020) describe frequentist inference and Bayesian inference in some detail. There are fundamental differences in the two approaches especially when it comes to their view of and use of probability. Because probability plays a central role in the analysis and interpretation of forensic evidence it is important to compare and contrast the two approaches to statistical inference, as well as other statistical perspectives. This chapter does so, focusing on issues most relevant to the analysis and interpretation of forensic evidence. Section 4.1 provides a short review of the different inferential philosophies. Section 4.2 compares the frequentist and Bayesian approaches to inference in general terms. Then Section 4.3 provides an overview of key issues in forensic science and the relevance of inferential philosophies in resolving these issues. The chapter includes a review of the three most common approaches to assessing the probative value of forensic evidence: expert opinion, two-stage procedures, and likelihood ratios and Bayes factors.

For notational and conceptual simplicity, we focus on data $Y = (Y_1, \ldots, Y_n)$ where the Y_i are independent, identically distributed random variables from a probability distribution $p(y|\theta)$ characterized by the (potentially multivariate) population parameter θ. The three most common inference tasks are point estimation, interval estimation and hypothesis tests. Without loss of generality we assume that the goal is point estimation, interval estimation, or tests of hypotheses regarding the value of θ, although some function of the parameters may be of interest in a given scientific context. Naturally, more complex models are possible, including additional forms of dependence (e.g., longitudinal or clustered data) and hierarchical structure (e.g., random effects). The discussion below applies to these as well. Some highlights of the earlier chapters describing frequentist inference and Bayesian inference are provided along with short descriptions of other approaches.

4.1.1 Frequentist Inference

The key feature of frequentist inference is that procedures are evaluated based on their calibration relative to a potentially infinite sequence of repetitions of the data collection from the probability distribution $p(y|\theta)$ with the same fixed but unknown values of the underlying parameter (Neyman, 1923, 1937). As the parameter θ is considered fixed, the only use of probability is to describe the distribution of the random sample. The relevant definition of probability relies on the long term frequency of an event (e.g., the proportion of interval estimates that contain the parameter value). There is no general strategy for how procedures (point estimates, interval estimates, tests) are to be developed, but there is a clear prescription for how to evaluate procedures. A point estimator for θ, say $T(Y)$, is evaluated based on its bias and variance calculated over the probability distribution of T generated by all possible samples. It is often taken as a goal to obtain unbiased estimators, so that $E(T(Y)|\theta) = \theta$, and within the class of unbiased estimators to prefer estimators with smaller variance (see, e.g., Lehmann and Casella, 2003). In the same manner, interval estimates are expected to provide at least the advertised degree of coverage (i.e., $100(1 - \alpha)\%$ confidence intervals should contain the estimand θ with probability at least $1 - \alpha$ averaging over repeated samples). Finally, tests of hypotheses are evaluated with respect to the specified type I error rate α with the expectation that true null hypotheses are rejected with probability less than or equal to α. Among tests that satisfy this constraint, those with smallest type II error rate (highest power) for relevant alternatives are preferred (see, e.g., Lehmann and Romano, 2006). There are many mathematical results that identify procedures with good frequentist properties. For example, maximum likelihood estimators have been shown to be asymptotically unbiased, and asymptotic confidence intervals with good properties have been developed. It is worth noting that the description of statistical tests provided above focused on the traditional Neyman-Pearson form of hypothesis testing rather than the significance testing approach that emphasizes p-values. The latter is also a frequentist concept for assessing the evidence against a null hypothesis; the p-value is the probability in repeated samples of obtaining results as or more extreme than the observed result.

4.1.2 Bayesian Inference

The Bayesian approach is characterized by the specificiation of a probability model to describe the distribution of all random quantities (this includes both Y and θ), followed by inference for unknown quantities (e.g., θ) conditional on observed data (Savage, 1954; Lindley, 1971; de Finetti, 1972) and an assessment of the fit of the model to the data. One important distinction between the two approaches to inference is that the starting point for Bayesian modeling is usually taken to be that the set of observations Y are exchangeable in the sense that the joint distribution of any subset is the same regardless of the specific subscripts (or specific observations). This is related to, but not the same as, the common assumption of independent identically distributed (iid) random variables. To be more specific, iid random variables are exchangeable but a set of exchangeable random variables need not be iid; fortunately a result of de Finetti (1972) establishes that an infinite set of exchangeable random variables are iid relative to some latent variable (which can be taken as the parameter θ).

The full probability model used in a Bayesian analysis describes the joint distribution of Y and θ. It is often specified via the factorization $p(Y, \theta) = p(Y|\theta)p(\theta)$, where $p(Y|\theta)$ is known as the sampling distribution or the data distribution and $p(\theta)$ is known as the prior distribution of θ. Both the data distribution and prior distribution are chosen by the statistician, but as the prior distribution is unique to the Bayesian approach, it tends to receive special attention as being a subjective element of the analysis. For this reason (and because the probabilities associated with the parameter values are understood as degrees of belief), we often talk about the Bayesian approach as relying on subjective probability. The choice of $p(Y|\theta)$ that is relevant to both frequentist and Bayesian inference is also a subjective choice, though often informed by scientific or mathematical considerations (e.g., means of large samples follow a normal distribution regardless of the underlying distribution of the individual measurements). There are a range of strategies for specifying the prior distribution (see, e.g., Gelman et al., 2013), which include the use of conjugate families, weakly informative prior distributions (i.e., those with large variances), and "objective" or reference prior distributions. Once the probability model is specified, inference for θ follows automatically from Bayes' Theorem, $p(\theta|Y) = p(Y|\theta)p(\theta)/p(Y)$. Point estimation is carried out through the specification of a loss function (a function of θ and the estimate $\hat{\theta}$) and the minimization of the expected loss averaged over the posterior distribution. As is well known, the posterior mean is the point estimate that minimizes squared-error loss and the posterior median is the point estimate that minimizes absolute value loss. Interval estimates for θ are also determined by the posterior distribution; it is common to use either highest posterior density intervals or central posterior intervals. Formal hypothesis tests are rare in the Bayesian approach. Instead, specific hypotheses about θ are generally addressed by examining the posterior distribution. Thus one might "reject" a hypothesis if the posterior interval for θ does not include the value. More formal approaches to model selection sometimes rely on the Bayes factor (see, e.g., Kass and Raftery, 1995), but we do not address this topic further here.

4.1.3 Other Approaches to Inference

The frequentist and Bayesian approaches are the most commonly applied. However, alternative approaches to inference have been proposed. These generally try to take advantage of the strengths of the two most common approaches while addressing their perceived weaknesses.

4.1.3.1 Fiducial Inference

One such approach is Fisher's (1935) notion of fiducial inference. There is not a unique description of fiducial inference (Seidenfeld, 1992); it is easiest to describe the goal. Fisher hoped to obtain probability statements about the parameter θ given the data Y without relying on a prior distribution. As with other critics of Bayesian methods it was not clear to Fisher where prior information about θ might come from in some problems. The fiducial argument is most clear in the simple setting of a univariate location family such as a single observation from the normal distribution with known scale parameter (perhaps equal to 1). The pivotal quantity $V = Y - \theta$ has a standard normal distribution under these assumptions. Fisher argued that the standard normal distribuiton for V would continue to hold conditional on Y and that this would allow probability statements about θ without having specified a prior distribution on θ. There is, in general, no guidance for when fiducial inference is possible. Many of the known cases where a fiducial argument seems to work correspond to settings where there is a possibly improper reference prior distribution that produces traditional frequentist results. For these and other reasons, fiducial inference does not receive much attention.

4.1.3.2 Likelihood Inference

The sampling or data distribution $p(Y|\theta)$ describes uncertainty about the random data Y given a fixed value of the parameters θ. When the distribution is viewed as a function of θ for fixed data Y, it is often called the likelihood function, $L(\theta|Y)$. Thus, it is often said that the posterior distribution is proportional to the product of the prior distribution for θ and the likelihood function for θ. Another approach to inference that has been proposed is sometimes called likelihood inference or likelihoodism (Edwards, 1992; Royall, 1997). It takes as its starting point the likelihood principle, a mathematical proposition that for a given statistical model, all of the evidence contained in the data Y about the model parameters θ is contained in the likelihood function. The Bayesian approach to inference satisfies the likelihood principle since it only involves the data through the likelihood function. For advocates of the likelihood approach, however, the prior distribution is viewed as not being well justified. Instead, inference proceeds only with reference to the likelihood function. Thus, for example, an interval estimate for θ might be obtained by incorporating all values of θ for which the value of the likelihood function is within a certain fraction of the maximum.

4.1.3.3 Confidence Distributions

There has been recent attention on confidence distributions (Singh et al., 2007), a frequentist concept that allows for a data(sample)-dependent function that can be used to derive confidence intervals for a parameter θ. Though a frequentist construct, the confidence distribution has been shown to be related to Bayesian and fiducial inference ideas as well. Indeed, a renewed focus on confidence distributions was a key motivation for the development of a series of conferences (begun in 2014) known as the Bayesian, Fiducial, and Frequentist (BFF) Conferences. In terms of inference, this is a frequentist idea so we do not discuss it further here.

4.2 Comparing the Approaches

The previous section defined the Bayesian and frequentist approaches to inerence without emphasizing their strengths and weaknesses. Frequentist inference is still the dominant mode of inference taught in the majority of introductory statistics courses and applied in many scientific disciplines. Bayesian inference, despite its rich history dating back to the 1700s, has become much more popular recently in part because advances in computation have made it feasible to learn about the posterior distribution in complex models. Some of the key advantages (and disadvantages) of the two approaches are reviewed here.

4.2.1 Planning Studies Using Frequentist Inference

One advantage of the frequentist approach to inference is that understanding the frequentist operating characteristics of procedures makes it possible to plan studies. Thus, the traditional approach to planning many studies relies on the fact that a Neyman-Pearson hypothesis testing framework allows one to relate the sample size, the hypothesized alternative, information about the values of nuisance parameters, and the operating characteristics (frequency of type I and type II errors) of the test. Given specified values for the hypothesized alternative and the type I/type II error rates of the test, it is possible to identify the sample size required to achieve those operating characteristics. Obtaining this information in advance of a study is enormously beneficial to avoid wasteful studies. In the same ways, it is possible to plan the sample size required to obtain a confidence interval with a desired degree of precision. The next paragraph describes some concerns associated with the interpretation of confidence intervals and hypothesis tests, but one reason for their popularity is the ability to use the operating characteristics of such procedures to plan studies.

4.2.2 Challenges for Frequentist Inference

A variety of well-established techniques with good frequentist properties such as the t-test and analysis of variance have been developed and applied broadly. There are, however, some conceptual difficulties associated with frequentist confidence intervals and hypothesis tests. One problem is that the interpretation of frequentist confidence intervals can be confusing and often doesn't match the expectations of users. The justification for the stated confidence level of a frequentist confidence interval is that the specified proportion of intervals constructed using the given procedure will contain the true (but unknown) parameter value. That is, if we use the same procedure to build a 95% confidence interval in a series of repeated studies containing data from the underlying probability model $p(Y|\theta)$, then 95% of these intervals will include the true value of θ. This does not mean that the probability associated with a given confidence interval is 0.95. Indeed, under the frequentist interpretation, it is not appropriate to talk about the probability associated with a single confidence interval. The parameter is thought of as a fixed unknown quantity, and once the sample is obtained, the data too are no longer random. Our particular interval either does or does not contain the unknown parameter. Our "confidence" in the interval derives from the known properties of the method used to construct the interval were it to be applied in many repeated samples. The same conceptual challenge holds for point estimates. Knowing that a particular estimator is unbiased or that it has the smallest possible varance among unbiased estimators does not provide any information about how well a particular point estimate has performed in a given sample.

A related challenge is associated with the use of frequentist hypothesis and significance tests. Neyman-Pearson hypothesis tests provide a decision rule for distinguishing between the null and alternative hypothesis for which the operating characteristics (e.g., proportion of true null hypotheses for which the null hypothesis would be rejected) are known. These characteristics do not reveal whether the current study is one of these false rejections. Making a series of decisions using the same procedure provides assurances about the operating characteristics over the series of decisions but not for an individual instance. In addition, some have expressed concerns about the asymmetry with which the two hypotheses are treated with the null hypothesis featured and type I errors controlled.

Bayesian inference allows one to make probability statements about parameters conditional on observed data. Thus a Bayesian posterior interval can be correctly interpreted as providing an interval that has the specified probability of containing the unknown parameter *given the specified probability model*. Because the model includes the subjective prior distribution, this is sometimes referred to as a subjective or personal confidence interval. In many simple settings, it is possible to define reference prior distributions such that Bayesian posterior intervals match traditional frequentist confidence intervals. For example, given a normal distribution model for Y, that is to say, $Y_i, i = 1, \ldots, n \sim$ iid $N(\mu, \sigma^2)$, and the non-informative (and improper) prior distribution $p(\mu, \sigma^2) \propto 1/\sigma^2$, one obtains the usual t_{n-1} confidence interval for μ. Under those assumptions the usual confidence interval admits its frequentist probability guarantees as well as the Bayesian probability interpretation. Of course, a Bayesian analysis with a different prior distribution will produce a different confidence interval.

4.2.3 Flexible Inference with Bayesian Methods

As described above, the Bayesian approach to inference provides results inviting a natural probabilistic interpretation through the posterior distribution. A Bayesian posterior interval provides an interval that matches our goal of providing a specified degree of certainty of containing the true value. If we have a specific hypothesis about the parameter, perhaps a concern as to whether θ is zero as opposed to an alternative that it is larger than zero, then the Bayesian approach does not provide a hypothesis test, but the posterior distribution does allow us to calculate the probability that θ is above any particular value that we may think is indicative of an important scientific effect. (There is a rich literature associated with addressing Bayesian interpretations of frequentist tests which we do not address here, see, e.g., Berger and Sellke, 1987.)

Perhaps more important than the fact that Bayesian procedures provide natural probabilistic interpretations for confidence intervals is their more general flexibility when it comes to statistical inference. For example, in problems with multiple related parameters (e.g., means of the response under several different treatments), the posterior distribution allows one to provide probabilities for a range of questions: what is the probability that the mean response under treatment A is greater than the mean response under treatment B? what is the probaiblity that treatment C yields the highest posterior mean? These questions are not well defined under the frequentist view of fixed unknown parameters. The explicit use of probability in the Bayesian approach enables them to be asked and answered.

4.2.4 Model Modifications and Adjustments

A related advantage of the Bayesian approach is that it provides a framework within which model modifications and adjustments can be accomodated easily. The frequentist approach tells us how to evluate procedures but not how to develop procedures or how to adjust them for complications that may arise. For example, it is easy to establish that the sample mean is the optimal estimate, in the frequentist sense of being unbiased with minimum variance, of the mean for a normal population distribution. Suppose, however, that we have reason to believe the population is not a single normal distribution but instead a mixture of two normal distributions, one of which corresponds to a heavy-tailed distribution of contaminated measurements. One can now try to develop new estimators and check their frequentist operating characteristics. If we were performing inference under the Bayesian approach, such a change to the model leads us to use a new choice of $p(Y|\theta)$ but the same approach (i.e., Bayes' Theorem) applies to obtaining posterior inferences. The computations required may be more complicated for the new model, but the target is straightforward. In the same manner, complications like missing data or censored data can be easily accommodated in the Bayesian framework.

4.2.5 The Prior Distribution and the Definition of Probability

The previous two subsections lay out some advantages of the Bayesian approach to inference. Why then are frequentist methods still more common in introductory statistics classes and in many scientific disciplines? The answer for many people involves the prior distribution. To many, the goal of science is to come up with objective assessment of the statistical evidence regarding unknown parameters. For such people, allowing prior information to enter the analysis increases the subjectivity of the analysis. There are mathematical results that guarantee the prior distribution becomes less important as we collect more data. But if an analyst chooses a strong prior distribution (i.e., has a strong prior opinion about the value of θ) then the resulting posterior distribution may still be far from the values suggested by the data. In the troubling extreme, an analyst might choose a prior distribution that assigns zero probability to the neighborhood containing the "true" value. In that case the analyst will never get the correct answer no matter how much data are collected. Related to this concern is the difficulty that many people have in combining a subjective prior distribution with a sampling distribution that may be well-justified based on theoretical or mathematical grounds (e.g., the normal distribution for a sample mean in a large sample). Such people might ask whether it is sensible to combine the two probability distributions, one subjective and one justified by theory.

The counter argument to all of these concerns emphasizes that probability is the mathematical language of uncertainty. Thus, probability should be used to describe the uncertainty associated with all random variables including both Y (before it is collected) and θ. Combining such distributions also makes sense in this view. And this view further makes clear that all of the choices in building a statistical model have some degree of subjectivitiy associated with them. Indeed, science is intrinsicially subjective at every step of a study – we decide which treatments to include in the study, how to design the study and what sampling distribution to use in building (or assessing) our procedure. If a particular analyst regularly chooses prior distributions that turn out not to be consistent with the data, then that analyst is still reflecting a personal state of knowledge given the analyst's prior distribution and the observed data. Learning that the analyst's inferences are poorly calibrated (i.e., the 95% posterior intervals never contain the correct value) may lead us not to use that statistician, but it does not invalidate the Bayesian approach.

There are also people who hold views in between the extremes of rejecting Bayesian methods because they rely on prior distributions and accepting Bayesian methods as inherently subjective. A compromise view argues for using the Bayesian approach to develop inferential procedures but suggests that there is a role for studying the frequentist operating characteristics of the resulting procedures (Rubin, 1984; Little, 2013).

4.3 Relevance to Forensic Statistics

What is the relevance of alternative inferential philosophies in the analysis and interpretation of forensic evidence? The 2009 report by the National Academies of Science,

Engineering and Medicine (NASEM, 2009) on the state of forensic science raised concerns about the underlying scientific foundation for a number of types of forensic evidence including the analysis of hair, shoe prints, and bitemarks. The aftermath of the report has seen increased attention to the collection of data to assess the strength of existing approaches to measuring and assessing forensic evidence and increased attention to the appropriate logic to use in assessing the evidence. This includes a renewed focus on measurement and uncertainty, the role of proficiency testing, and different approaches to the analysis and interpretation of evidence. As reflected in this volume, statisticians and the field of statistics have much to contribute to ongoing attempts to improve our understanding of forensic evidence and to advance the analysis of forensic evidence. Areas in which statistical philosophies play a role are reviewed here.

4.3.1 Likelihood Ratios and Bayes Factors

The scientific literature around the interpretation of forensic evidence relies heavily on Bayes' Theorem, which was introduced earlier, and can be expressed in odds form as

$$\frac{Pr(H_p|E)}{Pr(H_d|E)} = \frac{Pr(E|H_p)}{Pr(E|H_d)} \times \frac{Pr(H_p)}{Pr(H_d)},$$

where E denotes the evidence and H_p, H_d represent two alternative hypotheses or propositions regarding the evidence to be assessed with the subscript p denoting what might be the prosecution's view and d denoting what might be the defense's view (Aitken and Taroni, 2004; Stern, 2017). The evidence in a forensic examination often consists of different items or impressions such as glass fragments found at the crime scence and on the suspect, or a document of questioned authorship and a corpus known to have been written by a suspect. Probability models may be developed to describe the frequency distribution of features or measurements of the evidentiary items. These features or measurements we denote as E. To make things specific, we might suppose that the evidence E are measurements, perhaps chemical concentrations of various trace elements, of glass fragments found both at the crime scene and on a suspect's clothing. The prosecution's hypothesis might be that the suspect was at the scene and the defense hypothesis might be that the defendant was not there. The left-hand side represents the posterior (post-evidence) odds of the two hypotheses. The term on the far right is the prior (before the evidence) odds of the two hypotheses. The term in the middle $\frac{Pr(E|H_p)}{Pr(E|H_d)}$ is known as the Bayes factor or sometimes the likelihood ratio. Bayes' Theorem is a standard result in probability and statistics texts. It is well known as the mathematical result that allows one to reverse the order of the conditioning in a conditional probability. The application above shows that given a priori information about the relative likelihood of the two hypotheses and a probability model for the evidence conditional on each hypothesis we can reverse the order in the conditioning and learn about the posterior probabilities of the hypotheses given the evidence. Earlier in this chapter, Bayes' Theorem was used to combine a probability model for θ (the prior distribution) and a probaiblity model for Y conditional on θ (the sampling distribution) to obtain the posterior distribution of θ conditional on the data Y. In the forensic context, deliberations about the hypotheses, both the prior odds ratio and the posterior odds ratio, are the purview of the trier of fact (often the judge or jury in a trial). The forensic evidence can be assessed primarily in terms of the Bayes factor or likelihood ratio which considers the likelihood of observing evidence E under the two competing hypotheses.

Typically, the Bayes factor or likelihood ratio depends on a number of unknown parameters. In the glass example, under H_p we have two samples (one from the crime scene and one from the suspect) that represent the same glass source (i.e., the same population). The probability distribution of the evidence depends on the unknown mean and variance of the measurements obtained from the glass source. Under H_d we have two samples from different sources and hence even more parameters. Data from the case at hand are available for trying to estimate these parameters and may be supplemented by additional data such as from a survey of glass samples in the geographic area. The distinction between the use of the term likelihood ratio and the term Bayes factor relates to our discussion of inferential philosophies. Either a frequentist approach, in which one might identify good estimates of the needed parameters, or a Bayesian approach in which one would apply a Bayesian approach to inference for the parameters are possible. In the former case the term is usually referred to as the likelihood ratio, and in the latter case, as the Bayes factor. The discussion in the previous section regarding advantages and disadvantages of the approaches is relevant to understanding the likelihood ratio/Bayes factor approach and appying it in practice. At the moment the approach is applied most often with simple DNA evidence (DNA known to come from a single source or a simple mixture of two sources). The unknown parameters in this case are allele frequencies and are usually estimated using the frequentist approach, by plugging in allele frequencies in samples from certain populations.

4.3.2 Two-Stage Procedures in Forensic Science

An alternative approach to the assessment of forensic evidence is known as the two-stage process (Parker 1966, 1967; Parker and Holford 1968). The first stage compares measurements on a sample with unknown source (e.g., the glass fragments found on the suspect) to measurements from a sample with known source (e.g., the glass fragments known to have come from a window at the crime scene). A determination is made as to whether the two sets of measurements agree or match. Sometimes this determination is described as reflecting the notion that the samples can't be distinguished. The second stage of the analysis attempts to quantify the probative value of the identified agreement, perhaps by calculating the probability that a sample from a randomly chosen source from a relevant population would be found to match the sample from the known source.

The first stage of the process may be carried out through a traditional frequentist hypothesis test. (It is sometimes carried out using alternative procedures, but these are essentially equivalent to hypothesis tests.) As such, it inherits some of the issues associated with frequentist inference described above. The null hypothesis of the test is taken to be the hypothesis that the two samples come from sources or populations with a common mean; this is taken as evidence that the glass sources from which they are drawn are indistinguishable. The traditional type I error in this case refers to rejecting a true null hypothesis and thus exonerating a suspect for whom the evidence ought to have suggested possible guilt. It is common to choose the threshold for the test to control the type I (false exclusion) probability though this seems to run counter to the usual view of the American justice system in which the suspect should be presumed innocent (which would argue for controlling the probability of false inclusion or a false determination that the samples are indistinguishable), but see Kaye (2020). It is critical in this approach that the test procedure be developed with careful consideration of both type I (false exclusion) and type II (false inclusion) errors.

The second stage requires a calculation of the probability that a glass source randomly chosen from the relevant population of glass sources would yield a sample indistinguishable from the sample with known source. Defining the relevant population requires subjective input from the forensic examiner that can depend on details of the case that yielded the evidence. If a broken window occurred in an isolated neighborhood or subdivision full of homes built at the same time with windows provided by a single manufacturer and the suspect is from this neighborhood, then glass sources from just this neighborhood may be the relevant population. The notion in such a case is that many random individuals from the neighborhood may have trace glass fragments from such windows on their clothing. Alternatively, if the broken window is in a more heterogenous neighborhood and the suspect is from elsewhere in the area, then the relevant population may comprise all potential glass sources in the area. It is rare at present for the second stage to be carried out, primarily because there is a lack of data to describe the relevant population, which makes the calculation impossible. Case-specific data collection would be one approach to addressing this challenge.

4.3.3 Forensic Evidence as Expert Opinion and Error Rates

For many types of evidence, including fingerprints, shoe prints, firearms and tool marks, bloodstain pattern analysis and handwriting, current practice relies on the opinion of a forensic expert who analyzes the evidence based on training, experience and the standards of the field. Though research is underway to put such disciplines on a firmer scientific foundation, a current discussion focuses on what ought to be done to accurately reflect the uncertainty in these expert-based forensic conclusions. Statisticians have much to offer in the way of studies of reliability and validity of procedures as advocated in the 2016 report of the President's Council of Advisors on Science and Technology (PCAST, 2016).

4.4 Summary

The field of statistics, with its emphasis on understanding uncertainty and sources of variability, has much to contribute to the fields that analyze forensic evidence. Statistics' rich history in developing tools for data collection, experimental design, data analysis and inference can provide needed rigor to the assessment of evidence and thus contribute to the fair administration of justice. Though there are philosophical differences between statisticians when it comes to their approach to the data, the commonalities alluded to in this summary are likely to provide the biggest benefits to the forensic evidence community.

References

Aitken, C. and Taroni, F. *Statistics and the Evaluaton of Evidence for Forensic Scientists* (2nd ed.). Wiley, New York, 2004.

Banks, D. and Tackett, M. Bayesian methods and forensic inference. In D. Banks, K. Kafadar, D. Kaye, and M. Tackett, editors, *Handbook of Forensic Statistics*, CRC Press, Boca Raton, FL, 2020.

Berger, J.O. and Sellke, T. Testing a point null hypothesis: the irreconcilability of P values and evidence. *Journal of the American Statistical Association*, **82**, 112–122, 1987.

de Finetti, B. *Probability, Induction, and Statistics*. Wiley, New York, 1972.

Edwards, A.W.F. *Likelihood* (2nd ed.). Johns Hopkins University Press, Baltimore, 1992.

Fisher, R.A. The fiducial argument in statistical inference. *Annals of Eugenics*, **5**, 391–398, 1935.

Gelman, A., Carlin, J.B., Stern, H.S., Dunson, D.B., Vehtari, A., and Rubin, D.B. *Bayesian Data Analysis* (3rd ed.). Chapman and Hall/CRC Press, Boca Raton, 2013.

Kass, R.E. and Raftery, A.E. Bayes factors. *Journal of the American Statistical Association*, **90**, 773–795, 1995.

Kaye, D. Frequentist methods for statistical inference. In D. Banks, K. Kafadar, D. Kaye, and M. Tackett, editors, *Handbook of Forensic Statistics*, CRC Press, Boca Raton, FL, 2020.

Lehmann, E.L. and Casella, G. *Theory of Point Estimation* (2nd ed.). Springer, New York, 2003.

Lehmann, E.L. and Romano, J.P. *Testing Statistical Hypotheses* (3rd ed.). Springer, New York, 2006.

Lindley, D.V. *Bayesian Statistics: A Review*. SIAM, New York, 1971.

Little, R. In praise of simplicity not mathmatistry! Ten simple powerful ideas for the statistical scientist. *Journal of the American Statistical Association*, **108**, 359–369, 2013.

NASEM (National Academies of Science, Engineering and Medicine). *Strengthening Forensic Science in the United States: A Path Forward*. National Academy Press, Washington, 2009.

Neyman, J. On the application of probability theory to agricultral experiments. Essay on principles. *Translated into English in Statistical Science*, 1990, **5**, 463–472, 1923.

Neyman, J. Outline of a theory of statistical estimation based on the classical theory of probability. *Philosophical Transactions of the Royal Society of London, Series A*, Mathematical Physical Sciences, **236**, 333–380, 1937.

Parker, J. A statistical treatment of identification problems. *Journal of the Forensic Science Society*, **6**, 33–39, 1966.

Parker, J. The mathematical evaluation of numerical evidence. *Journal of the Forensic Science Society*, **7**, 134–144, 1967.

Parker, J. and Holford, A. Optimum test statistics with particular reference to a forensic science problem. *Journal of the Royal Statistical Society. Series C*, **17**, 237–251, 1968.

PCAST (President's Council of Advisors on Science and Technology). Report to the President: Forensic science in criminal courts: Ensuring scientific validity of feature-comparison methods, 2016. https://obamawhitehouse.archives.gov/sites/default/files/microsites/ostp/PCAST/pcast_forensic_science_report_final.pdf

Royall, R.M. *Statistical Evidence: A Likelihood Paradigm*. Chapman & Hall, London, 1997.

Rubin, D.B. Bayesianly justifiable and relevant frequency calculaitons for the applied statistician. *Annals of Statistics*, **12**, 1151–1172, 1984.

Savage, L.J. *The Foundation of Statistics*. Wiley, New York, 1954.

Seidenfeld, T.R.A. Fisher's fiducial argument and Bayes' theorem. *Statistical Science*, **7**, 358–368, 1992.

Singh, K., Xie, M., and Strawderman,W.E. Confidence distribution (CD) - distribution estimator of a parameter. *IMS Lecture Notes-Monograph Series* Vol. **54**, Complex Datasets and Inverse Problems: Tomography, Networks and Beyond, 132–150, 2007.

Stern, H.S. Statistical issues in forensic science. *Annual Review of Statistics and Its Application*, **4**, 225–244, 2017.

5

Decision Theory

Franco Taroni, Silvia Bozza, and Alex Biedermann

CONTENTS

5.1 Introduction 103
5.2 Concepts of Statistical Decision Theory 105
    5.2.1 Preliminaries: Basic Elements of Decision Problems 105
    5.2.2 Utility Theory 106
    5.2.3 Implications of the Expected Utility Maximisation Principle 108
    5.2.4 The Loss Function 108
    5.2.5 Particular Forms of the Expected Utility Maximisation Principle 109
    5.2.6 Likelihood Ratios in the Decision Framework 111
5.3 Decision Theory in the Law and Forensic Science 111
    5.3.1 Legal Applications 111
    5.3.2 Forensic Science Applications 113
5.4 Discussion and Conclusions 125
5.5 Further Readings 126
    5.5.1 Forensic Science 126
    5.5.2 General 127
Acknowledgments 127
References 127

5.1 Introduction

Forensic scientists, lawyers and other participants of the legal process are routinely faced with problems of making decisions under circumstances of uncertainty. Uncertainty relates to propositions of interest that are not completely known by the decision-maker at the time when a decision needs to be made. Propositions may relate to the source or nature of forensic traces, marks and objects. For example, with friction ridge marks, propositions of interest may be 'Does this fingermark come from the person of interest (POI) or from some unknown person?'. In forensic document examination, a scientist may ask 'Is this a genuine document or has it been modified (e.g., page substitution)?'. In forensic anthropology the question 'Are these human remains?' may arise, and so on. Replying in one way or another to such questions may be perceived as uncomfortable since knowledge about the relevant underlying truth-state of the world is incomplete to some extent. For example, in typical real-world applications of forensic science it is not known with certainty, when *deciding* to consider a POI as the source of a particular fingermark, whether the POI is in fact the source of the fingermark. Similarly, at an advanced stage of the legal process, the question of whether to convict or acquit a POI (i.e., the verdict) needs to be made in the presence of incomplete knowledge about whether or not the POI truly is the offender. There are analogies between the above questions, in terms of their logical underpinnings, that can be studied, analysed and described using formal methods, such as decision theory, which will be the main aim of this chapter.

Around the middle of the past century, discussions intensified and several fields of study emerged on decision-making concerning, for example, contexts where decisions have monetary consequences. These developments gravitated around questions such as how decisions should be made in order to be considered rational (Pratt et al., 1964). Though an important area, economics was not the only branch with strong interests in decision-making and decision analysis. Entire fields of study developed and interacted with each other in various ways, including psychology, mathematics and statistics, the law and philosophy of science, among others.

This chapter will primarily rely on statistical decision theory* as developed by Savage (1954) and in subsequent treatises (e.g., Lindley, 1985; Luce and Raiffa, 1958; Raiffa, 1968) as the framework for studying the formal structure of decision problems arising in forensic science and the law. Before proceeding with this presentation, an important preliminary needs to be considered. It deals with the question of how to understand decision analysis and the notion of theory of decision. To this point, the field of judgment and decision making, a branch of applied psychology, has contributed considerably by crystallizing three main perspectives and approaches, known as the descriptive, the normative and the prescriptive view (Baron, 2008; French et al., 2009). For a review of the history of these terms, see Baron (2006). Broadly speaking, the descriptive approach focuses on peoples' observable decision behaviour and extends to the development of psychological theories intended to explain how individuals make decisions. Such research is valuable in that it allows one to better understand the conditions under which decision behaviour departs towards incoherence or, worse, logical error. However, revealing such departures requires reference points against which observable decision behaviour can be compared. The provision of such reference points, also called normative standards, is the object of study of the normative approach. Decision theory and decision criteria (or, norms) derived from it, fall into this category of study. It is mainly pursued by mathematicians, statisticians and philosophers of science. The third perspective, the prescriptive approach, addresses the question of what recommendations ought to be derived from normative insights in order to improve practical decision making. For example, some strict normativists, such as Lindley (1985), consider that the normative concept of probability – that is, a standard for reasoning under uncertainty – and decision theory as its extension, are also prescriptive in the sense that they provide direct prescriptions on how to arrange one's reasoning and acting. Properly distinguishing the different intentions and goals of these kinds of decision science research is important for an informed discourse about notions of decision and decision analysis in forensic science applications (Biedermann et al., 2014).

This chapter is structured as follows. Section 5.2 outlines standard elements of statistical decision theory that will be exemplified in Section 5.3 for decision problems arising in the law in general (Section 5.3.1) and forensic science in particular (Section 5.3.2). This exposition will include examples such as decisions following forensic inference of source

---

* In later parts of this chapter, the discussion will be extended to the notion of *Bayesian* statistical decision theory, emphasizing the idea of using Bayesian inference procedures to inform decision makers, for example based on experimental information (Parmigiani, 2001).

(i.e., identification/individualization; Section 5.3.2.1). Discussion and conclusions will be presented in Section 5.4. Further readings on applications of decision theory in forensic science and treatments of decision theory in general are given in Section 5.5.

5.2 Concepts of Statistical Decision Theory

5.2.1 Preliminaries: Basic Elements of Decision Problems

Decision theory is a mathematical theory of how to make decisions when there is uncertainty about the true state of nature. The presence of uncertainty implies that a choice among the alternative courses of action leads to uncertainty regarding which consequences will effectively take place. In statistical terms, the states of nature may also be referred to as parameters, commonly denoted by θ, which may be discrete or continuous. The collection of all possible states of nature is denoted by Θ, the parameter space, and represents a first element of the formalization of decision problems. A second basic element is the feasible decisions (or courses of action), denoted by d. The space of all decisions, called the decision space, is denoted by $\mathcal{D}$. The third basic element is the consequences c. They are defined as the outcome following the combination of a decision d taken when the actual state of nature is θ, formally written $c(d, \theta)$. The space of all consequences is denoted by $\mathcal{C}$. Before proceeding, in Section 5.2.2, with presenting a formal approach to qualifying and quantifying the relative merit of rival courses of action, given the basic elements of the decision problem, it is useful to devote a few more comments to the description of the decision space and the parameter space.

Regarding the decision space, it is important for the decision maker to draw up an exhaustive list of m decisions that are available, say $d_1, d_2, \ldots, d_m \in \mathcal{D}$. As noted by Lindley: "(. . .) it would not be a properly defined decision problem in which the only decision was whether to go to the cinema, because if the decision were not made (that is, one did not go to the cinema) one would have to decide whether to stay at home and read, or go to the public-house, or indulge in other activities. All the possible decisions, or actions, must be included (. . .)" (Lindley, 1965, p. 63). Further, it is convenient to make the requirement of exclusivity, meaning that only one of the decisions can be selected. As noted by Lindley: "Hence, the decisions are both exclusive and exhaustive: one of them *has* to be taken, and at most one of them *can* be taken" (Lindley, 1985, p. 6).

The second task for a decision maker is to draw up a list of n exclusive and exhaustive events or states of nature, say $\Theta = \{\theta_1, \theta_2, \ldots, \theta_n\}$. Regarding the latter list, the decision maker may distinguish between situations of certainty and uncertainty. In the former case, certainty, the decision maker has complete knowledge about the states of nature. Hence, each alternative course of action leads to one and only one foreseeable consequence, and a choice among alternatives is equivalent to a choice among related consequences. In the latter case, uncertainty, the decision maker does not know which state of nature actually holds, or what the future will be. Consequently, each available course of action will have one of several consequences. It is possible, however, to measure uncertainty about the states of nature using a suitable probability distribution Pr over Θ. Note that in some fields, such as business decision analysis and operations research, this situation is called 'decision making under risk' and the expression 'decision making under uncertainty' is reserved to situations in which the decision maker is unable to provide a list of all possible outcomes and/or a probability distribution for the various outcomes. In this chapter, however, this interpretation will not be pursued.

5.2.2 Utility Theory

The principal issue in decision making under uncertainty is the selection of a member in the list of available decisions without knowing which state of nature is truly the case. The aim, therefore, is to create a framework that allows decision makers to assess the consequences of alternative courses of action in order to compare them and avoid irrational choices or behaviour.

The formulation of such a decision framework involves, first, the assumption that the decision maker can express preferences amongst possible consequences. It is in fact assumed that the space of consequences has a partial pre-ordering, denoted by $\preceq$, meaning that the decision maker must be able to specify, at any point, which consequence is suitable or whether they are equivalent (Piccinato, 1996). When comparing any pair of consequences $(c_1, c_2) \in \mathcal{C}$, $c_1 \prec c_2$ indicates that the consequence c_2 is strictly preferred to consequence c_1, $c_1 \sim c_2$ indicates that c_1 and c_2 are equivalent (or equally preferred), while $c_1 \preceq c_2$ indicates that c_1 is not preferred to c_2, that is either $c_1 \prec c_2$, or $c_1 \sim c_2$ holds. The measurement of preferences among decision outcomes is operated by a function, called a utility function, denoted by $U(\cdot)$ that associates a utility value $U(d, \theta)$ to each one of the possible consequences $c(d, \theta)$, also denoted $U(c)$; it specifies the desirability of each consequence on some numerical scale.

Second, the decision maker's uncertainty about the states of nature, when they are discrete, is expressed in terms of a probability mass function $\Pr(\theta \mid I)$, where I denotes the relevant information available at the time when the probability assessment is made. Combining the utilities $U(d, \theta)$ for decision consequences and the probabilities for states of nature leads to a measure of the desirability of alternative courses of action d in terms of their expected utility (EU)*:

$$EU(d) = \sum_{\theta \in \Theta} U(d, \theta) \Pr(\theta \mid I).$$

A standard decision rule, based on EU, instructs one to select the action with the maximum expected utility (see also Section 5.2.3). Hereafter, information I will be omitted to simplify the notation, though it is important to keep in mind that it conditions all probability assignments.

Some further conditions (axioms) must be imposed on the preference system in order for there to exist a function U, the utility function, such that for any pair $(c_1, c_2) \in \mathcal{C}$, the relationship $c_1 \preceq c_2$ holds if and only if $U(c_1) \leq U(c_2)$.

A.1 The first axiom requires that the preference system is *complete*. This amounts to assume that for any pair of consequences (c_1, c_2) of the space of consequences $\mathcal{C}$, it must always be possible to express a preference or indifference among them (one of the following relations must hold: $c_1 \prec c_2, c_2 \prec c_1, c_1 \sim c_2$).

A.2 The second axiom requires that the preference system is *transitive*. This means that for any $(c_1, c_2, c_3) \in \mathcal{C}$, if one prefers c_2 to c_1 ($c_1 \prec c_2$) and c_3 to c_2 ($c_2 \prec c_3$), then

---

* The same idea can be applied when θ is continuous and takes values in $\Theta_c, \theta \in \Theta_c$. The probability mass function $\Pr(\theta \mid I)$ is replaced by a probability density function $f(\theta \mid I)$ and the expected utility of decision d is:

$$EU(d) = \int_{\Theta_c} U(d, \theta) f(\theta \mid I) d\theta.$$

one prefers c_3 to c_1 ($c_1 \prec c_3$). In the same way, if one is indifferent between c_1 and c_2 ($c_1 \sim c_2$), and is indifferent between c_2 and c_3 ($c_2 \sim c_3$), then one is indifferent between c_1 and c_3 ($c_1 \sim c_3$). Not all the consequences are equivalent to each other, that is, for at least a pair of consequences (c_1, c_2), either $c_1 \prec c_2$ or $c_2 \prec c_1$ holds.

A.3 The third axiom requires that the ordering of preferences is invariant with respect to compound gambles. For any pair of consequences $(c_1, c_2) \in \mathcal{C}$, such that $c_1 \preceq c_2$, then, for any other consequence $c_3 \in \mathcal{C}$, and any probability α, the gamble that offers probability α of winning c_2, and probability $(1 - \alpha)$ of winning c_3 is preferred (or it is equivalent) to the gamble that offers probability α of winning c_1 and probability $(1 - \alpha)$ of winning c_3. Denote by $(c_i, c_j; \alpha, 1 - \alpha)$ the gamble offering c_i with probability α, and c_j with probability $(1 - \alpha)$, $i \neq j$. This axiom can then be formulated as follows: $c_1 \preceq c_2$ if and only if $(c_1, c_3; \alpha, 1 - \alpha) \preceq (c_2, c_3; \alpha, 1 - \alpha)$, for any $\alpha \in [0, 1]$ and any $c_3 \in \mathcal{C}$.

A.4 The fourth axiom requires that there are not (i) infinitely desirable or (ii) infinitely undesirable consequences. Let $(c_1, c_2, c_3) \in \mathcal{C}$ be any three consequences such that c_1 is preferred to c_2 and c_2 is preferred to c_3 ($c_3 \prec c_2 \prec c_1$). Then there exist probabilities α and β, such that (i) c_2 is preferred to the gamble $(c_1, c_3; \alpha, 1 - \alpha)$; (ii) the gamble $(c_1, c_3; \beta, 1 - \beta)$ is preferred to c_2.

If (i) does not hold, then one will always prefer the possibility of obtaining the best consequence c_1, no matter how small is the probability of obtaining it, to c_2; that is, one believes that c_1 is infinitely better than c_2 (and c_3). If (ii) does not hold, then one will prefer c_2, no matter how small is the probability of obtaining the worse consequence c_3 is; that is, one believes that c_3 is infinitely worse than c_2 (and c_1).

If these four conditions are satisfied, then one can prove the *expected utility theorem*, according to which there exists a function U on the space of consequences $\mathcal{C}$ such that for any d_i and d_k belonging to the decision space $\mathcal{D}$, d_i is preferred (or equivalent) to d_k if and only if the expected utility of d_i, $EU(d_i)$, is greater (or equal) than the expected utility of d_k, $EU(d_k)$, that is, assuming θ discrete, if

$$\sum_{\theta \in \Theta} U(d_i, \theta) \Pr(\theta) \geq \sum_{\theta \in \Theta} U(d_k, \theta) \Pr(\theta).$$

Consider, next, any consequence c and a pair $(c_1, c_2) \in \mathcal{C}$ such that $c_1 \prec c_2$, and $c_1 \preceq c \preceq c_2$. Following the stated conditions, it may be proved (see De Groot, 1970) that there exists a unique number $\alpha \in [0, 1]$ such that

$$c \sim [\alpha c_1 + (1 - \alpha) c_2], \tag{5.1}$$

and that

$$U(c) = \alpha U(c_1) + (1 - \alpha) U(c_2). \tag{5.2}$$

It can also be proved that the utility function is invariant under linear transformations. This means that if $U(c)$ is a utility function, then for any $a > 0$, $aU(c) + b$ is also a utility function preserving the same pattern of preferences.

Utility functions can be constructed in different ways. One possibility starts with a pair of non-equivalent consequences $(c_1, c_2) \in \mathcal{C}$ and assigns them a utility value. This will fix the origin and the scale of the utility function. The desirability of each consequence $c \in \mathcal{C}$

of interest will then be compared with those of c_1 and c_2. Given that utility functions are invariant under linear transformation, the choice of c_1 and c_2, and the choice of the scale of the utility, are not relevant. They are, however, generally identified with the worst and the best consequence, respectively. It is assumed, for example, that the utility of the worst consequence is zero, $U(c_1) = 0$, and the utility of the best consequence is one, $U(c_2) = 1$. The utilities of the remaining intermediate consequences are computed using Equation (5.2). This will be discussed further in Section 5.3.

5.2.3 Implications of the Expected Utility Maximisation Principle

Consider taking a decision d when the true state of nature is θ, so that the consequence is $c(d, \theta)$. It is possible to show, using relation (5.1), that there exists some α such that the consequence $c(d, \theta)$ is equivalent to a hypothetical gamble offering the worst consequence c_1 with probability α and the best consequence c_2 with probability $(1 - \alpha)$

$$c(d, \theta) \sim [\alpha c_1 + (1 - \alpha)c_2], \qquad c_1 \preceq c(d, \theta) \preceq c_2.$$

The utility $U(d, \theta)$ of the consequence $c(d, \theta)$ can then be calculated using Equation (5.2) as follows:

$$U(d, \theta) = \alpha \underbrace{U(c_1)}_{0} + (1 - \alpha) \underbrace{U(c_2)}_{1} = 1 - \alpha.$$

According to this, for any d and any θ, selecting decision d is equivalent to assigning a probability $U(d, \theta) = 1 - \alpha$ to the occurrence of the most favorable consequence. This hypothetical gamble can always be played. It can be played, in particular, after that decision d has been taken and it is known which state of nature θ holds. The term $U(d, \theta)$ can be understood as the conditional probability of obtaining the consequence c_2, given decision d has been taken and the state of nature θ occurred: $\Pr(c_2 \mid d, \theta) = U(d, \theta)$. Note that probability $\Pr(c_2 \mid d)$ can be written in extended form as

$$\Pr(c_2 \mid d) = \sum_{\theta \in \Theta} \Pr(c_2 \mid d, \theta) \Pr(\theta). \tag{5.3}$$

Therefore, (5.3) can be rewritten as

$$\Pr(c_2 \mid d) = \sum_{\theta \in \Theta} U(d, \theta) \Pr(\theta), \tag{5.4}$$

namely, the expected utility that quantifies the probability of obtaining the best consequence once decision d is taken (Lindley, 1985). The decision rule which instructs decision makers to select the decision which maximizes the expected utility (MEU criterion) is optimal because it is the decision which has associated with it the highest probability of obtaining the most favorable consequence.

5.2.4 The Loss Function

An alternative way to express preferences among decision consequences $c(d, \theta)$ is the use of non-negative loss functions. When a utility function is available, the loss function can be derived as follows (Lindley, 1985):

$$L(d, \theta) = \max_{d \in \mathcal{D}} U(d, \theta) - U(d, \theta) \tag{5.5}$$

The loss $L(d, \theta)$ for a given consequence $c(d, \theta)$ thus is defined as the difference between the utility of the best consequence under the state of nature at hand and the utility for the consequence of interest. That is, the loss measures the penalty for choosing a non-optimal action, also called opportunity loss (Press, 1989, p. 26–27): the difference between the utility of the best consequence that could have been obtained and the utility of the actual one received.

Note that following Equation (5.5), losses cannot, by definition, be negative because $U(d, \theta)$ will be smaller or at best equal to $\max_{d \in \mathcal{D}} U(d, \theta)$. The expected loss, $EL(d)$, thus characterises the undesirability of each possible decision, and can be quantified as follows:

$$EL(d) = \sum_{\theta \in \Theta} L(d, \theta) \Pr(\theta).$$

When using losses instead of utilities, the decision rule of maximising expected utility becomes the rule instructing the selection of the decision that minimizes the expected loss $EL(d)$. It might be objected that assuming a non-negative loss function is too restrictive. Note, however, that the loss function represents error due to an non-optimal choice. It thus makes sense to consider that even the most favorable decision will induce at best a zero loss.

5.2.5 Particular Forms of the Expected Utility Maximisation Principle

For the remainder of this chapter, it will be important to anticipate two particular forms in which the MEU principle may be formulated. Consider, first, the utility-based perspective of a two-action decision problem involving two states of nature, θ_1 and θ_2. The decision maker's probabilities for these states of nature are $\Pr(\theta_1 \mid \cdot)$ and $\Pr(\theta_2 \mid \cdot)$, respectively, such that $\Pr(\theta_1 \mid \cdot) + \Pr(\theta_2 \mid \cdot) = 1$. Note that $\mid \cdot$ is shorthand notation for the conditioning on any relevant evidence E or background information I. The two possible decisions are d_1 and d_2, representing the decision maker's acceptance of, respectively, θ_1 and θ_2 as the true states of nature. Hereafter, we write C_{ij} to denote the consequence $c(d_i, \theta_j)$ of taking decision d_i when θ_j is the actual state of nature and denote the corresponding utility by $U(C_{ij})$. The decision problem is summarized in Table 5.1.

According to the principle of maximization of expected utility, the decision maker should select decision d_1 rather than d_2 if $EU(d_1) > EU(d_2)$. This will be the case if

$$U(C_{11}) \Pr(\theta_1 \mid \cdot) + U(C_{12}) \Pr(\theta_2 \mid \cdot) > U(C_{21}) \Pr(\theta_1 \mid \cdot) + U(C_{22}) \Pr(\theta_2 \mid \cdot), \qquad (5.6)$$

TABLE 5.1

A Simple Decision Matrix with Two Decisions d_1 and d_2, Two States of Nature θ_1 and θ_2 and Corresponding Decision Consequences C_{ij} (for $i, j = \{1, 2\}$)

States of Nature	θ_1	θ_2
Decisions		
d_1	C_{11}	C_{12}
d_2	C_{21}	C_{22}

which can be rearranged to give

$$\frac{\Pr(\theta_1 \mid \cdot)}{\Pr(\theta_2 \mid \cdot)} > \frac{U(C_{22}) - U(C_{12})}{U(C_{11}) - U(C_{21})}. \tag{5.7}$$

The term $U(C_{22}) - U(C_{12})$ in the numerator on the right-hand side of (5.7) is the additional utility involved in making the correct decision when θ_2 turns out to be the correct state of nature. An alternative way to look at this term is to consider it as the potential regret: it is the potential loss in utility when erroneously deciding d_1 instead of d_2. The term $U(C_{11}) - U(C_{21})$ similarly deals with the potential regret of deciding d_2 when the true state of nature is θ_1. Relation (5.7) thus states that decision d_1 should only be taken if the odds in favour of θ_1 are sufficient to outweigh any extra potential regret associated with incorrectly deciding d_1 (Spiegelhalter et al., 2004).

Consider now the loss-based account. Recall, from Section 5.2.4, that the loss $L(d_i, \theta_j) = L(C_{ij})$ for a decision consequence C_{ij} is the difference between the utility of the outcome of the best decision under the state of nature at hand, and the utility of the outcome of the actual decision d_i under the same state of nature. Therefore, the decision that minimizes the expected loss is the same as the decision that maximizes the expected utility. Continuing the example introduced above, assume that there is a positive loss L incurred when falsely choosing a proposition that is not actually the case, that is $L(C_{ij}) > 0$ if $i \neq j$, and there is no loss when accepting a proposition that is actually the case, that is $L(C_{ij}) = 0$ if $i = j$. The loss can be symmetric, $L(C_{ij}) = L(C_{ji})$, or asymmetric, $L(C_{ij}) \neq L(C_{ji})$, $i \neq j$. The decision criterion depicted in Equation (5.6) will become to select d_1 rather then d_2 if $EL(d_1) < EL(d_2)$, that is if

$$L(C_{11}) \Pr(\theta_1 \mid \cdot) + L(C_{12}) \Pr(\theta_2 \mid \cdot) < L(C_{21}) \Pr(\theta_1 \mid \cdot) + L(C_{22}) \Pr(\theta_2 \mid \cdot), \tag{5.8}$$

and the expected loss of deciding d_i will be:

$$EL(d_i \mid \cdot) = \underbrace{L(C_{ii})}_{0} \Pr(\theta_i \mid \cdot) + L(C_{ij}) \Pr(\theta_j \mid \cdot) = L(C_{ij}) \Pr(\theta_j \mid \cdot), \qquad i \neq j.$$

Considering the principle of minimizing expected loss and given that

$$EL(d_1 \mid \cdot) < EL(d_2 \mid \cdot) \text{ if and only if } L(C_{12}) \Pr(\theta_2 \mid \cdot) < L(C_{21}) \Pr(\theta_1 \mid \cdot),$$

the decision problem involves a comparison of odds with the ratio of losses associated with erroneous decisions. Specifically, deciding d_1 rather than d_2 is optimal if and only if:

$$\frac{\Pr(\theta_1 \mid \cdot)}{\Pr(\theta_2 \mid \cdot)} > \frac{L(C_{12})}{L(C_{21})} \tag{5.9}$$

or, equivalently

$$\frac{\Pr(\theta_2 \mid \cdot)}{\Pr(\theta_1 \mid \cdot)} < \frac{L(C_{21})}{L(C_{12})}. \tag{5.10}$$

The loss ratio on the right-hand side in Equations (5.9) and (5.10) fixes a threshold for odds. The relation (5.9) specifies that if the odds in favour of θ_1 exceed the loss incurred from incorrectly choosing decision d_1 divided by the loss incurred from incorrectly choosing decision d_2, then the decision maker should take decision d_1.

5.2.6 Likelihood Ratios in the Decision Framework

So far it has been considered that the decision maker's probabilities for the state of nature are conditional probabilities written $\Pr(\theta_1 \mid \cdot)$ and $\Pr(\theta_2 \mid \cdot)$, incorporating all relevant evidence E and background information I available at the time when the decision needs to be made. The odds in Equation (5.9) can therefore be interpreted as *posterior* odds. It is useful to emphasize that likelihood ratios, commonly used in forensic science for quantifying the value of forensic results (e.g., Aitken and Taroni, 2004), play an important role in the inference process preceding the decision. Recalling that the posterior odds can be written as the product of the prior odds and the likelihood ratio for the forensic results E, the relation (5.9) can thus be rewritten as:

$$\frac{\Pr(\theta_1 \mid I, E)}{\Pr(\theta_2 \mid I, E)} = \underbrace{\frac{\Pr(\theta_1 \mid I)}{\Pr(\theta_2 \mid I)}}_{\text{prior odds}} \times \underbrace{\frac{\Pr(E \mid \theta_1, I)}{\Pr(E \mid \theta_2, I)}}_{\text{likelihood ratio}} > \underbrace{\frac{L(C_{12})}{L(C_{21})}}_{\text{loss ratio}}. \tag{5.11}$$

Relation (5.11) defines the conditions under which the decision d_1 is preferable to d_2, that is when the relative losses on the right are smaller than the product on the left, containing the likelihood ratio. Thus, it is now possible to reformulate the decision criterion, minimizing expected loss (Section 5.2.5), with an emphasis on the likelihood ratio, as follows:

> The decision d_1 is to be preferred to decision d_2 if the product of the likelihood ratio and the prior odds is larger than the ratio of the losses associated with adverse decision consequences.

A more intuitive form of (5.11) can be obtained when working with logarithms (e.g., Good, 1950):

$$\log\left[\frac{\Pr(\theta_1 \mid I)}{\Pr(\theta_2 \mid I)}\right] + \log\left[\frac{\Pr(E \mid \theta_1, I)}{\Pr(E \mid \theta_2, I)}\right] > \log\left[\frac{L(C_{12})}{L(C_{21})}\right]. \tag{5.12}$$

By re-arranging the terms one can isolate the log-likelihood ratio as follows:

$$\log\left[\frac{\Pr(E \mid \theta_1, I)}{\Pr(E \mid \theta_2, I)}\right] > \log\left[\frac{L(C_{12})}{L(C_{21})}\right] - \log\left[\frac{\Pr(\theta_1 \mid I)}{\Pr(\theta_2 \mid I)}\right]. \tag{5.13}$$

Note that following Good (1950), the logarithm of the likelihood ratio, the term on the left, is commonly referred to as the weight of evidence. The decision criterion minimizing expected loss (Section 5.2.5) thus becomes:

> The decision d_1 is to be preferred to decision d_2 if and only if the weight of evidence is *greater* than the difference between the logarithm of the ratio of the losses associated with adverse consequences and the logarithm of the prior odds in favour of proposition θ_1.

5.3 Decision Theory in the Law and Forensic Science

5.3.1 Legal Applications

Decision theory offers a formal framework for thinking analytically about decision problems, but this perspective – in particular the use of probability – is controversial both among some legal scholars and practitioners. Although formal decision theoretic discourses can be dated back about half a century ago (Kaplan, 1968), there has been a concentration of several article collections since the mid-1980s. See for example, Tillers and Green (1988), especially the paper by Kaye (1988), and the collection of articles in the *Internatinal Journal of Evidence & Proof* (Vol. 1, 1997) entitled 'Bayesianism and Juridical Proof', edited by R. J. Allen and M. Redmayne.

A generic outline of applying the decision-theoretic elements presented in Section 5.2 to legal decision problems proceeds among the following lines. Assume that in a case of interest there are only two possible decisions: decision d_1, finding for the plaintiff, and decision d_2, finding for the defense. Considering this decision problem in terms of expected utilities requires the specification of probabilities of the states of nature and utilities for decision consequences. Let θ_1 and θ_2 denote versions of the case wherein the plaintiff or defendant, respectively, is entitled to judgment. Let $\Pr(\theta_j)$ be the decision maker's probability, at a given time, for a given state of nature θ_j. Deciding in favour of, respectively, the plaintiff and the defendant may lead to accurate consequences, namely C_{11} and C_{22}, or adverse outcomes, namely C_{12} and C_{21}. Then, comparing the relative merit of decisions d_1 and d_2 comes down to criterion (5.6), that is deciding in favour of the plaintiff rather than for the defendant would require the expected utility of decision d_1 to be greater than the expected utility of decision d_2. The immediate question following this observation is 'When is this the case?'. As noted in Section 5.2.5, it can be helpful to illustrate the logic of the decision-theoretic result by formulating the MEU principle in an alternative form, such as relation (5.7), separating the thinking about probabilities from thinking about the utilities of the various decision consequences. For a discussion of relation (5.7) see, for example, Friedman (1997, 2017).

The decision-theoretic criterion may be more insightful if it is considered through a loss-based perspective (Section 5.2.5) using, for example, relation (5.9). Let $L(C_{12})$ denote the loss associated with wrongly deciding in favour of the plaintiff, and $L(C_{21})$ denote the loss associated with wrongly deciding in favour of the defendant. It is then clear to see that with a symmetric $0 - k$ loss function, that is with $L(C_{12}) = L(C_{21}) = k$ for adverse decision outcomes, and zero loss $L(C_{11}) = L(C_{22}) = 0$ for accurate decision outcomes, deciding in favour of, for example, the plaintiff is warranted if and only if the probability $\Pr(\theta_1 \mid \cdot)$ is greater than 0.5. This result is sometimes associated with the notions of 'balance of probabilities' and 'more probable than not' standards, translating common ideas in civil litigation according to which a correct judgment for the plaintiff is as preferable as a correct judgment for the defendant, and that erroneous verdicts for either side are equally undesirable (e.g., Kaye, 1999).

The expression (5.9) can also capture the logical structure of the decision problem of the typical criminal case where the prosecution has the burden of proving its case with respect to a particular standard. In such situations, $L(C_{12})$ is the loss of falsely declaring the defendant guilty, whereas $L(C_{21})$ is the loss associated with a false acquittal. A common viewpoint is the preference ordering $C_{12} \prec C_{21}$ according to which a false conviction is more undesirable than a false acquittal. Consequently, this amounts to consider the loss $L(C_{12})$ of a false conviction to be greater, often considerably greater, than the loss $L(C_{21})$ associated with a false acquittal. For any loss ratio thus defined, depending on the nature of the case and the stakes involved, the criterion (5.9) provides the minimal odds of liability required in order for a conviction, decision d_1, to be preferred to an acquittal, decision d_2. In this context, reference is sometimes made to Blackstone's 10 to 1 criterion according to which it is better that 10 truly liable defendants go free than 1 innocent defendant be wrongly convicted. It has been noted, however, that this criterion does not easily map to assignments of losses in a given singular case, but rather seems to refer to actual error rates across multiple distinct trials (Kaye, 1999).

It should be kept in mind that criterion (5.9) is an analytical result that can be thought through in two different ways: either starting from a preference structure (i.e., a loss ratio for adverse decision consequences) and derive the lower limit of the odds necessary to warrant a conviction, or starting with a given value for the odds of liability and then work out the loss ratio corresponding to these odds so that a given decision is warranted. Over the past decades, this normative account has stimulated considerable empirical research on, for example, what various subjects (e.g., judges, citizens, etc.) consider as required levels of probability before deciding in one way or another. See Dane (1985) and Simon and Mahan (1971), for example, and Hastie (1993) for a review. The quantitative values observed in such studies, using various elicitation procedures and methodologies, vary over broad ranges and depend largely on experimental conditions. Note, however, that this mismatch between, on the one hand, consistency requirements for utilities and probabilities implied by the theory, and, on the other hand, peoples' intuitive feelings about these assignments and their interdependency, does not invalidate the formal mathematical results. Similarly, arithmetics is not abandoned simply because practically operating individuals might reply, for example, with an answer other than 4 to the question of how much is $2 + 2$ (see, e.g., Lindley in de Finetti, 1974). This is an instance of the difference between the normative and the descriptive perspectives to decision mentioned in Section 5.1.

5.3.2 Forensic Science Applications

5.3.2.1 Forensic Identification

5.3.2.1.1 Preliminaries

One of the most well known (Champod, 2000) but also most widely challenged (Cole, 2014) notions in forensic science is 'individualization', or 'identification'. It is a conclusion following inference of source and involves the claim to reduce a pool of potential donors of a forensic trace to a single source. This section presents a decision-theoretic account of forensic individualization based on analyses previously given in Biedermann et al. (2008, 2016). For presentations in the wider context of Bayesian data analysis and Bayesian decision networks, see Taroni et al. (2010); Taroni et al. (2014). More generally, examiners' conclusions in forensic identification practice now are, increasingly often, referred to as decisions (Cole and Biedermann, 2020). See, for example, the reports issued by the President's Council of Advisors on Science and Technology (PCAST, 2016) and the AAAS (Thompson et al., 2017). There is an interest, thus, to devote attention to the ways in which decision may be understood and conceptualised from a scientific point of view. Decision theory provides a mathematically rigorous account for this.

Suppose extraneous material (e.g., blood) or a mark (e.g., finger- or toolmark) is collected at a crime scene and an individual is apprehended or a tool – called a potential source – is found. Similarly, one may imagine a litigation case in which a contested signature is present on a questioned document (e.g., a contract) and the question is whether or not the signature is from the POI. For the purpose of the current discussion, assume two uncertain events defined as 'The crime mark comes from the suspect' (θ_1), and 'The crime mark comes from an unknown person' (θ_2). These two states of nature are discrete and form the parameter space Θ. Assume further that 'identifying' (sometimes also called 'individualizing') an individual as being the source of a crime mark can be considered as a decision (d_1) made by a person authorized to do so. For the remainder of the analysis, it is not necessary

TABLE 5.2

Decision Matrix for a Forensic Identification Problem with d_i, $i = 1, 2, 3$, Denoting Decisions, θ_j, $j = 1, 2$, Denoting States of Nature and C_{ij} Denoting the Consequence of Taking Decision d_i When θ_j Turns Out to be the True State of Nature (Biedermann et al., 2008)

	States of Nature	
Decisions	θ_1: POI is Donor	θ_2: An Unknown Person is Donor
d_1: identification	C_{11}: correct identification	C_{12}: false identification
d_2: inconclusive	C_{21}: neutral	C_{22}: neutral
d_3: exclusion	C_{31}: false exclusion	C_{32}: correct exclusion

to specify whether this authorized person is a (forensic) scientist or some other participant in the legal process. As alternative decisions, consider the conclusions 'inconclusive' (d_2) and 'exclusion' (d_3). These forms of conclusion are currently used by many forensic practitioners. Combining these elements leads to the decision matrix shown in Table 5.2. The outcome of an 'identification' ('exclusion') conclusion, decision d_1 (d_3), is an accurate outcome if the POI is truly (is truly not) the origin of the crime mark. These consequences are referred to as 'correct identification' and 'correct exclusion', respectively. The outcome of an 'identification' ('exclusion') conclusion, decision d_1 (d_3), can be adverse if the POI is truly not (is truly) the origin of the crime mark. These consequences are listed as 'false identification' and 'false exclusion', respectively. Because the statement 'inconclusive', decision d_2, does not convey any information that tends to associate or otherwise the POI with the issue of the source of the crime mark, the consequences following d_2 are referred to as 'neutral'.

5.3.2.1.2 Preference Ordering and Construction of the Utility Function

Consider the following ordering of consequences:

$$C_{12} \prec C_{31} \prec C_{21} \sim C_{22} \prec C_{32} \sim C_{11}. \tag{5.14}$$

This preference ordering states that the most preferred consequences are a correct identification (C_{11}) and a correct exclusion (C_{32}), and the worst consequence is a false identification (C_{12}). To construct the utility function, after having chosen the scale, one starts by assigning the maximum utility value to the best consequence, in this case the couple C_{32} and C_{11}, and the minimum utility value to the worst consequence, C_{12}. Therefore, if a $(0, 1)$ scale is chosen, $U(C_{11}) = U(C_{32}) = 1$ and $U(C_{12}) = 0$.

The next steps consist in assigning an utility value to the intermediate consequences. Consider the consequence called 'neutral', C_{21}, and the above preference ranking

$$C_{12} \prec C_{21} \prec C_{11}.$$

In particular, it has been observed – see (5.1) and (5.2) – that if the preference system respects given conditions, there exists, for the decision maker, a unique number $0 \leq \alpha \leq 1$ such that the consequence C_{21} is equivalent to a hypothetical gamble where the worst consequence, C_{12}, is obtained with probability α, and the best consequence, C_{11} is obtained with probability $(1 - \alpha)$:

$$C_{21} \sim [\alpha C_{12} + (1 - \alpha)C_{11}], \tag{5.15}$$

and the utility of C_{21} can be computed as

$$U(C_{21}) = \alpha \underbrace{U(C_{12})}_{0} + (1 - \alpha) \underbrace{U(C_{11})}_{1} = 1 - \alpha.$$

Note that the utility of consequence C_{21} turns out to be the probability $(1 - \alpha)$ of finishing with the best consequence in the space of all possible consequences. In particular, note the equivalence of utility and probability in the latter sentence. Finding such an α is the most difficult part of the utility elicitation procedure. It involves answering the question what would make the decision maker indifferent between a neutral consequence, and a situation in which a false identification might occur. Specifically, the decision maker must specify the value α so that the sure consequence C_{21} appears equivalent to the gamble in which the worst consequence is obtained with probability α and the best consequence is obtained with probability $1 - \alpha$.

When thinking about the above question, one must be careful *not* to use as values for α the assigned probabilities $\Pr(\theta_1)$ and $\Pr(\theta_2)$ for the propositions of interest. In fact, the number α is a limiting value, asking decision makers to crystallize what would be, in general, their highest probability of running the risk of making the worse mistake they are willing to exchange with the consequence of rendering an 'inconclusive' statement. Suppose, for instance, as an extreme position, that the answer is zero, meaning that the decision maker never wants to run such a risk. This would mean, however, that no matter how high the probability of a correct identification is, the decision maker would consider that a neutral conclusion is as good as a correct identification. It is also worth noting that since $C_{32} \sim C_{11}$, one can substitute C_{32} for C_{11} in Equation (5.15), so obtaining another hypothetical gamble:

$$C_{21} \sim [\alpha C_{12} + (1 - \alpha)C_{32}].$$

To ensure coherence, the decision maker should again consider a neutral conclusion as much worth as a correct exclusion, no matter how high the probability of a correct exclusion is. Thus, if the decision maker would indeed consider that the highest probability for incurring the worst consequence, in exchange with the consequence of providing an 'inconclusive' statement, is strictly zero, then this belief cannot be coherent with the preference ranking (5.14). It can be coherent only with the following ordering:

$$C_{12} \prec C_{31} \prec C_{21} \sim C_{22} \sim C_{32} \sim C_{11}. \tag{5.16}$$

Note that there is a change in the fourth preference sign from the left.

Consider the logical implications of the preference ordering (5.16) through the decision matrix in Table 5.2. The preference ranking (5.16) implies that, if the proposition θ_1 is true, then decisions d_1 ('identification') and d_2 ('inconclusive') are equally preferred and both better than decision d_3 ('exclusion'). In turn, if the proposition θ_2 is true, then the decisions d_2 and d_3 are equally preferred and both better than decision d_1. Thus, decision d_2 is the best decision overall, because, if the proposition θ_1 is true, it is better than d_3, and, if the proposition θ_2 is true, it is better than d_1. Therefore, if the decision maker considers a strictly zero probability for the event of incurring the worst consequence, in exchange to the consequence of rendering an 'inconclusive' statement, then the decision maker should *always* take the decision 'inconclusive' (d_2).

This does not correspond, however, to the way in which decision makers behave in practice. Therefore, there *must exist* for these decision makers a unique value $0 < \alpha < 1$ such that the hypothetical gamble (5.15) does make sense for them, despite the inherent challenge to find the limiting value α. For the purpose of illustration, assume that the decision maker considers that $\alpha = 0.001$ is appropriate. Then

$$U(C_{21}) = \alpha U(C_{12}) + (1 - \alpha)U(C_{11}) = 1 - \alpha = 0.999.$$

Likewise, the utility of the consequence C_{31} can be elicited and quantified in comparison with $U(C_{12})$, the utility of the worst consequence, and $U(C_{11})$, the utility of the best consequence,

$$U(C_{31}) = \alpha^* U(C_{12}) + (1 - \alpha^*)U(C_{11}) = 1 - \alpha^*.$$

Here, α^* represents the decision maker's highest probability for incurring the worst consequence in exchange with the consequence of rendering a 'false exclusion' statement. For behaviour to be coherent, this limiting value α^* must necessarily be higher then the previous limiting value $\alpha = 0.001$. The reason for this is that the decision maker is facing, on the right-hand-side, a gamble with the same consequences as before and, on the left-hand-side, a less preferred consequence: recall the ranking $C_{31} \prec C_{21}$. Assume, for example, that $\alpha^* = 0.01$ is felt to be correct. Then $U(C_{31}) = 0.99$. This value means that the decision maker is indifferent between a false exclusion (C_{31}) and a gamble in which the worst consequence, a false identification (C_{12}), is obtained with probability $\alpha^* = 0.01$, and a correct identification with probability $1 - \alpha^* = 0.99$. Note that the order relation in the space of consequences is preserved. However, this is not the end of the matter. It is a good idea at this stage to check the appropriateness of the so-built utility function because there is no guarantee that the quantified utility values are coherent (Berger, 1988). This question can be examined by comparing different combinations of consequences as:

$$C_{31} \prec C_{21} \prec C_{11} \qquad \text{or} \qquad C_{12} \prec C_{31} \prec C_{21}$$

Consider the case on the left, for instance. There must exist, at this stage, a unique value α' such that

$$C_{21} \sim [\alpha' C_{31} + (1 - \alpha')C_{11}]. \tag{5.17}$$

According to the illustrated gambling scheme, and the quantified utilities,

$$U(C_{21}) = \alpha' U(C_{31}) + (1 - \alpha')U(C_{11})$$
$$\alpha' 0.99 + (1 - \alpha').$$

When solving this equation one obtains $\alpha' = 0.1$. Now, if one believes that this value is correct, in the sense that one is indifferent between a neutral consequence and a gamble where a false exclusion may occur with probability 0.1, then the utility function is coherent. Otherwise, one needs to go back and check previous assessments. For further discussion on such comparisons, see also Biedermann et al. (2008).

5.3.2.1.3 Computing Expected Utilities

Consider Table 5.3 for a summary of the utility values derived above. Assume that there is scientific evidence, denoted by E, available and used to inform the probabilities of the competing propositions θ_1 and θ_2, leading to posterior probabilities $\Pr(\theta_1 \mid E)$ and $\Pr(\theta_2 \mid E)$ at

TABLE 5.3

Illustrative Values for Utilities
$U(C_{ij}) = U(d_i, \theta_j)$, as Discussed in the
Text, for a Case of Forensic
Identification

Decisions	Uncertain Events	
	θ_1	θ_2
d_1: identification	1	0
d_2: inconclusive	0.999	0.999
d_3: exclusion	0.99	1

Note: The propositions of interest are θ_1 "The
crime stain comes from the POI" and θ_2 "The
crime stain comes from an unknown person"
(Taroni et al., 2010, 2014).

the time when the decision is made*. Start by considering the computation of the expected
utility of the decision d ("identification"),

$$EU(d_1) = U(C_{11})\Pr(\theta_1 \mid E) + U(C_{12})\Pr(\theta_2 \mid E).$$

Given the assigned utility values, it is immediately seen that the expected utility of decision
d_1 reduces to: $EU(d_1) = \Pr(\theta_1 \mid E)$. So, assuming a $(0, 1)$ utility function, and recalling result
(5.4) where the expected utility of a decision d is equated with the probability of obtaining
the best consequence once decision d is taken, it follows that

$$EU(d_1) = \Pr(\text{correct identification} \mid d_1) = \Pr(\theta_1 \mid E).$$

In the same way, one can compute the expected utilities of the decisions 'inconclusive' (d_2)
and 'exclusion' (d_3):

$$EU(d_2) = U(C_{21})\Pr(\theta_1 \mid E) + U(C_{22})[1 - \Pr(\theta_1 \mid E)]$$
$$= U(C_{21}) = U(C_{22}).$$
$$EU(d_3) = U(C_{31})\Pr(\theta_1 \mid E) + U(C_{32})[1 - \Pr(\theta_1 \mid E)]$$
$$= U(C_{31})\Pr(\theta_1 \mid E) + [1 - \Pr(\theta_1 \mid E)].$$

The decision with the highest expected utility, that is the optimal decision, depends on the
relative magnitude of $\Pr(\theta_1 \mid E)$, $U(C_{21})$ and $U(C_{31})$.

Recall, from the above analysis, that the probability α in Equation (5.1) is actually a limiting value. Intuitively, this implies that, if the probability of θ_2 is higher than this limiting
value, then 'identification' cannot be the best decision. To examine this aspect, consider the
utility values given in Table 5.3 and assume that the probability of θ_2 is 0.0011 (i.e., slightly
greater that α):

$$\Pr(\theta_2 \mid E) = 0.0011 > \alpha = 0.001.$$

---

* Note again that information I is omitted to simplify the notation, though it is important to keep in mind that it
  conditions all probability assignments.

The following expected utility values can then be calculated:

$$EU(d_1) = (1 \times 0.9989) = 0.9989$$

$$EU(d_2) = (0.999 \times 0.9989) + (0.999 \times 0.0011) = 0.999$$

$$EU(d_3) = (0.99 \times 0.9989) + (1 \times 0.0011) = 0.990011.$$

Thus, $EU(d_2) > EU(d_1) > EU(d_3)$ and the decision 'inconclusive' (d_2) is better than the decision 'identification' (d_1). 'Inconclusive' is the decision with the highest expected utility.

5.3.2.1.4 Comments

It is readily seen that the decision with the highest expected utility depends on the interplay between probabilities and utilities, though merely looking at formulae in isolation might not be helpful to get a sense of the interaction among the relevant factors. It may thus be useful to graphically display the expected utilities of the various decisions as a function of, for example, the probability of θ_1, the proposition according to which the crime mark comes from the POI. Following the above computations, it is clear that $EU(d_1)$ is an increasing linear function, corresponding to the probability of θ_1. In turn, $EU(d_2)$ is a constant, leading to a horizontal line at $y = U(C_{21}) = U(C_{22})$. Finally, $EU(d_3)$ is also a linear function, but decreasing. This is illustrated in Figure 5.1 (left). Note that, for illustrative purposes and improving readability, the computations plotted in Figure 5.1 are based on slightly modified values for α and α^*, that is 0.05 instead of 0.001 and 0.2 instead of 0.01, respectively. The bold solid lines in Figure 5.1 (right) highlight the decision with the maximum expected utility. Intersections between expected utility functions represent transition points, indicating a change in the decision with maximum expected utility when further increasing or

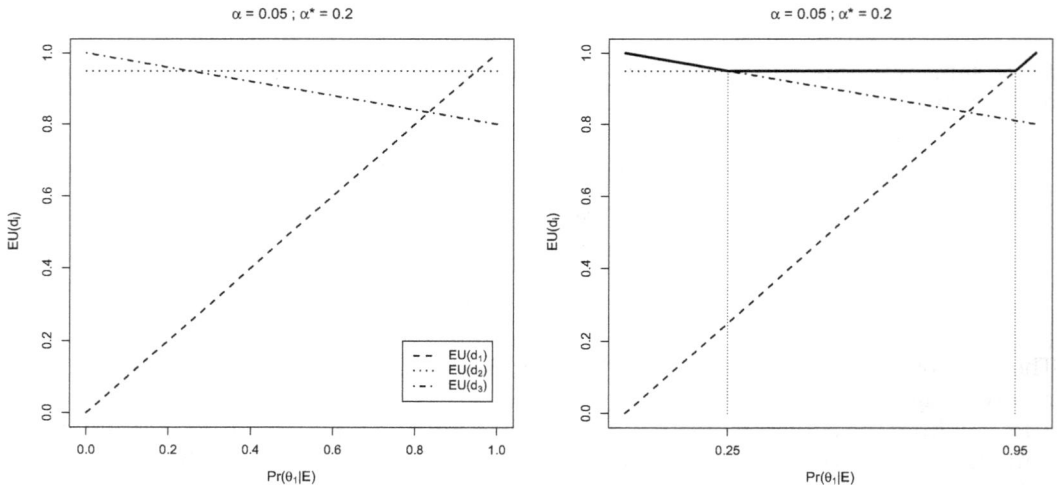

FIGURE 5.1
Illustrative expected utilities for a case of forensic identification with the propositions θ_1 'The crime stain comes from the POI' and θ_2 'The crime stain comes from an unknown person' (left). The available courses of action are 'identification' (d_1), 'inconclusive' (d_2), and 'exclusion' (d_3). The expected utilities are computed as a function of the probability of θ_1 (x-axis), using the utility values obtained by choosing $\alpha = 0.05$ and $\alpha^* = 0.2$. The bold solid lines (right) highlight, for each possible value of the probability of θ_1, the decision with the maximum expected utility.

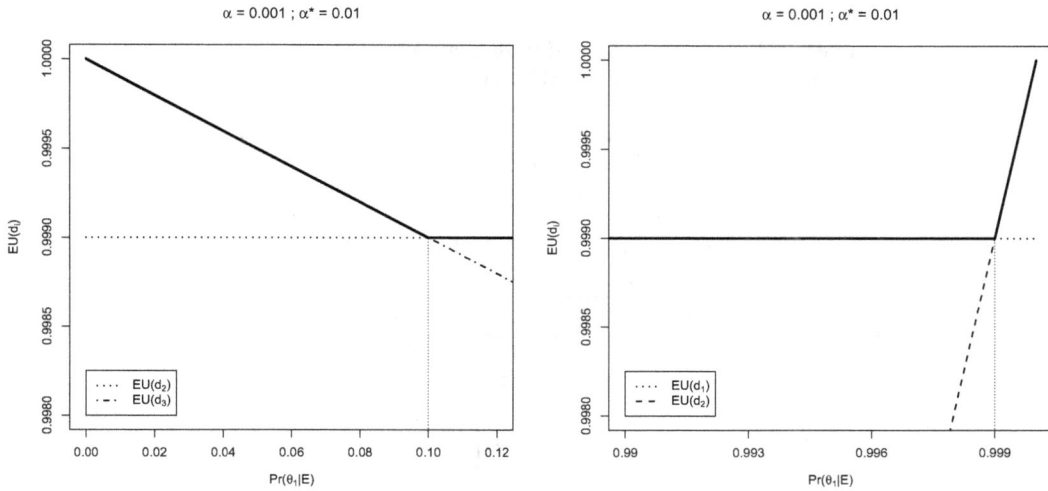

FIGURE 5.2
Illustrative expected utilities for a case of forensic identification with the propositions θ_1 'The crime stain comes from the POI' and θ_2 'The crime stain comes from an unknown person'. The available courses of action are 'identification' (d_1), 'inconclusive' (d_2), and 'exclusion' (d_3). The expected utilities are computed as a function of the probability of θ_1 ($x-$axis), using the utility values given in Table 5.3. The bold solid lines highlight, for each possible value of the probability of θ_1, the decision with the maximum expected utility.

decreasing the probability of θ_1. So, it can be easily observed that, by choosing $\alpha = 0.05$ and $\alpha^* = 0.2$,* for $\Pr(\theta_1 \mid E)$ smaller than 0.25 the optimal decision is d_3 (exclusion), and for $0.25 \leq \Pr(\theta_1 \mid E) \leq 0.95$ the optimal decision is d_2 (inconclusive). Finally, decision d_1 (identification) is optimal only whenever $\Pr(\theta_1 \mid E)$ is larger than 0.95.

It may be objected that these values may not be reasonable, and in fact decision makers will build their utility function coherently with their preference systems. Expected utility functions of decisions d_1, d_2 and d_3 obtained with the original choices $\alpha = 0.001$ and $\alpha^* = 0.01$ are plotted in Figure 5.2, where the $x-$ and $y-$ axes are focused on the transition points (i.e., on the left, the transition point from d_3 to d_2 is highlighted, fixed at $\Pr(\theta_1 \mid E) = 0.1$; on the right the transition point from d_2 to d_1 is shown, fixed at $\Pr(\theta_1 \mid E) = 0.999$). In particular, it is pointed out that according to this preference system, the probability of θ_1 must be at least equal to 0.999 to have d_1 (identification) as the optimal decision, or equivalently, the probability of θ_2 must be smaller than 0.001, the limiting value.

Note also that the decision-theoretic analysis of forensic identification can also be conducted with only two decisions, such as d_1 ('identification') and d_2 ('do not identify') and/or the quantification of decision consequences in terms of losses instead of utilities (Biedermann et al., 2016). The decision criterion of minimizing expected loss then comes down to criterion (5.9) with inferential properties as discussed previously in Section 5.2.5 and 5.3.1. For a discussion on the application of the decision-theoretic account to forensic identification in the particular area of fingermarks, see Champod et al. (2016). Identification decisions following database searches are discussed in Gittelson et al. (2012). For a formulation of the expected loss minimization decision criterion, and an explicit representation of the role of the likelihood ratio in the inference process preceding the decision,

---

* This amounts to have $U(C_{21}) = U(C_{22}) = 0.95$ and $U(C_{31}) = 0.8$.

TABLE 5.4

Numerical Examples, Presented in Biedermann et al. (2016), of
Minimum Likelihood Ratio (LR) Values for Satisfying the Expected
Loss Minimization Decision Criterion Necessary to Make the
Decision d_1 (Individualising the POI) Preferable to d_2 (Not
Individualizing the POI), Criteria (5.12) and (5.13), With PO
Denoting Prior Odds (Odds in Favour of the Proposition that the
POI of Interest is the Source of the Crime Stain) and RL Denoting
the Relative Losses (i.e., the Ratio of the Losses of Adverse
Decision Consequences)

PO $= \Pr(\theta_1 \mid I)/\Pr(\theta_2 \mid I)$	LR	RL	log(PO)	log(LR)	log(RL)
$1/10 = 0.1$	100	10	-1	2	1
$1/10 = 0.1$	1000	100	-1	3	2
$1/1000 = 0.001$	10^5	100	-3	5	2
$1/1000 = 0.001$	10^6	1000	-3	6	3

Note: The values in columns four to six are the logarithms (base 10) of the
values presented in the first three columns.

the development given in Section 5.2.6 applies, in particular (5.12) and (5.13). Consider, for
example, the following abbreviated form of the criterion (5.12):

$$\log(\text{PO}) + \log(\text{LR}) > \log(\text{RL}),$$

that is the requirement that the sum of the logarithm of the prior odds (PO) and the logarithm of the likelihood ratio (LR) must exceed the logarithm of the ratio of the losses
of adverse decision consequences (RL). Table 5.4 illustrates numerical examples of combinations of values, in particular limiting values that the likelihood ratio must exceed, in
order to make decision d_1 (individualization) preferable to decision d_2 (not individualizing
the POI).

5.3.2.2 Understanding Probability Assignment as a Decision: The Use of Proper Scoring Rules

So far in this section, probability has been encountered as a concept to express uncertainty
about the truth or otherwise of propositions or events that are not entirely known to the
decision maker at the time when a decision needs to be made. Interpreting probability
as a decision maker's personal degree of belief, it will be unreasonable then to say that
decision makers do not know their own states of mind. Throughout this chapter, it is tacitly
assumed that the decision makers have, each of them in their own way, and at any point
in time, their own probability, depending on their extent of knowledge and background
information. However, a legitimate question that may arise is what it means for a given
person to assign a so-called personal probability. In this context proper scoring rules allow
one to clarify the probability elicitation process, pointing out that it can itself be understood
as question of decision making

To illustrate the notion of proper scoring rule, consider a person who is asked to state the
probability that he or she assigns to a given event E. Assume further that part of the question is the information that the declared probability will be scored with respect to the actual
truth or falsity of E, denoted by, respectively, 1 and 0. The purpose of this added constraint
is to motivate people to report their actual beliefs, rather than a deliberately chosen value

(i.e., a value that is different from the one they have in mind), in a way that is made precise shortly below. Distorted probability assertions are sometimes encountered in the context of forensic identification, for example when experts round off small probabilities to zero, or high probabilities to 1. Another typical example is the notion of relevance of trace material which expresses the relationship between a given trace or stain and the offender. Typically, one cannot categorically assert relevance, as observed by Stoney, because relevance "(...) may range from very likely to practically nil (...)" (Stoney, 1994, at p. 18). Formally, relevance refers to a proposition of the kind 'the stain or mark comes from (or, was left by) the offender' and appears as a factor in various likelihood ratio developments (e.g., Evett et al., 1998). Uncertainty about propositions regarding evidential relevance is sometimes suppressed and evaluators declare a probability of relevance $p' = 1$, thus rounding up to 1 a probability actually smaller than one. Scoring rules allow one to show that this distortion of probabilities is not advisable.

The notion of score, in the decision-theoretic context, refers to the square of the difference between the probability p' for the event E, as stated by the person, and the actual truth-value of E, zero or one. Because of this squared difference the rule is also called 'quadratic scoring rule'. As an example, assume that the scientist, or any other person being asked to state their probability for event E, declares the value $p' = 0.8$. Thus, in the analysis here, statement $p' = 0.8$ is interpreted as a decision and the question is how to decide in an optimal way (i.e., what value p' to report). One can then distinguish two cases. In one case, E is true and thus has the truth value 1. This situation leads to the score $(1 - 0.8)^2 = 0.2^2$. In the other case, the event E is not true and thus assumes the truth value 0, leading to the score $(0 - 0.8)^2 = (-0.8)^2$. However, these score calculations, leading to expressions of *actual* penalty, are only hypothetical. Given that the scientist is uncertain about whether or not E is true, the scientist cannot know the actual score. At best the scientist can consider a prevision of the scoring and then seeking a way of proceeding that minimizes the expected penalty. This leads to the notions of 'prevision of the scoring', 'expected penalty' and 'expected loss'. These notions are based on the idea of combining the possible scores and the probabilities p and $(1 - p)$ with which the scores may be produced. Recall that the general concept of expected value was introduced in Section 5.2.2.

In general, the expected loss for reporting probability p' given the scientists' actual belief p is

$$\text{EL}(p_i') = \text{L}(p_i', E = true)\Pr(E = true) + \text{L}(p_i', E = false)\Pr(E = false),$$

or $(1 - p')^2 p + p'^2(1 - p)$ for short. One can readily see that the expected loss is minimal when the reported belief p' corresponds to the scientist's actual belief p. In the example considered here, with $p' = p = 0.8$, the expected loss is:

$$(1 - p')^2 p + p'^2(1 - p) = (1 - 0.8)^2 \times 0.8 + 0.8^2 \times (1 - 0.8)$$

$$= 0.16.$$

It is left as an exercise for readers to verify that for any reported value $p' \neq p$, the expected penalty would be larger. It is thus in the interest of scientists to report their actual belief. Note that the quadratic scoring rule can also be used for the elicitation of conditional probabilities (e.g., for an event E given another event H) when supposing that the penalties only apply if the conditioning event holds (e.g., de Finetti, 1972).

The quadratic scoring rule is a *proper scoring rule* because it implies that whatever one's true belief p, one should sincerely choose this value as one's reported probability; it is the choice with the minimal expected penalty. Other scoring rules may not necessarily exhibit this property. For example, the simple difference between the actual truth value of the event for which a probability is to be assessed and the reported probability p' does not imply optimality for sincerely reported beliefs. Historically, the quadratic scoring rule appears in many writings of the subjective probabilist (e.g., de Finetti, 1962, 1982). However, following a paper by Brier (1950) on an application in meteorology, the rule is also known as 'Brier's Rule'. The quadratic scoring rule is one of the most simple proper scoring rules, while others exist based on, for example, the logarithm (Good, 1952). See also Parmigiani and Inoue (2009) for further details on scoring rules and related concepts, such as calibration. Biedermann et al. (2013) discuss scoring rules in the context of forensic individualization and the use of influence diagrams (Bayesian decision networks) for implementation. The implications of this viewpoint for the subjectivist interpretation of probability is considered in Biedermann (2015), Biedermann et al. (2017b), and Biedermann and Vuille (2018).

5.3.2.3 Other Forensic Decision Problems: Consignment Inspection

Section 5.3.2.2 considered a problem in which the space of decisions covered any value in the interval between zero and one, including these endpoints, while the states of nature were binary. Consider now a situation in which the states of nature are continuous, but the space of decisions is binary. Suppose an unknown proportion, denoted by θ, such as the proportion of a consignment of individual items that are of a certain kind (e.g., contain an illegal substance) is the object of interest.

The proportion θ may take values in the range $[0, 1]$, including the endpoints. Suppose that the decision maker faces two available decisions, denoted by d_1 and d_2. The first decision, d_1, amounts to accepting the view that the proportion θ of 'positive' units (i.e., units with illegal content) in the consignment is not greater than some specified value θ^*, for example $\theta^* = 0.95$. The second decision, d_2, is the view according to which the proportion θ of positive units in the consignment is greater than the specified value θ^*. Formally, the two decisions d_1 and d_2 can be conceptualised as decisions to accept one of the two composite hypotheses $H_1 : \theta \leq \theta^*$ and $H_2 : \theta > \theta^*$. A decision d_i is accurate if the true value of the unknown proportion θ lies in the range of values defined by such a hypothesis H_i. Otherwise, it is an incorrect decision.

Assume that the undesirability of decision consequences is quantified in terms of losses. There is no positive loss associated with accurate decisions, though incorrect decisions have an associated positive loss. Formally, denote by $L(d_i, \Theta_j)$ the loss associated with the decision d_i, for $i = 1, 2$, while Θ_j, for $j = 1, 2$, is the true state of affairs. The term Θ_j is defined as follows: $\Theta_1 = [0, \theta^*]$ and $\Theta_2 = (\theta^*, 1]$. Thus, considering that accurate decisions consequences have zero loss is expressed by $L(d_i, \Theta_j) = 0$, for $i, j \in \{1, 2\}$ and $i = j$. Further, let l_i denote the loss $L(d_i, \Theta_j)$ associated with an erroneous conclusion when decision d_i is taken, $i, j \in \{1, 2\}$ and $i \neq j$. The so-built loss function is also called a $0 - l_i$ loss function.

Generally, there are different ways to implement a loss function considering both monetary and non-monetary components. Unlike a $0 - 1$ loss function, derived from the $0 - 1$ utility function presented above, the analysis here will interpret the losses l_i as purely monetary values, with l_1 representing the loss when the decision maker falsely regards a case as one in which $\theta \leq \theta^*$. Similarly, l_2 represents the loss from falsely considering the proportion θ of a consignment to be greater than θ^*. The monetary interpretation of losses is based on the following considerations. Suppose that the decision maker is a member of an investigative authority facing the practical problem of high workload. Thus, there may be an interest to focus primarily on cases where the proportion of seized items is above a certain threshold. The loss l_1, then, could consist of the funds or monetary value of property that could have been confiscated by the investigative authority as a penalty, and given to the public treasury. In turn, for assessing l_2, it is relevant to inquire about the consequence of pursuing a case that is not 'important' enough (i.e., falsely considering $\theta > \theta^*$). Here, the investigative authority might generate expenses which, when compared to the reduced funds that may be seized in a non-priority case with $\theta \leq \theta^*$, could represent a net loss. Also, the loss could represent the amount of compensation to be allocated to an erroneously pursued individual. As an example, consider $l_1 = l_2 = 100K$ USD, but readers may choose their own values, including asymmetric values $l_1 \neq l_2$.

To assign probabilities for the composite hypotheses $H_1 : \theta \in \Theta_1$ and $H_2 : \theta \in \Theta_2$, it is necessary to specify a probability distribution for θ. When the number of units in the consignment is large, it is possible to assume a continuous probability density function, such as a beta distribution with parameters α and β. Note that a beta distribution is chosen here because it allows one to readily incorporate sampling information regarding the number of positive units (i.e., units with illegal content), using a standard updating rule (Bernardo and Smith, 1994). For further discussion in forensic contexts see, for example, Aitken (1999), Aitken and Taroni (2004) and Taroni et al. (2010).

Initially, before inspecting any items of the consignment, suppose that all possible values the proportion θ may assume are considered equally probable, expressed in terms of a so-called uniform prior probability distribution. The probability of the hypothesis $H_1 : \theta \leq 0.95$ is then given by the integral of the uniform beta density with parameters $\alpha = \beta = 1$, with endpoints 0 and $\theta^* = 0.95$:

$$\Pr(\theta \leq 0.95) = \int_0^{0.95} f(\theta \mid \alpha = 1, \beta = 1)d\theta = 0.95 \tag{5.18}$$

From this result it follows that a priori $\Pr(\theta > 0.95) = 0.05$. On the basis of these probabilities one can calculate the expected loss of each decision d_i, for $i = 1, 2$. The result is written, for short, $\mathrm{EL}^0(d_i)$, where the superscript 0 refers to the 'initial' point of time, that is before any items are inspected. This loss is also sometimes called *prior expected loss*, whereas the posterior expected loss (i.e, after considering sampling information) is based on the posterior probability density for θ. In the case considered here, prior to considering sampling information, the assumed uniform probability distribution implies that the decision d_1 to accept the proposition $H_1: \theta \leq 0.95$ involves a 0.95 probability for a zero loss, and a 0.05 probability for a loss l_1. Therefore, the prior expected loss of decision d_1 is

$$\begin{aligned} \mathrm{EL}^0(d_1) &= \Pr(\theta \leq 0.95) \times \mathrm{L}(d_1, \Theta_1) + \Pr(\theta > 0.95) \times \mathrm{L}(d_1, \Theta_2) \\ &= 0.95 \times 0 + 0.05 \times l_1 = 0.05 \times l_1 \,. \end{aligned} \tag{5.19}$$

The prior expected loss of decision d_2 to accept the proposition $H_2: \theta > 0.95$ is obtained in the same way:

$$\begin{aligned} \mathrm{EL}^0(d_2) &= \Pr(\theta \leq 0.95) \times \mathrm{L}(d_2, \Theta_1) + \Pr(\theta > 0.95) \times \mathrm{L}(d_2, \Theta_2) \\ &= 0.95 \times l_2 + 0.1 \times 0 = 0.95 \times l_2 \,. \end{aligned} \tag{5.20}$$

The optimal decision d^0_{opt} will be the one which minimizes $\mathrm{EL}^0(d_i)$. The losses $l_1 = l_2 = 100K$ USD defined above thus lead to the following result:

$$\mathrm{EL}^0(d_1) = 5\,000, \ \mathrm{EL}^0(d_2) = 95\,000, \ \Rightarrow d^0_{opt} = d_1.$$

This result means that in a situation in which (i) the decision maker has not yet inspected any items of the consignment, (ii) prior beliefs about θ are based on an uniform prior distribution, and (iii) the loss function is symmetric (therefore the decision is based entirely on the prior beliefs about θ), it is preferable to conclude d_1, that is the proportion is smaller than 0.95.

It is important to note that the above result is crucially dependent on the decision maker's prior beliefs about the proportion θ and on the choice of the loss function. The optimal decision may change depending on the assigned probabilities and losses. To illustrate this dependency, suppose now that the initial beliefs of the decision maker (based on knowledge about previous consignments, domain expertise, and other sources of information) are represented by a beta$(3, 0.3)$ distribution, a distribution that places more density to high values of θ. Specifically, this distribution implies that approximately 60% of prior belief weight is given to values of θ that are greater than 0.95:

$$\Pr(\theta > 0.95) = \int_{0.95}^{1} f(\theta \mid 3, 0.3)d\theta = 0.6.$$

It follows from this that $\Pr(\theta \leq 0.95) = 0.4$ and the prior expected losses of decisions d_1 and d_2 become:

$$\mathrm{EL}^0(d_1) = 0.6 \times 100\mathrm{K} = 60\mathrm{K},$$

$$\mathrm{EL}^0(d_2) = 0.4 \times 100\mathrm{K} = 40\mathrm{K}, \Rightarrow d_{opt}^0 = d_2.$$

The result now is that it is advisable for the decision maker to decide d_2, that is the decision to consider $\theta > 0.95$, because the expected loss associated with this decision is lower than that for d_1.

A particular assumption in the above example is that the loss function is taken to be symmetric. There is no requirement for this, however, and it is possible to formulate the decision-theoretic criterion more generally. In particular, decision d_1 should be taken when the expected loss $\mathrm{EL}(d_1)$ is minimal, that is when

$$\Pr(\theta \in \Theta_2) \times l_1 < \Pr(\theta \in \Theta_1) \times l_2.$$

By rearranging terms, this criterion can be reformulated as follows: decide d_1 whenever $\Pr(\theta \in \Theta_2) < l_2/(l_1 + l_2)$. Thus, when the losses for adverse outcomes are considered equal, the action with the minimum expected loss is the one in which the associated parameter values have the higher probability.

Besides computing expected losses and determining optimal decisions, before or after taking into account sampling information, the above framework presents a starting point for a variety of further analyses and the development of additional concepts. For example, the expected loss of the a priori optimal action is also sometimes referred to as the *expected value of perfect information* (EVPI) about the true state of θ. Although, in many situations, it may not be possible to obtain perfect information about the true state of nature, the EVPI may be a useful measure to think about the decision problem. In particular, it allows one to indicate the maximum amount of money that one should be willing to pay for (expert) information that is such that it would allow one to determine the true state of nature with certainty. In the above example, where the a priori optimal decision d_{opt}^0 had an expected loss of 40K, the decision maker should accept additional information about the true proportion only if the cost for that information does not exceed 40K.

Another notion of interest in this context is the *value of sample information* (VSI), defined as the difference between the expected losses of the optimal actions before and after considering sampling information. Yet another concept are pre-posterior analyses that take into account the probabilities for various outcomes of item inspection (i.e., the proportion of inspected items that are of a certain kind). Such analyses may be conducted for fixed or variable sample sizes, and by taking into account or not the cost of inspecting items from the consignment. This leads to further notions, such as the *expected value of sample information* (EVSI) and the *expected net value of sample information* (ENVSI). See, for example Biedermann et al. (2012), Taroni et al. (2014) and Gittelson (2013) for a discussion of these concepts and the use of graphical models, such as decision trees and Bayesian decision networks (influence diagrams) for practically implementing these approaches. More generally, decision-theoretic approaches to sampling can be found, for example, in Schlaifer (1959) and Raiffa and Schlaifer (1961).

Note also that the discussion in this section concentrated on a binary decision among two composite hypotheses regarding a parameter space. A different decision problem is to consider the whole parameter space as the space of possible decisions. That is, given a parameter space and a posterior probability distribution over the possible values that the parameter may take, the question is which single value to select is, in some sense, optimal. This amounts to considering a problem of parameter estimation, that is inference, as a decision. In this case, other mathematically tractable non-constant loss functions, such as the quadratic loss or the piecewise linear loss function, can be implemented (Press, 2003). For general theory on this perspective, see Berger (1985), and for forensic applications Taroni et al. (2010).

5.4 Discussion and Conclusions

The core topic of this book, the application of statistics in forensic science, is primarily concerned with questions of inference – that is the reasonable reasoning in the face of uncertainty. This involves procedures for the coherent use of relevant data for revising and informing beliefs about competing propositions of interest. Propositions may be formulated at different levels regarding, for example, the source of particular (forensic) trace material (e.g., DNA, fibres, etc.), or at a more advanced stage, regarding alleged activities of particular individuals, such as the defendant (Cook et al., 1998). Two complications arise when taking this abstract account too literally. One is the occasionally raised claim that statistics can only be applied if data are available. This is not so. Reasoning methods, in particular the specification of conditions for coherence, reply to general questions of formal analysis that can be approached with whatever amount of information and new data – whereas the latter may be nil – there may be. The second complication is the flawed idea that inference is the end of the matter. Again, this is not so. As noted by Lindley (2000), inference is a preliminary to decision, an idea that started to expand more widely since the middle of the last century:

> Years ago a statistician might have claimed that statistics deals with the processing of data. As a result of relatively recent formulations of statistical theory, today's statistician will be more likely to say that statistics is concerned with decision making in the face of

uncertainty. Its applicability ranges from almost all inductive sciences to many situations that people face in everyday life when it is not perfectly obvious what they should do.

Chernoff and Moses, 1959, at p. 1

The neat connection between, on the one hand, informing beliefs through collected evidence and data, and addressing the question of how to decide, on the other hand, is also at the heart of applications in forensic science and the law. In these areas, decision theory has, over the past few decades, been considered by many authors as an analytical framework for studying selected decision problems. Though not as expanded as in economics and operations research, existing studies in the law are more numerous and also more controversially debated than applications to forensic science problems, which have a more recent history. It is widely acknowledged that decision theory provides a rigorous framework for thinking about decision problems, but it is also widely uncontested that the direct application in practice may not be immediate. The main reason for this is that additional argument needs to be invoked in order to interpret the formal elements of the theory with respect to the defining features of practical decision problems. This is a general challenge, however, that is equally encountered with other formal concepts, such as probability and Bayes' theorem. In the particular context of forensic science, decision theory is currently used to critically review understandings of traditional concepts, such as individualisation/identification, and practice thereof. Insight that is gained from such analyses helps better understand where current forensic practice makes assumptions that go above and beyond forensic examiners' areas of competence (Biedermann et al., 2016). Typical examples for this include assumptions about probabilities for propositions of interest and utilities/losses for decision consequences. The current role of decision theory in forensic science thus is advisory, by providing a reference point – in a normative sense (Section 5.1) – for delineating areas that require attention in ongoing reform efforts (Cole and Biedermann, 2020).

5.5 Further Readings

5.5.1 Forensic Science

Practising forensic scientists may encounter questions of decision making at various stages in their daily work. For example, scientists may need to decide about whether or not to use a particular analytical apparatus, apply a particular chemical substance or search for a particular type of transferred trace material. Several publications have addressed such questions, in particular in the context of forensic DNA analyses. Taroni et al. (2005) considered the question of whether or not to perform DNA analyses in a case of questioned kinship involving two individuals who are uncertain about whether they are full siblings or unrelated. Another question related to DNA analysis, addressed by Taroni et al. (2007), concerns the number of DNA loci that ought to be analyzed. Planning problems in forensic DNA analyses are also addressed in Mazumder (2010). Gittelson et al. (2014) used a decision-theoretic approach for the topic of genotype designation, that is a decision problem characterised by complications due to phenomena such as drop-in and drop-out. Decision-theoretic computations can readily become complex, in particular when computations need to be extended beyond one-stage analyses. An example for a staged decision analysis regarding the question of processing or not processing a fingermark is given in

Gittelson et al. (2013), focusing on the expected value of information (EVOI) and the cost of processing the fingermark. Other examples of sequential decisions are presented in Taroni et al. (2010); Taroni et al. (2014). General forensic applications of Bayesian decision theoretic criteria, for example in the context of kinship analyes and handwriting examinations, are presented in Biedermann et al. (2017a). A Bayesian classification criteria is presented in Bozza et al. (2014) to address the problem of determining plant's chemotype.

5.5.2 General

Ramsey (1980) is credited with pioneering expected utility, along with de Finetti (1937), followed by works of von Neumann and Morgenstern (1953) and Marschak (1950) on the axiomatization in terms of 'gambles'. Classical textbooks are Savage (1954), Luce and Raiffa (1958), Raiffa (1968) and De Groot (1970). More recent texbooks are Berger (1988), French (1988), Smith (1988, 2010), Parmigiani and Inoue (2009) and Robert (2007).

Acknowledgments

Franco Taroni and Alex Biedermann gratefully acknowledge the support of the Swiss National Science Foundation through grants No. IZSEZ0-19114 and No. BSSGI0-155809. Some of the writing of this chapter was carried out during a visiting research stay of Alex Biedermann at New York University School of Law.

References

Aitken, C.G.G. Sampling - How big a sample? *Journal of Forensic Sciences*, **44**, 750–760, 1999.

Aitken, C.G.G. and Taroni, F. *Statistics and the Evaluation of Evidence for Forensic Scientists* (2nd ed.). John Wiley & Sons, Chichester, 2004.

Baron, J. President's column: Normative, descriptive, and prescriptive. Newsletter of the Society for Judgment and Decision Making, December 2006. http://www.sjdm.org/newsletters/06-dec.pdf.

Baron, J. *Thinking and Deciding* (4th ed.). Cambridge University Press, New York, 2008.

Berger, J.O. *Statistical Decision Theory and Bayesian Analysis* (2nd ed.). Springer, New York, 1985.

Berger, J.O. *Statistical Decision Theory and Bayesian Analysis* (2nd ed.). Springer, New York, 1988.

Bernardo, J.M. and Smith, A.F.M. *Bayesian Theory*. John Wiley & Sons, Chichester, 1994.

Biedermann, A. The role of the subjectivist position in the probabilization of forensic science. *Journal of Forensic Science and Medicine*, **1**, 140–148, 2015.

Biedermann, A., Bozza, S., and Taroni, F. Decision theoretic properties of forensic identification: Underlying logic and argumentative implications. *Forensic Science International*, **177**, 120–132, 2008.

Biedermann, A., Bozza, S., Garbolino, P., and Taroni, F. Decision-theoretic analysis of forensic sampling criteria using Bayesian decision networks. *Forensic Science International*, **223**, 217–227, 2012.

Biedermann, A., Garbolino, P., and Taroni, F. The subjectivist interpretation of probability and the problem of individualisation in forensic science. *Science & Justice*, **53**, 192–200, 2013.

Biedermann, A., Taroni, F., and Aitken, C. Liberties and constraints of the normative approach to evaluation and decision in forensic science: A discussion towards overcoming some common misconceptions. *Law, Probability and Risk*, **13**, 181–191, 2014.

Biedermann, A., Bozza, S., and Taroni, F. The decisionalization of individualization. *Forensic Science International*, **266**, 29–38, 2016.

Biedermann, A., Bozza, S., and Taroni, F. Analysing and exemplifying forensic conclusion criteria in terms of Bayesian decision theory. *Science & Justice*, 2017a, in press.

Biedermann, A., Bozza, S., Taroni, F., and Aitken, C. The consequences of understanding expert probability reporting as a decision. *Science & Justice, Special Issue on Measuring and Reporting the Precision of Forensic Likelihood Ratios*, **57**, 80–85, 2017b.

Biedermann, A. and Vuille, J. The decisional nature of probability and plausibility assessments in juridical evidence and proof. *International Commentary on Evidence*, **16**, 1–30, 2018.

Bozza, S., Broséus, J., Esseiva, P., and Taroni, F. Bayesian classification criterion for forensic multivariate data. *Forensic Science International*, **244**, 295–301, 2014.

Brier, G.W. Verification of forecasts expressed in terms of probability. *Monthly Weather Review*, **78**, 1–3, 1950.

Champod, C. Identification/individualisation, overview and meaning of ID. In J.H. Siegel, P.J. Saukko, and G.C. Knupfer, editors, *Encyclopedia of Forensic Science*, pages 1077–1084. Academic Press, San Diego, 2000.

Champod, C., Lennard, C., Margot, P., and Stoilovic, M. *Fingerprints and Other Ridge Skin Impressions* (2nd ed.). CRC Press, Boca Raton, 2016.

Chernoff, H. and Moses, L.E. *Elementary Decision Theory*. John Wiley & Sons, New York, 1959.

Cole, S.A. Individualization is dead, long live individualization! Reforms of reporting practices for fingerprint analysis in the United States. *Law, Probability and Risk*, **13**, 117–150, 2014.

Cole, S.A. and Biedermann, A. How can a forensic result be a "decision"? A critical analysis of ongoing reforms of forensic reporting formats for federal examiners. *Houston Law Review*, **57**, 551–592, 2020.

Cook, R., Evett, I.W., Jackson, G., Jones, P.J., and Lambert, J.A. A hierarchy of propositions: Deciding which level to address in casework. *Science & Justice*, **38**, 231–239, 1998.

Dane, F.C. In search of reasonable doubt: A systematic examination of selected quantification approaches. *Law and Human Behavior*, **9**, 141–158, 1985.

de Finetti, B. La prévision: Ses lois logiques, ses sources subjectives. Annales de l'Institut Henri Poincaré, 7, 1–68 (English translation). In H.E. Kyburg and H.E. Smokler, editors, *Studies in Subjective Probability* (1980, 2nd ed.), pages 93–158. Dover, New York, 1937.

de Finetti, B. Does it make sense to speak of 'Good Probability Appraisers'? In I.J. Good, editor, *The Scientist Speculates: An Anthology of Partly-Baked Ideas*, pages 357–364. Basic Books, New York, 1962.

de Finetti, B. *Probability, Induction and Statistics, The Art of Guessing*. John Wiley & Sons, New York, 1972.

de Finetti, B. *Theory of Probability, A Critical Introductory Treatment*, Vol. **1**. John Wiley & Sons, London, 1974.

de Finetti, B. The proper approach to probability. In G. Koch and K. Spizzichino, editors, *Exchangeability in Probability and Statistics*, pages 1–6. North-Holland Publishing Company, Amsterdam, 1982.

De Groot, M.H. *Optimal Statistical Decisions* (Wiley Classics Library Edition Published 2004). John Wiley & Sons, Inc., Hoboken, New Jersey, 1970.

Evett, I.W., Lambert, J.A., and Buckleton, J.S. A Bayesian approach to interpreting footwear marks in forensic casework. *Science & Justice*, **38**, 241–247, 1998.

French, S. *Decision Theory, An Introduction to the Mathematics of Rationality*. Ellis Horwood Limited, Chichester, 1988.

French, S., Maule, J., and Papamichail, N. *Decision Behaviour, Analysis and Support*. Cambridge University Press, Cambridge, 2009.

Friedman, R.D. Answering the Bayesioskeptical challenge. *The International Journal of Evidence & Proof, Special Issue*, **1**, 276–291, 1997.

Friedman, R.D. *The Elements of Evidence* (4th ed.). West Academic Publishing, Saint Paul, Minnesota, 2017.

Gittelson, S. *Evolving from Inferences to Decisions in Forensic Science*. PhD thesis, The University of Lausanne, School of Criminal Justice, Lausanne, 2013.

Gittelson, S., Biedermann, A., Bozza, S., and Taroni, F. The database search problem: A question of rational decision making. *Forensic Science International*, **222**, 186–199, 2012.

Gittelson, S., Bozza, S., Biedermann, A., and Taroni, F. Decision-theoretic reflections on processing a fingermark. *Forensic Science International*, **226**, e42–e47, 2013.

Gittelson, S., Biedermann, A., Bozza, S., and Taroni, F. Decision analysis for the genotype designation in low-template-DNA profiles. *Forensic Science International: Genetics*, **9**, 118–133, 2014.

Good, I.J. *Probability and the Weighing of Evidence*. Griffin, London, 1950.

Good, I.J. Rational decisions. *Journal of the Royal Statistical Society. Series B (Methodological)*, **14**, 107–114, 1952.

Hastie, R. Algebraic models of juror decision making. In R. Hastie, editor, *Inside the Juror. The Psychology of Juror Decision Making*, pages 84–115. Cambridge University Press, New York, 1993.

Kaplan, J. Decision theory and the factfinding process. *Stanford Law Review*, **20**, 1065–1092, 1968.

Kaye, D.H. What is Bayesianism? In P. Tillers and E.D. Green, editors, *Probability and Inference in the Law of Evidence, The Uses and Limits of Bayesianism (Boston Studies in the Philosophy of Science)*, pages 1–19. Springer, Dordrecht, 1988.

Kaye, D.H. Clarifying the burden of persuasion: What Bayesian decision rules do and do not do. *The International Journal of Evidence & Proof*, **3**, 1–29, 1999.

Lindley, D.V. *Introduction to Probability and Statistics from a Bayesian Viewpoint: Part 2 Inference*. Cambridge University Press, Cambridge, UK, 1965.

Lindley, D.V. *Making Decisions* (2nd ed.). John Wiley & Sons, Chichester, 1985.

Lindley, D.V. The philosophy of statistics. *The Statistician*, **49**, 293–337, 2000.

Luce, R.D. and Raiffa, H. *Games and Decisions: Introduction and Critical Survey*. Wiley, New York, 1958.

Marschak, J. Rational behavior, uncertain prospects, and measurable utility. *Econometrica*, **18**, 111–141, 1950.

Mazumder, A. *Planning in Forensic DNA Identification using Probabilistic Expert Systems*. PhD thesis, University of Oxford, Department of Statistics, Oxford, 2010.

Parmigiani, G. Decision theory: Bayesian. In N. Smelser and P. Baltes, editors, *International Encyclopedia of Social and Behavioral Sciences*, pages 3327–3334. Elsevier, Oxford, 2001.

Parmigiani, G. and Inoue, L. *Decision Theory: Principles and Approaches*. John Wiley & Sons, Chichester, 2009.

Piccinato, L. *Metodi per le decisioni statistiche*. Springer-Verlag, Milano, 1996.

Pratt, J.W., Raiffa, H., and Schlaifer, R. The foundations of decisions under uncertainty: An elementary exposition. *Journal of the American Statistical Association*, **59**, 353–375, 1964.

President's Council of Advisors on Science and Technology (PCAST). *Forensic Science in Criminal Courts: Ensuring Scientific Validity of Feature-Comparison Methods*. Washington, DC, 2016. Executive Office of the President. The report is avaiblable at https://www.nitrd.gov/pubs/PCAST-NITRD-report-2015.pdf

Press, S.J. *Bayesian Statistics: Principles, Models, and Applications*. John Wiley & Sons, Ltd., New York, 1989.

Press, S.J. *Subjective and Objective Bayesian Statistics: Principles, Models, and Applications* (2nd ed.). Wiley-Interscience, Hoboken, 2003.

Raiffa, H. *Decision Analysis, Introductory Lectures on Choices under Uncertainty*. Addison-Wesley, Reading, Massachusetts, 1968.

Raiffa, H. and Schlaifer, R. *Applied Statistical Decision Theory*. The M.I.T. Press, Cambridge, Massachusetts, 1961.

Ramsey, F.P. Truth and probability. In R.B. Braithwaite, editor, *The Foundations of Mathematics and Other Logical Essays*, pages 156–198. Routledge & Kegan Paul Ltd., 1931. Reprinted in H.E. Kyburg and H.E. Smokler, editors, *Studies in Subjective Probability* (2nd ed.), pp. 61–92. Dover, New York, 1980.

Robert, C.P. *The Bayesian Choice, A Decision-Theoretic Motivation* (2nd ed.). Springer, New York, 2007.

Savage, L.J. *The Foundations of Statistics*. Wiley, New York, 1954.

Schlaifer, R. *Probability and Statistics for Business Decisions*. McGraw-Hill Book Company, Inc., New York, 1959.

Simon, R.J. and Mahan, L. Quantifying burdens of proof: A review from the bench, the jury, and the classroom. *Law and Society Review*, **5**, 319–330, 1971.

Smith, J.Q. *Decision Analysis: A Bayesian Approach*. Chapman and Hall, London, 1988.

Smith, J.Q. *Bayesian Decision Analysis, Principles and Practice*. Cambridge University Press, Cambridge, 2010.

Spiegelhalter, D.J., Abrams, K.R., and Myles, J.P. *Bayesian Approaches to Clinical Trials and Health-Care Evaluation*. John Wiley & Sons, Chichester, 2004.

Stoney, D.A. Relaxation of the assumption of relevance and an application to one-trace and two-trace problems. *Journal of the Forensic Science Society*, **34**, 17–21, 1994.

Taroni, F., Bozza, S., and Aitken, C.G.G. Decision analysis in forensic science. *Journal of Forensic Sciences*, **50**, 894–905, 2005.

Taroni, F., Bozza, S., Bernard, M., and Champod, C. Value DNA tests: A decision perspective. *Journal of Forensic Sciences*, **52**, 31–39, 2007.

Taroni, F., Bozza, S., Biedermann, A., Garbolino, G., and Aitken, C.G.G. Data analysis in forensic science: A Bayesian decision perspective. In *Statistics in Practice*, John Wiley & Sons, Chichester, 2010.

Taroni, F., Biedermann, A., Bozza, S., Garbolino, G., and Aitken, C.G.G. Bayesian networks for probabilistic inference and decision analysis in forensic science. In *Statistics in Practice* (2nd ed.). JohnWiley & Sons, Chichester, 2014.

Thompson, W.C., Black, J., Jain, A., and Kadane, J. Latent fingerprint examination. In *Forensic Science Assessments: A Quality and Gap Analysis*. American Association for the Advancement of Science, Washington, DC, 2017.

Tillers, P. and Green, E.D., editors, *Probability and Inference in the Law of Evidence. The Uses and Limits of Bayesianism*. Kluwer Academic Publishers, Dordrecht, 1988.

von Neumann, J. and Morgenstern, O. *Theory of Games and Economic Behavior* (3rd ed.). Princeton University Press, Princeton, 1953.

6

Association Does Not Imply Discrimination: Clarifying When Matches Are (and Are Not) Meaningful

Maria Cuellar, Lucas Mentch, and Cliff Spiegelman

CONTENTS

6.1   Introduction 132
6.2   Association and Discrimination 133
      6.2.1   Quality of Test: Sensitivity and Specificity 134
      6.2.2   Sources of Error 134
      6.2.3   Weight of Evidence: The Likelihood Ratio 135
      6.2.4   Useful Databases for Ascertaining Discriminatory Power 135
      6.2.5   Conflating Conditional Statements: The Prosecutor's Fallacy 136
6.3   Examples: The Discriminatory Power of Forensic Evidence 137
      6.3.1   Arson Investigation 137
      6.3.2   Other Types of Forensic Evidence: DNA, Fingerprints, and Shoe Prints . . 137
      6.3.3   Abusive Head Trauma 138
6.4   Conclusion 140
References 140

In determining the appropriate weight or strength of forensic evidence, one must be aware of the both the associative and discriminative power of that evidence. While much of traditional forensic science research has focused on best practices for establishing positive associations, for many forensic tools, the power to discriminate and uniquely identify individual sources is largely unknown and sometimes highly disputed. This chapter begins by defining the ideas of association and discrimination in a forensic science context and illustrates the importance of understanding both in order to properly weight the findings of forensic tests. We discuss the related notions of sensitivity and specificity of testing procedures and describe potential issues with reliance on databases alone for determining the discriminatory power of evidence. We go on to provide numerous examples from across various disciplines where particular forensic procedures with unknown or weak discriminatory power has led to inappropriate determinations and conclude by offering thoughts on how these issues can be addressed in ongoing research.

6.1 Introduction

Anyone who has taken an introductory statistics course—and likely even many who have not—will be familiar with the phrase "correlation does not imply causation". The co-occurrence of events does not necessarily imply a causal relationship between them. In this chapter, we will explore the related concepts of *association* and *discrimination* in forensic science. Though less often discussed than correlation and causation, these concepts are also often misunderstood. In this chapter we take *association* as used in the context of forensic-science identification evidence to refer to the degree of similarity between the characteristics of a crime-scene specimen and a proposed source (or control specimens from that source). Sufficient perceived or measured similarity of characteristics may be reported as a positive association, a match, or an identification. For example, the caliber of a cartridge case recovered at a crime-scene may match that of the ammunition used in a gun found in the suspect's possession, associating the gun with the expended cartridge case. *Discrimination* refers to how strongly a positive association points to one source instead of other possible sources. The caliber of ammunition is not very discriminating, making the positive association incapable of proving (by itself) the hypothesis that the cartridge case came from that particular gun.

Conflating the determination of an association with the assessment of discrimination can lead to anomalous results. A simple example will make this plain. Suppose that you are a police officer investigating a bank robbery and are told by employees working at the time that a single individual with a mask entered the bank, demanded money, and then fled. The employees noticed, and security footage confirms, that the man was wearing a red shirt. On your way home from work that day, you notice a man with a red shirt walking down the sidewalk. Should you arrest him for the robbery?

Of course not! Certainly many, many people have red shirts, yet almost none of them committed the robbery. In other words, although a red shirt is *associated* with the robber, this association has very low *discriminatory power*. An arrest made on this basis alone would almost certainly be deemed illegal for want of probable cause.

Unfortunately, many forensic evaluation procedures rely heavily, and sometimes exclusively, on matters of association when trying to determine whether a particular piece of evidence is connected to a given suspect or source. As the example above makes clear, the evidential strength of a positive association necessitates some understanding of the discriminatory power of that evidence. When conclusions are reached based on the observed values of variables whose true discriminatory power is unknown, untested, or overestimated, the consequences can be severe. In some cases, scientific tests have shown that some evidence types, though once highly regarded, may possess little more discriminatory power than random chance.

In Section 6.2, we describe how association and discrimination, as applied to the evaluation of trace evidence, relates to the common statistical notions of the quality of a test, sources of errors, weight of evidence via the likelihood ratio, and the prosecutor's fallacy (sometimes called the transposition of the conditional). In Section 6.3, we consider four examples from forensic science (arson, DNA, fingerprints, shoeprints, and abusive head trauma) in which the failure to attend properly to discriminatory power can lead to mistaken inferences or testimony. In the final section, we conclude by providing an overview of the problem in a statistical framework, making suggestions as to how the issue can be mitigated in practice.

6.2 Association and Discrimination

In statistics, association is a measure of the connection between two variables, more general than correlation. For example, exposure to a chemical could be associated with the occurrence of a type of cancer in that (for whatever reason) the disease occurs in a greater proportion of individuals exposed to the chemical than to those with no such exposure. The usual measures of association for categorical variables are the difference in proportions, the relative risk, and the odds ratio. The difference in proportions compares the relative frequency of important characteristic between two groups: $P(outcome\,|\,exposure)$ − $P(outcome\,|\,no\ exposure)$. The relative risk is the ratio of the two proportions: $P(outcome\,|\,exposure)/P(outcome\,|\,no\ exposure)$. The odds ratio is $Odds\ (outcome\,|\,exposure)/Odds\ (outcome\,|\,no\ exposure)$. These quantities often appear in civil litigation involving equal opportunity in employment (for a difference in proportions by race or gender) and toxic torts (for relative risk and odds ratios).

Association in forensics refers to the ability to eliminate alternative identifications based on measuring the similarity of the features. A forensic scientist who says a piece of evidence is associated with a putative source typically is only saying it exhibits *class characteristics* – so it may fall well short of being an "identification" or "individualization" (i.e., an association to one and only one possible source). Sometimes a limited identification is called a *positive association*.

In most pattern-matching forensic-science disciplines, analysts first attempt to identify the class characteristics of the trace evidence. Class characteristics, such as the brand and size of a shoe that left a shoeprint in a crime scene or the brand and make of a firearm from which a bullet was shot, are usually broad categories into which many objects are known to fall. Then analysts attempt to determine whether, within the class, additional, so-called "individual characteristics" that would distinguish a single object from every other one in the class are present. (Sometimes the analyst may make a one-to-one association with class characteristics in a small, closed population of possible candidates.) Identification or individuation occurs when the analyst decides that a single individual item must be the true (unique) source. Exclusion occurs when the analyst decides that the item could not have been the source. When there is not enough information for an identification or an exclusion, the evidence is inconclusive.

Whether individualization in open populations is possible in the different forensic disciplines is a source of debate. Traditionally, forensic feature-matching disciplines were based on the premise that a mark can be individualized to the specific object that produced it, to the exclusion of every other object of that kind that exists. Such terminology has fallen out of favor in several disciplines, but testimony of unique identification and individualization continues. Furthermore, there are many possible sources of error in reasoning from a class to an individual, including human error and errors in the premises of the discipline. Consequently, even if the theory of utterly individual features is true, the ability of analysts to apply it in practice is limited.

As noted above, discrimination refers to the ability to eliminate alternative identifications based narrowing the pool of candidates for individuation. For instance, an FBI article on fiber evidence states, "It is not possible to say positively that a fiber originated from a particular fabric," but that "the inability to positively associate a fiber with a source in no way diminishes the significance of a fiber association" (Deedrick, 2000). In other words, identifying a fiber associated with a fabric may have high discriminatory power, but in most cases, not enough to identify a specific unique piece of fabric to the exclusion of every other piece in existence.

Association and discrimination are therefore quite different concepts. Association, for example, can ask how similar a specific fiber is to the fabric of a t-shirt. Analysts can look at microscopic features of the fiber and t-shirt to see how similar they are to each other, according to some disciplinary guidelines. A high association would mean the fiber and t-shirt are very similar, and thus the fiber could have come from that t-shirt. Discrimination, on the other hand, asks how many other t-shirts also bear the same degree of similarity. If it turns out that a factory has made thousands of t-shirts that are indiscernible according to the disciplinary guidelines, then knowing that the fiber is associated to the t-shirt may not tell us much about individuation. Thus, despite having a high association, this analyst's conclusion would have low discrimination. Only if we knew that there are no other t-shirts in the world like this one would there be both high association and high discrimination, which would lead to a high likelihood of individuation.

6.2.1 Quality of Test: Sensitivity and Specificity

Suppose that a characteristic is associated with a hypothesis and that it has a certain amount of discriminatory power. How should we determine the strength of this evidence? Two useful concepts for assessing the quality of a test are sensitivity and specificity. In forensic-science evaluations, positive outcomes occur when an item or suspect "matches" a particular piece of evidence. (Analogously, in medical tests, a positive outcome denotes that a person being evaluated tests positive for a particular disease.) A test that accurately predicts the positive outcomes (registers "condition present" when the condition is present) is said to have high *sensitivity* while a test that does well at correctly identifying negative outcomes (registers "condition absent" when the condition is not present) has high *specificity*.

Ideally, effective tests require both high sensitivity and specificity, but in practice it is very often the case that one must trade one for the other. To see this, note that a test (or an examiner) that simply concludes that every comparison is a (positive) match will have perfect sensitivity since every "true" match will be properly identified as such. However, many of those claimed positive matches would in fact be negatives (i.e., false positives). Likewise, a test that always comes back negative will have perfect specificity but will miss the fact that many of the claimed negatives are in fact positives (i.e., false negatives). Well-designed tests must carefully balance the levels of sensitivity and specificity.

Sensitivity and specificity are related to association and discrimination. Tests with high sensitivity must utilize variables that have a strong association with the positive outcomes. On the other hand, if those variables fail to accurately discriminate between the outcomes, the model is likely to suffer from poor specificity. An effective test thus requires both high association and discrimination. Association is important, but it is not so important that one ought to sacrifice discrimination.

6.2.2 Sources of Error

In forensic science tests with binary outcomes, the analyst must make an identification (if possible) and then must assess the meaning of the reported match. What is the probability that the analyst has made the correct decision? It is possible that the method used to assess the match has flaws—perhaps the discipline has made untested assumptions that are too strong, or that the evidence was mislabeled or mixed up, or perhaps there was some other source of human error (Saks and Koehler, 2008).

To assess potential sources of error, groups of scientists and other academics have reviewed the current state of various forensic disciplines in terms of validity (see National Research Council, 2009 and PCAST, 2016). Validity is required for scientific statements to be admissible in court according to the rules of evidence of many U.S. jurisdictions (Kaye, 2020). The reports concluded that many procedures across forensic disciplines have not rigorously been shown to have scientific validity. Some of these conclusions were questioned by the forensic science community (see National District Attorneys Association, 2016). Although there is an ongoing debate about the reports' conclusions, some forensic laboratories and university researchers (the Center for Statistics and Applications in Forensic Evidence, for example) are currently working to assess the validity of numerous forensic tests and procedures.

While much of classical research in forensic science has dealt with whether analysts are warranted in making associations, current (and we hope, future) research is investigating not only the sources of error in association tests, but also identifying possible sources of error, and attempting to measure error rates for discrimination tasks.

6.2.3 Weight of Evidence: The Likelihood Ratio

One way to quantify the overall strength of a particular piece of evidence, is via the likelihood ratio (LR). In Bayesian decision theory, the LR is defined as a ratio of two probabilities: the probability of observing the data given that one hypothesis holds versus the probability of observing the data given an alternative hypothesis (see Lindley, 1977 and Lund and Iyer, 2016). These authors, among many others, suggest that, if a researcher could calculate a LR by using data, this value could be combined with the prior odds of the trier of fact (e.g., jurors, judge) by using Bayes' rule, and this would result in obtaining posterior odds, which would quantify the odds that the prosecution hypothesis holds versus the defense hypothesis. Forensic examiners could thus compute a LR as a summary of their analysis by using a database, and present its value as the weight of evidence to interested parties in a written report or to the triers-of-fact during testimony. Using LRs (in Bayesian and frequentist statistics) requires defining and estimating a probability model before use and thus care should be taken to ensure that the model selected offers an adequate and unbiased representation.

6.2.4 Useful Databases for Ascertaining Discriminatory Power

Currently in the field of forensic science there is an international debate regarding the proper role of databases in forensic evidence, and how these databases should be used to identify a suspect in a crime correctly, i.e., to ascertain discriminatory power. There are many databases created with the purpose of assisting law enforcement offers in linking a suspect to a crime. These databases include DNA, fingerprints, bullet casings, automobile paint, glass, and soil. However, many of the current forensic databases were developed on an ad hoc basis without careful consideration of the statistical and scientific principles associated with use.

Although it may seem straightforward to use a database to estimate discriminatory power, using a database requires careful considerations about its size, whether the data represent the population of interest, and what other information the researcher has besides the particular piece of evidence. Recall that the likelihood ratio requires comparing two quantities, each of which should be drawn from its corresponding population. One is the scenario in which the prosecution hypothesis holds, and the other is the scenario in which the defense hypothesis holds. Suppose, for example, a high-quality latent fingerprint was found at a crime scene. The scenario in which a suspect is guilty would result in a latent fingerprint from the evidence matching an individual's print on a database, or from a list of suspects and an examiner then concluding that the suspect is indeed a match. The error rate of the algorithms that find matches as well as that of the forensic analysts making the final determination are therefore crucial. Another relevant question is, what is the scenario in which the suspect would be considered innocent? How large of a database, or in how many databases, should a researcher check before determining whether the suspect is not guilty? There is currently no agreement about how to use forensic databases accurately.

Although likelihood ratios provide a quantitative measure of the weight that should be given to forensic evidence, the manner in which these should be calculated given the potential issues with databases discussed above is an issue that is not fully resolved. Ongoing research is attempting to establish requirements for a database to have the appropriate information to calculate a likelihood ratio, as well as the proper procedure for computing score-based likelihood ratios. It is essential to state, consider, and justify all assumptions carefully when using a likelihood ratio calculation. Calculating a likelihood ratio incorrectly can lead to decreased accuracy (false positives *and* false negatives) when matching crime scene evidence to a suspect.

6.2.5 Conflating Conditional Statements: The Prosecutor's Fallacy

The difference between association and discrimination plays a role in an important concept from law and psychology called the prosecutor's fallacy (Thompson and Schumann, 1987). Suppose a bank robbery is committed by an individual with blue eyes and red hair and that a man with blue eyes and red hair is later charged with the crime. For demonstration purposes, suppose that the city has a total population of 1 million people, of which only one out of every 10,000 has both blue eyes and red hair. The probability that a man would have blue eyes and red hair given that he is innocent is very small, $99/1,000,000 = 0.000099$ (0.0099%). So, the prosecutor says, "The chances that the suspect just happens to have blue eyes and red hair is so small that he (or she) must be guilty!"

But consider this: there is a 99% chance that a (randomly selected) person with red hair and blue eyes is innocent! Indeed, if the city contains only 100 such individuals and only one committed the crime, then we have $99/100 = 0.99$ (99%). Here we are seeing a confusion between the probability that an individual is innocent given the evidence and the probability of observing the evidence given that the individual is innocent. In the example above, the prosecutor's fallacy is a confusion between $P(blue\ eyes\ and\ red\ hair\ |\ innocent)$ and $P(innocent\ |\ blue\ eyes\ and\ red\ hair)$.

In the prosecutor's fallacy, any connection that can be made between a source or suspect and the crime scene is given a great deal of weight, but for the vast majority of forensic testing procedures there are no careful studies that firmly establish the discriminatory power of this association. The fallacy has been so prevalent in court that it is often taught to attorneys in training at law schools.

6.3 Examples: The Discriminatory Power of Forensic Evidence

Unfortunately, the problem of unknown or misunderstood discriminatory power is pervasive throughout many scientific fields. We begin this section by reviewing the particularly damning—and now largely discredited—field of arson investigation and then discuss the discrimination issue in the context of other areas of modern forensic science.

6.3.1 Arson Investigation

Parallels can be drawn between the ongoing search for informative disease-associated biomarkers and the development of arson science in the United States. The last several decades have seen drastic changes in the way fire scenes are processed and analyzed for arson indicators. Several indicators such as large *alligatoring* (deep patterns of severe scorching) and *crazed glass* (unusual fracture patterns in glass surfaces) were once thought to definitively indicate the presence of accelerants. Indicators such as these were even supported at the time by the National Bureau of Standards (NBS)—now the National Institute of Standards and Technology (NIST)—but are now recognized as no more than myth (NBS, 1980). These indicators were purely fanciful, lacking any power in either association or discrimination. Lentini (2006) gives a thorough history.

In addition to these incorrect interpretations, there are several other indicators that may often be present in cases of arson (high association) but can also routinely be found in scenes produced by accidental fires (poor discrimination). For example, *lines of demarcation*—well-defined patterns of differing char intensity—can be the result of liquid accelerants but may also be caused by clothing or falling drywall (Lentini, 2006). Since heat rises, the presence of *low burn areas* or *burn holes* in the floor is sometimes taken as proof that the fire started on the floor and likely involved accelerants (Lentini, 2006). However, NFPA (1992) now cautions that such effects may be the result of flashover—an instance of a fire spreading very rapidly through the air because of intense heat—an effect common in both intentional and accidental fires. Thus, while it may be the case that certain indicators are often present in many cases of arson, it is now understood that their presence does not clearly establish arson.

6.3.2 Other Types of Forensic Evidence: DNA, Fingerprints, and Shoe Prints

Unfortunately, the reliance on non-discriminative forensic evidence has not been limited to arson investigation. In recent years, the practice of bite mark analysis has come under pressure for much the same reasons. Dentitions once thought to uniquely identify an individual have been shown to in fact be far less discriminating (Saks et al., 2016).

On the other hand, a few forensic analysis procedures have been well-validated, and when sufficient evidence exists and the analyses are properly carried out, those methods can do an excellent job of discriminating the true source (suspect) from others. Nuclear DNA analysis ranks atop the list of these state-of-the-art identification procedures. When DNA evidence is found to "match" that of a given suspect (not from a database search, and using a full set of markers), it is extraordinarily likely that the suspect was the source of the crime-scene DNA (as compared to any unrelated person). Furthermore, suspects whose DNA does not match can be reliably ruled out as potential sources; see Kaye (2009) for a fuller discussion.

To a lesser extent, fingerprint analyses are also generally considered reliable procedures with substantial discriminatory power when quality latent prints are available. When matches are obtained through an automated database search, however, the examiner's threshold for declaring a match among the candidates may be too low. This issue arose in the infamous case of Brandon Mayfield. After the 2004 Madrid bombings, the Spanish authorities found fingerprints inside a bag that was believed to contain bomb detonating devices and shared the fingerprints with the FBI. The FBI performed a computerized search in a database of millions of fingerprints. The computer was programmed to generate a list of the 20 most similar prints. Three separate fingerprint examiners then concluded that one of them, which belonged to attorney Brandon Mayfield from Oregon, was a match. Mayfield was jailed for two weeks as a material witness before the Spanish police found that the print matched an Algerian man living in Spain. Mayfield received two million dollars to settle the lawsuit that he brought against the government. The U.S. Department of Justice conducted an extensive review of the FBI's handling of the case (Office of the Inspector General, 2006). Thompson and Cole (2005), Dror et al. (2006), and others wrote about the psychology involved in the misidentification and how such mistakes could be avoided in the future.

Unfortunately, the ability of most other forms of forensic evidence to discriminate and identify the true source effectively has yet to be rigorously investigated. Consider shoeprint evidence, for example. Suppose a high-quality latent shoeprint is obtained from a crime scene and a forensic examiner is given an exemplar from a suspect with which to make comparisons. Suppose further that the style, brand, and size of the shoe is easily identifiable and the exemplar shows very similar wear patterns to those observed on the latent print. Can the examiner conclude that the source of the exemplar is the same as the source of the latent with a high degree of certainty? Put differently, how well can we expect thorough analysis of a latent shoeprint—even one of exceptionally high quality—to discriminate the true source of that print from all shoes that could have made that print? Can we expect to identify the unique shoe, or, if the majority of shoes of a given kind can be expected to wear in a similar fashion, are we only safe in establishing the brand, style, and size of shoe? According to the 2009 report by the National Research Council, class characteristics can be ascertained, but there is no consensus and no standards regarding the number of individual characteristics or markers needed to make a positive identification with a particular degree of confidence. The discriminatory power of these types of forensic evidence remains largely unresolved (National Research Council, 2009).

6.3.3 Abusive Head Trauma

Trudy Muñoz, a nanny in Virginia, was taking care of four-month-old Noah Whitmer, when (she claimed) he stopped being responsive and his eyes rolled back. Muñoz called 911, and paramedics took Noah to the emergency room. At the hospital, the physicians noted that the child did not have external or visible injuries, but when they performed a CT scan and looked at his retinas, they realized he had retinal bleeding, brain swelling, and subdural hemorrhage. These symptoms together are sometimes called "the triad." They are considered to be characteristic of abusive head trauma (AHT), formerly called shaken baby syndrome. Muñoz was interrogated and placed under arrest. At trial, six physicians testified that nothing else could have caused this condition other than child abuse through shaking, and one physician claimed that the condition could have been caused by a spontaneous re-bleed. Muñoz was convicted of child abuse and is still serving her sentence in Virginia (Cenziper, 2015).

How did the physicians know for certain the child was abused? Abusive head trauma (AHT) is defined as "an injury to the skull of intracranial contents of an infant or young child due to inflicted blunt impact and/or violent shaking" (Parks et al., 2012). Could the condition have been caused by an underlying medical condition or an accident rather than a violent act? The University of Virginia Innocence Project filed a request for a pardon from the governor, pointing to doubts about the validity of AHT as proof of abuse. The President's Council of Advisors on Science and Technology (2016) noted that "Shaken Baby Syndrome requires urgent attention." Characteristics associated with AHT are often readily observable features like patterns of bruising or evidence of broken bones. Nonetheless, studies investigating AHT often overstate their discriminatory power (Cuellar, 2017), and controversy surrounds the proper diagnosis of AHT. Physicians often disagree about whether a child was abused or shaken.

Efforts to establish the factors that are diagnostic (discriminative) for AHT are complicated by the fact that in most cases, the only individual other than the infant victim who would know the truth about whether the child was abused or shaken is the possible abuser. Without known ground truth, the correlations derived from the datasets used to search for physical indicators of AHT are difficult to evaluate. They may contain data that is biased by the physician(s) making the diagnoses.

Furthermore, some AHT studies (Maguire et al., 2009, 2011; Cowley et al., 2015) determine that a child was abused only in cases satisfying at least one of the following criteria: (1) Abuse is confirmed at case conference or civil, family or criminal court proceedings or admitted by the perpetrator, or independently witnessed, or, (2) Abuse is confirmed by stated criteria including multi-disciplinary assessment. These criteria depend on the opinions of many individuals, from the nurse who first saw the child to the trier of fact in court. Thus, it is possible that some children are abused, but because that abuse could not be confirmed according to the above criteria, those cases are recorded as instances of non-abuse.

Court judgments have been reversed due to demonstrably wrongful convictions (see West and Meterko, 2015), false confessions have been shown to occur (Kassin et al., 1996), and in many cases, once a physician has made a diagnosis, it is quite possible that the multidisciplinary assessment is simply reinforcing the biases expressed by the original physician. Maguire et al. are aware that it is difficult to determine whether a child was abused—that is in fact their stated reason for writing their series of articles in the first place—but nevertheless, it is doubtful that a different physician, interdisciplinary team, or court proceed would consistently make the same decision with the same information given the dramatic variability in diagnoses by physician, hospital type, and geographic location.

In spite of the biases likely present within the available data, studies continue to utilize that data to construct models for predicting whether a given case is the result of abuse. The predictor variables found to be most informative in these models are then proposed as natural candidates for AHT markers, i.e., if these features are observed in a child, it is likely that the child was abused. Some recent articles claim to have identified features strongly associated with AHT that also have the power to discriminate this condition from other forms of head trauma. Berger et al. (2012) and Papa et al. (2013) claim that certain biomarkers are discriminatory of AHT. Much of the methodology and analysis is careful, but inaccuracies or biases present in the determination of the initial AHT diagnosis carry through to the construction of the statistical models and ultimately to the biomarkers determined to be most useful. Highly imprecise diagnoses of the target disease and lack of scientifically objective standards makes any formal assessment of the association and discrimination of proposed features extremely difficult.

It is important to keep the two concepts separate: certain brain (and other) conditions are associated with AHT, but simply observing these features does not mean one can discriminate and necessarily assert that the child has suffered these conditions as a result of AHT and not an accident or unrelated medical condition. If the diagnoses and thus studies of AHT are based on data that are biased and/or unreliable, then neither association nor the discriminatory power of a reported association is known.

6.4 Conclusion

This chapter has provided examples demonstrating that when the ability of certain indicators to discriminate is unknown or overestimated, the consequences can be severe. Features based only on association can lead to misunderstandings and potentially even wrongful convictions. Likewise, the same issue can lead investigators to incorrect conclusions as we saw was the case in arson science. Unlike many statistical issues, the problem of unknown discrimination is not one that can be solved by larger samples. In the case of abusive head trauma (AHT), for instance, the primary issue in uncovering accurate indicators is not the size of the sample, but the potential biases and inaccuracies embedded within the sample itself. A larger sample collected via a biased process would only serve to reinforce those biases.

This danger of reinforcing incorrect information is of particular concern in the practice of forensic science because of the use of precedents in the law. Consider a situation where a forensic examiner mistakenly believes that a certain type of evidence is capable of uniquely identifying an individual and in a particular trial, and he or she testifies to this effect. The defendant—due in part to that testimony—is found guilty, thereby reinforcing the misconception that this kind of evidence will only match those who are guilty. The next time that examiner finds that a suspect matches that kind of evidence, he or she may believe even more strongly in the guilt of that suspect because of the past experience.

Although some efforts are moving in the right direction, far more research is needed about the true discriminatory power of evidence. Further studies could mitigate these problems and begin to assign more reasonable weight to various forms of forensic evidence. Importantly, these studies must be rigorous and well designed. In the medical community, there is a large body of literature investigating improved study designs for biomarkers—see Diamandis (2010) and Rundle et al. (2012) for recent examples—but the majority of new studies often fail to utilize these improved methodologies. Analogous studies are far rarer in the forensic science literature, though with the renewed interest sparked by the (2009) National Research Council report, there is hope that the field will attract the kind of more rigorous scientific attention necessary to address these issues.

References

Berger, R.P., Hayes, R.L., Richichi, R., Beers, S.R. and Wang, K.K. Serum concentrations of ubiquitin C-terminal hydrolase-L1 and αII-spectrin breakdown product 145 kDa correlate with outcome after pediatric TBI. *Journal of Neurotrauma*, **29**(1), 162–167, 2012.

Brannigan, F.L., Bright, R.G. and Jason, N.H. *Fire Investigation Handbook (Vol. 134)*. US Department of Commerce, National Bureau of Standards, 1980.

Cenziper, D. Shaken science: A disputed diagnosis imprisons parents. *Washington Post*, March 20, 2015.

Cowley, L.E., Morris, C.B., Maguire, S.A., Farewell, D.M. and Kemp, A.M. Validation of a prediction tool for abusive head trauma. *Pediatrics*, 136(2), 290–298, 2015.

Cuellar, M. Causal reasoning and data analysis: Problems with the abusive head trauma diagnosis. *Law, Probability and Risk*, 16(4), 223–239, 2017.

Deedrick, D.W. *Hairs, Fibers, Crime, and Evidence: Part 2, Fiber Evidence*. US Department of Justice, Federal Bureau of Investigation, 2000.

Diamandis, E.P. Cancer biomarkers: Can we turn recent failures into success? *Journal of the National Cancer Institute*, 102(19), 1462–1467, 2010.

Dror, I.E., Charlton, D. and Péron, A.E. Contextual information renders experts vulnerable to making erroneous identifications. *Forensic Science International*, 156(1), 74–78, 2006.

Ioannidis, J.P. Why most published research findings are false. *PLoS Medicine*, 2(8), e124, 2005.

Kassin, S.M. and Kiechel, K.L. The social psychology of false confessions: Compliance, internalization, and confabulation. *Psychological Science*, 7(3), 125–128, 1996.

Kaye, D.H. Rounding up the usual suspects: A legal and logical analysis of DNA trawling cases. *North Carolina Law Review*, 87, 425, 2009.

Kaye, D.H. Forensic statistics in the courtroom. In Banks, D., Kafadar, K., Kaye, D.H., and Tackett, M., editors., *Handbook on Forensic Statistics*. CRC Press, Boca Raton, Florida, 2020.

Lentini, J.J. The mythology of arson investigation. In *ISFI 2006—Proceeding of 2nd International Symposium on Fire Investigation Science & Technology*, pages 301–312. NAFI, Sarasota FL, 2006.

Lindley, D.V. A problem in forensic science. *Biometrika*, 64(2), 207–213, 1977.

Lund, S.P. and Iyer, H. Likelihood Ratio as Weight of Forensic Evidence: A Metrological Perspective. *arXiv preprint arXiv*, 1608, 07598, 2016.

Maguire, S., Pickerd, N., Farewell, D., Mann, M., Tempest, V. and Kemp, A.M. Which clinical features distinguish inflicted from non-inflicted brain injury? A systematic review. *Archives of Disease in Childhood*, 94(11), 860–867, 2009.

Maguire, S.A., Kemp, A.M., Lumb, R.C. and Farewell, D.M. Estimating the probability of abusive head trauma: a pooled analysis. *Pediatrics*, 128(3), e550–e564, 2011.

National District Attorneys Association. Reference: Report Entitled 'Forensic Science in Criminal Courts: Ensuring Scientific Validity of Feature-Comparison Methods', 2016.

National Fire Protection Association. *Guide for Fire and Explosion Investigations (NFPA 921–92)*. NFPA, Quincy, MA, 1992.

National Research Council. *Strengthening Forensic Science in the United States: A PATH forward*. National Academies Press, 2009.

OIG A. *Review of the FBI's Handling of the Brandon Mayfield Case. Office of the Inspector General, Oversight and Review Division, US Department of Justice*, 1–330, 2006.

Papa, L., Ramia, M.M., Kelly, J.M., Burks, S.S., Pawlowicz, A. and Berger, R.P. Systematic review of clinical research on biomarkers for pediatric traumatic brain injury. *Journal of Neurotrauma*, 30(5), 324–338, 2013.

Parks, S.E., Annest, J.L., Hill, H.A. and Karch, D.L. Pediatric abusive head trauma: recommended definitions for public health surveillance and research, 2012.

President's Council of Advisors on Science and Technology (PCAST). *Forensic Science in Criminal Courts: Ensuring Scientific Validity of Feature-Comparison Methods*. Washington, DC, 2016.

Rundle, A., Ahsan, H. and Vineis, P. Better cancer biomarker discovery through better study design. *European Journal of Clinical Investigation*, 42(12), 1350–1359, 2012.

Saks, M.J., Albright, T., Bohan, T.L. et al. Forensic bitemark identification: Weak foundations, exaggerated claims. *Journal of Law and the Biosciences*, 3(3), 538–575, 2016.

Saks, M.J. and Koehler, J.J. The individualization fallacy in forensic science evidence. *Vanderbilt Law Review*, 61, 199, 2008.

Thompson, W.C. and Cole, S.A. Lessons from the Brandon Mayfield case. *The Champion*, **29**(3), 42–44, 2005.

Thompson, W.C. and Schumann, E.L. Interpretation of statistical evidence in criminal trials. *Law and Human Behavior*, **11**(3), 167–187, 1987.

West, E. and Meterko, V. Innocence project: DNA exonerations, 1989–2014: Review of data and findings from the first 25 years. *Albany Law Review*, **79**, 717, 2015.

7

Validation of Forensic Automatic Likelihood Ratio Methods

Daniel Ramos, Didier Meuwly, Rudolf Haraksim, and Charles E.H. Berger

CONTENTS

7.1 Introduction 143
    7.1.1 Scope 144
    7.1.2 Aim 144
    7.1.3 Structure 145
7.2 Validation Process 145
    7.2.1 Standardization 145
    7.2.2 Validation of Theoretical and Empirical Aspects 146
    7.2.3 Performance Characteristics for Automatic LR Methods 147
    7.2.4 Empirical Validation 147
    7.2.5 Validation Protocol 148
7.3 Primary Performance Characteristics 150
    7.3.1 Performance of Probabilities by Proper Scoring Rules 151
    7.3.2 Discrimination and Calibration of Probabilities 154
    7.3.3 Performance of Likelihood Ratios 155
    7.3.4 Properties of Well-Calibrated Likelihood Ratios 157
    7.3.5 Examples with Primary Performance Characteristics 158
7.4 Secondary Performance Characteristics 159
    7.4.1 Robustness 159
    7.4.2 Monotonicity 161
    7.4.3 Generalization 161
7.5 Conclusion 161
References 161

7.1 Introduction

Forensic practice is more and more under scrutiny, both from the general public and the scientific community. Press reports about forensic examination in criminal cases regularly question the scientific foundation of forensic science and challenge the results of its analysis and interpretation (Hsu, 2015). In 2016, a report from the President's Council of Advisors on Science and Technology (PCAST) entitled "Forensic Science in Criminal Courts: Ensuring Scientific Validity of Feature-Comparison Methods" (President's Council of Advisors on Science and Technology, 2016) was released. This report emphasizes the importance of the validity of methods for the credibility of forensic science. In their 2017 annual report, the UK forensic science regulator also stressed the importance of the validity of the methods in the accreditation process:

> The accreditation system is predicated on organisations being: a) accountable for the quality of their work and, b) able to demonstrate through regular audit and through evidence of staff competence and method validity that they are sustainably competent to produce reliable results.

Forensic Science Regulator
annual report, 2017

The forensic community is currently actively developing and implementing quality assurance by establishing worldwide harmonized minimum quality standards. Such standards can be used to demonstrate the scope of validity, the reliability and the adequacy of the methods applied to the data collected in forensic casework. The validation and accreditation of instrumental and automatic methods used for forensic analysis is well-studied and reflected in the scientific literature, and their harmonization and standardization are already in progress at the regional level in Europe (European Network of Forensic Science Institutes, ENFSI (De Baere et al., 2014)), in the US (Scientific Working Group for Forensic Analysis of Chemical Terrorism/Threats SWGFACT (Standards and Guidelines, 2005)), in Australia and New Zealand (Australian and New Zealand Policing Advisory Agency and National Institute of Forensic Science, ANZPAA NIFS and Standards Australia (Australian Standard, 2012)), and soon globally (International Organization for Standardization, ISO (ISO standard 21043)).

This chapter addresses the validation of automatic likelihood ratio methods for forensic evidence evaluation.

7.1.1 Scope

In forensic evidence evaluation practitioners assign a strength of evidence to forensic observations and analytical results, in order to address hypotheses at source or activity level (Cook et al., 1998). This assignment is based on the practitioner's assessment and increasingly on the computations of automatic likelihood ratio (LR) methods.

This chapter focuses on the validation of automatic methods* developed to assign a strength of evidence at source level to the analytical results originating from the comparison of distinctive features of two specimens: a trace or mark of an unknown source and a reference specimen of a known source (Meuwly 2006; Robertson et al., 2016). Usually, a trace or mark is produced under the uncontrolled conditions of a criminal activity while a reference specimen is produced under controlled and more ideal conditions.

We will review some of the performance characteristics needed to accomplish any validation process, and we will give special attention to the calibration of likelihood ratios, because of its importance and its relative novelty in forensic interpretation. Throughout this chapter, we will follow a Bayesian interpretation of probability (Lindley, 2013), and the recent guideline for evaluative reporting in forensic science in Europe (ENFSI, 2015).

7.1.2 Aim

The main aim of this chapter is to offer guidance to forensic practitioners assessing the scope of validity and applicability of automatic likelihood ratio methods, when

---

* Currently the validation of human-based interpretation methods focuses mainly on competence assessment. In the future it is desirable that the validation also addresses performance assessment, and the methods described in this chapter are also suitable for this purpose.

implementing a new and non-standard method in forensic practice. These are essential steps towards the demonstration that such a method provides results that are fit for their intended use and allow it to be accredited and used in forensic casework. The validation and accreditation of automatic forensic evaluation methods serves several purposes. Primarily, it enables the demonstration of compliance with the quality standards adopted globally (ISO/IEC standard 17025, 2017; International Laboratory Accreditation Cooperation), specifically the way in which specimens are handled, what methods are used and how the results are interpreted. Beyond that, scientific and transparent validation of new and non-standard forensic methods favors their acceptance within the forensic community. Accreditation enables the legal community to recognize the methods' merits and whether or not a method works reliably under forensic conditions.

Another aim of this chapter is to elaborate on the concept of calibration as a performance characteristic for likelihood ratios. We will justify its critical importance in the validation process.

7.1.3 Structure

This chapter is structured as follows. In Section 7.2 a review of the most important standards for validation is given, along with the concepts of performance characteristics, performance metrics and validation criteria, which constitute the validation process. The approach for the measurement of performance of the methods under validation is developed in Section 7.3. Sections 7.4 and 7.5 describe the primary and secondary characteristics used to assess the performance of automatic forensic evaluation methods. The chapter ends with our conclusions in Section 7.6.

7.2 Validation Process

7.2.1 Standardization

The ISO/IEC 17025:2017 standard (ISO/IEC standard 17025, 2017) is used worldwide as one of the main bases for the accreditation of forensic service providers carrying out laboratory activities, while some more specifically forensic ISO standards are currently in development (Wilson-Wilde, 2018). In Clause 7.2.2.1 the ISO/IEC 17025:2017 standard specifies that non-standard methods, laboratory-developed methods, and standard methods used outside their intended scope or otherwise modified need to be validated.

Likelihood ratio methods used for forensic evaluation can be considered as non-standard in two aspects. Firstly because they are laboratory-developed, and secondly because they address forensic evaluation from an automatic perspective, when forensic evaluation is generally only considered as an opinion formed by a practitioner.

In its Section 7.8.7.1 the ISO/IEC 17025:2017 standard specifies that:

> only personnel authorised for the expression of opinions and interpretations release the respective statement,

considering this step exclusively as a human competence.

A similar approach had already been pursued in 2010 by the Dutch Accreditation Council in its explanation of the ISO/IEC 17025:2005 standard. In its Section 3.2, the criteria to assess the competence of laboratories to express opinions and interpretations are listed as follows:

> (1) examining the implementation of the procedures and practices, (2) examining the adequacy of the competence criteria, (3) verifying qualifications, experience, training and knowledge of personnel, (4) examining the adequacy of mechanisms in place to monitor the competence of personnel, (5) examining reports where opinions and interpretations have been expressed, (6) examining records showing the basis on which opinions and interpretations are based, (7) using other appropriate assessment techniques.

A similar approach is also pursued in Section 4.8.3 of the ILAC- G19:08/2014 document (International Laboratory Accreditation Cooperation) *"Modules in a Forensic Science Process"*:

> personnel interpreting results shall have been assessed and deemed competent before reporting statements including interpretation and opinions of results and findings.

In Section 3.10 it also specifies that

> interpretations of results and findings shall be based on robust studies and documented procedures,

and in Section 3.12 it states that

> where software is used it shall be demonstrated as being fit for purpose. This may be a verification check of the software functionality, for example, the use of a spreadsheet to calculate values, or could be as part of the more wide reaching validation of the forensic science process in which the software is used, for example, the use of databases for matching specific characteristics.

But neither the ISO/IEC 17025:2017 standard nor the ILAC-G19:08/2014 document explicitly consider automatic interpretation methods for forensic evaluation, or the fact that these methods require a validation based on their performance, just as instrumental analytical methods require validation.

7.2.2 Validation of Theoretical and Empirical Aspects

Validation can address the theoretical or empirical aspects of the LR method. The validation of the theoretical aspects rests upon mathematical proof or falsification. The validation of the empirical aspects, on the other hand, rests upon the acceptance or rejection of validation criteria on the basis of experimental results. This requires the definition of a validation protocol and experiments, which are used to accept or reject the method's validity, based on the chosen validation criteria.

The theoretical validation is necessary and the literature regarding the theoretical grounds for using likelihood ratio methods for forensic evaluation is already abundant (Robertson et al., 2016). On the other hand, the empirical validation of automatic likelihood ratio methods is an emerging area, for which literature has been sparse to date (Haraksim et al., 2015; Meuwly et al., 2017; Ramos et al., 2017). Therefore, this chapter limits its focus to the empirical validation of automatic LR methods.

In essence, the approach for the empirical validation of automatic LR methods is analogous to the one described for the empirical validation of instrumental analytical methods in the ISO/IEC 17025:2017 standard. The aim is to establish the scope of validity of the method, and to determine the operational conditions under which it meets some performance requirements or validation criteria. In its Section 7.2.2, the ISO/IEC 17025:2017 standard mentions that validation can be one or a combination of measurements of several performance characteristics, such as:

1. The calibration or evaluation of bias, precision, a systematic assessment of the factors influencing the results;
2. The evaluation of the robustness for variation of controlled parameters;
3. The comparison of results achieved with other validated methods and the evaluation of measurement uncertainty of the results, based on an understanding of the theoretical principles of the method and practical experience with the method.

The note of its Section 7.2.2.3 also provides a definition of performance characteristics:

> performance characteristics can include, but are not limited to, measurement range, accuracy, measurement uncertainty of the results, limit of detection, limit of quantification, selectivity of the method, linearity, repeatability or reproducibility, robustness against external influences or cross-sensitivity against interference from the matrix of the sample or test object, and bias.

7.2.3 Performance Characteristics for Automatic LR Methods

Currently, accuracy, discrimination, calibration, robustness, monotonicity* and generalization have been identified as relevant to the validation performance characteristics for the assessment of automatic likelihood ratio methods (Meuwly et al., 2017). Performance metrics and graphical representations are associated with each performance characteristic for the measurement and representation of the method's performance.

Accuracy, discrimination and calibration have been defined as primary performance characteristics, as they relate directly to performance metrics and focus on desirable properties of the LR methods. They address the required behavior of the automatic LR method if it is intended to be fit for purpose. In Meuwly et al. (2017), their selection is based on the statistics literature on the evaluation of Bayesian probabilities, and in particular on the use of proper scoring rules.

Robustness, monotonicity and generalization have been identified as secondary performance characteristics. They describe how the primary characteristics behave in different conditions representing the extreme variability of forensic casework. Factors of variability are usually degrading, e.g., data sparsity, quality of the specimens or mismatch in the conditions between training data and operational data.

7.2.4 Empirical Validation

Empirical validation is strictly necessary before making use of a new method in practice, because of the variability and often low quality of the operational data analyzed, which may cause sound LR models to present undesirable behavior. Among the most common degrading factors are data sparsity, high variability of the quality of specimens, a shift between the conditions of the data used for LR model training, and the data captured in the different forensic scenarios.

---

* This was previously referred to as coherence (Haraksim et al., 2015), but the name was changed for the sake of clarity, and in order to avoid confusion with statistical coherence.

As a central procedure of the validation process, performance measurement requires careful definition. In particular, the performance characteristics must guarantee that the likelihood ratios are fit for purpose, and that they have desirable properties under operational conditions.

Some definitions are given here for better understanding of the rest of the chapter[*]:

- A *performance characteristic* represents the answer to the question *"What to measure?"* It is a characteristic of an LR method that is thought to have an influence on the desired or undesired behavior of a given interpretation method. For example, we want LR values that help the trier of fact to reach better decisions, and in that sense the LR values should possess the performance characteristic defined as *accuracy*[†].

- A *performance metric* represents the answer to the question *"How to measure?"* It gives a quantitative measure of a performance characteristic, usually as a scalar. For the performance characteristic defined above as accuracy, the performance metric can be implemented by the use of proper scoring rules (DeGroot and Fienberg, 1982; Gneiting and Raftery, 2007) on an empirical set of likelihood ratios (see Section 1.4.1). Thus, this performance metric will yield a single number that measures accuracy: the lower this number, the better the accuracy[‡], and *vice versa*.

- A *validation criterion* represents the answer to the question *"what performance is needed to regard a method as valid?"* It is defined as the decision rule to determine when a method is acceptable and fit for purpose according to a given performance characteristic. For the performance metric *accuracy* defined above (empirical average of a proper scoring rule), a possible validation criterion is a scalar threshold over the performance metric. When the metric is above the threshold, the method is not validated from the point of view of the accuracy, and *vice versa*.

7.2.5 Validation Protocol

The validation protocol begins with a validation plan describing the experiments. This plan lists the performance characteristics considered for validation of the method and the performance metrics and graphical representations used to assess those performance characteristics. It also describes the aim of the experiments, the data used and the validation criteria applicable. In order to get more insight into the expected performance of the method, a comparison with either the current state of the art or with a baseline method can be performed, which provides an initial set of validation criteria.

Experiments are performed in two stages, the first entails the development and validation of the method and the second the validation for varying conditions. The development and validation of the method uses a training dataset (with a known *ground truth*) to select the automatic LR method, and to refine the parameters of this method and the statistical models involved in it. The aim is to measure the primary performance characteristics of the method and to obtain the best performance with the most representative dataset for the widest possible range of conditions.

---

[*] As explained in Meuwly et al. (2017), these terms have been defined to be, as much as possible, in accordance with relevant ISO standards.

[†] Here, it can be seen that we define *accuracy* in terms of proper scoring rules, in contrast to its usual definition. See Section 1.4.1.

[‡] As we will see, the average of a proper scoring rule yields a penalty, which is lower when the accuracy is better.

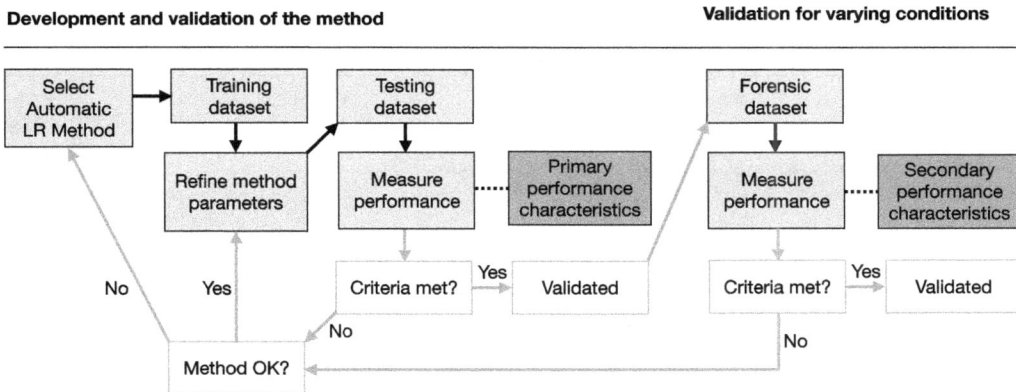

FIGURE 7.1
Diagram describing the development and validation stages of the validation process.

The validation of the developed method for varying conditions consists in measuring its performance on a previously unseen set of data captured under forensic conditions (with a known ground truth), using both the primary and secondary performance characteristics. The aim is to test the automatic LR method under conditions that are as similar as possible to conditions in forensic casework, and to arrive at the validation decision. If a dataset is used to assign the value of some hyperparameter, which is often the case in the method development stage, then the same dataset should not be used to estimate the performance in the validation stage. The reason is to avoid a possible inadequate generalization to new data in casework (overfitting). The validation experiments in two stages are summarized in the flowchart shown in Figure 7.1.

Finally, the results of the validation experiments are summarized in a validation report, recording the decision of acceptance or rejection, depending on whether the experimental results meet the validation criteria or not. A validation decision should always be linked to a specific set of experimental conditions determining the scope of validity of the method.

The protocol for the validation of an automatic LR method is summarized in the validation matrix as shown in Table 7.1. Note that all the validation processes, seen as columns of the validation matrix in Table 7.1 apply to each of the performance characteristics (i.e., all the rows in Table 7.2). This might mean that a validation process could end with a "pass" validation decision for some characteristics, and with a "fail" validation decision for some others. To apply the method in casework (or not) will be the decision of the forensic science institute, but the validation report should be transparent and made public.

The guideline for validation proposed in Meuwly et al. (2017) is the first initiative in a long-term effort. It will be improved in the future, considering suggestions from others (see, for example, Alberink et al., 2017).

An example of a validation report using development and forensic data can be found in Ramos et al. (2017). It is linked to the necessary data used to reproduce the results, in the form of empirical sets of likelihood ratios with corresponding ground-truth labels. Interested researchers can access the data and follow the set of steps presented in this report, which can help them to proceed with the empirical validation of their own methods.

Moreover, a toolbox for performance assessment is available with the main tools necessary to generate the performance metrics and graphical representations needed to validate

TABLE 7.1

Validation Matrix for Automatic Likelihood Ratio Methods

Performance Charac- teristic	Performance Metrics	Graphical Repre- sentation	Validation Criteria	Experiments	Data	Results	Validation Decision
For each listed character- istic	As appro- priate for character- istic	As appro- priate for character- istic	According to the defini- tion	Description of the exper- imental settings	Data used	$+/-$ [%] compared to the baseline	Pass/Fail

TABLE 7.2

Performance Characteristics and Examples of Performance Metrics and Graphical Representations

Performance Characteristic	Performance Metrics Examples	Graphical Representation Examples
Accuracy	Empirical average of a proper scoring rule for a given prior probability, such as C_{llr}.	Prior-dependent representation of a proper scoring rule, such as an ECE plot.
Discrimination	Discrimination component of the empirical average of a proper scoring rule for a given prior probability, such as C_{llr}^{min} or EER.	Discrimination component of a prior-dependent representation of a proper scoring rule, such as an ECE^{min} plot or a DET plot.
Calibration	Calibration component of the empirical average of a proper scoring rule for a given prior probability, such as C_{llr}^{cal}.	Calibration component of a prior-dependent representation of a proper scoring rule, such as an ECE^{cal} plot. Also visible in the symmetry of a Tippett plot (i.e., cumulative histograms).
Robustness, Monotonicity, Generalization	Variation of primary metrics such as C_{llr} or EER, range of LR values.	Variation of primary representations such as ECE, Tippett or DET plots.

an LR method from an empirical set of LR values. This toolbox is freely available online (https://sites.google.com/site/perfevtoolbox/).

7.3 Primary Performance Characteristics

The primary characteristics allow to define the minimum requirements that a LR method must satisfy empirically. This section begins with a description of how Bayesian probability theory has addressed this problem in other areas outside forensic science, leading to the concept of a *strictly proper scoring rule* as a function to assess the *accuracy* of probabilities. Then, the section describes the decomposition of the accuracy into discrimination and calibration, and their main properties.

The forensic evaluation of observations with regard to propositions is done by the practitioner assigning a likelihood ratio (ENFSI, 2015), while the prior probability of propositions is the province of the trier of fact. However, the framework proposed to measure performance focuses on validation and is based on proper scoring rules, which apply to posterior probabilities and thus also depend on the prior probabilities. In a validation process, the forensic scientist must demonstrate that the LR method is valid for a wide range of prior probabilities. The validation process therefore involves testing the LR method in such a range of prior probabilities. To measure the performance of LRs, the associated posterior probabilities are assessed by using proper scoring rules.

It does not suggest that forensic evaluation and reporting should involve assigning prior probabilities, but that the LR method should be tested for a wide range of prior probabilities, as described in Ramos and Gonzalez-Rodriguez (2013) and Meuwly et al. (2017).

7.3.1 Performance of Probabilities by Proper Scoring Rules

The assessment of the goodness of probabilistic opinions has been the focus of extensive research in statistics since long. According to a widely accepted interpretation of probability from a Bayesian perspective, probability is personal (Lindley, 2013), and therefore there is no such thing as a *true* probability or a *true* likelihood. One can assign a probability distribution for some uncertain observation, which might be different from the assignment made by someone else, but not necessarily better or worse. The Bayesian perspective also allows for subjective assignment of probabilities, as long as the rules of probability are respected, making it a coherent probability assignment.

Although this interpretation of probability proves to be flexible and useful, this does not mean that probabilities always lead to accurate actions when used to make decisions. For instance: a person might have a gut feeling that it will not rain the next day, which motivates an assignment of a probability of 1% to the event "rain next day". In accordance with this probability value, it is most probable that this person will decide not to take an umbrella the next day. However, if it rains the next day, the decision not to take the umbrella is not a successful one. Nothing in this example has violated the laws of probability, nor coherence, nor the logic of decision-making, but the outcome can rationally make us question the earlier probability assessment.

This fact has motivated the assessment of the goodness of probabilities. In fact, the earliest works on this topic addressed the problem of probabilistic weather forecasting (Brier, 1950), where an empirical criterion was proposed. Suppose a forecast on whether it will rain the next day or not is given by a forecaster every day. We denote $H \in \{H_r, H_{nr}\}$ the random variable that represents one of two propositions, taking two categorical values depending on whether it rains the next day or not. For a given day d_i, the forecaster assigns a probabilistic forecast $P_i \equiv P(H = H_r | d_i)$. Then, the next day, the true value of the random variable H for day d_i will be known, and will be denoted the ground-truth label for that day, as L_i, an observation drawn from random variable $L \in \{H_r, H_{nr}\}$. The empirical measurement of performance requires the availability of a database of past forecasts where the ground-truth labels are known. Empirical measurement has been proposed in the past by the use of a so-called *proper scoring rule* (PSR), in the following way:

$$S = \frac{1}{m} \sum_{i=1}^{m} R(P_i, L_i) \tag{7.1}$$

where S is the *average PSR score* of the forecaster, and $R(P_i, L_i)$ is the strictly proper scoring rule that assigns a *penalty* to the forecast P_i depending on the value of L_i. Useful examples of proper scoring rules are the quadratic rule and the Brier rule

(Brier, 1950):

$$R(P_i, L_i) = (1 - P_i)^2 \quad \text{if } L_i = H_r$$

$$R(P_i, L_i) = (P_i)^2 \quad \text{if } L_i = H_{nr}, \tag{7.2}$$

and also the logarithmic scoring rule:

$$R(P_i, L_i) = -\log(P_i)^2 \quad \text{if } L_i = H_r$$

$$R(P_i, L_i) = -\log(1 - P_i)^2 \quad \text{if } L_i = H_{nr}. \tag{7.3}$$

Figure 7.2 shows both examples of proper scoring rules. It can be seen that, if $L_i = H_r$, then it rained on day d_i, and therefore the proper scoring rule's penalty is lower if $P_i \equiv P(H = H_r | d_i)$ is closer to 1, and *vice-versa*. As a proper scoring rule is defined here as a penalty to a single probabilistic forecast, it penalizes forecasts P_i more when they are further from 1 while $L_i = H_r$, or further from 0 while $L_i = H_{nr}$. In other words, forecasters that will tend to assign probabilities that are closer to 1 when $L_i = H_r$ and closer to 0 when $L_i = H_{nr}$ will receive a better average PSR score.

This methodology to measure the performance of probabilistic forecasts also applies in the forensic evaluation context. We consider a trier of fact aiming to assign a posterior probability $P_i \equiv P(H = H_1 | E)$, where H_1 is the proposition that associates the suspect with some trace at a crime scene, and E are the observations to be evaluated by the forensic examiner. As is recommended in ENFSI (2015), the posterior probability is obtained from a prior probability, province of the trier of fact, and the likelihood ratio from the forensic examiner, using Bayes' theorem:

$$\frac{P(H_1 | E)}{P(H_2 | E)} = \frac{P(E | H_1)}{P(E | H_2)} \frac{P(H_1)}{P(H_2)} = LR \cdot \frac{P(H_1)}{P(H_2)}. \tag{7.4}$$

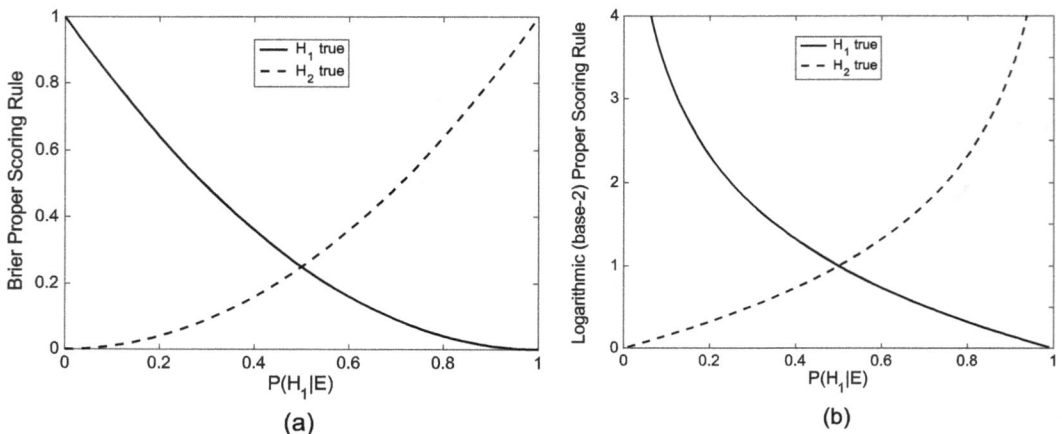

FIGURE 7.2
Examples of Proper Scoring Rules (PSR): (a) Brier, or quadratic, scoring rule. (b) Logarithmic scoring rule (base-2, although other bases could be considered, differing only by a scale factor).

If we have a database where the ground truth about the proposition for each of the observations in the database is known, we can replicate the weather forecasting performance measurement, and evaluate the posterior probabilities using proper scoring rules. The only relevant difference between the weather forecasting example and the forensic scenario is that, in the forensic evaluation scenario, we aim at measuring the performance of a method used by the forensic examiner, who only computes the likelihood ratio in Equation 7.4. In weather forecasting, the database contains the forecasts, which are equivalent to posterior probabilities in forensic science. However, in the validation of forensic methods, we cannot have a database of posterior probabilities, because the prior probabilities in Equation 7.4 are generally not known, and in any case, they are not the responsibility of the forensic examiner. Thus, in forensic science the equivalent of the database with forecasts is a database of likelihood ratios.

We denote $H \in \{H_1, H_2\}$ the random variable, taking one of two categorical values H_1 or H_2. For a given forensic case c_i, with findings E_i the forensic examiner will use their LR method to obtain LR_i. If the trier of fact would assign the prior probability, then the posterior probability $P_i \equiv P(H = H_1 \mid E_i)$ could be obtained by Equation 7.4. In the database of LR values, the true value of the random variable H for each case c_i is known, and will be denoted the *ground-truth* label for that case, as $L_i \in \{H_1, H_2\}$. The empirical measure of performance for that database of cases, where posterior probabilities and ground-truth labels are known, will then be given by the average PSR score as in Equation 7.1.

In order to compute the value of the average PSR score in forensic evaluation, a prior must be fixed, because proper scoring rules apply to posterior probabilities. However, it is well-known that the forensic examiner has no role in assigning the prior, since it concerns evidence and information that falls outside their area of expertise. Thus, our proposed validation methodology only uses prior probabilities for measuring performance. The performance will be dependent on those prior probabilities, and indeed it is observed in practice that an LR method, which presents a better average PSR score for a given prior probability, could present a bad average PSR score for another prior probability. Therefore, the performance of the average PSR score should be measured for a wide range of prior probabilities, in order to guarantee that the LR method will perform adequately in a range of diverse realistic scenarios.

Some popular proper scoring rules have been introduced in recent literature. Among them, those based on the logarithmic scoring rule have gained popularity, more specifically the Empirical Cross-Entropy (ECE) (Ramos et al., 2013, 2018), and the log-likelihood ratio cost C_{llr} (Brümmer and du Preez, 2006).

ECE is the average PSR score of the logarithmic strictly proper scoring rule, where the cases c_i, where H_1 or H_2 is true, are combined with the information of the prior probabilities $P(H = H_1)$ and $P(H = H_2)$, respectively. Thus, ECE is a measure of the average penalty that all the LRs in the set have. ECE is a prior-dependent measure, and therefore it must be represented as dependent on the prior probabilities in forensic science, since those are not the province of the forensic examiner. Therefore, ECE is better as it is lower, and the higher its value, the poorer the performance. C_{llr} is the value of ECE at the prior probabilities of 0.5, i.e., when $P(H = H_1) = P(H = H_2) = 0.5$, and it can be seen as a summarizing measure of ECE when the information about the propositions H_1 and H_2 is minimum (i.e., maximum prior uncertainty).

The ECE has been represented as a prior-dependent measure by means of ECE plots (Ramos et al., 2013; Ramos and Gonzalez-Rodriguez, 2013). It therefore measures the performance of LR values as a function of the prior probabilities. The C_{llr} also has an interpretation as the average cost for all possible prior probabilities, and can be seen as a summarizing measure of ECE in terms of information theory (Brümmer and du Preez, 2006; Ramos et al., 2013). Both performance measures have a relevant information-theoretical interpretation and have been scientifically justified (Ramos et al., 2018).

7.3.2 Discrimination and Calibration of Probabilities

An important property of proper scoring rules is that they allow a decomposition of the PSR score into two components: a refinement, discrimination, or sharpness component; and a calibration component (DeGroot and Fienberg, 1982; Murphy and Winkler, 1987). Because this decomposition is additive, both of these components should be minimized to optimize the performance of the system. This performance is measured for an empirical set of posterior probabilities P_i, obtained from a database of likelihood ratios and a value of the prior probability where the performance is to be measured. In general, we will refer to this decomposition using the following notation:

$$R(P_i, L_i) = R^{min}(P_i, L_i) + R^{cal}(P_i, L_i), \tag{7.5}$$

where R^{min} is the discrimination component of the penalty according to the proper scoring rule, and R^{cal} its component due to calibration. This decomposition can always be considered theoretically, and in some particular cases it allows a straightforward closed-form expression (see, for example, DeGroot and Fienberg (1982)). However, this is not the case in general and some algorithmic methods have been proposed in recent literature to separate both components, the most relevant one being the Pool Adjacent Violators (PAV) algorithm (Brümmer and du Preez, 2006; Fawcett and Niculescu-Mizil 2007).

In a forensic evaluation using a Bayesian decision framework, posterior probabilities must differentiate between different propositions, in the sense that the observations should lead to the true proposition by increasing its probability. This property of probabilities has been dubbed as *refinement* (DeGroot and Fienberg, 1982; Murphy and Winkler, 1987), *sharpness* (Gneiting et al., 2007) or *discrimination* (Brümmer and du Preez, 2006; Ramos and Gonzalez-Rodriguez, 2013). Roughly speaking, discrimination is the property that allows to separate the sets of posterior probabilities that are obtained when one or the other proposition in the case is true. In other words, with findings E for experiments where either H_1 or H_2 is true, $P(H_1 | E)$ should be higher when H_1 is true than where H_2 is true.

Thus, if we compute posterior probabilities for many cases in our development or validation database (i.e., by choosing a range of prior probabilities in our experimental set-up), the values of $P(H_1 | E)$ when H_1 is true should overlap as little as possible with the values of $P(H_2 | E)$ when H_2 is true. It is this relative overlap between probabilities which defines the discrimination. Popular measures of discrimination of probabilities are the Area under the Receiver Operating Characteristic curve (Area Under ROC, or AUC) (Fawcett and Niculescu-Mizil, 2007), the Equal Error Rate (Martin et al., 1997) and the C_{llr}^{min} (Brümmer and du Preez, 2006). Their corresponding graphical representations are Receiver Operating Characteristic (ROC) curves, Detection Error Tradeoff (DET) curves and minimum Empirical Cross-Entropy (ECE^{min}) curves.

But discrimination is not enough: posterior probabilities must also be *reliable* (Ramos and Gonzalez-Rodriguez, 2013), in the sense that triers of fact can rely on them to improve their decisions on average. This means they represent the findings that are being evaluated, and the prior probability for which the performance is evaluated. However, in this Bayesian context, *reliability* has a different meaning than the classical, frequentist one: in Bayesian statistics, as mentioned earlier, probability is personal (Lindley, 2013). Thus, reliability does not measure closeness to a given *true* probability distribution. The most common property that Bayesian statisticians have attributed to reliable probabilities is their *calibration* (Dawid, 1982; deGroot and Fienberg, 1982), and calibration has been used in many works as a synonym of reliability (Dawid, 1982). According to these works, posterior probabilities should not only be more discriminating, but also more reliable (in the sense of better-calibrated). If this is the case, probabilities lead the trier of fact to make better decisions on average (Ramos and Gonzalez-Rodriguez, 2013).

Calibration can be defined in many ways, but two definitions have been commonly accepted. The first definition is perhaps the most theoretical, and the most general: a method that assigns posterior probabilities is *perfectly calibrated* if, when a posterior probability is assigned using this method, and the method computes a posterior probability again once the previous posterior probability has been observed, it will yield the same posterior probability (van Leeuwen and Brümmer, 2013). This has the immediate implication that, if a posterior probability is perfectly calibrated, it will perfectly represent the probability of the propositions given the findings, and allows to arrive at better decisions (Robertson et al., 2016).

Another definition of calibration is more empirical, informal and practical. Imagine an empirical set of posterior probabilities $P(H_1 | E)$, with many cases when H_1 and H_2 are respectively true. Imagine also a subset of posterior probabilities that lay *close* to a value k. Then, k' is computed as the proportion of those probabilities in the subset where H_1 is actually true. The closer k is to k' for all values of k, the better the calibration. For instance, a weather forecaster will have a proportion k' of days with rain for all the days where s/he gave the posterior probability k of rain forecast. This definition motivates the use of performance representations like the so-called empirical calibration curves (Zadrozny and Elkan, 2002; Cohen and Goldszmidt, 2004), also known as reliability plots, where the values of k' (proportion of cases) are represented as a function of k (probabilities assigned in the empirical set) for a binning of the latter. In these representations, calibration will be better if the curve is closer to the diagonal of the plot. See, e.g., Figure 7.3d.

Of course, it follows from Equation 7.5 that perfect calibration does not imply certainty about the propositions, because the term due to discrimination still remains in the proper scoring rule (Brümmer and du Preez, 2006). But if the discrimination remains the same, then improving the calibration improves performance.

Following this approach, the performance of probabilities is measured as the average penalty, i.e., the empirical average of the proper scoring rule penalties (deGroot and Fienberg, 1982). Thus, the improvement in the accuracy of the decisions made by the trier of fact by a reduction of the proper scoring rule penalty may not manifest for a single case, where a single evaluation of findings is performed; but it will on average over a large number of cases. In fact, discrimination and calibration do not apply to single probabilities, or to single cases, but to averages over empirical sets.

7.3.3 Performance of Likelihood Ratios

As mentioned earlier, the validation process is aimed at the likelihood ratios given by a forensic method, not at posterior probabilities. The reason is that a forensic examiner cannot assign the prior probability, leaving that task to the trier of fact. However, LR methods that are going to be validated for their use in casework must be validated for a wide range of prior probabilities. This is because an LR method might pass the validation criterion for

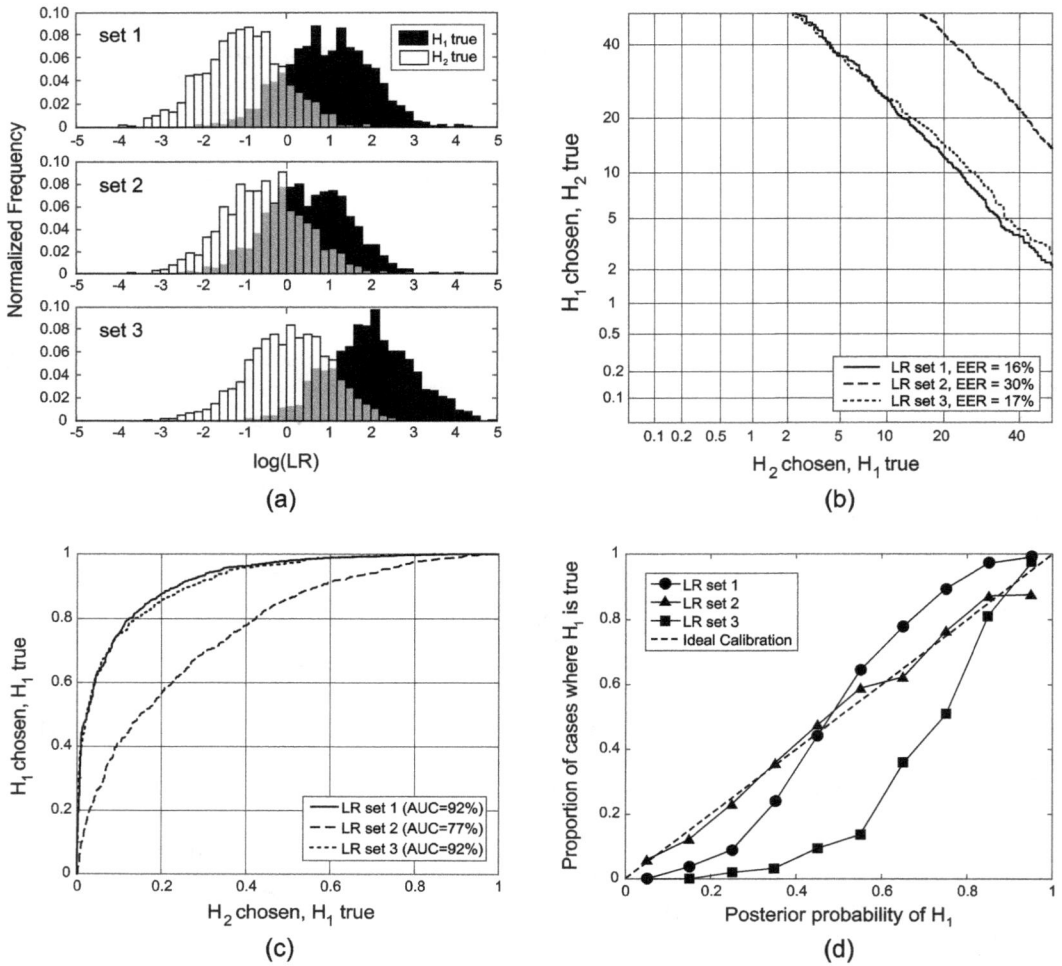

FIGURE 7.3
Some popular and useful performance metrics and graphical representations: (a) Histograms, (b) DET plots with EER, (c) ROC curves with AUC, (d) empirical calibration plots.

some prior probabilities, but not for others. There are several solutions to this issue, from the point of view of the validation process:

- The performance of likelihood ratios can be measured for a wide range of prior probabilities. Thus, LR methods that perform according to their validation criterion for the whole range of analyzed prior probabilities will be validated. This is done using prior-dependent performance representations of likelihood ratios, such as ECE plots (and the corresponding calibration and discrimination components, ECE^{min} and ECE^{cal}), ROC curves and DET plots.

- The performance of likelihood ratios can also be measured for a single, summarizing prior probability; or as an average over prior probabilities. Thus, those methods will give a single scalar measure that summarizes the performance for all the prior probabilities. This is done using e.g., the C_{llr} (with its corresponding discrimination and calibration components C_{llr}^{min} and C_{llr}^{cal}), the EER and the AUC.

The properties discussed above for probabilities easily extend to LRs, since the posterior probability is simply the product of the LR and a (fixed) prior probability. Therefore, the LR and the posterior probabilities share the same information if the prior is fixed. This can be proved using e.g., theorems of information theory (Cover and Thomas, 2006) related to the so-called *data-processing inequality*. Thus, an empirical set of LRs also presents the two properties determining the performance of a set of posterior probabilities: discrimination and calibration.

7.3.4 Properties of Well-Calibrated Likelihood Ratios

Calculating the LRs without mistake is necessary but, perhaps surprisingly, not sufficient for the numbers obtained to have the properties LRs should have. When used to update the prior probability ratio, on average LRs have to decrease the uncertainty about the hypotheses. If this property is observed on a large set of LRs, then the method is deemed to produce "well-calibrated" LRs. A method producing "well-calibrated" LRs would help, on average, the trier of fact to optimally reach a decision on the issue in a case. Numbers presented as LRs - from here on "reported LRs" - can have any degree of performance. Reported LRs are not always helpful for the trier of fact, whereas calibrated LRs are, on average.

The above should make clear that anyone considering using LRs should also be interested in the performance of those LRs. This not only shows whether a trier of fact can expect to benefit from the system at all, it can also help the forensic practitioner choose between different systems. For forensic scientists that attempt to improve a system's performance, it can help to assess the improvement achieved (Berger and Ramos, 2012).

The performance of an automatic system that generates LRs is limited for a variety of reasons. It can be due to blatant mistakes, but more often the laws of probability theory are followed and other factors are limiting performance. Limiting factors can be e.g., modeling assumptions, or databases that are never perfectly representing the population of interest because of their size or nature. For example, the database may be of studio recordings of speech but the system may be used to compare voices on tapped phone calls.

LRs have a specific probabilistic interpretation: they represent strength of evidence. Apart from discrimination (the low degree of overlap of LRs under either hypothesis), the values of the LRs themselves are important. We can consider the LR reported by a system (or person), as evidence E for our own evaluation. This evidence is interpreted, as always, by considering the probability of obtaining the evidence if either hypothesis is true:

$$\text{LR} = \frac{P(E|H_1)}{P(E|H_2)} = \frac{P(\text{reported LR}|H_1)}{P(\text{reported LR}|H_2)}. \tag{7.6}$$

If the reported LR is equal to LR, the result of our own trusted evaluation, then the calibration of the system is ideal. Ideal calibration means that our assessment of the evidential value of the reported LR agrees with the assessment of the evidence by the system. A system with ideal calibration thus reports LRs that fully capture the evidential strength of the observation, such that this original observation can be replaced by its LR. This means that the interpretation of the original observation gives the same result as the interpretation of the reported LR. Or, in other words, for ideal calibration the LR of the reported LR is equal to the reported LR. But if the calibration is not ideal, the reported LRs will be misleading more often than better-calibrated LRs (Dawid, 1982; Cohen and Goldszmidt, 2004; Brümmer and du Preez, 2006; Ramos et al., 2013, 2018).

An empirical observation is that, if the discrimination of an empirical set of LR values increases, the strength of evidence of well-calibrated LRs also tends to increase (Ramos and Gonzalez-Rodriguez, 2013). This agrees with common sense: an LR method with better discrimination (for instance, DNA analysis) should yield higher strength of evidence than an LR method with lower discrimination (for instance, forensic voice comparison). This happens if the set of LR values is well-calibrated, but not necessarily otherwise. In van Leeuwen and Brümmer (2013), a proof of this property is given for equal-variance Gaussian distributions of the scores of speaker recognition systems.

7.3.5 Examples with Primary Performance Characteristics

In this section several examples are given of performance metrics and graphical representations related to primary performance characteristics. The following performance metrics will be used for 3 sets of LR values produced by 3 simulated systems:

- Accuracy: C_{llr} (Brümmer and du Preez, 2006), ECE (Ramos et al., 2013, 2018).
- Discrimination: C_{llr}^{min} (Brümmer and du Preez, 2006), ECE^{min} (Ramos et al., 2013), AUC (Fawcett and Niculescu-Mizil, 2007), EER (Martin et al., 1997).
- Calibration: C_{llr}^{cal} (Brümmer and du Preez, 2006), ECE^{cal} (Ramos et al., 2013, 2018).

Also, the following graphical representations are shown:

- Accuracy: ECE plot (Ramos et al., 2013, 2018).
- Discrimination: ROC curve (Fawcett and Niculescu-Mizil, 2007), DET curve (Martin et al., 1997).
- Calibration: Empirical calibration plot (Zadrozny and Elkan, 2002).

It is out of the scope of this chapter to give a thorough interpretation of these performance measures, and interested readers can find further details in the indicated references. However, a brief description of their interpretation is given in relation with performance.

The histograms in Figure 7.3a show the explicit empirical distribution of log(LR) values for either hypothesis being true. These graphical representations do not explicitly measure performance, but show the degree of overlap as a measure of discrimination. They also give some indication of the calibration of the log(LR) values, since they should be centered around log(LR) = 0 if they are well-calibrated (vanLeeuwen and Brümmer, 2013). The cumulative version of these histograms is the Tippett plot (Meuwly, 2000; Lucena-Molina et al., 2015).

The DET and ROC graphs (Figure 7.3b and c respectively) are measures of discrimination. The closer the curves are to the top-left corner of the graph for ROC curves, and to the origin for the DET curves, the better the discrimination. The AUC is a summarizing measure of discrimination that integrates the whole ROC curve, the higher the better. Its equivalent is the C_{llr}^{min} (Brümmer and du Preez, 2006; Fawcett and Niculescu-Mizil, 2007), for which lower values represent better discrimination. The EER is a point-summary of the discriminating power, the lower the better. However, the calibration is not visible in these measures, as can be seen from the fact that Set 1 and Set 3 present similar DET and ROC curves (Figure 7.3b,c), but very different calibration (Figure 7.4).

The empirical calibration graph in Figure 7.3d gives a measure of calibration based on its empirical definition. It shows the relation between the proportion of cases where H_1 is true and the posterior probability of H_1, and takes the proportion of cases where H_1 is true as prior probability (named *empirical prior*). The set of LR values is well-calibrated when the data points approach the diagonal. This representation does not take into account discrimination. The example shows that Set 2 has the best-calibrated set of LR values, when the DET and ROC graphs show that it is also the least discriminating (Figure 7.3b,c). This example also shows that discrimination and calibration are two essential and complementary performance characteristics for the validation of LR methods.

The ECE plots in Figure 7.4 give the performance as a function of prior probabilities. The accuracy (solid curve, ECE) is plotted with the loss of accuracy due to imperfect discrimination (dashed curve, ECE^{min}) and to imperfect calibration (difference between both, ECE^{cal}). The lower the corresponding ECE curve, the better the performance. The plots also show the ECE of a non-informative system, such as the tossing of a coin (short-dash curve). They demonstrate that the LR method can produce results better than the non-informative system in some ranges of prior probability, while worse in other ranges. The scope of validity of the method increases with the range for which the method performs better than the level of chance. Finally, the values of C_{llr}, C_{llr}^{min} and C_{llr}^{cal} are the ECE values at the prior $\log_{10}$ (odds) value of 0 (i.e., y-axis).

These performance metrics and graphical representations are examples of the many that could be included in this validation framework based on proper scoring rules (Meuwly et al., 2017).

7.4 Secondary Performance Characteristics

The secondary characteristics describe how the primary metrics behave in different situations, such as typical forensic casework conditions (e.g., differing quality of the training data and the trace material). The aim of secondary performance characteristics is to assess the performance of the LR method in forensic casework. Therefore, secondary performance characteristics are mainly assessed at the stage of validation for varying conditions (see Figure 7.1). However, if necessary or if possible, they could also be used in the stage of development and validation of the method.

The secondary performance characteristics are related to a single primary performance metric or a single graphical representation. Thus, we talk about, for example, the robustness of the accuracy, of the discriminating power, or of the calibration.

We define the proposed secondary performance characteristics LR-based forensic evaluation methods as follows:

7.4.1 Robustness

The robustness of an LR method is the ability of the method to maintain the value of a primary performance metric when the data changes. For instance, Method A is more robust to a lack of data than Method B if, as the data gets sparser, the primary performance metric of Method A degrades less than the same metric for Method B. In the LR context, robustness

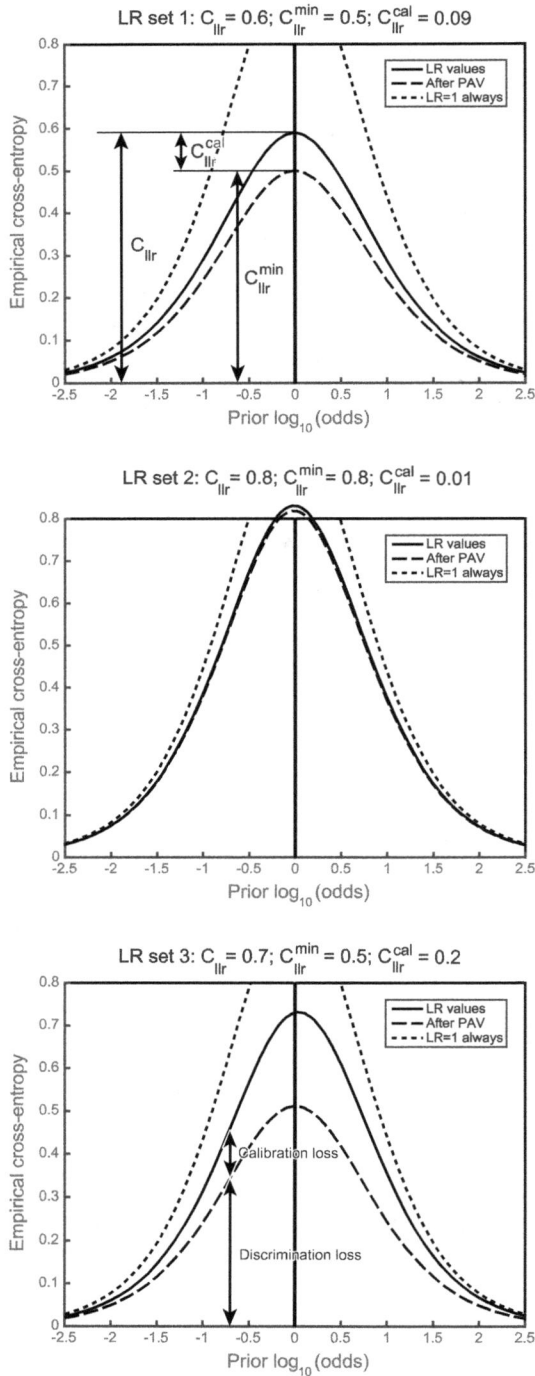

FIGURE 7.4
Empirical Cross-Entropy (ECE) plots with C_{llr}, C_{llr}^{min} and C_{llr}^{cal}, based on the same sets of LR values as Figure 7.3. The "After PAV" curve label refers to the ECE^{min} curve, since ECE^{min} is the ECE obtained after the application of the PAV algorithm.

usually refers to the stability of the performance of LR methods to varying conditions (e.g., quality/quantity of the data).

7.4.2 Monotonicity

The monotonicity* of an LR method is defined as the ability of the method to yield LR values with better performance when increasing the intrinsic quantity/quality of the data. Examples are the number of minutiae in a fingermark or the signal-to-noise ratio in a voice recording.

7.4.3 Generalization

The generalization of an LR method is the ability of the method to maintain its performance for previously unseen data (even when the quality/quantity of the data is the same). An LR system for speaker identification for example, could be used for recordings of the same quality and quantity but in a different language.

7.5 Conclusion

Many automatic methods have been developed to compute the strength of evidence, particularly for the forensic evaluation of genetic and biometric traces. But there is currently no standard for the validation of such forensic evaluation methods. This book chapter summarizes the first steps taken in the direction of the validation of forensic automatic likelihood ratio methods. Many more steps, and the involvement of the forensic community are necessary to further develop this multidisciplinary and complex matter.

References

Alberink, I., Bolck, A., Sjerps, M., and Vergeer, P. Comment to 'A guideline for the validation of likelihood ratio methods used for forensic evidence evaluation'. *Forensic Science International*, **276**, 154, 2017. https://doi.org/10.1016/j.forsciint.2017.03.011

Australian Standard AS 5388. Forensic analysis. 2012.

Berger, C. E. H., and Ramos, D. Objective paper structure comparison: Assessing comparison algorithms. *Forensic Science International*, **222**(1–3), 360–367, 2012. https://doi.org/10.1016/j.forsciint.2012.07.018

Brier, G. W. Verification of forecasts expressed in terms of probabilities. *Monthly Weather Review*, **78**(1), 1–3, 1950. https://doi.org/10.1175/1520-0493(1950)078<0001:VOFEIT>2.0.CO;2

Brümmer, N., and du Preez, J. Application-independent evaluation of speaker detection. *Computer Speech and Language*, **20**(2–3), 230–275, 2006. https://doi.org/10.1016/j.csl.2005.08.001

---

* In Haraksim et al. (2015), Meuwly et al. (2017), and Ramos et al. (2017), this property has been named *coherence*. However, it has been decided to change its name in order to not confound it with the statistical coherence of subjective probabilities.

Cohen, I., and Goldszmidt, M. Properties and benefits of calibrated classifiers. PKDD 2004: Knowledge Discovery in Databases: PKDD 2004 (LNCS, volume 3202), pages 125–136. Springer, 2004. https://doi.org/10.1007/978-3-540-30116-5_14

Cook, R., Evett, I. W., Jackson, G., Jones, P. J., and Lambert, J. A. A hierarchy of propositions: Deciding which level to address in casework. *Science and Justice*, **38**(4), 231–9, 1998.

Cover, T. M., and Thomas, J. A. *Elements of Information Theory*. John Wiley and Sons, 2006.

Dawid, A. P. The well-calibrated Bayesian. *Journal of the American Statistical Association*, **77**(379), 605–610, 1982. http://doi.org/10.2307/2287720

De Baere, T., Dmitruk, W., Magnusson, B., Meuwly, D., and O'Donnel, G. *Guideline for the Single Laboratory – Validation of Instrumental and Human Based Methods in Forensic Science*, ENFSI, 2014.

DeGroot, M. H., and Fienberg, S. E. The Comparison and Evaluation of Forecasters. *Journal of the Royal Statistical Society, Series D (The Statistician)* **32**(1–2), 1982. http://doi.org/10.2307/2987588

ENFSI. ENFSI Guideline for Evaluative Reporting in Forensic Science. Available at. 2015. http://enfsi.eu/news/enfsi-guideline-evaluative-reporting-forensic-science/ (last access: February 2018).

Fawcett, T., and Niculescu-Mizil, A. PAV and the ROC convex hull. *Machine Learning*, **68**(1), 97–106, 2007. https://doi.org/10.1007/s10994-007-5011-0

Forensic Science Regulator annual report. 2017. https://www.gov.uk/government/uploads/system/uploads/attachment_data/file/674761/FSRAnnual_Report_2017_v1_01.pdf

Gneiting, T., Balabdaoui, F., and Raftery, A. E. Probabilistic forecasts, calibration and sharpness. *Journal of the Royal Statistical Society*, Series B. **69**(2), 243–268, 2007. http://doi.org/10.1111/j.1467-9868.2007.00587.x

Gneiting, T., and Raftery, A. E. Strictly proper scoring rules, prediction, and estimation. *Journal of the American Statistical Association*, **102**(477), 359–378, 2007. https://doi.org/10.1198/016214506000001437

Haraksim, R., Ramos, D., Meuwly, D., Berger, C. E. H. Measuring coherence of computer-assisted likelihood ratio methods. *Forensic Science International*, **249**, 123–132, 2015. https://doi.org/10.1016/j.forsciint.2015.01.033

Hsu, S. S. FBI admits flaws in hair analysis over decades. *Washington Post*, April 18, 2015. https://www.washingtonpost.com/local/crime/fbi-overstated-forensic-hair-matches-in-nearly-all-criminal-trials-for-decades/2015/04/18/39c8d8c6-e515-11e4-b510-962fcfabc310_story.html

International Laboratory Accreditation Cooperation G19:08/2014, Modules in a Forensic Science Process.

ISO/IEC AWI 22842, Validation of automatic biometric methods for forensic purposes, Draft.

ISO/IEC standard 17025 *General Requirements for the Competence of Testing and Calibration Laboratories*, 3rd edition, 2017. Available at: https://www.iso.org/standard/66912.html

ISO standard 21043, Forensic Sciences, draft.

Lindley, D. *Understanding Uncertainty*. John Wiley and Sons, 2013. http://doi.org/10.1002/0470055480

Lucena-Molina, J-J., Ramos, D., and Gonzalez-Rodriguez, J. Performance of likelihood ratios considering bounds on the probability of observing misleading evidence. *Law, Probability and Risk*, **14**(3), 175–192, 2015. http://dx.doi.org/10.1093/lpr/mgu022

Martin, A., Doddington, G., Kamm, T., Ordowski, M., and Przybocki, M. The DET curve in assessment of detection task performance. In Proceedings of Eurospeech, pages 1895–1898, 1997.

Meuwly, D. Reconnaissance de Locuteurs en Sciences Forensiques: L'apport d'une Approche Automatique, (PhD thesis), 2000.

Meuwly, D. Forensic individualisation from biometric data. *Science and Justice*, Oct 1;**46**, 205–13, 2006.

Meuwly, D., Ramos, D., and Haraksim, R. A guideline for the validation of likelihood ratio methods used for forensic evidence evaluation, *Forensic Science International*, **276**, 142–153, 2017. https://doi.org/10.1016/j.forsciint.2016.03.048

Murphy, A.H., and Winkler, R.L. A General Framework for Forecast Verification. *Monthly Weather Review*, **115**, 1330–1338, 1987. https://doi.org/10.1175/1520-0493(1987)115<1330:AGFFFV>2.0.CO;2

President's Council of Advisors on Science and Technology (PCAST). Report to the president Forensic Science in Criminal Courts: Ensuring Scientific Validity of Feature-Comparison Methods, Washington DC, 2016. https://obamawhitehouse.archives.gov/sites/default/files/microsites/ostp/PCAST/pcast_forensic_science_report_final.pdf

Ramos, D., Franco-Pedroso, J., Lozano-Diez, A., and Gonzalez-Rodriguez, J. Deconstructing Cross-Entropy for Probabilistic Binary Classifiers. *Entropy*, **20**(3), 208, 2018.

Ramos, D., and Gonzalez-Rodriguez, J. Reliable Support: Measuring Calibration of Likelihood Ratios. *Forensic Science International*, **230**, 156–169, 2013. http://dx.doi.org/10.1016/j.forsciint.2013.04.014

Ramos, D., Gonzalez-Rodriguez, J., Zadora, G., and Aitken, C. Information-Theoretical Assessment of the Performance of Likelihood Ratio Models. *Journal of Forensic Sciences*, **58**(6), 1503–1518, 2013. http://dx.doi.org/10.1111/1556-4029.12233

Ramos, D., Haraksim, R., and Meuwly, D. Likelihood ratio data to report the validation of a forensic fingerprint evaluation method. *Data in Brief*, **10**, 75–92, 2017.

Robertson, B., Vignaux, G. A., and Berger, C. E. H. *Interpreting Evidence: Evaluating Forensic Science in the Courtroom* (2nd edition.), John Wiley and Sons, 2016. DOI: http://doi.org/10.1002/9781118492475

Standards and Guidelines. Validation Guidelines for Laboratories Performing Forensic Analysis of Chemical Terrorism. *FBI Law enforcement bulletin*, **7**(2), 2005.

van Leeuwen, D.A., and Brümmer, N. The distribution of calibrated likelihood-ratios in speaker recognition. In Proceedings of INTERSPEECH-2013, pages 1619–1623, 2013.

Wilson-Wilde, L. The International Development of Forensic Science Standards – A Review. *Forensic Science International*, **288**, 1–9, 2018.

Zadrozny, B., and Elkan, C. Transforming classifier scores into accurate multiclass probability estimates. In Proceedings of KDD'02, the eighth ACM SIGKDD international conference on Knowledge discovery and data mining, pages 649–699, 2002.

8

Bayesian Networks in Forensic Science

A. Philip Dawid and Julia Mortera

CONTENTS

8.1 Introduction 165
8.2 Probability Logic 166
8.3 Simple Bayesian Networks for Forensic Problems 168
8.4 Object-Oriented Bayesian Networks 176
8.5 Forensic Genetics 178
    8.5.1 Bayesian Networks for Simple Criminal Identification 178
    8.5.2 Simple Disputed Paternity 180
    8.5.3 Bayesian Networks for Complex Criminal Cases Involving Family
           Relationships 182
    8.5.4 Mutation 185
8.6 Bayesian Networks for Analysing Mixed DNA Profiles 185
    8.6.1 Discrete Features 186
    8.6.2 Continuous Features 187
8.7 Analysis of Sensitivity to Assumptions on Founder Genes 189
    8.7.1 Uncertainty in Allele Frequencies 190
    8.7.2 Heterogeneous Reference Population 191
8.8 Conclusions 192
Appendix 8A: Bayesian Network Basics 193
References 195

8.1 Introduction

Solving forensic identification problems frequently requires complex probabilistic argument and computation. The technology of *Bayesian Networks (BNs)*, also called *Probabilistic Expert Systems* (Cowell et al., 1999), supplies a valuable tool that can streamline and help solve such problems. Aitken and Gammerman (1989) gave one of the first applications of Bayesian networks in forensic science.

Building a Bayesian Network has two stages. We start by constructing a graphical representation of the problem, involving a node for each relevant variable or hypothesis, with connexions between these that encode the way in which they depend, probabilistically, on each other. We then add quantitative information about those probabilistic dependencies. Once the BN is set up in a suitable software environment, one enters the available evidence on the observed variables, and the system will compute the resulting remaining uncertainty about hypotheses of interest.

The aim of this chapter is to give an overview of the applications of BNs in forensic statistics with a particular emphasis on forensic genetics. This is largely done by example. Some of these are too complex to describe here in all their details, but the interested reader can find those details in the further references.

In Section 8.2 we recall the basic logical structure of probabilistic argument in the legal context. Section 8.3 introduces BNs by way of two illustrative examples: first, a specific problem of identification from eye witness evidence, and secondly a more generic structure for forensic identification. In Section 8.4 we describe a valuable extension of BNs as OOBNs (object-oriented Bayesian networks), whose key features are illustrated with a further elaboration of the problem of identification from eye witness testimony. In Section 8.5 we turn to consider the special features of forensic genetics identification problems, and how these can be incorporated into a BN. Illustrative examples include simple criminal identification, disputed paternity, and complex criminal cases involving family relationships. We also describe how a forensic BN can be easily modified to allow for additional complications, such as the possibility of mutation. In Section 8.6 we show how to construct BNs to analyse mixed DNA profiles, using discrete or continuous information, while in Section 8.7 we describe networks for representing uncertain allele frequencies and heterogeneous reference populations. Some concluding remarks are given in Section 8.8. The Appendix briefly outlines some of the essential underlying theory of Bayesian Networks.

8.2 Probability Logic

In a case at law, let E denote one or more items of evidence (perhaps its totality). We need to consider how this evidence affects the comparison of the hypotheses, H_0 and H_1 say, offered by either side. Thus in a criminal case with a single charge against a single defendant, the evidence might be that the defendant's DNA profile matches one found at the crime scene; hypothesis H_0, offered by the defence, is that the defendant is innocent ($\overline{G}$); the prosecution hypothesis, H_1, is that of guilt (G).

The adjudicator needs to assess his or her conditional probability for either hypothesis, *given* the evidence: $\Pr(H_0|E)$ and $\Pr(H_1|E)$. However, it will not usually be possible to assess these directly, and they will have to be constructed out of other, more basic, ingredients. In particular, it will often be reasonable to assess directly $\Pr(E|H_0)$ and $\Pr(E|H_1)$: the probability that the evidence would have arisen, under each of the competing scenarios.

The odds form of *Bayes's theorem*—a trivial consequence of the definition of conditional probability—tells us that

$$\frac{\Pr(H_1|E)}{\Pr(H_0|E)} = \frac{\Pr(H_1)}{\Pr(H_0)} \times \frac{\Pr(E|H_1)}{\Pr(E|H_0)}. \tag{8.1}$$

The left-hand side of (8.1) is the *posterior odds* for comparing H_1 and H_0, given the evidence E. When H_0 and H_1 are considered to exhaust all possibilities, this is a simple transformation of $\Pr(H_1|E)$, the *posterior probability* of H_1.

The second term on the right-hand side of (8.1) is constructed out of the directly assessed terms $\Pr(E|H_0)$ and $\Pr(E|H_1)$: it is the *likelihood ratio* $LR = \Pr(E|H_1)/\Pr(E|H_0)$ (for H_1, as against H_0) engendered by the evidence E. It is noteworthy that only the ratio of these terms enters, their absolute values being otherwise irrelevant.

To complete (8.1) we need the term $\Pr(H_1)/\Pr(H_0)$, the *prior odds* for comparing H_1 and H_0 (i.e., before the evidence E is incorporated). This might reasonably vary from one individual juror to another, so that it would not be appropriate to treat it as a subject for direct evidence. For this reason forensic experts are often instructed to give their evidence in the form of a likelihood ratio, it being left to the adjudicator to combine this appropriately with the prior assessment, using (8.1).

We can express (8.1) in words as:

POSTERIOR ODDS = PRIOR ODDS × LIKELIHOOD RATIO.

If we wish to compare more than two hypotheses, we require the full *likelihood function*, a function of the various hypotheses H being entertained (and of course the evidence E):

$$L(H) \propto \Pr(E|H). \tag{8.2}$$

The proportionality sign in (8.2) indicates that we have omitted a factor that does not depend on H, although it can depend on E. Such a factor is of no consequence and need not be specified, since it disappears on forming ratios of likelihoods for different hypotheses on the same evidence. Only such relative likelihoods are required, not absolute values.

We also now need to specify the prior probabilities, $\Pr(H)$, for the full range of hypotheses H. Then posterior probabilities in the light of the evidence are again obtained from Bayes's theorem, which can now be expressed as:

$$\Pr(H|E) \propto \Pr(H) \times L(H). \tag{8.3}$$

Again the omitted proportionality factor in (8.3) does not depend on H, although it might depend on E. It can be recovered, if desired, as the unique such factor for which the law of total probability, $\sum_H \Pr(H|E) = 1$, is satisfied.

When E denotes all the evidence in the case, all the probabilities in (8.1) or (8.2) are unconditional; in particular, the prior probabilities should be assessed on the basis that there is no evidence to distinguish the suspect from any other potential suspect—this can be regarded as one way of formalising the legal doctrine of "presumption of innocence" (which of course is not the same as an *assumption* of innocence). When E denotes a piece of evidence presented in mid-process, all the probabilities in (8.1) or (8.2) must be further conditioned on the evidence previously presented: in particular, the "prior" probabilities could themselves have been calculated using (8.1) or (8.2), as posterior probabilities based on earlier evidence.

In many cases the relationships between the various relevant items of evidence will be complex and subtle, and taking these relationships properly into account can be far from straightforward. It can then be helpful to structure these relationships in a more detailed way. The theory and technology of Bayesian Networks (BNs) (Cowell et al., 1999) can be useful to assist such structuring. In particular, for handling a variety of complex cases in forensic science, Bayesian networks, together with their associated computational methodology and technology, have been found extremely valuable, particularly in their "object-oriented" (OOBN) form, as implemented in commercial software such as HUGIN*. BNs can be used to assist police in their investigations, to aid a trial lawyer in developing an argument, and (in principle, if as yet rarely in practice) to assist the trier of fact in

---

* Obtainable from https://www.hugin.com

coming to a conclusion. Bayesian networks for evaluating DNA evidence were introduced by Dawid et al. (2002). Further description and developments can be found in Mortera (2003), Mortera et al. (2003), Dawid et al. (2006), Taroni et al. (2006), Dawid et al. (2007), Cowell et al. (2007b), and Vicard et al. (2008).

For some illustrative cases, we describe below how we can construct a suitable BN or OOBN representation of a complex forensic identification problem, incorporating all the individuals involved and the relationships between them.

8.3 Simple Bayesian Networks for Forensic Problems

A brief formal description of Bayesian Networks is provided in the Appendix. Here we introduce their main characteristics by means of examples.

Example 8.1: Eye Witness

We illustrate the simplest non-trivial Bayesian network with an example of a hit-and-run accident.

> A cab is involved in a hit-and-run accident at night. Only white and blue cabs operate in the city. 62% of the cabs in the city are white and 38% are blue. An elderly witness identifies the cab as blue. The reliability of the witness is tested under similar conditions: the witness correctly identifies a white cab 80% of the time, and correctly identifies a blue cab 70% of the time.

Qualitative BN

At this point we consider only qualitative, structural aspects, which are well displayed in a BN. As we shall see below, a BN can also incorporate quantitative information on probabilities, and provide a powerful machine for manipulating these.

The outline of this story is represented by the simple Bayesian network of Figure 8.1. There is a node for each of the relevant variables: (the actual) cab colour and (the report of the) eye witness. These each have two possible states, *white* or *blue* for cab colour, and *"white"* or *"blue"* for eye witness. The arrow indicates that the eye witness report is influenced, albeit non-deterministically, by the actual colour of the cab. We use a pseudo-genetic language to describe the dependencies between nodes: cab colour is a "parent" of eye witness, and eye witness is a "child" of cab colour.

An important aspect of this, as of any, BN representation is that the direction of an arrow indicates a stochastic relationship that can be interpreted as, in some intuitive

FIGURE 8.1
Bayesian network for hit-and-run accident. The eye witness report is probabilistically dependent on the actual cab colour.

sense, *causal* in nature. One useful way of thinking about this is that the nature of the dependence should be stable across different ways and contexts in which it might arise. Thus in this example the quoted success rates of the witness are assumed to be essentially the same, both in the test cases presented, and in the actual case at hand.

We will also be interested in *inferential* relationships; in this particular example, having heard the report of the eye witness, we would like to know what can be inferred about the actual colour of the cab? Some alternative representations, such as Wigmore charts (Wigmore, 1937; Anderson et al., 2005; Dawid et al., 2011), would reverse the direction of the arrow in Figure 8.1, so directly representing an inferential, non-causal, relationship. However, inferential relationships are typically not stable across different contexts, which is why we do not use them as basic ingredients. Nonetheless, as we shall see below, a BN with only causal arrows can readily compute such desired inferences.

Quantitative BN

We associate, with each node of the BN, a table describing the conditional probability distribution over its states, conditional on the state(s) of its parent node(s). In Figure 8.1, `cab colour` is a "founder" node, with no parents, so its table will just contain the unconditional (prior) probabilities for its states, as in Table 8.1.

As for the node `eye witness`, we need to specify the conditional probability for each of its states, given each state of its parent variable `cab colour`. This information is presented in Table 8.2, using the conditional probabilities as specified in the story.

The specification of the BN is now complete.

Computation and Inference

There exist powerful general algorithms (Lauritzen and Spiegelhalter, 1988; Cowell et al., 1999) for performing computations on a Bayesian network, which are embodied in a

TABLE 8.1

Probability Table for
`cab colour`

cab colour	**Probability**
white	0.62
blue	0.38

TABLE 8.2

Conditional Probability
Table for `eye witness`

	cab colour	
eye witness	*white*	*blue*
"*white*"	0.8	0.3
"*blue*"	0.2	0.7

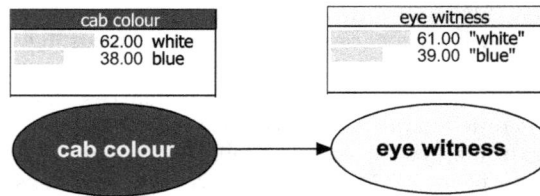

FIGURE 8.2
Bayesian network for hit-and-run accident showing prior marginal distributions.

number of specialist software systems, such as HUGIN*, GENIE†, and the $\mathcal{R}$ software gRain‡, among others. Here we have used HUGIN 8.3. The basic computational routine finds the marginal distribution at every node. The resulting probabilities are as displayed in Figure 8.2.

For cab colour this simply repeats the input prior probabilities, whereas the figures for eye witness show that, even without knowing the actual colour of the cab, we would expect the witness to report *"blue"* with probability 0.39. This computation automates the formula for "extension of the conversation" (generalised addition law):

$$\Pr(\text{eye witness} = \text{"blue"}) = \Pr(\text{eye witness} = \text{"blue"} \mid \text{cab colour} = \textit{white})$$
$$\times \Pr(\text{cab colour} = \textit{white})$$
$$+ \Pr(\text{eye witness} = \text{"blue"} \mid \text{cab colour} = \textit{blue})$$
$$\times \Pr(\text{cab colour} = \textit{blue})$$
$$= 0.2 \times 0.62 + 0.7 \times 0.38$$
$$= 0.39,$$

which yields the marginal probability.

By a straightforward variation on the basic computational algorithm, it is also possible to take account of new evidence about the state of one or more variables, thus executing "inference". In our example, we have obtained the evidence eye witness = *"blue"*. Entering this evidence and "propagating" using the software yields the results shown in Figure 8.3.

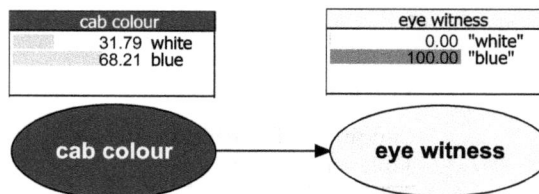

FIGURE 8.3
Bayesian network for hit-and-run accident, updated with eye witness evidence.

---

* https://www.hugin.com/
† https://www.bayesfusion.com/genie/
‡ https://cran.r-project.org/web/packages/gRain/index.html

Taking account of this fallible eye-witness's evidence, the posterior probability of cab colour = *blue* is 68%, as compared with its prior (unconditional) probability of 38%. The system has automated Bayes's Theorem:

$$\Pr(\text{cab colour} = blue \mid \text{eye witness} = \text{"}blue\text{"})$$

$$= \frac{\Pr(\text{cab colour} = blue) \times \Pr(\text{eye witness} = \text{"}blue\text{"} \mid \text{cab colour} = blue)}{\Pr(\text{eye witness} = \text{"}blue\text{"})} \quad (8.4)$$

$$= \frac{0.38 \times 0.7}{0.39} = 0.68. \quad (8.5)$$

Suppose however that the prior probabilities in Table 8.1 had been different, for example as in Table 8.3, while the conditional probabilities of Table 8.2, on account of their "causal" interpretation, were preserved.

We would then have computed different values:

$$\Pr(\text{eye witness} = \text{"}blue\text{"}) = 0.45$$

$$\Pr(\text{cab colour} = blue \mid \text{eye witness} = \text{"}blue\text{"}) = 0.78. \quad (8.6)$$

In particular, the difference between (8.5) and (8.6) demonstrates that, unlike "causal" probabilities, inferential probabilities are volatile and context-dependent—which is why we insist on using causal arrows and causal conditional probabilities as the basic ingredients when specifying a BN.

Likelihood Ratio

Even though the actual prior probabilities at the founder node cab colour are not equal, as in Table 8.3, it can be useful to proceed, formally, as if they were. For then we can use the odds form (8.1) of Bayes's Theorem:

$$\frac{\Pr(\text{cab colour} = blue \mid \text{eye witness} = \text{"}blue\text{"})}{\Pr(\text{cab colour} = white \mid \text{eye witness} = \text{"}blue\text{"})}$$

$$= \frac{\Pr(\text{cab colour} = blue)}{\Pr(\text{cab colour} = white)} \times \frac{\Pr(\text{eye witness} = \text{"}blue\text{"} \mid \text{cab colour} = blue)}{\Pr(\text{eye witness} = \text{"}blue\text{"} \mid \text{cab colour} = white)}$$

$$= \frac{\Pr(\text{cab colour} = blue)}{\Pr(\text{cab colour} = white)} \times \frac{0.7}{0.2}. \quad (8.7)$$

The final term in (8.7), $0.7/0.2 = 3.5$, is the likelihood ratio, constructed entirely of causal probabilities, and so relevant across different contexts.

We see from (8.7) that if we insert, purely formally, equal prior probabilities, $\Pr(\text{cab colour} = blue) = \Pr(\text{cab colour} = white) = 0.5$, the ratio of the resulting formal posteriors output by the software,

$$\frac{\Pr(\text{cab colour} = blue \mid \text{eye witness} = \text{"}blue\text{"})}{\Pr(\text{cab colour} = white \mid \text{eye witness} = \text{"}blue\text{"})},$$

TABLE 8.3

Alternative Probability Table for cab colour

cab colour	Probability
white	0.5
blue	0.5

will deliver the likelihood ratio. And once this stable element has been extracted, it can be used offline to construct a true posterior, again using (8.7), this time in conjunction with appropriately assessed prior odds,

$$\frac{\Pr(\texttt{cab colour} = blue)}{\Pr(\texttt{cab colour} = white)}.$$

Incorporating Additional Evidence

Now suppose that there is a further item of relevant evidence: a flake of paint was found at the scene, that can be assumed to have come from the offending cab. The flake is blue, but since the primarily blue cabs have some white areas, and the white cabs have some blue areas, this is not definitive evidence that the cab was blue. Rather, the conditional probabilities for the colour of the paint flake, given the predominant colour of the cab, are as in Table 8.4.

A BN incoporating the additional variable `paint flake` is shown in Figure 8.4. The absence of any arrow between `eye witness` and `paint flake` encodes an assumption that, conditional on the state of their common parent `cab colour`, these two variables are independent. Marginal prior probabilities for this extended BN are as in Figure 8.5, while Figure 8.6 shows the effect of incorporating both items of evidence, `eye witness` = *"blue"*, `paint flake` = *blue*. The posterior probability that the cab was blue has now gone up to 95%.

Which Fleet?

We now elaborate the story still further. There are two taxi fleets in the city, called *BLUE* and *WHITE*. The *BLUE* fleet operates 20% of the cabs, while the *WHITE* fleet has 80%, as in Table 8.5.

TABLE 8.4

Conditional Probability
Table for `paint flake`

	cab colour	
paint flake	*white*	*blue*
white	0.9	0.1
blue	0.1	0.9

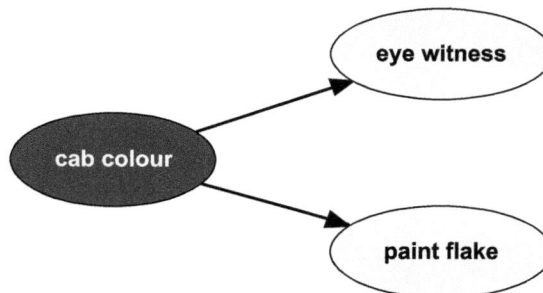

FIGURE 8.4
Bayesian network for hit-and-run accident, including paint-flake evidence. The two evidence items are independent, given the actual `cab colour`.

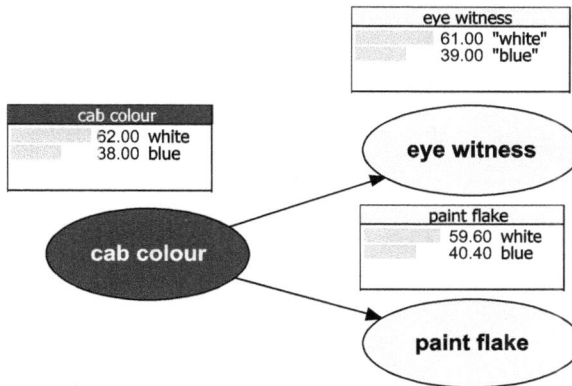

FIGURE 8.5
Bayesian network for hit-and-run accident, including paint flake, showing prior marginal distributions.

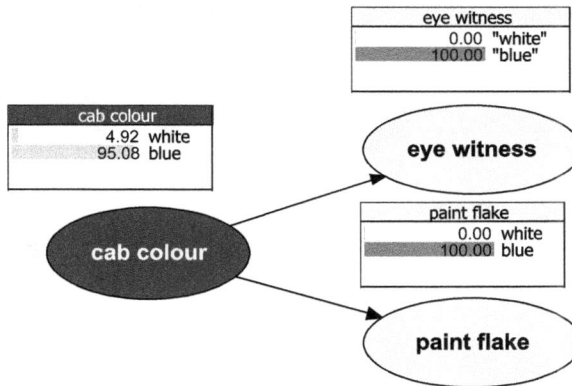

FIGURE 8.6
Bayesian network for hit-and-run accident, updated with eye witness and paint-flake evidence.

TABLE 8.5

Prior Probability Table
for `fleet`

`fleet`	Probability
WHITE	0.8
BLUE	0.2

In spite of their names, each fleet comprises only 70% cabs of the named colour, the remaining 30% being the other colour, as shown in Table 8.6.

A BN extended to incorporate information on the fleet owning the cab, showing the marginal prior probabilities, is shown in Figure 8.7. We have chosen the numbers in Tables 8.5 and 8.6 to make this extended story consistent with the more limited description in Figure 8.5. The absence of any arrow from `fleet` to `eye witness` or `paint flake`

TABLE 8.6

Conditional Probability
Table for `cab colour`, given
`fleet`

	fleet	
cab colour	*WHITE*	*BLUE*
white	0.7	0.3
blue	0.3	0.7

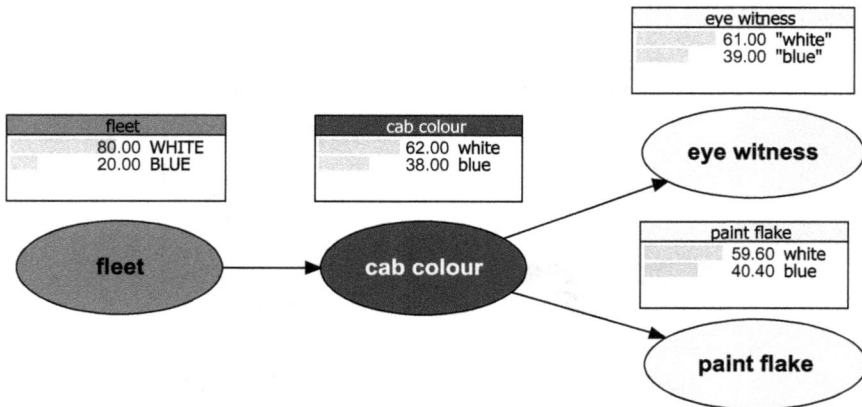

FIGURE 8.7
Bayesian network for hit-and-run accident including fleet information, showing prior marginal distributions.
Given the `cab colour`, the evidence items do not depend further on the `fleet` owning the cab.

in Figure 8.7 encodes the assumption that these two evidence variables are influenced only by the actual colour of the cab, and not further by which fleet it is owned by.

On again incorporating both items of evidence, we obtain the posterior distributions shown in Figure 8.8. The posterior probabilities that `cab colour` = *blue* is 95%, as before, but we now obtain additional information: the probability is 65% that the offending cab belongs to the *WHITE* fleet.

Example 8.2: General Forensic Identification

The general problem of identification of a suspect from forensic trace evidence gathered at the scene of a crime is extremely common, in a variety of forms. Examples include eye witness evidence (as in Example 8.1), handwriting, rifling marks on bullets, glass fragments, fibres, footprints, fingerprints, bitemarks, and, of especial importance and power, DNA profiles—this last is considered in detail in Section 8.5 below.

We wish to establish the evidential value of a comparison between the crime trace and a similar characteristic observed on a suspect. The comparison might be whether or not these are identical, a *match*, as in simple DNA profiling. In other cases, such as when comparing samples of handwriting, or a bullet with a gun, or DNA mixture samples, the

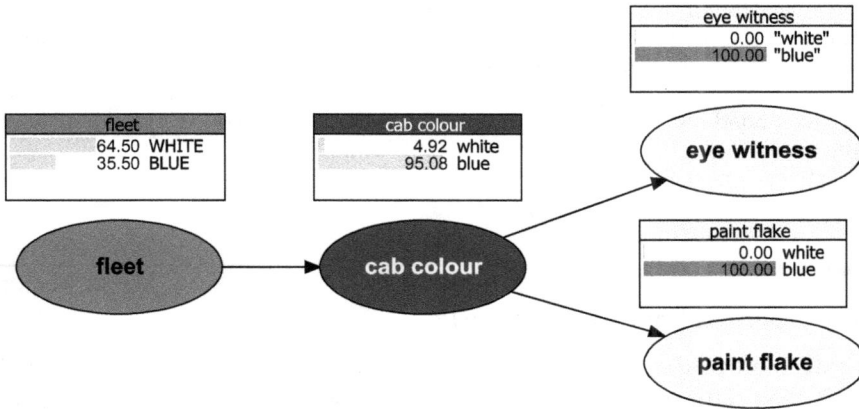

FIGURE 8.8
Bayesian network for hit-and-run accident including fleet information, showing posterior distributions based on eye witness and paint-flake evidence.

comparison may be less clearcut. An early use of probabilistic reasoning for forensic identification, dating back to 1865, was the Howland will case (Anon, 1870), which revolved around a possibly forged signature on a will. In another celebrated case, flawed probabilistic reasoning about handwriting identification resulted in the wrongful conviction of Alfred Dreyfus in 1894 (Reinach, 1901).

A generic BN for any simple forensic identification problem is displayed in Figure 8.9. Node evidence represents the evidence found at the crime scene, node suspect the characteristic of the suspect, and node other the characteristic of some alternative donor of the trace. The other individual could be an unidentified random individual in some specified population, or a specific known individual. Both suspect and other are initially populated with an appropriate prior distribution, typically estimated from a relevant data-base. The hypothesis variable, suspect left the evidence?, indicates whether the suspect or the other individual left the trace. It is populated with a formal prior distribution having equal probabilities for *true* and *false*. In a simple matching problem, such as for DNA evidence, evidence is logically determined by its parents, being equal to suspect or to other, according as suspect left the evidence? is *true* or *false*; other cases will require context-specific consideration.

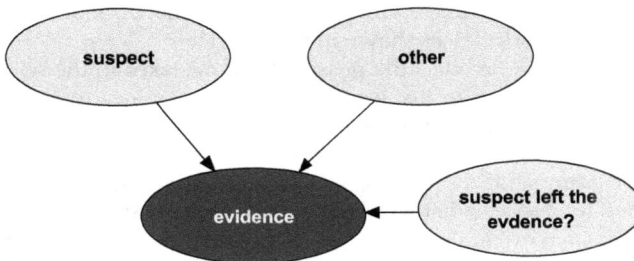

FIGURE 8.9
Bayesian network for general forensic identification. Was it the suspect, or some other person, who left the crime scence evidence?

One enters the observed trace evidence at `evidence`, the suspect's characteristic at `suspect`, and, if available, the alternative donor's characteristic at `other`. As seen in Example 8.1, after propagating this evidence through the system, the ratio of the formal posterior probabilities of *true* and *false* at node `suspect left the evidence?` will supply the likelihood ratio in favour of the suspect, rather than the other individual, having left the trace.

8.4 Object-Oriented Bayesian Networks

A useful extension of the idea of a Bayesian Network is the Object-Oriented Bayesian Network (OOBN). This organises the nodes into a hierarachy of subnetworks, which can greatly simplify specification and interpretation. (However, computation still takes place at the level of the basic nodes). A particularly valuable application of OOBNs is based on generic network modules, also termed classes, fragments, or idioms, that can be reused, both within and across higher-level networks (Laskey and Mahoney, 1997; Neil et al., 2000; Hepler et al., 2007; Fenton et al., 2013); we will typically use the description "(network) class". Thus the generic identification network of Figure 8.9 could form part of one or many other, more complex networks.

In what follows, we use **bold face** to indicate a network class, and `teletype face` to indicate a simple node or instance of a class. A class network is like a regular network, except that it can have interface—*input* and *output*—nodes, as well as internal nodes. Interface nodes are indicated by a grey outer ring, an input node having a dotted outline, and an output node a solid outline. Any network can have nodes that are themselves instances of other networks, in addition to regular nodes. Each instance of a class network within another network is displayed as a rounded rectangle, which can be expanded if desired to display its interface nodes. All instances of a class have identical probabilistic structure, except that an input node can be identified, by an incoming arrow, with a node in another network. A full description of forensic OOBN networks can be found in Dawid et al. (2006, 2007). We merely illustrate selected features here.

Example 8.3: Testimony

The generic network class **testimony** integrates several key attributes of eye witness testimony (Schum and Morris, 2007). This is conceived as involving three stages: `sensation`, `objectivity`, and `veracity`. These are combined in the network of Figure 8.10, which itself builds on the network classes shown in Figure 8.11.

The attribute `sensation` models the possibility of mistakes in the witness's perception of the event, due either to his sensory and general physical condition (leading to possible disagreement between the actual and the perceived features of the event), or to the conditions under which the observation is made. The latter aspect is termed `competence` (for example, if the witness was hiding under a table, he might not have been in a position to observe what was happening). These two processes are integrated in the node `sensation` of Figure 8.10, which is an instance of network class **sensation** shown in Figure 8.11a. Here the node `agreement` is itself an instance of the class **accuracy** of Figure 8.11b, which uses a random `Error` to determine whether or not its output reproduces its input. Only if this is so, and also `competent = true`, will the attribute `sensation` agree with the actual `event`.

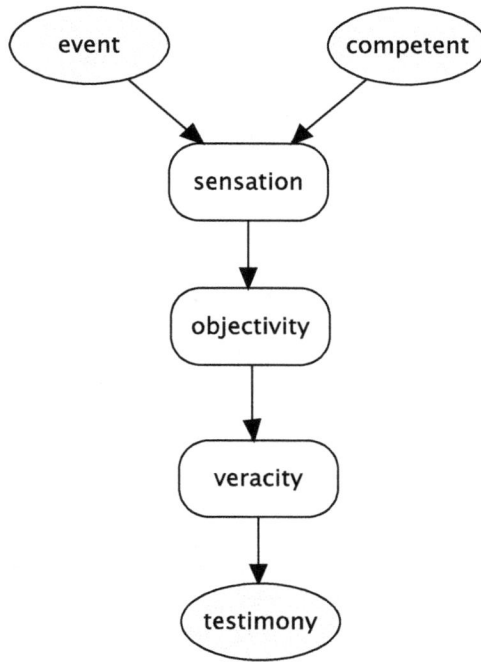

FIGURE 8.10
Network class **testimony**, showing the chain of error-prone processes intervening between an actual event and a witness's testimony of the event.

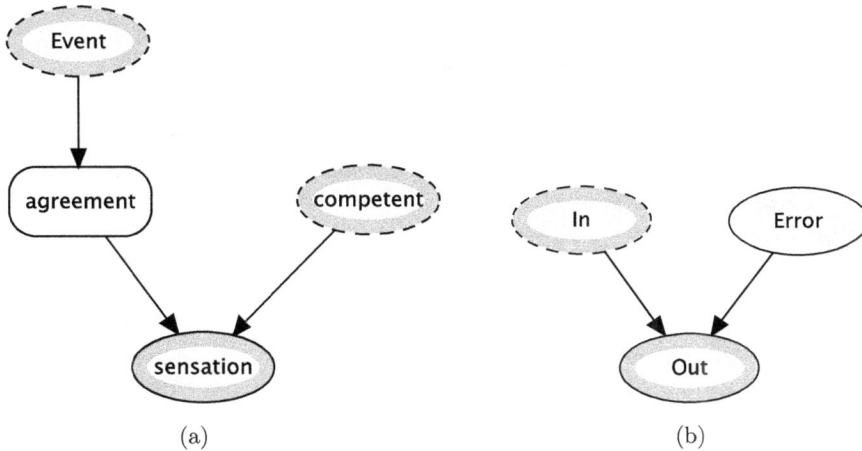

FIGURE 8.11
Testimony classes for use in Figure 8.10 (see text for details). (a) Sensation, (b) accuracy.

The attribute objectivity relates to whether or not the witness's belief is a correct interpretation of the evidence of his senses, while veracity relates to whether or not he truthfully reports his belief. The nodes objectivity and veracity in Figure 8.10 are likewise modelled as instances of the class **accuracy**.

8.5 Forensic Genetics

Forensic DNA evidence has some special features, principally owing to its pattern of inheritance from parent to child. These make it possible to use it to address queries such as the following:

> *Disputed paternity:* Is individual *A* the father of individual *B*?
> *Disputed inheritance:* Is *A* the son of deceased *B*?
> *Immigration:* Is *A* the mother of *B*? How is *A* related to *B*?
> *Criminal case – mixed trace:* Did *A* and *B* both contribute to a stain found at the scene of the crime? Who contributed to the stain?
> *Relationships and mixed trace:* Is *A* the son of a contributor to a mixture?
> *Disasters:* Was *A* among the individuals involved in a disaster? Who were those involved?

   In the simplest disputed paternity problem, the evidence *E* will comprise DNA profiles from mother, child and putative father. Hypothesis H_1 is that the putative father is the true father, while hypothesis H_0 might be that the true father is some other individual, whose DNA profile can be regarded as randomly drawn from the population. We can also entertain other hypotheses, such as that one of one or more other identified individuals is the father, or that the true father is the putative father's brother.

   In a complex criminal case, we might find a stain at the scene of the crime having the form of a *mixed trace*, containing DNA from more than one individual. DNA profiles are also taken from the victim and a suspect. We can entertain various hypotheses as to just who—victim?—suspect?—person or persons unknown?—contributed to the mixed stain.

8.5.1 Bayesian Networks for Simple Criminal Identification

In a simple criminal DNA identification case, the evidence is that the suspect's DNA profile matches a trace found at the scene of the crime. We are interested in comparing two mutually exclusive hypotheses: the *prosecution hypothesis* H_p: "the crime trace belongs to the suspect s" (loosely, "the suspect is guilty"), and the *defense hypothesis* H_d: "the crime trace belongs to another actor, o, *randomly drawn from the population*". Representation of such problems as BNs was first introduced by Dawid et al. (2002).

   In the current STR (short tandem repeat) technology, each DNA profile comprises around 16 to 20 forensic markers, with a genotype (unordered pair of alleles) measured for each marker. There is a fairly small repertory of allele values for each marker; the frequencies of these alleles in various reference populations can be estimated from research databases collected by forensic laboratories.

   The analysis proceeds separately for each marker *m*. The relevant OOBN is shown in Figure 8.12a; Figure 8.12b is an expanded version, showing the identification of nodes across subnetworks. Nodes s and o in Figure 8.12a are each instances of a class **founder**, as pictured in Figure 8.13. This contains nodes for paternal gene pg, maternal gene mg, and genotype gt. Each of the nodes pg and mg is identified with the sole node of an instance (pgin and mgin, respectively) of a simple class **gene**, that has as states the alleles for marker *m*, populated with their frequencies in a relevant reference population; while node gt of **founder** is an instance of a network class **genotype** (not shown) that simply forms the unordered combination of its input nodes pg and mg (since we cannot distinguish the

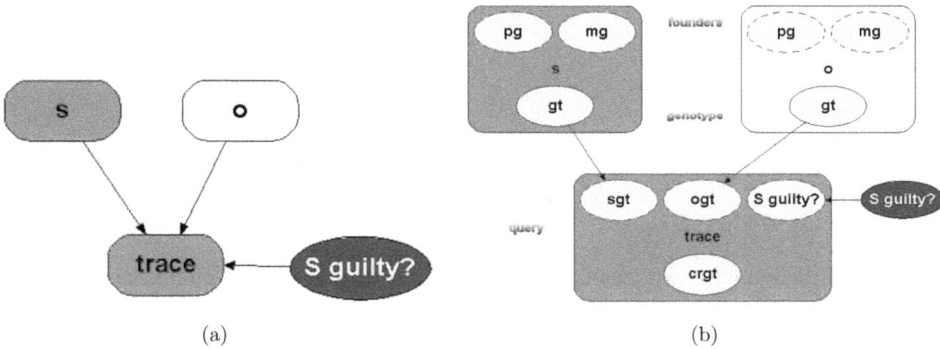

FIGURE 8.12
Network for criminal identification. (a) OOBN, showing network instances for alternative contributors s (suspect) and o (other) and crime trace, and hypothesis node. (b) Expanded network, showing identification of internal nodes across instance nodes.

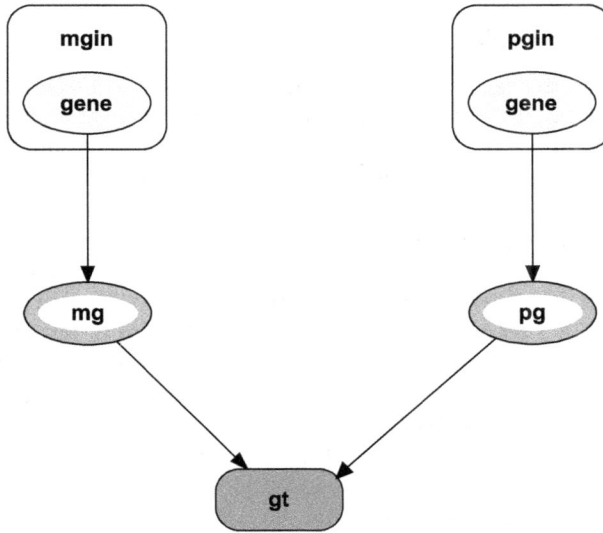

FIGURE 8.13
Network class **founder**. An individual's genotype gt is formed as the unordered pair of the individual's maternal gene mg and paternal gene pg.

paternal and maternal gene in a genotype). Node trace in Figure 8.12a is an instance of the **identification** network class shown in Figure 8.14, which is just a version of Figure 8.9: its output trace is modelled as equal to sgt or ogt, according as s guilty? is *true* or *false*, respectively. The full OOBN of Figure 8.12a elaborates this to account for the specific features of DNA profile identification, in particular the way in which a paternal and a maternal gene combine to form a genotype. As will be shown in Section 8.7, such an OOBN representation also allows us to include additional features, such as uncertain allele frequencies and heterogeneous populations.

Node s guilty? is assigned a formal prior probability 0.5 for *true*. The observed matching genotype at marker *m* is entered as evidence at gt in s, and again at crgt in trace, and

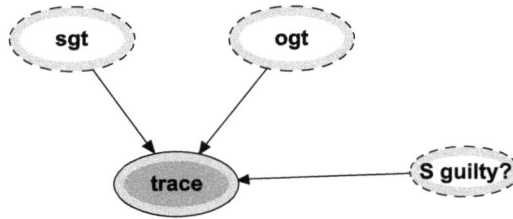

FIGURE 8.14
Network class **identification**: `trace` is identical to `sgt` or `ogt` according as `S guilty?` is *true* or *false*.

propagated through the network. The resulting computed odds on *true* at `S guilty?` can then be interpreted as the likelihood ratio in favour of H_p, based on the evidence of a match at marker m. Finally, under the assumption of independence across markers, multiplying these values across all markers delivers the overall likelihood ratio based on the full DNA evidence.

8.5.2 Simple Disputed Paternity

A man is alleged to be the father of a child, but disputes this. DNA profiles are obtained from the mother `m`, the child `c`, and the putative father `pf`. On the basis of these data, we wish to assess the likelihood ratio in favour of the hypothesis of *paternity*: H_1: `tf=pf`, the true father is the putative father; as against that of *non-paternity*: H_0: `tf=af`—where `af` denotes an unspecified alternative father, treated as unrelated to `pf` and randomly drawn from a suitable specified population.

Assuming independence across markers, we can again analyse them one at a time, before multiplying their associated likelihood ratio values together to obtain the overall likelihood ratio—which can finally be combined with the prior odds of paternity based on external background evidence, using (8.1), to obtain the posterior odds for paternity.

An OOBN representing the disputed pedigree, for a single forensic marker, is shown in Figure 8.15. We now describe the basic constituents of this OOBN. These will also be used in some of the more complex examples of forensic DNA inference that we shall consider below.

Each of `m`, `pf` and `af` is an instance of the class **founder** of Figure 8.13. Node `c` is an instance of a class **child**, whose structure is displayed in Figure 8.16a.

On the paternal (left-hand) side of **child**, the input nodes `fpg` and `fmg` represent the child's father's paternal and maternal genes. These are then copied into nodes `pg` and `mg` of an instance `fmeiosis` of network class **meiosis**, whose output node `cg` is obtained by flipping a fair coin (node `cg=pg?`) to choose between `pg` and `mg`; this is then copied to `pg` (child's paternal gene) in network **child**. A similar structure holds for the maternal (right-hand) side of **child**. Finally `pg` and `mg` are copied into an instance `gt` of the network class **genotype**, which forgets the information on parental origin. Any DNA evidence on the individual is entered here.

The hypothesis node `tf=pf?` embodies H_1 (`tf = pf`) when it takes the value *true* and H_0 (`tf = af`) when *false*; it feeds into the instance `tf` of class **query**, shown in Figure 8.17, to implement this selection.

We initially formally set both hypotheses as equally probable, so that, after propagation of evidence, the ratio of their posterior probabilities yields the likelihood ratio based on

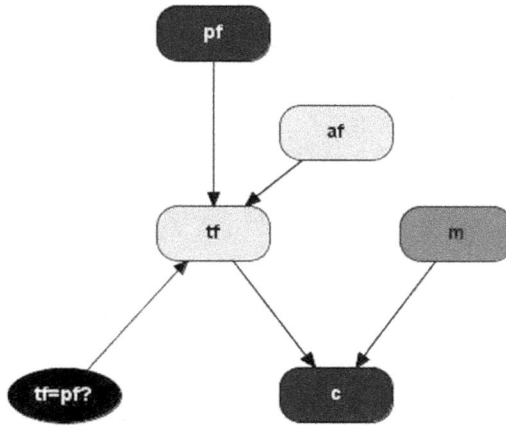

FIGURE 8.15
Pedigree for simple disputed paternity. Child c is the offspring of mother m and true father tf, who is either the putative father pf or an alternative father af.

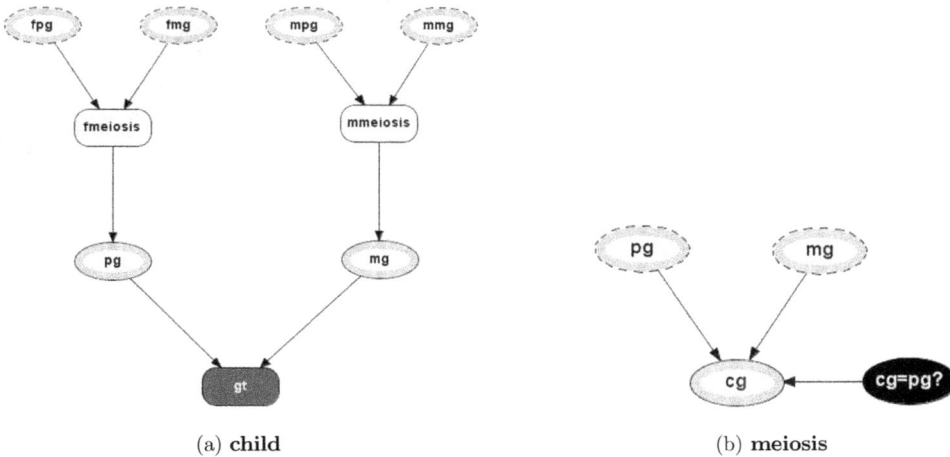

(a) **child**   (b) **meiosis**

FIGURE 8.16
Network classes (a) **child** and (b) **meiosis**. In **meiosis**, the gene cg a parent transmits to a child is chosen at random from that parent's paternal gene pg and maternal gene mg. In **child**, this process is repeated for both parents, the transmitted genes then being combined into the child's genotype gt.

this marker. By entering the data for each marker into the appropriate Bayesian network, we can thus easily calculate the overall likelihood ratio in favour of paternity.

As an illustrative example, suppose that the data, for marker D7S820, are child's genotype cgt= {12, 12}, mother's genotype mgt= {10, 12}, putative father's genotype pfgt= {10, 12}. The node gene in class **gene** is populated with the estimated population frequencies of the various alleles for this marker. In particular, the frequencies of the observed alleles 10 and 12 for the US Caucasian population (Butler et al., 2003) are, respectively, 0.243 and 0.166. Node tf=f1? in **query** is assigned formal equal prior probabilities for *yes*

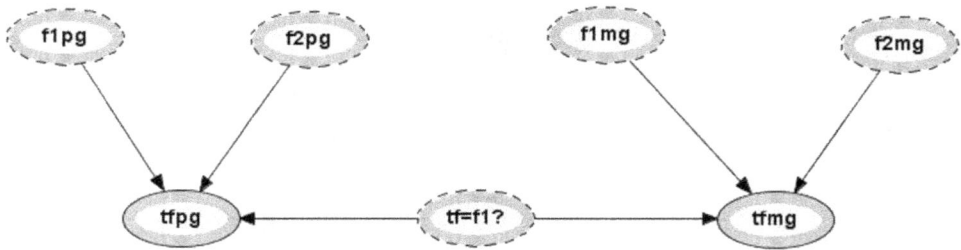

FIGURE 8.17
Network class **query** for disputed paternity testing. The query node `tf=f1?` determines, simultaneously for both genes `tfpg` and `tfmg` of the true father `tf`, whether these come from possible father `f1` or from `f2`.

and *no*. On inserting the genotype evidence and propagating, the posterior probability of *yes* is 0.751, so that the likelihood ratio in favour of paternity, based on marker D7S820 alone, is $0.751/0.249 = 3.01$.

The above procedure for the simple disputed paternity problem might be regarded as overkill, since it is easy to solve from first principles. Conditioning on the genotypes of mother and putative father (which will make no difference to the answer), we see that the child's genotype will be as observed if and only if both the mother and the true father contributed allele 12 to the child. Using Mendel's law of segregation, this event has probability $0.5 \times 0.5 = 0.25$ if the true father is the putative father, and probability $0.5 \times 0.166 = 0.083$ if the true father is, instead, some unrelated individual from the population. Thus the likelihood ratio in favour of paternity, based on marker D7S820 alone, is $0.25/0.083 = 3.01$. The OOBN analysis has merely automated this straightforward computation. The real power of the OOBN approach lies in its versatility and generalisability. Once supplied with the basic building blocks **founder**, **child** and **query**, we can connect them together in different ways, much like a child's construction set, to represent a wide range of problems that are not easy to solve by other means. Some illustrations are now given in Section 8.5.3.

8.5.3 Bayesian Networks for Complex Criminal Cases Involving Family Relationships

In cases of disputed paternity it commonly occurs that the DNA profiles of one or more of the principal actors in the pedigree is not available, but there is indirect evidence, in the form of DNA profiles of various known relatives. In Examples 8.4 and 8.5 we consider two examples of such a case. The analysis of all the data is clearly now much more complex.

Example 8.4: Complex Disputed Paternity for an Inheritance Case

The pedigree in Figure 8.18 represents a real case of disputed inheritance. It is essentially a more complicated uncertain paternity dispute.

An Australian woman `m1` was on holiday in Venice when she began to suffer from severe toothache. She went to the dentist a few times and they began an affair. When she returned to Australia she learned that the dentist had died. Some months later she gave birth to a boy `c1` and claimed that her son was the dentist's son and requested her son's share of the dentist's substantial inheritance. The dentist `d` was married to `m2` and had two children `c21` and `c22`. The case was brought as a civil law suit.

DNA profiles were taken of the undisputed family `m2`, `c21`, `c22` and the disputed family members `m1`, `c1`, and it is questioned whether the deceased father `d` is also the

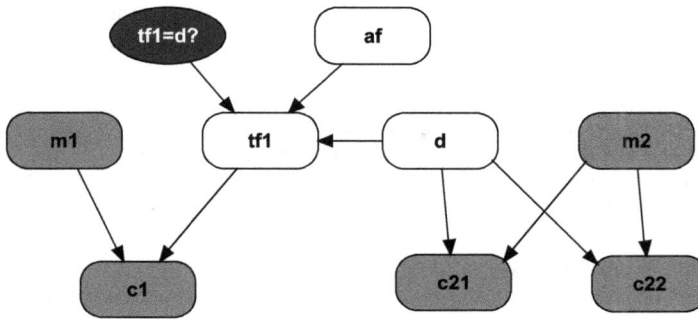

FIGURE 8.18
OOBN for disputed inheritance case (see text for details).

true father `tf1` of child `c1` of mother `m1`. The hypothesis of interest is represented by node `tf1=d?` which represents the two hypotheses: that `d` is the father of `c1`, or some other unspecified man `af` of Australian origin is the father. Here `c1`, `c21` and `c22` are all instances of class **child**, `tf1` is an instance of **query**, and `af`, `m1`, `m2` and `d` are all instances of class **founder**.

Example 8.5: Two Brothers

Figure 8.19 is a OOBN representation of a disputed paternity case where we have DNA profiles from a disputed child `c1` and from its mother `m1`, but not from the putative father `pf`. We do however have DNA from `c2`, an undisputed child of `pf` by a different, observed, mother `m2`, as well as from two undisputed full brothers `b1` and `b2` of `pf`.

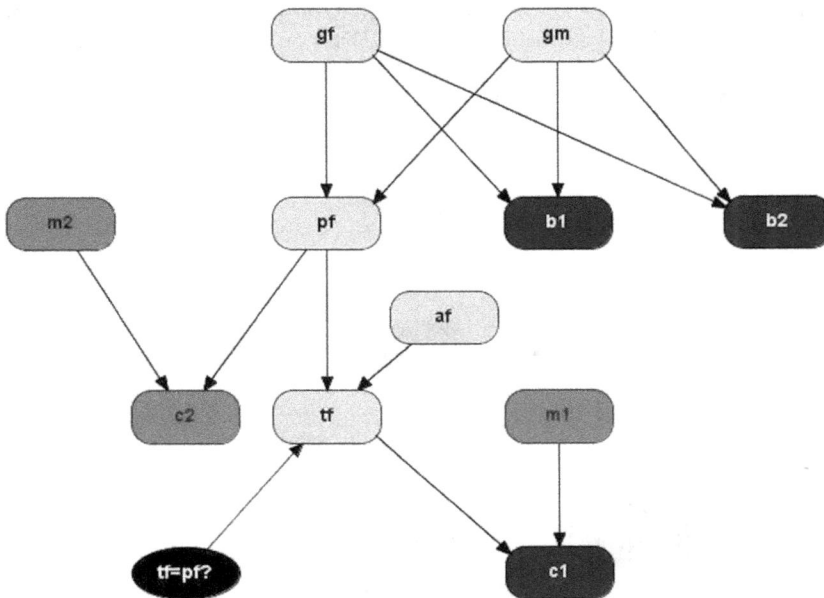

FIGURE 8.19
Pedigree for incomplete paternity case. DNA from the putative father `pf` of `c1` is unavailable, but indirect information on his genotype is provided by his two brothers `b1` and `b2`, as well as his undisputed child `c2` by `m2`.

The sibling relationship is made explicit by the incorporation of the unobserved grand-father gf and grandmother gm, parents of pf, b1 and b2. The "hypothesis node" tf=pf? again indicates whether the true father tf is pf, or is an alternative father af, treated as randomly drawn from the population.

Nodes gf, gm, m1, m2 and af are all instances of class **founder**; pf, b1, b2, c1 and c2 are instances of class **child**; tf is an instance of class **query**.

For this particular case the overall likelihood ratio evaluates to around 1300, meaning that the observed DNA evidence is 1300 times more probable on the hypothesis of pater-nity than it would be were we to assume non-paternity. According to Evett and Weir (1998, Chapter 9), such a value might be considered as offering "very strong support" to the hypothesis of paternity (although paternity applications such as this will never pro-duce the kind of likelihood ratio value, sometimes in the billions, that can occur when DNA profiling evidence is used to match a suspect to a crime). However it is important to remember, in all cases, that the likelihood ratio derived from the DNA evidence is only one element of the whole story, which also involves prior probabilities, and perhaps fur-ther likelihood ratios based on other evidence in the case. All these ingredients need to be combined appropriately, using Bayes's theorem, to produce the final probability of paternity.

Example 8.6: Hanratty

Another complex case is that of James Hanratty who, found guilty of rape and murder, was in 1962 the last person to be executed in the United Kingdom. In 1998, in an attempt to obtain a posthumous rehabilitation of Hanratty, his mother and brother underwent DNA testing. Their DNA profiles were compared with a DNA sample left at the scene of the crime. Figure 8.20 shows the OOBN for this case. Nodes father, mother and other are all instances of class **founder**, brother and Hanratty are instances of class **child**, and crime sample is an instance of class **query**.

Based on this familial evidence, the likelihood ratio in favour of Hanratty having been the source of the crime scene DNA was computed to be 440. This led to the exhumation of Hanratty's body and the extraction of his full DNA profile, which proved to be a full match with the crime sample, yielding a likelihood ratio of 2.5 million. Although the defence attempted to attribute the match to contamination during the many years for which the crime items had been stored, the Court reaffirmed the original guilty verdict.

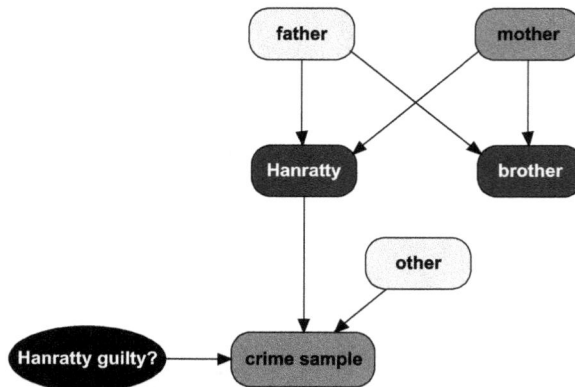

FIGURE 8.20
The case of James Hanratty. Indirect information on his DNA (for comparison with a crime sample) is supplied by his mother and brother.

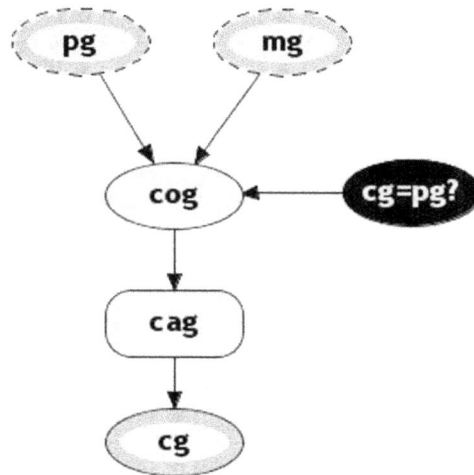

FIGURE 8.21
Revised network class **meiosis**, incorporating the possibility of mutation acting on an original gene to be transmitted from parent to child, cog, to transform it into the child's actually received gene, cg.

8.5.4 Mutation

Yet another advantage of the OOBN approach is that it is easy to modify the component networks to incorporate a variety of additional complications. One such is the possibility of *mutation* of genes in transmission from parent to child, which, in a simple case of disputed paternity, could lead to a true father appearing to be excluded (Dawid et al., 2001, 2003; Dawid, 2003; Vicard and Dawid, 2004; Vicard et al., 2008). We must now distinguish between a child's *original gene* cog, identical with one of the parent's own genes, and the *actual gene* cag available to the child, which may differ from cog because of mutation. We elaborate the network class **meiosis** of Figure 8.16b as shown in Figure 8.21, by passing its output cog through an instance cag of a new network class **mut**, constructed to implement whatever model is used to describe how the value of cog is stochastically altered by mutation. The output of cag is then copied to cg. Thus **meiosis** now represents the result of mutation acting on top of Mendelian segregation.

Once an appropriate network **mut** has been built, and **meiosis** modified as described above, pedigree networks constructed as in Section 8.5.2 or Section 8.5.3 will now automatically incorporate the additional possibility of mutation. Similar modifications to component networks can readily account for other complications, such as "silent" alleles, that fail to be picked up by the measuring apparatus.

8.6 Bayesian Networks for Analysing Mixed DNA Profiles

Bayesian networks have also been constructed to address the challenging problems that arise in the interpretation of mixed trace evidence. DNA samples that are found on crime scenes are often complex, as for example when the samples contain DNA from several individuals, or when the DNA samples are degraded. A sample found at a crime scene may contain biological material from the victim but also from individuals who might be involved in the crime. Identifying their DNA is fundamental to help solve the criminal investigation. Advanced DNA technology now extracts genetic material from a huge variety of surfaces and objects which may have been handled by several individuals, only some of whom are related to the specific crime. DNA mixtures with DNA from many individuals occur frequently in multiple-rape cases or in traces left by groups of perpetrators touching the same objects such as crowbars, guns, cigarette butts, *etc.* In many cases such mixed samples could contain DNA from a victim and a perpetrator. Typically one would be interested in testing whether the victim and suspect contributed to the mixture, H_0: $v \& s$, against the hypothesis that the victim and an unknown individual contributed to the mixture, H_1: $v \& u$. One might alternatively consider an additional unknown individual u_1 instead of the victim, with hypotheses H_0: $u_1 \& s$ versus H_1: $u_1 \& u_2$. Here we will firstly discuss the BNs used for analysing the discrete information from the alleles in the mixture and then briefly present the analysis for the continuous peak height/area.

8.6.1 Discrete Features

Figure 8.22 shows a top-level network which can be used for analysing a mixture with two contributors, *p*1 and *p*2. Nodes sgt, vgt, u1gt and u2gt are all instances of network class **founder**, and represent the suspect's, the victim's and two unknown individuals' genotypes. Boolean node p1=s? represents the hypothesis that contributor *p*1 is the suspect *s*. Node p1gt, the genotype of *p*1, is an instance of a network with similar structure to the **identification** class, which selects between the two genotypes sgt or u1gt according to the true/false state of the Boolean node p1=s?. A similar relationship holds between nodes p2gt, vgt, u2gt and p2=v?. Possible genotype information on the suspect and/or the victim is entered and propagated from nodes sgt and vgt. The target node is the logical combination of the two Boolean nodes p1=s? and p2=v? and represents the four different hypotheses described above. Ainmix? determines whether allele *A* is in the mixture: this will be so if at least one *A* allele is present in either p1gt or p2gt. Similarly for Binmix?,

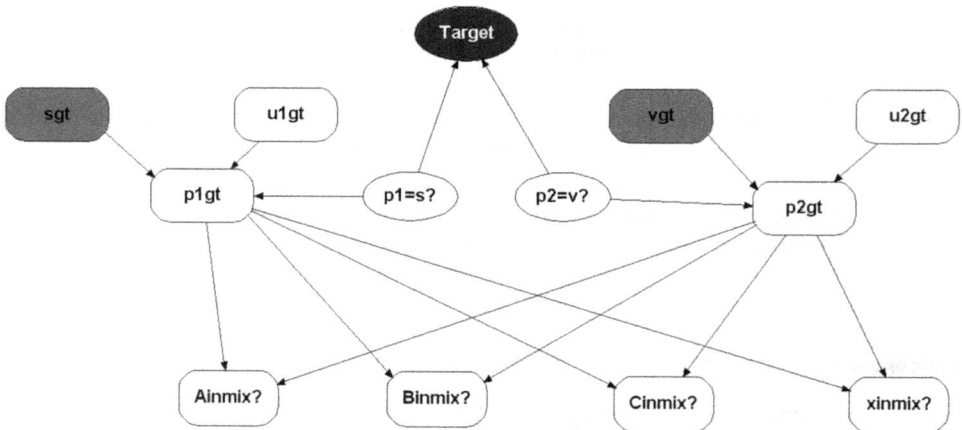

FIGURE 8.22
Bayesian network for DNA mixture from two contributors. The genotype of the first contributor p1gt can either be that of the suspect sgt or that of an unknown u1gt, according to the Boolean node p1=s?. Similarly p2gt, is either the victim vgt or a different unknown u2gt, according to p2=v? (see text for further details).

`Cinmix?`, `Dinmix?` and `xinmix?` (where x refers to all of the alleles that are not observed). Information on the alleles seen in the mixture is entered and propagated from these nodes.

The modular structure of Bayesian networks supports easy extension to mixtures with more contributors, as in cases where a rape victim declares that she has had one consensual partner in addition to the unidentified rapist, or that she has been victim of multiple rape. Simple modification of the network handles such scenarios, so long as the total number of contributors can be assumed known.

In general, however, although the evidence of the trace itself will determine a lower bound to this total, there is in principle no upper bound. Thus if in a trace we see that the maximum number of alleles in any marker is three, we know that the minimum number of contributors that could have produced this trace is two, but we can not be sure that there were only two. However Lauritzen and Mortera (2002) show that it is often possible to set a relatively low upper limit to the number it is reasonable to consider. Once it has been agreed to limit attention to some maximum total number of potential contributors, cases where the number of unknown contributors is itself uncertain can again be addressed using a Bayesian network, now including nodes for the number of unknown contributors and the total number of contributors (Mortera et al., 2003). This can be used for computing the posterior distribution of the total number of contributors to the mixture, as well as likelihood ratios for comparing all plausible hypotheses.

The modular structure of the Bayesian networks can be used to handle still further complex mixture problems. For example, we can consider together missing individuals, silent alleles and a mixed crime trace simply by piecing together instances of appropriate network classes.

8.6.2 Continuous Features

So far we have only used discrete information, namely which allele values are present in the mixture and the other profiles from the measured individuals. A more sensitive analysis additionally uses measured "peak areas" or "peak heights", which give quantitative information on the amounts of DNA involved. This requires much more detailed modelling, but again this can be effected by means of a Bayesian network (Cowell et al., 2007b). Figure 8.23 shows the top level network for two contributors, involving six markers, each an instance of a lower level network class **marker** as shown in Figure 8.24. Because the mixture proportion `frac` of DNA contributed by one of the parties is a common quantity across markers, we must now handle them all simultaneously within one "super-network". This network is an extended version of the one shown in Figure 8.22, incorporating additional structure to model the quantitative peak area information. In particular, the nodes `Aweight` *etc.* in **marker** are instances of a network class that models the quantitative information on the peak weight.

Cowell et al. (2007b) analyse the data shown in Table 8.7, taken from Evett et al. (1998), involving a 6-marker mixed profile with between 2 and 4 distinct observed bands per marker, and a suspect whose profile is contained in these. It is assumed that this profile is a mixture either of the suspect and one other unobserved contributor, or of two unknowns. Using only the repeat numbers as data, the likelihood ratio for the suspect being a contributor to the mixture is calculated to be around 25,000. On taking account of the peak areas also, this rises to about 170,000,000.

The appropriate extensions of the mixture models become relatively complex when the number of potential contributors to the mixture becomes large, or we want to allow for uncertainty in allele frequencies and/or population substructure.

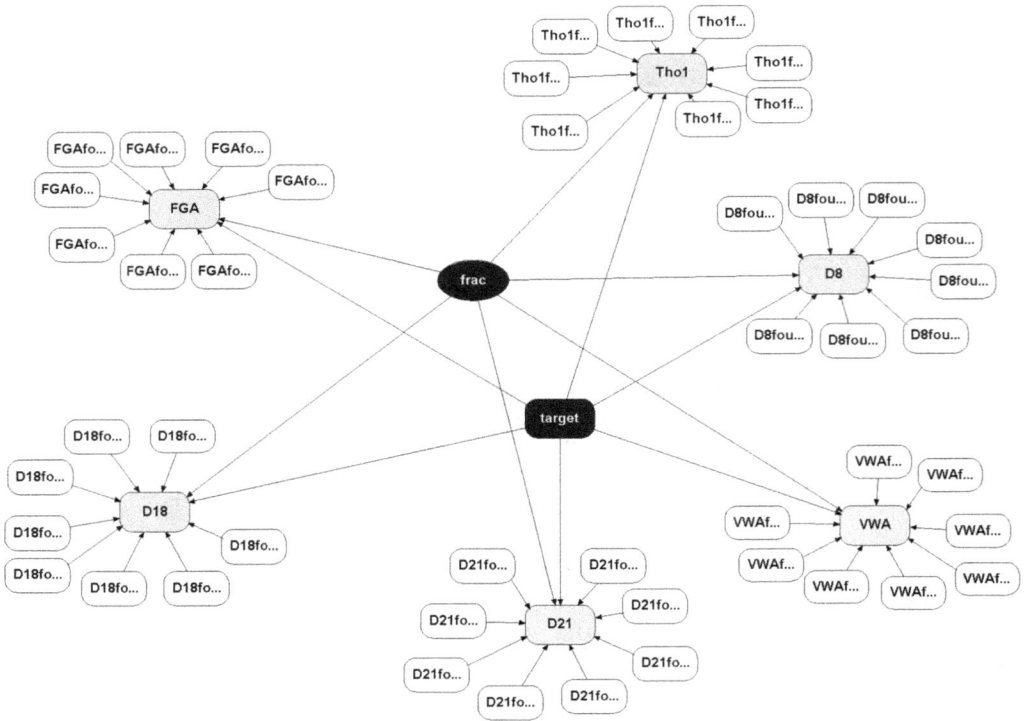

FIGURE 8.23
6-marker OOBN for a DNA mixture of two contributors, using peak areas (reproduced from Cowell et al., 2007b). The node `frac` represents the mixture proportion, while nodes D18, FGA, *etc.* are instances of the network class **marker** of Figure 8.24.

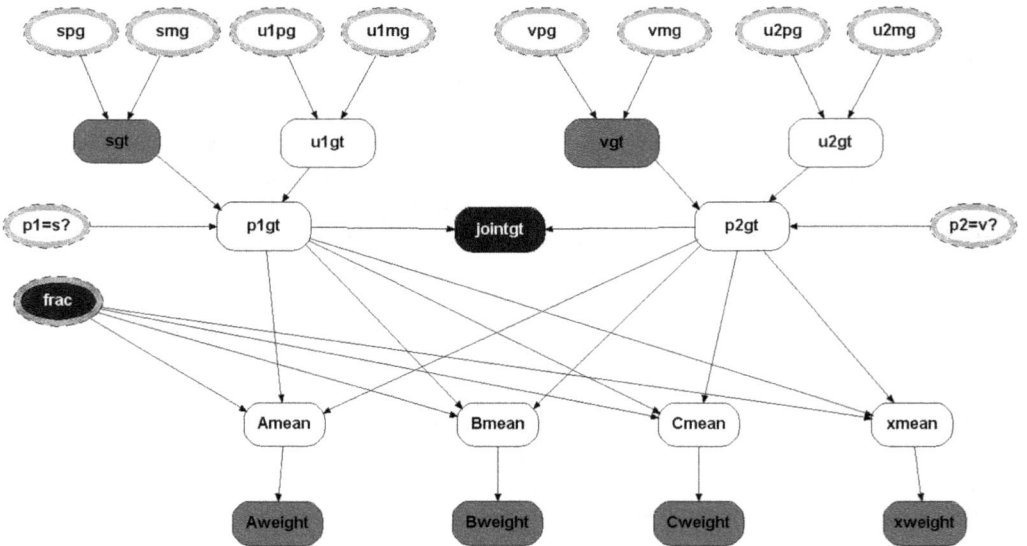

FIGURE 8.24
Network class **marker** with three observed allele peaks. For a description of the top three layers see Figure 8.22. The bottom two layers model the continuous peak area measurements.

TABLE 8.7

Data for Mixed Trace With Two Contributors. The Starred Values are the Suspect's Alleles

Marker	D8			D18			D21			
Alleles	10*	11	14*	13*	16	17	59	65	67*	70*
Peak area	6416	383	5659	38985	1914	1991	1226	1434	8816	8894
		FGA			THO1			VWA		
Alleles	21*	22*	23	8*	9.3*	16*	17	18*	19	
Peak area	16099	10538	1014	17441	22368	4669	931	4724	188	

Cowell et al. (2007a, 2011, 2015) extend the statistical model in Cowell et al. (2007b) for the quantitative peak information obtained from an electropherogram of a forensic DNA sample. A gamma model is used for the peak heights and the model further develops the modelling of various artefacts that can occur in the DNA amplification process.

The model can both find likelihood ratios for evidential calculations, and deconvolve a DNA mixture for the purpose of finding likely profiles of one or more unknown contributors to the mixture. Computation from this model relies on an efficient implementation of Bayesian network techniques. This allows for a ready extension to simultaneous analysis of more than one mixture trace. This modelling of peak height information provides for a very efficient mixture analysis.

Recently Mortera et al. (2016) applied this model to analyse a complex disputed paternity case, where the DNA of the putative father was extracted from his corpse, which had been buried for over 20 years. This DNA was contaminated and appeared to be a mixture of at least two individuals. This case, as well as a complex criminal DNA mixture case, was further analysed in Green and Mortera (2017), which presents general methods for inference about relationships between contributors to a DNA mixture and other individuals of known genotype. The model for relationship inference builds on the approach in Cowell et al. (2015), but makes more explicit use of the Bayesian networks in the modelling.

8.7 Analysis of Sensitivity to Assumptions on Founder Genes

Many forensic genetics problems, as we have shown, can be handled using structured systems of variables, for which Bayesian networks offer an appealing practical modelling framework, and allow inferences to be computed by probability propagation methods. However, when standard assumptions are violated—for example when allele frequencies are unknown, there is identity by descent or the population is heterogeneous—dependence is generated among founding genes, which makes exact calculation of conditional probabilities by propagation methods less straightforward. The standard assumptions that the allele frequencies are fixed and known, that the individual actors in the model are independent and that the allele frequency database is homogeneous can all be questioned (Green and Mortera, 2009). We now illustrate a couple of these issues.

8.7.1 Uncertainty in Allele Frequencies

In reality, the allele frequencies assumed when conducting probabilistic forensic inference are not known probabilities, but estimates based on empirical frequencies in a database.

For the criminal case of Section 8.5.1, the joint distribution of the founding genes is

$$\prod_m \{p(\mathrm{spg}_m)p(\mathrm{smg}_m)p(\mathrm{opg}_m)p(\mathrm{omg}_m)\}, \tag{8.8}$$

and all questions about sensitivity can be expressed through modifications to (8.8). Some generate dependence between founding genes. Following Green and Mortera (2009), assuming the idealisation of a Dirichlet prior and multinomial sampling, the posterior distribution of a set of probabilities is Dirichlet$(M\rho(1), M\rho(2), \ldots, M\rho(k))$, where M is the (posterior) sample size and the ρ's are essentially the database allele frequencies (posterior means). The founding genes (spg, smg, opg, omg) are drawn from this distribution, (conditionally) independently and identically across alleles. This corresponds to the standard set-up for a Dirichlet process model which, by marginalising over the Dirichlet distribution, can be represented in a BN using a Pólya urn scheme. This is represented by the network class **UGF** ("uncertain gene frequencies") shown in Figure 8.25: for further details see Green and Mortera (2009). For efficiency of the probability propagation, in order to create smaller clique tables this network is set up so that all choices are binary, following the "divorcing" procedure (Jensen, 1996), whereby auxiliary nodes are introduced in order to reduce the number of incoming edges of a selected node. An instance of this network class can then be incorporated as a building block in a higher level network that computes inference, for example, about a criminal identification case, a simple or complex paternity testing or a DNA mixture problem. Thus Figure 8.26 shows a network for criminal identification that integrates the network of Figure 8.12a with that of Figure 8.25. Similarly

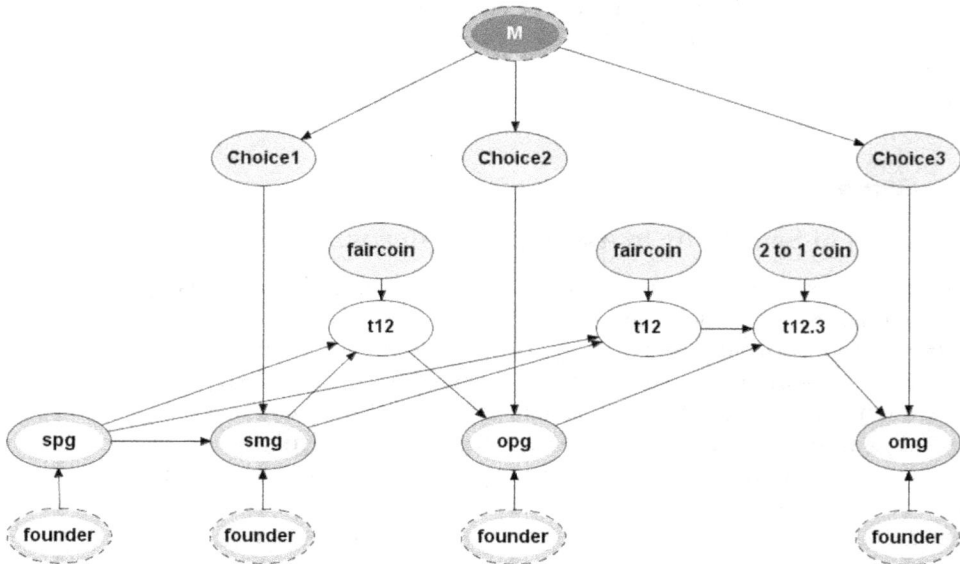

FIGURE 8.25
Network class **UGF** for modelling uncertain allele frequencies with the Pólya urn scheme (adapted with permission from Green and Mortera (2009), where details can be found).

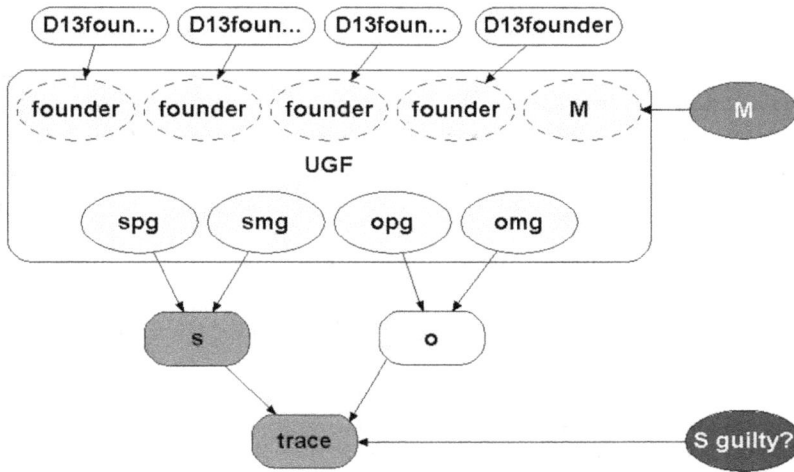

FIGURE 8.26
Network for criminal case with uncertain allele frequencies represented through the Pólya urn scheme. The parameter M can be interpreted as the effective size of the database.

instances of **UFG** of Figure 8.25, representing uncertain allele frequencies, can be integrated into the networks described in Sections 8.5.2, 8.5.3, 8.6. In this way, we can introduce uncertain allele frequencies for the reference population into any forensic identification problem.

8.7.2 Heterogeneous Reference Population

The assumption that the DNA reference population is homogeneous is questionable. The population is typically a mixture of subgroups.

Population heterogeneity raises two kinds of issues in the modelling. First, since unobserved actors are assumed to have genes drawn from a population, results can depend on which population (and correspondingly which allele frequency database) is used. Secondly, when there is uncertainty about which population is relevant, this can induce dependence between actors, observed or not. Additionally, when uncertainty about subpopulation relates to untyped actors, dependence between markers is induced.

The upper level network for sensitivity of inferences to population structure for criminal identification, based on a synthetic population that is a mixture of Afro-Caribbean, Hispanic and Caucasian subpopulations is shown in Figure 8.27.

Such problems are easily set up as Bayesian networks incorporating the structure shown in Figure 8.28. The variable s identifies the subpopulation, which may be dependent or independent between actors depending on the scenario of interest. Crucially, for each actor, s is the same for both genes for all markers, so that mixing across subpopulations is not the same as averaging the allele frequencies and assuming an undivided subpopulation. Note that, conditionally on subpopulation s, every gene at every marker is drawn independently from the appropriate subpopulation gene pool.

As before, instances of appropriate network classes, like that of Figure 8.28, can be integrated into the networks described in Sections 8.5.2, 8.5.3, 8.6. In this way, we can introduce both uncertain allele frequencies and heterogeneity into any forensic genetic identification problems.

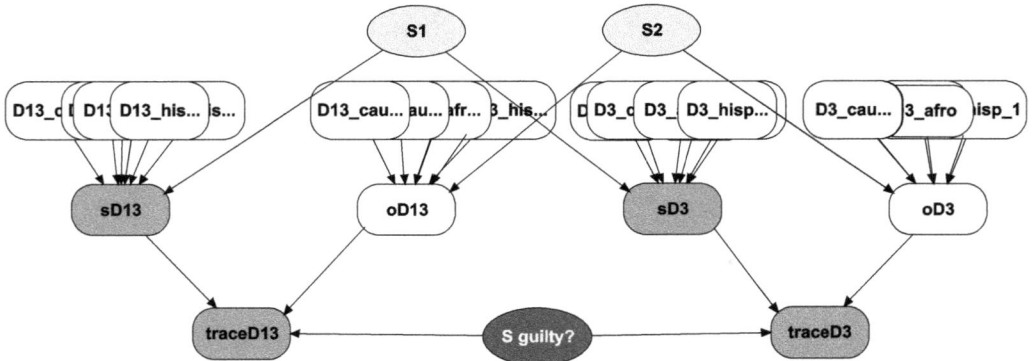

FIGURE 8.27
Network for two markers in a criminal identification allowing for subpopulation effect, where `S1` and `S2` are variables identifying the subpopulation. Nodes `sD13` [resp., `sD3`] for the suspect and `oD13` [resp., `oD3`] for an alternative suspect are instances of the network class shown in Figure 8.28, where the subpopulation nodes are populated with their associated allele frequencies for marker D13 [resp., D3].

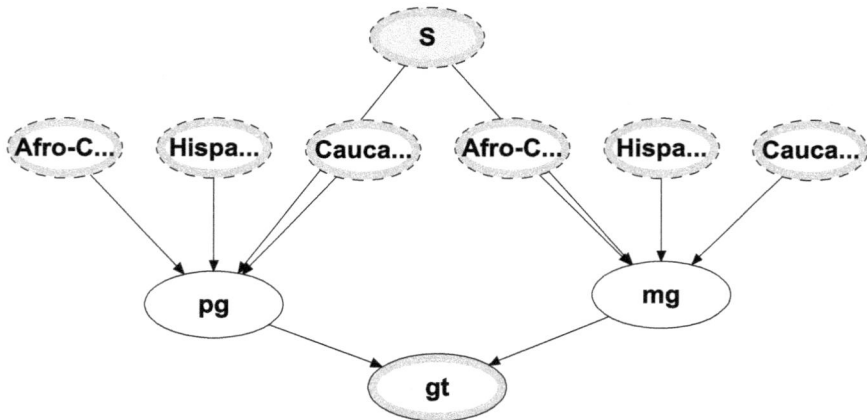

FIGURE 8.28
Network class for a genotype accounting for subpopulation effect, for Afro-Caribbean, Hispanic and Caucasian subpopulations.

8.8 Conclusions

We hope to have shown how useful BNs are for representing and solving a wide variety of complex forensic problems. The modularity and flexibility of BNs allows a complex problem to be broken down into simpler sub-networks that can then be pieced back together, so allowing one to build up a complex problem from simpler subproblems. Both genetic and non-genetic information can be represented in the same network. BNs can be used for a criminal case, both as a guide in the investigative phase, and to query the identification of an accused as the perpetrator. They can also be useful for combining different pieces of evidence pertaining to a specific case. The simplicity of their graphical representation makes them a useful tool for any legal scholar, for a judge, for the jury or for an investigator to understand the dependencies between different pieces of evidence. In particular, using OOBNs we can construct a flexible computational toolkit, and use it to analyse complex cases involving DNA profiles as well as other types of evidence. In this way one can address a wide range of forensic queries.

BNs and OOBNs can be helpfully applied in many branches of forensic analysis beyond those illustrated here. We hope we have stimulated the reader's interest in the use of BNs for modelling complex problems in forensic science.

Appendix 8A: Bayesian Network Basics

For a detailed account of the theory of Bayesian networks the reader is referred to Cowell et al. (1999).

8A.1 Qualitative Structure

A Bayesian network (BN) is a form of *directed acyclic graph (DAG)*, comprising a finite set *V* of *nodes*, with *arrows* between some of the nodes, in such a way that it is not possible, by following the arrows, to return to one's starting point. An example is shown in Figure 8A.1. The nodes having arrows pointing out from them and into a node $v \in V$ are called its *parents*, and denoted by pa(v); those with arrows pointing into them out from v are its *children*. This terminology is extended, in an obvious way, to *ancestor, descendant, etc.* Thus in Figure 8A.1 the black node has the horizontally striped node as its only child, and the two grey nodes as its parents.

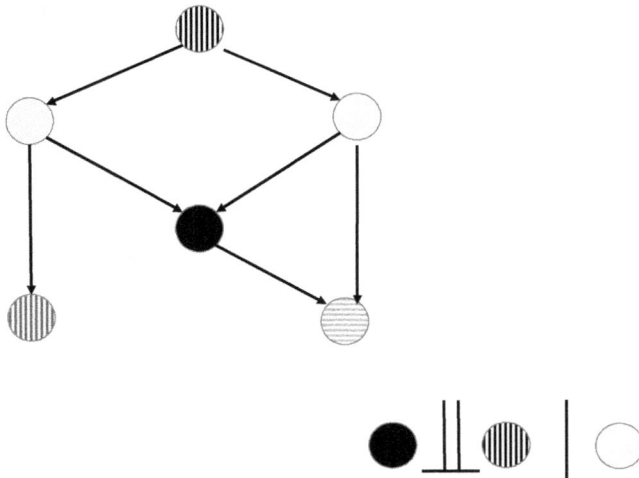

FIGURE 8A.1
A directed acyclic graph (DAG). The black node has the horizontally striped node as "child", and the two grey nodes as "parents". The black node is independent of the vertically striped nodes, conditional on the grey nodes.

8A.2 Independence Properties

Each node $v \in V$ is identified with a random variables X_v, and for $S \subseteq V$ we write X_S for $(X_v : v \in S)$. A BN *represents* a joint distribution P over the variables $(X_v : v \in V)$ when it is the case that, under P, each node in V is independent of all its non-descendants, conditional on its parents: in the conditional independence notation of Dawid (1979),

$$X_v \perp\!\!\!\perp X_{\mathrm{nd}(v)} \mid X_{\mathrm{pa}(v)}.$$

For example, in a distribution represented by Figure 8A.1 the black node is independent of the vertically striped nodes, conditional on the grey nodes.

For any such distribution P, we can deduce further implied probabilistic conditional independence properties, using the following entirely graphical routine.

Algorithm 8A.1: Moralisation

Suppose we wish to query the conditional independence property $X_A \perp\!\!\!\perp X_B \mid X_C$. We proceed by the following steps.

Step 1: Ancestral graph Delete from the DAG all nodes that are not in A, B, or C, or any of their ancestors.

Step 2: Moralisation Connect by an undirected arc any parents of a common child that are not already connected by an arrow. Then convert all arrows to undirected arcs.

Step 3: Separation In the resulting undirected graph, look for a path from a node in A to one in B that does not enter C.

If there is no such path, deduce that, under P, $X_A \perp\!\!\!\perp X_B \mid X_C$.   □

Using Algorithm 8A.1 we see that, for any distribution represented by Figure 8A.2, the black node is independent of the vertically striped nodes, conditional on the grey nodes.

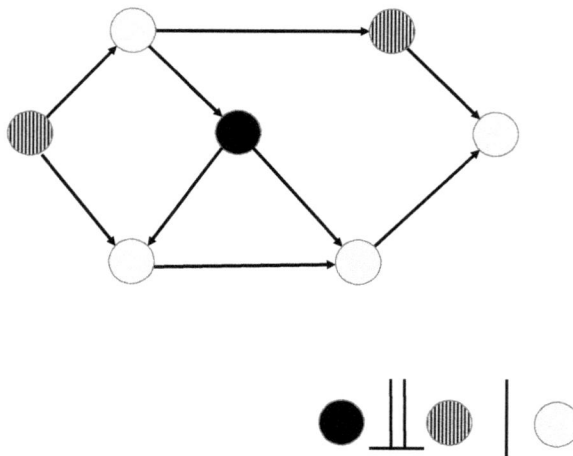

FIGURE 8A.2

Another DAG. The black node is independent of the vertically striped nodes, conditional on the grey nodes.

8A.3 Quantitative Structure

Let $\mathcal{X}_v$ denote the set of *states* (possible values) of X_v. For $S \subseteq V$ we write $\mathcal{X}_S$ for $\times_{i \in S} \mathcal{X}_i$. For $x = (x_i : i \in V) \in \mathcal{X}_V$, its projection (restriction), $(x_i : i \in S)$, to $\mathcal{X}_S$ is denoted by x_S.

We specify, for each node $v \in V$, a distribution over $\mathcal{X}_v$, conditional on each configuration $x_{pa(v)} \in \mathcal{X}_{pa(v)}$ of the states of its parents: the associated (discrete or continuous) density is denoted by $p(x_v \mid x_{pa(v)})$. For a distribution P represented by the BN, the joint density of all variables factorises as the product of its parent-child conditional densities:

$$p(x) = \prod_{i \in V} p(x_i | x_{pa(i)}).$$

Conversely any distribution with a density of this form will be represented by the BN.

When the variables are discrete, the parent-child conditional probability distributions can be represented as tables of the conditional probabilities $p(x_v | x_{pa(v)})$. The overall size of each table is the product of the sizes of the state-spaces of the variables concerned.

8A.4 Computation

The structure of a BN incorporates a degree of modularity, which makes it possible to execute complex computations by dividing them up into a sequence of "local" computations, each involving a subset of the variables, known as a clique. This is particularly useful when all cliques are relatively small. There exist elegant algorithms to identify the cliques and streamline the computations, and these have been implemented in a number of software packages. In particular, for a discrete distribution specified by its parent-child conditional probability tables, the Lauritzen–Spiegelhalter "probability propagation" algorithm (Lauritzen and Spiegelhalter, 1988) enables efficient computation of all marginal densities $\{p(x_i) : i \in V\}$. Moreover, for any $S \subseteq V$ and evidence $X_s = x_S^*$, essentially the same algorithm computes the prior probability of obtaining that evidence, $P(X_S = x_S^*)$, and the marginal posterior densities $\{p(x_i \mid X_S = x_S^*) : i \in V\}$. It is also straightforward to compute revised probabilities after incorporating external likelihood evidence relating to a subset of the variables.

References

Aitken, C. G. G. and Gammerman, A. J. Probabilistic reasoning in evidential assessment. *Science and Justice*, **29**, 303–316, 1989.

Anderson, T. J., Schum, D. A., and Twining, W. L. *Analysis of Evidence* (2nd ed.). Cambridge University Press, 2005. With on-line appendix on *Probability and Proof* by Philip Dawid at https://tinyurl.com/y9geusbb

Anon. The Howland will case. *American Law Review*, **4**, 562–581, 1870.

Butler, J. M., Schoske, R., Vallone, P. M., Redman, J.W., and Kline, M. C. Allele frequencies for 15 autosomal STR loci on U.S. Caucasian, African American and Hispanic populations. *Journal of Forensic Sciences*, **48**(4), 2003. doi:10.1520/JFS2003045

Cowell, R. G., Dawid, A. P., Lauritzen, S. L., and Spiegelhalter, D. J. *Probabilistic Networks and Expert Systems*. Springer, New York, 1999.

Cowell, R. G., Graversen, T., Lauritzen, S. L., and Mortera, J. Analysis of DNA mixtures with artefacts (with Discussion). *Journal of the Royal Statistical Society: Series C*, **64**(1), 1–48, 2015.

Cowell, R. G., Lauritzen, S. L., and Mortera, J. A gamma model for DNA mixture analyses. *Bayesian Analysis*, **2**(2), 333–348, 2007a.

Cowell, R. G., Lauritzen, S. L., and Mortera, J. Identification and separation of DNA mixtures using peak area information. *Forensic Science International*, **166**(1), 28–34, 2007b.

Cowell, R. G., Lauritzen, S. L., and Mortera, J. Probabilistic expert systems for handling artefacts in complex DNA mixtures. *Forensic Science International: Genetics*, **5**(3), 202–209, 2011.

Dawid, A. P. Conditional independence in statistical theory (with Discussion). *Journal of the Royal Statistical Society, Series B*, **41**(1), 1–31, 1979.

Dawid, A. P. An object-oriented Bayesian network for estimating mutation rates. In *Proceedings of the Ninth International Workshop on Artificial Intelligence and Statistics*, Jan 3–6 2003, KeyWest, Florida, editors, C. M. Bishop and B. J. Frey, 2003. http://tinyurl.com/y7pj6bwj

Dawid, A. P., Hepler, A. B., and Schum, D. A. Inference networks: Bayes and Wigmore. In *Evidence, Inference and Enquiry*, Proceedings of the British Academy, Vol. 171, editors, A.P. Dawid, W.L. Twining, and D. Vasilaki, pages 119–50. Oxford University Press, 2011.

Dawid, A. P., Mortera, J., Dobosz, M., and Pascali, V. L. Mutations and the probabilistic approach to incompatible paternity tests. In *Progress in Forensic Genetics 9*, International Congress Series, Vol. **1239**, editors, B. Brinkmann and A. Carracedo, pages 637–638. Elsevier Science, Amsterdam. International Society for Forensic Genetics, 2003. doi:10.1016/S0531-5131(02)00845-2

Dawid, A. P., Mortera, J., and Pascali, V. L. Non-fatherhood or mutation? A probabilistic approach to parental exclusion in paternity testing. *Forensic Science International*, **124**, 55–61, 2001.

Dawid, A. P., Mortera, J., Pascali, V. L., and van Boxel, D.W. Probabilistic expert systems for forensic inference from genetic markers. *Scandinavian Journal of Statistics*, **29**(4), 577–595, 2002.

Dawid, A. P., Mortera, J., and Vicard, P. Representing and solving complex DNA identification cases using Bayesian networks. In *Progress in Forensic Genetics 11*, International Congress Series, Vol. **1288**, editors, A. Amorim, F. Corte-Real, and N. Morling, pages 484–91. Elsevier Science, Amsterdam, 2006.

Dawid, A. P., Mortera, J., and Vicard, P. Object-oriented Bayesian networks for complex forensic DNA profiling problems. *Forensic Science International*, **169**(2–3), 195–205, 2007.

Evett, I.W., Gill, P. D., and Lambert, J. A. Taking account of peak areas when interpreting mixed DNA profiles. *Journal of Forensic Sciences*, **43**(1), 62–69, 1998.

Evett, I. W. and Weir, B. S. *Interpreting DNA Evidence*. Sinauer, Sunderland, MA, 1998.

Fenton, N., Neil, M., and Lagnado, D. A. A general structure for legal arguments about evidence using Bayesian networks. *Cognitive Science*, **37**(1), 61–102, 2013.

Green, P. J. and Mortera, J. Sensitivity of inferences in forensic genetics to assumptions about founder genes. *Annals of Applied Statistics*, **3**(2), 731–763, 2009.

Green, P. J. and Mortera, J. Paternity testing and other inferences about relationships from DNA mixtures. *Forensic Science International: Genetics*, **28**, 128–137, 2017.

Hepler, A. B., Dawid, A. P., and Leucari, V. Object-oriented graphical representations of complex patterns of evidence. *Law, Probability and Risk*, **6**(1–4), 275–293, 2007.

Jensen, F. V. *An Introduction to Bayesian Networks*. UCL Press and Springer Verlag, London, 1996.

Laṣkey, K. B. and Mahoney, S. M. Network fragments: representing knowledge for constructing probabilistic models. In *Proceedings of the Thirteenth Conference on Uncertainty in Artificial Intelligence, UAI97*, pages 334–341. Morgan Kaufmann Publishers Inc., San Francisco, CA, USA, 1997.

Lauritzen, S. L. and Mortera, J. Bounding the number of contributors to mixed DNA stains. *Forensic Science International*, **130**(2–3), 125–126, 2002.

Lauritzen, S. L. and Spiegelhalter, D. J. Local computations with probabilities on graphical structures and their application to expert systems. *Journal of the Royal Statistical Society Series B*, **50**, 157–224, 1988.

Mortera, J. Analysis of DNA mixtures using Bayesian networks. In *Highly Structured Stochastic Systems*, editors, P.J. Green, N.L. Hjort, and S. Richardson, chapter 1B, pages 39–44. Oxford University Press, 2003.

Mortera, J., Dawid, A. P., and Lauritzen, S. L. Probabilistic expert systems for DNA mixture profiling. *Theoretical Population Biology*, **63**(3), 191–205, 2003.

Mortera, J., Vecchiotti, C., Zoppis, S., and Merigioli, S. Paternity testing that involves a DNA mixture. *Forensic Science International: Genetics*, **23**, 50–54, 2016.

Neil, M., Fenton, N., and Nielson, L. Building large-scale Bayesian networks. *The Knowledge Engineering Review*, **15**(3), 257–284, 2000.

Reinach, J. *Histoire de L'Affaire Dreyfus*. Edition Fasquelle, Paris, 1901.

Schum, D. A. and Morris, J. R. Assessing the competence and credibility of human sources of intelligence evidence: Contributions from law and probability. *Law, Probability and Risk*, **6**(1–4), 247–274, 2007.

Taroni, F., Aitken, C., Garbolino, P., and Biedermann, A. *Bayesian Networks and Probabilistic Inference in Forensic Science*, Statistics in Practice. John Wiley and Sons, Chichester, 2006.

Vicard, P. and Dawid, A. P. A statistical treatment of biases affecting the estimation of mutation rates. *Mutation Research*, **547**, 19–33, 2004.

Vicard, P., Dawid, A. P., Mortera, J., and Lauritzen, S. L. Estimation of mutation rates from paternity casework. *Forensic Science International: Genetics*, **2**(1), 9–18, 2008.

Wigmore, J. H. *The Science of Judicial Proof* (3rd ed). Little, Brown, Boston, 1937.

Section III

Legal and Psychological Dimensions

9

How Well Do Lay People Comprehend Statistical Statements from Forensic Scientists?

Kristy A. Martire and Gary Edmond

CONTENTS

9.1 Methodological Overview 203
9.2 Consistency 203
    9.2.1 Framing 204
    9.2.2 Format 205
9.3 Sensitivity 206
9.4 (In)Coherence 209
    9.4.1 Prosecutor's Fallacy 209
    9.4.2 Defense Attorney's Fallacy 210
    9.4.3 Directional Errors 211
    9.4.4 Aggregation Errors 211
9.5 Ability 212
9.6 Orthodoxy 214
9.7 Discussion 216
9.8 Conclusion 221
References 221

> For the rational study of law the blackletter man may be the man of the present, but the man of the future is the man of statistics and the master of economics.
>
> **Holmes, 1897**

There are few legal rules or prescriptions that regulate how evidence should be evaluated during a trial (Twining, 1990; Edmond, 2015). For example, the trier of fact is not required to *accept* all of the evidence presented to it. It is free to choose between experts and even to reject opinions without providing any reasons for its preference. However, there is an expectation that the factfinder will *consider* all of the evidence, and that the evidence will be presented in a manner that can be understood by the decision-maker (*Kingswell v The Queen*, 1985: 301). This latter expectation poses a particular challenge for forensic scientists since they are frequently offering opinions in complex or technical domains about which lay people have limited knowledge (Hand, 1901; Jackson et al., 2006; Collins and Evans, 2007). There is also an expectation that the evaluation of the evidence will be (at least weakly) rational. In the sense that any verdict should accord with the burden and standard of proof, and not be inconsistent with evidence that is incontrovertible.

The statistical evidence presented by forensic scientists may include uncertainty statements, information about error, base rates, random match probabilities (RMP), false report probability (FRP), and likelihood ratios (LR) among others (Jackson, 2009; Thompson, 2018;

Thompson et al., 2018). These types of statements are important for the accurate, modest and transparent description of forensic science techniques and opinions (Edmond et al., 2016). The base or incident rate quantifies how commonly a characteristic appears in a population (Thompson, 1989); a random match probability describes the chance that a randomly selected person or object from a reference population will coincidentally "match" another person or object in terms of the specified characteristic (Koehler et al., 1995); a false report probability describes the chance of mistakenly misclassifying non-matching samples as matching, for example as a consequence of human error (Thompson et al., 2013); and a likelihood ratio communicates how many times more likely an observed match is under one hypothesis than another (Koehler, 1996). For examples of these types of statistical statements, see Table 9.1. Scholars of psychology and law have been examining lay comprehension of these types of statistical statements for decades.

TABLE 9.1

Statistical Statement Examples

Statement Type	*Example;* Source
Base/incident rate	*35% of the population had the blood type shared by the assailant and the defendant* (Smith et al., 1996)
Random match probability	*Only 1 in 5,072 Caucasian men have mtDNA types that match* (Kaye et al., 2007)
Exclusion probability	*99.98% of the population would be excluded as a possible source* (Kaye et al., 2007)
False report probability	*This type of error occurs in one out of 1,000 DNA tests* (Scurich and John, 2013)
Likelihood ratio	*100 times more likely if the partial profile came from the suspect than if it came from a randomly chosen person* (Thompson and Newman, 2015)

There has been heated debate about the potential for mathematical and statistical evidence to be misunderstood by lay people at least since 1970 (Eggleston, 1978). At that time Finkelstein and Fairley's (1970) suggestion that Bayes' theorem could potentially assist the trier of fact to appropriately combine and weight pieces of statistical evidence was met by opposition from those with concerns that statistical evidence would instead be overused in the context of the broader case (Tribe, 1971). Examination of the potential for over and underuse of forensic science evidence has only gained momentum since these debates began, partially as a response to the increasing statistical sophistication and critical analysis of forensic science evidence (National Research Council, 2009; President's Council of Advisors on Science and Technology, 2016; Stern, 2017). Where once the opinions of forensic practitioners were communicated in terms of unqualified categorical matches or "individualizations" (Saks and Koehler, 2008), now many forensic practitioners propose much more complex and nuanced formulations. For example, the interpretation of complex biological (DNA) samples in terms of the combined probability of inclusion (Bieber et al., 2016). Confronted with complexity, some forensic practitioners have attempted to translate statistical evidence, and impressions of the strength of evidence, into non-mathematical formulations and tables (Association of Forensic Science Providers, 2009).

Despite the evolving research literature and knowledge base around lay responses to statistical statements, it is fair to say that we are still seeking satisfactory answers to many fundamental questions: Do jurors understand statistical statements, and if so, how well? Which formats of expression or presentation facilitate optimal comprehension? And, how can expert reports and testimony be modified to improve lay performance?

The first of these questions "*do jurors understand statistical statements*" is deceptively simple because understanding of trial evidence is quite difficult to define and measure. What does it mean to "understand" a statistical statement presented in the context of a complex decision-making environment? And how do we know it when we see it? A survey of the literature examining lay responses to statistical statements shows that understanding or comprehension has implicitly and explicitly been conceptualised in a number of ways. For the purposes of this review we categorise these indicia as measures of: *Consistency, Sensitivity, Coherence, Ability* and *Orthodoxy* (abbreviated as "CASOC").

Consistency is giving equal weight to evidence with quantitatively equal strength. *Sensitivity* is assigning greater weight to evidence of greater value, and lesser weight to evidence of lesser value. *Coherence* is responding to evidence in a logical manner. *Ability* is being capable of applying statistical evidence or principles provided by a forensic scientist (or statistician) to the resolution of new problems. This is distinct from mere recognition or recollection of the statistical information. Ability requires an active application for the purposes of deriving information beyond what was originally provided. *Orthodoxy* is used here in the sense of compliance with or adherence to normative expectations. In this case orthodoxy is updating beliefs in a manner that is consistent with the normative expectations derived using Bayes' theorem. We provide a brief methodological overview of CASOC studies before we review the evidence in relation to each of the indicia below.

9.1 Methodological Overview

Various methodologies have been adopted by researchers to study lay comprehension of statistical statements. In the broadest terms these studies involve the recruitment of a sample of participants who are then presented with an expert opinion from a forensic scientist (e.g., DNA, serology or hair analyst) incorporating the focal statistical statement. Participants are then asked to answer questions that offer insights into the impact of the focal statement on judgements, beliefs and understanding of the evidence and case.

In practice these basic elements have resulted in a diverse literature incorporating responses from a range of lay populations such as university students and employees, venire persons, jury pool members and judges. Statistical statements are often presented to participants using brief written vignettes or summaries that include little more than some basic contextual information and the focal statement. However, several studies have attempted to increase the verisimilitude (i.e., ecological validity) of the experimental materials by instead using lengthy trial transcripts, written narrative summaries, or videotaped simulated trials that have included a deliberation phase.

Once participants have been presented with the statistical statement, a range of responses have been sought by researchers. These include binary verdict decisions (guilt or liability); estimates of the probability of guilt or probability that the defendant is the source - at times this information is elicited before and after the presentation of the focal statistical statement as a means of calculating change in beliefs; ratings of case strength, the expert (e.g., credibility, importance, likeability), the statistical statement (e.g., clarity and accessibility); as well as comprehension questions, recollection and recognition items. This diversity of measures reflects the plurality of definitions of comprehension and provides the basis for the CASOC indicators we now review in detail.

9.2 Consistency

There have been a number of studies examining the consistency of responses to evidence of equivalent statistical value. The main finding with respect to this dimension of comprehension is that mathematical equivalence does not guarantee psychological equivalence (Koehler, 2001). These results resonate with broader literatures identifying the role of heuristics and biases in human decision-making (Gigerenzer and Gaissmaier, 2011; Tversky and Kahneman, 1974).

9.2.1 Framing

Several studies have demonstrated that the interpretation of statistical statements can be affected by the context surrounding the evidence (i.e., the "frame"). Koehler (2001) examined the possible effects of the RMP target – either a single suspect or multiple people (in Washington, DC) – and format* of expression – either probability or frequency. In the first study single target probability formulation created a frame that resulted in higher estimates that the defendant was the source of recovered DNA than multiple target frequency frames (median 82.5% vs. 60.5% respectively). The same pattern of results was observed in the second study. This time target and format were independently varied to reveal main effects for both the estimates of the probability that the defendant was the source of the DNA, and the probability that the defendant was guilty of the crime. Single targets (suspect) and probability formats (RMP of 0.1%) resulted in significantly higher estimates than multiple targets (people in Houston) and frequency formats (RMP 1 in 1,000). The third study manipulated RMP (1/1,000; 1/1,000,000; 1/1,000,000,000) in addition to target and format. There was a significant three way interaction such that the framing effects were ameliorated by smaller RMPs to the point where there were no effects in the 1 in 1 billion condition. Koehler concluded that participants found it more difficult to imagine viable alternative suspects where the statistical statements were framed in terms of a probability about a single suspect, but that this was more difficult to achieve where the chances of a random match were smaller.

McQuiston-Surrett and Saks (2009), Exp. 1 also considered the role of single and multiple targets alongside subjective and objective RMP formulations. Judges and jurors were presented with forensic science evidence about the defendant's hair as either: a subjective estimate (stating a quantitative "guesstimate" that the profile occurred in 1 of every 1,000 people with an RMP of 0.001 or 0.1%); an objective single probability (0.001 or 0.1%); or an objective multi-frequency estimate (1 in 1,000 people, meaning 3,000 people in a city of 3,000,000). Estimates of how much the microscopic hair evidence contributed towards guilt in the case varied by frame and participant group. Judges were not affected by frame, however jurors considered the objective single-probability frame more persuasive (Mean (M) = 5.37) than either the subjective-probability (M = 4.42) or objective multiple-frequency frames (M = 4.51).

Wells (1992) used variants of the "Blue bus case" (Nesson, 1985) to explore the impact of statistical frame on the interpretation of "naked" statistical evidence in the context of liability verdict decisions. The Blue bus case describes a scenario where a bus is implicated by an eyewitness in a traffic accident. Depending on the specific formulation, the witness reports having seen a Blue bus in a town where only two bus companies service the area (e.g., Blue or Grey). The base rates for these bus companies are provided (e.g., 80% of buses are Blue and 20% are Grey) alongside other statistical information (e.g., regarding the accuracy of the eyewitness or log book entries that implicate the Blue bus). In this series of experiments Wells held the statistical values in the scenario constant (e.g., 80% Blue bus vs. 20% Grey bus) to examine the impact of: (1) the source of the uncertainty (eyewitness error

---

* This dimension was originally labelled 'frame' by the authors.

vs. incident frequency); (2) the causal relevance; (3) distributional fairness; (4) the type of reasoning supported (fact-to-evidence vs. evidence-to-fact); and (5) the verdict threshold (preponderance of evidence, more likely than not, or greater than 50%) on determinations of civil liability. While the subjective probabilities understood by participants matched the mathematical probabilities presented, judges, business students and psychology students *"react to probabilistic evidence in different ways, depending on how that evidence is framed, even under conditions in which subjective probabilities do not vary"* (p. 747).

9.2.2 Format

Many studies have examined the impact of presentation format (e.g., probability vs. frequency) on the interpretation of statistical statements. Presentation format is regularly seen to influence lay assessment of evidence. In Experiment 1, Thompson and Schumann (1987) presented university students with either the conditional probability of microscopic hair match evidence in a robbery scenario ("a two percent chance the defendant's hair would be indistinguishable from that of the perpetrator if he were innocent…" p. 173) or a percentage plus number of people (2% of people; 20,000 in a city of 1,000,000). Those in the conditional probability condition provided higher estimates of the probability of guilt than those in the percentage plus number condition.

Lindsey, Hertwig, and Gigerenzer (2003) and Goodman (1992) focused on the differences in interpretation resulting from the use of natural frequencies as compared to probabilities. Lindsey et al. (2003) gave law students and jurists a RMP of either "one in a million" or "0.0001%" in a rape case with DNA evidence. Goodman (1992), Exp. 2 gave potential jurors an RMP of either "one in 1,000" or "one-tenth of one percent" in an arson case involving chromatographic testing. Goodman found no significant effect of presentation format on verdicts although convictions were more common in the probability (59%) than frequency format (41%). The same pattern of results was observed by Lindsey et al. (2003) however, in this case the difference was statistically significant (probability: students 54.5%, jurists 44.5%; frequency: students 32.5%, jurists 32.0%).

Six papers have compared interpretations of a likelihood ratio (LR) format with frequencies (Koehler, 1996; Nance and Morris, 2002, 2005; Thompson and Newman, 2015), verbal quantification (Martire et al., 2013, 2014; Thompson and Newman, 2015) and visual or descriptive aids (Koehler, 1996; Nance and Morris, 2002, 2005; Martire et al., 2013, 2014). Overall, these studies suggest that there are differences between the interpretations of LRs and the other tested formats.

Koehler (1996) observed that jury-eligible respondents were more likely to assign lower probabilities to the hypothesis that the suspect was the source of DNA in a rape case if they received profile RMP frequencies (e.g., 1 person out of every 100) as compared to a LR (100 times more likely to see this match) or a posterior odds ratio (given this match the hypothesis is 100 times more likely). Those who received profile frequencies were also less likely to give a guilty verdict than those in the other conditions. In two studies Martire and colleagues (2013, 2014) compared numerically quantified likelihood ratios (e.g., 450 times more likely) with verbally quantified LRs (e.g., moderately strong support) in accordance with translation tables proposed by forensic science practitioners (Association of Forensic Science Providers, 2009). For larceny scenarios involving shoeprint (Martire et al., 2013) or fingerprint evidence (Martire et al., 2014), online respondents assigned statistically equivalent weight to verbal and numerically quantified LRs when the evidence was "moderate" or "strong" (Martire et al., 2014). However when the evidence was weak (e.g., 4.5 times more likely/"weak" or "limited support"), participants assigned significantly less weight to verbally quantified likelihood ratios compared to numerically quantified ratios (Martire et al., 2013, 2014), verbal/numerical translation tables, and visual scales locating the evidence on a continuum of support between the two hypotheses stated in the LR (Martire et al., 2014). Following on from these studies, Thompson and Newman (2015) presented participants with moderate or very strong DNA or shoeprint evidence in a RMP, LR or verbally quantified format. Those in the RMP condition were able to discriminate between moderate and very strong evidence, however this effect was only found for shoeprint evidence in the numerically and verbally quantified LR conditions.

Nance and Morris (2002, 2005) considered LRs (e.g., 25 times more likely), frequencies (1 in 1000), and an LR plus a chart illustrating how LRs relate a posterior probability to a prior probability (as determined by Bayes' theorem). In the first paper (2002) jury pool members presented with DNA evidence in a sexual assault scenario assigned probabilities of guilt in the frequency format condition that were statistically smaller than the probabilities of guilt in the LR plus chart condition. Responses in the LR condition fell between the two and did not differ significantly from either. In the second paper, also concerned with DNA evidence and sexual assault, there was a significant association between presentation format and guilty verdicts when the RMP and laboratory error rate were aggregated. Where there was aggregation, guilty verdicts were more likely in the LR plus chart format (64%) than either the frequency (35%) or LR formats (41%). Where there was no aggregation there was no difference between presentation formats.

In sum, the evidence regarding the consistency of lay responses to statistically equivalent statements generally shows that changes in contextual frame and presentation format affect evaluations. In some cases, such as where probabilities and frequencies differ (Lindsey et al., 2003), this does not seem consistent with comprehension or understanding. However, to characterise the evidence more broadly as demonstrating *mis*understanding overstates the scope of numerical equivalence and oversimplifies nuance in the case scenarios.

For example, although it is true that a 1 in 100 frequency is *statistically* equivalent to an LR of 100, participants are provided with more than this bare numerical information. A likelihood ratio necessarily involves the statement of two hypothetical alternatives because it is the ratio of two probabilities – the probability of the observations under a prosecution-consistent narrative compared to some alternate state of affairs. Thus, the fact that participants are more likely to believe the defendant is guilty in a frequency format that a LR format (Koehler et al., 1995) or a LR plus chart (Nance and Morris, 2002) may be a logical consequence of the fact that decision-makers in the LR conditions are given an alternative possibility to consider (e.g., "that the defendant is not the source of the semen" Koehler et al., 1995) whereas those in the frequency formats are not ("we would expect to see the DNA profiles observed in the semen and the defendant approximately 1 person out of every 100"). In short, it would be inappropriate to contend that a disconnect between mathematical and psychological equivalence is necessarily problematic in scenarios of this type. Rather, given the experimental scenarios used in the reported studies, it may be a mistake in some cases to assume that there should be equivalence in the first place.

9.3 Sensitivity

Sensitivity is defined here, and by Thompson and Newman (2015), in terms of responsiveness to evidence strength. Holding all else constant, the ability to assign greater weight to evidence of objectively greater strength is an important indicator of understanding. Despite the simplicity and clear logical value of this measure, the evidence for lay sensitivity is somewhat mixed.

Three studies have failed to identify effects of evidence strength manipulations (Goodman, 1992; Koehler, 1996; Nance and Morris, 2005). Goodman (1992) made blood type evidence appear more or less incriminating by varying the frequency of occurrence (Exp. 1: 10%, 5%, 1%, or 0.1%). Neither conviction rates nor culpability estimates differed according to the strength of the evidence. Participants hearing the strongest evidence (0.1%) convicted 31% of the time and estimated culpability at 47% whereas those given the weakest evidence (10%) convicted 14% of the time and estimated culpability at 34%. The author concluded that the closely clustered scores did *"not reflect much sensitivity to the differences in the statistical probabilities presented"* (p. 372). Similarly, Koehler (1996) observed no significant effect of DNA profile frequency (1 in 100 vs. 1 in 1,000) on estimates of the probability of guilt. Likewise Nance and Morris (2005) were unable to identify an effect of laboratory error rate (unquantified, 1 in 1,000, or aggregated with RMP) on guilty verdicts or estimates of the probability of guilt.

A number of studies have produced mixed or qualified evidence for lay sensitivity to evidence strength (Faigman and Baglioni, 1988; Martire et al., 2013, 2014; Scurich and John, 2013; Thompson et al., 2013; Thompson and Newman, 2015). Scurich and John (2013) obtained mixed evidence for sensitivity. The varied laboratory error rate (1 in 10 vs. 1 in 1,000) did not have a significant effect on verdict outcomes in their examination of the possible role of DNA database type. However there was a significant main effect of error rate on the likelihood of guilt. As error rate increased the likelihood of guilt decreased.

Thompson et al. (2013) conducted two experiments where students (Exp. 1) and jury pool members (Exp. 2) were presented with DNA evidence that had a random match probability of 1 in 1 trillion and was accompanied by a false report probability of either: zero, 1-in-10,000 or 1-in-100 (Exp. 1). The false report probability had a significant effect on both guilty verdicts and estimates of the probability of guilt. Post-deliberation conviction rates in the 1-in-100 and 1-in-10,000 conditions did not differ, but conviction rates in zero error condition were higher than the 1-in-100 condition. The chances of guilt in the zero error condition were higher than those in the 1-in-10,000 condition. These in turn were higher than the 1-in-100 condition. Together these results strongly suggest that lay decision-makers are sensitive to reported variations in evidence strength.

However, the second study by Thompson et al. (2013) provided weaker evidence of sensitivity. There were no significant differences between the conviction rates or chances of guilt in the zero, 1-in-10,000 and 1-in-100 error conditions. Following on from this, Martire et al. (2013), Exp. 1 also found evidence for sensitivity to evidence strength, but did not detect a clear linear relationship between strength and perceived evidence weight. There was a main effect of evidence strength (weak, moderate or strong), but beliefs about the guilt of the suspect did not differ markedly between LRs that were moderately strong (450 times more likely) and very strong (495,000 times more likely). Similarly, Faigman and Baglioni (1988) varied the frequency of blood grouping evidence (Type "A" – 40%; "O" – 20%; and "AB" – 5%). In this case there was a significant interaction between blood type and perceived weight of evidence, whereby those participants in the "AB" condition assigned significantly more weight to the blood typing evidence than did those in the "A" or "B" conditions, showing some sensitivity to evidence strength because the "AB" evidence was more probative. However, the weighting assigned to the blood evidence in the latter two conditions did not differ.

Finally, somewhat mixed evidence was obtained by Thompson and Newman (2015). As described above there was a significant interaction between evidence strength (moderate: 1 in 100 vs. strong: 1 in 1 million) and presentation format (RMP, numerical or verbal LR) such that when evidence was presented in a RMP format, convictions were more likely in a strong compared to a moderate case. There was no association between strength and conviction rates in the LR or VE conditions. Strength of evidence also had a significant effect on probability of guilt estimates elicited using a 17-point log scale and the odds measure (where odds of guilt were elicited before and after the presentation of the forensic science evidence). However, where there was a significant main effect of strength on the log scale, there was a three-way interaction on the odds measure. Those who received DNA evidence were sensitive to evidence strength irrespective of presentation format while those who received shoeprint evidence were only sensitive to evidence strength in the RMP format.

The following studies provide support for lay sensitivity to evidence strength (Smith et al., 1996; Koehler, 2001; Kaasa et al., 2007; Martire et al., 2014; de Keijser et al., 2016). Firstly, as already described, Koehler (2001) observed a significant interaction between incident frequency, target and format. Where the incident rate was 1/1,000 format and target effects were as previously observed (Koehler, 2001), Exp. 1; however, these effects were reduced for the 1/1,000,000 condition and were absent for the 1/1,000,000,000 condition. This suggests that participants were differentially sensitive to each level of the strength manipulation. Martire et al. (2014) also found a significant main effect of evidence strength such that participants presented with larger (more incriminating) LR (5,500 times more likely) assigned significantly more weight to the evidence than those presented with the smaller LR (5.5 times more likely). There was also a significant interaction effect, which has been described earlier (see Section 9.2.2).

de Keijser et al. (2016) presented law students with three reports about the results of DNA mixture analyses in a robbery case. The three reports were based on the same DNA samples but varied in a number of respects according to how three different DNA scientists chose to report their conclusions in the case. Of all the reports the "Dirksen" report arguably provides the strongest evidence, in particular, because Dirksen provides a LR of 5.7 million times more likely if the suspect is a contributor, as opposed to not being a contributor, for the nail dirt sample. The "Ten Cate" and "De Boer" reports indicate the sample is either "insufficient" or that "the suspect cannot be excluded". When participants were asked to rate how incriminating each of the three reports were, there were statistically significant differences. The Dirksen report was considered moderately incriminating overall, where the Ten Cate and De Boer reports were respectively considered to be mildly and moderately exculpatory.

Smith et al. (1996) presented students and community members with blood enzyme and blood type evidence in a rape trial modelled after the facts in *State v Kim* (1987). Half of participants were told that the assailant and defendant shared a blood type found in 35% of the population and a PGM enzyme found in 20% of the population (strong). The other half were told that the type was shared by 35% but the enzyme was shared by 80% (weak). Participants were more likely to believe that the defendant was guilty in the strong condition than in the weak condition.

Kaasa et al. (2007) varied the diagnosticity of bullet lead evidence. All student participants were told that the hit rate for two bullets coming from the same compositionally indistinguishable volume of lead was 90% and that the false positive rate was about 1 in 500. Those in the "strong" condition were further told that a random selection of 20 bullets were taken from the defendants' box of ammunition and that all 20 of these bullets match the murder bullet. Those in the "worthless" condition were told that just 2 of the 20 bullets matched the murder bullet and that 10 out of 100 bullets obtained from the community also matched the murder bullet. Participants in the "unknown" condition were given no information about the diagnosticity of the match. Post-deliberation ratings of case strength, probability of guilt and conviction rates were all significantly higher in the strong condition than those in the worthless condition.

On balance, the evidence suggests that lay people do understand and are sensitive to evidence strength, but they are perhaps not always as sensitive as statistical norms would dictate. In some cases participants are insensitive to evidence strength, and the weights assigned do not always accord with the magnitude of the manipulations.

9.4 (In)Coherence

As it is used here the term "coherence" describes decision-making that is logically sound.* This definition excludes a range of potentially "incoherent" lay responses to statistical statements that are incompatible with genuine comprehension such as the Prosecutor's and Defense Attorney's Fallacies (e.g., Thompson and Schumann, 1987), directional errors (e.g., Martire et al., 2013), and aggregation errors (e.g., Koehler et al., 1995).

9.4.1 Prosecutor's Fallacy

Thompson and Schumann (1987) coined the term "prosecutor's fallacy" to describe the mistaken inference that the random match probability (e.g., 1 in 1 million) equates to the probability that the defendant is innocent. This mistake is also described as an error of the "transposed conditional" because those who subscribe to the fallacy are presented with an RMP that describes the probability (P) of a "match" (M) given that the person is innocent (I): $P(M|I)$); but are instead making inferences about the probability of innocence given the match: $P(I|M)$ (Kaye et al., 2007).

In their seminal paper Thompson and Schumann (1987) observed that 13.2% of participants were victims of the prosecutor's fallacy, and that the fallacy occurred more often when the RMP evidence was presented as a conditional probability ("two percent chance") as opposed to a percentage plus number ("2% of people, 20,000 in a city of 1,000,000"). The fallacy occurred less frequently in the second experiment (3%), however, 28.8% of participants labelled statements consistent with the fallacy as "correct". A source probability error (also originating from transposed conditionals) was also commonly endorsed (63.59%) by participants surveyed by Thompson and Newman (2015).

Since 1987 researchers have regularly observed instances of the fallacy, although the rates vary from study to study. Goodman (1992) identified the fallacy among 2% of respondents in Experiment 1 and 8% of respondents in Experiment 2. Smith et al. (1996) presented participants with the fallacy in closing arguments. However, only 7 of 189 people made judgements consistent with the fallacy.

Nance and Morris (2002) reported the fallacy in 6% of cases and concluded that compared to a frequency or a LR plus chart format, the LR on its own may be the format

---

* This definition differs from that of "logical coherence" adopted by Thompson and Newman (2015). We use the term to describe failures of logic (e.g., fallacies) whereas Thompson and Newman used the term to refer to consistency with Bayesian normative expectations.

that generates the most fallacious interpretations. The authors elicited prior and posterior beliefs in order to follow up on the seemingly problematic role of the LR format and the possibility of mis-estimating the frequency of errors. The incidence of fallacies in the modified LR and modified LR plus chart formats dropped from 12% to 3%. The rate of observed fallacies declined further in their subsequent paper (Nance and Morris, 2005) where just one response out of 1,205 responses (0.3%) was clearly fallacious.

Kaye et al. (2007) found that nearly half of their participants (48%) agreed with a statement consistent with a prosecutor's fallacy (that mtDNA evidence shows that there is about a 1% chance that someone other than the defendant committed the crime), however, this was thought to overestimate the true fallacy rate. The rate was revised to 16% after considering the responses in control conditions. Thompson revisited the issue in 2013 and observed that 14% of post-deliberation judgments were consistent with the fallacy. Moreover, the rate of error increased along with the strength of evidence (False report probability condition: zero–30%; 1 in 10,000%–19%; 1 in 100%–11%). The fallacy was not evident in the second study reported in the paper.

Ultimately, this overview suggests that lay people can and do commit prosecutor's fallacies in response to statistical statements about forensic science evidence. While this error is generally made by a small proportion of decision-makers it is important to consider where the burden of this error falls. The risks associated with the prosecutor's fallacy are borne by the defendant whose chance of innocence is incorrectly inferred from the chance of a coincidental match. This must be taken into consideration when determining how problematic this type of incoherence might be in practice.

9.4.2 Defense Attorney's Fallacy

The "defense attorney's fallacy" was also identified by Thompson and Schumann (1987). In this case the fallacy is to infer that a piece of evidence has no incriminating value if it is possible that more than just the suspect (person or object) could be the source. For example if the scientist reports that 2% of people in a population will have a particular characteristic, it is a mistake to infer that because there might be 20,000 such people in a city of 1,000,000 that the evidence is therefore worthless. Such a reaction is inappropriate because the evidence reduces the pool of possible suspects and, in conjunction with other evidence, may help to link a person or object with the offense (Thompson and Schumann, 1987). Even so, Thompson and colleagues have identified this form of fallacious reasoning on a number of occasions: Thompson and Schumann (1987) observed the fallacy in 12.5% and 66% of participants (in Experiments 1 and 2 respectively); Thompson and Newman (2015) also identified high prevalence rates; 55.8% in a shoeprint case and 42.2% in a scenario involving DNA profiling, which varied according to the strength of the evidence. When the evidence was moderately strong 61.4% of participants assigned little or no probative value whereas when the evidence was very strong only 35.4% took the same approach. Furthermore, the participants who endorsed only the defense attorney fallacy had the lowest conviction rates and assigned the lowest chances of guilt.

Other studies support the existence of the defense attorney's fallacy. Nance and Morris (2002) observed the fallacy in 8% of those presented with evidence in a modified LR format and 6.5% of those presented with a modified LR plus chart format. Kaye et al. (2007) found that only half of their participants rejected a claim from the defense that was consistent with the defense attorney's fallacy, while 40% agreed that the DNA profile in the case was worthless because people other than the defendant could have contributed. Overall, the defense attorney's fallacy appears to be more prevalent than the prosecutorial equivalent and has been associated with verdict level effects in case vignette studies. Evidence of fallacious reasoning may undermine suggestions that lay decision-makers successfully comprehend statistical statements.

9.4.3 Directional Errors

In this context the term "directional error" is used to describe situations where objectively incriminating evidence is treated as though it is exculpatory. A weak evidence effect is a particular type of directional error that occurs where the evidence in favour of a proposition is quite weak (Martire et al., 2013). Martire and colleagues (2013, 2014) observed weak evidence effects when participants were presented with a likelihood ratio about a shoeprint (2013) or fingerprint (2014) "match" that gave "weak or limited support" for a hypothesis consistent with the guilt of the accused. Given this evidence, a majority (62%) of participants considered the defendant less likely to be guilty than they had prior to hearing the evidence even though the evidence implicated the defendant. These results were replicated in their 2014 paper where 64% of participants receiving weakly incriminating evidence revised down their beliefs in the guilt of the defendant. The same downward revisions were not apparent to the same extent when the strength of the evidence was communicated numerically (2013: 13%; 2014: 13%) rather than using a verbal equivalent.

While weak evidence effects in particular (Fernbach et al., 2011) and more general directional effects (Lopes, 1987) have been observed fairly regularly in other contexts, these effects have been given relatively little attention in the context of statistical statements from forensic scientists. Smith et al. (1996) observed that a small number of participants reduced their guilt assessments in light of incriminating statistical statements and Thompson and Newman (2015) noted that a small number of participants (7.6% in the log scale condition and 6.4% in the odds conditions) treated incriminating evidence as exculpatory. Although there was no "weak" evidence in this study, the directional error was more likely to occur when the evidence was moderate (9.9%) rather than strong (3.7%). Overall, it is not yet clear how robust directional errors like the weak evidence effect are.

9.4.4 Aggregation Errors

Consideration has also been given to the lay ability to correctly combine and evaluate statistical elements of forensic science evidence such as the random match probability and laboratory error rates. This became particularly important as RMPs shrank rapidly in the DNA context (at that time 1 in 1 million or 1 billion) while the much larger and therefore limiting probability of a laboratory error rate was estimated at around 1 in 1,000 (Koehler et al., 1995). An example of an aggregation error is where participants fail to realize that when there are two possible outcomes with different chances of occurring, the larger probability defines how likely it is that *either* of those outcomes has taken place. For example, if a decision-maker is provided information about DNA evidence suggesting that: (1) the crime scene and suspect DNA samples would match 1 in 1,000 people in a specified population by coincidence alone (i.e., a RMP); and (2) that DNA analysts mistakenly declare a "match" (i.e., a laboratory error (LE)) in 1 in 100 examinations; then the chance that *either* a random match error or a laboratory error in this case is 1 in 100 because that is the most likely outcome. It is does not matter how unlikely the second outcome is – it could be 1 in 1 million or 1 in 1 billion and it wouldn't change the situation – in this case a laboratory error is more likely than a random match and it is that chance that should be considered when evaluating the possibility of an error in the case.

Koehler et al. (1995) completed the seminal studies in this area. In two experiments jury-eligible students (Exp. 1) and jury pool members (Exp. 2) were asked to reach verdicts in a murder case. The RMP and LE for the DNA match were presented either separately (e.g., Exp. 1: RMP 0.000000001, LE 0.02) or in an aggregated format (0.02). In the first experiment the introduction of a highly diagnostic LE had little impact on juror verdicts suggesting that participants were insensitive to the fact that the probability of error is set by the largest estimate. However, when the two probabilities were combined for participants, conviction rates roughly halved showing that decision-makers do not automatically aggregate. This effect of aggregation on verdicts was not evident in the second study where those in the LR and aggregated RMP plus LE conditions performed similarly.

The effect of aggregation was pursued by Schklar and Diamond (1999). Jury-eligible students were presented with DNA evidence in a sexual assault scenario. Participants either received: a single estimate of a LE or an RMP; two separate estimates (one with a large RMP or one with a large LR); two separate estimates plus and instruction about how to combine them (to reach "about 2 in 100"); or a single combined estimates (a 2 in 100 chance of either error occurring). The provision of an aggregation instruction did not affect guilty verdicts (84%) when compared to those provided with two separate estimates (88%). Although not discussed, the conviction rate in the single combined estimate condition was approximately 60%. If the difference is statistically significant it would suggest that an aggregated format does alter perceptions and that lay people are not able to correctly combine RMP and LEs either aided or unaided. Finally, as described previously, Nance and Morris (2005) observed a significant interaction between aggregated RMP and LE and presentation format. This further suggests that the aggregation of statistical estimates made the differences in presentation format more salient to decision-makers.

Aggregation errors, directional errors, prosecutor's and defense attorney's fallacies all run counter to sound logical interpretations of statistical statements. These reasoning errors must therefore threaten lay understanding and comprehension, but it is unclear how often and by how much. On one hand the evidence quite strongly suggests that lay decision-makers have trouble appropriately aggregating disparate probability estimates, and suggests that the defense attorney's fallacy might be both prevalent and important. This is a cause for concern. On the other hand, the evidence for directional errors and the prosecutor's fallacies is more varied and remains open to interpretation. However, it must be acknowledged that the risks associated with the prosecutor's fallacy fall to the defendant and therefore may be more problematic than those associated with the defense attorney's fallacy – which fall to society. We do have some clues about how these failures of coherence might be avoided, for example through the use of numerical rather than verbal quantifications of evidence strength (Martire et al., 2013, 2014), or providing information about the application of Bayes' theorem (Smith et al. 1996; Nance and Morris, 2002). However, there is a need for more research focused on the coherence of lay responses to uncertain evidence.

9.5 Ability

As mentioned above, manipulation checks and recall or restatement of the testimony provided by the forensic scientist do not constitute ability as we have defined it. Here we are interested in whether there is evidence that decision-makers can take the information provided and complete some mathematical or conceptual transformation. That is, can they successfully apply or adapt what they have been told? Ability has been examined relatively infrequently in the literature.

Both Koehler (2001) and McQuiston-Surrett and Saks (2009) were interested in whether lay people presented with a RMP pertaining to either DNA or microscopic hair comparison, could use that information to calculate the number of people not implicated – in a city of 500,000 people – who would give a match with the recovered evidence. Koehler (2001) Exp. 2 reported that 58.1% of participants were able to calculate the right answer (500 people) given the RMP. Jury-eligible students presented with the RMP in a frequency format were significantly more likely to produce the correct answer than those presented with a probability format.

McQuiston-Surrett and Saks (2009), Exp. 1 observed a similar level of accuracy among the judges and venire persons. Approximately half of the participants provided the correct answer. Again, performance varied according to the presentation format. Participants who were presented with a subjective probability, or an objective multiple-frequency were significantly more likely to answer correctly (41% and 46.5% respectively) than those in the single probability condition (25.3%). However, this is unsurprising given that the calculation in the single probability condition was objectively more difficult.

A similar approach was taken by Goodman (1992), Exp. 1. Student participants were presented with a base rate and were asked to compute the odds that the blood samples gathered at the scene of the crime would match the defendant's blood type. Participants were given a choice of 10 possible alternatives. Only 27% selected the correct option given the evidence presented.

Lindsey et al. (2003) asked law students and jurist to combine the population frequency information with the laboratory error rate to derive the conditional probability that someone would be the source of the incriminating DNA evidence. While only a handful of participants were able to reach the correct answer when the RMP and LE were presented in a probability format, nearly a half (40%–50%) of the students and roughly three-quarters of jurists (70%–75%) answered correctly when the information was presented in a natural frequency format.

Finally, Kaye et al. (2007) sought to explore conceptual rather than mathematical comprehension. Firstly, jury pool members who were presented with a RMP and an exclusion probability were asked to reconcile competing arguments about the role of heteroplasmy in mtDNA evidence. After deliberation, roughly two-thirds of participants were correctly able to indicate that it was necessary to consider heteroplasmy even though the defendant was not heteroplasmic because of the way the FBI defines an "exclusion". Participants were also asked to agree or disagree with the statement that "the mtDNA evidence in this case excludes at least 99% of the population as the source of the hairs" (p. 809). This fact had to be inferred from the forensic science testimony, and roughly two-thirds of participants (69%) agreed.

Other studies have either labelled recall questions as measures of understanding (Thompson et al., 2013) or vaguely referred to measures of understanding without providing enough information to establish whether any demonstration of ability was required (Faigman and Baglioni, 1988). Therefore, given the available evidence about demonstrated ability as an indicator of comprehension, the picture is somewhat mixed. Response accuracy has ranged from 27% (Goodman, 1992) to approximately 75% depending on the type of respondent and the question asked. In most cases performance was far from optimal, clustering around 50% accuracy. Performance also seems to improve somewhat with frequency formats as compared to probabilities, particularly where mathematical calculations are required. However, there may be a mediating role for individual mathematical competence which may constrain demonstrations of ability.

9.6 Orthodoxy

Normative expectations obtained through the application of Bayes' theorem provide a standard for assessing comprehension of statistical statements (Nance and Morris, 2002). Bayes' rule provides a formula for combining initial (prior) beliefs about a proposition with new evidence in order to reach a final updated belief (posterior). Within this framework the statistical statements provided by forensic scientists are the evidence lay decision-makers must use to update their beliefs either about the probability that the defendant is the source of the questioned sample (e.g., the fingerprint or blood sample), or about the probability that the defendant is guilty as accused. While few people suggest that the psychological processes involved in belief updating are accurately described by Bayes' theorem (Kahneman and Tversky, 1973) there is general acceptance that the theorem can be used to set an appropriate norm (Thompson and Newman, 2015). We have used the term Orthodoxy to describe lay belief updating that conforms with these normative expectations. Thompson and Newman (2015) called this "logical coherence".

Comparisons between lay beliefs and Bayesian norms initially suggested poor correspondence and therefore poor or limited comprehension. Thompson and Schumann (1987) observed that participants tended to be conservative and underused associative evidence in two experiments. For example in Experiment 2, where the Bayesian posterior estimate for probable guilt was 0.92, participants generated mean posteriors of either 0.28 or 0.43 depending on whether participants falling victim to the defense attorney's fallacy were included in the analysis or not (respectively). This result was replicated by Goodman (1992) and Faigman and Baglioni (1988) who both observed underuse of statistical information compared to Bayesian norms.

Smith et al. (1996) were the first to focus on Bayesian orthodoxy in earnest. Not only were participants in some conditions provided with an instruction about how to update beliefs in line with Bayes' theorem, individual beliefs were extensively documented and explored. Overall, there was a slight tendency to underuse the statistical evidence (by 3 percentage points). However there was substantial variability at the individual level. In particular, when participants were presented with more incriminating enzyme evidence, those with very low prior beliefs slightly overused the evidence and as priors increased overuse tended to decrease.

However, the authors suggested that undervaluing is more likely to occur than overvaluing for several reasons that are not necessarily indicative of misunderstanding. Firstly, the case scenarios chosen by researchers often involve either high prior beliefs (i.e., a strong case) or powerful statistical evidence, or both. Under these circumstances Bayesian expectations will be high and therefore difficult to exceed. Secondly, participants with high priors or those receiving strong evidence will be required to provide extreme probability estimates (over 99%), but ceiling effects may prevent them from doing so. Thirdly, the models for deriving normative expectations may be overly simplistic and fail to take into account relevant factors that might rationally reduce the perceived strength of the evidence (operationalised via the likelihood ratio); for example the false report or random match probabilities. Unless these things are also included in the model, Bayesian expectations will be inappropriately inflated. Consequently they argued that it was important for researchers to consider weaker cases or evidence as well as more sophisticated normative models if they hope to use Bayesian orthodoxy as a meaningful indicator of comprehension.

Schklar and Diamond (1999) seemingly responded to the challenge and included participants' expectancies and beliefs about intentional tampering into their Bayesian normative estimates (reducing the likelihood ratio for the scientific evidence). By refining the model in this way the gap between lay and normative guilt beliefs was reduced, but undervaluing was still evident. Nance and Morris (2002, 2005) also explored more sophisticated models by asking participants to provide estimates of the probability of a false report, a random match and any other reason consistent with the innocence of the defendant. In both papers two models were constructed. The first assumed that the false report and random match probabilities provided by the scientist were believed by the participants, the second model used participants own beliefs about these possibilities to moderate the strength of the evidence. Irrespective of which model was used, participants undervalued the scientific evidence. Responses did approach Bayesian norms when participants using a probability of 0.5 to express uncertainty were removed from the sample, but only in one condition (Nance and Morris, 2005). In general, across both papers, Bayesian orthodoxy increased as decision-makers were provided with more information about the meaning of the DNA match. Frequency formats resulted in the least orthodox results, while the likelihood ratio plus explanatory chart resulted in the most orthodox.

Subsequent studies by Martire and colleagues (2013, 2014) addressed concerns about potential ceiling effects and evidence strength. Participants were presented with evidence of varying strength (e.g., from "weak or limited"/LR 4.5 to "very strong"/LR 495000; 2013) and were asked to provide prior and posterior beliefs in the defendants' guilt using an odds elicitation method. Importantly unlike probabilities, which are bounded at 0 and 1, odds are unbounded and a participant is free to use any positive number to express how many times more likely they think it is that the defendant is guilty than not guilty (or vice versa). Even so, the results were consistent with earlier research. Participants showed extreme undervaluing of the scientific evidence in both studies, irrespective of evidence strength.

Thompson and colleagues (2013, 2015) used a log interval response scale to make it easy for participants to express extreme values without encountering ceiling effects. When using this measure to elicit prior beliefs (from the control condition) and posterior beliefs (from the experimental conditions) responses "were generally consistent" with Bayesian expectations (Thompson et al., 2013, p. 388). Indeed, many participants *overvalued* the evidence, and only those with low prior beliefs undervalued the evidence.

Thompson and Newman (2015) built on this contribution by comparing their log-scale with the odds elicitation method used by Martire et al. (2013, 2014), and by using a Bayesian network incorporating expectancies of coincidental match, false report match due to laboratory error and the possibility of a frame up. The log-scale and odds elicitation methods resulted in very different performance compared to Bayesian norms. Where the odds method again resulted in extreme undervaluing, the log-scale suggested that participants underused shoeprint evidence but not DNA evidence.

Overall, while researchers have regularly found evidence suggesting that lay decision-makers undervalue statistical statements when compared to Bayesian norms, we would argue that the evidence is more mixed than generally thought. Not only it is difficult to build normative models that incorporate all of the relevant factors a diligent reasoner might take into account when evaluating statistical evidence, it is also difficult to access these beliefs without distorting them in the process – as illustrated by the differences between responses on the log scale as compared to an odds elicitation method. Under these circumstances it is more likely that a participant will undervalue than overvalue the statistical evidence and it is not yet clear how much can be attributed to imperfect research methodologies, and how much of it is due to genuine lay misunderstandings of the evidence.

9.7 Discussion

The evidence regarding lay comprehension of statistical statements is mixed, however, perhaps not as damning as generally thought. There is undoubtedly evidence of weaknesses and limitations but there are also positive signs. Overall, consistency is lacking but this is not clearly a negative sign since the assumption of psychological equivalence may be unfounded in some instances. Absence of Bayesian orthodoxy where previously thought to be unassailable evidence of misunderstanding is now more questionable given an increasing appreciation for the nuances and challenges associated with eliciting and appropriately modelling beliefs and belief updating. Sensitivity to evidence strength is often apparent, although not always commensurate with evidence strength. Failures of coherence are regularly observed but when they do occur they tend to threaten the ability of the State to convict, rather than the defendant's right to be presumed innocent. Ability is largely untested but seems to vary considerably from study to study, and perhaps with individual differences. All this begs the question, what can we say about lay comprehension of statistical statements in the forensic sciences?

To answer this question it is worth first considering whether we have ever *explicitly* assessed lay *comprehension* of statistical statements. Despite the substantial number of studies, we have seen that comprehension can be operationalised in many different ways. Yet, on their own, none of these indicia adequately capture what it means to comprehend a statistical statement. Being able to update one's beliefs in an orthodox manner does not entail the ability to apply statistical information for the purpose of deriving new information. Being sensitive to evidence strength does not prevent someone from adopting a prosecutor's or defense attorney's fallacy (i.e., (in)coherence). And, treating equal evidence equally (i.e., consistency) does not mean that you will do so in line with normative expectations.

Instead, given the range of indicators measured in the literature, comprehension should arguably be conceived of as the sum of numerous parts. It is a multidimensional construct comprising a range of behaviours. The more of these behaviours an individual displays, the more confident we can be that they have understood the *meaning* of a statistical statement. Yet there may be a great deal more to the construct of comprehension than what has previously been measured. Moreover, our research has not considered individual performance against multiple dimensions of comprehension. To date, we have not calculated an individual's "score" or summed performance across the CASOC indicators of comprehension. As a consequence, we know almost nothing about either the extent to which individual decision-makers broadly understand statistical statements (e.g., the aggregation of performance across a range of comprehension indicators), or how this multidimensional construct of comprehension might be optimised or impaired by the use of different forms of statistical statements. We have also given limited attention to how comprehension might be augmented or impaired by processes of deliberation.

A second issue is whether comprehension as defined here and in the literature aligns well with legal considerations. When framing their studies about lay understanding of statistical statements researchers regularly report motivation to facilitate communication in trials and improve the fairness of justice procedures (Faigman and Baglioni, 1988; Kaye and Koehler, 1991; Goodman, 1992; Smith et al., 1996; Koehler, 2001, 2012; Nance and Morris, 2002, 2005; Scurich and John, 2013; Thompson et al., 2013). But it is not clear that observed or potential CASOC failures are necessarily problematic from a legal perspective. Indeed, notwithstanding the increasing prevalence of statistical evidence, particularly in relation to DNA profiles, there is surprisingly little jurisprudence on the comprehension (by jurors) of statistical evidence in criminal proceedings.

In general, courts tend to be disinterested in precise statistical formulations and actual demonstrations of competence from jurors. With the appearance of DNA evidence, a few courts initially excluded or qualified the admission of statistical evidence, partly due to concerns about comprehensibility (Aronson, 2007). As procedures and standards for reporting have evolved, concerns about lay comprehension in criminal jurisdiction have largely subsided. Now, apart from the validity and reliability of new DNA procedures (such as the algorithms used to evaluate mixed DNA samples), evaluations of the admissibility and unfair prejudice associated with forensic science evidence in criminal trials rarely incorporate comprehension concerns. Instead, courts have drawn on judicial anxieties and impressions to determine how statistical statements will be regarded and managed, rather than scientific evaluation (Cole, 2011). This has led to a range of responses that are inconsistent across jurisdictions and unprincipled in form. For example the High Court of Australia has gone so far as to conclude that the prosecutor could use any of a range of statistical expressions, provided they were mathematically equivalent (*Aytugrul v the Queen*, 2012) – despite scientific research suggesting that decision-makers may not treat them equally.

Other courts have embraced scientific advice and require that DNA profiling evidence be expressed as a RMP or likelihood ratio (National Research Council, 1996; Kaye, 2010), whereas for other procedures – often relying on relatively weaker scientific foundations (e.g., latent fingerprints and ballistics) – the same critical scholarship has often been overlooked. While these responses may not be representative of civil litigation, where the resources available to corporate defendants (and some plaintiffs) are often substantial, it is fair to say that courts are not particularly concerned with the comprehensibility of statistical statements. Perhaps the main exceptions involve causation evidence in mass torts and product liability litigation, and suits involving discrimination and economic evidence (Cranor, 2016).

As for adopting an orthodox or statistically normative approach, courts and some commentators have questioned the necessary application to criminal proof. In *R v Adams* (1996, 1998), the English Court of Appeal was concerned about the ability of the jury to apply Bayesian belief updating and ultimately the defense was prevented from presenting a Bayesian analysis of the evidence in the case (Lynch et al., 2010). For the court, the jury were not required to generate explicit (i.e., mathematical) values for a range of events and possibilities or combine them according to Bayesian rules. Thus some courts are actively opposed to any formalised expectation of orthodoxy.

Indeed, some commentators (e.g., Allen, 1996; Ligertwood and Edmond, 2012) have questioned whether Bayesian approaches clearly articulate with the accusatorial criminal trial, where the burden of proof rests on the state and the standard is beyond reasonable doubt. These approaches draw on legal principle, before even considering lay ability to respond to complex and technical forms of evidence. Even if a Bayesian approach were the only logical way to update and assess evidence, in the end it is not clear that legal institutions would require its use or that it would be workable in current criminal justice systems, especially those retaining lay jurors. *Legal orthodoxy* does not encourage, and certainly does not expect, judges and juries to adhere to the normative expectations held by statisticians, particularly the updating of beliefs in a manner consistent with the application of Bayes' theorem.

Although perhaps surprising, this disinterestedness in lay comprehension of statistical statements is not irrational from a legal perspective.Beliefs about the protective power of the fair trial arguably focus critical evaluation on trial procedures and away from the individual case-specific performances of decision-makers. According to common law traditions and accusatorial trial values it is the responsibility of parties to adduce relevant evidence, present it in a manner that can reasonably be expected to be understood, and to effectively test opposing perspectives through cross examination and rebuttal witnesses (e.g., "*Daubert v. Merrel Dow Pharmaceuticals, Inc.,*" 1993). Beyond this, risks or limitations that might arise from complex statistical statements can be managed through admissibility rules (e.g., The U.S. Federal Rule of Evidence 702), judicial discretions (e.g., The U.S. Federal Rule of Evidence 403 and the *English Police and Criminal Evidence Act* (1984) s 78), opening and closing addresses, and judicial instructions. Thus, where all the parties involved are playing their part to ensure fair trial procedures are followed, lay decision-makers are trusted to comprehend and respond appropriately. This is despite the fact that many of the trial safeguards have proven ineffective in restricting the admission and reliance on expert evidence that is not demonstrably reliable (McQuiston-Surrett and Saks, 2009; Edmond and Roberts, 2011; Edmond and San Roque, 2012; Edmond, 2015).

Furthermore, demonstrations of lay comprehension are also very difficult to elicit and observe in real world legal contexts. When judges make decisions there is an expectation that they will articulate their reasoning process, thereby allowing some assessment of their comprehension. Although, even in this circumstance, engagement with evidence and signs of comprehension are frequently complicated by interacting strands of evidence, standards of proof, and other legal issues. In contrast to those of judges, the decisions of jurors are generally inscrutable; particularly outside the United States where it is often an offense to disclose jury discussions. Jurors are not required to divulge their deliberative processes or to state the reasons for their verdict. Consequently, scope for the supervision and evaluation of the reasoning and decision-making of jurors is seriously constrained. This makes it easier for courts to be disinterested in the quality or characteristics of evidence evaluation, than to focus on its weaknesses and possible avenues for improvement.

Moreover, even where some insight into jury reasoning is available, miscomprehension can be difficult to identify. Trials are unquestionably complex decision-making environments. Where experimental examinations of lay responses to statistical statements are constructed to detect causal relations by reducing and restricting complexity, matters that actually come to trial tend to be among the most unclear and complex. In most cases responses to statistical evidence are confounded by the interdependence of evidence, the dynamics of adversarial procedures, the effects of (in)competent cross examination, post hoc summaries of evidence, repetition and group deliberation processes. Each of these trial elements may dilute or distort statistical statements and compound the difficulties associated with evidence evaluation and integration. This in turn makes it very difficult to detect instances of evidence misuse, misunderstanding or irrationality. These same considerations pose challenges for the realistic specification of Bayesian normative models and expectations. This means that appropriate normative expectations are difficult to set and decision-making problems are difficult to identify.

These difficulties detecting comprehension are further exacerbated by trial selection and attrition. Pre-trial filtering (e.g., deciding whether a prosecution has a reasonable chance of success and is in the public interest) means that, in the vast majority of cases, almost any verdict (i.e., guilty, not guilty, hung, not proven in Scotland) might appear or be defended as reasonable given the evidence presented (Jasanoff, 1998, 2002). On rare occasions a verdict might appear inconsistent with the evidence, or some combination of verdicts against one or more defendants might seem to be inconsistent. This might lead to judicial intervention, but this is the exception and not the rule. Only a tiny proportion of cases are reversed on the basis of jurors misusing expert evidence (Medwed, 2012).

Ultimately, courts can be disinterested in comprehension because it is difficult to observe and is presumed where fair trial procedures have been followed. Furthermore, however it is defined from a legal perspective, comprehension only rises to notice when *mis*comprehension threatens to undermine the rationality of decision-making (Gans, 2017). Jury rationality and the rationality attributed to their decisions is a longstanding legal presupposition with foundations in the common law before scientific and technical forms of evidence (including complex statistical statements) were admitted. While indicia of comprehension (i.e., CASOC) are arguably consistent with rationality, the rational assessment of statistically complex evidence requires more than just consistency, sensitivity, coherence, ability and orthodoxy.

A basic model of rationality requires that decision-makers are at least given access to information about the validity and reliability of a technique as well as the proficiency of the analyst who applied it (Edmond, 2015). Each of the elements of rationality may need to be expressed using some form of statistical statement, for example the true positive rate (validity), a random match probability (reliability) or a laboratory error rate (proficiency). Consequently the rational evaluation of evidence will require the integration of multiple statistical statements to form a broad impression of the legal construct reliability or *trustworthiness* of the field, the technique and the practitioner (Martire et al., 2017). Although this basic model seems both sensible and consistent with scientific assessments of trustworthiness, legal models of reliability also include a range of additional factors, which may or may not contribute to sound evaluations of probative value.

It is a central function of the jury to determine the probative value of a piece of evidence. Moreover, it is legally uncontroversial for this assessment to incorporate heuristic perceptions of witness credibility, demeanor, experience, and ability to withstand cross examination or rebuttal evidence (*Daubert*, 1993). Although these latter cues are not necessarily good indicators of the scientific reliability of a piece of evidence, courts do not distinguish between potentially invalid assessments of the expert based on heuristics (e.g., Rachlinski, 2000; Saks and Kidd, 1980) and more valid assessments of the evidence based on studies and associated statistics. Indeed, in many cases confidence in the experience of the witness, long use of a technique and/or legal recognition (in the guise of authoritative precedent) might transcend the lack of validation. While in *Daubert*, the US Supreme Court tried to enforce a distinction between these two types of indicators and place emphasis on scientifically accepted indicators of reliability (e.g., peer review, falsifiability and error rate) this has not changed the fact that legal perceptions of reliability are still based on impressions, historically accepted practices, previous admissions, confidence in individuals and their institutions, and safeguards (Edmond et al., 2013). Thus, if a juror or judge does not consider the expert to be trustworthy for almost any reason, they are not required to rely on their evidence when forming a verdict, irrespective of the strength of the accompanying statistical statements. They are permitted to simply ignore it.

The broad conceptualization of reliability adopted in law contrasts with the narrow approach adopted in cognitive experimental research. At times, researchers have employed relatively sophisticated Bayesian models that vary evidence strength while also attempting to account for broader beliefs that might affect evidence evaluation (e.g., by including estimates of intentional tampering or laboratory error rate, etc., see Schklar and Diamond, 1999; Nance and Morris, 2002, 2005; Thompson and Newman, 2015). But most often experimenters have constructed scenarios where the statistical statement of the random match probability, laboratory error rate or likelihood ratio are meant to completely encapsulate the reliability of the evidence. When reliability is defined narrowly in this way the failure to assign more weight to evidence with a larger likelihood ratio is taken as evidence of miscomprehension. The same can be said of the failure to update beliefs in an orthodox manner, or failing to update ones beliefs at all. Yet from a legal perspective, if these failures are accompanied by an evaluation of the evidence or the expert as unreliable, or less reliable than another piece of evidence or expert, then these failures do not indicate miscomprehension or misunderstanding. They might be considered rational responses to unreliable evidence.

For example, giving more weight to DNA evidence than shoeprint evidence of the same statistical value may not be an indicator of misunderstanding of a statistical statement. Rather it may reflect differences in the perceived trustworthiness of two different types of evidence that is not otherwise captured (or not perceived to be captured) in the bare statistical statements of evidence strength (Martire et al., 2017). Under these circumstances it may be rational to attribute less weight to shoe print than DNA evidence despite equivalent likelihood ratios. Attributing different weights to the evidence may not signal misunderstanding. If decision-makers do not attribute twice as much weight to evidence of double the strength because of general skepticism about expertise or forensic science (or even the credibility of a witness or the trustworthiness of investigative institutions), this might be defended as rational. Similarly, if decision-makers treat inculpatory evidence as exculpatory evidence (i.e., a directional error) because they did not receive enough evidence to convict, and so tend to believe the evidence supports innocence rather than guilt, this might also be rational, or defensible in terms of social and institutional values. To date our experimental approaches have not really been attuned to these possibilities.

If reliability is not measured or manipulated, the indicators of comprehension we have examined might not be as meaningful as hoped. If the perceived reliability of the evidence is undermined by reasonable beliefs and inference, and this influences performance on the CASOC measures, our understanding of comprehension is potentially confounded. From a legal perspective, a study that does not allow for decision-making to be influenced by broad assessments of reliability, that might extend to witness credibility, demeanour and performance in cross examination, falls short. Consistency, sensitivity, coherence, ability and orthodoxy do not matter, and would not be expected if the evidence is primarily understood to be unreliable. Thus it may be more important to understand how reliability and comprehension interact. When evidence is seen as reliable, do jurors show comprehension of statistical statements? Are some forms of expressing statistical statements better than others for facilitating this? The evidence we have thus far suggests that the orthodoxy of belief updating is improved when broader perceptions of reliability are considered (e.g., Schklar and Diamond, 1999; Nance and Morris, 2002, 2005; Thompson and Newman, 2015). However we need more research designed to explore how lay people evaluate the *expert* as well as the statistical statements they provide if we hope to better understand comprehension in the context of legal reliability assessments.

9.8 Conclusion

Ultimately, the question of whether lay decision-makers appropriately comprehend statistical statements from forensic scientists is an important one, but its answer remains elusive. Irrespective of legal interest in empirically established understanding, cognitive and forensic scientists interested in just trial outcomes can and should continue collaborating to optimize the communication of uncertainty in the high-stakes forensic decision-making environment. The evidence gathered is valuable for fueling discussion and debate surrounding justice reforms and is necessary for improved communication between practitioners and decision makers. Knowledge means little if it is unable to be shared (Howes, 2015).

The preceding review also provides tantalizing glimpses of productive avenues for future research, for example examining comprehension as a multidimensional construct likely to vary based on individual differences. Even though our research has tended to adopt a limited conceptualization of reliability, this is something we can change so that we can better differentiate legally-orthodox disregard from undesirable miscomprehension. By using more ecologically valid and complex scenarios that capture evaluations of the expert, their field and their stated opinion, we can better understand the relationship between reliability and comprehension. These insights have the potential to improve the communication of statistical statements and contribute positively to the administration of justice and fair trial outcomes.

References

Allen, R. J. Rationality, algorithms and juridical proof: A preliminary inquiry. *International Journal of Evidence and Proof*, **1**, 254, 1996.

Aronson, J. *Genetic Witness: Science, Law, and Controversy in the Making of DNA Profiling*. Rutgers University Press, 2007.

Association of Forensic Science Providers. Standards for the formulation of evaluative forensic science expert opinion. *Science and Justice*, **49**, 161–164, 2009.

Aytugrul v The Queen (2012) 247 CLR 170 (Australia).

Bieber, F. R., Buckleton, J. S., Budowle, B., Butler, J. M., and Coble, M. D. Evaluation of forensic DNA mixture evidence: Protocol for evaluation, interpretation, and statistical calculations using the combined probability of inclusion. *BMC Genetics*, **17**(1), 125–140, 2016.

Cole, S. A. Splitting Hairs-Evaluating Split Testimony as an Approach to the Problem of Forensic Expert Evidence. *Sydney Law Review*, **33**, 459, 2011.

Collins, H., and Evans, R. *Rethinking Expertise*. University of Chicago Press, Chicago and London, 2007.

Cranor, C. F. *Toxic Torts: Science, Law, and the Possibility of Justice* (2nd ed.). Cambridge University Press, Cambridge, 2016.

Daubert v. Merrel Dow Pharmaceuticals, Inc., 509 US 579 (1993) (United States).

de Keijser, J. W., Malsch, M., Luining, E. T., Kranenbarg, M. W., and Lenssen, D. J. Differential reporting of mixed DNA profiles and its impact on jurists' evaluation of evidence. An international analysis. *Forensic Science International: Genetics*, **23**, 71–82, 2016.

Edmond, G. Forensic science evidence and the conditions for rational (jury) evaluation. *Melbourne University Law Review*, **39**(1), 77–121, 2015.

Edmond, G., Cole, S., Cunliffe, E., and Roberts, A. Admissibility compared: The reception of incriminating expert evidence (ie, forensic science) in four adversarial jurisdictions. *University of Denver Criminal Law Review*, **3**, 31, 2013.

Edmond, G., Found, B., Martire, K., Ballantyne, K., Hamer, D., Searston, R., and San Roque, M. Model forensic science. *Australian Journal of Forensic Sciences*, **48**(5), 496–537, 2016.

Edmond, G., and Roberts, A. Procedural fairness, the criminal trial and forensic science and medicine. *Sydney Law Review*, **33**, 359, 2011.

Edmond, G., and San Roque, M. The cool crucible: Forensic science and the frailty of the criminal trial. *Current Issues in Criminal Justice*, **24**, 51, 2012.

Eggleston, R. *Evidence, Proof and Probability* Vol. 2: Weidenfeld and Nicolson, 1978.

Faigman, D. L., and Baglioni, A. Bayes' theorem in the trial process: Instructing jurors on the value of statistical evidence. *Law and Human Behavior*, **12**(1), 1–17, 1988.

Fernbach, P. M., Darlow, A., and Sloman, S. A. Asymmetries in predictive and diagnostic reasoning. *Journal of Experimental Psychology: General*, **140**(2), 168, 2011.

Finkelstein, M. O., and Fairley, W. B. A Bayesian approach to identification evidence. *Harvard Law Review*, **83**, 489–517, 1970.

Gans, J. *The Ouija Board Jurors: Mystery, Mischief and Misery in the Jury System*. Waterside Press, 2017.

Gigerenzer, G., and Gaissmaier, W. Heuristic decision making. *Annual Review of Psychology*, **62**, 451–482, 2011.

Goodman, J. Jurors' comprehension and assessment of probabilistic evidence. *American Journal of Trial Advocacy*, **16**, 361–390, 1992.

Hand, L. Historical and practical considerations regarding expert testimony. *Harvard Law Review*, **15**, 40–58, 1901.

Holmes, O. W. The path of the law. *Harvard Law Review*, **110**(5), 991–1009, 1897.

Howes, L. M. The communication of forensic science in the criminal justice system: A review of theory and proposed directions for research. *Science and Justice*, **55**(2), 145–154, 2015.

Jackson, G. Understanding forensic science opinions. In Fraser, J. and Williams, R. (eds), *Handbook of Forensic Science*. Willan Publishing Ltd, pp. 419–455, 2009.

Jackson, G., Jones, S., Booth, G., Champod, C., and Evett, I. The nature of forensic science opinion—a possible framework to guide thinking and practice in investigation and in court proceedings. *Science and Justice*, **46**(1), 33–44, 2006.

Jasanoff, S. The eye of everyman: Witnessing DNA in the Simpson trial. *Social Studies of Science*, **28**(5–6), 713–740, 1998.

Jasanoff, S. Science and the statistical victim: Modernizing knowledge in breast implant litigation. *Social Studies of Science*, **32**(1), 37–69, 2002.

Kaasa, S. O., Peterson, T., Morris, E. K., and Thompson, W. C. Statistical inference and forensic evidence: Evaluating a bullet lead match. *Law and Human Behavior*, **31**(5), 433–447, 2007.

Kahneman, D., and Tversky, A. On the psychology of prediction. *Psychological Review*, **80**(4), 237, 1973.

Kaye, D. H. *The Double Helix and the Law of Evidence*. Harvard University Press, 2010.

Kaye, D. H., Hans, V. P., Dann, B. M., Farley, E., and Albertson, S. Statistics in the jury box: How jurors respond to Mitochondrial DNA match probabilities. *Journal of Empirical Legal Studies*, **4**(4), 797–834, 2007.

Kaye, D. H., and Koehler, J. J. Can jurors understand probabilistic evidence? *Journal of the Royal Statistical Society. Series A (Statistics in Society)*, **154**, 75–81, 1991.

Kingswell v The Queen (1985) 159 CLR 264 (Australia).

Koehler, J. J. On conveying the probative value of DNA evidence: Frequencies, likelihood ratios, and error rates. *University of Colorado Law Review*, **67**, 859–886, 1996.

Koehler, J. J. When are people persuaded by DNA match statistics? *Law and Human Behavior*, **25**(5), 493–513, 2001.

Koehler, J. J. Linguistic confusion in court: Evidence from the forensic sciences. *Journal of Law and Policy*, **21**, 515, 2012.

Koehler, J. J., Chia, A., and Lindsey, S. The random match probability in DNA evidence: Irrelevant and prejudicial. *Juriemetrics*, **35**(2), 201–219, 1995.

Ligertwood, A., and Edmond, G. Expressing evaluative forensic science opinions in a court of law. *Law, Probability and Risk*, **11**(4), 289–302, 2012.

Lindsey, S., Hertwig, R., and Gigerenzer, G. Communicating statistical DNA evidence. *Jurimetrics*, **43**, 147–163, 2003.

Lopes, L. L. Procedural debiasing. *Acta Psychologica*, **64**(2), 167–185, 1987.

Lynch, M., Cole, S. A., McNally, R., and Jordan, K. *Truth machine: The Contentious History of DNA Fingerprinting.* University of Chicago Press, 2010.

Martire, K., Edmond, G., Navarro, D. J., and Newell, B. R. On the likelihood of "encapsulating all uncertainty". *Science and Justice*, **57**(1), 76–79, 2017.

Martire, K., Kemp, R., Sayle, M., and Newell, B. On the interpretation of likelihood ratios in forensic science evidence: Presentation formats and the weak evidence effect. *Forensic Science International*, **240**, 61–68, 2014.

Martire, K., Kemp, R., Watkins, I., Sayle, M., and Newell, B. The expression and interpretation of uncertain forensic science evidence: Verbal equivalence, evidence strength, and the weak evidence effect. *Law and Human Behavior*, **37**(3), 197, 2013.

McQuiston-Surrett, D., and Saks, M. J. The testimony of forensic identification science: What expert witnesses say and what factfinders hear. *Law and Human Behavior*, **33**(5), 436–453, 2009.

Medwed, D. S. *Prosecution Complex: America's Race to Convict and its Impact on the Innocent.* NYU Press, 2012.

Nance, D. A., and Morris, S. B. An empirical assessment of presentation formats for trace evidence with a relatively large and quantifiable random match probability. *Jurimetrics*, **42**, 403–448, 2002.

Nance, D. A., and Morris, S. B. Juror understanding of DNA evidence: An empirical assessment of presentation formats for trace evidence with a relatively small random-match probability. *The Journal of Legal Studies*, **34**(2), 395–444, 2005.

National Research Council. *The Evaluation of Forensic DNA Evidence: An Update.* National Academy Press, Washington, DC, 1996.

National Research Council. *Strengthening Forensic Science in the United States: A Path Forward.* National Academies Press, 2009.

Nesson, C. The evidence or the event? On judicial proof and the acceptability of verdicts. *Harvard Law Review*, **98**, 1357–1392, 1985.

Police and Criminal Evidence Act 1984 (s.66), Codes of Practice. London: H.M.S.O, 1985. Print.

President's Council of Advisors on Science and Technology. *Forensic Science in Criminal Courts: Ensuring Scientific Validity of Feature-Comparison Methods.* Executive Office of the President of the United States, Washington, DC, 2016.

R v Adams (1996) 2 Cr App R 467 (England).

R v Adams (1998) 1 Cr App R 377 (England).

Rachlinski, J. J. Heuristics and biases in the courts: Ignorance or adaptation. *Oregon Law Review*, **79**, 61, 2000.

Saks, M. J., and Kidd, R. F. Human information processing and adjudication: Trial by heuristics. *Law and Society Review*, **15**, 123–160, 1980.

Saks, M. J., and Koehler, J. J. The individualization fallacy in forensic science evidence. *Vanderbilt Law Review*, **61**(1), 199–219, 2008.

Schklar, J., and Diamond, S. S. Juror reactions to DNA evidence: Errors and expectancies. *Law and Human Behavior*, **23**(2), 159–184, 1999.

Scurich, N., and John, R. S. Mock jurors' use of error rates in DNA database trawls. *Law and Human Behavior*, **37**(6), 424–431, 2013.

Smith, B. C., Penrod, S. D., Otto, A. L., and Park, R. C. Jurors' use of probabilistic evidence. *Law and Human Behavior*, **20**(1), 49–82, 1996.

State v Kim 398 N.W.2d 544 (1987) (United States).

Stern, H. S. Statistical Issues in Forensic Science. *Annual Review of Statistics and Its Application*, **4**, 225–244, 2017.

Thompson, W. C. Are juries competent to evaluate statistical evidence? *Law and Contemporary Problems*, **52**(4), 9–41, 1989.

Thompson, W.C. How should forensic scientists present source conclusions? *Seton Hall Law Review*, **48**, 773–813, 2018.

Thompson, W.C., Hofstein Grady, R., Lai, E. and Stern, H.L. Perceived strength of forensic scientists' reporting statements about source conclusions. *Law, Probability and Risk*, **17**, 133–155, 2018.

Thompson, W. C., Kaasa, S. O., and Peterson, T. Do jurors give appropriate weight to forensic identification evidence? *Journal of Empirical Legal Studies*, **10**(2), 359–397, 2013.

Thompson, W. C., and Newman, E. J. Lay understanding of forensic statistics: Evaluation of random match probabilities, likelihood ratios, and verbal equivalents. *Law and Human Behavior*, **39**(4), 332–349, 2015.

Thompson, W. C., and Schumann, E. L. Interpretation of statistical evidence in criminal trials: The prosecutor's fallacy and the defense attorney's fallacy. *Law and Human Behavior*, **11**(3), 167–187, 1987.

Tribe, L. H. Trial by mathematics: Precision and ritual in the legal process. *Harvard Law Review*, **84**, 1329–1393, 1971.

Tversky, A., and Kahneman, D. Judgment under uncertainty: Heuristics and biases. *Science*, **185**(4157), 1124–1131, 1974.

Twining, W. *Rethinking Evidence: Exploratory Essays*. Northwestern University Press, 1990.

U.S. Federal Rules of Evidence, Pub. L. No. 93-595, January 2, 1975.

Wells, G. L. Naked statistical evidence of liability: Is subjective probability enough? *Journal of Personality and Social Psychology*, **62**(5), 739–752, 1992.

10

Forensic Statistics in the Courtroom

David H. Kaye

CONTENTS

10.1 The Purpose, Form, and Prerequisites of Expert Testimony 225
    10.1.1 Lay and Expert Testimony 225
    10.1.2 Qualifications for Statistical Experts (and Experts Who Use Statistics) . . 226
    10.1.3 Forms of Statistical Expert Testimony 228
    10.1.4 Reasonable Scientific or Statistical Certainty 229
10.2 Special Rules for Scientific Expert Testimony 229
    10.2.1 The General-Acceptance Standard 230
    10.2.2 The Scientific-Validity Standard 231
10.3 Selected Evidentiary Issues in Forensic Statistics 234
    10.3.1 Two Uses of Statistical Analysis as Evidence 234
    10.3.2 Theory and Application 235
    10.3.3 Error Rates in Determining Admissibility 238
    10.3.4 Error Rates, Likelihood Ratios, and Bayes Factors for Quantifying
        Probative Value 241
10.4 Conclusion 244
References 245
Cases and Rules 248

10.1 The Purpose, Form, and Prerequisites of Expert Testimony

10.1.1 Lay and Expert Testimony

In the Anglo-American legal system, rules of evidence developed incrementally via court opinions resting on the inherent power of courts to define their own procedures. Today, most of the rules in the United States are contained in comprehensive codes, usually enacted by legislatures. The Federal Rules of Evidence (FRE) govern factfinding in the federal courts, and most states have similar rules. The rules treat all relevant testimony, documents, and real (physical) evidence as admissible unless: (1) a categorical rule (such as the rule against hearsay)* or a procedural or constitutional rule (such as the Sixth Amendment right to confront one's accusers in a criminal case) excludes it (FRE 402); (2) it would be unduly time consuming, confusing, or prejudicial in light of its probative value

---

* Hearsay is an out-of-court statement that a party offers to prove that the substance of the statement is true (FRE 801(c)). The rule against hearsay, however, is subject to many exceptions (FRE 802–807).

in the particular case (FRE 403); or (3) the documents or real evidence are not properly authenticated (FRE 901).*

Trial evidence normally is introduced through the testimony of witnesses sworn to tell the truth, but the law distinguishes between two types of witnesses. Lay or percipient witnesses generally are expected to testify to personal observations rather than to offer opinions or high-level conclusions about what they have witnessed (FRE 602, 701). For example, a witness might testify that a document contains a given individual's signature because the witness saw the individual sign it or is well acquainted with how the individual signs her name. Because jurors have their own experience and understanding of such matters, they can decide for themselves how much to credit the witness's testimony.

In contrast, observations or conclusions arrived at through esoteric processes or unusual skills require testimony from an expert witness (FRE 701). If qualified "by knowledge, skill, experience, training, or education," an expert witness may testify about matters outside the range of common knowledge if "the expert's scientific, technical, or other specialized knowledge will help the trier of fact to understand the evidence or to determine a fact in issue" (FRE 702). Appropriate statistical data and assessments are generally admissible under Rule 702.

10.1.2 Qualifications for Statistical Experts (and Experts Who Use Statistics)

Trial courts have great leeway in deciding whether an expert has sufficient knowledge, skill, experience, training, or education to testify about a particular matter, and they generally will allow a witness with only minimal qualifications to testify as an expert. Thus, extensive publications and distinguished academic credentials are not necessary.[†] Even an individual who declines to call himself an "expert" in the relevant field is not necessarily barred from testifying, because the standard for expertise in a professional community may diverge from the very liberal standard for being qualified as an expert in court. As long as a witness meets the minimal requirement of specialized knowledge or skill not likely to be possessed by the judge or a jury member, limitations in his or her credentials go to the weight the trier of fact will give to the expert's testimony, not to its admissibility.

Nevertheless, no matter how well credentialed experts are on a particular issue, they may not testify beyond their expertise. As a classic treatise emphasized,

> The capacity is in every case a relative one, i.e. relative to the topic about which the person is asked to make his statement. The object is to be sure that the question to the witness will be answered by a person who is fitted to answer it. His fitness, then, is a fitness to answer on that point. He may be fitted to answer about countless other matters, but that does not justify accepting his views in the matter in hand (Wigmore, 1940, vol. 2 §555).

Unfortunately, this standard does not translate into a simple rule for determining what level of specialization and experience a statistician or scientist needs in order to testify

---

* There are other limiting rules or as well. For examine, the "best evidence rule" requires proof of the contents of a writing by means of the original or a suitable copy (FRE 1001–1003).

† A court also has the power to appoint experts. This procedure is unusual, especially in criminal cases, but a court employing it normally would seek an expert with strong qualifications. Experts appointed pursuant to FRE 706 can assist with the admissibility inquiry and may be called to testify in the trial itself. For an account of the contributions of an appointed expert epidemiologist and other specialists in a medical product liability case, see Kaye and Sanders (1997).

about particular methods and applications. A guide to statistics for judges suggests that

> For convenience, the field of statistics may be divided into three subfields: probability theory, theoretical statistics, and applied statistics. Probability theory is the mathematical study of outcomes that are governed, at least in part, by chance. Theoretical statistics is about the properties of statistical procedures, including error rates; probability theory plays a key role in this endeavor. Applied statistics draws on both of these fields to develop techniques for collecting or analyzing particular types of data (Kaye and Freedman, 2011, p. 214).

Clearly, a degree in statistics is not required to perform particular statistical analyses. Because statistical reasoning underlies all empirical research, researchers in many fields are exposed to statistical ideas and computations. Experts with advanced degrees in the physical, medical, and social sciences—and some of the humanities—may receive formal training in statistics. Such specializations as biostatistics, epidemiology, chemometrics, econometrics, and psychometrics are primarily statistical, with an emphasis on methods and problems most important to the related substantive discipline.

There is no hard-and-fast requirement that the statistical expert must be a subspecialist with respect to the method or subject area. A common set of basic procedures is taught in all the statistical-subject-area specialties, and a generalist or a subspecialist in a related field of statistics may know enough to be qualified as an expert regarding a subspecialty. The key point is that individuals "who specialize in using statistical methods—and whose professional careers demonstrate this orientation—are most likely to apply appropriate procedures and correctly interpret the results" (ibid., p. 214).

At the same time, the choice of which data to examine, or how best to model a particular process could require subject-matter expertise that a statistician might lack. Statisticians often advise experts in substantive fields on the procedures for collecting data and often analyze data collected by others. As a result, cases involving statistical evidence often are (or should be) "two-expert" cases of interlocking testimony. A chemist or biologist, for example, may use capillary electrophoresis to measure the lengths of DNA fragments, and, as discussed in Chapters 11 (Weir, 2021a) and 12 (Weir, 2021b), a statistical geneticist may compute the probability that a half-sibling's DNA, or the DNA of an unrelated individual, would have very similar measurements.

Whether forensic scientists and technicians are qualified to testify to probabilities or statistics is not always obvious. In contrast to statisticians or subject-area statistical experts, some of these testifying experts have limited exposure to statistical methods and reasoning. They may lack the training or knowledge required to understand the statistical studies or results to which they testify. *State v. Garrison*, a murder prosecution involving bite-mark evidence, is a dated but unusually stark illustration of the problem. In that case, a dentist was allowed to testify that "the probability factor of two sets of teeth being identical in a case similar to this is, approximately, eight in one million," even though "he was unaware of the formula utilized to arrive at that figure other than that it was 'computerized'" (*State v. Garrison* 585 P. 2d at 566).

Of course, forensic-science practitioners sometimes use pencil-and-paper algorithms or sophisticated computer programs specifically designed to analyze and interpret their data. The programs may be the product of extensive study by knowledgeable statistical experts. If the algorithms have been tested and their limitations are known to the practitioners, then these witnesses could be considered qualified to present the output. Thus, it has been said that

> When generally accepted formulas are available, when it is reasonably clear (from authoritative texts) that they apply to a given situation, and when the use of the selected formulas is straightforward, credentials in a statistical subject area should not be necessary. After all, the defense can produce a statistically trained specialist to counter testimony that turns out to be too simple-minded (Kaye et al., 2011, §3.2 p.99).*

10.1.3 Forms of Statistical Expert Testimony

Unlike lay witnesses, statistical and other expert witnesses are not ordinarily confined to reporting personal observations. Their testimony also can take the form of opinions, explanations, or answers to hypothetical questions. For example, in the 2002 trial of a bank robbery case, an image analyst at the FBI testified about the similarities between photographs of a handbag carried by a bank robber taken by a surveillance camera in the bank and a handbag seized in the defendant's home (*United States v. McKreith*, 2005). He described "four specific characteristics": (1) "the alignment of the black and silver stripes from the back side of the bag with the end of the bag"; (2) "the location of the snaps at the top on a silver line"; (3) "the very small silver line at the top of the back piece"; and (4) "the silver line at the very top of the back piece." These were personal observations. He also testified, that with respect to all these features, the bag was "indistinguishable" from the picture of it in the bank. This was an opinion (based on unarticulated criteria for the degree of similarity that would make the features distinguishable).[†] In addition, the analyst explained the significance of the similarities with a probability model of the manufacturing process in which every feature being indistinguishable has a probability of one-half and is independent of every other feature. On that basis, he opined that the random-match probability was $1/2^4$, so that one can "eliminate 15 out of 16 silver bags that would be coming off the manufacturing process from whatever company made those bags."

Whether the expert should be allowed to testify to the calculations based on the probability model he posited is one issue, which we will consider in § 10.2. As far as the source of the data goes, the expert's opinions were based on admissible personal observations of common features of handbags.

Complications can arise when a different individual provides the data to the testifying expert—a common situation when statistical analysts are asked to testify. The data may constitute hearsay. If they are a public or business record, an exception to the rule against hearsay applies (FRE 803(6), 803(8)). Furthermore, the expert interpreting the case-specific data can use hearsay or other inadmissible evidence when it is of the type that "experts in the particular field would reasonably rely on ... in forming an opinion on the subject"

---

* For a comprehensive analysis of examining machine-generated evidence of all kinds in the litigation process, see Roth, 2017.

† Eyewitnesses often give similar opinions when they say things like "Yes, the man who demanded the money from me is the man in the photograph." These opinions are admissible without a precise set of criteria for identifying images because factfinders understand how ordinary witnesses reach these conclusions. The eyewitness is not claiming to be an expert at matching features in photographic images to those of physical objects, and a point-by-point specification of similarities from this witness in lieu of the witness's conclusion as to identity would not necessarily help the jury. In contrast, if an image analyst is relying on specialized training and experience to decide what is "indistinguishable," more of a foundation for the testimony is necessary. This foundation is the subject of § 10.2 below.

(FRE 703). Even so, the Sixth Amendment right of a criminal defendant "to confront the witnesses against him" may require the prosecution to make that individual available for cross-examination. The Supreme Court has construed this constitutional requirement of confrontation to apply to virtually all "testimonial" hearsay, but the meaning of "testimonial," especially in the context of scientific data, is highly contested (*Williams v. Illinois*, 2012; Mnookin and Kaye, 2013).

10.1.4 Reasonable Scientific or Statistical Certainty

It is common in some jurisdictions to have medical experts testify to "a reasonable degree of medical certainty." By extension, some judges demand that scientists testify to "a reasonable degree of scientific certainty." For example, in *United States v. Natson*, the government sought to prove that defendant murdered his girlfriend and hid her body in the woods because she would not abort her pregnancy. A DNA expert found that DNA from the defendant, the woman, and the bones of the fetus indicated defendant's paternity with a likelihood ratio of 26 (known in the parentage-testing field as a paternity index). But this expert declined to opine that this finding amounted to "a reasonable degree of certainty" for parentage testing. The federal district court excluded testimony about the likelihood ratio, reasoning (oddly) that even if the probability of paternity was 96% (presuming a prior probability of 50%), "it does not rise to any reasonable level of scientific certainty."

Nevertheless, it is clear that the rules of evidence do not require experts to present conclusions "to a reasonable degree of scientific (or other) certainty" (Kaye et al., 2019, § 1.3.2). Sheer speculation is inadmissible, of course, but if a reasonably estimated probability would assist the judge or jury to determine a fact in issue, the expert may testify about it (assuming it satisfies the other rules for admissibility).

10.2 Special Rules for Scientific Expert Testimony*

The evidentiary doctrines summarized in § 10.1 apply to all sorts of expert testimony—from accounting to zoology. For centuries, Anglo-American law did not distinguish one type of expert testimony from another. On the surface, a uniform standard governed the admission of the testimony of all qualified experts. As indicated above, the evidence simply had to be relevant and not too prejudicial or time-consuming, and it had to deal with matters not comprehensible to ordinary jurors without the assistance of an expert. A pristine statement of the position is that

> Any relevant conclusions which are supported by a qualified expert should be received unless there are other reasons for exclusion. Particularly, its probative value may be overborne by the familiar dangers of prejudicing or misleading the jury, unfair surprise, and undue consumption of time (McCormick, 1954).

Although this relevance-helpfulness requirement applies to all expert testimony, scientific and nonscientific alike, it need not have the same impact on all types of expert testimony.

---

* This section is adapted and updated from Kaye et al. (2011).

Scientific evidence tends to be time-consuming or difficult to understand. Courts fear that it comes cloaked in an aura of infallibility that leads jurors to give it more credence than it deserves. Consequently, ad hoc balancing of probative value and its counterweights can operate to exclude scientific evidence, especially if the science is not well established. Nevertheless, in practice, the relevance-helpfulness standard promotes "a generally laissez-faire approach to the admissibility of expert evidence" (Law Commission, 2011, p. 16).

Given the perception that scientific evidence poses special problems, courts in the United States came to supplement the relevance-helpfulness standard with more exacting rules that attend to the special features of scientific evidence. Two forms of additional scrutiny—a "general-acceptance standard" and a "scientific-validity standard" typically are used.

10.2.1 The General-Acceptance Standard

The general acceptance standard made its debut in 1923, in the now celebrated case of *Frye v. United States*. James Alphonso Frye was charged with murder. He sought to introduce the testimony of a psychologist who had administered a systolic blood pressure test and was prepared to testify that Frye was truthful when he denied committing the murder. The trial judge refused to allow the jury to hear this testimony. In affirming the trial court's rulings, the court of appeals wrote that exclusion was proper because other psychologists had yet to accept the expert's claim that he could verify honesty by measuring the speaker's blood pressure. The court wrote:

> Somewhere in this twilight zone [between the "experimental" and the "demonstrable"] the evidential force of the principle must be recognized, and while courts will go a long way in admitting expert testimony deduced from a well-recognized scientific principle or discovery, the thing from which the deduction is made must be sufficiently established to have gained general acceptance in the particular field in which it belongs (293 F. at 1014).

Concluding that the deception test lacked the requisite "standing and scientific recognition among physiological and psychological authorities," the court of appeals upheld the exclusion of the psychologist's testimony.

This *Frye* standard was not widely adopted at first, but in succeeding decades, it became the dominant rule in state and federal courts. It has been invoked to exclude from the courtroom polygraphy, graphology, hypnotic and drug-induced testimony, voice stress analysis, voice spectrograms, various forms of spectroscopy, infrared sensing of aircraft, blood alcohol tests, retesting of breath samples for alcohol content, polarized light microscopy, psychological profiles of battered women and child abusers, post traumatic stress disorder as indicating rape, penile plethysmography as indicating sexual deviancy, astronomical calculations, blood group typing, DNA testing, therapy to recover repressed memories, and identifications based on ear prints and knife marks.

It may seem that several of these techniques should be admissible, and today a number of them are, while others remain pseudoscience or insufficiently validated. Because the *Frye* test requires sufficient time for a method to be widely accepted as valid, it is fundamentally a conservative standard. The major advantage of the rule is that by looking to the views of the scientific community, it avoids having judges who are ill-equipped for the task act like independent scientists in ascertaining whether the science in question is good enough for forensic use.

10.2.2 The Scientific-Validity Standard

Since the early 1970s, the *Frye* standard was subjected to increasingly critical analysis, limitation, modification, and finally, outright rejection. The adoption of the Federal Rules of Evidence in 1974 intensified the retreat from *Frye*. These rules do not explicitly distinguish between scientific and other forms of expert testimony, and they do not mention general acceptance. As originally enacted, Federal Rule 702 simply provided that "[i]f scientific, technical, or other specialized knowledge will assist the trier of fact to understand the evidence or to determine a fact in issue, a witness qualified as an expert by knowledge, skill, experience, training, or education, may testify thereto ...." By 1990, a strong minority of jurisdictions had expressly repudiated *Frye* in favor of a "relevancy-plus" analysis that required a certain extra trustworthiness, accuracy, or fit beyond that needed to admit nonscientific testimony.

The trend culminated in a series of Supreme Court cases on scientific and expert testimony. The seminal case is *Daubert v. Merrell Dow Pharmaceuticals*. In *Daubert*, two young children were born with deformed limbs, and their parents sought damages against the manufacturer of Bendectin, a prescription drug taken by the boys' mothers to treat nausea and vomiting during pregnancy. The plaintiffs' case foundered when they were unable to point to any published epidemiological studies concluding that Bendectin causes limb reduction defects. The federal district court excluded the plaintiffs' evidence of causation—so-called structure-activity studies, *in vitro* animal cell experiments, *in vivo* animal research, and an unpublished reanalysis of the epidemiological data. The court of appeals affirmed.

The Supreme Court held that in looking solely to general acceptance as indicated by peer-reviewed publications, the lower courts had applied the wrong standard. The Court proclaimed that the "austere [general acceptance] standard, absent from and incompatible with the Federal Rules of Evidence, should not be applied in federal trials" (509 U.S. at 589). Having jettisoned general acceptance as "the exclusive test for admitting expert scientific testimony," the Court announced that as the gatekeeper of evidence, "the trial judge must ensure that any and all scientific testimony or evidence admitted is not only relevant, but reliable." This *"evidentiary* reliability" presumes "scientific knowledge"—"the proffered testimony must be "ground[ed] in the methods and procedures of science." In a further elaboration, the Court suggested that this "reliability" determination "entails a preliminary assessment of whether the reasoning or methodology underlying the testimony is scientifically valid" and "properly can be applied to the facts in issue." This, in turn, depends on such things as (1) "whether it can be (and has been) tested," (2) "whether the theory or technique has been subjected to peer review and publication," (3) "the known or potential rate of error," (4) "the existence and maintenance of standards controlling the technique's operation," and, as in *Frye*, (5) the "degree of acceptance within [a relevant scientific] community."

In addition, the Court noted the value of ensuring that "expert testimony proffered in the case is sufficiently tied to the facts of the case. This "fit," the Court added, "is not always obvious, and scientific validity for one purpose is not necessarily scientific validity for other, unrelated purposes." As a putative example of valid scientific knowledge that should be excluded for lack of fit, Justice Blackmun wrote that the

> study of the phases of the moon, for example, may provide valid scientific "knowledge" about whether a certain night was dark, and if darkness is a fact in issue, the knowledge will assist the trier of fact. However (absent creditable grounds supporting such a link), evidence that the moon was full on a certain night will not assist the trier of fact in determining whether an individual was unusually likely to have behaved irrationally on that night.

This lexicon is another example of different meanings of terms in law and science. As used in statistics, "validity" already incorporates the use to which the measurements are put. Lunar phase is a valid predictor of darkness at night but not of instances of lunacy.

In the wake of *Daubert*, some state courts observed that they already used this version of the relevancy-plus standard, and a large number agreed to relax *Frye*'s insistence on general acceptance. Still others adhered to *Frye*. But *Daubert* left the courts that sought to apply it with a series of questions. The most obvious was how to apply the loose multifactor test in which no factor was said to be determinative and none were fully defined. In 2009, a seventeen-member committee convened by the National Academy of Sciences at the request of Congress pointedly observed that "[i]n a number of forensic science disciplines, forensic science professionals have yet to establish either the validity of their approach or the accuracy of their conclusions, and the courts have been utterly ineffective in addressing this problem" (National Research Council Committee on Identifying the Needs of the Forensic Sciences Community, 2009, p. 53). The committee also observed that "[f]ederal appellate courts have not with any consistency or clarity imposed standards ensuring the application of scientifically valid reasoning and reliable methodology in criminal cases involving *Daubert* questions" (ibid., p. 96). This situation, it added, was "not surprising" given that *Daubert* is so "flexible" (ibid.). Even after this report, courts managed to avoid faithfully or rigorously applying the "*Daubert* factors" to many forms of forensic-science testimony (Kaye, 2018a).

More subtly, whether a showing of "evidentiary reliability" as a condition for admitting evidence applied only to the methods or tests used to reach conclusions in particular cases, or whether it also extended to conclusions derived from those methods became a central issue in many cases. On its face, *Daubert* maintained the well-established distinction. The Court provided general guidance for trial courts on how to conduct "a preliminary assessment of whether the reasoning or methodology underlying the testimony is scientifically valid." The criteria pertained to "the principles that underlie a proposed submission," and the majority cautioned that "[t]he focus, of course, must be solely on principles and methodology, not on the conclusions that they generate."

But in *General Electric Co. v. Joiner*, the Court shied away from the labels "methodology" and "conclusion." *Joiner* is another toxic-tort case, this time involving the association between polychlorinated biphenyls (PCBs) and lung cancer. Finding that the epidemiological and animal studies on which the plaintiffs' experts relied were too weak to justify the conclusion that PCBs can promote cancers, the district court granted summary judgment to defendants. The Court of Appeals for the Eleventh Circuit reversed. The Supreme Court reviewed the research literature on whether PCBs promote cancers and held that excluding the experts' opinions about the actions of PCBs was within the district court's discretion.

This disposition required the Court to confront the argument that because the District Court's disagreement was with the conclusion that the experts drew from the studies, the District Court committed legal error and was properly reversed by the Court of Appeals. Rather than analyze whether the "weight of the evidence" assessment that plaintiffs' experts used was a scientific methodology subject to special scrutiny under *Daubert* or instead a fairly debatable conclusion from scientific studies and thus admissible unless unfairly prejudicial or time-consuming, the Court blithely stated that

> [C]onclusions and methodology are not entirely distinct from one another. Trained experts commonly extrapolate from existing data. But nothing in either *Daubert* or the

Federal Rules of Evidence requires a district court to admit opinion evidence which is connected to existing data only by the ipse dixit of the expert. A court may conclude that there is simply too great an analytical gap between the data and the opinion proffered (522 U.S. at 146).

Not all states that have adopted the *Daubert* standard have jettisoned the methodology-conclusion distinction. Some continue to insist that "once the validity of the expert's reasoning or methodology has been satisfactorily established, any remaining questions regarding the manner in which that methodology was applied in a particular case will generally go to the weight of such evidence" (Farm and Garden Center v. Kennedy, 2018, p. 594). A number of federal courts also continue to place great weight on the methodology-conclusion distinction, and the distinction is unavoidable when assessing the evidence under the *Frye*-like general acceptance factor in *Daubert*. After all, it makes little sense to ask whether case-specific data or conclusions are generally accepted, for such acceptance can only extend to the method of arriving at the result in the particular case, leaving the question of infirmities in the application of the method to the factfinder. Or, the court could exclude the evidence if the application is so flawed as to make the result inadmissible under the balancing test for probative value and possible prejudice and time-consumption.

Following *Joiner*, the Supreme Court continued to blur the distinction between methodology and conclusion. In *Kumho Tire Co., Ltd. v. Carmichael*, the Court observed that "[t]he objective of [*Daubert*] is to ... make certain that an expert, whether basing testimony upon professional studies or personal experience, employs in the courtroom the same level of intellectual rigor that characterizes the practice of an expert in the relevant field." Various lower federal courts had drawn the same lesson from *Daubert*, and several have spoken of a departure from the level of professional care normally observed outside of litigation as a reason to exclude statistical testimony.

Federal Rule 702 was amended in 2000 to reflect the holdings of *Daubert*, *Joiner*, and *Kumho Tire*. The rule now contains an additional proviso requiring that "(1) the testimony is based upon sufficient facts or data, (2) the testimony is the product of reliable principles and methods, and (3) the witness has applied the principles and methods reliably to the facts of the case." Nonetheless, this supplemental language does not authorize the trial court to exclude evidence just because it finds an expert's opinion unpersuasive. As the Advisory Committee that drafted the amendment cautioned, "[w]hen a trial court, applying this amendment, rules that an expert's testimony is reliable, this does not necessarily mean that contradictory expert testimony is unreliable. The amendment is broad enough to permit testimony that is the product of competing principles or methods in the same field of expertise" (Federal Rules of Evidence Advisory Committee, 2000).

The tension between the desire to exclude unreliable opinions and the need to leave the resolution of ostensibly scientific conclusions that are fairly debatable to the jury is palpable, but Rule 702's articulation of how to do so leaves a great deal unsaid. In 2016, the President's Council of Advisors on Science and Technology released a report on pattern-matching evidence that proposed detailed criteria for the validation of largely subjective determinations of when the feature sets of different specimens are sufficiently similar to conclude that the specimens come from the same source (PCAST, 2016). This group maintained that scientific validation had to include multiple tests, conducted by independent researchers, in which analysts succeeded in correctly classifying pairs of same-source and different-specimens representative of case work with a false-positive probability of no more than five percent. Some critics of the report attacked this specification as novel *ipse dixit* (Hunt, 2018).

10.3 Selected Evidentiary Issues in Forensic Statistics

Section 10.2 outlined some of the history and nature of the leading standards for judicial evaluations of the admissibility of scientific evidence generally and the division of authority between, on the one hand, the judge as gatekeeper of the evidence that the finder of fact receives* and, on the other hand, the trier of fact as the arbiter of whether the evidence proves the essential elements of the criminal charge (or civil complaint) against the defendant. This section sharpens the focus on the admissibility standards as they apply to some forms of statistical proof, and it discusses the probative value of scientific and statistical evidence.

10.3.1 Two Uses of Statistical Analysis as Evidence

There are two ways in which statistical analysis can become part of a party's proof. First, statistical analysis can be a component of the process that leads to the scientific conclusions. For example, a statistical rule can be used to declare that the physical or chemical properties of glass fragments are indistinguishable (Kaye, 2015). The current standard (ASTM International, 2017) for comparing elemental concentrations with micro-X-ray fluorescence spectrometry uses nine (or even fewer) replicate measurements to estimate a standard deviation (s) as a basis for the conclusion that questioned and known specimens "are not from the same source" when the mean for a questioned specimen is more than $\pm 3s$ from the mean for the known specimen. It suggests that this is a useful decision rule because "This range corresponds to 99.7% of a normally distributed population." However, the mean for such small samples is not normally distributed (even if the concentrations are), and the $\pm 3s$ procedure may not account correctly for the variability of the mean for questioned specimens.[†] Thus, a party might object that the reasoning in the ASTM standard does not satisfy *Daubert* or *Frye* (with respect to the relevant statistical community).

Second, statistical or probabilistic analysis can be offered to assist the judge or juror in understanding the significance of other facts. The infamous California case of *People v. Collins* is one of many vivid examples of this use. In 1964, a seventy-one-year-old woman was walking, cane in hand and groceries in tow, down an alley in Los Angeles late one morning. Her purse sat atop the packages in her wicker-basket carryall. Suddenly, she found herself knocked to the ground with a dislocated shoulder and fractured arm. All she saw was the fleeing woman's blond hair. Her screams attracted the attention of a man who was watering his lawn nearby. He saw a bearded and mustached African-American man in a yellow car snatch up a blond woman with a ponytail and speed off. Officers arrested a Malcolm Ricardo Collins and his wife, Janet Louise Collins, on the strength of these descriptions. But there was little else to prove their guilt. To convince the jury that the characteristics of the couple were highly significant, the prosecution had a newly hired assistant professor of mathematics at California State College at Long Beach testify to the "product rule" for computing "the probability of the joint occurrence of a number of mutually independent events." The prosecutor later proposed figures for the frequencies of an

---

* In criminal cases, the Sixth Amendment ensures the defendant the right to a trial by jury. A defendant may waive that right in favor of a bench trial (one in which the judge also is the trier of fact tasked with reaching a verdict of guilt or innocence based on the admissible evidence introduced at the trial).

† If the measurements of the concentration of an element are normally distributed around the true value, then a t-test would apply for a small sample of measurements. That test entails a smaller rejection region to achieve the false-rejection probability (such as 0.003 for $\pm 3\,s$) that applies to the z-test suggested by the ASTM standard.

interracial couple in a car (1/1,000), a girl with a ponytail (1/10), a partly yellow automobile (1/10), a man with a mustache (1/4), and a black man with a beard (1/10). The prosecutor applied the product rule to his "conservative" estimates to conclude "that there was but one chance in 12 million that any couple possessed the distinctive characteristics of the defendants." He argued that "the chances of anyone else besides these defendants being there, ... having every similarity, ... is something like one in a billion." The California Supreme Court reversed the resulting conviction. Its opinion condemned the introduction of this testimony about the product rule as "fundamental error" because of at least two "glaring defects" in the calculation—"an inadequate evidentiary foundation" for the numbers that were multiplied and "an inadequate proof of statistical independence" (438 P.2d at 38).

For our purposes, the point is not that the prosecution's statistical analysis was silly. It is that the expert testimony was used to assess, however poorly, the probative value of the features shared by the robbers and the defendants. To that extent, *Collins* is like thousands of DNA-evidence cases in which a much better-founded statistical calculation of the probability of a random match is introduced as part of the scientific evidence (Kaye, 2010). The probability computation produces a p-value, with the null hypothesis being that the defendant is unrelated to the individual whose DNA was found at crime scene.

The image-comparison case, *United States v. McKreith* (§ 10.1.3), is another such application of probability theory. That the features in the photograph from the bank robbery were similar to those of the handbag was not disputed. The p-value of 1/16 was used to suggest that the photograph was pretty good proof of the defendant's involvement. Indeed, in *McKreith*, the expert also computed p-values down to $1/(6.5 \times 10^{11})$ for measurements of the position of lines in a plaid shirt owned by McKreith and those seen on videos from the bank robberies. The expert propounded a theory of the manufacturing process that gave rise to the probability model he used, and the district court found it all acceptable under *Daubert* (Kaye, 2019). Although that outcome can be questioned, it was understood that estimating the probability that the observed features would match if they were the result of randomness in the manufacturing process was a step that also had to be shown to be scientifically valid.

We turn, therefore, to the issues that arise in making such a showing. In addition, we will consider a third use of statistical or probabilistic analysis—to inform a decision as to the admissibility of nonstatistically based forensic-science findings.

10.3.2 Theory and Application

The fact that the statistical or probabilistic analysis draws on existing theory and tools presented in academic journal articles and textbooks may suggest that general acceptance or scientific validity is not much of an issue with such evidence. That view, however, is dangerously superficial. As emphasized in one treatise

> *Daubert*'s reference to "fit" serves as an important reminder that the "scientific validity" of a technique or instrument depends on the use to which it is put. It will not do, for example, to assert that because regression analysis always is a valid technique for examining the association between two variables, bivariate regression is valid, in itself, for drawing causal inferences about these variables. The statistical technique is valid for some purposes, but not for all. As a result, it may fit one case but not another (Kaye et al., 2011, § 7.3.1(c)(3)).

Thus, the normal distribution plays a central role in statistics, but that does not mean it fits the problem that the ASTM micro-X-ray standard tries to solve. The ASTM theory is that a one-sample z-test with a sample standard deviation based on only nine observations is appropriate for comparing the means of two independent samples to decide whether to reject the hypothesis that the specimens have a common source.* As a first impression, most statisticians would not be inclined to establish the rejection region this way (compare National Research Council, 2004, p. 171), although sufficient empirical research might show that the unusual procedure performs adequately in distinguishing between same-source and different-source glass fragments (Kaye, 2015).

Likewise, the binomial probability model has a storied place in probability theory, but in a case like *McKreith*, a court asked to exclude testimony of the probability of the four matching features needs to know the grounds for believing that the model applies to the manufacture of handbags. In the case itself, the grounds were more theoretical than empirical.

The concern also can be framed without reference to legal "fit" by asking whether it is *valid* to use the binomial model with a success (match) probability of one-half for individual characteristics as a description of what is expected under the hypothesis of coincidence. A court applying the *Daubert* factors could reason and inquire as follows:

1. The model could be falsified. The distribution of features in the relevant population of handbags could have been studied. The process for producing the components of the handbags and fastening them together could have been studied. But they weren't.

2. Has the theory that handbags (or similarly manufactured items) have features that occur as described by the probability model been subject to peer review and publication? The government relied on one case report describing its expert's comparison of denim trousers to photographs in a different series of bank robberies on the basis of what he presumed were "individual characteristics that differentiate the item from all other similar items" (Vorder Bruegge, 1999, p. 613).

3. What is the likely range of error in using the theory to compute a probability of 1/16?

4. There are no formal standards that control when an analyst adopts a particular probability model. Considering the lack of data, has the witness taken the degree of care expected for professional statistical work that has the "intellectual rigor" called for in *Kumho Tire*?

5. Is the theory that handbags (or similarly manufactured items) have features described by the model generally accepted in the statistical community?

In the above list, the "theory" has been defined as the claim that "handbags (or similarly manufactured items) have features that occur as described by the model" rather than as the theory that the binomial model gives the probability of x out of n successes for a Bernoulli process. That the binomial model has been known for centuries does not prove validity

---

* In actuality, the hypothesis being tested is that the two specimens have the same elemental composition. The estimated standard error s determines the rejection region (and hence the match window). How often "matches" within a window such as $\pm 3s$ occur in samples from different sources is a further question. As a result, forensic statisticians and criminalists sometimes speak of the use of the $\pm 3s$ windows as the first part of a two-stage procedure. The ASTM standard ignores what Parker, 1996, p. 38, and many later writers call the second stage (see Chapters 1 (Aitken and Taroni, 2021) and 4 (Stern, 2021)).

or general acceptance of the application to the task at hand. Nonetheless, even in cases in which it is obvious that a statistical analyst is presenting a new and unpublished method for computing quantities of interest for the specific case, the party offering the evidence may well try to overcome a *Daubert* or *Frye* objection by pointing to the impeccable pedigree of some component of the calculations. Consider these questions and answers like these from *Commonwealth v. McNair*:

Q. Is the use of likelihood ratios generally accepted in the biostatistical community?

A. Yes, it is.

Q. Is there any controversy that you are aware of about the validity of the use of likelihood ratios in biostatistics?

A. No, the likelihood ratio has been used, for example, in forensic genetics for decades. There had been ongoing discussion about the kind of data that are used to calculate likelihood ratios, but the concept of likelihood ratio has been used in forensic genetics for, as I said 20, 30 years at least.

Q. And, Doctor –

A. And there's currently no controversy about the likelihood ratio being the best way of summarizing the evidence that is included in the data in a single numerical figure. (Transcript of Hearing, 2017, 21–22)

Does the acceptance of the general concept of a likelihood ratio for some purposes mean that the choices of the probability model and the parameters used to arrive at the numerator and denominator of the ratio are generally accepted or valid? Surely not.

Many similar questions about the line between methodology and conclusions of statistical assessments can be posed. Is Neyman-Pearson hypothesis testing with $\alpha = 0.05$ a method and the multitudinous applications just case-specific matters for the factfinder to ponder? As with the computation of a likelihood ratio, that procedure leaves open too many nontrivial modes of implementation to wall off those choices from heightened scrutiny. Is "Bayesian inference" a single "complex statistical analysis" (Forensic Science Regulator, 2019, 14.2.1) or a broad approach to statistical problems that encompasses a large number of distinct methods? Again, the answer seems obvious, but at what point does the more specific method that must pass muster become so detailed as to be the mere application of a generally accepted method? It may be that only some algebra or arithmetic is so case specific as to be exempt from strict pre-trial scrutiny.

In *McNair*, the biostatistician characterized his application of probability theory to genetic events as mathematically "very straightforward." These computations yielded a likelihood ratio ranging from 320 to 2×10^9 or so, depending on certain assumptions (Memorandum of Decision and Order, 2017). But the arithmetic emerged from a previously unknown method for quantifying the probability that mutations could have caused several base pairs in the DNA from semen from a rape to differ from those of saliva samples from two twin brothers. The cross examination produced the following exchange:

Q. But there is no established formula among forensic DNA statisticians as to the methodology to use to calculate the likelihood ratio of one twin being more likely than another to be the source of a trace evidence sample based on the saliva genotype of the two twins, correct?

A. That is correct.

Q. Can we agree that the methodology that you created in this case, and that you had to create because there was no other methodology, is not generally accepted in the forensic DNA statistics community?

A. That is correct because it has not been put to the test of peer review, and that's correct, yeah.

After the court excluded the DNA sequencing evidence, this biostatistician and several other prosecution experts published a paper describing a related (and more complex) procedure for computing a likelihood ratio in such cases (Krawczak et al., 2018).

10.3.3 Error Rates in Determining Admissibility

The risk of error is a key concept in inferential statistics. In *Daubert*, the Court invoked the notion as follows: "in the case of a particular scientific technique, the court ordinarily should consider the known or potential rate of error, see, e.g., *United States v. Smith*, 869 F.2d 348, 353–354 (CA7 1989) (surveying studies of the error rate of spectrographic voice identification technique)." In *Smith*, unlike *McNair*, the issue was not whether some statistical method was adequate for courtroom use. Rather, the statistical analysis was introduced to judge whether other evidence—forensic-science practitioners' classifications of voice exemplars—is accurate enough to be legally acceptable. The *Smith* court found this evidence to be admissible in part because the witness, who was then with the Los Angeles County sheriff's department and later became a senior scientist with the FBI, testified that studies showed small proportions of "false identifications" (false positives) and "false eliminations" (false negatives) in experiments (2.4% and 6%, respectively)* and in practice (0.31% and 0.53%, respectively).[†]

In using error rates as a factor in ascertaining admissibility of classifications, a series of issues arise: (1) How should the rates of misclassifications be determined? (2) What rates are small enough to warrant admission of the classifications? (3) Are there other statistics besides false-positive or false-negative probabilities that courts can or should use in deciding admissibility questions?

Determining misclassification rates. To begin with, the crucial question addressed by error rates in cases like *Smith* is whether experts examining the features of pairs of items can correctly classify the specimens as coming from the true source and coming from a different source. Relevant data could come from experiments or observational studies. The problem with observational studies in this context is the lack of a suitable criterion for validity. A laboratory that uses blind verification of all positive and negative conclusions could ascertain inter-examiner reliability (also called reproducibility) within the one laboratory from historical data. Or, researchers could pick a sample of past cases and submit specimens as if they were parts of new cases. High reproducibility argues in favor of admission, although courts do not inquire into statistical reliability as much as they should.

Yet, reproducibility does not ensure low error rates. When there are discrepancies, how do we know which examiner is correct—or whether either analyst is extracting

---

* The court referred to 35,000 comparisons (of voice exemplars from 250 students). There are a number of reasons to think that the error proportions, which pertain only to the subset of identifications that the analysts did not consider "uncertain," underestimate the conditional error probabilities in realistic conditions. See United States v. Angleton, 269 F.Supp.2d 892, 899-900 (S.D. Tex. 2003). The *Smith* court also observed that in a "follow-up … study … involving actual cases," the same researcher "found no errors whatsoever."

† On cross examination, he acknowledged that two "studies had higher error rates … of 62.7 percent and 83.33 percent respectively."

and applying useful information to come to a conclusion? Even so, statistical reliability is related to error rates. Low reliability (within or across examiners) limits accuracy because inconsistent results for the same specimens cannot all be correct. Furthermore, *Daubert* spoke of "the existence and maintenance of standards controlling the technique's operation" in the same sentence as "the known or potential rate of error." Inconsistent outcomes are an indication that the conclusions are largely discretionary and not effectively channeled by controlling standards.

To obtain more direct data on the accuracy of the classifications, we need to know the true state of affairs.* Several types of studies can make use of that information to determine the error rates. First, one can take a sample of cases in which analysts have reached conclusions and check those against a test that is known to be highly accurate. Because the latter, "gold standard" test makes almost no errors, the false-positive error rate for the test being validated is approximately the proportion of the specimens with negatives on the gold-standard test that have positives on the test in question; likewise, the false-negative error rate is approximately the proportion of the specimens registering positive on the gold-standard test that are classified as negative on the test in question. Such an approach was used with microscopic hair comparisons followed (years later, for the purpose of the study) by mitochondrial DNA analysis of the hairs (Houck and Budowle, 2002). The false-positive proportion was 9/46 = 20%;[†] the false-negative proportion was 0/69.

Second, rather than taking a sample of specimens that have been analyzed in practice, one can create specimens from known sources and present them to analysts in an experimental setting. Estimates for latent fingerprint examinations and firearms toolmarks using this methodology are discussed in the PCAST report. Although analysts do not know the prevalence of true positives in the test materials, they are volunteers who know they are being tested. That reduces the realism of the experiments. In addition, arguments can arise as to whether the test materials are truly representative of case work.

Another type of experiment overcomes the volunteer and knowledge problems by injecting known test specimens into the flow of case work. If the test specimens and accompanying materials are indistinguishable from actual case items, and if there are enough of them to produce reasonable sample sizes, then the observed error rates will be informative.

The magnitude of the error rates. The PCAST report announced a greatest lower bound for deciding when error rates are too high to admit an analyst's conclusion that specimens originated from a common source. The Council declared that

> Methods with a high [false positive rate] are scientifically unreliable for making important judgments in court about the source of a sample. To be considered reliable, the [rate] should certainly be less than 5 percent and it may be appropriate that it be considerably lower, depending on the intended application (PCAST, 2016, 104).

PCAST did not explain what applications would require error rates below a smaller number. Perhaps the advisors meant that the same forensic-science identification method that is "scientifically reliable" in a prosecution for a minor crime could lack sufficient validity for a capital case. If so, it seems to follow that in civil cases, a false-positive error rate in

---

* However, courts have been impressed with anecdotal reports that no one discovered any errors in an analyst's casework. In *Smith*, the court of appeals placed in the same category as experimental studies the speech scientist's report that in the 150 earlier instances in which he made (but did not testify to) spectrographic voice identifications, no one informed him that he had made a mistake.

† If one excludes cases in which the microscopic comparisons were inconclusive from the denominator, the false-positive rate increases to 9/26 = 35%.

excess of five percent might be tolerable. The report also suggests that a large false-positive error probability should not keep the defendant from introducing evidence even though the prosecution cannot present it. But the usual rules for admitting expert testimony do not depend on the type of case or the party proffering the evidence. To be sure, only a "preponderance of the evidence" is needed for most civil verdicts, whereas criminal convictions demand proof beyond a reasonable doubt. A preponderance usually is equated with a probability of just over 50%, whereas "beyond a reasonable doubt" might mean a probability of 95% or more. But whatever these posterior probabilities may be, they apply to the totality of the evidence—not to each item in isolation. The law insists on a high burden of persuasion in criminal cases because false convictions entail greater social disutility than false acquittals. It does not demand the exclusion of probative evidence that would contribute to the determination of a better informed posterior probability (Kaye, 2015, p. 181; Memorandum from the Legal Research Committee, 2016).

Furthermore, PCAST's dictum does not seem consistent either with *Daubert's* reference to the "rate of error" as one part of a "flexible" five-factor test or with the definition of validity as sufficient accuracy to make a method fit for the purpose for which it is used. Suppose that a test has a 20% false-positive probability. By definition, $P(+|\sim S) = 0.20$, where $+$ is a positive test result and $\sim S$ is the different-source hypothesis. Suppose also that the false-negative probability is $P(-|S) = 0\%$, where $-$ is a negative test result and S is the same-source hypothesis. (These are the point estimates from the hair-DNA study.) Then the probability of a positive test result is larger when the specimens are from the same source than when they are from different sources, and the test has probative value. Specifically, the likelihood ratio (with a positive test result) is

$$L = \frac{P(+|S)}{P(+|\sim S)} = \frac{1 - P(-|S)}{P(+|\sim S)} \tag{10.1}$$

Substituting the values, we conclude that a positive classification has a likelihood ratio of $1/0.20 = 5$. So a positive result is slightly useful in discriminating between S and $\sim S$. If presented to the factfinder without leading to an exaggerated sense of probative value, it is fit for the Rule 702 purpose of assisting the trier of fact "determine a fact in issue."

The upshot is that the scientific-validity standard of *Daubert* and Rule 702 does not lend itself to a simple, numerical rule like the PCAST 0.05 rule for dismissing as scientifically invalid a forensic-science test on the basis of its false-positive probability. A likelihood ratio is a more appropriate statistic for binary classifications. When that figure is close to one, the classification test has low validity. Presenting the classification as definitive in those circumstances could be called scientifically invalid and should not be acceptable, but presenting it for what it is worth is potentially permissible.

Consistent with this analysis, courts looking at error rates for forensic-science classifications have not attempted to draw sharp boundaries. Unfortunately, they have sometimes confused the false-positive probability with the complement of the positive predictive value of a test* and have dismissed false-negative probability as irrelevant despite its potential impact on the likelihood ratio. In *United States v. Love*, for instance, a federal

---

* The positive predictive value (PPV) is the posterior probability $P(S|+)$ when the prior probability is the prevalence of S in the set of samples used to estimate the false-positive probability, $P(+|\sim S)$. The probability that a report is incorrect when it is positive is $P(\sim S|+) = 1 - P(S|+) = 1 - \text{PPV}$. For discussion of an opinion describing a false-positive probability as if it were the probability of an incorrectly reported false-positive, see Kaye (2016).

district court anchored its understanding of a false-positive error rate on an FBI supervisor's earlier claim that the FBI Latent Print Unit's "error rate was 1 in 11 million cases." The supervisor based the number on the absence of revelations of any false identifications in casework over the years (before the unit's notorious misidentification of an Oregon lawyer's latent print on a fragment of an explosive device used in the 2004 Madrid train bombing). The judge described "an overall false positive rate of 0.1%" from a far more rigorous 2011 study as only "marginally higher" and the "somewhat higher" 7.5% false-negative rate in the experiment (Ulery et al., 2011) as "not relevant here." As the next subsection shows, however, both error rates jointly determine the probative value of a positive test result.

Error rates like the ones in *Smith* are useful in studying the validity of a system that makes "matches" between specimens. Declaring a "match" versus an "exclusion" is appropriate for features such as blood types or DNA alleles in single-source specimens, that are measured as discrete variables. But this "two-stage" procedure does not make full use of the information when the variables are continuous ones such as breath alcohol concentration, refractive index, elemental concentrations, or the size of bones. The validation of those measurements requires other statistical methods.

The same is true for likelihood ratios that arise without an artificial match/no-match step, as occurs for objective similarity scores for multiple features, for likelihood ratios based on an examiner's sense of the degree of correspondence between feature sets, and for "probabilistic genotyping" of ambiguous genotypes in complex DNA mixtures. Internationally, most forensic statisticians advocate skipping the match/no-match step in favor of a more nuanced and transparent presentation and explanation of a likelihood ratio or Bayes factor (e.g., Aitken et al., 2011; Nordgaard et al., 2012; Hicks et al., 2016; Evett et al., 2017; Gittelson et al., 2018; Morrison et al., 2017).

The courts have yet to address what analog of simple error rates for classifications would be appropriate in judging validity in these situations. Bounds on the expected rates of "misleading evidence" could be helpful here. If X has density f_1, then for any alternative density f_2 and any $k > l$, the probability of misleading evidence, $P(f_2/f_1 \geq k; H_1)$, can never exceed $1/k$. Similarly, $P(f_1/f_2 > k; H_2) \leq 1/k$ (Royall, 2000, p. 762). Thus, a bound on the expected rate of evidence that seems to support a false hypothesis at least as strongly as the likelihood ratio L states it does is the reciprocal of L (Taylor et al., 2015, p. 166). But this bound presupposes that the likelihood ratio is correct. Ideas on the validation and calibration of likelihood ratios are provided in Chapter 7 (Ramos et al., 2021) and Chapter 13 (Neumann et al., 2021).

Better estimates of a tail-end probability may be derived from simulations. For example, the distribution of the likelihood ratio for a suspect's DNA being present in a complex DNA mixture as opposed to that of an unrelated individual *when the suspect is not a source* can be estimated by computing likelihood ratios using a large enough number of genotypes for a simulated population of unrelated individuals (Moretti et al., 2017; Taylor et al., 2017; compare Kaye et al., 1991). In this way, an analog to the false-positive probability for a classification test can be obtained for reported likelihood ratios.

10.3.4 Error Rates, Likelihood Ratios, and Bayes Factors for Quantifying Probative Value

Error rates like the ones in *Smith* have a second role in the trial process. Not only are they important if not vital to the admissibility decision, but they can help reveal to the finder of fact the probative value of the expert's conclusion. An analyst's judgment that two specimens have the same source can be presented as indicative rather than apodictic. The error rates seen in experiments with similar analysts performing the same classification task are relevant to the metaphorical strength or weight that the analyst's classification should have in reaching a verdict. Equation 10.1 expressed the likelihood ratio L (with a positive classification) in terms of the error probabilities. For the error rates in the accompanying example, an expert could testify that he or she regards the specimens as positively associated or "matching" and that analysts make such categorical matches five times more often when similar specimens come from the same source than when they originate from different sources. To give context to this number, one might add that the positive match has the same strength of association as discovering a matching blood type in a suspect and a crime-scene stain, where the blood type in question occurs in 1 person out of every 5 (and the blood test always identifies the blood type correctly).* In general, if the value of the likelihood ratio from the actual test is some number L from Equation 10.1, then the frequency that generates the value with an unambiguous discrete-trait test is just $1/L$.

Of course, the point estimates for the error probabilities from validation experiments are subject to sampling error. If the counts in the classification table are binomial random variables, confidence intervals for the error probabilities as well as the resulting likelihood ratio (or other interval estimates) can be computed (Simel et al., 1991). The PCAST report proposes informing the jury of the upper bound of a one-sided 95% confidence interval for the false-positive probability. Other methods for expressing the likely extent of sampling error also should be legally acceptable.†

The "error rates" mentioned in *Daubert* presuppose what I have called "conclusion-centric" as opposed to "evidence-centric" testimony (Kaye, 2018b). As shown in Figure 10.1, when working in the traditional conclusion-centric classification mode, criminalists tell the factfinder not merely that features match (or cannot be distinguished) and that

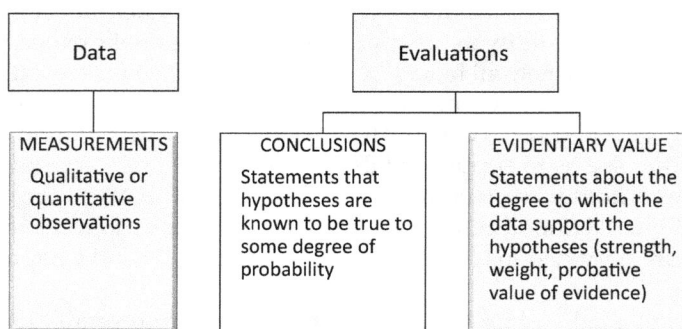

FIGURE 10.1
Distinguishing data from evaluations and conclusions from expressions of evidentiary value.

---

* A positive association on a blood test that is certain to be positive for S and has probability $1/5$ of being positive for $\sim S$ entails a likelihood ratio $L = 1/(1/5) = 5$.

† PCAST goes so far as to say that "[t]he use of lower values may rightly be viewed with suspicion as an attempt at obfuscation" (PCAST, 2016, p. 153). It is hard to imagine what is suspicious about the common procedure of reporting the entire range of a two-sided confidence interval or how that procedure is more obfuscatory than reporting only the upper end of an interval (if that is what the report means when it states that "[w]hen reporting a false positive rate to a jury, it is scientifically important to state the 'upper 95 percent one-sided confidence bound' to reflect the fact that the actual false positive rate could reasonably be as high as this value" (ibid., p. 51)). For criticism of some of PCAST's "upper bound" estimates, see Chapter 13 (Neumann et al., 2021).

the same-source hypothesis "could be" true, but also that the hypotheses is true (to some probability or even to a certainty). An evidentiary-value approach avoids these classifications by using relative likelihoods or degrees of support to indicate the weight or probative value of evidence. For trace evidence, it essentially consists of describing the noteworthy features being compared and explaining how strongly the observations of these features support competing conclusions.

Conclusion-centric testimony goes a step further (often without making information about relative support explicit). From a Bayesian perspective, the expert's conclusion is a statement of the analyst's posterior probability for a source attribution or some other hypothesis, H. If we let D stand for data, the expert making a positive source attribution believes that $P(H|D)$—the probability of the hypothesis given the data (and ascertained in light of the analyst's knowledge of the world)—is very large. So the difference between conclusion-centric and evidence-centric testimony is the difference between $P(H|D)$ and $P(D|H)$. In the first expression, the probability is attached to the conclusion H (conditional on the data); in the second, the probability is attached to the data—the expert measurements or observations of the specimens (conditional on the hypothesis).

As noted in § 10.3.3, the dominant view among forensic statisticians is that evaluations should be cast in the evidence-centric mold—as statements about the extent to which the evidence supports competing hypotheses rather than opinions about the hypotheses themselves.[*] Legal theorists have taken the same position (e.g., Robertson and Vignaux, 1995; Cheng, 2017; Kaye, 2018b; Kaye et al., 2011, § 15.6). The legal phrase "probative value" refers to the extent to which evidence tends to prove the hypothesis that it is offered to prove. Likelihoodists contend that the likelihood ratio (or its logarithm) expresses this idea quantitatively (Kaye, 2017). Moreover, for two simple, collectively exhaustive hypotheses, the ratio is a Bayes factor because it is then the ratio of posterior to prior odds on the proponent's hypothesis.[†] The logarithm of the Bayes factor has properties that make it a convenient metric for the weight of the evidence (Good, 1960, 1985, 1991). Of course, the term "weight" is metaphorical—it is not a quantity like a kilogram that has metrological traceability (compare Lund and Iyer, 2017, with Aitken et al., 2018).

Given the division of responsibility between the expert witness and the trier of fact described in § 1, the legal commentators maintain that the trier of fact's task is to adjust personal prior odds (odds formed without regard to the forensic-science evidence) and to return a verdict according to whether that probability satisfies the applicable burden of persuasion. The expert's role is more limited, but still important. It is to advise the judge or jury of a reasonably derived Bayes factor that factfinders can use (or not, as they see fit) to reach the posterior odds. Figure 10.2 summarizes these ideas.

By "confin[ing] their evaluative statements to expressions of support for stated hypotheses" (ASA, 2019), expert witnesses avoid venturing beyond their expertise and using their own, unstated prior probabilities to reach conclusions (Kaye et al., 2011, § 15.6; Thompson et al., 2018).

---

[*] This view is reflected in American Statistical Association Position on Statistical Statements for Forensic Evidence (2019), which "strongly advise[s] forensic science practitioners to confine their evaluative statements to expressions of support for stated hypotheses: e.g., the support for the hypothesis that the samples originate from a common source and support for the hypothesis that they originate from different sources."

[†] The Bayes factor when $\sim H$ is a disjunction of distinct alternative hypotheses (or parameter values in a composite hypothesis) will involve a weighted average of likelihoods, where the weights come from the prior probability distribution (Lee, 1989, pp. 126–127). Statistical experts still could provide the likelihood ratios for pairs of hypotheses, but they could not provide a single Bayes factor for the relative support provided solely by the data.

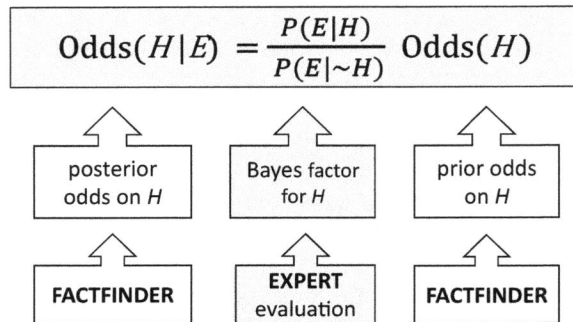

$$\text{Odds}(H|E) = \frac{P(E|H)}{P(E|\sim H)} \, \text{Odds}(H)$$

posterior odds on H	Bayes factor for H	prior odds on H
FACTFINDER	EXPERT evaluation	FACTFINDER

FIGURE 10.2
Bayesian view of the expert's role. In the simplest case, the expert can help the factfinder evaluate the strength of evidence by presenting the likelihood ratio as a Bayes factor.

The courts have not discussed likelihood ratios and Bayes factors extensively. American courts have been open to them in genetic tests for parentage, DNA-mixture cases, and a few other situations (Kaye et al., 2019). Testimony founded on a likelihood ratio informed by sample data and personal impressions of population frequencies for shoe print characteristics did not fare well in a prominent English Court of Appeal opinion (*R. v. T.*, 2010, 2011); however, the opinion was widely criticized (e.g., Berger et al., 2011; Kaye, 2012), and the court soon deemed similar testimony admissible (*R. v. South*, 2011).

Attempts to encourage jurors to perform their own computations of posterior probabilities via Bayes' rule were overruled on appeal in England (Donnelly, 2005), but in the United States, a state supreme court required parentage testers to describe the impact of a likelihood ratio across a range of prior probabilities (*Plemel v. Walter*, 1987; Kaye, 1988), and the Uniform Parentage Act makes an unchallenged likelihood ratio of 100 or more determinative of the biological fact of paternity or maternity (National Conference of Commissioners on Uniform State Laws, 2017).

10.4 Conclusion

Statistical analysis can contribute to the proof of disputed facts in litigation and to the judicial screening of forensic-science methods to ensure that they have sufficient "evidentiary reliability" to produce the "scientific knowledge" that the law requires (*Daubert v. Merrell Dow Pharmaceuticals*, 1993). For such analyses to be admissible, the expert needs to be qualified to testify by virtue of education, training, and knowledge of the statistical methodology. Furthermore, the statistical expert (or the substantive expert with statistical training) should be able to demonstrate, at a suitable level of particularity, that the methodology has been applied appropriately.

For forensic-science classification tasks, it is not always clear to courts that two types of conditional "error rates" from suitably designed and executed studies are usually needed to establish the reliability and validity of the classifications. The conditional error rates can be combined to derive estimates of the likelihood ratio for classifications. It is this quantity that helps answer the legal question of whether the expert classifications are generally good enough for presentation in the courtroom. An important legal consideration in the determination of admissibility is the expert witness's claimed degree of certainty. Extraordinary claims of certainty require extraordinarily low error probabilities. Good estimates of error probabilities and resulting likelihood ratios can give factfinders a sense of the probative value of forensic-science classifications, making the expert conclusions more useful to factfinders (and buttressing arguments for their admission).

Traditional classification procedures can be replaced with statements of the likelihoods or relative support that the forensic-science examination provides for the propositions of interest in light of data on measurement precision and the rarity of features. These statements help shift the expert's role from decisionmaker ("in my expert opinion, this is the correct conclusion") to advisor ("the data support the hypothesis that much more than the hypothesis that ... "). Validating qualitative or quantitative statements of relative support is more complex, but, if that project proves successful, this framework for interpretation fits well into the legal framework that governs expert statistical testimony.

References

American Statistical Association Position on Statistical Statements for Forensic Evidence Jan. 2, 2019. https://www.amstat.org/asa/files/pdfs/POL-ForensicScience.pdf

Aitken, C.G.G., Berger, C.E.H., Buckleton, J.S. et al. Expressing evaluative opinions: A position statement. *Science & Justice*, **51**, 1–2, 2011, http://dx.doi.org/10.1016/j.scijus.2011.01.002

Aitken, C., and Taroni, F. The history of forensic inference and statistics: A thematic perspective. In *Handbook of Forensic Statistics*, D. Banks, K. Kafadar, D.H. Kaye, and M. Tackett, editors. CRC Press, Boca Raton Florida, 2021.

Aitken C., Nordgaard, A., Taroni, F., and Biedermann, A. Commentary: Likelihood ratio as weight of forensic evi-dence: A closer look. *Frontiers in Genetics*, **9**, 1–2, 2018.

ASTM International. *ASTM E2926–17, Standard Test Method for Forensic Comparison of Glass Using Micro X-ray Fluorescence (μ-XRF) Spectrometry*. ASTM International, West Conshohocken, PA, 2017.

Berger, C.E., Buckleton, J., Champod, C., Evett, I.W., and Jackson, G. Evidence evaluation: A response to the court of appeal judgment in R v T. *Science & Justice*, **51**, 43–49, 2011. doi:10.1016/j.scijus.2011.03.005

Cheng, E.K. The burden of proof and the presentation of forensic results. *Harvard Law Review Forum*, **130**, 154–162, 2017.

Donnelly, P. Appealing statistics. *Significance*, **2**, 46–48, 2005. doi:10.1111/j.1740--9713.2005.00089.x

Evett, I.W., Berger, C.E.H., Buckleton, J.S., Champod C., and Jackson, G. Finding the way forward for forensic science in the US—A commentary on the PCAST Report. *Forensic Science International*, **278**, 16–23, 2017.

Forensic Science Regulator. Guidance on the Content of Reports Issued by Expert Witnesses in the Criminal Justice System in England and Wales, 2019. Birmingham, available at https://www.gov.uk/government/publications/expert-report-content-issue-3

Gittelson, S., Berger C.E.H., Jackson, G. et al. A response to "Likelihood ratio as weight of evidence: A closer look" by Lund and Iyer. *Forensic Science International*, **288**, e15–e19, 2018. doi:10.1016/j.forsciint.2018.05.025, PMID: 29857959

Good, I.J. Weight of evidence, corroboration, explanatory power, information and the utility of experiments. *Journal of the Royal Statistical Society, Series B*, **22**, 319–331, 1960.

Good, I.J. Weight of evidence: A brief survey. In *Bayesian Statistics 2*, J.M. Bernardo, M.H. Degroot, D.V. Lindley, and A.F.M. Smith, editors, pages 249–270. North-Holland, Amsterdam, 1985.

Good, I.J. Weight of evidence and the Bayesian likelihood ratio. In *The Use of Statistics in Forensic Science*, C.G.G. Aitken and David A. Stoney, editors, pages 85–106. Ellis Horwood, Chichester, 1991.

Hicks, T., Buckleton, J.S., Bright, J., and Taylor, D. A framework for interpreting evidence. In *Forensic DNA Interpretation*, J.S. Buckleton, J. Bright, and D. Taylor, editors, pages 37–86 (2nd ed.). Chapman & Hall/CRC, Boca Raton, FL, 2016.

Houck, M., and Budowle, B. Correlation of microscopic and mitochondrial DNA hair comparisons. *Journal of Fo-rensic Science*, **47**, 1–4, 2002.

Hunt, T.R. Scientific validity and error rates: A short response to the PCAST report. *Fordham Law Review Online*, **86**, 24–39, 2018. https://ir.lawnet.fordham.edu/flro/vol86/iss1/14/

National Research Council. 2004. *Forensic Analysis: Weighing Bullet Lead Evidence*. National Academies Press, Washington D.C., 2004. https://doi.org/10.17226/10924.

Kaye, D. Plemel as a primer on proving paternity. *Willamette Law Journal*, **24**, 867–883, 1988.

Kaye, D.H. *The Double Helix and the Law of Evidence*. Harvard University Press, Cambridge, MA, 2010.

Kaye, D.H. Likelihoodism, Bayesianism, and a pair of shoes. *Jurimetrics Journal*, **53**, 1–9, 2012.

Kaye, D.H. Reflections on glass standards: Statistical tests and legal hypotheses. *Statistica Applicata—Italian Journal of Applied Statistics*, **27**, 173–186, 2015.

Kaye, D.H. The false-positive fallacy in the first opinion to discuss the PCAST report. *Forensic Science, Statistics and the Law*, Nov. 3, 2016, http://for-sci-law.blogspot.com/2016/11/the-false-positive-fallacy-in-first.html

Kaye, D.H. Digging into the foundations of evidence law. *Michigan Law Review*, **115**, 915–934, 2017.

Kaye, D.H. How *Daubert* and its progeny have failed criminalistics evidence, and a few things the judiciary could do about it. *Fordham Law Review*, **86**, 1639–59, 2018a.

Kaye, D.H. The Nikumaroro bones: How can forensic scientists assist factfinders? *Virginia Journal of Criminal Law*, **6**, 101–118, 2018b.

Kaye, D.H. Propublica's picture of photographic analysis at the FBI laboratory (pt. 2). Forensic Science, Statistics, and the Law, Mar. 19, 2019, http://for-sci-law.blogspot.com/2019/03/propublicas-picture-of-photographic_19.html

Kaye, D.H., Bernstein, D.E., and Mnookin, J.L. *The New Wigmore: A Treatise on Evidence: Expert Evidence* (2nd ed.). Aspen, New York, 2011.

Kaye, D.H., Bernstein, D.E., and Mnookin, J.L. *Cumulative Supplement to The New Wigmore: A Treatise on Evidence: Expert Evidence* (2nd ed.). Aspen, New York, 2019.

Kaye, D.H., and Freedman, D.A. Reference guide on statistics. In *Reference Manual on Scientific Evidence*, pages 241–302 (3rd ed.). Committee on the Development of the third edition of the Reference Manual on Scientific Evidence, National Academy Press, Washington, DC, 2011.

Kaye, D.H., and Sanders, J. Expert advice on silicone gel breast implants: Hall v. Baxter Healthcare Corp. *Jurimetrics Journal*, **37**, 113–128, 1997.

Kaye, D.H., Vyvial, T.M., and Young, D.L. Validating the probability of paternity. *Transfusion*, **31**, 823–828, 1991. https://www.ssrn.com/abstract=2705941

Krawczak, M., Budowle, B., Weber-Lehmann, J., and Rolf, B. Distinguishing genetically between the germlines of male monozygotic twins. *PLOS Genetics*, **14**, e1007756, 1991. https://doi.org/10.1371/journal.pgen.1007756

Law Commission. *Expert Evidence in Criminal Proceedings in England and Wales*. Stationery Office, London, 2011.

Lee, P.M. *Bayesian Statistics: An Introduction*. Oxford University Press, New York, 1989.

Legal Research Committee. Memorandum to the Organization of Scientific Area Committees for Forensic Science, National Institute of Standards and Technology (rev. ed. Oct. 7, 2016), reprinted, Harvard Law Review Forum 130:137–144, 2016.

Lund, S. and Iyer, H. Likelihood ratio as a weight of forensic evidence: A closer look. *Journal of Research of National Institute of Standards and Technology*, **122**, 1–32, 2017.

McCormick, C.T. *Handbook of the Law of Evidence*. West, St. Paul, 1954.

Mnookin, J.L., and Kaye, D.H. Confronting science: Expert evidence and the Confrontation Clause. *Supreme Court Review*, **2012**, 99–159, 2013.

Moretti, T.R., Just, R.S., Kehl, S.C., Willis, L.E., Buckleton, J.S., Bright, J., Taylor, D.A., and Onorato, A.J. Internal Validation of STRmixTM for the Interpretation of Single Source and Mixed DNA Profiles. *Forensic Science International: Genetics*, **29**, 126–144, 2017.

Morrison, G.S., Kaye, D.H., Balding, D.J. et al. A comment on the PCAST report: Skip the "match"/"non-match" stage. *Forensic Science International* **272**:e7–e9, 2017.

National Conference of Commissioners on Uniform State Laws. Uniform Parentage Act, 2017. https://www.uniformlaws.org/committees/community-home/librarydocuments/viewdocument?DocumentKey=3fd1d24b-11e7-4e09-b82e-f2b2ef9956af

National Research Council Committee on Identifying the Needs of the Forensic Sciences Community. *Strengthening Forensic Science in the United States: A Path Forward*. National Academy Press, Washington D.C, 2009.

Neumann, C., Hendricks, J., and Madeline Ausdemore. Statistical support for conclusions in Fingerprint examinations. In *Handbook of Forensic Statistics*, D. Banks, K. Kafadar, D.H. Kaye, and M. Tackett, editors. CRC Press, Boca Raton, 2021.

Nordgaard, A., Ansell, R., Drotz, W., and Jaeger, L. Scale of conclusions for the value of evidence. *Law, Probability and Risk*, **11**, 1–24, 2012.

Parker, J.B. A statistical treatment of identification problems. *Journal of the Forensic Science Society*, **6**, 33–39, 1966.

President's Council of Advisors on Science and Technology. Forensic Science in Criminal Courts: Ensuring Scientific Validity of Feature-comparison Methods, 2016. https://obamawhitehouse.archives.gov/sites/default/files/microsites/ostp/PCAST/pcast_forensic_science_report_final [https://perma.cc/VJB4-5JVQ]

Ramos, D., Meuwly, D., Haraksim, R., and Berger, C.E.H. Validation of forensic automatic likelihood ratio methods. In *Handbook of Forensic Statistics*, D. Banks, K. Kafadar, D.H. Kaye, and M. Tackett, editors. CRC Press, Boca Raton Florida, 2021.

Robertson, B., and Vignaux, G.A. *Interpreting Evidence: Evaluating Forensic Science in the Courtroom*. John Wiley and Sons, Chichester, 1995.

Roth, A. Machine testimony, *Yale Law Journal*, **126**, 1972–2259, 2017.

Royall, R. On the probability of observing misleading statistical evidence. *Journal of the American Statistical Association*, **95**, 760–768, 2000, https://doi.org/10.1080/01621459.2000.10474264

Simel, D.L., Samsa, G.P., and Matchar, D.B. Likelihood ratios with confidence: Sample size estimation for diagnostic test studies. *Journal of Clinical Epidemiology*, **44**, 763–770, 1991.

Stern, H. Comparing philosophies of statistical inference. In *Handbook of Forensic Statistics*, D. Banks, K. Kafadar, D.H. Kaye, and M. Tackett, editors. CRC Press, Boca Raton Florida, 2021.

Taylor, D., Buckleton, J., and Evett, I.W. Testing Likelihood ratios produced from complex DNA profiles. *Forensic Science International: Genetics*, **16**, 165–171, 2015

Taylor, D., Curran, J.M., and Buckleton, J. Importance sampling allows H_d true tests of highly discriminating DNA profiles. *Forensic Science International: Genetics*, **27**, 74–81, 2017.

Thompson, W.C., Vuille, J., Taroni, F., and Biedermann, A. After uniqueness: The evolution of forensic-science opinions. *Judicature*, **102**, 19–27, 2018. https://judicialstudies.duke.edu/wp-content/uploads/2018/04/JUDICATURE102.1-THOMPSON-etal-1.pdf

Ulery, B.T., Hicklin, R.A., Buscaglia, J., and Roberts, M.A. Accuracy and reliability of forensic latent fingerprint decisions. *Proceedings of the National Academy of Sciences (USA)*, **108**, 7733–7738, 2011.

Vorder Bruegge, R.W. Photographic identification of denim trousers from bank surveillance film. *Journal of Forensic Sciences*, **44**(3), 613–622, 1999.

Weir, B.S. DNA frequencies and probabilities. In *Handbook of Forensic Statistics*, D. Banks, K. Kafadar, D.H. Kaye, and M. Tackett, editors. CRC Press, Boca Raton Florida, 2021a.

Weir, B.S. Kinship. In *Handbook of Forensic Statistics*, D. Banks, K. Kafadar, D.H. Kaye, and M. Tackett, editors. CRC Press, Boca Raton Florida, 2021b.

Wigmore, J.H. *A Treatise on the Anglo-American System of Evidence in Trials at Common Law* (3rd ed.). Little Brown, Boston, 1940.

Cases and Rules

Commonwealth v. McNair, No. 8414CR10768 (Mass. Super. Ct., Suffolk County, Jan. 12, 2018).

Daubert v. Merrell Dow Pharmaceuticals, 509 U.S. 579 (1993).

Farm & Garden Center v. Kennedy, 921 N.W.2d 615 (Neb. Ct. App. 2018).

Federal Rule of Evidence 702 advisory committee note to 2000 amendment.

Frye v. United States, 293 F. 1013 (D.C. Cir. 1923).

General Electric Co. v. Joiner, 522 U.S. 136 (1997).

Kumho Tire Co., Ltd. v. Carmichael, 526 U.S. 137 (1999).

Memorandum of Decision and Order on Defendant's Motion in Limine to Exclude All Results of "Ultra-deep next Generation Sequencing" for Distinguishing DNA from Identical Twins with Samples from Different Cell Types, Commonwealth v. McNair, No. 8414CR10768 (Mass. Super. Ct., Suffolk County, Apr. 11, 2017), available at http://for-sci-law.blogspot.com/2019/03/the-gordian-knot-in-commonwealth-v.html

People v. Collins, 438 P.2d 33 (Cal. 1968).

Plemel v. Walter, 735 P.2d 1209 (Or. 1987).

R. v. South [2011] EWCA (Crim) 754.

R. v T [2010] EWCA (Crim) 2439; [2011] 1 Cr. App. R. 9.

State v. Garrison, 585 P.2d 563 (Ariz. 1978).

Transcript of Hearing, Commonwealth v. McNair, No. 1484CR10768 (Mass. Super. Ct., Suffolk County, Feb. 15, 2017).

United States v. Angleton, 269 F.Supp.2d 892 (S.D. Tex. 2003).

United States v. Love, No. 10cr2418–MMM, 2011 WL 2173644 (S.D. Cal. June 1, 2011).

United States v. McKreith, 140 Fed. Appx. 112, 2005 WL 1600471 (11th Cir. 2005) (per curiam).

United States v. Natson, 469 F.Supp.2d 1253 (M.D. Ga. 2007).

United States v. Smith, 869 F.2d 348 (7th Cir. 1989).

Williams v. Illinois, 567 U.S. 50 (2012).

Section IV

Applications of Statistics to Particular Fields in Forensic Science

11

DNA Frequencies and Probabilities

Bruce S. Weir

CONTENTS

11.1 Introduction 251
11.2 Likelihood Ratios 252
11.3 Population Genetics 254
    11.3.1 Single Loci 255
    11.3.2 Multiple Loci 255
    11.3.3 Population Structure 256
    11.3.4 Lineage Markers 257
11.4 Mixtures 258
    11.4.1 Semi-Continuous Model......................... 259
    11.4.2 Continuous Model 259
11.5 DNA Sequence Data 260
11.6 Future Directions 260
11.7 Conclusions 261
References 262

11.1 Introduction

In many ways, DNA profiling has served as the poster child for attaching statistical calculations to forensic evidence. This may reflect the early involvement of statistical geneticists who were experienced in performing calculations for genetic data, and it may reflect the absence of a culture in which a forensic analyst would examine DNA evidence, declare it to have the same type as that for a person of interest (POI), and then claim the POI was the source of the evidence. There was widespread acceptance of Mendel's Laws, by which a parent is equally likely to pass on the maternal or paternal copy of a genetic element to a child, with these transmissions being independent for different genetic elements. Early statistical treatments, however, did not pay attention to the field of evidence interpretation, and apparently simple concepts such as "random match probability" or "random man not excluded" may have led to unsatisfactory analyses in the early days of forensic DNA profiling.

Thirty years after the pioneering work of Alec Jeffreys and the UK Forensic Science Service (Jeffreys et al., 1985a,b), both the nature of DNA profiling and the methods used to quantify the strength of DNA evidence have evolved. There is now general acceptance of the value of such evidence. Science is not static of course, and forensic DNA typing is in the process of changing from the examination of a small number of discrete genetic elements or markers to the sequencing of increasingly large regions of the genome, and it is adopting probabilistic genotyping analyses whereby the "exclusive, inconclusive or inclusive" characterization of a POI for an item of evidence is replaced by probabilities associated with the involvement of the POI.

This chapter reviews the basic elements of DNA evidence interpretation, describes current areas of development, and speculates on future trends. The chapter shows that these are challenging and rewarding times for forensic genetic statisticians.

11.2 Likelihood Ratios

To introduce concepts and notation, suppose a biological sample is typed and found to have DNA profile G_C. A POI has profile G_S and is not excluded from being a contributor to the sample if the two profiles are the same, $G_C = G_S$. Clearly this matching between the sample and the POI will be more probative for profiles that are less common in a population, but approaching questions about strength of evidence based only on population proportions is not fully informative and can be misleading. A coherent approach can be extended to apply to more complex situations such as multi-contributor samples, the possible involvement of relatives of the POI, the genetic structure of the relevant population, and the possibility of erroneous typing or typing that has been affected by allelic drop out or drop in. Such situations call for an approach that addresses a basic question: does the sample profile represent DNA from the POI? Even this simple question, however, is best considered after articulation of alternative hypotheses about the sample (Evett and Weir, 1998), further refined into a "hierarchy of propositions." As given in Cook et al. (1998), three levels of hypotheses about the matching profile can be considered. If the sample is a bloodstain found on a broken window at the scene of a burglary, these three levels are

- Source level: H_0: The blood on the broken window is from the POI, versus H_1 : The blood on the broken window is from some other person.
- Activity level: H_0: The window was broken by the POI, versus H_1 : The window was broken by some other person.
- Offense level: H_0: The burglary was committed by the POI, versus H_1 : The burglary was committed by some other person.

It is the offense level propositions that are of interest to a court, whereas it is the source level that can be addressed by the forensic analyst. The source level propositions have been further divided, in Taylor et al. (2018) for example, into source and sub-source, with sub-source referring particularly to the DNA profile. In the burglary example, the origin of the DNA profile could have been either blood resulting from a cut while breaking the window or it could have been from a fingerprint from touching the window at some other time. It is only sub-source propositions that will be considered here, but it is worth stressing that strong statements about whose DNA is in a sample need not imply the POI whose type matches the sample committed a crime, and there have been situations (Gill, 2014), where the presence of DNA from a POI has an explanation unrelated to the crime being investigated.

With sub-source hypotheses formulated for DNA evidence E, attention is appropriately focussed on a non-excluded POI with the likelihood ratio LR, as described in other chapters of this volume:

$$LR = \frac{Pr(E|H_0)}{Pr(E|H_1)}$$

Probabilities are attached to evidence, not to propositions. In the present example of the evidence being the two profiles G_C, G_S, standard probability arguments allow the LR to be re-written as

$$LR = \frac{Pr(G_C|G_S, H_0)}{Pr(G_C|G_S, H_1)} \tag{11.1}$$

The numerator is the probability of the person whose blood is on the window having profile G_C, if that person is the POI whose type is known to be $G_S = G_C$. If the DNA typing procedure is free of errors and ambiguities, this conditional probability is 1 and the LR numerical value reduces to the reciprocal of the match probability $Pr(G_C|G_S, H_1)$: the probability of an untyped person, not the POI, having the same profile as the POI given the observation of the POI profile.

The LR is a statement describing the strength of the evidence against a POI whose DNA profile matches that in the crime sample. It is the ratio of two probabilities: it is not itself a probability and it is not the proportion or number of people in any population with that profile. In other words, it is not the reciprocal of the profile probability $Pr(G_C)$, the probability of an untyped person having the crime sample profile. It would be confusing to term $Pr(G_C)$ the "random match probability" and also confusing to express an LR of, say, one million as "1 in a million." This last expression leads to the possibility of the Defense Attorney's Fallacy (Thompson and Schumann, 1987) where a proportion 1 in a million is interpreted as there being 10 people in a city of 10 million people with the particular profile and the POI being regarded as being chosen from 10 people. Apart from confusing expected and actual numbers of times an event occurs, and supposing all people with the particular profile should be considered as equally likely contributors to the sample regardless of their age or gender, this line of argument ignores the LR being the ratio of two probabilities. A much more common misinterpretation of the LR, the Prosecutor's Fallacy (Thompson and Schumann, 1987), is to equate $Pr(E|H_0)$ to $Pr(H_0|E)$ and interpret a low match probability as a low probability of the proposition H_0. A low probability for H_0, in turn, may suggest a high probability the POI is not the source of the DNA in a sample whereas the forensic analyst can say only that, if the POI is not the source, there is a low probability of a match between POI and sample profiles.

The LR approach brings in the match probability $Pr(G_C|G_S, H_1)$ as a statement about two profiles, one from each of two people. The early "random match probability" (Koehler et al., 1995) was often less explicit in being about two people and was unnecessarily restrictive in requiring G_C to be from a person, not the POI, being chosen randomly from a population. Match probabilities can be calculated for any degree of dependence between the two people, as shown below. The other early quantity, the probability of a "random man not excluded" is the chance a person selected randomly from a population does not have alleles not detected in the evidence profile. A person homozygous for allele A, for example, would not be excluded as the source of single-contributor evidence that was heterozygous for alleles A and B. It does not take account of the profile of the POI, and it can substantially understate inclusion of a POI profile in multi-contributor evidence.

Although the likelihood ratio is an appropriate quantity to express the probative value of matching DNA profiles, it does not have the intuitive appeal of probability statements about the source of a crime-scene profile after examination of evidence. In other words, posterior probabilities for hypotheses given the evidence are desired, but Bayes' Theorem states that

$$\frac{\Pr(H_0|E)}{\Pr(H_1|E)} = \frac{\Pr(E|H_0)}{\Pr(E|H_1)} \times \frac{\Pr(H_0)}{\Pr(H_1)}$$

Posterior odds for H_0 = Likelihood Ratio × Prior odds for H_0

so posterior probabilities require prior probabilities, and these are beyond the province of the forensic analyst. However, there has been recent work to attach some sense to the magnitude of LRs. Should an LR of a billion convince triers-of-fact? What about an LR of a thousand? Or ten? A narrow answer was provided in Beecham and Weir (2011) with attention to the uncertainty in LRs introduced by the use of sample allele proportions for allele probabilities. A similar treatment in Gittelson et al. (2017) accounted for the choice of database on calculated LR values: if the unknown contributor to the evidence was of European ancestry, for example, but allele frequencies from a database of profiles from people not of European ancestry were used to calculate the match probability, then the result would not be a good estimate of the appropriate value. Different databases will provide different match probability estimates, and some idea of the variation among resulting LRs can be given by using a series of different databases. Alternatively, population genetic theory can be used as described in Section 11.3, on population structure.

Another approach was reviewed in Swaminathan et al. (2016) where the distribution of LRs was used to assess robustness: for hypotheses that the POI is or is not a contributor to a profile, how often will the LR be greater than 1, and so support the POI being a contributor, when in fact he is not? This approach was extended in Gill and Haned (2013) with the introduction of a "*p*-value" or the probability a randomly chosen individual results in a LR greater than the LR obtained for a POI. This may be interpreted as the false positive rate resulting from a hypothesis test between the prosecution and defense hypotheses. A continuous model was used in Swaminathan et al. (2016) to show results such as *p*-values less than 10^{-9} occurring when the LR is greater than 10^8. This concept of a *p*-value has been criticized (Kruijver et al., 2015) because the *p*-value, defined as the probability of an LR at least as large as the LR for the POI when the POI is not a contributor, is bounded above by the reciprocal of the LR. Moreover, the *p*-value does not address the alternative hypotheses concerning the POI for which the LR was calculated.

There does seem to be value in simulating DNA profiles under alternative hypotheses and comparing the empirical distributions of the likelihood ratios when either H_0 or H_1 is true. For a particular number of genetic markers and a particular database of DNA profiles used for the simulations, this can suggest thresholds above which LR values occur much more often when H_0 is true than when H_1 is true. Ultimately though, the inherent uncertainty about the truth of any hypothesis must be acknowledged.

11.3 Population Genetics

Most current forensic DNA profiles consist of the genetic constitution, or genotype, of a sample or a person at up to 20 individual positions in the human genome. At each of these positions a person has material, or an allele, from each of their two parents so that the profile is a set of 20 pairs of allele types. The types reflect the number of repeat units, or very short stretches of DNA, at each Short Tandem Repeat (STR) genetic marker (Butler, 2009). The 20 markers in common use are referred to as the CODIS expanded set, where:

> CODIS is the acronym for the Combined DNA Index System and is the generic term used to describe the FBI's program of support for criminal justice DNA databases as well as the software used to run these databases. The National DNA Index System or NDIS is considered one part of CODIS, the national level, containing the DNA profiles contributed by federal, state, and local participating forensic laboratories. FBI (2020)

There are currently 18 million profiles in NDIS.

11.3.1 Single Loci

To a first approximation, the two alleles at a single STR locus carried by an individual are independent, as expected if the individual's parents are unrelated. Formally, allelic independence is known as Hardy-Weinberg Equilibrium (HWE), and if this situation holds then the probability for any single-locus genotype can be expressed as the product of the corresponding allele probabilities. If P_{uv} is the probability of a random person having alleles of types u and v, and if p_u is the probability of a single allele being of type u,

$$P_{uv} = \begin{cases} p_u^2, u = v & \text{Homozygote} \\ 2p_u p_v, u \neq v & \text{Heterozygote} \end{cases} \tag{11.2}$$

In practice, HWE is assumed and population genotype probabilities are estimated from database allele proportions. Following an early claim (Lander, 1989), refuted in Devlin et al. (1990), of HWE departures in a Lifecodes database, thousands of HWE tests in forensic databases have been conducted and generally non-significant results published. Estimation of a genotype probability can proceed by multiplying together sample allelic proportions, even from FBI frequency databases with sample sizes around 200 (Moretti et al., 2016) where many of the possible genotypes for an STR marker are not observed. Although the need for the HWE assumption has largely been circumvented by the population-structure modifications to match probabilities approach (Balding and Nichols, 1994) described next, HWE testing still has a role as a quality control measure (Balding and Steele, 2015) even though it is not as important as was once thought. HWE testing, if performed, should follow best practice (Graffelman and Weir, 2018).

11.3.2 Multiple Loci

The CODIS markers are located on the autosomes, the chromosomes other than X and Y that are not associated with sex determination. These markers are generally regarded as being independent, so that probabilities calculated for each marker can be multiplied together across all 20 loci, and the logarithms of such products are approximately normally distributed since the Lindeberg conditions hold for the Central Limit Theorem. There is growing evidence, however, that there are dependencies among single-locus match probabilities and that these may have an appreciable effect for large numbers of loci. The issue of dependencies arising from relatedness were discussed in Donnelly (1995): "after the observation of matches at some loci, it is relatively much more likely that the individuals involved are related (precisely because matches between unrelated individuals are unusual) in which case matches observed at subsequent loci will be less surprising. That is, knowledge of matches at some loci will increase the chances of matches at subsequent loci, in contrast to the independence assumption." A theoretical prediction of dependencies in match probabilities, focusing on the joint effects of mutation and genetic drift, was given in Laurie and Weir (2003). An empirical study of matching at up to six loci was given in Weir (2004): among 15,000 forensic STR profiles, the ratios of multi-locus match proportions to products of single-locus proportions were 1.000, 1.000, 1.008, 1.034 and 1.041 for two, three, four, five and six loci. This observation suggests that attempting to test for multi-locus dependencies is not necessary as dependencies are expected, and it also suggests the need to question the extent to which increasing the number of loci will decrease the probability of matching between pairs of profiles. Work continues in this area, with empirical support for the hope that modifications to single-locus probabilities to account for population structure, as discussed next, will accommodate dependencies among loci.

11.3.3 Population Structure

Even if assumptions of allelic independence within and between STR loci are reasonable for individuals within the relevant population for a particular crime, there are the problems of describing that population with any precision and then in collecting data to estimate its allele probabilities. Instead, use is made of frequency databases from a population thought to encompass the relevant population. Such databases are typically described by major continental ancestries, and a forensic laboratory report may include statements such as

> Statistical approximations have been generated and are reported using the 2015 FBI STR population data and allele frequencies.
> The DNA profile from Item Number 1 [sample] matches the DNA profile from Item Number 2 [POI] at 9 STR loci. The approximate incidence of this profile is 1 in 1.0 trillion Caucasians, 1 in 18 trillion African Americans and 1 in 3.2 billion Hispanics.

The impression given by this report is that it is unlikely anyone but the POI would have the particular 9-locus STR profile in the sample, regardless of the continental ancestry of that person. Is the FBI Caucasian database described in Moretti et al. (2016) relevant for the crime addressed in this report? For people of European ancestry living in the US, for example, allele frequency differences are expected among groups whose recent ancestors were from different European countries, and the relevant population for this crime may be enriched for people whose ancestry was in one of these countries. Variation among allele proportions for subpopulations of the population represented by a database is described by the parameter θ, and the allele probabilities p_{iu} for allele type u in subpopulation i might be supposed to follow the Beta distribution with parameters $(1 - \theta_i)\pi_u/\theta_i$ and $(1 - \theta_i)(1 - \pi_u)/\theta_i$. The allele probabilities π_u apply to the whole population. Each p_{iu} and π_u are averages over replicates of the evolutionary processes leading to subpopulation i or the whole population. There is a population-genetic literature (Wright, 1951) supporting this model and more recent notice (Balding and Nichols, 1994) that there are simple consequent expressions for probabilities of sets of alleles in the subpopulation. The probability of drawing an allele of type u from subpopulation i after having seen n_u copies of that type in the previously drawn n alleles is

$$\Pr(u|n_u, n) = \frac{n_u\theta_i + (1 - \theta_i)\pi_u}{1 + (n - 1)\theta_i}$$

Applications of this result gives the Balding-Nichols match probabilities for homozygotes and heterozygotes (Balding and Nichols, 1994):

$$\Pr(uu|uu) = \frac{[2\theta_i + (1-\theta_i)\pi_u][3\theta_i + (1-\theta_i)\pi_u]}{(1+\theta_i)(1+2\theta_i)}$$

(11.3)

$$\Pr(uv|uv) = \frac{2[\theta_i + (1-\theta_i)\pi_u][\theta_i + (1-\theta_i)\pi_v]}{(1+\theta_i)(1+2\theta_i)}, \quad u \neq v$$

The match probabilities in Equations 11.3 are greater than the profile probabilities $\Pr(uu)$ and $\Pr(uv)$, meaning that the chance of seeing a profile increases after it has already been seen to exist in the population. The theory leading to these equations has assumed HWE in each subpopulation but the equations explicitly show that HWE does not hold in the whole population represented by the database used to furnish estimates of the allele probabilities π_u. There are published recommendations for θ values (Buckleton et al., 2016; National Research Council, 1996). Buckleton et al. (2016) base their recommendations on published STR allele proportions from many populations within the major human ethnic groups. The θ value estimates range from 0.008 for variation among African populations to 0.037 for variation among Native American populations. These θ estimates are the best available guides for use within any one country, such as the US, even though they are not based on observed variation among actual populations of African or Native American ancestry within that country. Using values in the range 0.01 - 0.05 along with national database allele proportions seems to be appropriate. These recommended values are to be used with population allele proportion estimates from a database, and the match probability estimates are intended to apply to any subpopulation of that population. There is no need for the subpopulations to all have the same ancestry: providing the database represents all ancestries, then Equation 11.3 would apply to any ancestry in the database population.

To the extent that data are available, empirical profile match proportions can be compared to those predicted by Equations 11.3. As matches are rarely observed beyond ten loci, these empirical demonstrations are limited, but the equations seem to be appropriately conservative with θ in the recommended range. Moreover, this conservatism (predicted match probabilities exceeding observed match proportions) seems to accommodate the between-locus dependencies referred to earlier (Weir, 2004).

11.3.4 Lineage Markers

In addition to the CODIS set of autosomal STR markers, forensic scientists make use of mitochondrial DNA sequences and STR profiles for the Y chromosome. Mitochondria are small molecules that exist outside the nucleus of cells and are distinct from the autosomes and the X and Y chromosomes within the nuclei. They are carried by males and females although they are transmitted only by females to their children. Because of their small size and large number of copies per cell they are less susceptible to degradation than STR markers and they find great use in typing bones, even those from ancient remains. Mitochondrial typing proceeds by sequencing the hypervariable region of the mitochondrial genome, and identifying the few positions in the sequence where there are differences from a published reference sequence (Bar et al., 2000). The process has been used extensively to identify war remains (Holland et al., 1993).

The Y chromosome is carried only by males, and so is transmitted only by men to their sons. Up to 30 STR markers on the Y chromosome are used, primarily in sexual assault cases where the evidence sample may contain DNA from both a female victim and a male perpetrator. Typing proceeds as for autosomal STR markers, but it produces only one allele per man for each typed marker. There is an international database, YHRD (https://yhrd.org/) of over 200,000 Y-STR profiles with national and continental ancestry identifiers. Each YHRD profile has alleles for up to 27 markers.

The absence of recombination in the mitochondrion and in the non-recombining portion of the Y chromosome argues against independence of single-locus mitochondrial variant or Y-STR match proportions, although the independence of mutational events at different loci results in low levels of linkage disequilibrium (Hall, 2016).

The simplest alternative to multiplying match proportions over loci is to determine mitochondrial or Y-STR haplotype proportions in a database. The limitation of these databases for giving profile proportions is that a profile not observed at some number of DNA sequence positions or STR loci will also not be observed at an increased number of loci, even though the strength of the evidence would seem to increase with the number of markers. Just as with autosomal profiles, however, there are diminishing returns from adding loci past some optimal number of loci. A match at 10 or more loci suggests two profiles are from the same mitochondrial female lineage or the same Y chromosome male lineage and further matching is expected at additional loci apart from some mutations.

The match probability formulation of Balding and Nichols (1994) is also relevant for Y-haplotypes. The probability for haplotype A, given that it has already been seen, is

$$\Pr(A|A) = \theta + (1 - \theta)\pi_A \tag{11.4}$$

where θ is the probability of identity by descent for two haplotypes drawn randomly from a population and π_A is the probability a single haplotype is of type A. The value of θ decreases with the number of loci, and for 20 loci (Hall, 2016) estimated values are of the order of 10^{-5}. For very rare haplotypes, the match probability reduces to θ for the relevant population.

For either mitochondrial or Y-STR profiles, a high degree of matching between profiles suggests relatedness, and therefore some degree of autosomal STR matching as well. There have been investigations of dependencies between matching proportions among the three systems, autosomal, Y chromosome and mitochondrial (Walsh et al., 2008) and although the dependencies are low, they do exist. Combining match probability estimates over systems is not generally recommended.

11.4 Mixtures

Evidentiary samples may contain material from more than one contributor. One situation is for evidence collected in sexual assault cases where material from the victim, possible consensual partners, and the perpetrator(s) may all be present. Another situation is "touch DNA" where swabbing a surface may collect DNA from several people who have touched that surface. Even if some of the contributors have provided only a small proportion of the DNA detected in the sample, improved technology has made it easier to detect their alleles in the mixed profile. Interpretation problems arise when not all the alleles of all contributors, especially minor contributors, are detected. There is allele drop-out and, once this is recognized, account should also be taken of possible allele drop-in. Probabilities for the set of alleles constituting the genetic evidence E are statements about alleles carried by typed people known to have contributed as well as by unknown people, but now the earlier binary model with an allele being declared absent or present in a sample profile needs to be modified. Semi-Continuous and Continuous models will now be described.

11.4.1 Semi-Continuous Model

Situations in which some of the alleles in the evidence profile may be masked by typing artifacts such as stutter or may have dropped out completely and are not detected have been discussed in Gill et al. (2006). There may also be the possibility of sporadic alleles from fragmented genomes dropping in to the evidentiary profile, a different situation from whole genomes contaminating the evidence profile. A complete analysis needs to take into account the relative amounts of DNA inferred to be present at each of the alleles detected in the mixture. Having to allow for not-detected alleles reduces the possibility of being able to exclude a potential contributor to the mixture simply because that person's alleles are not detected. Great care needs to be taken to avoid prejudicial conclusions if it is decided to ignore those loci in a profile for which interpretation is difficult or alleles are suspected of not being detected. A straightforward approach that is based only on the alleles detected in the mixture, without regard to their relative amounts, was given in Gill and Haned (2013).

There is a set E of alleles detected in the evidence sample, a set of individuals specified to be contributors under proposition H, a set of individuals specified to not be contributors under H, and a set of unknown contributors U. The numbers of known contributors (likely to include a suspect under a prosecution hypothesis) and known non-contributors (likely to include a suspect under a defense hypothesis) are known, along with their genotypes. The number of unknown contributors is not known, but may be inferred if the analyst specifies how many contributors there are to the evidence, and the probabilities of E under each hypothesis require summation over all values of U. The probability for each U depends on the population structure formulation discussed above. The sets of alleles carried by the known and unknown contributors are compared to the alleles detected in the evidence and drop-out and drop-in events invoked to make these sets match.

Once the probabilities of the evidence mixture profile have been determined for each of two alternative hypotheses, the likelihood ratio can be calculated. The need to specify the drop-out rate is avoided by the approach in Slooten (2017) where the likelihood ratio is integrated over the range of drop-out rates.

11.4.2 Continuous Model

In the continuous model STR allelic signals are treated as continuous variables. The probabilities for likelihood ratios are replaced by probability densities based on models that are fit using training data from mixtures with contributors of known genotype in known proportions (Perlin et al., 2011; Taylor et al., 2013). An overview of continuous-model software was given in Moretti et al. (2017): The software weighs potential genotypic solutions for a mixture by utilizing more DNA typing information, such as the amount of DNA detected for an allele and the molecular weight of an allele. Models for allelic drop out and drop in are specified. It is still the case that likelihood ratios are generated to express the weight of the DNA evidence given two user-defined propositions. Such probabilistic genotyping software has been demonstrated to reduce subjectivity in the interpretation of DNA typing results and, compared to binary interpretation methods, to provide a more powerful tool supporting the inclusion of contributors to a DNA sample and the exclusion of non-contributors. There are now both commercial and open-source software packages available, and guidelines have been published for validating them (Coble et al., 2016; SWGDAM, 2015).

11.5 DNA Sequence Data

The emergence of sophisticated statistical models and software packages for existing capillary electrophoretic detection of STR variants represents considerable effort by statisticians, statistical geneticists and forensic scientists. This continues a tradition of attention to the error structure in forensic genetic data. For example. at the time when DNA fragment lengths were being estimated from gel electrophoresis data, a pair of papers using Bayesian techniques (Berry et al., 1992; Devlin et al., 1992) modeled the correlation structure of observations on the variants in a profile. A quarter-century later, forensic scientists are gaining access to other new genomic data: SNPs (single nucleotide polymorphisms) at a targeted number of positions, or at all the positions in a targeted region revealed by DNA sequencing. The latter has been called NGS (next-generation sequencing) or MPS (massively parallel sequencing) by forensic scientists (Borsting and Morling, 2015). Not only are these sequence-based profiles highly-discriminatory, but also they may contain information about physical appearance, ethnicity and even health status (Kidd et al., 2015).

The massive amounts of SNP data being collected for biomedical research and by direct-to-consumer testing makes it inevitable that forensic applications for such data will be explored, as illustrated by the recent "Golden State Killer" case (Phillips, 2018) whereby SNP profiles of an unknown person can identify relatives of that person in a genealogy database, and hence the person. Some of those issues are discussed in the next chapter in this handbook.

For the forensic science community to embrace SNP typing, there would need to be compatibility with large STR-profile databases such as NDIS mentioned earlier. The forensic applications of DNA sequencing currently rest on the sequencing of relatively short regions of DNA including the current set of STR markers. There are software packages to convert these sequence data to STR profiles (Woerner et al., 2017), making the sequence profiles backwards compatible with STR profile databases, and also to reveal single nucleotide variation (SNV) in the sequenced region to augment the length variation among STR variants.

In addition, other omic data are being introduced to forensic science. Amino acid variation in protein sequences has been used to infer SNP profiles as a way to avoid problems with degradation of DNA (Parker et al., 2016), and to be able to generate profiles from hair shafts. There is considerable activity (Chong et al., 2015) in using RNA assays to identify the source organs for biological tissue or body fluids. Methylation patterns seen in samples can allow estimation of the age of the person who was the source of the sample (Aliferi et al., 2018).

11.6 Future Directions

The recent past has seen substantial changes in forensic genetics. Increased sensitivity of STR typing has led both to an increased number of evidence samples that are mixtures and to the increased need to allow for allelic drop out and drop in. This has led, in turn, to the introduction of probabilistic genotyping. These new statistical approaches will be extended to DNA sequence data, with read depth providing quantitative units of observation (Ricke et al., 2019). Omic data, including proteomic, RNA and methylation data will play an increasing role in forensic science.

Statistical analyses will reflect more of a big-data approach with use of machine learning and other techniques to mine DNA sequence databanks and provide new methods of matching profiles. It is expected that the likelihood ratio framework will prove resilient to advances in typing technologies and computational power. Current uncertainty over the best way to interpret lineage marker profiles is expected to be resolved (Andersen and Balding, 2017; Brenner, 2013; Hall, 2016).

The embrace of direct-to-consumer SNP typing by participants curious about their ancestry or future health status may well swing public opinion towards universal genetic profiling with provision for forensic use of those data. The guidance of legal and ethics scholars will be crucial in any such move (Murphy, 2018).

Interpreting forensic genetic profiles in the future will offer new challenges for statistical genetics.

11.7 Conclusions

The forensic use of genetic information, such as the ABO blood group and other proteins in blood, dates back to the early 1900s, but the introduction of DNA profiles 35 years ago has had a profound impact on forensic science. The potential to use these profiles to associate a person with forensic evidence, to distinguish between any two profiles or to associate a profile with a family, have wide-ranging implications for decisions aided by statistical analyses. Quantifying the strength of DNA evidence rests on the Mendelian principles of genetics, including the randomness with which maternal or paternal copies of autosomal genetic elements are transmitted from parent to child and the field of statistical genetics had a rich history before Alec Jeffreys described DNA fingerprints.

Statistical geneticists already had the machinery to accommodate dependencies among genetic elements, such as those arising from population structure or family membership, and to incorporate mutation that can change the state of an element. They were also experienced in taking account of genetic typing technologies that may provide false or ambiguous results. Forensic DNA typing required the additional aspect of being sensitive to legal issues. As noted in Feinberg (1989)

> The very goals of science and law differ. Science searches for the truth and seeks to increase knowledge by formulating and testing theories. Law seeks justice by resolving individual conflicts, although this search often coincides with one for truth. . . . Rules of decision that are not tailored to individual cases, such as those that turn on statistical reasoning, are often viewed as suspect.

One response to this apparent conflict has been to adopt analyses that tend to understate the strength of a match, such as using estimates of profile probabilities that are larger than unbiased estimates (National Research Council, 1996). A better approach combines forensic science and statistical thinking in the calculation of a likelihood ratio: the probability of

DNA evidence under one hypothesis divided by the probability under an alternative. This approach leads naturally to an emphasis on a match probability, the probability an untyped individual has a particular DNA profile given that a typed person of interest has been seen to have that profile. Match probabilities incorporate genetic features and aspects of typing technology. Formulation of alternative hypotheses and calculation of corresponding likelihood ratios reduces the chance of misstating the strength of DNA evidence.

Advances in DNA characterization technology are being accompanied by advances in statistical analyses. Probabilistic genotyping (Taylor et al., 2013) uses Markov chain Monte Carlo methods to attach probability-based weights to many sets of profiles that collectively match a multi-contributor profile, and these weights used for likelihood ratios where a POI belongs to one set. These calculations incorporate many of the features of the capillary electrophoresis technology used to generate STR profiles, but the models underlying probabilistic genotyping are avoided in machine-learning approaches (Ricke et al., 2019). Other statistical advances are being used to interpret profiles generated by DNA sequencing (Woerner et al., 2017), and non-DNA evidence such as mass spectrometry-based characterization of protein variants (Parker et al., 2016) are bringing new statistical challenges.

References

Aliferi, A., Ballard, D., Gallidabino, M.D., Thurtle, H., Barron, L., and Court, D.S. DNA methylation-based age prediction using massively parallel sequencing data and multiple machine learning models. *Forensic Science International: Genetics*, **37**, 215–226, 2018.

Andersen, M.M. and Balding, D.J. How convincing is a matching Y-chromosome profile? *PLoS Genetics*, **13**(11), 2017. e1007028. https://doi.org/10.1371/journal.pgen.1007028

Balding, D.J. and Nichols, R.A. DNA match probability calculation: how to allow for population stratification, relatedness, database selection and single bands. *Forensic Science International*, **64**, 125–140, 1994.

Balding, D.J. and Steele, C.D. *Weight-of-Evidence for Forensic DNA Profiles*. Wiley, New York, 2015.

Bar, W., Brinkmann, B., Budowle, B., Carracedo, A., Gill, P., Holland, M., Lincoln, P.J., Mayr, W., Morling, N., Olaisen, B., Schneider, P.M., Tully, G., and Wilson, M. DNA commission of the international society for forensic genetics: guidelines for mitochondrial DNA typing. *International Journal of Legal Medicine*, **113**, 193–196, 2000.

Beecham, G.W. and Weir, B.S. Confidence intervals for DNA evidence likelihood ratios. *Journal of Forensic Sciences Supplement Series*, **1**, S166.S171, 2011.

Berry, D.A., Evett, I.W., and Pinchin, R. Statistical-inference in crime investigation using deoxyribonucleotideacid profiling. *Journal of the Royal Statistical Society Series C – Applied Statistics*, **41**, 499–531, 1992.

Borsting, C. and Morling, N. Next generation sequencing and its applications in forensic genetics. *Forensic Science International: Genetics*, **18**, 78–89, 2015.

Brenner, C.H. Understanding Y haplotype matching probability. *Forensic Science International: Genetics*, **8**, 233–243, 2013.

Buckleton, J.S., Curran, J.M., Goudet, J., Taylor, D., Thiery, A., and Weir, B.S. Population-specific *FST* values: a worldwide survey. *Forensic Science International: Genetics*, **23**, 91–100, 2016.

Butler, J.M. *Fundamentals of Forensic DNA Typing*. Academic Press, New York, 2009.

Chong, K.W.Y., Wong, Y.X., Ng, B.K., Thong, Z.H., and Syn, C.K.C. Development of a RNA profiling assay for biological tissue and body fluid identification. *Forensic Science International: Genetics Supplement Series*, **5**, E196.E198, 2015.

Coble, M.D., Buckleton, J., Butler, J.M., Egeland, T., Fimmers, R., Gill, P., Gusma, L., Guttman, B., Krawczak, M., Morling, N., Parson, W., Pinto, N., Schneider, P.M., Sherry, S.T., Willuweit, S., and Prinz, M. DNA commission of the international society for forensic genetics: recommendations on the validation of software programs performing biostatistical calculations for forensic genetics applications. *Forensic Science International: Genetics*, **25**, 191–197, 2016.

Cook, R., Evett, I.W., Jackson, G., Jones, P.J., and Lambert, J.A. A hierarchy of propositions: deciding which level to address in casework. *Science and Justice*, **38**, 231–239, 1998.

Curran, J., Triggs, C.M., Buckleton, J., and Weir, B.S. Interpreting DNA mixtures in structured populations. *Journal of Forensic Sciences*, **44**, 987–995, 1999.

Devlin, B., Risch, N., and Roeder, K. No excess of homozygosity at loci used for DNA fingerprinting. *Science*, **249**, 1416–1420, 1990.

Devlin, B., Risch, N., and Roeder, K. Forensic inference from DNA fingerprints. *Journal of the American Statistical Association*, **87**, 337–350, 1992.

Donnelly, P. Nonindependence of matches at different loci in DNA profiles – quantifying the effect of close relatives on the match probability. *Heredity*, **75**, 26–34, 1995.

Evett, I.W. and Weir, B.S. *Interpreting DNA Evidence*. Sinauer, Sunderland, MA, 1998.

FBI. Frequently Asked Questions on CODIS and NDIS. https://www.fbi.gov/services/laboratory/biometric-analysis/codis/codis-and-ndis-fact-sheet

Feinberg, S.E. (Editor). *The Evolving Role of Statistical Assessments as Evidence in the Courts*. Springer, New York, 1989.

Gill, P., Brenner, H., Buckleton, J.S., Carracedo, A., Krawczak, M., Mayr, W.R., Morling, N., Prinz, M., Schneider, P.M. and Weir, B.S. DNA commission of the international society of forensic genetics (ISFG): recommendations on the interpretation of mixtures. *International Journal of Legal Medicine*, **160**, 90–101, 2006.

Gill, P. and Haned, H. A new methodological framework to interpret complex DNA profiles using likelihood ratios. *Forensic Science International: Genetics*, **7**, 251–263, 2013.

Gittelson, S., Moretti, T.R., Onorato, A., Budowle, B., Weir, B.S. and Buckleton, J. The factor of 10 in forensic DNA match probabilities. *Forensic Science International: Genetics*, **28**, 178–187, 2017.

Gill, P. *Misleading DNA Evidence. Reasons for Miscarriages of Justice*. Academic Press, New York, 2014.

Graffelman, J. and Weir, B.S. Multi-allelic exact tests for Hardy-Weinberg equilibrium that account for gender. *Molecular Ecology Resources*, **18**, 461–473, 2018.

Hall, T.O. *The Y-Chromosome in Forensic and Public Health Genetics*. PhD Dissertation, University of Washington, Seattle, WA, 2016.

Holland, M., Fisher, D., Mitchell, L., Rodriquez, W., Canik, J., Merril, C., and Weedn, V. Mitochondrial DNA sequence analysis of human skeletal remains: identification of remains from the Vietnam War. *Journal of Forensic Sciences*, **38**, 542–553, 1993.

Jeffreys, A.J., Wilson, V., and Thein, S.L. Individual-specific 'fingerprints' of human DNA. *Nature*, **316**, 76–79, 1985a.

Jeffreys, A.J., Brookfield, J.F.Y., and Semeonoff. Positive identification of an immigration test-case using human DNA fingerprints. *Nature*, **317**, 818–819, 1985b.

Kidd, K.K., Speed, W.C., Wootton, S., Lagace, R., Langit, R., Haigh, E., Chang, J., and Pakstis, A.J. Genetic markers for massively parallel sequencing in forensics. *Forensic Science International: Genetics Supplement Series*, **5**, E677.E679, 2015.

Koehler, J.J., Chia, A., and Lindsey, S. The random match probability in DNA evidence: irrelevant and prejudical? *Jurimetrics*, **15**, 201–220, 1995.

Kruijver, M., Meester, R., and Slooten, K. p-values should not be used for evaluating the strength of DNA evidence. *Forensic Science International: Genetics*, **16**, 226–231, 2015.

Lander, E.S. DNA fingerprinting on trial. *Nature*, **330**, 501–505, 1989.

Laurie, C. and Weir, B.S. Dependency effects in multi-locus match probabilities. *Theoretical Population Biology*, **63**, 207–219, 2003.

Moretti, T.R., Moreno, L.I., Smerick, J.B., Pignone, M.L., Hizon, R., Buckleton, J.S., Bright, J.A., and Onorato, A.J. Population data on the expanded CODIS core STR loci for eleven populations of significance for forensic DNA analyses in the United States. *Forensic Science International: Genetics*, **25**, 175–181, 2016.

Moretti, T.R., Just, R.S., Kehl, S.C., Willis, L.E., Buckleton, J.S., Bright, J.A., Taylor, D.A., and Onorato, A.J. Internal validation of STRmix (TM) for the interpretation of single source and mixed DNA profiles. *Forensic Science International: Genetics*, **29**, 126–144, 2017.

Murphy, E. Law and policy oversight of familial searches in recreational genealogy databases. *Forensic Science International*, **292**, e5.e9, 2018.

National Research Council. *The Evaluation of Forensic DNA Evidence*. National Academy Press, Washington, DC, 1996.

Parker, G.J., Leppert, T., Anex, D.S., Hilmer, J.R., Matsunami, M., Baird, L., Stevens, J., Parsawar, K., Durbin-Johnson, B.F., Johnson, B.F., Rocke, D.M., et al. Demonstration of protein-based human identification using the hair shaft proteome. *PLoS One*, **11**, 2016. Article e0160653.

Perlin, M.W., Legler, M.M., Spencer, C.E., Smith, J.L., Allan, W.P., Belrose, J.L., and Duceman, B.W. Validating TrueAllele (R) DNA mixture interpretation. *Journal of Forensic Sciences*, **56**, 1430–1447, 2011.

Phillips, C. The Golden State Killer investigation and the nascent field of forensic genealogy. *Forensic Science International: Genetics*, **36**, 185–188, 2018.

Ricke, D.O., Fremont-Smith, P., Watkins, J., Stankiewicz, S., Boettcher, B.S., and Schwoebel, E. Estimating individual contributions to complex DNA SNP mixtures. *Journal of Forensic Sciences*, **64**, 1468–1474, 2019.

Slooten, K. Accurate assessment of the weight of evidence for DNA mixtures by integrating the likelihood ratio. *Forensic Science International: Genetics*, **27**, 1–16, 2017.

Swaminathan, H., Garg, A., Grgicak, C.M., Medard, M., and Lun, D.S. CEESIt: a computational tool for the interpretation of STR mixtures. *Forensic Science International: Genetics*, **22**, 149–160, 2016.

SWGDAM. Guidelines for the validation of probabilistic genotyping systems, 2015. http://www.swgdam.org/publications

Taylor, D., Bright, J.A., and Buckleton, J. The interpretation of single source and mixed DNA profiles. *Forensic Science International: Genetics*, **7**, 516–528, 2013.

Taylor, D., Kokshoorn, B., and Biedermann, A. Evaluation of forensic genetics findings given activity level propositions: a review. *Forensic Science International: Genetics*, **36**, 34–49, 2018.

Thompson, W.C. and Schumann, E.L. Interpretation of statistical evidence in criminal trials – the prosecutor's fallacy and the defense attorney's fallacy. *Law and Human Behavior*, **11**, 167–187, 1987.

Walsh, B., Redd, A.J., and Hammer, M.F. Joint match probabilities for Y chromosomal and autosomal markers. *Forensic Science International*, **174**, 234–238, 2008.

Weir, B.S. Matching and partially-matching DNA profiles. *Journal of Forensic Sciences*, **49**, 1009–1014, 2004.

Woerner, A.E., King, J.L., and Budowle, B. Fast STR allele identification with STRait Razor 3.0. *Forensic Science International: Genetics*, **30**, 18–23, 2017.

Wright, S. The genetical structure of populations. *Annals of Eugenics*, **15**, 323–354, 1951.

12

Kinship

Bruce S. Weir

CONTENTS

12.1 Introduction 265
12.2 Genetic Models for Allele and Genotype Frequencies 266
    12.2.1 Population Structure 266
    12.2.2 Relatedness 268
    12.2.3 SNP Data 272
    12.2.4 Genealogy Data 272
12.3 Conclusions 273
References 274

12.1 Introduction

DNA profiling of biological material has had an enormous impact on forensic science because of its potential role in associating an individual with a crime. As two examples, the biological material may be taken from the clothing or body of a victim and found to have the same profile as that of a person suspected of assaulting the victim, or it may be from a stain in a vehicle and found to have the same profile as that of a missing person thought to have been abducted by the owner of the vehicle. Such direct matches between an evidence sample and a individual, however, are by no means the only use for DNA profiles; because they are genetic they have been shaped by evolutionary processes and they may contain information about both distant ancestry and recent family membership of the profiled people. Most recently, for example, the use of genealogical databases has provided leads in long-cold cases such as that of the "Golden State Killer" (Phillips, 2018).

This chapter considers two ways in which forensic DNA profiles provide information about kinship. The current set of 20 short tandem repeat (STR) markers in common use do not contain sufficient information to estimate the degree of relatedness between two individuals from their DNA profiles; at least 50 markers would be needed (Anderson and Weir, 2007). The current panels do, however, allow probabilities of the profiles to be compared under alternative relatedness scenarios, and this has proven helpful for many years in parentage disputes and remains identification. The emerging use of single nucleotide polymorphism (SNP) profiles from DNA sequencing is bringing forensic science closer to relatedness determination at the level of precision reported in genealogical studies (Henn et al., 2012).

12.2 Genetic Models for Allele and Genotype Frequencies

A body of population genetic theory predicts the DNA profile types for people of known relationship, and this theory makes use of the probabilities of the allelic components of these profiles. There are now several databases with estimates of allele probabilities for different ethnic groups, such as those assembled by the National Institute for Standards and Technology (NIST) and accessible at https://strbase.nist.gov/. The degree of relationship is described in this theory with sets of kinship parameters, and the theory can be expressed as probabilities of profiles given relationship: Pr(Profiles|Relationship). The forensic interest, however, is almost always in making statements about relationship given the observed profiles. It is the probability Pr(Relationship|Profiles) that is likely to be of interest.

In the absence of prior information about the degree of relationship in a particular situation, which would allow the transposition of Pr(Profiles|Relationship) to Pr(Relationship|Profiles) via Bayes' Theorem, the population genetic theory is used to form the likelihood ratio

$$\text{LR} = \frac{\text{Pr(Profiles|One Relationship)}}{\text{Pr(Profiles|Another Relationship)}}$$

The numerator relationship may be of full-sibship for a set of remains thought to be from a missing person whose brother has been genotyped, and the denominator for no relationship between remains and family member. A high LR value would support sibship.

The LR depends on a genetic model reflecting common ancestral origins of alleles in different individuals or populations. If these ancestral alleles were in recent generations, the individuals carrying the current alleles are said to be related. More distant ancestral alleles could be considered as leading to "evolutionary relatedness." The key point is the explicit recognition that present genotype proportions are shaped by events that have taken place in previous generations.

12.2.1 Population Structure

For a single genetic marker with alleles u having proportions p_u in a population, the proportions P_{uv} for genotypes uv within that population may be expressed as

$$P_{uv} = \begin{cases} p_u^2 + f p_u(1 - p_u), & u = v, \\ 2p_u p_v - 2f p_u p_v, & u \neq v. \end{cases} \tag{12.1}$$

This formulation may be used to introduce the within-population inbreeding coefficient f (also written as F_{IS}). Setting $f = 0$ gives the classic Hardy-Weinberg Law linking genotype proportions to squares and products of allele proportions. As the loci used for forensic typing are unlikely to be under the influence of natural selection, the use of a single coefficient f for all genotypes may be reasonable, although genotype-specific coefficients should strictly be used for loci with allele-specific mutation processes (Graham et al., 2000). Expressing genotype proportions as functions only of allele proportions, i.e., $f = 0$, is used for highly variable loci, when many genotypes are not seen in a sample and their population proportions are difficult to estimate as sample proportions. For example, the NIST databases used for calculations in the US have sample sizes between 97 (Asian) and 361 (Caucasian) (Ruitberg et al., 2001), and many of the possible genotypes at a locus have observed counts of zero. An STR locus with 10 different alleles has 55 different genotypes, for example, and not all of these are expected to be seen in 300 or so profiles.

From an evolutionary perspective, an extant population can be regarded as just one realization of a process over time involving the stochastic processes of genetic drift, mutation, migration and so on. Predictions of quantities such as genotype probabilities are expectations over replicates of the population history. The allele proportions p_u in a specific population have expected values over these replicates of π_u and genotype probabilities Pr(uv) are expectations of proportions in that population

$$\text{Pr}(uv) = \begin{cases} \pi_u^2 + F\pi_u(1 - \pi_u), & u = v, \\ 2\pi_u\pi_v - 2F\pi_u\pi_v, & u \neq v. \end{cases} \tag{12.2}$$

where F is the "total inbreeding coefficient" (also written as F_{IT}). Balding and Nichols (1994) assumed the population-specific proportions p_u have Beta distributions with parameters $(1 - \theta)\pi_u/\theta$ and $(1 - \theta)(1 - \pi_u)/\theta$. The "population-structure parameter" θ (or F_{ST}) can be regarded as a measure of relationship for pairs of alleles within a population where the measure is the probability of allelic identity by descent. The trio F_{IS}, F_{IT}, F_{ST} are referred to as "Wright's F-statistics" (Wright, 1951), and are related as $F_{IS} = (F_{IT} - F_{ST})/(1 - F_{ST})$. A random-mating population is expected to be in Hardy-Weinberg equilibrium ($F_{IS} = 0$) but this does not imply allelic independence ($F_{IT} = 0$) in an evolutionary sense. Hardy-Weinberg equilibrium does imply $F_{IT} = F_{ST}$: pairs of alleles within populations have the same relationship whether they are carried by the same or different individuals in that population.

Related individuals carry alleles from recent common ancestors and these alleles can consequently be identical through their descent from that ancestor. The genetic sampling process inherent in the selection of which of a parent's two alleles at one locus is transmitted to their child means that relatives' genotype probabilities involve the total inbreeding coefficient and its analogs for alleles in different individuals. The usual kinship parameter for relatives X and Y, θ_{XY}, is the identity probability for an allele taken randomly from relative X with one taken randomly from Y, and it equals the inbreeding coefficient F of any child they may have.

The Beta distribution leads to probabilities of sets of alleles, nicely summarized by a "sampling formula" for alleles (Balding and Steele, 2015). Suppose n alleles have been sampled from a population and n_u are of type u. The probability the next allele sampled is also of type u is

$$\text{Pr}(u|n_u \text{ of type } u \text{ in } n) = \frac{n_u\theta + (1 - \theta)\pi_u}{1 + (n - 1)\theta} \tag{12.3}$$

Here θ is for a random pair of alleles in the population under study. This result allows probabilities for sets of alleles, in kinship or parentage calculations, to be computed very easily.

Equation 12.3 requires knowledge of the parameters π_u and θ. Allele probabilities π_u are replaced by sample proportions $\tilde{p}_u$ in databases such as those reported in Moretti et al. (2016) and Ruitberg et al. (2001) and categorized by ancestral origin: "Caucasian", "African-American" and so on. It is recognized that the relevant population for a particular situation, even if the origins of that population are known, is unlikely to be the same as the population from which the database was constructed. Variation among populations within the broad group represented by the database is analogous to variation among evolutionary replicates of a single population, and θ also represents variation among (sub)populations within the database population. For this reason, θ can be referred to as a population-structure parameter. Values such as $\theta = 0.01$ for major ethnic groups, and higher values, such as $\theta = 0.03$, for smaller groups like Native Americans have been suggested (National Research Council, 1996). From estimating θ in worldwide national populations relative to continental-scale databases, (Steele et al., 2014) recommended $\theta = 0.03$ as being almost always conservative, even if the source of the DNA is from a different continent than the suspected source.

Results consistent with those in Steele et al. (2014) were found in another comprehensive survey (Buckleton et al., 2016) of forensic STR marker allelic data. Those authors extracted allelic sample proportions from 250 publications and estimated θ values within and between continental-ancestry groups of populations. This survey included Native American populations, and for those populations the relevant θ values can be as high as 0.10 relative to world-wide populations. The survey employed an estimation procedure based on allelic matching and described in more detail in Goudet et al. (2018). There the sample proportion of matching pairs of alleles sampled from population i, without regard to which individuals carry those alleles, was written as $\tilde{M}_i$, and the sample proportion of matching pairs of alleles, one taken from each of two populations, and averaged over pairs of populations, as $\tilde{M}_B$. An estimate of θ_i, the relationship for pairs of alleles within population i, relative to the relationship of pairs of alleles from pairs of populations, was $\hat{\beta}_i = (\tilde{M}_i - \tilde{M}_B)/(1 - \tilde{M}_B)$. When database sample proportions $\tilde{p}_u$ are used in place of population proportions π_u in Equation 12.3, it is appropriate to use the average $\hat{\beta}_W$ of the population-specific values $\hat{\beta}_i$ in place of θ_i, and probability estimates will apply to any population within the group of populations represented by the database. The work in Goudet et al. (2018) applies to haploid (e.g., Y-haplotypes) and diploid data.

12.2.2 Relatedness

Equation 12.3 refers to the probabilities of sets of alleles with a shared evolutionary history. The expressions have an implicit assumption of Hardy-Weinberg equilibrium within a single replicate population ($F_{IS} = 0$) and the equivalence of pairs of alleles within individuals with those between individuals within the same population ($F_{IT} = F_{ST}$). Related individuals, however, may share alleles from recent common ancestors: if neither is inbred, meaning each has unrelated parents, the probabilities they share 0, 1 or 2 pairs of alleles identical by descent are written as k_0, k_1, k_2 respectively. Values for these probabilities for common pairs or relatives are shown in Table 12.1, and expressions for joint genotypic probabilities are shown in Table 12.2. They are often summarized by the kinship coefficient $\theta = k_2/2 + k_1/4$.

In theory, the expressions in Table 12.2 would allow the k's to be estimated from the observed profiles for a pair of individuals, assuming the k's were the same for every locus contributing to the profile and the genotypes were regarded as being independent over loci. Maximum likelihood and method of moments approaches can be used. In practice, however, there is insufficient information in a pair of 20-locus profiles to furnish reliable estimates. There is also the substantial problem that the allele probabilities are not known, and relatedness can be estimated only relative to the population from which these probabilities are estimated (Goudet et al., 2018; Weir and Goudet, 2017). This issue is discussed below in the section on SNPs. Instead of estimating the relatedness parameters, the expressions in Table 12.2 are used to form likelihood ratios for one hypothesized relationship

TABLE 12.1

Relationship Probabilities for Common Relatives

Relationship	k_2	k_1	k_0	θ
Identical twins	1	0	0	0.5
Full sibs	0.25	0.5	0.25	0.25
Parent-child	0	1	0	0.25
Three-quarter sibs	0.125	0.5	0.375	0.1875
Double first cousins	0.0625	0.375	0.5625	0.125
Half sibs[a]	0	0.5	0.5	0.125
First cousins	0	0.25	0.75	0.0625
Unrelated	0	0	1	0

[a] Also grandparent-grandchild and avuncular (e.g., uncle-niece).

TABLE 12.2

Joint Genotypic Probabilities for Pairs of Relatives

Genotypes[a]	Probability
uu, uu	$k_0\pi_u^4 + k_1\pi_u^3 + k_2\pi_u^2$
uu, vv	$k_0\pi_u^2\pi_v^2$
uu, uv	$2k_0\pi_u^3\pi_v + k_1\pi_u^2\pi_v$
uu, vw	$2k_0\pi_u^2\pi_v\pi_w$
uv, uv	$4k_0\pi_u^2\pi_v^2 + k_1\pi_u\pi_v(\pi_u + \pi_v) + 2k_2\pi_u\pi_v$
uv, uw	$4k_0\pi_u^2\pi_v\pi_k + k_1\pi_u\pi_v\pi_k$
uv, wz	$4k_0\pi_u\pi_v\pi_w\pi_z$

[a] $u \neq v \neq w \neq z$.

versus another, with Table 12.1 providing the k values for the probabilities of the observed profiles for each relationship.

In the special situation where H_0 is a specified relationship, with specified k's, and H_1 is that the individuals are unrelated ($k_2 = k_0 = 0; k_0 = 1$), the likelihood ratios for each possible pair of genotypes are shown in Table 12.3. These values may be greater than or less than 1, depending on the genotype pairs at each locus, and they are multiplied over loci.

The expressions in Tables 12.2 and 12.3 involve kinship coefficients k, specified by the hypotheses being considered, and the unknown allele probabilities π. These probabilities can be estimated as sample proportions $\tilde{p}$ in a database. If the continental ancestry is known for the two people whose relationship is in question, then the database would ideally represent people of like ancestry. There is still an issue (Goudet et al., 2018; Weir and Goudet, 2017) of relatedness among individuals whose profiles are in the database as that will affect the extent to which terms like $1/\tilde{p}$ represent $1/\pi$. The representation will be better for databases constructed with profiles from unrelated individuals.

The expressions in Tables 12.1 and 12.2 also allow the alternative to the hypothesis H_1: The sample profile is from a particular typed person ($k_2 = 1$) to be H_2: The sample profile is from a relative of the typed person (e.g., $k_2 = 1/4, k_1 = 1/2, k_0 = 1.4$ for a full sibling). For example, if the POI and sample profiles are both homozygous uu at a single locus, and

TABLE 12.3

Likelihood Ratio Values for H_1: Related Versus H_2: Unrelated

Genotypes[a]	Probability
uu, uu	$k_0 + k_1/\pi_u + k_2/\pi_u^2$
uu, vv	k_0
uu, uv	$k_0 + k_1/(2\pi_u)$
uu, vw	k_0
uv, uv	$k_0 + k_1(\pi_u + \pi_v)/(4\pi_u\pi_v) + k_2/(2\pi_u\pi_v)$
uv, uw	$k_0 + k_1/(4\pi_u)$
uv, wz	k_0

[a] $u \neq v \neq w \neq z$.

ignoring population structure, the likelihood for the POI versus an untyped brother of the POI being the source of the sample is

$$LR = \frac{4}{(1 + \pi_u)^2}$$

and this can be substantially smaller than the $1/\pi_u^2$ for unrelated alternative sources.

12.2.2.1 Relatedness and Population Structure

To account for both population structure and relatedness, the four alleles carried by two individuals are reduced to sets of two, three or four alleles identical by descent from family relatedness and then the allelic sampling formula, Equation 12.3, used for evolutionary relatedness of the alleles in those sets. For two related homozygotes or heterozygotes

$$\Pr(uu, uu) = k_0 \Pr(uuuu) + k_1 \Pr(uuu) + k_2 \Pr(uu)$$

$$\Pr(uv, uv) = 4k_0 \Pr(uuvv) + k_1[\Pr(uuv) + \Pr(uvv)] + 2k_2 \Pr(uv)$$

so the match probabilities are

$$\Pr(uu|uu) = \frac{[2\theta + (1 - \theta)\pi_u][3\theta + (1 - \theta)\pi_u]k_0}{(1 + \theta)(1 + 2\theta)} + \frac{[2\theta + (1 - \theta)\pi_u]k_1}{1 + \theta} + k_2$$

$$\Pr(uv|uv) = \frac{2[\theta + (1 - \theta)\pi_u][\theta + (1 - \theta)\pi_v]k_0}{(1 + \theta)(1 + 2\theta)} + \frac{[2\theta + (1 - \theta)(\pi_u + \pi_v)]k_1}{2(1 + \theta)} + k_2$$

Parameters π_u and θ are assumed to have the same value in successive generations. Setting $k_0 = 1, k_1 = k_2 = 0$ reduces these expressions to the match probabilities for unrelated homozygotes and heterozygotes, respectively.

12.2.2.2 Parentage Calculations for Structured Populations

The previous section is also the basis for parentage calculations when population structure is taken into account. A typical situation is when a mother M and her child C are typed,

TABLE 12.4

Paternity Indices for Structured Populations

G_M	G_C	G_{AF}	PI
uu	uu	uu	$(1+3\theta)/[4\theta+(1-\theta)\pi_u]$
		uv	$(1+3\theta)/2[3\theta+(1-\theta)\pi_u]$
	uv	vv	$(1+3\theta)/[2\theta+(1-\theta)\pi_v]$
		uv	$(1+3\theta)/2[\theta+(1-\theta)\pi_v]$
		vw	$(1+3\theta)/2[\theta+(1-\theta)\pi_v]$
uv	uu	uu	$(1+3\theta)/[3\theta+(1-\theta)\pi_u]$
		uv	$(1+3\theta)/2[2\theta+(1-\theta)\pi_v]$
	uw	ww	$(1+3\theta)/[2\theta+(1-\theta)\pi_w]$
		uw	$(1+3\theta)/2[\theta+(1-\theta)\pi_w]$
		wx	$(1+3\theta)/2[\theta+(1-\theta)\pi_w]$

along with an alleged father AF. The relevant hypotheses are

H_1: AF is the father of C.
H_2: AF is not the father of C.

The likelihood ratio, often called the Paternity Index PI, compares the probabilities of all three profiles G_M, G_C, G_{AF} under each hypothesis and may be expressed as

$$PI = \frac{Pr(G_C|G_M, G_{AF}, H_1)}{Pr(G_C|G_M, G_{AF}, H_2)}$$

The numerator just applies Mendelian rules for the possible genotypes from known parents, whereas the denominator requires the probability of the paternal allele after having seen the alleles carried by the mother and the alleged father. The most common scenarios are shown in Table 12.4 (Evett and Weir, 1998).

12.2.2.3 Identifying Remains

There are many situations in which remains need to be identified. For remains of a single individual, the profile from a reference sample from tissue, such as cells on a toothbrush, known to be from a missing person can be compared to the profile from the remains and the likelihood ratio calculated for the hypotheses that the remains are, or are not, those of the missing person. In the absence of direct reference samples, the profiles from relatives of a missing person may be used in ways similar to parentage disputes. There are software packages, such as Egeland et al. (2016), that allow quite complicated scenarios to be addressed.

For mass disasters, such as airplane crashes, tidal waves and acts of war, there may be large numbers of missing people and their family members to associate with large numbers of remains. For the 2001 World Trade Center disaster (Brenner and Weir, 2003) described an effective screening procedure for every comparison of a remains sample with a direct reference sample or a sample from a family member of a missing person: the profiles from these two sources were considered to be from the same person, or from parent and child, or from full siblings versus being from unrelated people in each case.

12.2.3 SNP Data

With large numbers of SNPs, as might be revealed by sequencing STR regions (Woerner et al., 2017), an estimation method with good properties for the kinship parameter θ has been described in Goudet et al. (2018) and Weir and Goudet (2017) where it was also stressed that relatedness is not an absolute quantity; pairs of individuals are considered to be more or less related than a random pair in some reference population. In particular, two individuals in an African population are less related, on average, than a pair of individuals from anywhere in the world, although they show only the average degree of relatedness compared to individuals drawn from Africa. A reference population is needed, and must be specified, in estimating relatedness.

At SNP l, the allele dosage X_{jl} is the number (0,1,2) of one of the alleles for individual j, and the observed allelic matching proportion for individuals j, j' is $\tilde{M}_{jj'} = [1 + (X_{jl} - 1)(X_{j'l} - 1)]/2$. These proportions have the values $0, 0.5$ and 1, and their average over all pairs of individuals taken from the reference population is written as $\tilde{M}_B$. The estimated kinship $\theta_{jj'}$ for the two individuals is $\hat{\beta}_{jj'} = (\tilde{M}_{jj'} - \tilde{M}_B)/(1 - \tilde{M}_B)$ and information from multiple SNPs is obtained by summing the numerator and denominator separately over SNPs. These estimates will be negative for individuals less related than the average pair in the reference group, although there could be an adjustment made to give estimates of zero for the least related individuals in a study (Weir and Goudet, 2017).

12.2.4 Genealogy Data

Relatedness analyses based on a single STR or a single SNP, even when combined over STRs or SNPs, are unlikely to distinguish among classes of distant relationships. Although an individual receives two copies of each autosomal chromosome, one from each parent, he or she transmits to a child a single chromosome representing non-overlapping sections of the two parental copies. These sections are formed by the process of genetic recombination with the result that the child's two chromsomes are generally different from all four of those carried by the parents. Over generations, recombination means that the section of an individual's chromosome remaining intact and present in a descendant becomes increasingly smaller. Relatives can receive identical chromosome sections from a common ancestor, and the length of these identical sections contains information about their degree of relatedness. In the forensic setting this observation has recently been used to identify the source of an evidence profile by looking for the same person among the descendants of different distant relatives of that source, where these relatives have been identified by submitting the evidence profile to a genealogy database (Phillips, 2018). A process for finding relatives in a database from a SNP profile has been described by Henn et al. (2012). Those authors simulated pairs of individuals with relatedness up to 9th cousins. They related the pedigree relationships of these pairs to values of a statistic they termed IBD$_{\text{half}}$. This metric looked at sections of a chromosome bounded by SNPs with the opposite homozygotes: one individual homozygous for the minor allele and the other homozygous for the major allele. Only sections that contained over 400 genotyped SNPs homozygous in at least one of the two individuals and that are at least 5cM in length were considered. (One centiMorgam, cM, is approximately one million base pairs of DNA sequence and the human genome is over three billion base pairs in length.) The total length of such regions, provided the largest of them was at least 7cM in length, is IBD$_{\text{half}}$. There were some additional features added to this definition, but the statistic indicates the portion of the genomes of two individuals that is identical by descent (IBD) through having derived from the same ancestor. The number of IBD segments was also considered, and good success was found in identifying up to 5th cousins—considerably beyond what is possible with methods based on single SNP analyses (Goudet et al., 2018), and not requiring phased data, in which the set of alleles an individual receives from each parent is identified, as is used in Browning and Browning (2010). With two or more relatives of the source of an evidence profile identified, public records are searched to find their descendants in the hope of identifying descendants common to both pedigrees who may then be subject to investigation. Aspects such as geography, continental ancestry, age and gender help refine these searches. The Parabon-Nanolabs company, https://parabon-nanolabs.com, had reported 55 successful forensic genealogy searches by May, 2019. In September, 2019, the U.S. Department of Justice promulgated guidelines (Department of Justice, 2019) limiting use of the method.

12.3 Conclusions

DNA profiles have been used to link suspects to forensic evidence since the Pitchfork case (Wambaugh, 1989) in 1986, but they have been used to link relatives for much longer. Alec Jeffreys, whose typing linked two murders, exonerated a suspect who had falsely confessed to one of them, and then confirmed that Colin Pitchfork had committed both, first used his DNA fingerprinting methods in an immigration case (Jeffreys et al., 1985). Jeffreys and his colleagues showed that a Ghanaian boy was the son, and not the nephew, of a woman living in England and was therefore entitled to enter that country. This finding rested on profiles from the boy, the woman and her other son and two daughters. Genetic information, such as ABO blood groups, had been used in parentage disputes for much longer but the amount of information in current STR profiles, and the newer SNP profiles, allow for more sensitive detection of even distant relationships among the sources of forensic genetic patterns.

There is a continuum in the degree of relatedness between two people, ranging from common membership in a continental ancestry group to membership in the same nuclear family, and statistical calculations of the strength of DNA evidence supporting relatedness between, say, a POI and an item of evidence, should take account of both these evolutionary and recent family time scales. The key element of these analyses is that kinship is a relative concept. Relative to a hypothetical closed community where every pair of people is first cousins, no two people are related and their children would not be inbred, but relative to the rest of the population any two people in the community have a kinship coefficient of 1/16 and their children would have an inbreeding coefficient of 1/16. Both these numbers would be higher if account is taken of the shared continental ancestry of people in the population.

One application of kinship is "familial searching." A law-enforcement database such as the FBI's National DNA Index System (FBI, 2019) can be queried for exact matches to an evidence profile when a POI has not already been identified as a possible source for that profile. Failure to find a match could lead to a query for profiles sufficiently similar to the evidence profile to suggest the database entry is from a relative of the evidence source. The current STR profiles make this feasible only for first degree relative pairs: parent-offspring and full-sibs. More distant relationships lead to many false positives (LR values for relatedness versus unrelated greater than one for truly unrelated pairs) or false negatives (LR values less than one for true relatives). Even then, there have been sufficient concerns raised about familial searching, including those related to privacy, that it is not universally used and may be prohibited in some jurisdictions (Murphy, 2010). Even when familial searching is allowed, it may be subject to the requirement of strong preliminary evidence of relatedness, such as a Y-chromosome profile match in crimes with a male perpetrator (Myers et al., 2011).

The population structure effect is generally ignored in kinship calculations, including familial searching, on the grounds that it has a smaller effect on profile probabilities for two people than their family membership. From Equation 12.3 and Table 12.3, the LR for related versus unrelated individuals when the two profiles are uv and uw, for example, is

$$\mathrm{LR} = k_0 + \frac{(1+2\theta)k_i}{4[\theta + (1-\theta)\pi_u]}$$

When $\theta = 0$ this reduces to the values in Table 12.3. For all relative types, and all non-zero values of θ, the LR will be smaller when θ is included unless the allele probability π_u has the high value of $1/3$ or more, with the change in LR being greater for smaller π_u. Given that statistical calculations are computer-based, it is prudent to include population structure effects in all kinship-related calculations, including those involving parentage disputes.

DNA typing is a powerful tool for forensic science, not least because of the information it reveals about kinship among typed individuals. This aspect of forensic science mirrors the wider societal interest in genealogical research spurred by the availability of direct-to-consumer genetic testing. Statisticians have an important role in translating genetic profiles to valid statements about kinship.

References

Anderson, A.D. and Weir, B.S. A maximum-likelihood method for the estimation of pairwise relatedness in structured populations. *Genetics*, **176**, 421–440, 2007.

Balding, D.J. and Nichols, R.A. DNA match probability calculation: how to allow for population stratification, relatedness, database selection and single bands. *Forensic Science International*, **64**, 125–140, 1994.

Balding, D.J. and Steele, C.D. *Weight-of-Evidence for Forensic DNA Profiles*. Wiley, New York, 2015.

Browning, S.R. and Browning, B.L. High-resolution detection of identity by descent in unrelated individuals. *American Journal of Human Genetics*, **86**, 526–539, 2010.

Brenner, C.H. and Weir, B.S. Issues and strategies in the DNA identification of World Trade Center victims. *Theoretical Population Biology*, **63**, 173–178, 2003.

Buckleton, J.S., Curran, J.M., Goudet, J., Taylor, D., Thiery, A., and Weir, B.S. Population-specific *FST* values: a worldwide survey. *Forensic Science International: Genetics*, **23**, 91–100, 2016.

FBI, Frequently Asked Questions on CODIS and NDIS, 2019. https://www.fbi.gov/services/laboratory/biometric-analysis/codis/codis-and-ndis-fact-sheet

Department of Justice. *Interim Policy on Forensic Genetic Genealogical DNA Analysis and Searching*, 2019. https://www.justice.gov/olp/page/file/1204386/download

Evett, I.W. and Weir, B.S. *Interpreting DNA Evidence*. Sinauer, Sunderland, MA, 1998.

Egeland, T., Kling, D., and Mostad, P. *Relationship Inference with Familias and R. Statistical Methods in Forensic Genetics*. Academic Press, London, 2016.

Goudet, J., Kay, T., and Weir, B.S. How to estimate kinship. *Molecular Ecology*, **27**, 4121–4135, 2018.

Graham, J., Curran, J., and Weir, B.S. Conditional genotypic probabilities for microsatellite loci. *Genetics*, **155**, 1973–1980, 2000.

Henn, B.M., Hon, L., Macpherson, J.M., Eriksson, N., Saxonov, S., Pe'er, I., and Mountain, J.L. Cryptic distant relatives are common in both isolated and cosmopolitan genetic samples. *PLoS One*, **7**, 2012. Article e34267.

Jeffreys, A.J., Brookfield, J.F.Y., and Semeonoff, R. Positive identification of an immigration test-case using human DNA fingerprints. *Nature*, **317**, 818–819, 1985.

Moretti, T.R., Moreno, L.I., Smerick, J.B., Pignone, M.L., Hizon, R., Buckleton, J.S., Bright, J.A., and Onorato, A.J. Population data on the expanded CODIS core STR loci for eleven populations of significance for forensic DNA analyses in the United States. *Forensic Science International: Genetics*, **25**, 175–181, 2016.

Murphy, E. Relative doubt: Familial searches of DNA databases. *Michigan Law Review*, **109**, 291–348, 2010.

Myers, S.P., Timken, M.D., Piucci, M.L., Sims, G.A., Greenwald, M.A., Weigand, J.J., Konzak, K.C., and Buoncristiani, M.R. Searching for first-degree familial relationships in California's offender DNA database: Validation of a likelihood ratio-based approach. *Forensic Science International: Genetics*, **5**, 493–500, 2011.

National Research Council. *The Evaluation of Forensic DNA Evidence*. National Academy Press, Washington, DC, 1996.

Phillips, C. The Golden State Killer investigation and the nascent field of forensic genealogy. *Forensic Science International:Genetics*, **36**, 185–188, 2018.

Ruitberg, C.M., Reeder, D.J., and Butler, J.M. STRBase: a short tandem repeat DNA database for the human identity testing community. *Nucleic Acids Research*, **29**, 320–322, 2001.

Steele, C.D., Court, D.S., and Balding, D.J. Worldwide F_{ST} estimates relative to five continental-scale populations. *Annals of Human Genetics*, **78**, 468–477, 2014.

Wambaugh, J. *The Blooding: The True Story of the Narborough Village Murders*. Bantam, New York, 1989.

Weir, B.S. and Goudet, J. A unified characterization for population structure and relatedness. *Genetics*, **206**, 2085–2103, 2017.

Woerner, A.E., King, J.L., and Budowle, B. Fast STR allele identification with STRait Razor 3.0. *Forensic Science International: Genetics*, **30**, 18–23, 2017.

Wright, S. The genetical structure of populations. *Annals of Eugenics*, **15**, 323–354, 1951.

13

Statistical Support for Conclusions in Fingerprint Examinations

Cedric Neumann, Jessie Hendricks, and Madeline Ausdemore

CONTENTS

13.1 Introduction 278
13.2 Fingerprint Examination Framework (ACE-V) 280
    13.2.1 Analysis Stage 280
    13.2.2 Comparison Stage 282
    13.2.3 Evaluation Stage 282
    13.2.4 Verification Stage 282
    13.2.5 Decision Making in the ACE-V Framework..................... 283
    13.2.6 Summary 287
13.3 Common Source vs. Specific Source Scenarios 287
    13.3.1 Common Source Scenario 288
    13.3.2 Specific Source Scenario 288
    13.3.3 Simulations 289
    13.3.4 Summary 291
13.4 Weight of Fingerprint Evidence 293
    13.4.1 Similarity Metrics and Kernel Functions 293
    13.4.2 Score-Based Likelihood Ratios 294
    13.4.3 Approximate Bayes Factor for Fingerprint Evidence 301
    13.4.4 Summary 304
13.5 Factors Affecting Examiners' Decision Making and Error Rates 304
    13.5.1 Decision Making During ACE 304
    13.5.2 PCAST Controversy 308
    13.5.3 Statistical Analysis of Experiments Designed to Study the ACE-V
          Framework 310
    13.5.4 Summary 312
13.6 U.S. Defense Forensic Science Center's *FRStat* 312
    13.6.1 Summary 316
13.7 Conclusions... 317
References 318

Fingerprint evidence has been used to support the identification of criminals for more than a century. Statistical analysis of information present on finger impressions is not a new endeavour. Sir Francis Galton is credited with the first extensive analysis of fingerprint variability (Galton, 1892). However, with the increased use of statistics and probability theory to support and report inferences on the source of biological material recovered at crime scenes in the 1980s and 1990s, there has been a demand from scientific and legal scholars for the application of similar methods to handle the uncertainty in the determination of the source of other types of evidence, and in particular, fingerprint evidence. This chapter investigates the two main approaches that are currently advocated for supporting the conclusions of fingerprint examinations: Bayesian inference and error rates. The findings of this chapter show that most ad hoc methods offered to support Bayesian inference in fingerprint examination may have some merits as deterministic decision tools; however, the use of these methods within a Bayesian paradigm is not appropriate. As a result, so-called score-based likelihood ratios cannot be used to update prior beliefs on the source of a finger impression as part of Bayesian reasoning. The error rate studies performed during the past decade inform the community on the magnitude of the expected rates of erroneous identifications and exclusions. Unfortunately, it is very difficult to relate these community-wide expected error rates to the risk of error in a specific case since that risk will depend on the quality of the impression, the appropriateness of the examination and documentation procedures established by a specific laboratory, as well as the competency of the examiner performing the examination. In conclusion, we find that the foundations of fingerprint examination are much stronger now than a decade ago, but that much is left to be done in terms of providing tools for data-driven decision-making in individual cases.

13.1 Introduction

The historical intent behind the use of finger impressions was to verify the identity of an individual for administrative or legal purposes.* In this context, one or more finger impressions from the individual, whose identity has yet to be verified, are compared to impressions previously acquired and associated with the known identity of their donors. All impressions are acquired under controlled conditions and are usually of good quality. These impressions are sometimes referred to as *control impressions* (right panel of Figure 13.1).

Today, this verification process still represents the vast majority of the fingerprint comparisons performed worldwide. For example, the largest biometric database in the world, known as the *Aadhaar* program, which contains biometric data from more than one billion individuals, was initiated as an attempt to streamline government processes for residents of India (Safi, 2018).

In the U.S., the Next Generation Identification (NGI) system of the federal government receives slightly less than 5,000,000 requests for fingerprint searches monthly (as of February 2019). Out of these 5,000,000 requests, approximately 2.9 million are related to civil applications (e.g., teacher background checks), 1 million are requests for criminal records, and 1.1 million are requests related to arrestees. Only 25,000 requests involve the submission of trace impressions recovered in connection with the commission of a crime (U.S. Federal Bureau of Investigation, 2019).

The verification process relies primarily on good quality impressions. This allows for most of the requests for verification (>99%) to be processed entirely automatically, with a response time of a few minutes, if not seconds. Furthermore, the quality of the impressions

---

* For an historical introduction to fingerprint examination and other background information on fingerprints, readers may refer to Barnes (2014), Maceo (2014) and Wertheim (2014).

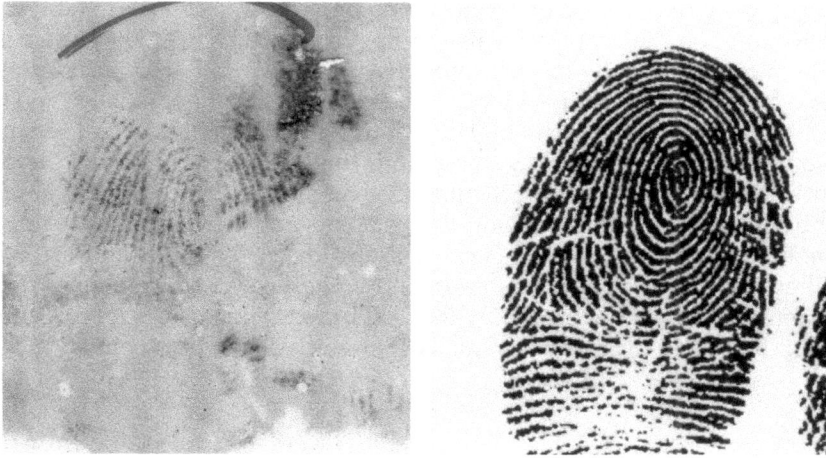

FIGURE 13.1
Left panel: latent print. Right panel: control print. Ridges appear darker than background. Both impressions were made by the same finger. Their comparison shows that both ridge flows are affected by different distortion and degradation effects.

and the amount of information available are usually such that associations of individuals with previously recorded identities are virtually error free and undisputed.

In contrast, trace impressions inadvertently left at crime scenes, or on objects connected with criminal activities, usually referred to as *latent prints* in the U.S. or *fingermarks* in Europe, are often partial, degraded or distorted (left panel of Figure 13.1). The search of these latent prints in large databases, or their direct side-by-side comparisons with control impressions from known individuals, involves a fair amount of manual labour. Due to the quality of latent prints, and the human involvement in the examination process, the uncertainty associated with the determination of the identity of the donor of a latent print is much higher than the determination of individuals based on control prints. This uncertainty needs to be handled as part of a formal inference process and conveyed to customers of fingerprint services.

This chapter focuses on statistical data and models proposed to support the inference of the identity of the donor of a trace impression recovered in connection with a crime. This chapter does not address the topic of the selection of candidate donors through the search of latent prints in large databases such as NGI.* From the perspective of this chapter, database searches are only used to propose donors. This chapter considers that the process of inferring the identity of the donor of a trace impression is the same, regardless of whether the potential donors were generated by a database or by police investigation.[†]

---

* Readers interested in the testing and performance of fingerprint database search engines can refer to Indovina et al. (2012) and Watson et al. (2015).

† There is an ongoing controversy on the weight associated to forensic DNA evidence when the potential donor has been detected through a cold search in a DNA database. The reader is encouraged to consult Balding and Donnelly (1996), Dawid (2001), Balding (2002), Meester and Sjerps (2003), Sjerps and Meester (2009), Berger et al. (2015), Wixted et al. (2019), and Chapter 14 (Sjerps, 2020). Presumably, the outcome of this controversy will be applicable to the inference of the identity of the source of latent prints.

This chapter does not review the majority of previously published statistical models aimed at informing on the probative value of fingerprint evidence. The reader is directed to the excellent review of early models by Stoney (2001). A more recent discussion of statistical data related to fingerprint evidence can be found in Neumann (2012) and Champod et al. (2016). Instead, this chapter mainly concentrates on the main current research trends aimed at supporting conclusions resulting from fingerprint comparisons,* and explores their potential and limitations. In particular, this chapter focuses on the quantification of the weight of fingerprint evidence using Bayes factors and the use of so-called *score-based likelihood ratios* as ad hoc proxies. This chapter also discusses the use of error rate studies as an alternative to Bayesian reasoning and some related controversies. Finally, this chapter explores a method developed by the U.S. Defense Forensic Science Center laboratory (DFSC) called *FRStat* (Swofford et al., 2018), which sits in between Bayes factors and error rates.

13.2 Fingerprint Examination Framework (ACE-V)

Fingerprint examiners form opinions on the identity of the donor of a trace impression (left panel of Figure 13.1) through its comparison with a control impression (right panel of Figure 13.1) using a framework known as ACE-V, which stands for Analysis, Comparison, Evaluation and Verification. This framework and its related processes are sufficiently inclusive that they can be used to compare almost anything. The Scientific Working Group on Friction Ridge Analysis, Study and Technology (SWGFAST) has produced several documents that attempt to describe the implementation of the ACE-V framework in the context of fingerprint examination (Scientific Working Group on Friction Ridge Analysis, Study and Technology, 2012a,b,c, 2013a,b). The following sections give an overview. More detailed descriptions can be found in Langenburg (2012) and Champod et al. (2016).

13.2.1 Analysis Stage

During the analysis stage of the ACE-V framework, examiners detect and characterize the features of the latent print that will be used in the subsequent stages of the examination. Traditionally, fingerprint examiners characterize finger impressions according to three levels of friction ridge detail. The first level is the general pattern of the ridge flow. Figure 13.2 shows the three main pattern types: loop, whorl and arch. The left panel displays a loop: the ridges (in black) flow in from one side of the impression, turn around, and flow out through the same side. The middle panel displays a whorl: the ridges appear mostly as concentric ellipses. Finally, the right panel displays an arch: the ridges flow from left to right across the impression. Many other patterns exist and can be related to a combination of these three primary patterns.

---

* The reader will note that, although the material covered in this chapter focuses on impressions of the skin of the tip of the finger (distal phalange), its content also applies to other friction ridge skin impressions, such as intermediate and proximal phalanges, palms, soles and toes.

FIGURE 13.2
First level of details. Ridges appear darker than background. Left panel: loop pattern type. Middle panel: whorl pattern type. Right panel: arch pattern type.

FIGURE 13.3
Second and third levels of detail. Ridges appear darker than background. Second level of detail: ridges can end (*ridge endings*), or split (*bifurcations*); ridges can create enclosures (*lakes* - upper left corner), or can be very short (*islands* - upper right corner); ridges can be so short that they appear as *dots*. Short, narrow, and non-continuous ridges that appear between two parallel ridges are called *incipient ridges*. Third level of detail: white dots within the ridges are sweat pores.

The second level of detail relates to ridge events. In Figure 13.3, we observe that ridges do not necessarily flow continuously. Ridges may end (or begin) or split (or merge). Ridges may form *lakes* (upper left corner of Figure 13.3) or *islands* (upper right corner of Figure 13.3). The ridge flow may include *dots* or *incipient ridges* (short, narrow and non-continuous ridges in between two parallel ridges). These events are often called *minutiae*. Examiners do not only consider the presence and type of minutiae, but, more importantly, they also note their spatial relationships and directions with respect to the ridge flow.

Finally, the third level of detail is associated with the shape of the ridges. Elements such as the position and spacing of sweat pores (small white dots on the ridges) or the shape of ridge edges are usually considered when present.

At the end of the analysis stage, the quality and quantity of detected features are assessed. If they are deemed *sufficient*, the examiner proceeds with the comparison of the latent print to control impressions from one or more donors.

13.2.2 Comparison Stage

Latent prints that have been deemed suitable for further examination are then compared to control impressions from known donors. During the comparison stage of the ACE-V framework, examiners must determine if the features that they observed on a latent print during the analysis stage can be detected on a control impression, and if they correspond in terms of their types and spatial relationships.

In principle, this comparison process is meant to be unidirectional: only features detected during the analysis stage should be used and searched for in the control impression. In most practical situations, examiners go back and forth between the trace and control impressions. This circular process can lead examiners to see similarities where there are none. This effect is called *confirmation bias* (U.S. Department of Justice, Office of the Inspector General, 2006; Kassin et al., 2013; Ulery et al., 2015). It occurs when examiners observe a clear characteristic on the ridge flow of the control impression and associate it to a previously undetected feature in the image of the trace impression. Unfortunately, in many cases, it is highly questionable whether the previously undetected feature is even part of the ridge flow, is part of the background noise of the surface on which the impression was laid, or results from the chemical or physical development of the latent print during its recovery. Confirmation bias has been shown to result in erroneous conclusions (Kassin et al., 2013; Neumann et al., 2013; Ulery et al., 2015) and is believed to have played a significant role in the FBI misidentification of a print in relation to the 2004 terrorist attack in Madrid, Spain (U.S. Department of Justice, Office of the Inspector General, 2006).

13.2.3 Evaluation Stage

During the evaluation stage of the ACE-V framework, examiners must assess (1) whether the features observed on the trace impression correspond, within some tolerance level, to the features observed on the control impression; and (2) the specificity (or rarity) of these features. When there are clear dissimilarities between the trace and control impression, the source of the control impression is *excluded* as a possible source for the trace impression. When the level of similarity between the latent and control prints is weak or when the features observed on the trace impression are not very discriminative, examiners may deem the examination to be *inconclusive*. Finally, when there is a high level of similarity between the prints and the features observed on the trace impression are deemed to be discriminative, examiners may conclude that the donor of the trace impression has been *identified* as the donor of the control impression.

13.2.4 Verification Stage

Finally, during the verification stage, a second examiner attempts to confirm the conclusion reached by the first examiner. Different types of verification processes exist: during a *technical review*, the second examiner has access to the notes and conclusions of the first examiner; during a *verification*, the second examiner does not have access to the notes of the first examiner, but may or may not know the initial conclusions; during a *blind verification*, the second examiner has access to limited information regarding the initial examination and the circumstances of the criminal case in order to prevent *contextual bias*; finally, during a *double blind verification*, the second examiner does not know that an initial examination was performed. The choice of which verification process is used usually depends on case circumstances, quality of the latent print, and laboratory policy.

13.2.5 Decision Making in the ACE-V Framework

The ACE-V framework has three critical decision points. At the moment, each individual decision relies on the training and experience of the fingerprint examiner involved and is not directly supported by data.* Such subjectivity can result in lack of transparency, lack of coherence, and overall irreproducible results (see Langenburg (2004, 2009, 2012), Langenburg et al. (2004), and Ulery et al. (2011, 2012, 2013, 2014, 2015, 2016, 2017)). Some of this work is discussed in Section 13.5.

The first critical decision occurs at the end of the analysis stage. Besides the determination of the amount and types of features present on latent prints, the analysis stage serves to filter out latent prints that display so few features that their comparison to control impressions would be pointless. This filtering process can be understood from a quality assurance perspective, as low quality latent prints may have a higher risk of resulting in erroneous conclusions, or from an efficiency perspective, as low quality latent prints may take significantly more time to compare. At present, the decision that a latent print has *sufficient* information that it is worth comparing to a control impression is subjective and does not rely on well defined criteria. Several research projects have demonstrated the variability that exists between examiners who have to decide whether the examination of a given latent print should proceed (Langenburg, 2004, 2012; Ulery et al., 2013). For example, Figure 13.4 shows data obtained by Neumann et al. (2013) who asked approximately 150 examiners to analyse the same 15 latent impressions.[†] Although there is a strong consensus that it is appropriate to proceed with the examination of some impressions (e.g., trials 01, 02, 03, 07, 08, 09, 10, 12, 13 and 14), the same cannot be said for other impressions (e.g., trials 04, 05, 06, 11 and 15).

Researchers have attempted to propose models that predict some measure of usefulness of a latent print based on its quality, and on the number and types of features that it displays. None of these models are currently used in casework, despite their strong potential for improving the efficiency of the examination workflow and for improving quality control (Yoon et al., 2013; Chugh et al., 2018; Cao et al., 2016; Hicklin et al., 2011, 2013; Neumann et al., 2016; Langenburg and Champod, 2011; Langenburg et al., 2012).

The second critical decision arises during the evaluation stage. Traditionally, fingerprint examination can result in three possible conclusions: *identification*, which implies that the donor of the control impression is deemed to be the source of the trace impression; *exclusion*, which implies that the donor of the control impression is excluded as the source of

---

* We do not claim that subjectivity is intrinsically bad. We can think of several daily activities where human subjective decision-making currently outperforms objective algorithms. For example, driving is an activity that involves a great deal of subjectivity, relies on training and experience, and at which human beings presently are more qualified and less error-prone than machines.

[†] These latent prints were selected for their complexity and were not meant to be representative of casework.

the trace impression; or *inconclusive* when it is not possible to reach one of the other two decisions.*

Subjectivity plays a critical role in the conclusions resulting from fingerprint examinations. Different countries have different rules regarding the quality and quantity of features that need to correspond between the trace and control impressions. Two main trends coexist in the community: the point standard (see Champod et al. (2016) for a comprehensive review) and the so-called *holistic* approach (Ashbaugh, 1999). On the one hand, the point-standard method, which consists of defining a fixed number of features that serves as a minimum threshold to reach a conclusion of identification (Champod et al., 2016; see Bertillon, 1912 and Locard, 1914 for historical references), provides a clear decision rule, but has some flaws. For one, proponents of this approach never clearly define the type of features (level 2 or level 3 details) that can be considered or the minimum quality that they need to be counted towards the threshold. Second, Evett and Williams (1996) have demonstrated the arbitrariness in the count of the features. On the other hand, the holistic approach is more flexible and encompassing. By definition, it does not require a specific number, type or quality of features. However, it completely lacks any guidance on how and when decisions should be made (see SWGFAST (2013a) for the best implementation of decision points so far).

As with the analysis stage, the conclusions resulting from the evaluation stage are highly variable (Langenburg, 2009, 2012; Ulery et al., 2011, 2012, 2014). Figures 13.4 and 13.5 reproduce data obtained by Neumann et al. (2013) that illustrate that, while some examinations result in a strong consensus towards an identification, an exclusion or even an inconclusive examination (e.g., trials 01, 02, 03, 04, 09, 10, 11), other examinations result in more split decisions (e.g., trials 05, 06, 07, 08, 12, 13, 14, 15). Some examinations even result in an alarming number of erroneous conclusions (i.e., trial 08 with 19 erroneous exclusions, trial 12 with 11 erroneous identifications, and trial 14 with 18 erroneous exclusions).

Statistical inference frameworks have been associated with forensic DNA analysis since its inception. In particular, many scientific and legal scholars have promoted the use of Bayesian reasoning to support the inference process (see Evett (1998) for an historical reference, Aitken and Taroni (2004) for a general introduction, and Chapters 3 (Banks, 2020), 4 (Stern, 2020), 5 (Taroni et al., 2020), and 10 (Kaye, 2020)). This approach is centered on the Bayes factor, which represents the relative support of the scientific evidence for two mutually exclusive propositions related to the source of the evidence. In Bayesian reasoning, the

---

* The implications of reaching a conclusion of *identification* are not clear. On February 19, 2009, the day after the release of the report from the U.S. National Academies on the state of forensic science in the U.S. (National Research Council, 2009), fingerprint examiners were advised by a memo (Garrett, 2009) from Robert Garrett, former President of the International Association for Identification, that a conclusion of *identification* does not imply that all other possible sources in the world have been excluded. This recommendation is confusing. Indeed, from a logical perspective, it seems that the decision to identify a source *de facto* excludes all other *possible* sources. However, some examiners believe that the decision to identify a source only excludes all other *potential* sources from a *relevant* population of sources using some inductive inference framework. They believe that to exclude all *possible* sources, all of these sources should be physically compared to the latent impression, which is not practically feasible.

The implications of reaching a conclusion of *exclusion* are similarly confusing. In many cases, an *exclusion* implies that the donor of the control impression is not the individual who left the latent print; however, in some cases, an *exclusion* only implies that the specific area of the skin represented in the control impression is not the source of the latent print. In these cases, examiners do not exclude the possibility that another, unobserved, area of the skin of the same individual could be the source of the latent print. In general, fingerprint examination reports are not explicit enough that a lay reader would realise the nuance. This lack of transparency can lead to great misunderstandings.

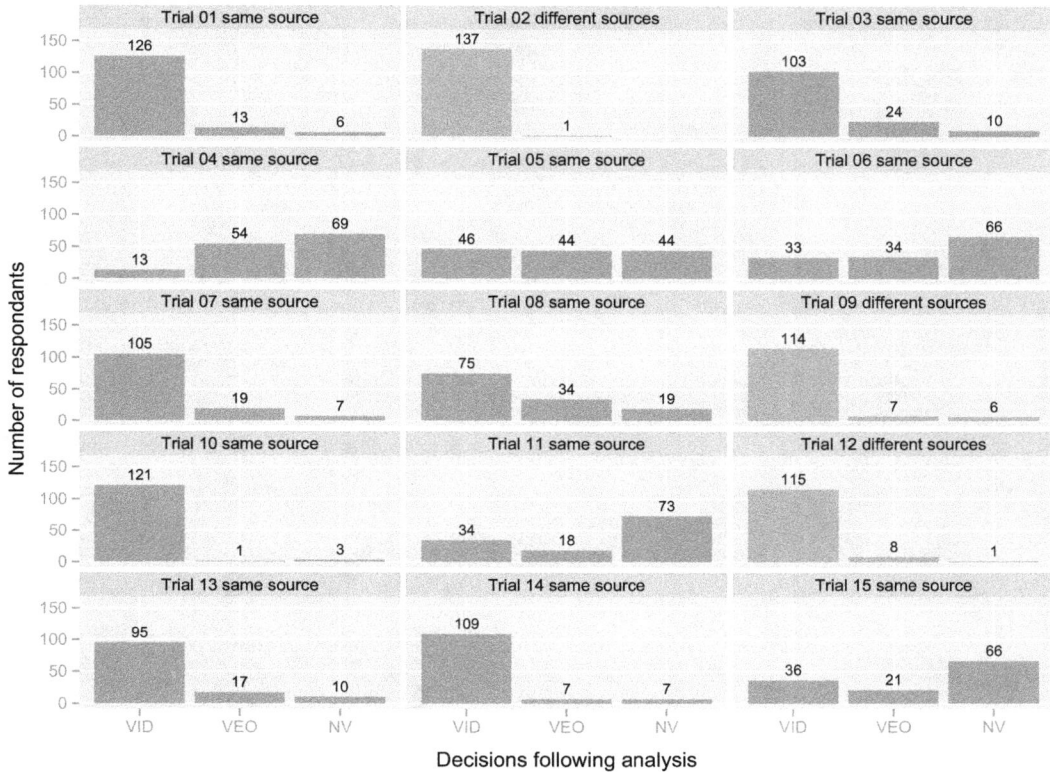

FIGURE 13.4
Decisions after the analysis stage of ACE-V provided by approximately 150 examiners who were asked to determine the quality and quantity of features present on the same 15 latent prints. The possible decisions were: (VID) value for identification, which implies that an impression contains enough information that it could result in the identification of the true source; (VEO) value for exclusion only, which implies that an impression only contains enough information to exclude sources but not identify a source; and (NV) no value, which implies that an impression has so little information that it should not be examined further. Note that the 15 trials were selected to explore the decision-making process and are not representative of casework.

Bayes factor is used to update prior beliefs about the two propositions. The updated beliefs are often called posterior beliefs. Posterior beliefs are probabilities and do not equate to categorical decisions. Hence, they are not directly compatible with the conclusion scheme currently used in fingerprint examinations. The path leading from a posterior probability to a decision involves the use of loss functions and has been described, in the forensic context, by Biedermann et al. (2008). Proponents of the use of Bayesian reasoning argue that it is the only coherent and logical manner to perform inferences in forensic science (see the work by Champod (Champod, 1995; Champod and Evett, 2001; Champod, 2008) for discussions specifically related to fingerprint evidence and Chapters 3 (Banks, 2020), 4 (Stern, 2020), and 5 (Taroni et al., 2020) for a more general discussion). They further argue that, in casework, forensic scientists do not possess the information that would allow them to assign prior beliefs to the propositions that they are considering. Thus, they should limit themselves to reporting Bayes factors and let others (e.g., fact-finders, jurors, judges) complete the

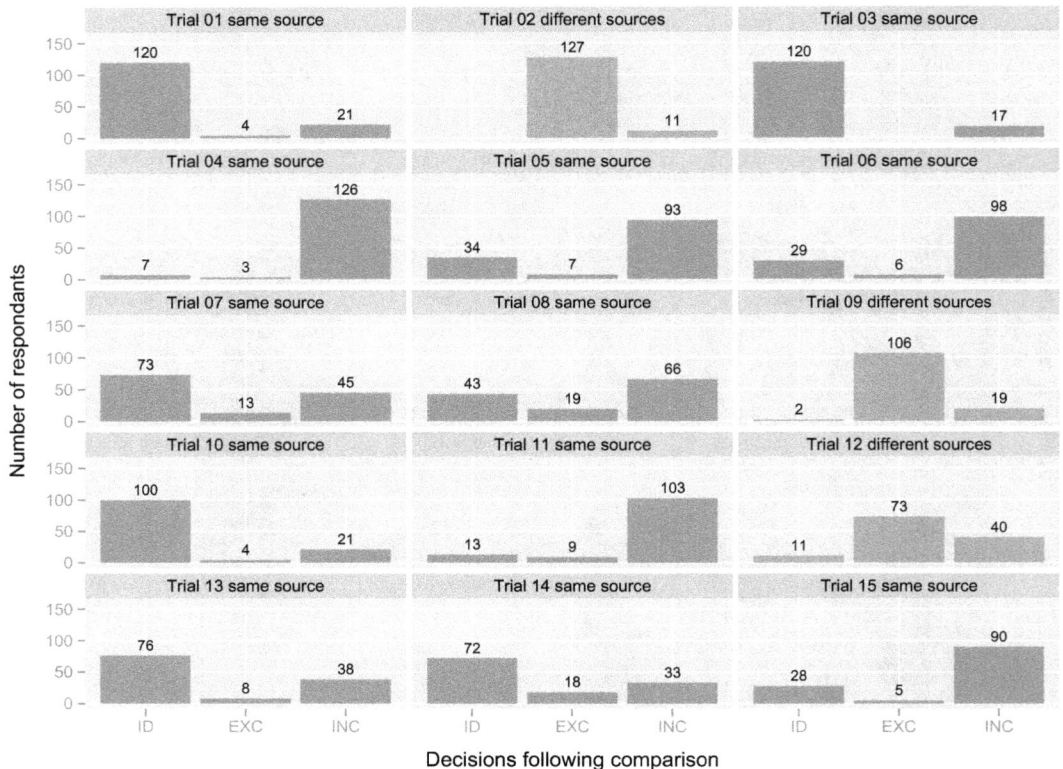

FIGURE 13.5
Decisions after the evaluation stage of ACE-V provided by approximately 150 examiners asked to compare the 15 latent prints from Figure 13.4 with a control impression. Trials 01, 03-08, 10, 11, and 13-15 were same source comparisons, while trials 02, 09 and 12 were different source comparisons. The possible conclusions were: (ID) identification, which implies that the donor of the latent print was identified to be the donor of the control print; (EXC) exclusion, which implies that the control print was deemed to be from a different finger than the latent print; and (INC) inconclusive, which implies that neither of the other two conclusions could be reached based on the material provided. Note that the 15 trials were selected to explore the decision-making process and not to be representative of casework.

inference process. Confronted with the difficulties inherent to the development of models required to assign Bayes factors to fingerprint evidence, researchers have proposed many ad hoc solution, such as so-called *score-based likelihood ratios*, to act as proxies for the Bayes factor. Bayes factors and these ad hoc decision models are explored further in Sections 13.3 and 13.4.

Bayesian reasoning may require a significant paradigm shift in what examiners would be allowed to report and how they would report it. In addition, Bayes factors may be misconstrued by fact-finders (Neumann et al., 2016; Thompson and Schumann, 1987; McQuiston-Surrett and Saks, 2008, 2009; Thompson et al., 2013; Thompson and Newman, 2015 and Chapter 9 (Martire and Edmond, 2020)). Thus, several scholars, mainly in the U.S., have focused on the use of error rates to accompany opinions formed by examiners. This inference framework is very close to the frequentist inference framework. Accordingly, examiners are allowed to communicate categorical conclusions, but they should ideally report the error rates associated with these conclusions. These error rates are estimated by considering the examiners as black boxes, which can be tested using pairs of impressions of known origin. Limitations of this approach include the design of the error rate studies and their generalization to casework. These limitations are discussed in Section 13.5. A recent alternative to large scale error rate studies has been proposed by the Latent Print Branch of the U.S. Defense Forensic Science Center laboratory (DFSC) (Swofford et al., 2018). This alternative, called *FRStat*, essentially allows for calculating error rates that are somewhat more case specific than the generic error rate estimates resulting from the black-box studies. This alternative is reviewed in Section 13.6.

Finally, the third critical decision is made during the verification stage of the ACE-V framework. Technically, this decision is very similar to the one made during the comparison stage, except that it is made by another examiner. Given this similarity, we will not expand on the decisions made during the verification stage beyond stating that conflicting conclusions between two examiners should be documented and trigger a conflict resolution process.

13.2.6 Summary

The fingerprint examination process is often described by the acronym ACE-V. This process includes several decision points. In current practice, none of these decision points are rigorously standardized and data-driven. This results in a significant lack of between-examiners reproducibility and within-examiner repeatability of the decisions, in particular when low quality or complex trace impressions are considered. Some have argued that Bayesian reasoning should be used to support the decision-making in fingerprint examination (and in forensic science in general). Others have argued that it is appropriate to consider fingerprint examiners as black boxes as long as error rates can be associated with the conclusions reached at the end of the examination process.

13.3 Common Source vs. Specific Source Scenarios

In general, the inference of the identity of the donor of a trace impression from its comparison with a control impression from a known donor requires considering two mutually exclusive hypotheses* (Aitken and Taroni, 2004). Over the years, and perhaps because of the legacy of the inference frameworks developed to report forensic DNA evidence (see comment in Section 13.3.3), these hypotheses have not been formulated particularly rigorously. Unfortunately, this lack of formalism has resulted in the development of models and the collection of data that do not properly address the hypotheses that are most often of interest.

The next sections briefly develop two formal scenarios that frame the inference of the source of latent prints: the *common-source scenario* and the *specific-source scenario* (Ommen

---

* The first hypothesis, denoted H_0 below, is commonly called the *prosecution hypothesis*. The alternative hypothesis, denoted H_1 is commonly called the *defense hypothesis*. This often represents an abuse of language since, in most cases, neither the prosecutor nor the defense attorney explicitly states these propositions.

et al., 2017). These scenarios are often confused with one another. This results in models developed under one scenario that answer the question considered by the other one. Thus, understanding their differences is important and helps assess the potential and limitations of the different inference frameworks for fingerprint evidence.

13.3.1 Common Source Scenario

The common source scenario considers that two impressions are *from the same source* without formally specifying who that source is. This scenario typically relates to the comparison of two latent prints, e_{u_1} and e_{u_2} (e.g., two impressions recovered on two different crime scenes or even on the same crime scene), with the goal of determining if they were made by the same person when that person is unknown (e.g., determining whether the two scenes are linked or the number of perpetrators). The hypotheses considered in the common source scenario can be stated as follows*:

$H_{0_{CS}}$: e_{u_1} and e_{u_2} were left by the same, unknown, person;

$H_{1_{CS}}$: e_{u_1} and e_{u_2} were left by two different, unknown, individuals.

In this scenario, the true donor of each impression is considered to be a random individual from a population of potential donors. Under $H_{0_{CS}}$, the donor of the two impressions is the same random individual, while two different random donors from the population left the trace impressions under $H_{1_{CS}}$.

13.3.2 Specific Source Scenario

The specific-source scenario typically involves the comparison of a trace impression, e_u, with a control impression from a known individual, e_s, with the goal of determining if the trace was made by the considered individual. The hypotheses considered in the specific source scenario can be stated as follows:

$H_{0_{SS}}$: e_u and e_s were made by Mr. X.;

$H_{1_{SS}}$: e_u was made by somebody other than Mr. X.

In this scenario, Mr. X is a known donor. He can be considered fixed. Under $H_{1_{SS}}$, the true donor of e_u is unknown and is considered to be a random individual from a population of potential donors, while Mr. X remains the undisputed donor of e_s.

The distinction between both scenarios is not merely theoretical. Each scenario results in different likelihood functions for the same information, and, as we discuss below, in different interpretations of the results of fingerprint examinations.

In the vast majority of cases, the inference questions of greatest interest to the criminal justice system fall under the umbrella of the specific-source scenario. Nevertheless, the determination that two latent prints were made by the same unknown person may

---

* Note that the hypotheses are stated as a function of the donor(s) of the impressions, since they are the ones of interest to the criminal justice system. In practice, fingerprint examiners may consider specific fingers or areas of friction ridge skin. Most of the models and data discussed in this chapter focus on finger-related hypotheses. Neumann et al. (2011) discuss some of the assumptions and implications of moving from finger-based hypotheses to person-based hypotheses.

be relevant to some investigations (e.g., for forensic intelligence-led investigations). Since these two scenarios are different and consider two radically different pairs of hypotheses, they should not be interchanged. Unfortunately, they are often confused.

13.3.3 Simulations

This chapter explores the convergence of different models and inference frameworks partly through simulations. The simulations rely on generative models that give simplified representations of how the data arise under the different hypotheses laid out in Sections 13.3.1 and 13.3.2. We consider a simple univariate setting to explore the different models in the common and specific source scenarios. In the common source scenario, the generative models under both hypotheses can be represented by two hierarchical random effects models:

$$e_{u_1} = \mu + d + u_1, \text{ where } d \sim N(0, \sigma_d^2) \text{ and } u_1 \sim N(0, \sigma_{u_1}^2);$$
$$e_{u_2} = \mu + d + u_2, \text{ where } d \sim N(0, \sigma_d^2) \text{ and } u_2 \sim N(0, \sigma_{u_2}^2);$$

where μ is the mean of the population of sources, d is a random effect due to sources, and u_1 and u_2 are random effects due to objects within sources.*

Under $H_{0_{CS}}$, both impressions originate from the same donor and, thus, have the same value for d (but not necessarily the same σ_u^2 if both traces were left in different conditions). Under $H_{1_{CS}}$, both impressions originate from two different donors and are therefore independent. Thus, the joint distributions of e_{u_1} and e_{u_2} are

$$\begin{pmatrix} e_{u_1} \\ e_{u_2} \end{pmatrix} | H_{0_{CS}} \sim MVN\left(\begin{pmatrix} \mu \\ \mu \end{pmatrix}, \begin{pmatrix} \sigma_d^2 + \sigma_{u_1}^2 & \sigma_d^2 \\ \sigma_d^2 & \sigma_d^2 + \sigma_{u_2}^2 \end{pmatrix}\right);$$

$$\begin{pmatrix} e_{u_1} \\ e_{u_2} \end{pmatrix} | H_{1_{CS}} \sim MVN\left(\begin{pmatrix} \mu \\ \mu \end{pmatrix}, \begin{pmatrix} \sigma_d^2 + \sigma_{u_1}^2 & 0 \\ 0 & \sigma_d^2 + \sigma_{u_2}^2 \end{pmatrix}\right). \tag{13.1}$$

The generative models in the specific source scenario differ depending on whether $H_{0_{SS}}$ or $H_{1_{SS}}$ is considered. Under $H_{0_{SS}}$, when both impressions originate from the same donor, the models are two simple random effects models:

$$e_u = \mu_d + u, \text{ where } u \sim N(0, \sigma_u^2);$$
$$e_s = \mu_d + s, \text{ where } s \sim N(0, \sigma_s^2);$$

where μ_d represents the mean for the considered specific donor, and u and s are random effects respectively corresponding to trace and control samples.

Under $H_{1_{SS}}$, the generative model for the control impressions from the specific donor is the same as under $H_{0_{SS}}$ (indeed, there is no dispute that e_s originates from its donor).

---

* In terms of fingerprints, μ is the mean of the distribution of the characteristics of all friction ridge skin in a population; d represents the distance between the characteristics of the friction ridge skin on various random individuals and the mean of the population; and u_1 and u_2 are random effects that affect the final appearance (after development, transfer, photography, etc.) of fingerprints resulting from different impressions of the fingers represented by d on various surfaces. Note that u_1 and u_2 may be different as two different impressions may be affected by different sets of factors.

However, the model for the trace impression, e_u, is a hierarchical random effects model to reflect that its true donor is an unknown individual from a population of donors:

$$e_u = \mu + d + u, \text{ where } d \sim N(0,\sigma_d^2) \text{ and } u \sim N(0,\sigma_u^2);$$
$$e_s = \mu_d + s, \text{ where } s \sim N(0,\sigma_s^2);$$

and where μ, μ_d, d, u and s are defined as above.[*]

Under $H_{0_{SS}}$, trace and control observations are independent given μ_d, and their joint distribution is multivariate normal. Under $H_{1_{SS}}$, trace and control observations are independent since they are not from the same donor, and their joint distribution is also multivariate normal. We have:

$$\begin{pmatrix} e_u \\ e_s \end{pmatrix} | H_{0_{SS}} \sim MVN\left(\begin{pmatrix} \mu_d \\ \mu_d \end{pmatrix}, \begin{pmatrix} \sigma_u^2 & 0 \\ 0 & \sigma_s^2 \end{pmatrix}\right);$$

$$\begin{pmatrix} e_u \\ e_s \end{pmatrix} | H_{1_{SS}} \sim MVN\left(\begin{pmatrix} \mu \\ \mu_d \end{pmatrix}, \begin{pmatrix} \sigma_d^2 + \sigma_u^2 & 0 \\ 0 & \sigma_s^2 \end{pmatrix}\right).$$

(13.2)

If we take the view that forensic evidence has to be evaluated within a Bayesian paradigm, then we are interested in quantifying the weight of the evidence using Bayes factors (or, when the parameters are known, likelihood ratios). In the common source framework, the likelihood ratio for e_{u_1} and e_{u_2} is:

$$LR_{CS} = \frac{f(e_u, e_s|H_{0_{CS}})}{f(e_u, e_s|H_{1_{CS}})} = \frac{f(e_u, e_s|H_{0_{CS}})}{f(e_u|H_{1_{CS}})f(e_s|H_{1_{CS}})},$$

(13.3)

while the likelihood ratio for e_u and e_s in the specific source framework is

$$LR_{SS} = \frac{f(e_u, e_s|H_{0_{SS}})}{f(e_u, e_s|H_{1_{SS}})} = \frac{f(e_u|H_{0_{SS}})}{f(e_u|H_{1_{SS}})}.$$

(13.4)

We already mentioned that, in most cases, fingerprint examiners are working under the specific source scenario. They are provided with a latent impression and they have to infer whether it was left by the same known person who provided a set of control impressions. Using the toy examples in (13.1) and (13.2), we can study the convergence of the common source likelihood ratio in (13.3) to the specific source likelihood ratio in (13.4) that should be used to quantify appropriately the weight of the evidence.

To compare these likelihood ratios, we consider pairs of e_u and e_s generated by model (13.2) under $H_{0_{SS}}$ or $H_{1_{SS}}$ and we calculate the likelihood ratios (13.3) and (13.4). To calculate the common source likelihood ratio using the data generated under the specific source model, we set $e_{u_1} = e_u$, $e_{u_2} = e_s$, $\sigma_{u_1}^2 = \sigma_u^2$ and $\sigma_{u_2}^2 = \sigma_s^2$.

---

[*] A similar analogy to the one made in the previous footnote can be made here, with μ_d representing the characteristics of the friction ridge skin of a specific finger on a known individual (e.g., a suspect); μ representing the mean of the distribution of the characteristics of all friction ridge skin in a population; d representing the distance between the characteristics of the friction ridge skin of a specific finger from an unknown individual (e.g., the true donor of the latent print) and the mean of the population; and u and s representing random effects that affect the final appearance (after development, transfer, photography, etc.) of fingerprints resulting from different impressions of the fingers represented by μ_d and d on various surfaces. As before, u and s may be different, as latent and control prints are acquired under different sets of conditions.

Figure 13.6 presents the results of three experiments in which, $\mu = 10$, $\sigma_d^2 = 10$ and $\sigma_u^2 = 2$. All simulations were repeated 1,000 times. In the first experiment, the characteristics of the donor of e_s were chosen to be relatively common in the population of donors ($\mu_d = 9$) but also quite variable ($\sigma_s^2 = 1$). In the second experiment, the characteristics of the donor of e_s were chosen to be rare in the population of donors ($\mu_d = 0$) and equally variable ($\sigma_s^2 = 1$). In the last experiment, the variability of the characteristics of the known donor of the control impressions was chosen to be negligible ($\sigma_s^2 = 10^{-5}$).*

The results of the experiments show that likelihood ratios for the common and the specific source scenarios do not converge unless the variability of the donor of the control material is negligible.[†,‡] Importantly, the results for the first two experiments in Fig. 13.6 show that the common source likelihood ratio unpredictably over- or underestimates the value of the specific source likelihood ratio. That said, although common-source likelihood ratios that are assigned when $H_{1_{SS}}$ is true may underestimate the corresponding specific-source likelihood ratios, they have a marked tendency to overestimate their counterparts (we have not found a situation where common-source likelihood ratios consistently underestimate specific-source likelihood ratios).

The lack of convergence raises the issue of whether $H_{0_{SS}}$ or $H_{1_{SS}}$ is true. When $H_{0_{SS}}$ is true, underestimating the value of the specific-source likelihood ratio may result in the erroneous exclusion of the donor of the control impressions as the source of the trace impression. The criminal justice system considers this to be a better outcome than the erroneous identification of an innocent individual. Furthermore, when $H_{0_{SS}}$ is true, overestimating the value of the specific-source likelihood ratio only results in being overconfident in the support of the correct conclusion that the donor of the control impressions is also the donor of the trace impressions; thus, the impact of the overestimation may be considered minimal. Unfortunately, when $H_{1_{SS}}$ is true, overestimating the value of the specific-source likelihood ratio may result in exculpatory evidence not being given the appropriate weight in favour of an innocent, yet suspected, donor. In fact, Figure 13.6a,b show that some pieces of evidence result in values of their specific-source likelihood ratios that are less than one and in values of their common source likelihood ratios that are greater than one. This may result in serious miscarriages of justice if the common-source likelihood ratio is used instead of the specific-source one.

13.3.4 Summary

There are two different scenarios under which forensic evidence can be evaluated: the common-source scenario and the specific-source scenario. The former scenario focuses on evaluating whether two trace objects originate from the same unknown donor, while the latter focuses on evaluating whether a single trace object originates from a specific known donor. Forensic evidence is considered differently under these two scenarios and does not

---

* It can be argued that finger impressions collected from a known donor under controlled conditions have virtually no variability. This argument is reasonable. For example, control prints have more variability than forensic DNA profiles; however, they have much less variability than trace impressions.

† This can also be seen directly from the analytical forms of the joint distributions of e_{u_1} and e_{u_2}, and e_u and e_s, as $\sigma_{u_2}^2 = \sigma_s^2 \to 0$.

‡ This is typically the case for forensic DNA analysis when single full DNA profiles are considered. Since the allelic designation of a full DNA profile is extremely reproducible, the inference of the identity of source of a pair of full DNA profiles will be the same under both common and specific source scenarios. This may explain why the distinction between common and specific source scenarios was not discussed until recently by Ommen et al. (2017).

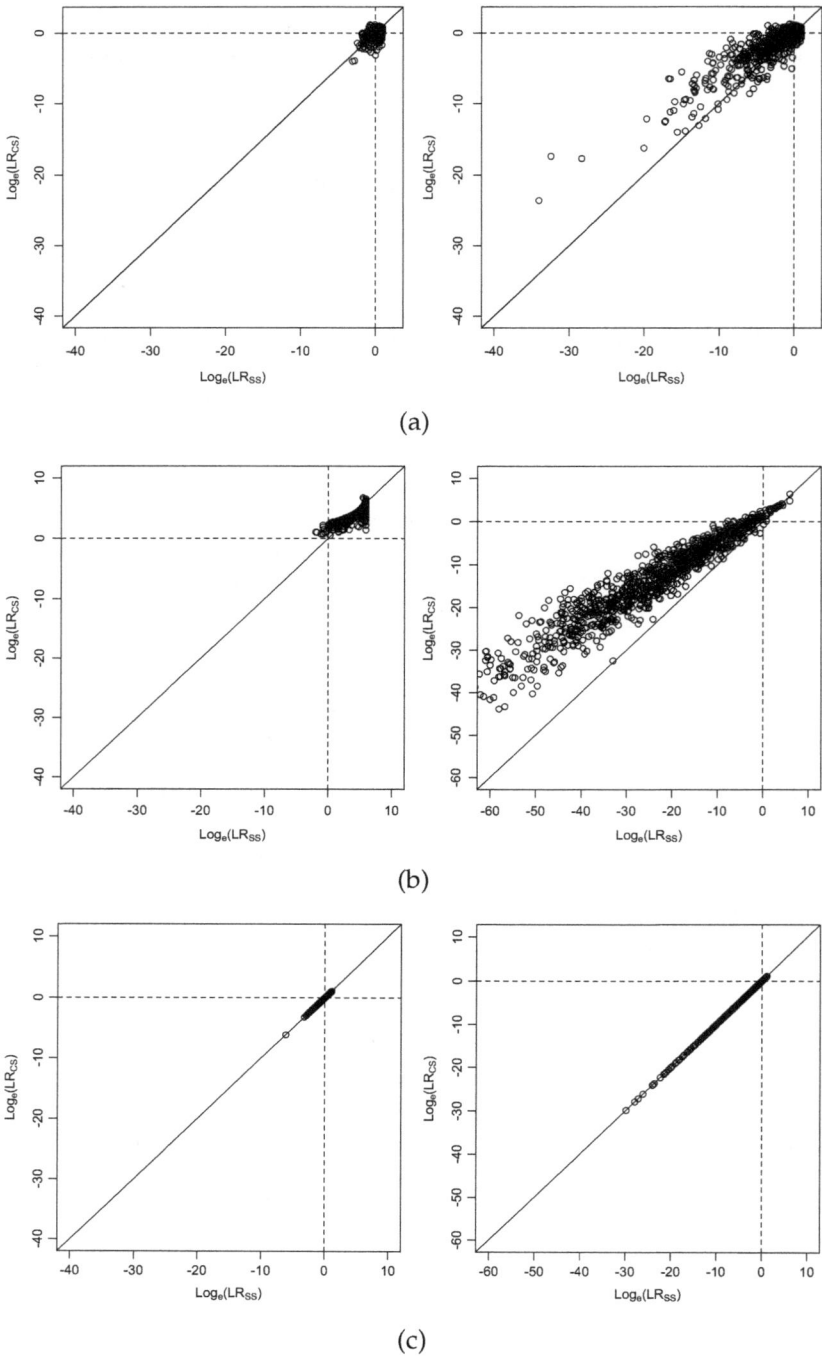

FIGURE 13.6
Comparisons between LRs in the common source and specific source scenarios. Columns: the left column reports the results when e_u and e_s have been sampled under $H_{0_{SS}}$; the right column reports the results under $H_{1_{SS}}$. Rows: (a) the source of the control impression is common; (b) the source of the control impression is rare; (c) the source of the control impression is common, but it has virtually no variance.

have the same weight. Both scenarios converge only when the variability of the evidence is negligible (e.g., single source DNA profile, finger impression taken under controlled conditions). In all other circumstances, misspecifying the interpretation framework can lead to dramatic results.

13.4 Weight of Fingerprint Evidence

Anecdotal attempts have been made to assign Bayes factors to fingerprint evidence by directly modelling fingerprint features (Forbes et al., 2014; Tackett, 2018; Neumann et al., 2015). These feature-based models all suffer from the difficulty of defining reasonable likelihood functions to represent the joint distributions of heterogenous and high-dimensional feature vectors. For example, a single minutia can be represented by its Euclidean coordinates (bivariate continuous variable), its type (nominal variable) and its direction in the ridge flow (circular variable). An impression where n minutiae are observed is then represented by a $4n$-long vector, which contains three different types of variables. To bypass the need to work with intractable likelihood functions, most researchers concentrate on the use of *(dis)similarity metrics* or *kernel functions* to map finger impressions from their natural space to $\mathbb{R}$. But working in a continuous space is not without limitations, which are explored below.

13.4.1 Similarity Metrics and Kernel Functions

The natural proximity of the fields of forensic fingerprint examination and biometry have led researchers to quickly realize the benefits of modelling the (dis)similarity between pairs of impressions, rather than modelling fingerprint features in their original space. This strategy offers great flexibility to researchers. First, they can design algorithms that measure the distance between two impressions, such that the *score* representing that distance is minimised when the two impressions originate from the same finger and is maximised when they originate from different fingers.* Second, the level of (dis)similarity between pairs of impressions can be expressed as a univariate continuous random variable, which is significantly more convenient to model than the original vectors representing the observations made on the impressions.

A score can have two interpretations: it can be seen as a summary statistic resulting from the comparison of two impressions, or it can be seen as the scalar projection resulting from the inner product of two vectors. In the first case, we can discuss the sampling distributions of the score under various situations; in the second case, the score has a geometric interpretation. Formally, a score interpreted as a summary statistic of the (dis)similarity between two objects e_i and e_j can be defined as $\delta(e_i, e_j)$, where δ is any function with a real-valued output. A score interpreted as the inner product of two vectors can be similarly defined as $\kappa(e_i, e_j) = \langle \eta(e_i), \eta(e_j) \rangle$, where κ is a kernel function and η is a set of basis expansions. The main difference between δ and κ is that κ has to be a positive semi-definite symmetric function, while there is no requirement for the construction of δ.

---

* The score representing the similarity between a pair of impressions is maximized when these impressions originate from the same finger, while it is minimized when the impressions originate from different fingers. The upcoming discussion on the use of scores is not affected by their interpretation as similarity measures or distances.

These two different perspectives are used to investigate the different score-based models in the next section, as well as the Approximate Bayesian Computation Bayes factors discussed in Section 13.4.3. The generative models described in Section 13.3.3 are used to discuss the convergence of these models to the specific-source likelihood ratio of interest in (13.4). To perform these simulations, both δ and κ are defined as the squared Euclidean distance, which is both a summary statistic and a valid kernel function, and has tractable distributions for the chosen generative models.

13.4.2 Score-Based Likelihood Ratios

The use of scores to calculate score-based likelihood ratios can be tracked back to the late 1990s and early 2000s and the field of speaker recognition (Meuwly and Drygajlo, 2001; Gonzalez-Rodriguez et al., 2003, 2006). In the same period, several authors (Gonzalez-Rodriguez et al., 2005; Egli et al., 2006; Neumann et al., 2006, 2007) proposed to use scores from biometric systems to evaluate the probative value of fingerprint and other types of evidence. Different constructions of score-based likelihood ratios have been proposed over the years.* Despite their limitations, their use in casework is advocated (at least in Europe (European Network of Forensic Science Institutes, 2016)). The concept behind most models rests on the comparison of the likelihood of the score calculated between a latent print, e_u, and a control print from a known individual, e_s, evaluated in two different functions based on the sampling distributions of the score under two alternative propositions (left panel of Figure 13.7). These models are best introduced through the sampling experiments that can be imagined to study the sampling distributions under the two alternative propositions.

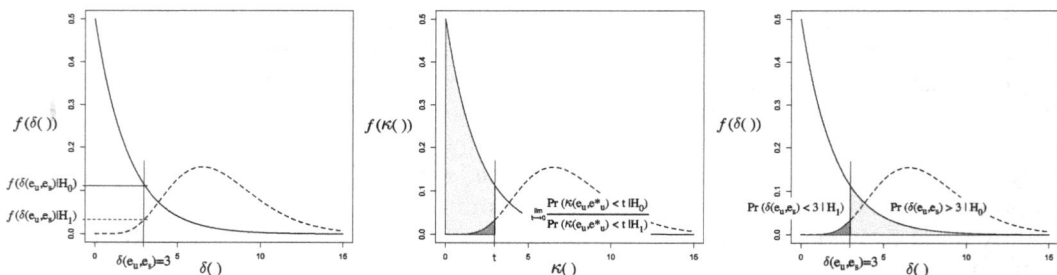

FIGURE 13.7
Three different concepts for the use of summary statistics/kernel functions to provide information on the probative value of fingerprint evidence. Left panel: ratio of the density of a summary statistic for an observed pair of trace and control impressions, $\delta(e_u, e_s)$, in its sampling distribution under the first proposition and its density in its sampling distribution under the second proposition (Section 13.4.2). Middle panel: limit as $t \to 0$ of the ratio of two tail probabilities, where t is a level of tolerance for the dissimilarity of the latent print observed at the crime scene, e_u, and latent prints generated under the two alternative propositions, e_u^* (Section 13.4.3). Right panel: α- and β-error types associated with a decision of identification or exclusion at an observed level of dissimilarity of a pair of trace and control impressions, $\delta(e_u, e_s)$, using the two sampling distributions for $\delta(e_i, e_j)$ under the two propositions (Section 13.6).

---

* In some cases (van Es et al., 2017; Chapter 20 (Morrison et al., 2020)), the score itself is a proper likelihood ratio. In these cases, the score-based likelihood ratio is the "likelihood ratio of a likelihood ratio". These models have different properties than the ones discussed below. These types of models are not encountered in fingerprint or other pattern evidence and are not discussed in this chapter.

13.4.2.1 Common Source Score-based Models

In the first type of score-based model (Meuwly and Drygajlo, 2001; Gonzalez-Rodriguez et al., 2005), the score, $\delta(e_u, e_s)$, is evaluated using sampling distributions based on the following thought experiments:

1. When the prosecutor proposition is correct, $\delta(e_u, e_s)$ is a score that is calculated by comparing trace and control impressions from the same, random, individual. The sampling distribution of $\delta(e_u, e_s)$ under H_0 can be studied by considering a sample of individuals from a relevant population, and by sampling and comparing a single trace and a single control impression from each individual.

2. When the defense proposition is correct, $\delta(e_u, e_s)$ is a score that is calculated by comparing trace and control impressions from different individuals. The sampling distribution of $\delta(e_u, e_s)$ under H_1 can be studied by sampling independent pairs of individuals from a relevant population, and by comparing a trace impression from the first individual to a control impression from the second individual.

This type of model has two main limitations. First, it is only reporting the *average* density of $\delta(e_u, e_s)$ under both propositions. Neither sampling distribution is specific to the donor of e_s. This type of model clearly addresses the common-source pair of propositions and is not relevant to a case involving the comparison of a latent to a control print from a known donor. Second, Bayes factors can roughly be viewed as the ratio between some measure of similarity between the characteristics of the trace and control objects, and some measure of the rarity of the characteristics of the trace. However, the model described above does not account for the rarity of the trace characteristics at all. This type of model only accounts for the rarity of the level of similarity.*

To compare this type of model to the specific-source likelihood ratio in (13.4), we use the generative models proposed in (13.1). By defining $\delta(e_u, e_s) = (e_{u_1} - e_{u_2})^2$, we have

$$(e_{u_1} - e_{u_2})^2 | H_{0_{CS}} \sim \frac{1}{\sigma_{u_1}^2 + \sigma_{u_2}^2} \chi^2 \left(\frac{(e_{u_1} - e_{u_2})^2}{\sigma_{u_1}^2 + \sigma_{u_2}^2} \right),$$

$$(e_{u_1} - e_{u_2})^2 | H_{1_{CS}} \sim \frac{1}{\sigma_{u_1}^2 + \sigma_{u_2}^2 + 2\sigma_d^2} \chi^2 \left(\frac{(e_{u_1} - e_{u_2})^2}{\sigma_{u_1}^2 + \sigma_{u_2}^2 + 2\sigma_d^2} \right),$$

(13.5)

and

$$SLR_{CS} = \frac{f((e_{u_1} - e_{u_2})^2 | H_{0_{CS}})}{f((e_{u_1} - e_{u_2})^2 | H_{1_{CS}})}.$$

(13.6)

The results of the comparison of (13.4) and (13.6) using our toy example are presented in Figure 13.8. To study the convergence of LR_{SS} and SLR_{CS}, we set $e_{u_1} = e_u, e_{u_2} = e_s, \sigma_{u_1}^2 = \sigma_u^2$ and $\sigma_{u_2}^2 = \sigma_s^2$. In both models, $\mu = 10, \sigma_d^2 = 10$ and $\sigma_u^2 = 2$. All simulations were repeated

---

* For example, consider a bloodstain recovered at a crime scene. A suspect, who has blood of the same type as the bloodstain, is considered. Clearly, the information that the type of the blood recovered at a crime scene is the same as the one of the suspect will be a lot more helpful to support the inference that the blood comes from the suspect if the blood type is AB⁻ (less than 1% of the population) than if the blood type is O⁺ (approx. 40% of the population). Yet, under the defence proposition, the model described above only assigns a probability to the event that the two blood types correspond by chance without accounting for the specific type of the blood at the crime scene.

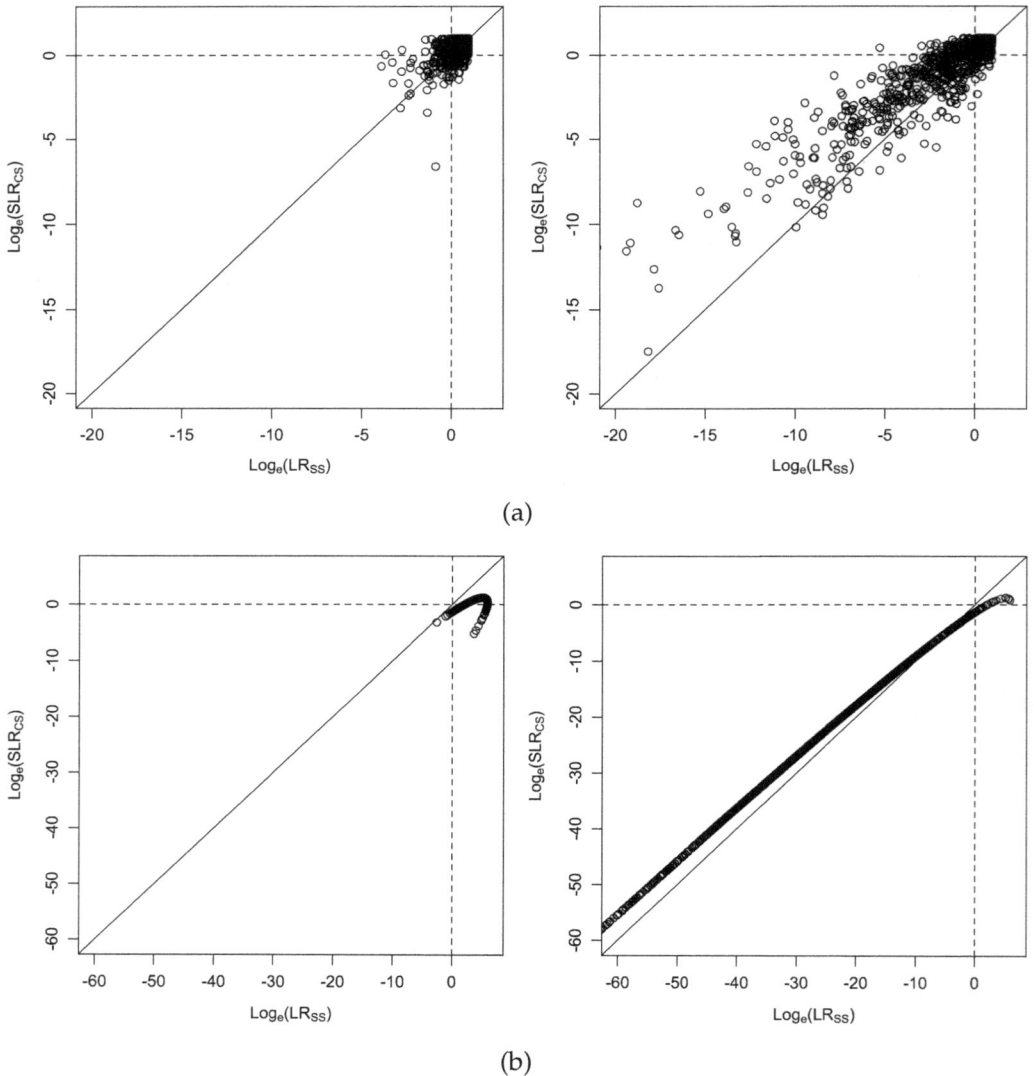

(a)

(b)

FIGURE 13.8
Comparisons between SLRs in the common-source scenario with the LR in the specific-source scenario. Columns: the left column reports the results when e_u and e_s have been sampled under $H_{0_{SS}}$; the right column reports the results under $H_{1_{SS}}$. Rows: (a) the source of the control impression is common and has some variance; (b) the source of the control impression is rare and has virtually no variance.

1,000 times. In the first experiment, the characteristics of the donor of e_s were relatively common with respect to the population of donors ($\mu_d = 9$) but also quite variable ($\sigma_s^2 = 1$). In the second experiment, the characteristics of the donor of e_s were rare with respect to the population of donors ($\mu_d = 0$) with a negligible variability ($\sigma_s^2 = 10^{-5}$).

The same general conclusions drawn from the data presented in Figure 13.6 can be reached when observing the data presented in Figure 13.8. Figure 13.8 shows that the SLR_{CS} have a marked tendency to overestimate their LR_{SS} counterparts, which may not necessarily be a problem when $H_{0_{SS}}$ is true; however, the use of SLR_{CS} to report fingerprint evidence in court may be very detrimental to innocent suspects. We also note the particular behaviour of the relationship between LR_{SS} and SLR_{CS} when the variance of the control impressions is very small.

Overall, common-source score-based models may be convenient to implement, but are not relevant to most fingerprint comparisons of forensic interest, and do not converge to the weight of fingerprint evidence. The lack of convergence between LR_{SS} and SLR_{CS} is not only a by-product of the use of a potentially non-sufficient summary statistic as the score between a pair of impressions, but also results from the inadequacy of (13.6) to account for the rarity of the characteristics observed on the trace impression.

13.4.2.2 Suspect-Centred Score-based Models

A second type of score-based model was proposed to be more relevant to the case at hand (Helper et al., 2012; Alberink et al., 2014). According to this type of model, the sampling distributions of $\delta(e_u, e_s)$ can be represented by the following thought experiments:

1. When the prosecutor proposition is correct, $\delta(e_u, e_s)$ is a score calculated by comparing a latent print and a control impression that have been both obtained from the donor of e_s. Thus, the sampling distribution of $\delta(e_u, e_s)$ under H_0 can be studied by sampling, and comparing, independent pairs of trace and control impressions from the donor of e_s.

2. When the defense proposition is correct, $\delta(e_u, e_s)$ is a score calculated by comparing trace impressions sampled from random individuals in a relevant population to control impressions from the donor of e_s.*

While this type of score-based model has been designed to address the specific-source pair of propositions, it does not address the issue of the rarity of the trace characteristics. Furthermore, it may be not be trivial to repeatedly sample latent and control prints from an uncooperative donor. To overcome the latter issue, authors have proposed to generate pseudo-fingerprints (Neumann et al., 2012; Rodriguez et al., 2012) or to use parametric models for the score distributions (Egli et al., 2006; Egli-Anthonioz and Champod, 2014).

To avoid repeatedly sampling control impressions from the donor of e_s, some authors have proposed to condition the score-based model on e_s. The difference between the unconditioned suspect-centred score-based model described in the previous paragraph and the conditioned model is that $\delta(e_u, e_s)$ and both sampling distributions use the same fixed e_s. Mathematically:

$$
\begin{aligned}
SLR_{SS|e_s} &= \frac{f(\delta(e_u, e_s), e_s | H_{0_{SS}})}{f(\delta(e_u, e_s), e_s | H_{1_{SS}})} = \frac{f(\delta(e_u, e_s) | e_s, H_{0_{SS}}) f(e_s | H_{0_{SS}})}{f(\delta(e_u, e_s) | e_s, H_{1_{SS}}) f(e_s | H_{1_{SS}})} \\
&= \frac{f(\delta(e_u, e_s) | e_s, H_{0_{SS}})}{f(\delta(e_u, e_s) | e_s, H_{1_{SS}})}.
\end{aligned}
\tag{13.7}
$$

The second ratio in (13.7) cancels out since the characteristics observed on the control impressions have the same density irrespective of whether the donor of e_s is also the donor

---

* This may seem counterintuitive, and the reader may wonder why trace impressions, rather than control impressions, are randomly sampled from donors in the relevant population. This sampling model is rooted in the definition of the generative model in (13.2): indeed, in the specific-source scenario, there is no dispute that e_s originates from the suspected donor.

of the trace impression. From the generative models proposed in (13.2), and with $\delta(e_u, e_s) = (e_{u_1} - e_{u_2})^2$, we obtain the following sampling distributions for $\delta(e_u, e_s)$:

$$(e_u - e_s)^2 | e_s, H_{0_{SS}} \sim \frac{1}{\sigma_u^2} \chi^2 \left(\frac{(e_u - e_s)^2}{\sigma_u^2}, \lambda = \frac{(\mu_d - e_s)^2}{\sigma_u^2} \right)$$

$$(e_u - e_s)^2 | e_s, H_{1_{SS}} \sim \frac{1}{\sigma_u^2 + \sigma_d^2} \chi^2 \left(\frac{(e_u - e_s)^2}{\sigma_u^2 + \sigma_d^2}, \lambda = \frac{(\mu - e_s)^2}{\sigma_u^2 + \sigma_d^2} \right).$$

$$(13.8)$$

These sampling distributions enable us to compare the likelihood ratio of interest, LR_{SS}, with its proxy, $SLR_{SS|e_s}$. This comparison is reported in Figure 13.9 using the same parameter values as those used to generate the results presented in Figures 13.6 and 13.8.*

The model proposed in (13.7) certainly seems to be a reasonable ad hoc solution: it is specific to the suspected donor; the required sampling-simulation of trace impressions from relevant donors can be achieved by using a suitable fingerprint distortion model or a parametric model of the score distributions; furthermore, under the reasonable assumption that control fingerprints have very limited variance, it will mostly converge to the specific source likelihood ratio of interest (Figure 13.9c). Unfortunately, this type of model is plagued by a fundamental lack of coherence: indeed, with these models, a given piece of evidence can provide support for either of the alternative propositions, depending on which proposition is considered first.

To demonstrate this lack of coherence, consider a model designed to address the two following specific-source propositions:

H_A: e_u was made by Mr. A;

H_B: e_u was made by Mr. B.

We observe the trace impression as well as two control impressions, one from Mr. A and one from Mr. B. The specific-source generative models under H_A and H_B are, under H_A,

$$e_a = \mu_a + a, \text{where } a \sim N(0, \sigma_a^2); \qquad e_u = \mu_a + u, \text{where } u \sim N(0, \sigma_u^2);$$

$$e_b = \mu_b + b, \text{where } b \sim N(0, \sigma_b^2),$$

and, under H_B:

$$e_a = \mu_a + a, \text{where } a \sim N(0, \sigma_a^2);$$

$$e_b = \mu_b + b, \text{where } b \sim N(0, \sigma_b^2); \qquad e_u = \mu_b + u, \text{where } u \sim N(0, \sigma_u^2).$$

The specific-source likelihood ratio that addresses H_A and H_B is:

$$LR_{SS} = \frac{f(e_u, e_a, e_b | H_A)}{f(e_u, e_a, e_b | H_B)} = \frac{f(e_u | H_A)}{f(e_u | H_B)} = \left(\frac{f(e_u | H_B)}{f(e_u | H_A)} \right)^{-1}.$$

$$(13.9)$$

Thus, LR_{SS} coherently supports the same proposition irrespective of which one is considered first. However, the specific-source SLR_{SS} conditioned on the control impression of the

---

* Note that the results presented in Figure 13.9 are highly dependent on the value chosen for e_s. In particular, the patterns in Figure 13.9a,b are very sensitive to the value of e_s.

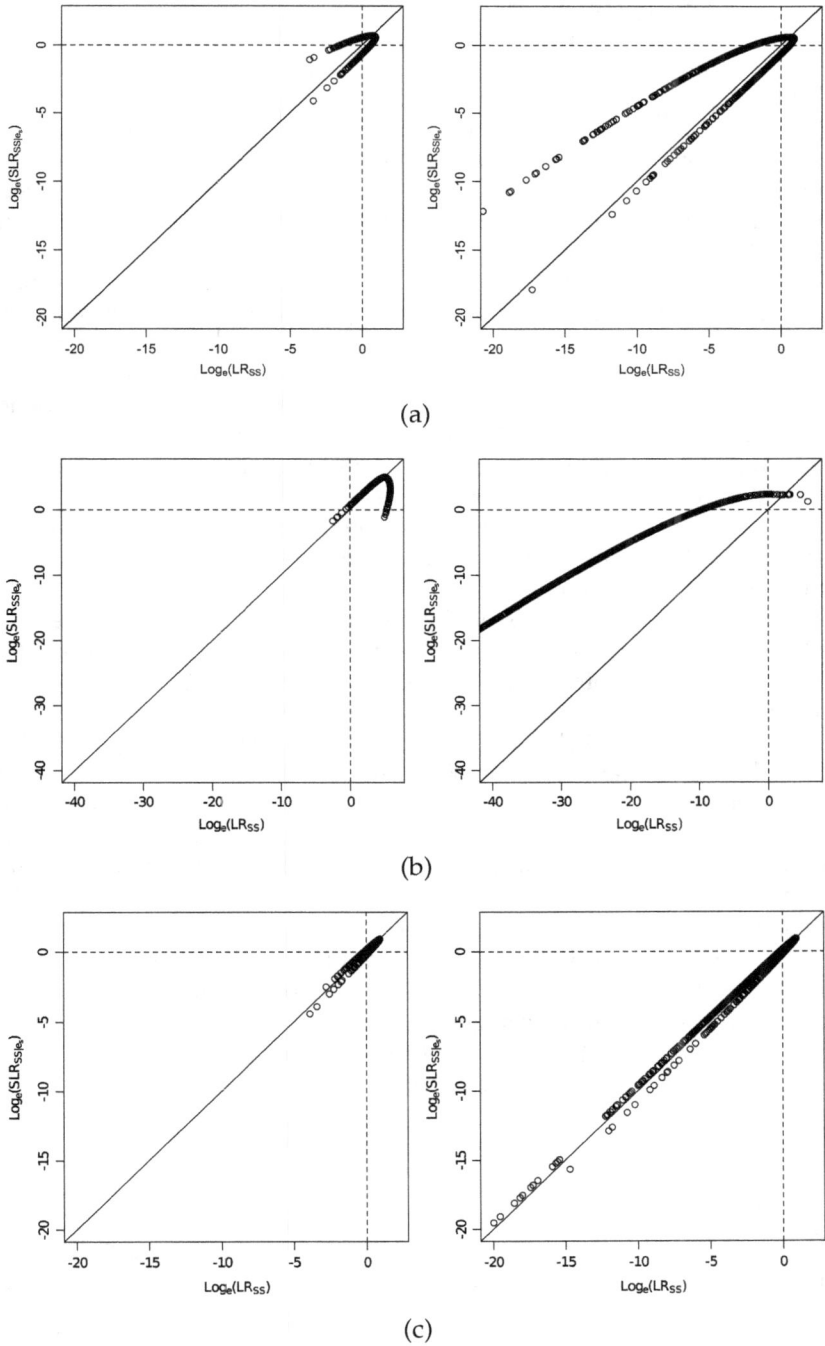

FIGURE 13.9
Comparisons between SLRs conditioned on e_s in the specific source scenario with the LR in the specific source scenario. Columns: the left column reports the results when e_u and e_s have been sampled under $H_{0_{SS}}$; the right column reports the results under $H_{1_{SS}}$. Rows: (a) the source of the control impression is common and has some variance; (b) the source of the control impression is rare and has some variance; (c) the source of the control impression is common and has virtually no variance.

first proposition considered by the model is

$$SLR_{SS|e_a} = \frac{f(\delta(e_u, e_a)|e_a, H_A)}{f(\delta(e_u, e_a)|e_a, H_B)} \neq \left(SLR_{SS|e_b}\right)^{-1} = \left(\frac{f(\delta(e_u, e_b)|e_b, H_B)}{f(\delta(e_u, e_b)|e_b, H_A)}\right)^{-1}. \tag{13.10}$$

Equation (13.10) shows that $SLR_{SS|e_s}$ is not coherent in general since it potentially does not support the same proposition depending on which one is considered first. This lack of coherence can similarly be demonstrated for the unconditioned suspect-centred score-based likelihood ratio.

The conditioning of $\delta(e_u, e_s)$ on e_s has an interesting geometric interpretation. When e_s is fixed, all trace impressions from the donor considered under $H_{0_{SS}}$ and from the individuals from the population considered under $H_{1_{SS}}$ are compared to the same control impression. It is thus possible to consider that all scores considered in our thought experiment are equivalent to the scalar projections of the vectors representing all trace impressions onto a vector space defined by e_s. Figure 13.10 illustrates this interpretation.

In Figure 13.10, a vector $\mathbf{e}_u$, representing a trace impression recovered in connection with a crime, is compared to impressions from donors A and B, represented by two mean vectors, $\mathbf{e}_a$ and $\mathbf{e}_b$. The left panel shows the orthogonal projection of $\mathbf{e}_u$ onto $\mathbf{e}_a$ and $\mathbf{e}_b$. The resulting scalars are the scores calculated by $\delta(\mathbf{e}_u, \mathbf{e}_a)$ and $\delta(\mathbf{e}_u, \mathbf{e}_b)$, which are, in this case, equivalent to $\kappa(\mathbf{e}_u, \mathbf{e}_a)$ and $\kappa(\mathbf{e}_u, \mathbf{e}_b)$.

The middle panel shows the orthogonal projections of (1) $\mathbf{e}_u$, (2) multiple trace impressions sampled from the donor of $\mathbf{e}_b$ (represented by dots near the tip of $\mathbf{e}_b$), and (3) multiple trace impressions sampled from the donor of $\mathbf{e}_a$ (represented by dots near the tip of $\mathbf{e}_a$) onto $\mathbf{e}_a$. The two density functions represent the distributions of the scalar projections of these vectors onto $\mathbf{e}_a$. According to the middle panel, $SLR_{SS|e_a}$, in (13.10), is equivalent to the ratio of the likelihoods of the scalar projection of $\mathbf{e}_u$ onto $\mathbf{e}_a$ evaluated using the two distributions of the scalar projections of the trace impressions from both suspects onto $\mathbf{e}_a$. We see that, in this case, $SLR_{SS|e_a}$ would support the proposition that $\mathbf{e}_u$ was made by donor B. The right panel shows the same information as in the middle panel, but this time, projected onto $\mathbf{e}_b$. We see that in this case, $SLR_{SS|e_b}$ would support the proposition that $\mathbf{e}_u$ was made by donor A. This geometric interpretation holds in the general case, when the alternative hypothesis is not specific to a single donor, but considers a population of donors as in Section 13.3.2. In the general case, all trace impressions from all donors of the relevant population are projected onto the vector representing a single donor under $H_{0_{SS}}$ or under $H_{1_{SS}}$. This results in the same lack of coherence in the support of the evidence for the alternative propositions when the donor defining the vector space changes.

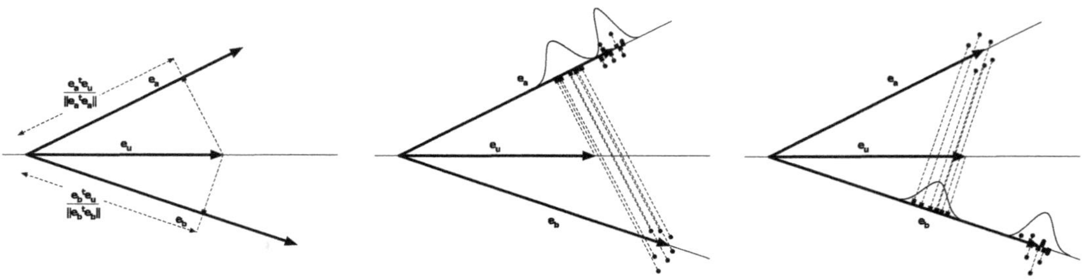

FIGURE 13.10
Left panel: vector projection of $\mathbf{e}_u$ onto $\mathbf{e}_a$ and $\mathbf{e}_b$. Middle panel: vector projection of all trace impressions, including $\mathbf{e}_u$, onto $\mathbf{e}_a$. Right panel: vector projection of all trace impressions, including $\mathbf{e}_u$, onto $\mathbf{e}_b$.

13.4.2.3 Trace-Centred Score-based Models

The two types of models presented above lack the ability to account for the rarity of the characteristics of the trace impression, which is crucial to properly quantify the weight of fingerprint evidence. To remedy this shortcoming, the use of trace-centred score-based models has been proposed (Alberink et al., 2014). This type of model is somewhat similar to the family of suspect-centred models. However, since it is not possible to sample more trace and control impressions from the true donor of e_u (since he is unknown), these models must be conditioned on the observed trace, e_u. Mathematically, we have:

$$SLR_{SS|e_u} = \frac{f(\delta(e_u, e_s), e_u | H_{0_{SS}})}{f(\delta(e_u, e_s), e_u | H_{1_{SS}})} = \frac{f(\delta(e_u, e_s) | e_u, H_{0_{SS}}) f(e_u | H_{0_{SS}})}{f(\delta(e_u, e_s) | e_u, H_{1_{SS}}) f(e_u | H_{1_{SS}})}$$
$$= \frac{f(e_u | H_{0_{SS}})}{f(e_u | H_{1_{SS}})}.$$
(13.11)

Interestingly, the second ratio in (13.11) does not cancel out. Indeed, the likelihood of observing the trace impression, e_u, is very different under $H_{0_{SS}}$ and $H_{1_{SS}}$, and this ratio corresponds exactly to the likelihood ratio in (13.4). In contrast, the first ratio, which includes the score, does cancel since e_u is fixed under both propositions due to conditioning, and e_s has the same distribution under both propositions in the specific-source scenario. This can be seen when using the generative model in (13.2) with $\delta(e_u, e_s) = (e_{u_1} - e_{u_2})^2$, which results in the same sampling distributions under both propositions:

$$(e_u - e_s)^2 | e_u, H_{0_{SS}} \sim \frac{1}{\sigma_s^2} \chi^2 \left(\frac{(e_u - e_s)^2}{\sigma_s^2}, \lambda = \frac{(e_u - \mu_d)^2}{\sigma_s^2} \right);$$
$$(e_u - e_s)^2 | e_u, H_{1_{SS}} \sim \frac{1}{\sigma_s^2} \chi^2 \left(\frac{(e_u - e_s)^2}{\sigma_s^2}, \lambda = \frac{(e_u - \mu_d)^2}{\sigma_s^2} \right).$$
(13.12)

The results in (13.11) and (13.12) may seem suspicious at first. Some readers may think that under $H_{1_{SS}}$, the sampling distribution should involve control impressions from donors in the relevant population. However, $H_{1_{SS}}$ is very clear on the origins of e_s: its donor is undisputed, and it is the same specific donor considered in $H_{0_{SS}}$ (see Section 13.3.2 and Equation (13.2)). Geometrically, this type of model has a similar interpretation as the suspect-centred models. Trace-centred models can be understood as the projection of all control impressions onto a vector space defined by e_u. However, as mentioned above, the only control impressions available in this type of model are control impressions of the same donor (i.e., the donor of e_u) under both alternative propositions. Therefore, the first ratio in $SLR_{SS|e_u}$ will always be one. In summary, it appears that trace-centred score-based likelihood ratios are not very useful.

13.4.3 Approximate Bayes Factor for Fingerprint Evidence

Suspect-centred score-based likelihood ratios do not account for the rarity of the characteristics of the trace. Trace-centred score-based likelihood ratios only account for the impressions of one known donor. Neumann et al. (2012) proposed an algorithm that accounts for the rarity of the trace characteristics and leverages impressions from all individuals considered under both alternative propositions. The concept of this model is illustrated in the middle panel of Figure 13.7. The results obtained in Neumann et al. (2012)

were used to support the admissibility of fingerprint evidence in U.S. courts in several landmark cases (State v. Hull, 2008; State v. Dixon, 2011), that contributed to reducing legal challenges to the scientific foundations of fingerprint evidence. In particular, Neumann et al.'s data (Neumann et al., 2012) show that:

1. Fingerprint evidence is highly discriminative;
2. Fingerprint evidence carries a lot of evidentiary value, which increases with the number of concordant features that can be observed between trace and control impressions;
3. Pairs of impressions sharing few rare features can carry more weight than pairs of impressions sharing many common features. Thus, there is no scientific justification for a numerical point standard and all comparisons need to be evaluated based on their own merit.

Neumann et al. (2012) did not recognise the similarities between their algorithm and the Approximate Bayesian Computation (ABC) model selection family of algorithms (Marin et al., 2019), which allows for quantifying the support of a set of observations for one of several alternative propositions in settings where the likelihood functions are intractable. Fortunately, the work presented in Neumann et al. (2012) was revisited by Hendricks et al. (2019), who recently provided a theoretical justification for Neumann et al.'s initial attempt and further developed it. In particular, Hendricks et al. (2019) show that, under certain circumstances, the ABC Bayes factor converges to the Bayes factor for the specific scenario of interest.

Mathematically, the ABC Bayes factor for the specific source scenario is given by:

$$BF_{abc} = \lim_{N \to \infty} \frac{\sum_{i=1}^{N} \mathbb{I}(H^{(i)} = H_{0_{SS}}) \cdot \mathbb{I}(\kappa(\eta(e_u), \eta(e_u^{*(i)})) \leq t)}{\sum_{i=1}^{N} \mathbb{I}(H^{(i)} = H_{1_{SS}}) \cdot \mathbb{I}(\kappa(\eta(e_u), \eta(e_u^{*(i)})) \leq t)} \cdot \frac{\pi(H^{(i)} = H_{1_{SS}})}{\pi(H^{(i)} = H_{0_{SS}})}, \quad (13.13)$$

where $\mathbb{I}(\cdot)$ is the indicator function, η is a summary statistic of the observation made on a finger impression, π is a probability measure, $H^{(i)}$ is the proposition chosen at the ith iteration of the algorithm, $e_u^{*(i)}$ is a pseudo-trace impression generated under the proposition chosen during the ith iteration of the algorithm, and t is a tolerance threshold. The ABC algorithm for comparing two models is summarised in Algorithm 1 (Hendricks et al., 2019).

Algorithm 1: ABC model selection algorithm

Data: Observed trace impression e_u.
Results: ABC Bayes factor, BF_{abc}.
for $i = 1$ *to* N **do**

> Sample a model index $H^{(i)}$ from the model prior, $\pi(H^{(i)} = h)$, where $h = H_{0_{SS}}, H_{1_{SS}}$;
> Sample a vector of parameters, $\boldsymbol{\theta}^{(i)}$, from the prior density, $\pi(\boldsymbol{\theta}|H^{(i)})$;
> Generate a pseudo-trace impression, $e_u^{*(i)}$, from the assumed likelihood, $f(e_u^{*(i)}|\boldsymbol{\theta}^{(i)})$;
> Compute $\kappa(\eta(e_u), \eta(e_u^{*(i)}))$.

Assign BF_{abc} as in (13.13).

The Bayes factor in Equation 13.13 is trace-centred since each iteration involves e_u. However, it differs from the trace-centred score-based likelihood ratio described in the previous section in three fundamental ways:

1. The observed trace, e_u, is compared to other traces generated under the two alternative propositions, and not to control impressions;
2. The observed trace is compared to traces generated by all potential donors under the two alternative propositions, and not only by the donor considered under $H_{0_{SS}}$;
3. More importantly, the Bayes factor is not obtained by evaluating the likelihood of a single score between e_u and e_s using two different functions. The Bayes factor is obtained by counting how many scores obtained under each proposition are smaller than a tolerance threshold at the limit when this threshold tends to zero. Doing so accounts for the rarity of the features observed on the latent impression.

Geometrically, the ABC Bayes factor can be represented as in Figure 13.11, which shows that all trace impressions left by two donors (the extension to a population of donors is trivial) are projected on a vector space defined by e_u (or in the case of the ABC algorithm by $\eta(e_u)$). Figure 13.11 shows that the concern regarding the lack of coherence associated with the use of score-based likelihood ratios does not apply to the ABC Bayes factor. Indeed, all projections are made with respect to the vector space defined by $\eta(e_u)$, which will always remain fixed in a given case, irrespective of who is suspected to have left the trace impression.

It has been shown that the ABC Bayes factor only converges to the Bayes factor if η is a sufficient summary statistic across both $H_{0_{SS}}$ and $H_{1_{SS}}$ (Didelot et al., 2011; Robert et al., 2011), which is a condition that needs to be investigated in the context of fingerprint evidence. The main contribution of Hendricks et al. (2019) is to show that:

$$\lim_{t \to 0^+} BF_{abc} \approx \lim_{p \to 0^+} \frac{\mathrm{ROC}(p)}{p}, \tag{13.14}$$

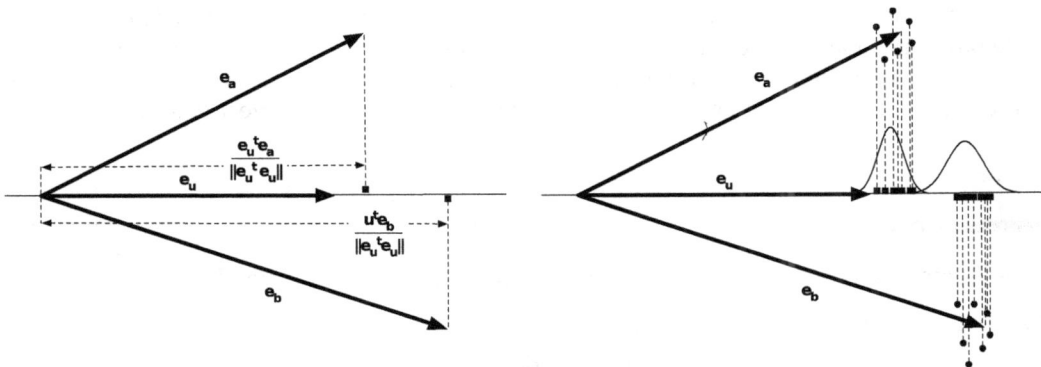

FIGURE 13.11
Left panel: vector projection of $\mathbf{e}_a$ and $\mathbf{e}_b$ onto $\mathbf{e}_u$. Right panel: vector projection of all trace impressions generated by the donors of $\mathbf{e}_a$ and $\mathbf{e}_b$ onto $\mathbf{e}_u$.

where p is the rate of false positives (i.e., the rate of $\kappa(\eta(e_u), \eta(e_u^*))|H_{1_{SS}} \leq t)$, and $ROC(p)$ is the well-known Receiver Operating Characteristic curve. We note that, contrary to score-based models, the ROC-ABC model described in (13.14) is assigned by calculating the ratio of two tail probabilities (as the limit of one of them tends to zero), while score-based like-lihood ratios generally consider the ratio of two likelihoods.* The convolution of the ABC algorithm with the ROC curve allows for monitoring the convergence as $p \to 0^+$, which alleviates some of the issues related to the curse of dimensionality (Blum and Francois, 2010) and the convergence at $t \to 0^+$.

The ROC-ABC algorithm proposed by Hendricks et al. (2019) was applied to large datasets of finger impressions. Figure 13.12 presents some of the results published in Hendricks et al. (2019). These results are similar to those of Neumann et al. (2012). The data show a good separation between the Bayes factors assigned to same source pairs of impressions (Figure 13.12a) and Bayes factors assigned to pairs of impressions where the control impressions originate from random sources that are different from the donors of the trace impressions (Figure 13.12c). Furthermore, Figure 13.12b shows that Bayes factors assigned to the comparisons of trace impressions with control impressions selected for their high level of similarity overwhelmingly support the proposition that the trace and control impressions were made by different donors.

13.4.4 Summary

Various attempts have been made to quantify the weight of fingerprint evidence. Most of them suffer from severe shortcomings, which result in unpredictable bias with respect to the Bayesian inference framework. Some of these shortcomings include addressing the common-source scenario instead of being specific to a suspected donor, failing to account for the rarity of the features observed on the latent impression, or providing incoherent evidence that may support both of two mutually exclusive propositions. Several guidelines have been proposed to "validate" these ad hoc methods (Ramos and Gonzalez-Rodriguez, 2013; Ramos et al., 2013; Haraksim et al., 2015; Leegwater et al., 2017; Meuwly et al., 2017). Unfortunately, these guidelines focus on the accuracy of the models (i.e., does a model support the correct proposition on average?) and not on the coherence or the magnitude of support in specific cases. Overall, the ad hoc models resulting from these various attempts appear to be valuable deterministic data-driven decision-making tools, but are not appropriate to support Bayesian inference. One attempt, based on Approximate Bayesian Computation, may converge to the desired Bayes factor under conditions of sufficiency and infinite number of samples. These conditions are not necessarily reasonable in the context of fingerprint examination and are currently being investigated.

13.5 Factors Affecting Examiners' Decision Making and Error Rates

13.5.1 Decision Making During ACE

The description of the ACE-V examination framework in Section 13.2 highlights that conclusions from fingerprint examinations are traditionally reported according to a tripartite

---

* In addition, we note that the two tails are two left tails, which is significantly different from FRStat (Swofford et al., 2018) (see the illustration in Figure 13.7 and the description of FRStat in Section 13.6).

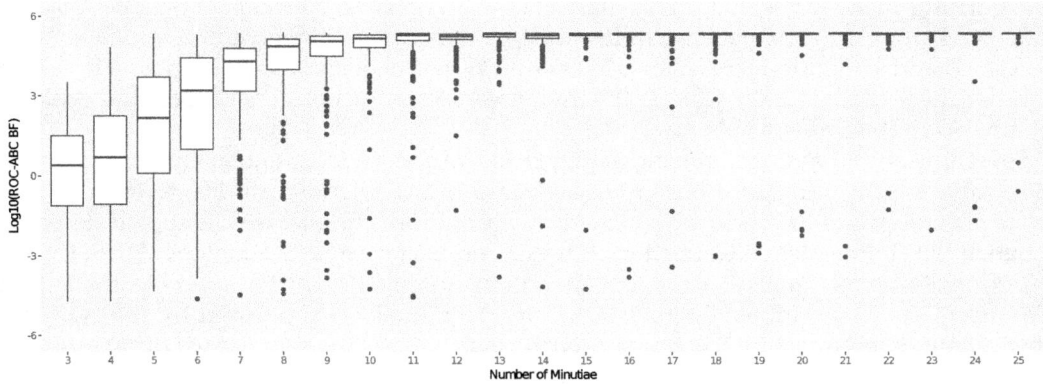

(a)

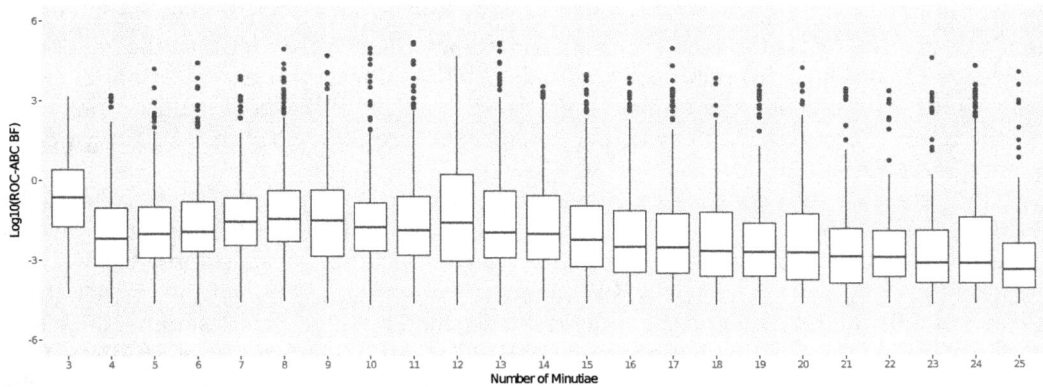

(b)

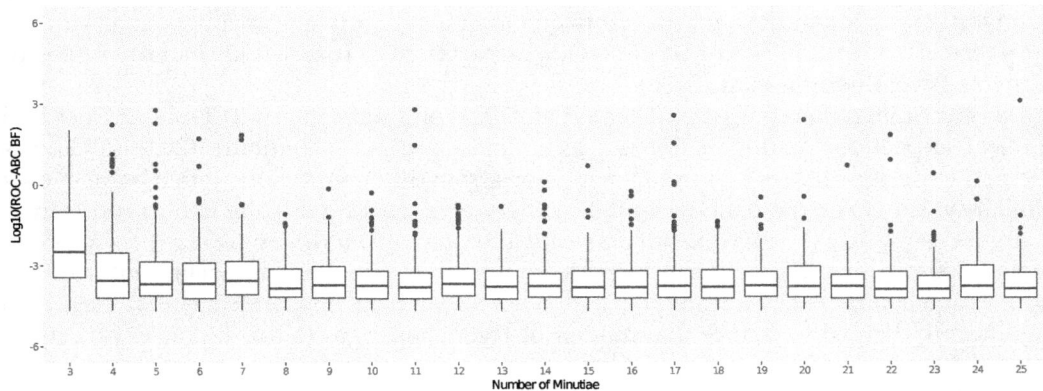

(c)

FIGURE 13.12
ABC Bayes factors calculated for configurations of 3 to 25 minutiae (approximately 200 configurations each) based on their spatial relationships, types, and directions in the ridge flow. Rows: (a) trace and control impressions originating from the same source; (b) control impressions selected out of 7,000,000 fingerprints for their similarity with the trace impressions; (c) control impressions randomly selected out of 7,000,000 fingerprints.

categorical conclusion scheme. This implies that, sometimes, conclusions will be misclassified, resulting in erroneous identifications or exclusions. Fingerprint examiners have received, and continue to receive, criticism (Saks and Koehler, 2005; National Research Council, 2009; Zabell, 2005; Cole, 2005) due in part to erroneous conclusions that occur at an unknown (but presumably low) rate.

Several reports have recommended the study of the validity and reliability of forensic analyses, which includes designing and performing scientific studies to estimate the rates of error present in the conclusions drawn from forensic analyses (National Research Council, 2009; President's Council of Advisors on Science and Technology, 2016; Expert Working Group on Human Factors in Latent Print Analysis, 2012). For example, the Expert Working Group on Human Factors in Latent Print Analysis sponsored by the National Institute of Standards and Technology (NIST) emphasised the importance of quantifying the rate of erroneous conclusions resulting from fingerprint examinations (Expert Working Group on Human Factors in Latent Print Analysis, 2012). They assert that knowledge about error rates enables the assessment of the reliability of forensic analyses and allows for appropriate confidence to be placed on a given evidence type. More recently, the President's Council of Advisors on Science and Technology (PCAST) argued that the validity of subjective feature-based comparison methods (such as fingerprint examination, firearms analysis, or footwear analysis) must be assessed through "black box" studies that analyse the rates of occurrence of erroneous conclusions (President's Council of Advisors on Science and Technology, 2016).

Many researchers did not wait for these reports to propose the use of studies to assess the accuracy and reliability of fingerprint examinations and to explore the impact of various subjective aspects of the examination process on the resulting conclusions. We want to mention the early work by Evett and Williams (1996) who investigated the impact of the use of a numerical threshold on the conclusions reached by fingerprint examiners. We also want to mention the work by Langenburg (2004, 2009, 2012) who conducted studies on the performances of fingerprint examiners as early as 2004. Finally, we want to mention the extensive work by a team of scientists from Noblis and the FBI (Ulery et al., 2011, 2012, 2013, 2014, 2015, 2016, 2017) that spawned from the review of the U.S. Federal Bureau of Investigation's handling for an erroneous identification made in connection with the 2004 terrorist attacks in Madrid, Spain (U.S. Department of Justice, Office of the Inspector General, 2006; Budowle et al., 2006).

The error rate data reported by Ulery et al. (2011) and reproduced in Table 13.1 has had a major impact on the fingerprint and legal communities. Langenburg (2009, 2012) was the first to design studies that used impressions from different donors with a high level of similarity. However, the data presented in Ulery et al. (2011) were the first to result from a larger-scale study (i.e., 169 examiners) with a reasonably large sample size (i.e., approximately 100 examinations per participant), commissioned by a well-known law enforcement agency and conducted by a team of experienced researchers. It is fair to say that these data significantly helped to reduce the number of legal challenges to the scientific validity of fingerprint examination.

The other papers in the series provide a wealth of information on the factors affecting the entire chain of decisions made by examiners within the ACE framework. For instance, Ulery et al. (2012) found that identification conclusions are reproduced in 55% (between-examiners variability) and repeated in 69% (within-examiner variability) of the cases where complex latent impressions are considered (similar values are observed for exclusion conclusions). These observations rejoin the data from Neumann et al. (2013) presented in Section 13.2 and indicate that there is very little consistency in the decision-making process supporting ACE-V. When attempting to study the decision-making process in more

TABLE 13.1

Reproduction of Table S5 from Ulery et al.'s Supplemental Information

Comparison Decision	Latent Value	Total	Mates	Non-mates	% of Mated Pairs			% of Non-mated Pairs		
					PRES	CMP	VID	PRES	CMP	VID
(not compared)	NV	3,947	3,389	558	29.3%			10.1%		
Exclusion	VEO	486	161	325	1.4%	2.0%		5.9%	6.5%	
Exclusion	VID	4,072	450	3,622	3.9%	5.5%	7.5%	65.3%	72.7%	88.7%
Inconclusive	VEO	2,596	2,019	577	17.4%	24.7%		10.4%	11.6%	
Inconclusive	VID	2,311	1,856	455	16.0%	22.7%	31.1%	8.2%	9.1%	11.1%
Individualisation	VEO	40	40	0	0.3%	0.5%		0.0%	0.0%	
Individualisation	VID	3,669	3,663	6	31.6%	44.7%	61.4%	0.1%	0.1%	0.1%
Totals		17,121	11,578	5,543	100.0%	100.0%	100.0%	100.0%	100.0%	100.0%
Total comparisons	Value (either)	13,174	8,189	4,985						
Total comparisons	VID	10,052	5,969	4,083						

Note: NV (no value): The Latent Print Is Deemed of No Value. VEO (Value for Exclusion Only): The Characteristics Observed on the Latent Print can be Used to Exclude Potential Sources. VID (Value for Identification): The Characteristics Observed on the Latent Print Are Such That an Identification Conclusion is Possible. PRES: Latent Prints Submitted for Examination. CMP: Latent Prints Actually Compared.

detail, Ulery et al. (2015) found that examiners added on average 21% more features on trace impressions after having observed control impressions from potential donors (i.e., at the comparison stage of ACE-V) when compared to the number of features that they observed on the latent impressions in isolation (at the analysis stage of ACE-V). This type of circular reasoning and confirmation bias (Kassin et al., 2013) is one of the root causes of erroneous identifications: Ulery et al. (2015) observed that all features reported during the analysis were removed (16 features) or moved (1 feature) and that 13 new features were added after the observation of the control impression in the only erroneous identification that they observed.

In Ulery et al. (2016), Ulery et al. noted high between-examiner variability in minutiae annotation: between 47% (low quality impressions) and 69% (higher quality impressions) of the features observed by an examiner were also used by another examiner. Ulery et al. (2014, 2017, 2013) found that minutiae count is the main force behind the various decisions made during the examination process. Their studies show the direct relationship between the lack of standardization of the examination process, which leads to bias and to an extreme variability in the selection of the features that are used to form opinions, and ultimately to the large variability in the resulting conclusions observed in Ulery et al. (2014, 2017, 2013) and Neumann et al. (2013).

Interestingly, in all of these studies, a single factor is found to systematically affect the variability within and between examiners. Although many factors seem to influence examiners' decision-making, the quality of the latent impression drives the number of features that can be detected as well as the level of confidence associated with these features. This, in turn, drives the confidence that can be associated with their final conclusion on the source of the latent print. The past decade has seen research projects shifting from attempting to collect data supporting the final conclusions to attempting to support, render more transparent, and document the entire sequence of decisions made during the examination process. In light of the data collected by Langenburg, the Noblis-FBI team and others, we can only support this shift and the recommendations recently made by Champod (2015):

1. The inference process needs to be better anchored in formal decision theory as described in Biedermann et al. (2008);

2. The inference process needs to be more transparent. Case files, reports and testimonies needs to articulate which features have been considered and how their combined weight justifies the conclusions that have been reached;

3. The inference process needs to be data-driven and the fingerprint community needs to move away from purely opinion-driven conclusions;

4. The risks of bias need to be managed throughout the examination process. A more formal decision-making process supplemented by transparent documentation will help identify the potential risks and allow for mitigating them.

13.5.2 PCAST Controversy

The PCAST report only found two black-box studies worth mentioning: the well-known study by Ulery et al. (2011) discussed in the previous section and a study by the Miami Dade Police Department (MDPD) (Pacheco et al., 2014) that had been fairly unnoticed up to that point. The PCAST report relies heavily on these results to discuss the estimated rate of erroneous identification associated with fingerprint examination. In fact, the PCAST report uses the MDPD study to consider that, using the upper bound of a 95% confidence interval, the true rate of erroneous identification of fingerprint examination could be as high as 1 in 18 (or approximately 5.4%) cases (President's Council of Advisors on Science and Technology, 2016).

The extravagance of the upper bound of 1 erroneous identification in 18 examinations claimed by PCAST raised major concerns among fingerprint examiners regarding PCAST's interpretation of the MDPD error rates and the subsequent calculation of the confidence intervals.* The MDPD experiment was designed such that participants would not receive pairs of trace and control impressions, as in Langenburg (2009) or Ulery et al. (2011), but would receive trace impressions associated with the control impressions of all 10 fingers of three individuals. The test cases were designed so that, in some cases, the 10 control impressions from the true donor of the latent impression were provided, while in other cases, the impressions from three random individuals were provided. The difficulties in analysing the MDPD data arise as a result of the possibility for participants to associate a trace impression with a control impression from the correct donor, even though the chosen control impression is not an impression from the finger that made the trace impression.

Table 13.2 reproduces the data collected by the MDPD research team and shows that, in 35 instances, participants erroneously identified the wrong finger of the correct person. Furthermore, the participants erroneously identified 4 latent prints even though impressions from the true donor were provided among the control impressions. Finally, participants erroneously identified 3 latent prints when impressions from the true donor were not provided as control material. Based on their data, the MDPD research team estimated that the rate of erroneous identifications is 3.0%. This estimate is obtained by relating the 42

---

* In fact, the PCAST committee merely reproduced the error that the original MDPD research team made during the analysis of their data. Nonetheless, the prestige associated with the PCAST committee contributed to the spread of the erroneous result in the legal and scientific communities.

TABLE 13.2

A Modified Version of Table 11 in the Miami-Dade Report Where the Erroneous ID Column has been Divided into Erroneous Identifications in which the Correct Individual was Identified, but the Incorrect Finger was Reported and Erroneous Identifications in which the Wrong Individual was Identified

Source Present (Y/N)	# of Latent Prints	# of Deci-sions	Correct ID	Erroneous ID		Inc.	Correct Excl.	Erroneous Excl.
				Correct Person, wrong Finger	Incorrect Person			
Yes	56	3177	2457	35	4	446	N/A	235
No	24	1359	N/A	N/A	3	403	953	N/A
Totals	80	4536	2457	35	7	849	953	235

erroneous identifications observed during the study to a total of 1398 decisions (including inconclusive examinations). This estimate raises to 4.2% if the inconclusive examinations are not accounted for ($1398 - 403 = 995$ conclusions).

Usually, false positive rates are estimated by the ratio between the number of false positives and the total number of opportunities to reach a false positive decision. In the MDPD case, it was possible to erroneously identify all test latent prints; thus, a possible denominator could have been 4536. However, for their denominator, MDPD used the 1359 test cases that did not have the true donor as part of the control material and added the 39 test cases that had the true donor as part of the control material but resulted in an error ($1359 + 39 = 1398$) (Table 13.2).

The inconsistency between the numerator and the denominator for calculating the rate of erroneous identifications was pointed out by the Friction Ridge Subcommittee of the Organization of Scientific Area Committees for Forensic Science (OSAC FRS) (OSAC, 2016). The OSAC FRS proposed that the rate of erroneous identifications should consider all 42 erroneous identifications out of the total number of decisions made, excluding inconclusive decisions ($4536 - 849 = 3687$). This calculation lowers the false positive rate to approximately 1.1%, with a 95% confidence interval upper bound of approximately 1.5%.

Ausdemore et al. (2019) recreated, by simulation, the MDPD experiment and tested different error rates, including the 3.0% proposed error rate. These authors noted that the main interpretation challenge resides with the 35 erroneous identifications. Table 13.3 reports the results from one of their simulations when the rate of erroneous identifications is set to 3.0%. The data in Table 13.3 show that a 3.0% false positive rate should result in a much larger number of erroneous identifications than the 42 observed by the MDPD research team (95% credible interval: 109 to 157). These results, therefore, imply that 3.0% is much larger than the true rate of erroneous identification.

Ausdemore et al. (2019) propose two distinct interpretations of the MDPD results:

1. Fingerprint examiners have different rates of erroneous identification depending on whether the true donor is part of the comparison material. This is consistent with the observation that examiners also have significantly different rates of inconclusive conclusions depending on whether impressions from the true donor are part of the control material.

2. Fingerprint examiners have a unique rate of erroneous identification; however, they are prone to clerical errors and lack of attention to details, which inflates

TABLE 13.3

Average Number of Conclusions in each Category Using Miami-Dade Rates of Erroneous
Identification (3.0%) and Erroneous Exclusion (7.5%)

	Source Present	Source Not Present	Totals
# of latent prints	56	24	80
# of decisions	3187.07 [3003, 3392]	1364.26 [1163, 1533]	4551.33 [4481, 4612]
Correct IDs	2405 [2252, 2554]	N/A	2405.07 [2252, 2554]
Erroneous IDs			
Correct person, wrong finger	N/A	N/A	N/A
Incorrect person	95.85 [76, 116]	40.74 [27, 54]	136.59 [109, 157]
Inconclusive examination	447.54 [402, 494]	403.73 [339, 468]	851.28 [785, 902]
Correct exclusions	N/A	919.79 [791, 1051]	919.79 [791, 1051]
Erroneous exclusions	238.61 [209, 270]	N/A	238.61 [209, 270]

Note: The Numbers in [ ] Represent the Bounds of Maximum Density 95% Credible Intervals from 1,000
Simulations.

their rate of erroneous identification when impressions from the true donor are
available, and has no consequence when impressions from the true donor are not
available.

Overall, Ausdemore et al.'s analysis rejoins the results proposed by Langenburg (2009)
and Ulery et al. (2011) with a worst-case scenario rate of erroneous identifications of
approximately 1.7%, which includes clerical errors but does not account for the verification
stage.

13.5.3 Statistical Analysis of Experiments Designed to Study the ACE-V Framework

The PCAST controversy (Section 13.5.2) highlights another issue related to the analysis
and interpretation of data collected during studies designed to explore various aspects of
the ACE-V framework, such as the ones reported in Langenburg (2009); Ulery et al. (2011);
Pacheco et al. (2014). In general, these studies rely on the goodwill of many benevolent participants who agree to perform a certain number of examinations outside of their regular
work hours or work duties. Thus, even though the initial design of an experiment may be
rigorously structured to render the different effects orthogonal, it rarely survives its initial
contact with the real world. Therefore, it is usually the case that:

1. The various examinations are not independent:
   a. The examination of several pairs of images by each participant introduces
      dependencies between the resulting conclusions for that participant.
   b. The examination of each test case by multiple participants introduces depen-
      dencies between the conclusions reached for a given test case.
   c. Dependencies between participants may also be created by common training
      and operating procedures.
2. The experiment is unbalanced:
   a. Some participants examine more test cases than others.

    b. Examiners are not assigned an equal number of mated and non-mated image pairs, nor are they assigned the same number of mated and non-mated pairs as each other.

    c. Some test cases are examined by more participants than others. Test cases are not assigned in a systematic manner.

    d. Not all test cases result in the same number of decisions on the source of the latent print since participants have the possibility to determine that a latent print is not suitable for further examination.

3. There are missing data:

    a. Some data can be removed due to data entry error.

    b. In other studies, many participants do not finish the assigned tasks and only return results for a portion of the latent prints they were asked to examine.

Error rate studies, such as the ones considered by PCAST and reported in Ulery et al. (2011) and Pacheco et al. (2014), present point estimates, calculated using the plug-in principle, associated with confidence intervals. The dependencies in the design of the experiment, and the unbalanced and missing data make it inappropriate to use a binomial likelihood and, thus, to rely on plug-in estimates. In other words, not accounting for the various dependencies results in underestimating the variance of the estimates, which then results in confidence or credible intervals that do not include the true error rates.* Over-all, likelihood-based methods for the type of experiments designed to study error rates are impractical. For example, error rate data could be described by a generalised linear model with a fully specified covariance matrix. However, in the case of Ulery et al.'s study, this would require defining all dependencies in a 17,121 by 17,121 covariance matrix.

Traditionally, after using one of these methods to calculate parameter estimates, $(1 - \alpha)100\%$ confidence intervals are usually constructed. Aside from the lack of independence of the data discussed above, interpretation of confidence intervals relies on ad aeternem theoretical repetition of the experiment and does not provide measures of uncertainty for the data at hand. Confidence intervals are not coherent measures of the uncertainty associated with the estimates, and are confusing and misleading, even for statistical audiences. For example, we mentioned in Section 13.5.2 that PCAST makes the dubious statement that "the [false positive] rate could be as high as 1 error in 18 cases". Indeed, the rate *could* be as high as a certain value, but this statement gives little information about the probability that this event would happen.

Hendricks and Neumann (2020) propose a simulation-based approach that can account for the dependencies that arise from complex experimental designs where the likelihood function is intractable. This approach, based on Approximate Bayesian Computation (see Section 13.4.3 for a model selection version of ABC) provides a potential solution that assigns posterior distributions to the parameters of interest. Some of the results obtained by Hendricks and Neumann (2020) when re-analysing the results from Ulery et al. (2011) are reported in Table 13.4. Overall, the data are in agreement with the original estimates proposed by Ulery et al. (2011). As expected, the credible intervals obtained by Hendricks and Neumann (2020) are much wider than the confidence intervals calculated with the Agresti-Coull method used by Ulery et al. (2011). For example, according to Hendricks and

---

* Note that Ulery et al. (2011) recognised this point regarding their experiment but decided to move forward with confidence intervals based on the assumption of independent data (see Section 2.1 in the appendix of Ulery et al. (2011)).

TABLE 13.4

ABC Population Rate Estimates and 95% Highest Posterior Density Intervals
for each of the 14 Possible Decisions in Ulery et al. (2011)

	Mated Est.	Mated HDI	Non-mated Est.	Non-mated HDI
NV	29.61%	[18.04%, 40.36%]	9.35%	[2.98%, 15.30%]
Exc. VEO	0.95%	[0.00%, 3.26%]	5.06%	[0.63%, 8.50%]
Exc. VID	3.18%	[0.00%, 6.03%]	68.09%	[58.98%, 77.11%]
Inc. VEO	17.67%	[7.58%, 27.25%]	9.71%	[3.72%, 16.28%]
Inc. VID	15.69%	[6.99%, 25.35%]	7.75%	[1.67%, 14.02%]
Ind. VEO	0.46%	[0.00%, 1.80%]	0.04%	[0.00%, 0.13%]
Ind. VID	31.81%	[20.55%, 43.02%]	0.15%	[0.00%, 0.48%]

Note: Values are Presented as Percents of Total Mated or Non-mated Presented Pairs and
are Rounded to Two Decimal Places. Note that Medians of the Posterior Distributions are
Used as Point Estimates; Thus, They do not Sum to 100% Across the Seven Categories
for Mated Comparisons and the Seven Categories for Non-mated Comparisons. For a
Description of NV, VEO and VID, See the Caption of Table 13.1 Above.

Neumann (2020), approximately 0.15% of the non-mated latent prints were erroneously
identified to the donor of the associated control impressions with a 95% credible interval
ranging from 0.00% to 0.48%. The corresponding plug-in estimate from Ulery et al. (2011)
is 0.108% with a confidence interval ranging from 0.04% to 0.24%.

13.5.4 Summary

A great deal of data measuring various aspects of the fingerprint examination process have
been acquired over the past decade. The data show the variability in the application of the
ACE-V process by different examiners and the variability in the resulting conclusions. The
papers by Ulery and his colleagues identify key steps of the examination process that need
to be strengthened, such as features definition and selection, latent quality determination,
bias mitigation and data-driven decision-making.

While the magnitude of the expected rate of erroneous conclusions is reasonably
low (between 0.1% and 1% expected erroneous identifications and approximately 7.6%
expected erroneous exclusions), care needs to be taken when designing and analysing error
rate studies. Furthermore, error rates are community-wide measures and cannot directly
be used to communicate the risk that an error occurred in a specific case without factoring
in the quality of the trace impression and the competency of the examiner assigned to the
case.

13.6 U.S. Defense Forensic Science Center's *FRStat*

The Latent Print Branch of the U.S. Defense Forensic Science Center laboratory (DFSC) has
gone through significant changes in their operating procedures to account for the criticisms
made over the past decades. In 2015, the Latent Print Branch issued an information paper
(U.S. Department of the Army, Defense Forensic Science Center, 2015) announcing its decision to move away from the categorical decision scheme of identification, exclusion and

inconclusive examination. Instead, the Latent Print Branch proposed more probabilistic language:

> The latent print on Exhibit ## and the record finger/palm prints bearing the name XXXX have corresponding ridge detail. The likelihood of observing this amount of correspondence when two impressions are made by different sources is considered extremely low (U.S. Department of the Army, Defense Forensic Science Center, 2015).

This language was modified in 2017 to reflect the use of two alternative propositions rather than a single likelihood:

> The latent print on Exhibit ## and the standards bearing the name XXXX have corresponding ridge detail. The probability of observing this amount of correspondence is approximately ## times greater when impressions are made by the same source rather than by different sources (U.S. Department of the Army, Defense Forensic Science Center, 2017).

To assign these probabilities, the Latent Print Branch developed its own statistical model named FRStat (Swofford et al., 2018). In Swofford et al. (2018), the authors propose to calculate the ratio of the risks of *erroneous exclusion* and *erroneous identification* for such a categorical decision based on a given value for $\delta(e_u, e_s)$ (right panel of Figure 13.7). The two rates of erroneous conclusions are estimated for the two common-source alternative propositions discussed in Section 13.3.1, using the same sampling experiments described in Section 13.4.2. In other words, FRStat calculates the ratio between the rates of Type II and Type I errors at any particular observed value of $\delta(e_u, e_s)$.

The strategy is interesting and novel, at least in areas of forensic science concerned with the examination of pattern evidence. The argument seems to be that if an observed value of $\delta(e_u, e_s)$ would result in a risk of erroneous exclusion that is overwhelmingly larger than the corresponding risk of erroneous identification, then it would be safe to report that the trace and control impressions were made by the same donor. This strategy diverges from the popular trend of attempting to assign Bayes factors to fingerprint evidence. In fact, the authors are explicit that they are not attempting to do so (Swofford et al., 2018). Importantly, this strategy somewhat addresses the criticisms raised regarding the use of error rate studies to support casework: instead of providing generic estimates for the rates of erroneous conclusions that would apply to the entire population of latent print examiners and to all examinations, FRStat provides rate estimates that apply to each specific case based on the level of similarity between the trace and control impressions observed in the case.

Nonetheless, FRStat has some serious limitations:

1. The model does not address the specific source propositions, and, therefore, does not address the issue of interest (in most cases) to the criminal justice system;
2. FRStat does not enable fingerprint examiners to account for the rarity of the characteristics observed on the trace impressions. As discussed above, the mere information that the characteristics observed on a pair of trace and control impressions are within some tolerance level does not allow for a full interpretation of the probative value of a piece of evidence;
3. FRStat does not allow for assigning a value to the likelihood mentioned in the 2015 information paper published by the DFSC or to the probability mentioned

in the 2017 information paper. The discrepancy between the proposed wording of conclusions resulting from fingerprint examinations performed by the Latent Print Branch and the tool they may be using to support this language is concerning;

4. Despite Swofford et al.'s clear statement that they are not attempting to assign Bayes factors (Swofford et al., 2018), the expression of FRStat as a ratio with respect to two alternative hypotheses will result in major confusion in the community. This confusion is even present in FRStat's foundation paper, where the authors go back and forth between likelihoods and tail probabilities, and claim that:

> values greater than 1 indicate a higher probability of the observed similarity statistic value among mated sources compared to non-mated sources and values less than 1 indicate a higher probability of the observed similarity statistic values among non-mated sources compared to mated sources. Values equal to 1 indicate equal probability of the observed similarity statistic value among mated and non-mated sources (Swofford et al., 2018).

First, this statement corresponds exactly to the type of statement that would be communicated as a result of the calculation of a common-source score-based likelihood ratio in Section 13.4.2. Second, this statement does not refer to the tail probabilities that FRStat is calculating.* And last, but not least, there is no guarantee that the ratio of Type II and Type I probabilities has the behaviour described in this statement. In fact, we can test the relationship between FRStat ratios and LR_{SS} using our toy example. For these simulations, we consider the same sampling distributions as in the case of the common-source score-based likelihood ratio in Section 13.4.2. We set $e_{u_1} = e_u$, $e_{u_2} = e_s$, $\sigma^2_{u_1} = \sigma^2_u$ and $\sigma^2_{u_2} = \sigma^2_s$. In both models, $\mu = 10$, $\sigma^2_d = 10$ and $\sigma^2_u = 2$. All simulations were repeated 1,000 times. In the first experiment, the characteristics of the donor of e_s were chosen to be relatively common with respect to the population of donors ($\mu_d = 9$) but also quite variable ($\sigma^2_s = 1$). In the second experiment, the characteristics of the donor of e_s were chosen to be rare ($\mu_d = 0$) and variable ($\sigma^2_s = 1$). In the last experiment, the characteristics of the donor of e_s were chosen to be common with respect to the population of donors ($\mu_d = 9$) with a negligible variability ($\sigma^2_s = 10^{-5}$).

The data presented in Figure 13.13 show that FRStat does not have the same behaviour as a Bayes factor, or even as a score-based likelihood ratio. In particular, we note the severe bias of FRStat values relative to the specific-source likelihood ratios when $H_{1_{SS}}$ is true. This can result in grave misrepresentation of the contribution of the fingerprint evidence to a case, if FRStat's output is perceived as a number supporting one of the alternative propositions rather than the ratio between two error rates;

5. FRStat is meant to be used following the determination by an examiner that a latent and control impressions have enough features in agreement that they can be associated. FRStat is meant to qualify the strength of this association. Thus, according to this operating procedure, FRStat is only meant to be used for pairs of impressions that truly originate from the same finger, or for pairs of impressions that do not

---

* We believe that the probabilities mentioned in this statement are in fact meant to be likelihoods or densities since the summary statistic is a continuous random variable. The same comment can be made with respect to the language proposed by the DFSC in 2017.

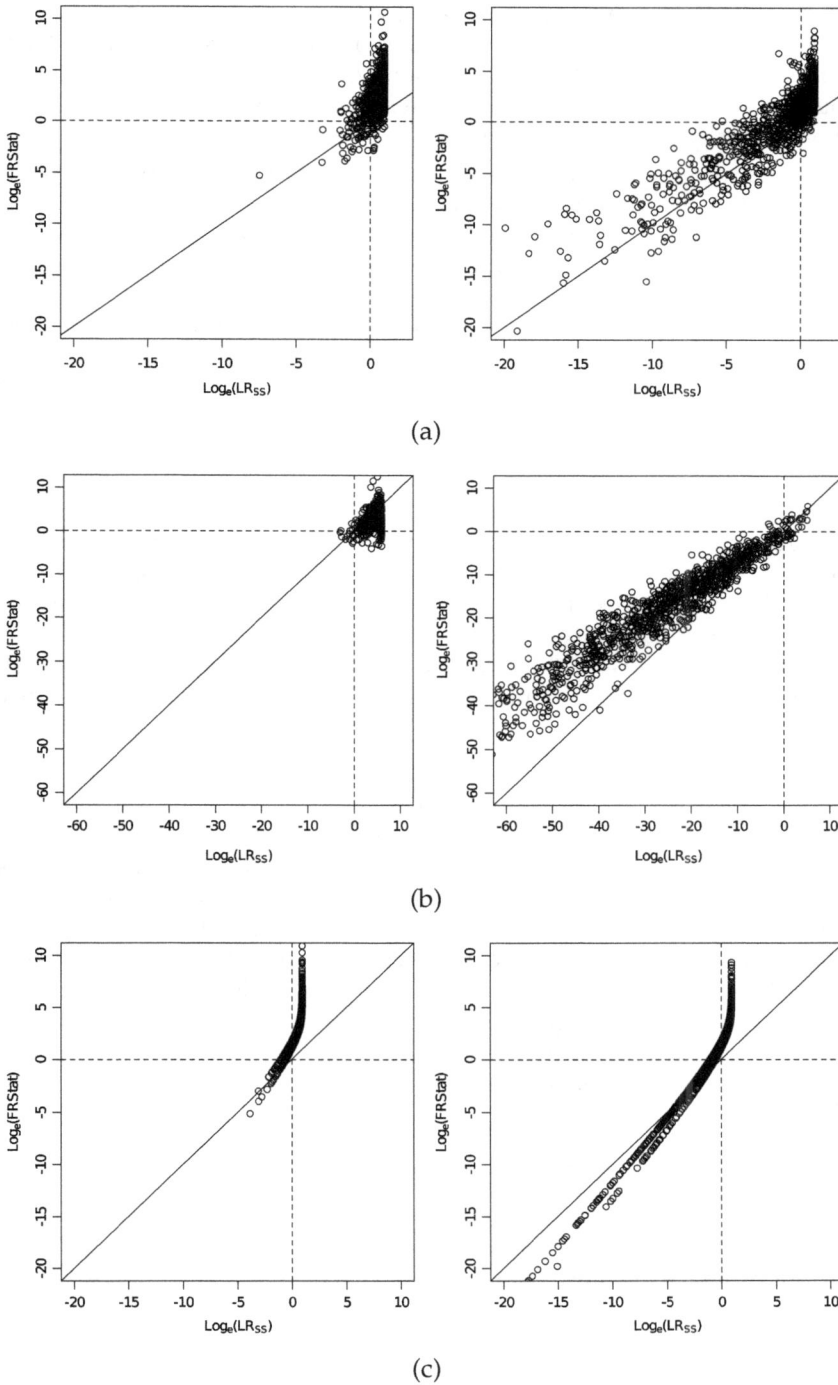

FIGURE 13.13
Comparisons of FRStat-like values with the LR in the specific source scenario. Columns: the left column reports the results when e_u and e_s have been sampled under $H_{0_{SS}}$; the right column reports the results under $H_{1_{SS}}$. Rows: (a) the source of the control impression is common and has some variance; (b) the source of the control impression is rare and has some variance; (c) the source of the control impression is common and has virtually no variance.

originate from the same finger, but whose levels of similarity are fooling examiners. Yet, the probability distributions underlying FRStat are built on data acquired from randomly selected pairs of impressions. Given the planned operating context, it would have been more appropriate to rely on impressions selected for their high level of similarity;

6. While FRStat proposes error rate estimates that are more case specific than generic error rate estimates resulting from large scale studies, it only addresses half of the concerns. FRStat certainly provides estimates that are based on the level of similarity observed between the specific pair of impressions considered in the case; in that respect, the estimates are not meant to represent an average risk for all possible pairs of impressions that examiners might encounter in casework. However, the estimates remain based on experiments involving many examiners, and, therefore, are average estimates for a population of examiners, as opposed to estimates for the specific examiner handling the case.

FRStat also suffers from some technical issues. For example, the authors of Swofford et al. (2018) have collected data to study the sampling distributions of their test statistic under both propositions. These score distributions are based on a very small sample and are quite well separated (see, for example, minutiae 9 to 15 in Figure 5 in Swofford et al. (2018)). Hence, the authors decided to use parametric models to assign tail probabilities in their model and chose to use a Kolmogorov-Smirnov test to measure the adequacy of their proposed parametric models for their data. The Kolmogorov-Smirnov test statistic for the proposed cumulative distribution function, $F(x)$, of a random variable x is defined as:

$$D_n = \sup_x |F_n(x) - F(x)|, \tag{13.15}$$

where $F_n(x)$ is the empirical distribution function for n observations. Equation (13.15) shows that the Kolmogorov-Smirnov test is based on absolute distances. It focuses on discrepancies between the proposed model and the observed data in the body of the distribution, and, by construction, is not powerful enough to detect discrepancies in the tails. The crux of the matter is that FRStat exclusively cares about tail probabilities in remote areas of the distributions. Thus, the use of the Kolmogorov-Smirnov test to choose the distributions that underpin FRStat irremediably results in unreliable tail probabilities, and thus, in misleading ratios. A solution would be to rely on more powerful tests based on relative distances, such as the Anderson-Darling test statistic:

$$A^2 = n \int_{-\infty}^{\infty} \frac{(F_n(x) - F(x))^2}{F(x)(1 - F(x))} dF(x), \tag{13.16}$$

to ensure that differences between the tails of the empirical and proposed distributions are appropriately weighted, and that the risks of erroneous conclusions are accurately represented.

13.6.1 Summary

FRStat is designed to provide some case-specific measure of the probative value of fingerprint evidence. FRStat adopts a novel strategy that departs from the commonly advocated Bayesian paradigm. FRStat is a valuable tool that could be used to study the impact of examiner variability on the resulting probative value, to discuss data-driven operating procedures and workflows, or to evaluate the impact of data-driven reports and testimonies on fact-finders. Nevertheless, a rapid overview of the scientific foundations of FRStat shows that this tool suffers from too many fundamental issues to be reliably used in casework.

13.7 Conclusions

The general framework used by fingerprint examiners relies on a series of subjective decisions that are neither reproducible from one examiner to another, nor repeatable by a given examiner. Traditionally, the goal of fingerprint examination is to form a categorical opinion on the source of a particular latent print. Two general approaches have been developed to support these opinions with data. The first approach consists in quantifying the weight of evidence within a Bayesian paradigm. Unfortunately, we found that the foundations of this paradigm are not well defined and that two scenarios (Section 13.3), which lead to different models, are often used interchangeably. Furthermore, assigning probability distributions to fingerprint features has proven to be extremely difficult due to the heterogeneity of the information that is traditionally used to characterised friction ridge patterns. Thus, Section 13.4 focused on the use of scores to simplify the problem of defining tractable likelihood functions to evaluate fingerprint evidence.

At the moment, the use of scores appears to be the only viable method to reduce the complexity inherent in modelling fingerprint features. Unfortunately, it also appears that none of the score-based models proposed to date can be considered suitable proxies for the Bayes factor of interest. We are not arguing that these ad hoc methods are not useful in their own way. Indeed, score-based likelihood ratios can be calibrated to minimise false positives or false negatives (Ramos and Gonzalez-Rodriguez, 2013; Ramos et al., 2013; Haraksim et al., 2015; Leegwater et al., 2017; Meuwly et al., 2017), and thus act as deterministic decision-making engines with known error rates. However, these score-based models are not valid likelihood ratios or Bayes factors for the propositions of interest: (1) they do not account for the rarity of the trace's characteristics; and (2) they are not coherent in the sense that a given piece of evidence may support either of two alternative propositions depending on which one is considered first. Our main argument is that if one wants to abide by the idea that forensic evidence should be reported within a Bayesian paradigm, then one cannot use score-based likelihood ratios.

Fortunately, scores can also be used within the context of Approximate Bayesian Computation. In the ABC paradigm, an ABC Bayes factor can converge to the desired Bayes factor (1) under the questionable assumption that fingerprint features observed on trace impressions are represented by a sufficient summary statistic and (2) at the limit when the threshold, t, in Equation 13.13 tends to zero. Further research is needed on the sufficiency of the representation of observed fingerprint features. For example, it may be possible to use the strategy proposed by Pudlo et al. (2016) and rely on a large collection of summary statistics, with the hope that, collectively, they will be sufficient across models.

The second approach arises from several committees in the U.S., backed by legal cases that argue that the reliability and validity of fingerprint examination should be demonstrated through the estimation of error rates. Section 13.5 discussed several studies of the factors influencing the decision-making process and the (lack of) reproducibility of conclusions. Two series of research projects conducted by two different research teams show the major impact of the quality of the latent prints on the reliability of the resulting conclusions. These findings lead us to agree with the recommendations proposed by Champod (2015) for structuring the inference process and providing corresponding training to examiners, increasing the level of documentation and transparency of the examination process, and overall making data-driven decisions. We also agree with other authors (Champod, 2015; Evett et al., 2017; Morrison et al., 2017) that error rates are only an average measure of performance over a population and do not provide information regarding the reliability of conclusions in individual cases.

Lastly, Section 13.6 described a hybrid approach that provides case-specific error rates. The approach, proposed by Swofford et al. (2018), is interesting and has merit. Its output may prove easier to communicate to lay individuals. That said, this approach is not mature enough to be used in casework. More specifically, there are several discrepancies between what the method calculates, what it claims to calculate, and how its results are conveyed. In addition, the method entirely relies on tail probabilities; yet, these tails are not well represented by the parametric models underpinning the algorithm, leading to unreliable results with potentially dramatic consequences for the criminal justice system.

In conclusion, much remains to be done when it comes to the rigorous application of statistical techniques to fingerprint examination. Decades of criticisms and challenges have pushed the fingerprint community to strengthen the foundations of its field. Fingerprint examiners and researchers have succeeded in many repects; however, they have yet to provide rigorously developed and validated tools to support the examination process in individual cases.

References

Abraham, J., Champod, C., Lennard, C., and Roux, C. Modern statistical models for forensic fingerprint examinations: a critical review. *Forensic Science International*, **232**, 131–150, 2013.

Aitken, C.G.G. and Taroni, F. *Evaluation of Evidence for Forensic Scientists*. Wiley & Sons Ltd, Chichester, 2nd edition, 2004.

Alberink, I., de Jongh, A., and Rodriguez, C. Fingermark evidence evaluation based on automated fingerprint identification system matching scores: the effect of different types of conditioning on likelihood ratios. *Journal of Forensic Sciences*, **59**, 70–81, 2014.

Ashbaugh, D.R. *Quantitative-Qualitative Friction Ridge Analysis: an Introduction to Basic and Advanced Ridgeology*. Taylor & Francis, 1st edition, 1999.

Ausdemore, M.A., Hendricks, J.H., and Neumann, C. Review of several false positive error rate estimates for latent fingerprint examination proposed based on the 2014 Miami Dade Police Department study. *Journal of Forensic Identification*, **69**(1), 59–81, 2019.

Balding, D.J. The DNA database search controversy. *Biometrics*, **58**, 241–244, 2002.

Balding, D.J. and Donnelly, P. Evaluating DNA profile evidence when the suspect is identified through a database search. *Journal of Forensic Science*, **41**, 603–607, 1996.

Banks, D. Bayesian inference for forensic science. In D. Banks, K. Kafadar, D. Kaye, and M. Tackett, editors, *Handbook of Forensic Statistics*, chapter 5. CRC Press, Boca Raton, Florida, 2020.

Barnes, J.G. History. In *The Fingerprint Sourcebook*, chapter 1, pages 7–24. National Institute of Justice, Office of Justice Programs, U.S. Department of Justice, 2014.

Berger, C.E.H., Vergeer, P., and Buckleton, J.S. A more straightforward derivation of the LR for a database search. *Forensic Science International: Genetics*, **14**, 156–160, 2015.

Bertillon, M.A. Les empreintes digitales. *Archives d'Anthropologie Criminelle de Lacassagne*, **27**, 36–52, j1912.

Biedermann, A., Bozza, S., and Taroni, F. Decision theoretic properties of forensic identification: underlying logic and argumentative implications. *Forensic Science International*, **177**, 120–132, 2008.

Blum, M. and Francois, O. Non-linear regression models for approximate bayesian computation. *Statistics and Computing*, **20**, 63–73, 2010.

Bozza, S., Taroni, F., and Biederman, A. Decision theory. In D. Banks, K. Kafadar, D. Kaye, and M. Tackett, editors, *Handbook of Forensic Statistics*, chapter 7. CRC Press, Boca Raton, Florida, 2020.

Budowle, B., Buscaglia, J., and Schwartz-Perlman, R. Review of the scientific basis for friction ridge comparisons as a means of identification: committee findings and recommendations. *Forensic Science Communications*, **8**(1), 2006.

Cao, K., Chugh, T., Zhou, J., Tabassi, E., and Jain, A.K. Automatic latent value determination. *2016 International Conference on Biometrics*, pages 1–8, 2016.

Champod, C. Fingerprint identification: advances since the 2009 National Research Council report. *Philosophical Transactions of the Royal Society B*, **370**, 123–132, 2015. http://dx.doi.org/10.1098/rstb.2014.0259

Champod, C. *Reconnaissance Automatique et Analyse Statistique des Minuties sur les Empreintes Digitales*. PhD thesis, University of Lausanne, 1995.

Champod, C. Fingerprint examination: towards more transparency. *Law, Probability and Risk*, **7**, 111–118, 2008.

Champod, C. and Evett, I.W. A probabilistic approach to fingerprint evidence. *Journal of Forensic Identification*, **51**, 101–122, 2001.

Champod, C., Lennard, C.J., Margot, P., and Stoilovic, M. *Fingerprints and Other Ridge Skin Impressions*. CRC Press, Boca Raton, 2nd ed. 2016.

Chugh, T., Cao, K., Zhou, J., Tabassi, E., and Jain, A.K. Latent fingerprint value prediction: crowd-based learning. *IEEE Transactions on Information Forensics and Security*, **13**, 20–34, 2018.

Cole, S.A. More than zero, accounting for error in latent fingerprint identification. *Journal of Criminal Law and Criminology*, **95**(3), 985–1078, 2005.

Dawid, A.P. Comment on Stockmarr's Likelihood ratios for evaluating DNA evidence when the suspect is found through a database search. *Biometrics*, **57**, 976–978, 2001.

Didelot, X., Everitt, R.G., Johansen, A.M., and Lawson, D.J. Likelihood-free estimation of model evidence. *Bayesian Analysis*, **6**(1), 49–76, 2011.

Egli, N.M., Champod, C., and Margot, P. Evidence evaluation in fingerprint comparison and automated fingerprint identification systems - modelling within finger variability. *Forensic Science International*, **176**, 189–195, 2006.

Egli-Anthonioz, N.M. and Champod, C. Evidence evaluation in fingerprint comparison and automated fingerprint identification systems - modeling between finger variability. *Forensic Science International*, **235**, 86–101, 2014.

European Network of Forensic Science Institutes (ENFSI). *ENFSI Guideline for Evaluative Reporting in Forensic Science*, 2016. http://enfsi.eu/wp-content/uploads/2016/09/m1_guideline.pdf

Evett, I.W. Towards a uniform framework for reporting opinions in forensic science casework. *Science & Justice*, **38**(3), 1198–1202, 1998.

Evett, I.W., Berger, C.E.H., Buckleton, J.S., Champod, C., and Jackson, G. Finding the way forward for forensic science in the USA: commentary on the PCAST report. *Forensic Science International*, **278**, 16–23, 2017.

Evett, I.W. and Williams, R. A review of the sixteen points fingerprint standard in England and Wales. *Journal of Forensic Identification*, **46**, 49–73, 1996.

Expert Working Group on Human Factors in Latent Print Analysis. Latent Print Analysis and Human Factors: Improving the Process Through a Systems Approach, 2012. Kaye, D.H. Editor. National Institute of Standards and Technology: Washington, D.C.

Forbes, P.G.M., Lauritzen, S., and Moller, J. Fingerprint analysis with marked point processes. Technical report, 2014. https://arxiv.org/abs/1407.5809

Galton, F. *Finger Prints*. Macmillan & Co., London, UK, 1892.

Garrett, R. Letter to the IAI membership. Document on file with the authors, 2009.

Gonzalez-Rodriguez, J., Drygajlo, A., Ramos-Castro, D., Garcia-Gomar, M., and Ortega-Garcia, J. Robust estimation, interpretation and assessment of likelihood ratios in forensic speaker recognition. *Computer Speech and Language*, **20**, 331–355, 2006.

Gonzalez-Rodriguez, J., Fierrez-Aguilar, J., and Ortega-Garcia, J. Forensic identification reporting using automatic speaker recognition systems. *2003 IEEE International Conference on Acoustics, Speech, and Signal Processing, 2003. Proceedings. (ICASSP '03)*, pages 11–93, 2003.

Gonzalez-Rodriguez, J., Fierrez-Aguilar, J., Ramos-Castro, D., and Ortega-Garcia, J. Bayesian analysis of fingerprint, face and signature evidences with automatic biometric systems. *Forensic Science International*, **155**, 126–140, 2005.

Haraksim, R., Ramos-Castro, D., Meuwly, D., and Berger, C.E.H. Measuring coherence of computer-assisted likelihood ratio methods. *Forensic Science International*, **249**, 123–132, 2015.

Hendricks, J.H. and Neumann, C. A Bayesian approach for the analysis of error rate studies in forensic science. *Forensic Science International*, 306, 2020.

Hendricks, J.H., Neumann, C., and Saunders, C.P. A ROC-based Approximate Bayesian Computation algorithm for model selection: application to fingerprint comparisons in forensic science. Technical report, 2019. https://arxiv.org/abs/1803.10121

Hepler, A.B., Saunders, C.P., Davis, L.J., and Buscaglia, J. Score-based likelihood ratios for handwriting evidence. *Forensic Science International*, **219**, 129–140, 2012.

Hicklin, R.A., Buscaglia, J., and Roberts, M.A. Assessing the clarity of friction ridge impressions. *Forensic Science International*, **226**, 106–117, 2013.

Hicklin, R.A., Buscaglia, J., Roberts, M.A., Meagher, S.B., Fellner, W., Burge, M.J., Monaco, M., Vera, D., Pantzer, L.R., Yeung, C.C., and Unnikumaran, T.N. Latent fingerprint quality: a survey of examiners. *Journal of Forensic Identification*, **61**, 385–419, 2011.

Indovina, M.D., Dvornychenko, V.N., Hicklin, R.A., and Kiebuzinski, G.I. *ELFT-EFS Evaluation of Latent Fingerprint Technologies: Extended Feature Sets [Evaluation 2]*. National Institute of Standards and Technology, 2012. https://doi.org/10.6028/NIST.IR.7859

Kassin, S.M., Dror, I.E., and Kukucka, J. The forensic confirmation bias: problems, perspectives, and proposed solutions. *Journal of Applied Research in Memory and Cognition*, **2**, 42–52, 2013.

Kaye, D.H. Forensic statistics in the courtroom. In D. Banks, K. Kafadar, D.H. Kaye, and M. Tackett, editors, *Handbook of Forensic Statistics*, chapter 10. CRC Press, Boca Raton, Florida, 2020.

Langenburg, G. Pilot study: a statistical analysis of the ACE-V methodology – Analysis stage. *Journal of Forensic Identification*, **54**(1), 64–79, 2004.

Langenburg, G. A performance study of the ACE-V process: a pilot study to measure the accuracy, precision, reproducibility, repeatability, and biasability of conclusions resulting from the ACE-V process. *Journal of Forensic Identification*, **59**(2), 219–257, 2009.

Langenburg, G. *A Critical Analysis and Study of the ACE-V Process*. PhD thesis, University of Lausanne, 2012.

Langenburg, G. and Champod, C. The GYRO system – A recommended approach to more transparent documentation. *Journal of Forensic Identification*, **61**(4), 373–384, 2011.

Langenburg, G., Champod, C., and Genessay, T. Informing the judgments of fingerprint analysts using quality metric and statistical assessment tools. *Forensic Science International*, **219**, 183–198, 2012.

Langenburg, G., Champod, C., and Wertheim, P. Testing for potential contextual bias effects during the Verification stage of the ACE-V methodology when conducting fingerprint comparisons. *Journal of Forensic Sciences*, **54**(3), 571–582, 2004.

Leegwater, A.J., Meuwly, D., Sjerps, M., Vergeer, P., and Alberink, I. Performance study of a score-based likelihood ratio system for forensic fingermark comparison. *Journal of Forensic Sciences*, **62**(3), 626–640, 2017.

Locard, E. La prevue judiciaire par les empreintes digitales. *Archives d'Anthropologie Criminelle, de Médecine Légale et de Psychologie Normale et Pathologique*, **28**, 312–348, 1914.

Maceo, A. Anatomy and physiology of adult friction ridge skin. In *The Fingerprint Sourcebook*, chapter 2, pages 25–50. National Institute of Justice, Office of Justice Programs, U.S. Department of Justice, 2014.

Marin, J.-M., Pudlo, P., Estoup, A., and Robert, C. Likelihood-free model choice. In S.A. Sisson, Y. Fan, and M.A. Beaumont, editors, *Handbook of Approximate BAyesian Computation*, chapter 6, pages 153–178. CRC Press, 2019.

Martire, K.A. and Edmond, G. How well do lay people comprehend statistical statements from forensic scientists? In D. Banks, K. Kafadar, D. Kaye, and M. Tackett, editors, *Handbook of Forensic Statistics*, chapter 10. CRC Press LLC, Boca Raton, Florida, 2020.

McQuiston-Surrett, D. and Saks, M.J. Communicating opinion evidence in the forensic identification sciences: accuracy and impact. *Hastings Law Journal*, **59**, 1159, 2008.

McQuiston-Surrett, D. and Saks, M.J. The testimony of forensic identification science: what expert witnesses say and what fact-finders hear. *Law and Human Behavior*, **33**(5), 436–453, 2009.

Meester, R. and Sjerps, M. The evidential value in the DNA database search controversy and the two-stain problem. *Biometrics*, **59**, 727–732, 2003.

Meuwly, D. and Drygajlo, A. Forensic speaker recognition based on a bayesian framework and gaussian mixture modeling. *Proc. of Odyssey 2001 Speaker Recognition Workshop, Crete (Greece)*, 2001.

Meuwly, D., Ramos, D., and Haraksim, R. A guideline for the validation of likelihood ratio methods used for forensic evidence evaluation. *Forensic Science International*, **276**, 142–153, 2017.

Morrison, G.S., Enzinger, E., Ramos, D., Gonzalez-Rodriguez, J., and Lozano-Diez, A. Statistical models in forensic voice comparison. In D. Banks, K. Kafadar, D. Kaye, and M. Tackett, editors, *Handbook of Forensic Statistics*, chapter 21. CRC Press LLC, Boca Raton, Florida, 2020.

Morrison, G.S., Kaye, D.H., Balding, D.J., Taylor, D., Dawid, P., Aitken, C.G.G., Gittelson, S., Zadora, G., Robertson, B., Willis, S., Pope, S., Neil, M., Martire, K.A., Hepler, A., Gill, R.D., Jamieson, A., de Zoete, J., Ostrum, R.B., and Caliebe, A. A comment on the PCAST report: skip the match/non-match stage. *Forensic Science International*, **272**, e7–e9, 2017.

National Research Council (NRC) of the National Academies. *Strengthening Forensic Science in the United States: A Path Forward*. The National Academies Press, Washington, DC, 2009.

Neumann, C. Statistics and probabilities as a means to support fingerprint examination. In R. Ramotowski, editor, *Advances in Fingerprint Technology, Lee and Gaensslen's Advances in Fingerprint Technology*, chapter 15, pages 419–466. CRC Press LLC, Boca Raton, Florida, 3rd edition, 2012.

Neumann, C., Armstrong, D.E., and Wu, T. Determination of AFIS sufficiency in friction ridge examination. *Forensic Science International*, **263**, 114–125, 2016.

Neumann, C., Champod, C., Puch-Solis, R., Egli, N., Anthonioz, A., and Bromage-Griffiths, A. Computation of likelihood ratios in fingerprint identification for configurations of any number of minutiae. *Journal of Forensic Sciences*, **52**, 54–64, 2007.

Neumann, C., Champod, C., Puch-Solis, R., Meuwly, D., Egli, N., and Anthonioz, A. Computation of likelihood ratios in fingerprint identification for configurations of three minutiae. *Journal of Forensic Sciences*, **51**, 1255–1266, 2006.

Neumann, C., Champod, C., Yoo, M., Genessay, T., and Langenburg, G. Improving the understanding and the reliability of the concept of sufficiency in friction ridge examination. Technical report, National Institute of Justice, 2013. Award 2010-DN-BX-K267, https://www.ncjrs.gov/App/Publications/abstract.aspx?ID=266312

Neumann, C., Champod, C., Yoo, M., Genessay, T., and Langenburg, G. Quantifying the weight of fingerprint evidence through the spatial relationship, directions and types of minutiae observed on fingermarks. *Forensic Science International*, **248**, 154–171, 2015.

Neumann, C., Evett, I.W., and Skerrett, J.E. Quantifying the weight of evidence from a forensic fingerprint comparison: a new paradigm. *Journal of the Royal Statistical Society (Series A)*, **175**, 1–26, 2012.

Neumann, C., Evett, I.W., Skerrett, J.E., and Mateos-Garcia, I. Quantitative assessment of evidential weight for a fingerprint comparison I: generalisation to the comparison of a mark with set of ten prints from a suspect. *Forensic Science International*, **207**, 101–105, 2011.

Neumann, C., Kaye, D., Jackson, G., Reyna, V., and Ranadive, A. Presenting quantitative and qualitative information on forensic science evidence in the courtroom. *CHANCE*, **29**, 37–43, 2016.

Ommen, D.M., Saunders, C.P., and Neumann, C. The characterization of Monte Carlo errors for the quantification of the value of forensic evidence. *Journal of Statistical Computation and Simulation*, **87**(8), 1608–1643, 2017.

OSAC. *Organization of Scientific Area Committees (OSAC) Friction Ridge Subcommittee's Response to the President's Council of Advisors on Science and Technology's Request for Additional References*. https://www.nist.gov/topics/forensic-science/friction-ridge-subcommittee, December 2016.

Pacheco, I., Cerchiai, B., and Stoiloff, S. Miami-Dade research study for the reliability of the ACE-V process: Accuracy and precision in latent fingerprint examinations. Technical report, National Institute of Justice, 2014. Award 2010-DN-BX-K268, https://www.ncjrs.gov/pdffiles1/nij/grants/248534.pdf

President's Council of Advisors on Science and Technology (PCAST). *Forensic Science in Criminal Courts: Ensuring Scientific Validity of Feature-Comparison Methods*. Executive Office of the President's Council of Advisors on Science and Technology, Washington, DC, 2016.

Pudlo, P., Marin, J.-M., Estoup, A., Cornuet, J.-M., Gautier, M., and Robert, C.P. Reliable ABC model choice via random forests. *Bioinformatics*, **32**(6), 859–866, 2016.

Ramos, D. and Gonzalez-Rodriguez, J. Reliable support: measuring calibration of likelihood ratios. *Forensic Science International*, **230**, 156–169, 2013.

Ramos, D., Gonzalez-Rodriguez, J., Zadora, G., and Aitken, C. Information-theoretical assessment of the performance of likelihood ratio computation methods. *Journal of Forensic Sciences*, **58**(6), 1503–1518, 2013.

Robert, C.P., Cornuet, J.-M., Marin, J.-M., and Pillai, N.S. Lack of confidence in ABC model choice. *Proceedings of the National Academy of Sciences of the United States of America*, **108**(37), 5112–5117, 2011.

Rodriguez, C., de Jongh, A., and Meuwly, D. Introducing a quick and simple approach to simulate large numbers of crime scene fingermarks for research on fingerprint identification. *Journal of Forensic Sciences*, **57**, 334–342, 2012.

Safi, M. Indian court upholds legality of world's largest biometric database. *The Guardian*, 2018. https://www.theguardian.com/world/2018/sep/26/indian-court-upholds-legality-of-worlds-largest-biometric-database

Saks, M.J. and Koehler, J.J. The coming paradigm shift in forensic identification science. *Science*, **309**(892), 893–894, 2005.

Scientific Working Group on Friction Ridge Analysis, Study and Technology (SWGFAST). *Standard for the Application of Blind Verification of Friction Ridge Examinations (Latent/Tenprint)*, 2012a. version 2–0, http://clpex.com/swgfast/documents/blind-verification/121124_Blind-Verification_2.0.pdf

Scientific Working Group on Friction Ridge Analysis, Study and Technology (SWGFAST). *Standard for the Documentation of Analysis, Comparison, Evaluation and Verification (ACE-V) (Latent)*, 2012b. version 2–0, http://clpex.com/swgfast/documents/documentation/121124_Standard-Documentation-ACE-V_2.0.pdf

Scientific Working Group on Friction Ridge Analysis, Study and Technology (SWGFAST). *Standard for the Technical Review of Friction Ridge Examinations (Latent/Tenprint)*, 2012c. version 2–0, http://clpex.com/swgfast/documents/technical-review/121124_Technical-Review_2.0.pdf

Scientific Working Group on Friction Ridge Analysis, Study and Technology (SWGFAST). *Guideline for the Articulation of the DecisionMaking Process for the Individualization in Friction Ridge Examination*, 2013a. version 1–0, http://clpex.com/swgfast/documents/articulation/130427_Articulation_1.0.pdf

Scientific Working Group on Friction Ridge Analysis, Study and Technology (SWGFAST). *Standards for Examining Friction Ridge Impressions and Resulting Conclusions (Latent/Tenprint)*, 2013b. version 2–1, http://clpex.com/swgfast/documents/examinations-conclusions/130427_Examinations-Conclusions_2.1.pdf

Sjerps, M.J. Probabilistic considerations when interpreting database search and selection effects. In *Handbook of Forensic Statistics*, D. Banks, K. Kafadar, D.H. Kaye, and M. Tackett, editors. CRC Press, Boca Raton, Florida, 2020.

Sjerps, M. and Meester, R. Selection effects and database screening in forensic science. *Forensic Science International*, **192**, 56–61, 2009.

State v. Dixon, No. 27-CR-10-3378 (D. Ct. Cty. Hennepin, Minn.). 2011.

State v. Hull, No. 48-CR-07-2336 (Minn. D. Ct. Cty. of Mille Lacs). 2008.

Stern, H. Comparing philosophies of statistics inference. In D. Banks, K. Kafadar, D. Kaye, and M. Tackett, editors, *Handbook of Forensic Statistics*, chapter 6. CRC Press LLC, Boca Raton, Florida, 2020.

Stoney, D. Measurement of fingerprint individuality. In H.C. Lee and R.E. Gaensslen, editors, *Advances in Fingerprint Technology*, chapter 9, pages 327–388. CRC Press LLC, Boca Raton, Florida, 2nd edition, 2001.

Swofford, H.J., Koertner, A.J., Zemp, F., Ausdemore, M.A., Liu, A., and Salyards, M.J. A method for the statistical interpretation of friction ridge skin impression evidence: Method development and validation. *Forensic Science International*, **287**, 113–126, 2018.

Tackett, M. *Creating Fingerprint Databases and a Bayesian Approach to Quantify Dependencies in Evidence*. PhD thesis, University of Virginia, 2018.

Thompson, W., Kaasa, S.O., and Peterson, T. Do jurors give appropriate weight to forensic identification evidence? *Journal of Empirical Legal Studies*, **10**(2), 359–397, 2013.

Thompson, W. and Newman, E.J. Lay understanding of forensic statistics: evaluation of random match probabilities, likelihood ratios, and verbal equivalents. *Law and Human Behavior*, **39**(4), 332–349, 2015.

Thompson, W. and Schumann, E.L. Interpretation of statistical evidence in criminal trials: the prosecutor's fallacy and the defense attorney's fallacy. *Law and Human Behavior*, **11**(3), 167, 1987.

Ulery, B.T., Hicklin, R.A., Buscaglia, J., and Roberts, M.A. Accuracy and reliability of forensic latent fingerprint decisions. *Proceedings of the National Academy of Sciences*, **108**(19), 2011.

Ulery, B.T., Hicklin, R.A., Buscaglia, J., and Roberts, M.A. Repeatability and reproducibility of decisions by latent fingerprint examiners. *PLoS ONE*, **7**(3), 2012.

Ulery, B.T., Hicklin, R.A., Kiebuzinski, G.I., Buscaglia, J., and Roberts, M.A. Understanding the sufficiency of information for latent fingerprint value determinations. *Forensic Science International*, **230**, 99–106, 2013.

Ulery, B.T., Hicklin, R.A., Roberts, M.A., and Buscaglia, J. Measuring what latent fingerprint examiners consider sufficient information for individualization determinations. *PLoS ONE*, **9**(11), 2014.

Ulery, B.T., Hicklin, R.A., Roberts, M.A., and Buscaglia, J. Changes in latent fingerprint examiners' markup between Analysis and Comparison. *Forensic Science International*, **247**, 54–61, 2015.

Ulery, B.T., Hicklin, R.A., Roberts, M.A., and Buscaglia, J. Interexaminer variation of minutia markup on latent fingerprints. *Forensic Science International*, **264**, 89–99, 2016.

Ulery, B.T., Hicklin, R.A., Roberts, M.A., and Buscaglia, J. Factors associated with latent fingerprint exclusion determinations. *Forensic Science International*, **275**, 65–75, 2017.

U.S. Department of Justice, Office of the Inspector General. *A Review of the FBI's Handling of the Brandon Mayfield Case*. https://oig.justice.gov/special/s0601/final.pdf, 2006.

U.S. Department of the Army, Defense Forensic Science Center. Information paper: use of the term.identification. in latent print technical reports. Document on file with the authors, November 2015.

U.S. Department of the Army, Defense Forensic Science Center. Information paper: modification of latent print technical reports to include statistical calculations. Document on file with the authors, March 2017.

U.S. Federal Bureau of Investigation. *February 2019 Next Generation Identification (NGI) System Fact Sheet*, February 2019. https://www.fbi.gov/services/cjis/fingerprints-and-other-biometrics/ngi

van Es, A., Wiarda, W., Hordijk, M., Alberink, I., and Vergeer, P. Implementation and assessment of a likelihood ratio approach for the evaluation of LA-ICP-MS evidence in forensic glass analysis. *Science & Justice*, **57**(3), 181–192, 2017.

Watson, C.I., Fiumara, G.P., Tabassi, E., Cheng, S.L., Flanagan, P.A., and Salamon, W.J. *Fingerprint Vendor Technology Evaluation*. National Institute of Standards and Technology, 2015. https://doi.org/10.6028/NIST.IR.8034

Wertheim, K. Embryology and morphology of friction ridge skin. In *The Fingerprint Sourcebook*, chapter 3, pages 51–76. National Institute of Justice, Office of Justice Programs, U.S. Department of Justice, 2014.

Wixted, J.T., Christenfeld, N.J.S., and Rouder, J.N. Calculating the posterior odds from a single-match DNA database search. *Law, Probability and Risk*, **18**, 1–23, 2019.

Yoon, S., Cao, K., Liu, E., and Jain, A.K. LFIQ: Latent Fingerprint Image Quality. *IEEE International Conference on Biometrics: Theory, Applications and Systems*, 2013.

Zabell, S.L. Fingerprint evidence. *Journal of Law and Policy*, **13**, 143–170, 2005.

14

Probabilistic Considerations When Interpreting Database Search and Selection Effects

M.J. Sjerps

CONTENTS

14.1 Introduction 325
14.2 Probabilistic Analysis 326
14.3 Scientific Debate: The Database Search Controversy 328
14.4 The Number of Adventitious Matches 331
14.5 Reporting Database Matches 332
    14.5.1 Comparing a "Probable Cause" and a "Database Search" Match 332
    14.5.2 The Other Evidence in the Case 333
    14.5.3 Guidance on Interpreting Database Matches 333
14.6 Dealing with "Selection Effects" More Generally 335
14.7 Interesting Cases 336
14.8 Conclusion 338
References 338

14.1 Introduction

When a crime is committed, and traces are gathered at the scene or otherwise obtained, comparing these traces to a database is now a standard means of identifying a suspect or suspected item. Traditionally, forensic scientists used fingerprints and shoeprints in this way. More recently, DNA databases have become very prominent, and currently a variety of other trace types also are used for database searches national and internationally (Walsh, 2016). For example, biometric features like fingerprints, chemical features of glass, physical features of cartridges, and digital features such as pixel defect patterns in digital photos are used, but DNA databases feature most prominently in current police investigations and prosecution.

With DNA databases quickly increasing in size (e.g., the UK National DNA Database contained about 5.5 million profiles of individuals in March, 2019) and the legal possibility to search in other databases nationally and internationally (in Europe, for example, the EU Prüm treaty (2008) allows for an automatic way to do this), we find ourselves in the following situation. When a DNA profile is derived from some trace, and a suspect is not yet available or is excluded because his profile does not match, the trace DNA profile is compared with those in a number of databases until either a match is found or until there are no more databases of reference profiles left to compare. In the latter case, the profile will be added to a database of unmatched DNA traces to be compared with new reference profiles as they become available. Typically, the expected frequencies of the profiles analyzed with recent techniques is much smaller than 1 in a billion (10^{-9}). However, we also search with partial profiles from DNA samples of poor quality or made with an old technique, with profiles from a mixture of DNA from several persons, and we can perform familial searches as well.

Consequently, billions of comparisons are made every day with all kinds of profiles. This process generates an ever-increasing stream of matches with non-donors (adventitious or random matches), along with a far bigger increasing stream of matches with donors ("true" matches). In the EU international exchange DNA database, many instances of adventitious 6- and 7-locus matches have been identified. In a joint investigation with the UK where 137 6-locus matches and 394 7-locus matches were verified by additional DNA testing by the Dutch DNA database staff, 85 (62%) of the 6-locus matches and 23 (6%) of the 7-locus matches proved to be false matches. This chapter explains that these numbers are not unexpected from a statistical point of view. Nevertheless, they may come as a surprise to lay persons. Simultaneously, many cases are solved using these intelligence databases. For example, in 2017 the Dutch DNA Database reported 1754 international matches to the investigative authorities obtained in the EU database (NLDNAD, 2017). The national DNA database in the UK reported about 31,000 subject-to-crime links, including about 1500 murders or rape cases, for the years 2017-2018 (UK National DNA Database, 2019).

Thus, the general public is faced with both success stories and with alarming news reports on false matches leading to serious problems for innocent people (e.g., Porter, 2009; Geddes, 2011). The scientific community has debated for quite a long time how to interpret and report DNA database matches. This debate has become known as the "DNA database search controversy" and spans several decades (see, e.g., NRC, 1992; Wixted et al., 2019 with commentary).

In this chapter, the main positions and the associated probabilistic analyses in this debate are explained. It is concluded that from a mathematical point of view the controversy is solved. It is also argued that similar reasoning can be used to answer puzzling questions in a much wider range of "search and selection" situations where persons or items become the focus of interest after multiple comparisons are made. However, it is far from clear how the results of a database search (or some other form of screening) should be reported so that a lay person interprets the result appropriately. Several suggestions are presented, but there is no consensus on this important topic, and more research on it is greatly needed. Finally, some interesting legal cases are discussed.

14.2 Probabilistic Analysis

From a probabilistic point of view, the problem is quite simple and the solution has been presented in various forms in the scientific literature (see next section). Here, we follow the analysis of Dawid (2001).

Suppose that a sample of a crime stain yields a DNA profile G which is assumed to originate from a single person. There is a population of N unrelated persons who could be the donor of this DNA. Each of these persons has a known probability p to have the profile G, independently from each other*. We call this the "profile probability". Note that p is directly related to the rarity of G, because it is the expected relative frequency of G in the population. We have a DNA database (D) containing the DNA profiles of n different persons, and we search this database. Suppose that the profile of a single person, Smith, is observed to match (i.e., his DNA profile is also G) and all others do not match. We define this outcome of the search as the evidence E, and the hypothesis that Smith is the donor of the DNA in the sample as H_1. A judge or juror will be interested in $Pr(H_1 \mid E)$. This problem is one variant of a so-called "island problem", where we picture the population as living on an island. The simplest "island problem" is when the database contains only Smith.

From a Bayesian point of view, this probability is defined as a personal degree of belief. From a frequentist point of view, this probability is difficult to interpret and a probabilistic model is required for proper interpretation. For example, we model the donor of the crime stain as a red ball in an urn with only $N - 1$ white balls. We know that each ball in the urn has a probability p to have a black dot on it (determined by a toss of an (unfair) coin), and we know that the red ball has a black dot. We now take at random n balls out of the urn, and observe the evidence E that only one of these has a black dot. We are interested in the probability of event H_1 that this is the red ball, $Pr(H_1 \mid E)$. With this interpretation, the following probability calculus can be interpreted in a Bayesian or in a frequentist way.

To calculate $Pr(H_1 \mid E)$, we define H_2 as the hypothesis that Smith is not the donor of the DNA in the sample, and use Bayes' rule (in odds form) to find that

$$\frac{\Pr(H_1 \mid E)}{\Pr(H_2 \mid E)} = \frac{\Pr(H_1)}{\Pr(H_2)} \cdot \frac{\Pr(E \mid H_1)}{\Pr(E \mid H_2)}.$$

Let's consider the terms on the right-hand side. The first ratio is the odds on H_1 prior to considering E. The second ratio is the likelihood ratio (LR). For the numerator of the LR, it is assumed that Smith is the crime stain donor. In that case his DNA profile will match for sure with that of the crime stain (unless an error occurred, but we will ignore this possibility here). Thus, the numerator can be calculated as the probability that all $n - 1$ other persons in the database do not match

$$\Pr(E \mid H_1) = (1 - p)^{n-1}.$$

The denominator of the LR depends on the probability that the donor of the crime stain has a DNA profile in the database D (prior to knowing E). We call this probability δ. The only way to obtain E when H_2 is true is when (a) the crime stain donor is not in the database, (b) Smith matches by chance, and (c) all others do not match. The probability of (a) given H_2 depends on the prior probability that Smith is the donor, which we denote as π. The probability of (a) given H_2 then equals $(1 - \delta)/(1 - \pi)$. Thus, we obtain

$$\Pr(E \mid H_2) = \frac{1 - \delta}{1 - \pi} \cdot p \cdot (1 - p)^{n-1}.$$

---

* This simple sentence hides a world of subtle issues. In practice, p is unknown and must be estimated from data. One usually calculates the "random match probability" (Pr(the crime stain has profile G | Smith has G, H_2), see below for a description of Smith and H_2) that is related to the rarity of the matching profile. This differs from the unconditional "profile probability". Furthermore, a common interpretation of probability is that it is personal; hence different people may assign different probabilities. Finally, people may be related, and so their profiles are dependent on each other. For simplicity, we ignore in this chapter all these subtleties and assume that the profile probability is given to us as a fixed and known number p.

Therefore, the LR of E for H_1 versus H_2 is the ratio of the two probabilities[*]

$$LR_{H_1:H_2}(E) = \frac{1-\pi}{1-\delta} \cdot \frac{1}{p}.$$

Multiplying this by the prior odds $\pi/(1-\pi)$, we obtain the posterior odds as

$$\frac{\Pr(H_1 \mid E)}{\Pr(H_2 \mid E)} = \frac{\pi}{1-\delta} \cdot \frac{1}{p}.$$

Thus, the odds increase with the prior probability that Smith is the donor (π), and decrease with the profile probability (p), as one would expect. Interestingly, they increase with the prior probability that the donor is in the database (δ). This is because if we are surer that the donor is in the database, and Smith is the only match, we become surer that Smith is the donor.

If we assume that *a priori* each of N persons in the population of interest is equally likely to be the crime stain donor, we have $\pi = 1/N$, and if we furthermore assume that the DNA database D is a random sample of n of these persons, we have $\delta = n/N$. This leads to prior odds of $1/(N-1)$ and

$$LR_{H_1:H_2}(E) = \frac{N-1}{N-n} \cdot \frac{1}{p},$$

and

$$\frac{\Pr(H_1 \mid E)}{\Pr(H_2 \mid E)} = \frac{1}{(N-n)} \cdot \frac{1}{p}.$$

Intuitively, this makes sense. There are $N-n$ persons not in the database, and we expect a fraction p of them to match the crime stain profile. So, knowing that Smith is the only person in the database who matches, either Smith is the donor or one of the other people in the population who has the same DNA profile is, and we expect $(N-n)p$ of those. Furthermore, we see that the probability that Smith is the donor increases when the database size n grows. This is because with increasing database size more people are being excluded as a possible stain donor, and consequently the expected number of matching persons other than Smith decreases. Ultimately, when all persons are in the database ($n = N$), we are sure that Smith is the donor, corresponding to infinite odds.

14.3 Scientific Debate: The Database Search Controversy

The interpretation of database matches has been the subject of a very intense scientific debate (see e.g., Aitken and Taroni (2004), or Walsh et al. (2005, 2016) for an overview including a large number of references). Here, some key issues and papers are discussed.

The first report on DNA evidence of the influential US National Research Council (1992) showed concern about the fact that the probability of obtaining a match when comparing to a database of profiles is much larger than when comparing to a single person. Methods

---

[*] Dawid (2001) equation (3) differs due to a typo which was later corrected. Also we switched the notation H and H'.

to correct for multiple comparisons were dismissed because the NRC committee deemed the estimation of profile frequencies too unreliable, due to small reference databases at the time. They advised using a subset of DNA loci to search the database for a matching suspect, and then to doing additional DNA testing with other loci to confirm the match. The reported evidential strength of the match should then be calculated only from the additional loci.

This was criticized by the forensic community as being too conservative (e.g., Balding and Donnelly, 1996). Often, the DNA material in the crime stain was depleted so that additional typing yielded no or only a few additional loci. Also, with growing databases many loci were needed for identification of a single matching person, and none or few loci were left for calculating evidential strength. The second report of the US National Research Council (1996) instead advised multiplying the random match probability p of the profile by the number of persons in the database n and reporting this value np. The argument was based on the "frequentist" idea of hypothesis testing. Under the hypothesis that the donor is not in the database, the probability of finding at least one match in the database is at most the sum of the probabilities that each individual person matches, np. Thus, when this probability is small, this hypothesis can be rejected in favor of the hypothesis that the matching individual is the donor.

The second NRC report conflicted with the Bayesian analysis of Dawid (1994), Dawid and Mortera (1996) and Balding and Donnelly (1995a), who analyzed various versions of the "island problem" mentioned earlier. The latter argued that the probability of finding at least one match is irrelevant, and that the suggestion that a DNA match in the case of database search is n times less convincing than had the suspect been the only individual examined is misleading. All papers showed that ignoring the fact that the suspect was identified through database search leads to a conservative evaluation of the posterior probability of donorship. Balding and Donnelly (1995b, 1996) summarized these arguments for a general scientific audience. Evett and Weir (1998) provided a similar analysis in a leading textbook for forensic DNA scientists. The fact that the second NRC report ignored the arguments from these Bayesian analyses and focused instead on a frequentist approach of hypothesis testing was remarkable, and was criticized by Balding (1997) and others. It ignited a scientific discussion between those who were in favor of reporting an uncorrected random match probability "p", basically ignoring the search, and those who were in favor of reporting a corrected random match probability "np" as suggested by the second NRC report.

Stockmarr (1999) refueled the debate with the argument that the hypotheses H_1 and H_2 above are "data-dependent", i.e., they depend on the outcome of the database search. In other words, we can only focus on the hypothesis that Smith is the donor *after* we know that Smith is the only match in the database. He proposed to use other hypotheses which can be specified beforehand: the donor of the crime stain is in the database (H_1') or he is not (H_2'). The probability calculus then runs as follows.

The prior odds of hypotheses H_1' versus H_2' are $\delta/(1-\delta)$. The evidence E is defined as before, Smith is the only person in the database matching the profile G of the crime stain. The numerator of the LR is

$$\Pr(E \mid H_1') = \frac{\pi}{\delta} \left(1-p\right)^{n-1}$$

since we can only obtain E if Smith is the donor (given H_1' this happens with probability π/δ) and all others do not match. The denominator is

$$\Pr(E \mid H_2') = p \cdot (1-p)^{n-1}.$$

This yields a LR of

$$LR_{H_1':H_2'}(E) = \frac{\pi}{\delta} \cdot \frac{1}{p}.$$

In case of uniform priors, the LR is simply

$$LR_{H_1':H_2'}(E) = \frac{1}{n} \cdot \frac{1}{p}.$$

Thus, Stockmarr concluded that the evidential strength of the database match decreases when the database becomes larger. Also, he noted that his formula is in concordance with the recommendation in the second NRC report. This reasoning is in line with the intuition of many statisticians that in case of multiple comparisons the probability of an adventitious match increases and therefore it is important to correct the conclusion for the number of comparisons made.

Stockmarr's paper sparked more discussion in the scientific literature, which mainly focused on arguments why hypotheses set $H_1 : H_2$ should be used (this is the set that it relevant in court; the database is not on trial) or why hypotheses set $H_1' : H_2'$ should be used (this set is data-independent and takes proper account of multiple comparisons).

In an important paper (Donnelly and Friedman, 1999) and commentary Dawid (2001), it was shown that, from a mathematical point of view, the two approaches are simply equivalent, and there is no reason to prefer one over the other. After we know that Smith is the only match in the DNA-database, the hypotheses become logically equivalent

$$H_1 \,|\, E \Leftrightarrow H_1' \,|\, E, \text{ and } H_2 \,|\, E \Leftrightarrow H_2' \,|\, E.$$

This implies that the posterior odds of the two hypotheses pairs are the same. Thus, we have:

$$\frac{\Pr(H_1' \mid E)}{\Pr(H_2' \mid E)} = \frac{\pi}{1 - \delta} \cdot \frac{1}{p}.$$

The LR follows as the ratio of these posterior odds and the prior odds $\delta/(1 - \delta)$

$$LR_{H_1':H_2'}(E) = \frac{\pi}{\delta} \cdot \frac{1}{p}.$$

Assuming uniform prior probabilities as before, these equations become

$$\frac{\Pr(H_1' \mid E)}{\Pr(H_2' \mid E)} = \frac{1}{N - n} \cdot \frac{1}{p},$$

and

$$LR_{H_1':H_2'}(E) = \frac{1}{n} \cdot \frac{1}{p}.$$

So, in the end all we did when moving from the "data-dependent" pair H_1 and H_2 to the "data-independent" pair H_1' and H_2' is to move a factor of

$$\frac{\pi/(1 - \pi)}{\delta/(1 - \delta)}, \text{ or in case of uniform priors, } \frac{N - n}{N - 1} \cdot \frac{1}{n}.$$

from the prior odds to the LR (Dawid 2001). In other words, the prior odds are divided by this factor, but the LR is multiplied by it. Since this factor lies in the interval [0,1], the net result of the move is larger prior odds, smaller LR, and equal posterior odds.

Dawid (2001) warned that misunderstandings arise when important details are omitted in the communication. When evaluating a match obtained through database search, people consider the LR as a measure of the strength of the evidence E, without specifying which hypotheses pair they are considering. Those who consider implicitly H_1 and H_2 will claim that the evidence gets *stronger* when the database grows larger. Those who consider implicitly H_1' and H_2' will claim that the evidence gets *weaker* when the database grows larger. Both are correct. As we saw above, they are simply talking about different hypotheses. But when the hypotheses are left implicit, and both talk about *the* LR, it seems that they are in conflict.

Numerous papers have been published after Donnelly and Friedman (1999) and Dawid (2001), exploring the controversy with, e.g., Bayesian networks, or rewriting formulas in a more convenient way, reaching for the most part the same conclusion. For example, by comparing frequentist p-values and Bayesian posterior probabilities on hypotheses, Storvik and Egeland (2007) concluded that the gap between the frequentist and Bayesian reasoning was bridged. Thus, from a mathematical point of view, the database search controversy has been solved.

14.4 The Number of Adventitious Matches

How often may we expect to find adventitious matches? Basic probability calculus tells us that if we consider a DNA profile of a crime stain, where each person has this profile with known probability p, and we compare it with a database of n unrelated (independent) profiles, then the number of matching profiles will be binomially distributed (with parameters n and p). The expected number of matches with non-donors is thus np for a database search with a single profile. Furthermore, the probability to find at least one match in a database of all non-donors is $1 - (1 - p)^n$, i.e., $1 - q$, where q is the probability that nobody in the database matches. For example, given a database with $n = 100,000$ persons and $p = 10^{-9}$ (1 in a billion), we expect to find no matches with non-donors, and the probability of finding at least one such match is about 0.1%. However, this is when we consider a search with a single rare profile.

When we compare each profile in the database with n profiles to each other, we have $n(n-1)/2$ pairwise comparisons. This amounts to 5 billion comparisons if $n = 100,000$. The probability of finding at least one matching pair in a database of independent profiles, each with a profile probability p, can now be calculated according to the solution of the "birthday problem" (Weir, 2007). It amounts to $1 - [(1 - p)(1 - 2p)(1 - 3p) \ldots (1 - (n-1)p)]$, and this is about $1 - e^{-n^2 p/2}$ (where e is Euler's number, 2.71828...).

For our example, with $n = 100,000$, and $p = 10^{-9}$, we have a probability of 99% to find at least one matching pair. Furthermore, we see that if $n = 1/\sqrt{p}$, then this probability is $1 - e^{-1/2} = 40\%$. This calculation is under the assumption that all profiles are independent of each other, and all profiles have the same profile probability. Weir (2004, 2007) worked out formulas and calculations which take dependencies due to (distant) relatedness into account, as well as formulas for the number of profiles that match only for a part of the profile. He showed that in large databases, we may expect to find matches and partially matching profiles between two different persons, even if p is quite small. Considering the current worldwide practice of database search (see introduction), we see that adventitious matches are a realistic possibility.

14.5 Reporting Database Matches

The key issue when interpreting database search results is to distinguish between the strength of the *case* and the strength of the *evidence* (Balding and Donnelly 1995b, 1996; Robertson and Vignaux 1995; Balding 2002). Translated to the question considered here, the source of the DNA in the crime stain: it is essential to distinguish between the posterior odds and the LR. It is the posterior odds that are ultimately of interest to the court. As we saw above, these are the same for the pair H_1 versus H_2 and for the pair H_1' versus H_2'.

Unfortunately, forensic scientists who write the reports cannot calculate these odds. This is because they contain the prior odds, which are for the court to decide. The expert can calculate the LR, but as we saw above, the LR differs between the two hypotheses pairs. Furthermore, in some jurisdictions the jury is not allowed to be informed about the fact that the suspect's DNA profile is in the database, as the jury members may be prejudiced by this information (Meester and Sjerps, 2004). In these jurisdictions the pair H_1' versus H_2' is problematic. Thus, the expert has to choose a way to report on the evidential strength of the search result within legal constraints, and such that lay persons will arrive at their posterior odds correctly and in an intuitive way. This is far from easy. Hence, while the database search controversy may be trivial from a mathematical point of view, it is far from trivial from the point of view of a forensic scientist who has to write a report or testify in court.

14.5.1 Comparing a "Probable Cause" and a "Database Search" Match

When reporting DNA database matches, the following simple insight is key. In cases where the suspect is first identified by some other evidence such as an eye witness, and for this reason his DNA profile is compared to that of the crime stain ("probable cause cases"), we know that there is other evidence that is strong enough to single out this person for DNA comparison. We do not have this knowledge in cases where the suspect is identified through a database match. This simple fact is the crucial difference between a "probable cause match" and a "database match".

This difference is reflected in the prior odds of the hypotheses H_1 and H_2. In a database search case, it is possible that besides the DNA match there is (almost) no other evidence supporting H_1. In fact, if we are dealing with an adventitious match, this situation is the one we may expect. This means that in a database search case, it is possible that the prior odds for H_1 versus H_2 are extremely small. They may be so small that they effectively counterbalance the LR concerning the DNA match, resulting in posterior odds that leave quite some doubt on Smith being the donor. For example, when the LR is 10^9 and the prior odds are 10^{-9}, we end up with a probability of only 50% that Smith is the donor of the crime stain. This counterbalancing effect will be counterintuitive to many people, and thus needs to be explained. In a probable cause case, we cannot have this situation of (almost) no other evidence supporting H_1, since the probable cause constitutes evidence supporting H_1 strongly enough to order a DNA test for Smith (Balding and Donnelly 1995b, 1996; Donnelly and Friedman 1999).

14.5.2 The Other Evidence in the Case

When Smith is identified through a database search, and his name is passed on to the police, they may or may not find other evidence. If Smith is in fact the donor, and it is a recent and important case, we may expect to find some other evidence supporting Smith as the donor. For example, an eye witness may be found who recognized Smith as the offender, so that the strength of this other evidence becomes comparable to a probable cause case. In this situation, the court should conclude that there is a large probability that Smith is the donor, depending on the other evidence produced. If other evidence is found strongly supporting that Smith is not a donor, such as a reliable alibi, it will be clear that the court needs to weigh this with the DNA evidence.

In database search, there can also be several situations in which we expect to find (almost) no other evidence supporting donorship. We may expect this if Smith's DNA profile matches by chance. Furthermore, in quite a number of cases the police will produce little or no other evidence for economic reasons, because the case is deemed not important enough to warrant further investigation. Finally, for a cold case that happened many years ago, it may be difficult to find other evidence altogether. In these situations, we just know that Smith is in the database for some reason, plus some additional information such as his place of residence and his age, but this may be considered as weak evidence by the court. The court may then conclude that the prior odds on donorship are small. The posterior odds follow from the multiplication of these small prior odds and the large $LR_{H1:H2}$ of the DNA evidence. Depending on the outcome the court should conclude that Smith probably is the donor, or the opposite, Smith probably is *not* the donor. The latter conclusion may feel very counterintuitive.

The intuition of most lay people is that strong DNA evidence implies a large probability of donorship. This intuition leads to a correct conclusion in many cases, except in those where the prior probability of donorship is so small that it counterbalances the DNA evidence. This counterbalancing effect needs to be explained very clearly as lay persons are not likely to recognize it. This has very serious consequences in those cases; i.e., innocent persons may be prosecuted and convicted while the real offender may go on committing other crimes.

14.5.3 Guidance on Interpreting Database Matches

The forensic scientist can observe a match but not whether it was an adventitious or a true match. The distinction must be made by the lay people involved in the investigative or legal process starting after the match is identified. It is therefore crucial that they receive information and guidance that allows them to do this properly. It is the task of the DNA expert reporting the match to provide this information and guidance. The message that we need to get across consists of the following elements:

1. A DNA match between Smith and a crime stain typically is very strong evidence supporting that Smith is the donor of the stain.

2. However, it is possible that Smith is not the donor and matches by chance.

3. The probability that Smith is in fact the donor can vary from small to large, depending also on the other evidence in the case. If the other evidence supports donorship, the probability is large. If the other evidence supports non-donorship, or if there is (almost) no other evidence, the probability can be small.

4. In case of a probable cause match we know that there is some other evidence supporting donorship. In case of a database match we do not know this, and there may even be (almost) no other evidence.

5. If multiple comparisons are made with the crime stain profile (as in database search) and all except Smith have been excluded as the donor, this increases the probability that Smith is the donor.

6. If more certainty is required it may be possible to do additional DNA analysis.

Reporting np has been advised by some (NRC 1996; Stockmarr 1999), but, as we saw above, has been criticized by many. It may misleadingly convey the opposite of element (1), i.e., that the DNA evidence is only weak evidence that Smith is the donor of the stain. Reporting the profile probability p and ignoring the search has been recommended as being appropriate and conservative (e.g., Donnelly and Friedman, 1999). Balding (2002) emphasizes that the expert only needs to provide p without any further guidance. Kaye (2009) endorses this. Thus, the literature seems to have reached consensus by now that p is the relevant number that should be reported to the court, and the gap between the frequentist and Bayesian reasoning is bridged (Storvik and Egeland, 2007). However, reporting p in isolation will only convey element (1). The possible counterbalancing effect of the prior odds (element (3)) is not something that lay persons will intuitively recognize (Meester and Sjerps, 2004). In fact, it has been shown that lay persons tend to fall prey to "base rate neglect", i.e., ignoring the prior odds, and so merely providing p may lead to serious misunderstandings (Wixted et al., 2019).

To provide guidance of element (3), reporting a table relating prior probabilities of donorship with posterior probabilities of donorship has been proposed (NRC 1996; Meester and Sjerps 2003, 2004).

Recently, Wixted et al. (2019) have expressed their doubts whether this works in practice, as they fear that a jury would not be able to come up with a sensible prior probability. Instead, they suggested a "solution" based on reporting the posterior odds that Smith is the offender (assuming that the crime stain was left by the offender). To be able to do this, they estimate the number of individuals who are plausible candidates for having their DNA profiles in the database. Meester and Slooten (2019) expose this as an unnecessary detour, and show that it can be avoided by estimating the probability that the stain donor C is in the database D directly from database search match rates. Sjerps (2019) notes that either approach to estimate the posterior odds requires questionable assumptions. She also warns that reporting posterior odds may be a recipe for new disaster and calls for more psychological research. Finally, Neumann and Ausdemore (2019) express several other objections against reporting posterior odds. Recently, Meester and Slooten (2020) propose to report posterior odds in some situations, but not in others.

Concerning element (5) above, Walsh et al. (2016), referring to personal communication with Buckleton and Curran) provide some "fables" to explain in court that excluding other suspects merely increases the probability that Smith is the donor.

Sjerps and Meester (2009), based on suggestions by experts of the Netherlands Forensic Institute (NFI), proposed another way of conveying essential guidance, that can be used to convey all elements (1)–(6). This is to report p, accompanied with a special textbox that explicitly conveys these elements. This has been adopted by the experts of the NFI and possibly by other European DNA laboratories (recommendation 25 and 26 with an example in Appendix 3 in ENFSI DNA working group (2017)).

In conclusion, from a mathematical point of view, the main issues in the "database search controversy" have been solved. However, effectively communicating database search matches remains a challenge. Currently, laboratories have different ways to report database matches, and there is no apparent consensus in the literature. Several suggestions have been made, but more research is needed to find the optimal way to communicate database search results to lay persons.

14.6 Dealing with "Selection Effects" More Generally

The database search controversy is an example of a general situation in which suspects or suspected items become the focus of interest as a result of some operation on a database that records certain features of a number of persons or items. We currently see an explosion of such possibilities, which are very interesting for both intelligence and subsequent prosecution. For example, in forensic DNA we perform such activities to identify a suspect in a specific case through "normal" DNA search in (inter)national DNA databases, but also in the context of crime linkage, population mass screens, familial searches, disaster victim identification (DVI) and missing persons identification. Very similar kinds of activities are performed on other kinds of traces, such as biometric features (e.g., fingerprints, voice, face), chemical features (glass; drug profiling), physical features (bullets and cartridges, footwear), and digital features (e.g., pixel defects patterns* in digital photos, and mobile phone co-locations). The operations used vary from simple match comparison to more advanced algorithms such as social network analysis.

There are also situations in which a large number of comparisons are made in order to select a suspect or suspected item without use of a database. For example, a surveillance tape with a robber is shown on internet or television, and all people watching compare the robber to the people they know. As another example, the police may search the house, car, office, caravan, et cetera, of several suspects for matching fibres, or for objects that could cause a certain type of wound. Robertson and Vignaux (1995, p. 95) present an example in which the police search the street for someone with bloodstains on his clothing. These ways of selecting persons or objects for comparison by an expert are very similar to a database search; many comparisons are made and the "matching" one (or the one with the highest score) becomes the focus of interest.

As in the DNA database search, people's intuitions may differ on how his selection procedure affects the evidential value of the result of the comparison. The following questions are examples where we can expect different answers:

1. Suppose that an observation is first used for identification and subsequently as evidence. Is this double use of observations allowed?
2. Hypotheses are formulated after looking at the observations. Are such data-dependent hypotheses allowed?

---

* The technical term is Photo Response Non-Uniformity (PRNU) patterns.

3. Suppose that features of a trace are compared to features of a large set of potential sources (this is often called a search or a screening). Subsequently one zooms in on the source that is the most alike and reports about the comparison of the features. Should we correct for the fact that multiple comparisons were made?

4. We know at the start that the selection procedure results in a comparison of the trace with a source that has similar features (a "look-alike"). Should we correct for such "selection bias"?

In current forensic practice, many experts feel uneasy because they recognize the difficulties and questions above but are unsure about the answer. The forensic literature does not provide much guidance (to the best of my knowledge).

From a mathematical point of view, Sjerps and Meester (2009) have argued that the solution lies in the same kind of Bayesian reasoning as applied in the database search. Also Evett et al. (1998) referred to Bayesian reasoning in the context of searching a database of shoe soles, to argue that the search does not weaken the evidential value but merely affects the prior odds. Especially the general weight-of-evidence-formula by Balding (2005, see also Balding and Steele, 2015) is quite useful for analyzing this kind of problems. For example, it can be used to show that there is no problem in "double use" of observations (see also Robertson and Vignaux, 1995 p. 95). It can also be used to show that data-dependent hypotheses and selection bias are corrected through prior odds, and that more comparisons just add information. However, as in the database search controversy, the challenge lies in the communication, and we need more research to find how we can report results in such a way that lay people will intuitively reason with it in a sound way.

14.7 Interesting Cases

Some cases where the database search aspect played a prominent role have been reported in the literature or elsewhere. Kaye (2009) describes how U.S. courts struggled with the database search controversy in the cases *People v. Jenkins* (2005) and *People v. Nelson* (2006): "Oddly—from the statistical standpoint—both opinions suggest that the jury might be given two match probabilities and left to decide for themselves which one to use in assessing the significance of the fact that the defendant's DNA profile matches a sample from the crime scene." He also describes a newspaper article on the *Puckett* case (Felch and Dolan, 2008), which stated that the jury was told that the random match probability was about 1 in 1.1 million—but not that Puckett was identified by a search in a database of 338,000 profiles, so that np was about 1 in 3.

In the Netherlands, the database controversy did not receive much attention either from the media or in the court. In a case concerning both rape and burglary (Dutch case 1, 2010), the defense argued that the forensic expert should have multiplied the random match probability with the number of people in the database, i.e., np. The defense used a paper in the Dutch legal literature (Sjerps et al., 2010) explaining the controversy. The appeals court did not accept the defense's reasoning, a decision which was upheld by the Supreme Court of the Netherlands, based on advice by the Prosecutor General who recognized that the paper also argued that p is the number relevant in court (not np), and how the court can combine p with the other evidence in the case. In a robbery case and subsequent DNA database search (Dutch case 2, 2011), a Dutch court explicitly stated that a DNA match with a very large evidential strength, in combination with a large prior probability based on other available evidence, and without any alternative explanation by the defense, led them to conclude that the suspect was guilty. A Dutch appeal court released a previously convicted person in a theft case (Dutch case 3, 2011), stating that based on only a partial DNA database match, and without any other evidence, there was not enough evidence to convict. In 2015, the Bayesian method for interpreting evidence (a combination of a DNA database match and physical end matching* of a piece and a roll of tape) in a violent theft case (Dutch case 4, 2015) was contested by the defence as "unreliable and controversial"; however the court dismissed this argument, stating that the method is a generally accepted scientific method and hence a reliable method. In 2018, a Dutch court convicted seven persons for part of a series of violent attacks on cash machines and for being part of a criminal organization. The evidence included a combination of DNA database matches and glass database matches. The forensic report on the glass matches provided guidance on the interpretation of the numerous comparisons of glass samples (Dutch case 5, 2018). The court used the glass evidence and explained for each attack how the glass evidence was combined with the other evidence.

There are several case reports on adventitious DNA matches. Goodwin et al. (2007) describe one of the first cases in the UK. In 1999, Raymond Easton's profile, with a random match probability of 1 in 37 million, was reported to match the DNA profile found at a burglary. At the time, Easton suffered from Parkinson's disease and could not walk well. Nevertheless, he was accused of the crime. Geddes (2011) describes the case of Peter Hamkin, an adventitious hit in the international database of the EU (Prüm-database). Hamkin was arrested in 2003 for the murder of an Italian woman. "The Italian police had requested a search of the UK DNA database and claimed he was a perfect match, and that he fitted witness descriptions of the murderer. After a 20-day ordeal, a second DNA test ruled Hamkin out and he was released without charge." Sjerps et al. (2010) describe two instances of adventitious matches with DNA mixture profiles.

Finally, there are many cases where persons were wrongly associated by a DNA database match with a crime because of an error. Kloosterman et al. (2014) provide an overview of such errors, their frequencies at NFI, and their consequences. They mention a case where a contamination of a sample in a rape case and subsequent database search with this sample led to the accusation of an innocent suspect in 2004. Besides contamination, swapping of crime samples and of reference samples is a relatively common error that can lead to false matches. Reference sample swaps may even lead to a wrong association with a series of crimes, for example in the case of a Dutch man who was convicted for three car burglaries (Trouw, 2014). Dodd et al. (2012) report on the UK case of Adam Scott, whose profile matched a sperm sample taken from a rape victim because of a serious contamination error.

Currently, forensic scientists, statisticians and some legal scholars are knowledgeable in (forensic) evidence interpretation and understand many important and sometimes subtle or counterintuitive issues. However, the decision makers using this evidence are policemen, prosecutors, defense lawyers, juries and judges, and persons working in intelligence agencies. Politicians decide on the development of (inter)national databases containing forensic information for intelligence purposes. To ensure that these lay decision makers use the evidence in an appropriate way, it is key that the scientific community pays more attention to the communication of their knowledge. Some of the cases mentioned above illustrate the rather urgent need for this.

---

* Comparing the physical fit between the ends of two pieces of tape (or the end of a roll of tape).

14.8 Conclusion

From a mathematical point of view, the database search controversy has been solved. Furthermore, the insights obtained in this scientific debate are useful to understand "selection effects" more generally. The challenge now lies in communicating database search results to lay persons in such a way that they intuitively evaluate and combine all relevant information appropriately. This requires more research.

References

Aitken, C.G.G. Population and samples. In *The Use of Statistics in Forensic Science*, C.G.G. Aitken and D.A. Stoney, editors. Ellis Horwood Lmt, Chichester, 1991.

Aitken, C.G.G., and Taroni, F. *Statistics and the Evaluation of Evidence for Forensic Scientists* (2nd ed.). John Wiley & Sons, Chichester, 2004.

Balding D.J. Errors and misunderstandings in the Second NRC Report. *Journal of Jurimetrics*, 37, 469–476, 1997.

Balding, D.J. The DNA database search controversy. *Biometrics*, 58, 241–244, 2002.

Balding, D.J. *Weight-of-Evidence for Forensic DNA Profiles*. Wiley & Sons, Chichester UK, 2005.

Balding, D.J., and Donnelly, P.J. Inference in Forensic Identification. *Journal of the Royal Statistical Society. Series A (Statistics in Society)*, 158(1), 21–53, 1995a.

Balding, D.J., and Donnelly, P. Inferring identity from DNA profile evidence. *Proceedings of the National Academy of Sciences of the United States of America*, 92, 11741–11745, 1995b. Medical Sciences.

Balding, D.J., and Donnelly, P.J. DNA profile evidence when the suspect is identified through a database search. *J. Forensic Sci.*, 41, 603–607, 1996.

Balding, D.J., and Steele, C.D. *Weight-of-evidence for forensic DNA profiles* (2nd ed.). John Wiley & Sons, Chichester UK, 2015.

Buckleton, B., Triggs, C.M., and Walsh, S.J., editors. *Forensic DNA Evidence Interpretation*. CRC Press, Boca Raton, 2005.

Cook, R., Evett, I.W., Jackson, G., Jones, P.J., and Lambert, J.A. A hierarchy of propositions: Deciding which level to address in casework. *Science and Justice*, 38, 231–239, 1998.

Dawid, A.P. The island problem: Coherent use of identification evidence. In *Aspects of Uncertainty: A Tribute to D. V. Lindley*, P. R. Freeman, and A. F. M. Smith, editors, ch. 11. Wiley, Chichester, 1994.

Dawid, A.P. Comment on Stockmarr's "Likelihood ratios for evaluating DNA evidence when the suspect is found through a database search". *Biometrics*, 976–980, 2001.

Dawid, A.P. and Mortera, J. Coherent analysis of forensic identification evidence. *Journal of the Royal Statistical Society, Series B*, 58, 425–443, 1996.

Dodd, V., Bowcott, O., Laville, S., and Malik, S. Forensics firm investigated over DNA blunder in rape case. *The Guardian* 9 March, 2012.

Donnelly, P. and Friedman, R.D. DNA database searches and the legal consumption of scientific evidence. *Michigan Law Review*, 97, 931–984, 1999.

Dutch case 1: 2010 available in Dutch on the internet on, http://deeplink.rechtspraak.nl/uitspraak?id=ECLI:NL:PHR:2010:BM9128

Dutch case 2: 2011. available in Dutch on the internet on, http://deeplink.rechtspraak.nl/uitspraak?id=ECLI:NL:RBALK:2011:BO9753

Dutch case 3: 2011. available in Dutch on the internet on, http://deeplink.rechtspraak.nl/uitspraak?id=ECLI:NL:GHLEE:2011:BP4646

Dutch case 4: 2015. available in Dutch on the internet on, http://deeplink.rechtspraak.nl/uitspraak?id=ECLI:NL:RBLIM:2015:1588

Dutch case 5: 2018. available in Dutch on the internet on, http://deeplink.rechtspraak.nl/uitspraak?id=ECLI:NL:RBMNE:2018:6343

ENFSI DNA working group. DNA Databasemanagement-Review and recommendations, 2017. download from website European Network of Forensic Science Institutes (ENFSI): http://enfsi.eu/wp-content/uploads/2017/09/DNA-databasemanagement-review-and-recommendatations-april-2017.pdf

EU Prüm treaty. Council Decision 2008/615/JHA of 23 June 2008 on the stepping up of cross-border cooperation, particularly in combating terrorism and cross-border crime, 2008. available at http://data.europa.eu/eli/dec/2008/615/oj

Evett, I.W., Lambert, J.A., and Buckleton, J.S. A Bayesian approach to interpreting footwear in forensic casework. *Science & Justice*, **38**, 241–247, 1998.

Evett, I.W., and Weir, B.S. *Interpreting DNA evidence*. Sinauer, Sunderland MA, 1998.

Felch, J., and Dolan, M. "DNA Matches Aren't Always a Lock." *Los Angeles Times*: *May* 3, 2008. available at http://www.latimes.com/news/local/la-me-dna4-2008may04,0,6156934,full.story

Fung, W.K., and Hu, Y-Q. *Statistical DNA Forensics- Theory, Methods and Computation*. Wiley & Sons, Chichester UK, 2008.

Geddes, L. DNA super-network increases risk of mix-ups. *New Scientist*, 2828, 2011.

Goodwin, W., Linacre, A., and Hadi, S. *An Introduction to Forensic Genetics*. John Wiley & Sons, Chichester UK, 2007.

Kaye, D.H. Rounding up the usual suspects: A legal and logical analysis of DNA trawling cases. *North Carolina Law Review*, **87**, 425–503, 2009.

Kloosterman, A., Sjerps, M., and Quak, A. Error rates in forensic DNA analysis: Definition, numbers, impact and communication, *Forensic Science International: Genetics*, **12**, 77–85, 2014.

Meester, R., and Sjerps, M. The evidential value in the DNA database search controversy and the two-stain problem. *Biometrics*, **59**, 727–732, 2003.

Meester, R., and Sjerps, M. Why the effect of prior odds should accompany the likelihood ratio when reporting DNA evidence. *Law, Probability and Risk*, **3**, 51–62, 2004. (see also the discussion and response pp. 63–86)

Meester, R., and Slooten, K. Ne bis in idem - a commentary on "Calculating the Posterior Odds from a Single-Match DNA Database Search". *Law, Probability & Risk*, **18**, 35–38, 2019.

Meester, R., and Slooten, K. DNA database matches: a p versus np problem. Forensic Science International: Genetics, **46**, 102229, 2020.

National DNA database. National DNA Database Strategy Board Annual Report 2017/18, 2019. available from (visited at April 27 2019) https://assets.publishing.service.gov.uk/government/uploads/system/uploads/attachment_data/file/778065/National_DNA_Database_anuual_report_2017-18_print.pdf

National Research Council. *DNA technology in Forensic Science*. National Academy Press, Washington, D.C., 1992.

National Research Council. *The Evaluation of Forensic DNA Evidence*. National Academy Press, Washington, D.C., 1996.

Neumann, C., and Ausdemore, M. Communicating forensic evidence: Is it appropriate to report posterior beliefs when DNA evidence is obtained through a database search? *Law, Probability, and Risk*, **18**, 25–34, 2019.

NLDNAD. Jaarverslag 2017 Nederlandse DNA databank, published 11 Dec 2018, 2017. available at https://dnadatabank.forensischinstituut.nl/binaries/nederlandse-dna-databank/documenten/publicaties/2018/11/01/jaarverslag-2017/DNADB-jaarverslag-2017.pdf

People v. Jenkins. 2005. 887 A.2d 1013 (D.C. 2005).

People v. Nelson. 2006. 78 Cal.Rptr.3d 69 (Cal. 2008), aff'g, 48 Cal.Rptr.3d 399 (Ct. App. 2006).

Porter, H. The rising odds of false DNA matches. *The Guardian* May 25 2009.

Robertson, B., and Vignaux, G.A. *Interpreting Evidence-Evaluating Forensic Science in the Courtroom*. Wiley & Sons, Chichester UK, 1995.

Sjerps, M. Reporting DNA database matches: We need more research. *Law, Probability & Risk*, **18**, 39–41, 2019.

Sjerps, M., Kloosterman, A., and Van der Beek, K. De interpretatie van een DNA-databankmatch. *Delikt en Delinkwent*, **40**, 138–155, 2010 (in Dutch).

Sjerps, M., and Meester, R. Selection effects and database screening in forensic science. *Forensic Science International*, **192**, 56–61, 2009.

Stockmarr, A. Likelihood ratios for evaluating DNA evidence when the suspect is found through a database search. *Biometrics*, **55**, 671–677, 1999.

Storvik, G., and Egeland, T. The DNA Database Search Controversy Revisited: Bridging the Bayesian–Frequentist Gap. *Biometrics* **63**, 922, 2007.

Trouw [newspaper name]. Onterechte veroordeling na verwisseling DNA, Nov 4, 2014 (in Dutch).

Walsh, S.J., Bright, J-A., and Buckleton, J.S. "DNA intelligence databases", chapter 13. In *Forensic DNA evidence interpretation* (2nd ed.), J.S. Buckleton , J-A. Bright, and D. Taylor, editors. CRC Press, Boca Raton, 2016.

Walsh, S.J., and Buckleton J.S. DNA intelligence databases. In *Forensic DNA Evidence Interpretation*, J.S. Buckleton, C.M. Triggs, and S.J. Walsh, editors. CRC Press, Boca Raton, 2005.

Weir, B.S. Matching and Partially-Matching DNA Profiles. *Journal of forensic sciences*, **49**, 1009–1014, 2004.

Weir, B.S. The rarity of DNA profiles, *The Annals of Applied Statistics*, **1**(2), 358–370, 2007.

Wixted, J.T., Christenfeld, N.J.S., and Rouder, J.N. Calculating the posterior odds from a single-match DNA database search. *Law, Probability & Risk*, **18**, 1–23, 2019.

15

Comparing Handwriting in Questioned Documents

Alan Julian Izenman

CONTENTS

15.1 Introduction 341
    15.1.1 Outline of This Chapter 342
15.2 Historical Background 343
    15.2.1 Roman Law 343
    15.2.2 English Law 344
    15.2.3 United States Law 348
15.3 Challenges in Studying Questioned Documents 350
15.4 The Art of Conducting a Handwriting Examination 351
    15.4.1 Principles of Handwriting Analysis 352
    15.4.2 The ACE-V Process of Comparing Handwriting 353
    15.4.3 A Probability Scale for Expressing Conclusions 354
15.5 Statistical Approaches to Handwriting Analysis 355
    15.5.1 Comparing Howland Will Signatures 355
    15.5.2 A Bayesian Approach to Comparing Signatures 356
    15.5.3 A Dynamic Model for Handwriting 357
15.6 Proficiency Studies and Error Rates 359
15.7 Discussion 360
Acknowledgments 361
References 361
Cases and Statutes 363

15.1 Introduction

Handwriting and the analysis of questioned documents have had a contentious legal history throughout the centuries. Comparing handwriting in a genuinely written document with handwriting in a possible forged document has been treated in various ways in the courts of law: sometimes, such handwriting comparisons have been banned, while at other times, they have been allowed. In fact, as the knowledge and practice of handwriting has proliferated, so has the incidence of forgery.

What are questioned documents? They are communications between individuals that have been challenged, in part or in whole, as to their authenticity. Almost any type of document may be disputed. The word "document" has a very broad meaning that could involve sheets of paper with handwriting on them, or any marks, signs, or symbols intended to mean something to someone. Examples of such documents are checks, wills, contracts, ransom notes, handwritten letters, and perhaps even gas station receipts or concert tickets.

More often than not, disputed documents can help link a suspect to a crime, including financial extortion, identity theft, medical malpractice, insurance fraud, elder abuse, and contract disputes in business. Forensic document examiners (who are usually referred to as FDEs) prepare reports in which disputed documents are compared to documents of known origin to determine whether the disputed document is a forgery. Lawyers may then call a forensic document examiner as an expert witness in court so that information in the report may be used to prosecute a case of possible forgery.

Issues that arise when dealing with questioned documents include determining the authorship of an anonymous letter (possibly a threatening letter), the identity of the individual who signed the reverse side of a check endorsing the amount of the check, the source of a computer printout, reconstructing damaged or altered documents, and investigating photocopy manipulation.

If the disputed document involves a handwritten signature, there are two types of forgeries (Gaborini et al., 2017). If the forger knows the name of the victim, but signs a document using the forger's own handwriting style, this is called "blind" forgery. If, however, the forger also has genuine samples of the victim's handwriting and signature, and signs the document by trying to mimic the victim's signature, this is "skilled" forgery.

Several famous court cases have involved questioned documents. They included the 1868 Howland will trial (involving a disputed will for a $2 million estate, where the signatures of Sylvia Ann Howland of Bedford, Massachusetts were found to have been traced from original signatures); the 1934 Charles Lindbergh baby kidnapping trial (involving alleged handwriting and spelling similarities of Richard Hauptmann to those found in ransom notes), the 1948 treason trial of Alger Hiss (involving documents copied and typed on his typewriter and turned over to the Soviets), and the 1978 Mormon will trial (involving Melvin Dummar's claim to $156 million of Howard Hughes's estate based upon a disputed handwritten will).

Other famous questioned documents included the 1971 forged autobiography of Howard Hughes (written by Clifford Irving and purchased by *Life* magazine to run excerpts), the 1983 Hitler diaries (involving a massive number of forgeries of diaries, paintings, and other memorabilia), and the 2004 forged memos ostensibly written in the 1970s by Lt. Col. Jerry B. Killian (involving six forged memoranda that were critical of President George W. Bush's service in the Air National Guard during 1972–73; four of these memos were presented uncritically as authentic by Dan Rather in the CBS TV program *60 Minutes II*).

15.1.1 Outline of This Chapter

In Section 15.2, we provide some historical background into identification and comparison of handwriting specimens. We describe decisions that led to the present acceptance of handwriting comparisons in the law. We focus on how Roman, English, and United States legal systems treat the comparison of questioned documents, noting how each legal system learned from the other systems. In Section 15.3, we discuss a number of unusual challenges in studying questioned documents. In Section 15.4, we describe how a handwriting examination is conducted. In Section 15.5, we discuss three different statistical approaches to analyzing handwriting comparisons. In Section 15.6, we present the results of published experiments that attempt to determine error rates for handwriting comparisons by FDEs and non-FDEs. In Section 15.7, we provide some concluding remarks.

15.2 Historical Background

Forging a handwritten document to deceive people is not new. Apparently, handwriting forgery has been found everywhere handwriting was an important medium of communication. Some of the historical discussion on handwriting in this section is based on Clayton (2013) and Trubek (2016), both of whom describe the history of writing and of the written word in great detail. Historical details regarding questioned documents are based on Osborn (1910, 1929), Carlson (1920), Baker (1971), Costain (1977), Hilton (1979, 1982), Norwitch and Seiden (2003), and Koppenhaver (2007).

The whole idea of examining a handwritten document for possible forgery makes several assumptions: that paper was readily available and that people were literate and that they knew how to write. Breakthroughs in writing occurred at two separate times: when vellum books took over for papyrus scrolls and when Johannes Gutenberg invented the printing press in the early-1450s using movable type.

15.2.1 Roman Law

The first and second centuries CE saw the population of Rome grow to over one million residents. As the Roman empire expanded, so did the financial, economic, and administrative sectors of the empire. This expansion led to an increased need for more systematic document handling and record keeping. To keep track of these documents, the Emperor Augustus created a new imperial bureaucracy staffed by slaves and freedmen who reported directly to him. These slave administrators kept all written records involving legal and financial transactions; they managed businesses and tended to the property of their masters, and they oversaw the writing, production, and copying of literary works.

Augustus died in 14 CE and was succeeded by Tiberius, Caligula, and then Claudius. During the mid-first century CE, the Emperor Claudius started replacing his slave administrators by salaried staff. The citizenry interacted with the empire's bureaucracy mainly through the local town councils, which maintained local written records. According to Roman law, documents such as land agreements and wills were placed on the council's rolls, which then formed the official record of those documents.

Writing changed as new cursive methods, which involved the rounding of certain letters, were used to speed up the writing process. Books started to appear with covers, which contained a number of parchment (or vellum) pages made from dried animal skins (usually calf or sheep skins) stitched together at the spine. The great advantage of such books was that parchment pages could be written on both sides, which replaced the one-sided scrolls from previous Roman, Greek, and Egyptian periods. Scrolls were very awkward to read, and the compact book format proved to be simple by comparison. Rather than having to unroll and roll the scrolls with two hands, one could open a book to the desired page with a single hand. This new writing style spread over the Roman empire as demand for these books increased.

As far back as 80 BCE, a Roman law had prohibited the falsification of documents that dealt with the inheritance of land. This seems to be the first instance in Roman law that acknowledged that forgery was taking place. In the 3rd century CE, the Romans established protocols for determining instances of forged documents. Two of the biggest culprits were Titus (described as "the most skillful forger of his time") and Anthony (who was accused of fraud by Cicero), both of whom became quite wealthy by forging documents related to property rights (Baker, 1971).

The investigation and procedures for the detection of forged handwriting in a disputed document were permitted under Roman Law. In 539 CE, the Code of the Emperor Justinian established additional guidelines for the comparison of handwritten documents in Roman courts. Chapter II of TITLE IV of the 49TH NEW CONSTITUTION reads as follows:

> COMPARISON OF HANDWRITING SHALL ONLY BE MADE IN THE CASE OF PUBLIC DOCU-
> MENTS, AND IN THE CASE OF PRIVATE INSTRUMENTS WHERE THE ADVERSE PARTY CAN
> USE THEM TO HIS OWN ADVANTAGE.
>
> We have decided that the following addition should be made. We have, sometimes since, drawn up a constitution forbidding the comparison of handwriting in the case of private instruments, and only authorizing this to be done with public documents; but experience has convinced Us that this law should be amended, and, as this is the case, We are going to proceed in accordance with the custom observed by litigants.
>
> For as We entertain hatred for the crime of forgery, We order that experts charged with the comparison of the handwriting of public documents shall be sworn before any private instruments are placed in their hands for this purpose. Wherefore this law, as well as the present modification of the same, shall remain in full force, and experts aforesaid shall by all means be sworn.

Under these guidelines, the judge would appoint individuals, who were skilled at writing, as examiners of the disputed handwritten documents. They would testify as to the document's authenticity based solely upon the visual similarity (or not) of the questioned handwriting to examples of known, genuine handwriting of the person concerned. As one can imagine, this method was probably not very accurate.

The Justinian procedures for investigating disputed documents were accepted and practiced throughout Europe.

15.2.2 English Law

English law followed Roman law in that it used charters to confirm land rights, privileges, and tax orders. The Domesday Book, compiled during 1085–86, brought about a huge increase in administrative documents in its detailed survey of the state of the realm. This and later surveys encouraged writing at all levels of society. By the 13th century, around 8 million charters that defined property rights and obligations had been produced in England.

The founding of universities (e.g., Oxford University in 1167) changed the appeal of the writings to the general populus. The universities introduced the study of mathematics, music, astronomy, philosophy, and theology to an academic curriculum, and readers were encouraged to offer criticisms and opinions on the texts other than those set out in books. These advances in scholarly pursuits led, in turn, to smaller handwriting due to using a quill pen rather than a reed, to the use of a smooth and thin vellum surface rather than a slightly grainy papyrus surface, and, as a result, to scaled-down book sizes.

By the middle of the 15th century, there were hundreds of thousands of manuscript books written before Gutenberg's printing press was invented. By the end of the century, forty thousand printed books were added to those manuscript books already in circulation. Some of the difficulties involved in reading such books were that there were numerous styles of writing (typically, class-based styles), not all of them legible to all readers. Furthermore, there was no standardized spelling or punctuation in England until the 18th century, and words were often written as the writer sounded them out. Even Shakespeare used different spellings for his name.

For many years, English common law seemed to pay little attention to acts of forgery. This was aided by an extensive period of general illiteracy in England, which encouraged a "mystique" connected with the written word. Forgery was first treated as a crime in England in 1562, when a law prohibited the forgery of publicly recorded, officially sealed documents relating to land titles.

For many generations and in many trials, including those of libel and disputed wills, the concept of "comparison of hands" (i.e., comparing two writings with each other with the goal of learning whether they were both written by the same person) by witnesses who had neither seen the defendant write the disputed documents nor corresponded with him had been rejected in the English courts as highly improper and extremely dangerous.

The issue of "comparison of hands" appeared at the 1683 "Rye House Plot" trial of Colonel Algeron Sidney, who was an English politician, elected to the Long Parliament as Member of Parliament for Cardiff, Wales. Sidney was arrested in London and charged with participating in a conspiracy to assassinate King Charles II and his brother James, Duke of York. Sidney was arraigned for high treason before the King's Bench, which was presided over by the Lord Chief Justice, Sir George Jeffries. Jeffries was described by the biographer of Sidney, Van Santvoord (1851, Chapter VIII), as "infamous," "loathsome," "utterly unfit and unworthy," "a violent, shameless, cruel, vindictive man, a renegade and traitor to every principle," "a debauchee and a common drunkard, a demon incarnate," and "a corrupt and wicked judge." Jeffries's friendship with the Duke of York led to him being appointed Lord Chancellor of England. Jeffries repeatedly refused to provide Sidney with a copy of the indictment and packed the jury with men of "ruined character and fortune." Moreover, under English law at the time, defendants were not allowed to have the benefit of counsel.*

At Sidney's trial, the prosecution introduced certain papers found in his closet at home. The papers formed an unfinished manuscript entitled *Discourses Concerning Government*,[†] which was entered as evidence (treated by the court as equivalent to a "second witness") against Sidney. Three witnesses claimed that they were familiar with Sidney's handwriting. Two said that although they had never seen him write, they had seen his endorsements on several bills of exchange or had seen notes with Sidney's name endorsed on them. The third witness had actually seen him write endorsements on several bills (which were never produced as evidence at trial) and testified that the handwriting in the book was like that of Sidney. Sidney actually complained after the trial that "similitude of hands can be no evidence," and that "there is nothing but the similitude of hands offered for proof." Following a bullying of the jury by Jeffries to return a guilty verdict, Sidney was convicted of high treason and executed in 1683. In 1688, an Act of Parliament repealed and reversed Sidney's conviction and declared it null and void because the disputed handwriting, which the jury was directed to believe was his when compared with his other writings, had not been proved in his trial to have been written by Sidney. In his 1689 commentary on the trial, Sir John Hawles noted that

> And as this indictment was an original in one part, ... the evidence on it was an original in another part, which was proving the book produced to be Col. Sidney's writing, because the hand was like what some of the witnesses had seen him write; an evidence never permitted in a criminal matter before.

---

* This and other "severe and unjust" rules of the common law were abolished after the Glorious Revolution of 1688 that overthrew James II and placed William and Mary jointly on the throne of England.

† This book, apparently written twenty years prior to the trial, was published in 1698, and was later considered to be a primary source for the American Revolution.

In 1726, the forgery laws were expanded to cover the false endorsement of an unsealed private document, which was henceforth deemed a capital crime punishable by death. This is the first time that English law recognized individual differences in handwriting. Geoffrey Gilbert, a jurist, made the following claim (Clayton, 2013, p. 200) regarding legal evidence:

> Men are distinguished by their handwriting as much as by their faces, for it is seldom that the shape of their letters agree any more than the shape of their bodies.

Gilbert went on to assert that forged writing is different from the real thing, implying that this fact should be recognized and accepted by the law courts. However, little changed regarding the comparison of handwriting in the courts of law.

Testimony consisting of opinions of handwriting experts had been allowed in certain cases in English civil courts since the late 18th century. There was, however, the curious case of Lord Kenyon, an English judge, who flip-flopped on whether he would allow comparison of handwriting at trials of civil cases. In 1792, 1795, and 1797 cases, he allowed it, but in 1793, 1799, and 1801 cases, he did not. The admissibility of comparison of hands was allowed when no other evidence existed of proving the ownership of the handwriting in a questioned document. So, there was never a total ban on expert opinion of handwriting in English civil courts.

The late 18th century saw a series of important and world-shaking events: the Industrial Revolution (which led to division of labor, economies of scale, mass production, steam power, iron-making, and the invention of textile machinery), the 1776 American Revolution, and the 1789 French Revolution. But whatever influence the Industrial Revolution had on English society, it had little impact on writing and reading. In fact, literacy rates dropped because poor children, who were working as unpaid apprentices in coal-mining, iron-smelting, and cotton-spinning factories, were too tired to learn to write in evening schools.* An 1807 proposal to make elementary education available also to the poor (the *Parochial Schools Bill*) was defeated in the House of Lords, which introduced a class-based system into education.† The lack of an educated citizenry during the 19th century probably contributed to the origins of the common-law rule that handwriting samples that were not part of the case were excluded and, hence, could not be used for handwriting comparison by witnesses.

The many reasons quoted in various cases for the common-law rule of exclusion were ignorance of jurors, and their inability to make intelligent comparisons; danger of unfairness and fraud in the selection of specimens, with no sufficient opportunity for the opposing party to investigate and expose; collateral issues as to the genuineness of the specimens presented; the writings offered in evidence as specimens might be manufactured for the occasion; the opponent might be surprised by the introduction of documents otherwise foreign to the case; and the handwriting of a person might be changed by age, health, state of mind, position, haste, penmanship, and writing materials (*University of*

---

* There was also a movement to restrict poor children from learning to write so that they would not rise above their station in life.
† It was not until the late-19th century, when a series of three *Elementary Education Acts* were passed by Parliament (the 1870 Act, which made elementary education universally available, at least in principle, for all children aged 5–13; the 1880 Act, in which education up to the age of ten was made compulsory; and the 1891 Act, in which it became freely available), that England had a national system of elementary education (Gillard, 2011).

Illinois v. Spalding, 1901; Masington, 1960, p. 492 n. 7). Harris (1892, Chapter XI, No. 479, p. 332) argued that

> [T]o exclude the comparison of the two papers or signatures, and then permit the witness to compare one paper with his recollection of another paper (previously seen by him), is a glaring absurdity in the very nature of things. And to exclude such comparison from a jury is equally absurd, unless it be upon the untenable position taken by the early English cases that the jurors were illiterate. Even then the witness may be equally illiterate. What then?

In *Doe on the demise of (Mudd v. Suckermore,* 1836–1837), regarding the trial of a disputed will, the trial judges were evenly divided as to whether handwriting comparisons should be admitted into evidence, and so, according to protocol, admittance was denied. One of the judges, Sir John Taylor Coleridge, wrote that (Coleridge, 1838):

> The test of genuineness ought to be the resemblance, not to the formation of the letters in some other specimen or specimens, but to the general character of writing, which is impressed on it as the involuntary and unconscious result of constitution, habit, or other permanent cause, and is therefore itself permanent. And we best acquire a knowledge of this character by seeing the individual write at times when his manner of writing is not in question, or by engaging with him in correspondence; either supposition giving reason to believe that he writes at the time, not constrainedly, but in his natural manner.

Judge Coleridge's comments were controversial, with many judges and legal scholars disagreeing with him (see Osborn, 1910, pp. 33–35). In fact, during that time, English law took the position that handwriting could not be identified and that "expert" witnesses were not to be trusted.

At the start of the 19th century, writing implements had changed from quill pens, which were so unreliable that they were viewed as disposable, to steel nibs, which were brought from France. Of particular interest to English society during the mid-19th century were signatures and writings of famous men and women, and the collection of large numbers of autographs (including personal papers) was becoming very popular as a new hobby. By the end of the 19th century, more people had become literate, and writing styles were less class-based. The general increase in literacy in England was now changing general attitudes regarding legal situations, and so a fresh look at the common-law rule was inevitable.

As noted earlier, *Doe on the demise of Mudd v. Suckermore* held that handwriting comparison was inadmissible at trial. This state of affairs lasted until 1854, when Parliament passed the *Common Law Procedure Act*. Section XXVII of the Act provided:

> Comparison of a disputed writing with any writing proved to the satisfaction of the judge to be genuine, shall be permitted to be made by witnesses; and such writings and the evidence of witnesses respecting the same, may be submitted to the court and jury as evidence of the genuineness, or otherwise of the writing in dispute.

This Act, which abolished the common-law rule, applied only to civil proceedings. Note that this law did not require that the genuine document be relevant to the issue in the case and could well be completely irrelevant except for the purpose of comparison. In 1865, Parliament passed the *Criminal Procedure Act*, of which Section 15.8 extended the rule further to criminal cases.

15.2.3 United States Law

The United States adopted much of England's system of jurisprudence, including the common-law rule that a comparison of handwritings could not be used to prove that certain documents were forgeries. Proof of a forgery had to be achieved by using other types of evidence. This practice was followed in the federal courts and some state courts even though England had abolished that practice in 1854.

The field of questioned documents is due to Albert S. Osborn, who created the basic principles in Osborn (1910) that are still used today. As Osborn states (in Chapter III), "The average man is perfectly astounded to learn that a writing cannot be proved by going out and getting undoubtedly genuine writing to compare it with."

The English common-law rule regarding the ban on handwriting comparisons was deeply unsatisfying. Using the 1854 Act of Parliament in England as motivation, a bill was passed in 1894 by the House of Representatives that would change by statute the rule of evidence so that handwriting comparisons would be authorized by courts and juries in disputed document cases. Yet it took almost twenty years before a law was passed that changed the way that federal courts dealt with handwriting comparisons. Meanwhile, individual states moved away from the English common-law rule on handwriting comparison, either through statute (as in England) or by judicial decision.*

In 1913, the Attorney General George W. Wickersham wrote to the Judiciary Committee of the United States Senate that the English common-law rule had resulted in

> many miscarriages of justice, particularly in cases against persons charged with sending obscene matter through the mails ... [A]doption of a new law on handwriting comparison would remove the obstacles which now stand in the way of a successful presecution of many Federal criminal cases, and would in no wise (sic) prejudice any right of the accused.

He argued that the U.S. Congress should pass a bill changing the law. Many federal judges had urged the passage of such legislation, and the Judiciary Committee was unanimous in its support. Accordingly, Congress passed *The General Comparison Statute* (1913), which provided:

> [I]n any proceeding before a court or judicial officer of the United States where the genuineness of the handwriting of any person may be involved, any admitted or proved handwriting of any person shall be competent evidence as a basis for comparison by witnesses, or by the jury, court, or officer conducting such proceeding, to prove or disprove such genuineness.

Thus, the new law allowed exemplar handwriting to be admitted as evidence, even if that handwriting had no bearing on the case other than the need for it to be used as a standard of comparison. This change in court procedure amounted to a revolution (Osborn, 1933).

Handwriting experts were now being officially referred to as "questioned document examiners," and handwriting became the first example of forensics to be declared admissible in a court of law. Prior to that time, there were few handwriting experts in the United States because there was very little need for them. As the states broke away from the English rule, however, there was greater emphasis on questioned documents, and so a document examiner became a valued professional (Carlson, 1920).

There followed a series of legal decisions and rulings that had a direct impact upon the admissibility of expert testimony on the courts. *Frye v. United States* (1923) required that

---

* See Risinger et al. (1986, Appendix 3) for a list of dates and cases for each state. See also Osborn (1910, p. 17, n. 1), whose list of states and dates differs from Risinger et al.

scientific testimony have gained general acceptance in the particular discipline to which it belongs. Forensic handwriting examination was admitted under the *Frye* rule because of its longstanding acceptance in U.S. courts. In fact, it was used to great success in the 1925 Lindbergh kidnapping trial when "forensic document examiners" (FDEs), as they were now called, including Albert S. Osborn, testified on the similarities between the handwriting, misspellings, and other language issues of Richard Hauptmann and those on the 14 ransom notes. By 1933, questioned document examiners were in great demand and were being hired by the FBI and state crime laboratories.

The Federal Rules of Evidence (FRE), signed into law in 1975, provided that

> If scientific, technical, or other specialized knowledge will assist the trier of fact to understand the evidence or to determine a fact in issue, a witness qualified as an expert by knowledge, skill, experience, training, or education, may testify thereto in the form of an opinion or otherwise (FRE 702, 1975).

Various courts interpreted this rule as overturning the requirement of general acceptance, but most continued to apply the *Frye* standard. In 1993, the Supreme Court held in *Daubert v. Merrell Dow Pharmaceuticals* that the *Frye* test had not survived. Instead, the Court determined that federal judges could admit scientific testimony only if the "reasoning or methodology" were "scientifically valid." *Daubert* set out four "factors," in addition to general acceptance to help trial judges determine whether scientific testimony was reliable:

1. Could the theory or technique be tested, and has it been tested?
2. Are there standards that control the technique's operation?
3. Was the theory or technique subjected to peer review and publication?
4. Is there a known or potential error rate?

Under *Daubert*, trial judges serve as "gatekeepers" to make sure that unreliable scientific testimony is excluded.

In *United States v. Starzecpyzel* (1995), a federal district court ruled that questioned document examination did not meet the *Daubert* criteria, but nevertheless was still admissible as a "technical skill" rather than as scientific knowledge. This ruling shocked the forensic document community. Its impact led to an increase in peer-reviewed empirical studies to prove the individuality of handwriting.

Extensions of *Daubert* were given by the Supreme Court in *General Electric v. Joiner* (1997), which allowed trial courts to reject opinion testimony based upon flagrant error (e.g., unsupported facts), and *Kumho Tire Co. v. Carmichael* (1999), which extended *Daubert*'s gatekeeping function to all expert opinion testimony (not just to scientific testimony, which was the nature of the expertise at issue in *Daubert*) as in FRE 702. In *Kumho*, the Court noted that the *Daubert* factors "may or may not be pertinent"; it will all depend upon "the nature of the issue, the expert's particular expertise, and the subject of his testimony." The *Daubert*, *Joiner*, and *Kumho Tire* cases are generally referred to as the "*Daubert* trilogy." In 2000, FRE 702 was amended to reflect the *Daubert* trilogy.

There were several cases, however, in which the courts cited *Starzepyzel's* holding that even though handwriting analysis did not satisfy the *Daubert-Kumho* criteria, it was admissible under FRE 702. Courts also found that the prosecution failed to show peer review, publication, error rates, and general acceptance of the techniques (Page et al., 2011). Furthermore, there is no fixed number of characteristics that an FDE has to find in the handwriting comparison to declare a positive match. As a result, the judges either limited the expert to pointing out similarities and differences or excluded the testimony completely (*United States v. McVeigh* (1997), *United States v. Hines* (1999), *United States v. Rutherford* (2000), and *United States v. Oskowitz* (2003)). For commentary on these cases and others, see Risinger (2007).

These cases proved to be a wake-up call to document examiners, who, because of all the *Daubert* challenges, suddenly realized that they had to standardize their work and practices. Although judges gave greater scrutiny to testimony by experts, judges tended to be reluctant to reject established areas of forensic science, such as handwriting comparison.

Enter the FBI, which, in 1997, created an organization composed of members from various laboratories called the Technical Working Group for Document Examination (TWGDOC) to expedite the drafting of questioned document standards. This group was later renamed SWGDOC, with the word 'scientific' replacing 'technical.' SWGDOC and ASTM (the American Society for Testing and Materials) have published many standards for document examiners (see, e.g., SWGDOC E2290, 2016), and several federal appellate courts have declared the *Daubert* factors satisfied by FDEs.

The American Board of Forensic Document Examiners (ABFDE, founded in 1977) recognized that the profession had to have a well-defined strategy for dealing with *Daubert* challenges. In particular, they were motivated by the cases *United States v. Fujii* (2000) and *United States v. Saelee* (2001) in which FDE testimony was completely excluded. By 2002, the ABFDE had organized the *Daubert* Group of three FDEs to track all state and federal *Daubert* decisions and provide suitable responses. Some states (e.g., Florida) also set up local *Daubert* groups.

The *Daubert* rules as they apply to handwriting expert testimony were examined in detail in the Ninth Circuit case *United States v. Prime* (2005), in which the introduction of handwriting expert testimony was affirmed. The judges emphasized that *Daubert* was not intended to apply to well-established forensic investigation procedures, such as questioned document examination, which had long been accepted in U.S. courts.

Recently, the Advisory Committee on Rules of Evidence became concerned that courts were applying FRE 702 incorrectly, especially on challenges to forensic expert testimony, such as handwriting identification. A symposium was held in October 2017 to consider possible changes to FRE 702.* Presentations were made by scientists, jurists, academics, and practitioners. A suggestion was made either to add an extra section to FRE 702 or to introduce a new rule to govern forensic expert testimony. The committee has not moved forward with revised language for forensic-science evidence.

15.3 Challenges in Studying Questioned Documents

FDEs often find certain types of questioned documents challenging, some of which can be illustrated by the following examples.

The study of questioned documents usually involves a comparison of signatures or handwriting records when one of the writers is known and another individual is suspected of being the other writer. A different type of case involves identifying the writer of an anonymous letter, especially when the item is typed rather than handwritten. Sometimes, the FDE's job is simplified if a clue to the writer's identity becomes available. For example,

---

* See www.uscourts.gov/sites/default/files/a3_0.pdf for the symposium agenda.

the person designated by the news media as the "Unabomber," who carried out a 20-year letter-bombing campaign directed at universities and airlines starting in 1978, and who was also author of an anonymous manifesto (a 35,000 word essay typed on a 1930s-era manual typewriter), was identified only in 1996 as Theodore ("Ted") Kaczynski when his brother recognized the message and the writing style of the manifesto. After his brother provided the FBI with a collection of Ted Kaczynski's typed letters, a linguistic expert noted instances of idiosyncratic spelling and hyphenation in the various writings and agreed that Kaczynski's prior writings almost certainly matched that of the manifesto. Without the brother's help, it would have been difficult to identify Kaczynski as the Unabomber.

In situations in which the value of an ancient document, such as a will or land deed, depends upon its date of origin, fraud can often be proved using technical information. For example, fraud can be established if it is determined that the paper on which the document was written was manufactured several years after the date listed on the document. An example of such a mistake occurred with the 62 forged Hitler diaries (covering 1932–1945), where the forger, Konrad Kujau, was found to have used modern stationary to create letterheads. Furthermore, West Germany's Federal Archives announced that chemical analyses showed that the paper, ink, and glue that held the book bindings together were definitely manufactured after the end of World War II and Hitler's death! In a similar vein, the Killian memos, ostensibly written during 1972 and 1973, were shown in 2004, when they were brought to the public's attention, to have been written on a computer using a word processor with proportional-spaced font, an anachronism that demonstrated that they could not have been written on a manual typewriter in the early 1970s.

Many instances of forgery involve attempts at concealment. These include the falsification of items such as birth certificates, medical records, or land deeds with the goal of either establishing a fraudulent claim or ensuring the existence of an alibi. Concealment could involve the disfiguring of the questioned document by folding, tearing, discoloring, soiling, or otherwise mutilating it so that all evidence of fraud is hidden. In the case of the Hitler diaries, the forger poured tea over them to make them look old and hit the diaries hard against his desk to age them. Similarly, the Killian memos had been photocopied over and over again 15 times with the intent of making them look old.

15.4 The Art of Conducting a Handwriting Examination

We now describe the current state-of-the-art process by which an FDE is trained to carry out a detailed examination of a questioned document examination.

The goal of a handwriting examination is to confirm whether a specific person is the writer of a questioned document or to exclude that person as the writer. Examples of such questioned documents include wills, checks, anonymous letters (where the handwriting may be disguised), letters of recommendation, diplomas, certificates, and autographs of famous people.

An FDE is trained to recognize the gross features as well as the subtle details of a particular piece of handwriting so that it may be differentiated from the writing of another person. A handwriting examination is based upon a set of basic principles that govern handwriting analysis, together with the ACE-V methodology used in handwriting comparisons and how the FDE's conclusions are expressed in probability terms for reporting the likely source of the questioned handwriting.

15.4.1 Principles of Handwriting Analysis

There are three basic principles that enable FDEs to carry out handwriting identification:*

1. No two people share the same combination of handwriting characteristics, given an extended sample of their writings. This principle, like that of fingerprints (no two people have yet been found with identical fingerprints), has not yet been refuted. Handwriting, especially a signature, has unique peculiarities that are shared by no one else. So far, no two persons have been shown to have identical handwriting characteristics.

    There have also been several studies regarding handwriting comparisons of twins or other individuals of multiple births. Because twins tend to share the same upbringing, schools, and, for identical twins, DNA, it is of interest to know how similar are their handwritings. Studies have shown that twins (of all types: monozygotic, dizygotic, identical, fraternal) are highly likely to have very similar, but not identical, handwriting (Srihari et al., 2008; Dziedzic et al., 2007; Boot, 1998; Gamble, 1980; Beacom, 1960).

2. No person has yet been found who can duplicate exactly his or her own writing when repeating. Handwriting is a very complicated activity involving a large number of characteristics that are not easy to duplicate. This, in turn, contributes to the idea of a natural range of variation of a person's handwriting that can be differentiated from the writing of a different person.

3. No person can suddenly change his or her writing skill. In other words, over the years, a writer, with a given "skill level," should be able to achieve a certain level of "uniformity" or "consistency" in the way letters are constructed, shaped, and aligned with the rest of the text. The many influences that affect the way a person writes may be considered to be a person's unique writing "style," which cannot be duplicated by a different person.

There are several factors, however, that may cause difficulty in determining the authorship of a given sample of handwriting. These include the possibility of a person deliberately concealing his or her natural writing style by disguising the writing, perhaps by using the non-dominant hand, or suffering the effects of a medical condition, such as a stroke or the use of drugs, that would prevent easy identification of the person's handwriting.

It is important to note that expert opinions on handwriting similarities and differences depend upon certain limitations: the quality of the samples provided (e.g., the original handwritten document, or a photocopy or fax of the original, or a copy of a copy, or even a forged document presented as the original), the state of health (e.g., physical handicaps, medications, substance abuse), age, and other conditions of the writer, the writing conditions, writing instruments, the quality of the paper, and the number of writing samples presented at trial or at an investigation (Kelly and Lindblom, 2006). A photocopy or fax of a document may introduce distortions and defects, and, therefore, cannot be relied upon as an exact copy of the original document.

---

* See, for example, the supporting documentation for *Department of Justice Proposed Uniform Language for Testimony and Reports for the Forensic Handwriting Analysis Discipline* published in July 2016 and available at the website www.justice.gov/archives/dag/forensic-science. The FBI viewpoint is provided by Harrison et al. (2009).

15.4.2 The ACE-V Process of Comparing Handwriting

Document examiners use a methodology for comparing handwriting known as ACE-V, which is an acronym for Analysis, Comparison, Evaluation, and Verification.* The four steps in this process are:[†]

> **Analysis.** In this step, both the questioned writing and the known writing are examined, independently, in detail by the FDE. The FDE determines whether the writing is original (e.g., ink on paper) and whether it was written as natural writing. Characteristics of interest include slant and size of the letters, thickness of letters as direction changes, and beginning and ending strokes of words. The FDE also looks for internal consistency and variability of the writing, and whether there are any personal characteristics.

> **Comparison.** When the analysis step is completed and it is determined that the questioned writing and the known writing are each available for comparison, the FDE does a side-by-side comparison of them. The FDE looks for similarities and differences, and any subtle peculiarities that would distinguish one from the other. Any known limitations on the writer are taken into consideration during the comparison step.

> **Evaluation.** In this step, the FDE has to evaluate the significance of the similarities and differences found in the comparison step. It must be determined whether the person who wrote the known writing is the same person who wrote the questioned writing. For the two persons to be the same writer, the FDE should not find any significant differences in the writings, other than taking into consideration natural writing variability.

> **Verification.** After the ACE steps have been completed, the process and conclusions undergo verification by an independent FDE, who repeats the ACE process and explains any exceptions found.

The National Research Council (2009, pp. 163–167) Report on *Strengthening Forensic Science in the United States*, recommended that

> The scientific basis for handwriting comparisons needs to be strengthened. Recent studies have increased our understanding of the individuality and consistency of handwriting ... and suggest that there may be a scientific basis for handwriting comparison, at least in the absence of intentional obfuscation or forgery.

The NRC Report recommended research "to quantify the reliability and replicability" of the ACE-V methodology used by FDEs and concluded that "there may be some value in handwriting analysis."

Although the ACE-V methodology leaves room for subjectivity, the method is recognized as passing the *Frye* test for admissibility. Furthermore, a federal judge in the Eastern

---

* The ACE-V method is similar in spirit to the FBI's method for comparing latent fingerprints found at a crime scene with a database of known fingerprints. See Chapter 13 (Neumann et al., 2020).

[†] The ACE-V process is described in the supporting documentation for *Department of Justice Proposed Uniform Language for Testimony and Reports for the Forensic Handwriting Analysis Discipline* published in July 2016.

District of Pennsylvania commented in a sentencing hearing in *United States v. Christopher McDaniels* (2014) that the ACE-V methodology is "generally reliable and scientifically sound, as required under Rule 702 and *Daubert*."

15.4.3 A Probability Scale for Expressing Conclusions

Following the completion of the ACE part of the process, an FDE makes a conclusion regarding the comparison of the handwriting samples. This conclusion is stated as an assessment of the amount of confidence the FDE has in the result of the analysis using a nine-point scale suggested by McAlexander et al. (1991). The ASTM approved this scale in a standard it adopted in 1995. It is given by:*

1. **Identification**: The conclusion here is that the questioned writings "match" the known writings, and the writer is judged to be the same person in both sets of writings. No significant differences are observed between the questioned and known writings, other than the normal range of variation.

2. **Strong Probability Did Write**: The FDE concludes that the evidence is persuadable and that although it is a virtual certainty that the same person wrote both documents, some critical quality is missing.

3. **Probably May Have Written**: This conclusion allows the FDE to give a higher probability that the person wrote both documents than simply saying "may" have written both documents.

4. **Indications May Have Written**: Although the two sets of writings are observed to have many characteristics in common and that one person may indeed be the writer for both sets, there are still indications of doubt regarding certain details or limitations in the questioned writings that cannot be seen in the known writings.

5. **No Conclusion**: The writing samples cannot be determined to have come from the same writer (or from different writers) because of a lack of legibility and detail in the writings, or because possible inconsistencies or unexplained characteristics are observed in the writings.

6. **Indications May Not Have Written**: There are indications that the questioned writing and the known writing were not written by the same person. There are too many differences in the writings. However, these differences are not thought to be definitive, resulting in a certain amount of uncertainty, either because some characteristics may be seen as being common to the two sets of writings, or because some limitations or inconsistencies are observed in the writings.

7. **Probably May Not Have Written**: This conclusion allows the FDE to give a higher probability that the person did not write both documents than simply saying "may not" have written both documents.

8. **Strong Probability Did Not Write**: The FDE concludes that the evidence is persuadable and that although it is a virtual certainty that the same person did not write both documents, some critical quality is missing.

---

* Some forensic laboratories (e.g., the New York Police Department's Forensic Investigations Division) use a seven-point scale, omitting the categories of Strong Probability Did Write (#2) and Strong Probability Did Not Write (#8).

   The Questioned Documents Unit of the FBI Laboratory (and some state forensic laboratories) use a five-point scale, omitting also Probably May Have Written (#3) and Probably May Not Have Written (#7) (Harrison et al., 2009).

9. **Elimination**: Because of too many differences in characteristics observed in the two sets of writings, it is clear that they were not written by the same person. Any limitations are outweighed by the observed differences in the writings.

See also National Research Council (2009, p. 166). This nine-point scale provides a discrete set of ordered recommendation categories that gives decreasing levels of the FDE's confidence that the writer of the questioned document is the same person as the writer of the known document.

15.5 Statistical Approaches to Handwriting Analysis

Although there have been many qualitative studies of handwriting, there have been very few formal statistical studies, especially of signatures. We describe three very different statistical attempts at analyzing handwriting. First, we describe the reassessment by Meier and Zabell (1980) of Benjamin Pierce's analysis in the Howland will trial; second, we describe a Bayesian approach by Gaborini et al. (2017) to comparing signatures, which built upon several previous studies in the forensic literature; and, third, we describe the approach to handwriting analysis by Ramsay (2000) and Ramsay and Silverman (1997, 2002), who modeled printed and cursive handwriting as linear differential equations, and then estimated the parameters of those equations from multiple handwriting samples.

15.5.1 Comparing Howland Will Signatures

At the heart of this case was a question of forgery: was the signature on a will of Sylvia Ann Howland, who died on July 2, 1865 with an estate valued at over \$2 million, forged by her niece, Hetty H. Robinson, a wealthy woman in her own right? Robinson contended that an earlier (1862) will that left the entire estate to her should be controlling and that no later will should be honored. The Executor, Thomas Mandell, refused to recognize the earlier will. Robinson sued the Executor.

An important witness at trial was Benjamin Pierce, Professor of Mathematics at Harvard University, who was one of the first to develop the study of mathematical statistics in the United States. Pierce testified for the Executor and he sought to show that the two signatures on the second page of the disputed will were forgeries. His testimony focused on the number of downstrokes that matched in the different signatures. The evidence consisted of photographic copies of 42 genuine signatures, plus one original, and one of the two disputed signatures.

All possible pairings of the 42 signatures ($\binom{42}{2} = 861$) were compared using the number of downstroke matches (out of the 30 possible). They were found to have an average agreement in downstrokes of close to six with a standard deviation of about 2.5. However, when the disputed signature was compared with the original signature, Pierce found that all 30 downstrokes matched exactly. Under his assumption of a binomial model for the number of downstroke agreements plus the product rule for probabilities, agreement in all 30 had an infinitesimal probability (one in 9.31×10^{20}), and, hence, the disputed signature had to be a forgery, most likely traced. He also used a graphical goodness-of-fit test for his model. At the end, the evidence was judged to be insufficient to support Hetty's claim, and the court found for the defendant.

Further details can be found in Meier and Zabell (1980). They show that, although Pierce's arguments were novel and ingenious, at least given the almost-nonexistent state of statistical theory in that era, those arguments failed the test of time. Specifically, they note, first, that the positions of the 30 downstrokes could not be equally likely, and second, that the 861 pairs of signatures could not be statistically independent. Hence, the product rule for independent random variables would not hold, and the number of agreements in the different signatures could not be binomial. They also computed a chi-squared goodness-of-fit test of the distribution of matching downstrokes, a modern alternative to Pierce's graphical display. The chi-squared test yielded a value of over 170 on 12 degrees-of-freedom, showing that Pierce's binomial model was an "exceedingly poor fit" to the data. A more sophisticated test found further support for that conclusion. They also criticized a lack of attention by Pierce to the effect of possible time dependence between signatures.

15.5.2 A Bayesian Approach to Comparing Signatures

It is important, when dealing with questioned handwriting, to be able to identify features of the writing that will help an FDE to distinguish between genuine and disputed writing. An FDE will set up two competing hypotheses, one for the prosecution, $\mathcal{H}_p$, and the other for the defense, $\mathcal{H}_d$. For example, in the Howland will trial, $\mathcal{H}_p$ would state that Sylvia Ann Howland wrote the disputed signature (it is given that she wrote the known signature), and $\mathcal{H}_d$ would state that the signatures on the two wills were written by different individuals, which implies a forged will.

The probabilities for each of these hypotheses depend upon many factors that may exhibit the similarities and differences between the questioned (or disputed) writing and the known (or reference) writing. These similarities and differences constitute the evidence at trial. The role of an FDE is to evaluate the probability of seeing the evidence under each of the two hypotheses $\mathcal{H}_p$ and $\mathcal{H}_d$, assuming the evidence can be viewed as having been drawn from a specific population of such writers. If E denotes the evidence, then the two (conditional) probabilities are $P\{E|\mathcal{H}_p\}$ and $P\{E|\mathcal{H}_d\}$, respectively, where other background information is subsumed into the conditioning but not explicitly shown. The rarer are the similarities and differences, the easier it is for the FDE to make statements regarding the two hypotheses with high probability.

Gaborini et al. (2017) explored a Bayesian approach to the signature comparison problem. They report on experiments involving 143 signatures made by individual A plus a total of 96 signatures that forged A's signature made by four people other than A for a total of $n = 239$ signatures. All signatures were made using a black ballpoint pen on unruled white paper while in a normal sitting position. For the forgers, tracing an original signature was forbidden. The authors handpicked 19 keypoints in the signature to identify features that would be of interest to an FDE. These keypoints would then be used to measure distances, angles, and ratios of distances. For convenience the authors reduced these 19 to two keypoints, the width and height of the signatures. One of the issues with this type of approach is that of selecting the characteristics on which to base the handwriting analysis.

For the defense hypothesis, $\mathcal{H}_d$, the number of potential forgers is assumed to be unknown. So, the four forgers, who are considered to be indistinguishable from each other, are grouped together under an umbrella population $\mathcal{F}$. The comparison problem is, thus, a two-class comparison problem, "authentic" vs. "forged," where $\mathcal{H}_d$ states that the disputed signature was made by "someone in $\mathcal{F}$." For the prosecution hypothesis, $\mathcal{H}_p$, the genuine signatures are used to assess their variability.

Based on scatterplots of the width (S_1) and height (S_2) of signatures, the writings were treated as n realizations from two (number of writers) independent bivariate Gaussian distributions. The writer of the ith specimen was chosen from $\{A\} \cup \mathcal{F}$, $i = 1, 2, \ldots, n$, where A is the writer of the genuine signature. Set $\mathbf{S} = (S_1, S_2)$. The authors model the ith realization of $\mathbf{S}$, given an hypothesis and parameters, as distributed as bivariate Gaussian with a mean vector and covariance matrix, the values of which depend only upon the hypothesis in question.

The authors used a simplistic Bayesian approach for their analysis of signature comparison. They proposed that the posterior odds (i.e., likelihood ratio times prior odds) should be used to decide whether a particular person wrote the questioned signature. They then reduced the Bayesian problem to just the likelihood ratio (by ignoring the prior probabilities) computed under each of the competing hypotheses, with the prosecutor's bivariate Gaussian distribution in the numerator and that of the defendant in the denominator. The parameters of each distribution are estimated by maximum likelihood, and then the estimates are used as if they were the known values of the parameters. The authors sum up their model and how it was used with the following cautionary remarks:

> A fully Bayesian analysis would heavily reduce this problem by leveraging on past data (represented by priors) as well as hypotheses on population parameters. However, the development of a fully Bayesian method is not straightforward.

15.5.3 A Dynamic Model for Handwriting

A very different approach to analyzing handwriting was presented by Ramsay (2000) and Ramsay and Silverman (1997, Chapter 5; 2002, Chapter 11). They viewed handwriting as a complex set of neural events and a biomechanical process that activates muscle contractions and finger, wrist, and forearm motion. They then argued that, as a biomechanical process, handwriting can be described in terms of derivatives, and relationships between derivatives can be described by a differential equation model. In particular, a linear differential equation model permits a more interesting error structure than the more standard type of statistical model.

Because handwriting is a dynamic activity, a sequence of written characters, such as words, phrases, and signatures, can be viewed as a curve drawn over time. So, changes in curve values over time can be described by one or more of the derivatives of the curve with respect to time. Let $x(t)$ be a handwritten curve evaluated at time t. Denote by $D^m x(t)$ the mth derivative of the curve $x(t)$, where $D^m = d^m/dt^m$. Specifically, $Dx(t)$ represents velocity and $D^2 x(t)$ represents acceleration of the curve. A linear differential equation model of order m has the general form,

$$D^m x(t) = \alpha(t) + \sum_{j=0}^{m-1} \beta_j(t) D^j x(t), \tag{15.1}$$

where $\beta_j(t)$ are *coefficient functions* that vary with time, $j = 0, 1, 2, \ldots, m - 1$, and $\alpha(t)$ is the intercept, often called the *forcing function*. If $\alpha(t) = 0$, the equation is referred to as being *homogeneous*; otherwise, it is *nonhomogeneous*.

Examples of this method for determining handwriting variability were obtained through repetitions of the same writing exercise by a single individual. To illustrate this aspect of handwriting, three data sets were created, each one of which had one person write the same words many times. In Ramsay and Silverman (1997, Chapter 5), the characters "fda" were handwritten 20 times by one person in cursive (or script) fashion; in Ramsay and Silverman (2002, Chapter 11), the characters "fda" were printed 20 times by one person; and in Ramsay (2000), the words "statistical science" were written in Chinese characters 50 times by a single person.

For the different sets of writings, two or three coordinate functions were recorded, depending upon the type of writing: for both cursive and printed writing, the two coordinates were lateral movement (left-to-right), X, and vertical movement (up-and-down), Y, of the tip of the pen on the writing surface; in the case of printed writing, a third coordinate was recorded, the movement, Z, of lifting the pen away from the surface.

For handwriting analysis, Ramsay and Silverman (2002) set $m = 3$ and $\beta_0(t) = 0$. Suppose there are n handwriting samples of the same "word." Then, a second-order nonhomogeneous linear differential equation model of the X-coordinate for the ith writing sample is given by

$$D^3 x_i(t) = \alpha_x(t) + \beta_{x1}(t)Dx_i(t) + \beta_{x2}(t)D^2 x_i(t) + e_{xi}(t), \tag{15.2}$$

$i = 1, 2, \ldots, n$, where $e_{xi}(t)$ is the error function for the ith writing sample. Similar models for X and Y can also be set up.

Ramsay and Silverman (1997) and Ramsay (2000) computed smooth nonparametric estimates of the derivative functions, $D^j x_i(t)$, $j = 1, 2, 3$, using penalized regression splines. This was followed by a data-registration step that aligns each coordinate function, X and Y (and Z), so that clock times are nonlinearly transformed to take into account temporal variation across handwriting samples. Once this is done, the coefficient functions $\beta_{x1}(t)$ and $\beta_{x2}(t)$, and the forcing function $\alpha_x(t)$ are estimated by minimizing the error sum of squares integrated over all $t \in [0, T]$. This method was introduced by Ramsay (1996) under the name of *principal differential analysis*.

To view the amount of variability in these functions, Ramsay and Silverman recommended that the estimates $\widehat{\beta}_{x1}(t)$, $\widehat{\beta}_{x2}(t)$, and $\widehat{\alpha}_x(t)$ should each be plotted, separately, against time t, together with their averages. Also, just as is done with multiple regression analysis, a plot of the ith residual process, $\widehat{e}_{xi}(t)$, $i = 1, 2, \ldots, n$, against time t should be made. If this method yields good estimates, then the residual process should be relatively small over time for all writing samples. These plots should display any idiosyncratic behavior of the estimated coefficient functions, the estimated forcing function, and the residual process if such behavior exists.

They also proposed that a measure of the goodness-of-fit of the estimated equation can be computed by a version of the squared multiple correlation coefficient:

$$R_X^2 = 1 - \frac{\sum_{i=1}^n \int_0^T \widehat{e}_{xi}^2 dt}{\sum_{i=1}^n \int_0^T (D^3 x_i(t))^2 dt}. \tag{15.3}$$

This measure compares the overall level of the residual functions against the overall level of the third derivatives.

The dynamic model and estimation techniques proposed by Ramsay and Silverman can also be used to compare handwriting samples from different individuals. Parallel box plots of the root-mean-square magnitudes of the residual process can be drawn by applying the differential operators to each data set from the different individuals.

15.6 Proficiency Studies and Error Rates

Some of the most helpful proficiency studies of handwriting comparison by FDEs were undertaken by Dr. Moshe Kam and his colleagues. His initial foray into this field took place in 1991 when the FBI asked him to design computer software for proficiency testing of scanned handwriting images. After searching the literature, Kam could not find any controlled proficiency studies in which the expertise of FDEs were proved to be superior to that of non-FDEs. Consequently, he also viewed statistics reported from professional proficiency tests to be untrustworthy.

He conducted a pilot study (Kam et al., 1994) to determine whether or not there were any differences in skills between FDEs and laypersons in handwriting recognition. His pilot study involved seven FDEs (then with the FBI) and ten graduate students (playing the role of laypersons). Each participant was shown 86 original handwritten documents and were asked to sort the documents into piles in such a way that each pile consisted of the writings of a single person. It was considered an error if writings by two different persons were assigned to the same pile, or if writings by the same person were assigned to different piles. Kam found the results "striking." Five of the seven FDEs had perfect performances, the other two each made two errors. The graduate students made 7 to 45 errors.

In a second study (Kam et al., 1997), responding to critiques of his pilot study, Kam conducted a much larger study of handwriting, with over 100 FDEs, a control group of 41 non-experts, and eight subjects who were training to become FDEs. Each person was tested separately using the same six questioned and 24 known documents. They were each asked to carry out $6 \times 24 = 144$ pairwise comparisons, and decide whether a "match" was detected (i.e., whether two documents were written by the same person). The FDEs' error rate was 6.5%, the non-experts' error rate was 38.3%, and the trainees' error rate was 8.3%. So, the non-experts have a greater than six-to-one larger error rate than the FDEs, which suggested that FDEs have a skill that the non-experts lack in handwriting comparison. In a third study (Kam et al., 1998), Kam and colleagues showed that this expertise did not derive from a remuneration difference.

A fourth study (Kam et al., 2001) showed a significant difference in the error rates between FDEs and laypersons when they had to compare genuine (G) as opposed to non-genuine (NG) signatures. The error rate for FDEs was 7.05% for designating a G signature as NG, and 0.49% for designating an NG signature as G. The error rate for the laypersons was 26.1% for designating a G signature as NG, and 6.47% for designating an NG signature as G.

These results by Kam and colleagues also proved to be a crucial piece of information that could be used to defend against *Daubert* challenges. In fact, the court in *United States v. Prime* (2005) noted that the studies of Dr. Kam provided error rates for FDEs and showed how a well-designed statistical study superseded all studies of previously flawed proficiency tests that were carried out in the 1980s.

A more recent study (Kam et al., 2015) looked at how well 19 FDEs and 26 laypersons were at identifying simulated handwriting (but not signatures). The sample of laypersons was drawn from a wide population, so that it resulted in a sample similar in composition to a typical jury pool. They showed that there was a significant statistical difference between FDEs and laypersons in differentiating between genuine and simulated handwriting.

Another study (Sita et al., 2002), which also focused upon the genuineness of a person's signature, found that FDEs' error rate in judgment was 3.4%.

Criticisms of all these studies and more can be found in Risinger (2007).

15.7 Discussion

In this chapter, we discussed the current status of handwriting comparisons in the courts. Such comparisons are necessary when dealing with disputed wills and checks, identifying the writer of ransom notes, and the falsification of birth certificates, medical records, or land deeds with the purpose of inserting a name or altering a name to establish a claim or an alibi.

We showed how long it took handwriting comparisons to be accepted in the Roman, English, and United States legal systems. It is surprising how resistant to change England and the United States were when it came to providing a legal foundation for handwriting comparisons in the courts. Some of it can be blamed on poverty and the prevalence of illiteracy in England during the 19th century. These conditions, in turn, led to a lack of trust by the English courts in the ability of juries to comprehend and compare handwritten documents. This was all made moot by the 1854 and 1865 Acts of Parliament so that handwriting comparisons became admissible at trial. In the United States, resistance to admitting handwriting comparisons in federal court lasted until 1913, almost sixty years after England had overcome that hurdle. Even then, it took several major cases to define the conditions for the federal courts to allow scientific or technical testimony at trial, especially testimony by FDEs on handwriting comparisons. Most states had, by 1913, already admitted such testimony.

We then discussed some of the challenges that FDEs face, including the examination of anonymous letters and forgeries of dated materials. Often, an attempt is made to conceal the true age of a questioned document by hiding the evidence behind an alleged accident so that the document is torn, discolored, or otherwise mutilated. In such cases, an FDE has to identify (or exclude, if possible) a known writer as the source of a handwritten item.

The basic principles of handwriting analysis hold that handwriting is unique to the individual, cannot be duplicated exactly even by the same individual, and each person has a certain consistent writing style or skill that is unique to that individual. These principles enable FDEs to declare a particular writing, such as a signature, as authentic or a forgery.

The ACE-V methodology plays a significant role in the FDE's handwriting comparison process. At the completion of the ACE part of the process, the FDE makes a recommendation using a probability scale that quantifies how certain the FDE is in the conclusions. However, there are still courts in the United States that limit an FDE at trial to explain only similarities and dissimilarities in handwriting samples without specifying an opinion as to who actually wrote the questioned document.

There have only been a few attempts at constructing a statistical model of handwriting. A very early attempt was made by Benjamin Pierce as an expert witness in the 1868 Howland will trial, when he counted the number of downstrokes that matched in the different handwriting samples. Although this idea was considered novel at the time, it contained faulty statistical reasoning in computing the odds of the two signatures being written by the same person. In a second approach, a Bayesian model for handwriting comparison uses the posterior odds to distinguish whether a specific person wrote the questioned document. However, complications in computing the likelihood ratio and the prior odds led the authors to recognize that a fully Bayesian model would be difficult to construct. A third approach used functional data analysis to construct a linear differential equation that modeled the curves of the signature, and was used to provide visual displays for comparing different handwriting samples. More research along these lines should be encouraged by those statisticians involved in forensic science.

Finally, because known error rates are one of the conditions set out in *Daubert* that would allow an FDE's full testimony at trial, we describe the differences between error rates of FDEs and non-FDEs obtained from several experiments on handwriting comparisons. Various studies have been conducted on FDEs and non-FDEs, comparing their abilities to distinguish authentic from forged documents and signatures. Statistical results from these experiments show that FDEs have a skill that makes them significantly more accurate than non-FDEs in determining whether handwritten documents derive from the writing of the same or different persons.

Acknowledgments

Many thanks go to David Banks for the invitation to write this chapter and for editorial suggestions, and David H. Kaye and Michael D. Risinger for helpful correspondence and references on the *Daubert* Group. Thanks also to Eliot (Elyah) Springer, Deputy Director, Police Laboratory, New York Police Department, for drawing attention to the Gaborini et al. (2017) article. Conversations on the same article with Marc Sobel were also useful. Thanks also go to Betty-Ann S. Izenman, Esq., for help in editing a draft of this chapter and for help with legal questions.

References

Baker, J.N. *Law of Disputed and Forged Documents*, Michie Co., Charlottesville, VA, 1971.

Beacom, M.S. A study of handwriting by twins and other persons of multiple births, *Journal of Forensic Sciences*, **5**, 121–131, 1960.

Boot, D. An investigation into the degree of similarity in the handwriting of identical and fraternal twins in New Zealand, *Journal of the American Society of Questioned Document Examiners*, **1**, 70–81, 1998.

Carlson, M. Handwriting testimony, *Virginia Law Register, New Series*, **6**, 175–178, 1920.

Clayton, E. *The Golden Thread: The Story of Writing*, Counterpoint, Berkeley, CA, 2013.

Coleridge, J.T. Doe on the demise of *Mudd v Suckermore*. In J.L. Adolphus and T.F. Ellis, editors, *Reports of Cases Argued and Determined on the Court King's Bench*, Vol. 5, Saunders and Benning, 1838.

Costain, J.E. Questioned documents and the law: handwriting evidence in the Federal court system, *Journal of Forensic Sciences*, **22**, 799–806, 1977.

Dziedzic, T., Fabiansa, E., and Toeplitz, Z. Handwriting of monozygotic and dizygotic twins, *Problems of Forensic Science*, **69**, 30–34, 2007.

Gaborini, L., Biedermann, A., and Taroni, F. Towards a Bayesian evaluation of features in questioned handwriting signatures, *Science and Justice*, **57**, 209–220, 2017.

Gamble, D. The handwriting of identical twins, *Canadian Journal of Forensic Science Journal*, **13**, 11–30, 1980.

Gillard, D. *Education in England: A Brief History*, 2011. educationengland.org.uk/history

Harris, G.E. *A Treatise on the Law of Identification: A Separate Branch of the Law of Evidence*, H.B. Parsons, Albany, NY, 1892.

Harrison, D., Burkes, T.M., and Seiger, D.F. Handwriting examination: meeting the challenges of science and the law, *Federal Bureau of Investigation (FBI), Forensic Science Communications*, **11**(4), 2009.

Hilton, O. History of questioned document examination in the United States, *Journal of Forensic Sciences*, **24**, 890–897, 1979.

Hilton, O. *Scientific Examination of Questioned Documents*, Elsevier Science Publishing Co., New York, 1982.

Kam, M., Abichandani, P., and Hewett, T. Simulation detection in handwritten documents by forensic document examiners, *Journal of Forensic Sciences*, **60**, 936–941, 2015.

Kam, M., Fielding, G., and Conn, R. Writer identification by professional document examiners, *Journal of Forensic Sciences*, **42**, 778–786, 1997.

Kam, M., Fielding, G., and Conn, R. Effects of monetary incentives on performance of nonprofessionals document-examination proficiency tests, *Journal of Forensic Sciences*, **43**, 1000–1005, 1998.

Kam, M., Gummadidala, K., Fielding, G., and Conn, R. Signature authentication by forensic document examiners, *Journal of Forensic Sciences*, **46**, 884–888, 2001.

Kam, M., Wetstein, J., and Conn, R. Proficiency of professional document examiners in writer identification, *Journal of Forensic Sciences*, **39**, 5–14, 1994.

Kelly, J.S. and Lindblom, B.S. (Eds). *Examination of Questioned Documents*, CRC Press, Boca Raton, FL, 2006.

Koppenhaver, K.M. *Forensic Document Examination: Principles and Practice*, Humana Press, Totowa, NJ, 2007.

Masington, R.S. Evidence – Jury comparison of handwriting without aid of experts, *University of Miami Law Review*, **14**, 491–495, 1960.

McAlexander, T.V., Beck, J., and Dick, R. The standardization of handwriting opinion terminology, *Journal of Forensic Sciences*, **36**, 313, 1991.

Meier, P. and Zabell, S. Benjamin Peirce and the Howland Will, *Journal of the American Statistical Association*, **75**, 497–506, 1980.

National Research Council. *Strengthening Forensic Science in the United States: A Path Forward*, National Academies Press, Washington, DC, 2009.

Neumann, C., Hendricks, J., and Ausdemore, M. Statistical support for conclusions in fingerprint examinations. In *Handbook of Forensic Statistics*, D. Banks, K. Kafadar, D.H. Kaye, and M. Tackett, editors. CRC Press, Boca Raton Florida, 2020.

Norwitch, F.H. and Seiden, H. Questioned documents, In S.H. James and J.J. Hordby, editors, *Forensic Science*, pages 357–373. Chapman & Hall/CRC, Boca Raton, FL, 2003.

Osborn, A.S. *Questioned Documents*, Boyd Printing Co., Albany, NY, 1910.

Osborn, A.S. *Questioned Documents*, second edition, Boyd Printing Co., Albany, NY, 1929.

Osborn, A.S. Progress in proof of handwriting and documents, *Journal of Criminal Law and Criminology*, **24**, 118–124, 1933.

Page, M., Taylor, J., and Blenkin, M. Forensic identification science evidence since *Daubert*: Part I, A quantitative analysis of the exclusion of forensic identification science evidence, *Journal of Forensic Sciences*, **56**, 1180–1184, 2011.

Ramsay, J.O. Principal differenial analysis: data reduction by differential operators, *Journal of the Royal Statistical Society, Series B*, **60**, 365–375, 1996.

Ramsay, J.O. Functional components of variation in handwriting, *Journal of the American Statistical Association*, **95**, 9–15, 2000.

Ramsay, J.O. and Silverman, B.W. *Functional Data Analysis*, Springer, New York, 1997.

Ramsay, J.O. and Silverman, B.W. *Applied Functional Data Analysis: Methods and Case Studies*, Springer, New York, 2002.

Risinger, M.D. Cases involving the reliability of handwriting identification expertise since the decision in *Daubert*, *Tulsa Law Review*, **43**, 477–595, 2007.

Risinger, D.M., Denbeaux, M.P., and Saks, M.J. Exorcism of ignorance as a proxy for rational knowledge: the lessons of handwriting identification – expertise., *University of Pennsylvania Law Review*, **137**, 731–792, 1986.

Sita, J., Found, B., and Rogers, D. Forensic handwriting examiners' expertise for signature comparison, *Journal of Forensic Sciences*, **47**, 1117, 2002.

Srihari, S., Huang, C., and Srinivasan, H. On the discriminability of the handwriting of twins, *Journal of Forensic Sciences*, **53**, 430–446, 2008.
SWGDOC E2290. *Standard for Examination of Handwritten Items*, 2016.
Trubek, A. *The History and Uncertain Future of Handwriting*, Bloomsbury Publishing Plc, New York, NY, 2016.
Van Santvoord, G. *Life of Algeron Sidney With Sketches of Some of His Contemporaries and Extracts From His Correspondence and Political Writings*, Charles Scribner, New York, 1851.

Cases and Statutes

Daubert v. Merrell Dow Pharmaceuticals, 509 U.S. 579 (1993).
Frye v. United States, 293 F 1013 (D.C. Cir. 1923).
The General Comparison Statute, 28 U.S.C. 638 (1913).
General Electric v. Joiner, 522 U.S. 136 (1997).
Kumho Tire Co. v. Carmichael, 526 U.S. 137 (1999).
United States v. Fujii, 152 F.Supp.2d 939 (N.D. Ill. 2000).
United States v. Hines, 55 F.Supp.2d 62 (D. Mass. 1999).
United States v. McDaniels, Docket No. 12-393-01, (E.D.PA 2014) (unpublished).
United States v. McVeigh, 964 F.Supp. 313 (D. Colo. 1997).
United States v. Oskowitz, 294 F.Supp.2d 379 (E.D.N.Y. 2003).
United States v. Prime, 431 F.3d 1147 (9th Cir. 2005).
United States v. Rutherford, 104 F.Supp.2d 1190 (D.Neb. 2000).
United States v. Saelee, 162 F.Supp.2d 1097 (D. Alaska 2001).
United States v. Starzecpyzel, 880 F.Suppl. 1027 (S.D.N.Y. 1995).
University of Illinois v. Spalding, 71 N.H. 163 (N.H. 1901).

16

An Introduction to Firearms Examination for Researchers in Statistics

Susan VanderPlas, Alicia Carriquiry, Heike Hofmann, James Hamby, and Xiao Hui Tai

CONTENTS

16.1 Introduction 365
16.2 The Anatomy of Guns and of Ammunition 368
16.3 A Brief History of Firearms Examination 370
      16.3.1   Microscopic Imperfections 372
16.4 Comparison of Cartridge Case Marks 373
16.5 Comparison of Marks on Land Engraved Areas of Bullets 377
      16.5.1   Pairwise Comparisons 378
16.6 Revisiting the Question of Source 384
16.7 Some Final Thoughts 386
References 387

16.1 Introduction

In the United States, where firearms are readily accessible, the annual number of gun-related crimes is in the hundreds of thousands, and about two thirds of all murders are committed with a gun (Federal Bureau of Investigation, 2017). Almost 400 million guns of all types are owned by civilians, meaning that in the United States, there are about 1.2 guns for every man, woman and child. Therefore, most crime labs employ one or more experts in firearms examination, who can visually compare striations on bullets or cartridge cases found in a crime scene with those on test shots obtained from the suspect's gun, or on samples recovered from a different crime scene.

Visual comparisons are problematic for several reasons, including the fact that the assessment of similarity is typically subjective, and consequently, estimation of error rates is difficult. This said, establishing a more formal, data-based protocol of the comparison is challenging. A typical image of a bullet or a cartridge case has tens of thousands of pixels, and a pixel-by-pixel comparison is not robust to changes in scale, rotation, and translation. To add to the challenge, there is no generative model to represent the process by which striations occur that would permit reducing the dimensionality of the images and establishing a formal testing approach. Finally, it is not clear whether the information contained in the image can be summarized into a set of measurements to enable a data-based comparison.

Consequently, at the present time, forensic practitioners rely on subjective methods for evaluating and interpreting most types of pattern evidence including striations on bullets and cartridge cases. In a typical evaluation, the examiner compares two or more samples side by side, and concludes that the samples may have a common origin (or source) if the number of common features in the samples warrants "sufficient agreement". Examiners may rely on instruments such as a comparison microscope and limited measurements of characteristics such as rifling pitch, or examine national databases such as NIBIN (US Department of Justice and Bureau of Alcohol Tobacco Firearms and Explosives, 2018), but the final decision is almost entirely subjective and greatly depends on the training and experience of the examiner. The final decision is typically one of identification if the samples are sufficiently similar, exclusion if there are significant dissimilarities, or inconclusive if there are not sufficient features to make a determination.

As noted in Carriquiry et al. (2019), the fact that the evaluation of pattern evidence continues to be subjective is problematic. Except for fingerprint examination, there is no agreed upon threshold to declare that two samples are "similar enough". Even among fingerprint examiners, the set of specific minutiae used for comparison is common, but there is no rule that establishes the number of matching minutiae that must be exceeded to conclude that two prints have the same origin. Another challenge that subjective methods make it is impossible to estimate the rate of errors incurred by individual examiners or by the discipline as a whole. Experience is no substitute for experimentation, where ground truth is known (President's Council of Advisors on Science and Technology, 2016), and claims by practitioners that the rate of error is zero simply reflect the fact that their conclusions have never been challenged in court or elsewhere. Finally, when assessments are subjective, it is difficult to estimate the *repeatabilty* and *reproducibility* of firearm evaluations. Here, we say that an approach is repeatable if the examiner reaches the same conclusion when presented with the same evidence at two different times. A method is reproducible if two examiners presented with the same evidence, reach the same conclusion.

Because of these and several other concerns, the National Research Council (2009) published a report that was strongly critical of all forensic disciplines with the exception of DNA analysis of single-donor or simple mixture samples. The report singled out pattern evidence as particularly lacking in scientific validity and rife with subjectivity, and called for an immediate and sustained research effort to shore up pattern evidence by—among other recommendations—developing the scientific and statistical framework that underpin any scientific discipline. Of particular concern for firearms and toolmark analysis is that the process of individualization be "precise and repeatable," in contrast to the current methodological guidelines.

In the last few decades, there have been some attempts to develop methods to quantify the similarity between two bullets or two cartridge cases. An early attempt by Biasotti (1959) consisted of counting the number of consecutively matching striae (CMS) on two bullets. The idea here is that when two bullets are fired from the same barrel, the number of consecutively matching striations is expected to be high. While this proposition is likely to be true, the method was not widely applied in practice because the discriminating power of CMS is limited. Hamby et al. (2009) were the first to assemble an experimental set of bullets that could be used to investigate the reproducibility of firearms examination. They created 240 replicate sets of 20 reference bullets fired from 10 consecutively manufactured barrels and 15 questioned bullets that were known to have been fired by the barrels in the study. By now, the 240 sets have been examined by over 700 participating firearms examiners worldwide, 14 of whom have used some form of ballistic imaging technology. While the design of this experimental dataset would have been improved by, for example, including questioned bullets fired by barrels other than the 10 barrels considered, the database is nonetheless a good resource for researchers. Figure 16.1 was drawn using one of the Hamby sets of bullets (Hamby et al., 2009). The set of 20 bullets was fired from

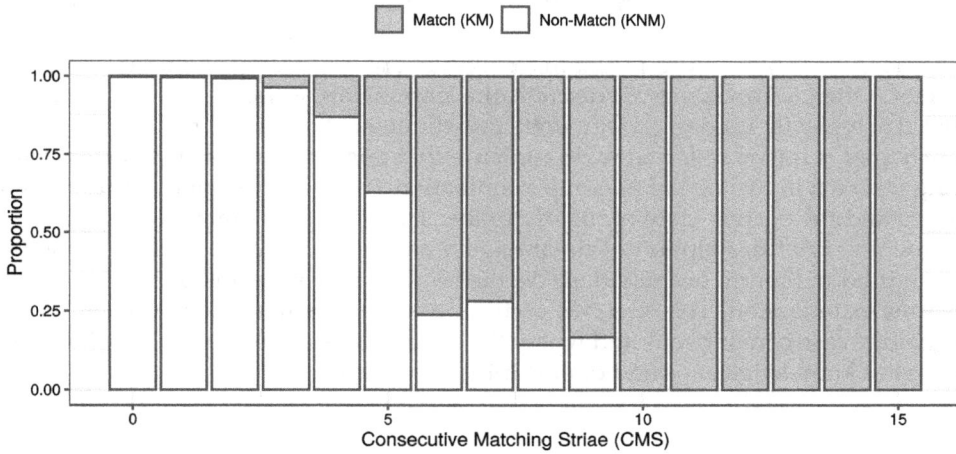

FIGURE 16.1
Proportion of pairs of bullets with same (gray) or different (white) source given observed number of CMS.

10 consecutively rifled barrels, and the provenance of each bullet (same or different barrel) is known. We constructed all possible pairs of bullets, some of which came from the same barrel and some of which did not. The values along the horizontal axis represent the number of CMS, which for this dataset varied from 0 to 25. For each value of CMS, the two colors on each column represent the proportion of pairs of bullets that had a common (gray) or a different (white) source. For example, when CMS is equal to 2, 100% of pairs with 2 CMS had a different source. When CMS is 10 or higher, 100% of all pairs of bullets have a common source.

From the figure, it seems that – at least for this particular dataset – when the observed number of CMS is between 4 and 10, a binary decision of same/different source has a non-negligible probability of being incorrect.

In recent years, there has been a push to develop more robust methods to address the question of source. Much of the effort has been devoted to cartridge cases: breech face impressions, firing pin impressions, firing pin aperture. We describe some of these approaches later in this chapter. Less attention has been devoted to bullets, perhaps because bullets sometimes get damaged on impact and this makes the comparison more difficult. In both cases, methods based on machine learning and computer vision have gained in popularity among researchers and appear to be promising for use in case work.

Machine learning can be used to augment the subjective perceptions of examiners, providing a quantitative foundation for the assessment of questions of source in firearms examination. In order to leverage machine learning techniques for firearm identification, researchers mostly use supervised learning algorithms and large amounts of labeled training data, called a training set, to assess how features from the data relate to the labels. From the training set, the algorithm "learns" how the features relate to the labels, or in other words, produces an estimate of the function that describes the association between features and label. Once the algorithm has been trained, it can be used to examine new data outside the original training set.

A good algorithm strikes a balance between the competing objectives of optimizing its performance on the training data and minimizing classification errors when presented with new units. This is known as the variance-bias tradeoff; more flexible models, that fit the training data very well, have high variance, because even a small change in the data might result in a large change in the fitted model. On the other hand, models that fit the training data less closely, may be more robust to changes in the observations, but tend to exhibit a larger bias. Often, when labels are discrete, the learning algorithm is known as a classifier because it is typically used to classify units into different classes.

This chapter is intended for an wide audience that might include statisticians interested in doing research in the area of firearms examination, or for lawyers and judges who wish to understand the current state of the discipline. The chapter is organized as follows. We first provide a brief description of firearms and ammunition and introduce some of the language used in firearm examination. We then briefly revisit the history of the discipline of firearms examination. The next two sections describe the new methodology for bullet and cartridge case comparisons and where the science is likely to go next. We finish with a partial list of knowledge gaps and opportunities for research.

16.2 The Anatomy of Guns and of Ammunition

About 12.5 million guns were manufactured in the United States in 2016, of which approximately 50% were pistols or revolvers. In the same year, the United States imported slightly over five million firearms, the vast majority of which were handheld pistols and revolvers. Here we briefly describe the components and operation of pistols and revolvers, since according to the FBI (Federal Bureau of Investigation, 2017) over 70% of all gun-related crimes involve one of those types of firearms.

The main difference between a revolver and a pistol is the mechanism to load ammunition into the chamber. In a revolver, cartridges are loaded into their own chamber in a cylinder that rotates when the hammer is cocked and the trigger is pulled (in single-action revolvers) or advances with just the pull of the trigger (double-action). The capacity of the cylinders tends to be limited to six cartridges, and cartridges are not ejected from the cylinder each time the revolver is fired. In a pistol, cartridges are loaded into a magazine that can contain anywhere between 5 and 18 rounds, depending on caliber and type of magazine. Pistols are self-loading firearms because a fresh cartridge from the magazine is automatically moved into the single chamber when the slide of the gun is pulled rearward. The spring-loaded slide then returns forward, feeding the first cartridge into the chamber for firing. The pistol is fired with each new pull of the trigger, and the next cartridge in the magazine is moved into the chamber. This is why this type of firearm is known as a semi-automatic weapon. In contrast to what happens when a revolver is fired, cartridge cases are ejected from the barrel of a pistol at the time of firing (Heard, 1997). Figure 16.2 shows a Luger 1901 pistol cutaway with important parts labeled.

When the firearm is fired, several things occur. First, the firing pin hits the primer at the base of the cartridge, causing a small explosion that ignites the propellant. As it burns, the propellant creates gas inside the cartridge case. The tremendous change in pressure inside the cartridge case separates the bullet from the cartridge and pushes it down the barrel at a speed anywhere between 200 to 650 m/s, depending on caliber. Most barrels are *rifled*, meaning that their interior has grooves and elevated surfaces between the grooves called *lands* (typically five or six of each) at a left or right angle, and this imparts rotation to the

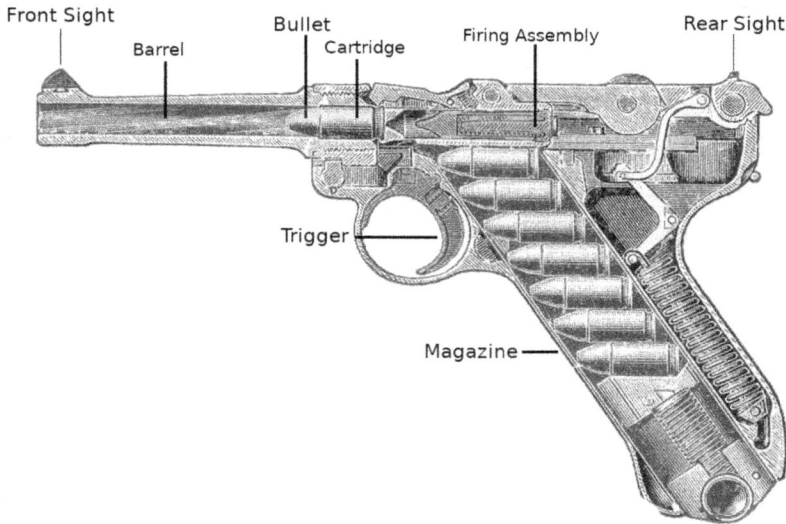

FIGURE 16.2
A 1901 Luger pistol. Image modified from Rock Island Auction Co. [CC0], via Wikimedia Commons (https:// commons.wikimedia.org/wiki/File:Luger_cutaway.jpg).

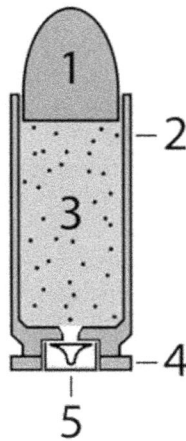

FIGURE 16.3
Diagram of a cartridge with its components: (1) bullet, (2) cartridge case, (3) propellant or powder, (4) base of the cartridge case, and (5) primer. Image downloaded from Wikipedia (https://en.wikipedia.org/wiki/Bullet).

bullet as it travels down the barrel. This improves stability in the flight of the bullet. Consequently, fired bullets also show a pattern of land impressions and grooves that correspond to the rifling of the barrel. In addition to separating the bullet from the cartridge, the explosion in the cartridge case makes the base of the cartridge case slam against the breech face of the pistol. Cartridge cases in semi-automatic pistols are typically ejected once the bullet is separated. Figure 16.3 shows a diagram of a cartridge.

16.3 A Brief History of Firearms Examination

Firearms examination has been practiced for a long time. The discipline is sometimes incorrectly referred to as *ballistics*, but the term is more appropriately associated with flight and impact properties of projectiles once they are fired. Its first recorded use around 1835 is attributed to the Bow Street Runners, the first police-like force in London, precursor of the Metropolitan police (Beattie, 2012). Wilson and Wilson (2003) tell the story of Confederate General Jackson, who was killed in battle in 1863, during the Civil War. The fired bullet recovered from his body was examined and found to be a 0.67-caliber ball, the type of ammunition used by Confederate, not Union soldiers, leading to the surprising conclusion that General Jackson had been killed by one of his own men.

In 1900, a paper by Hall (1900) in the *Buffalo Medical Journal* detailed methods for test firing weapons (into a bag of meal) for the purposes of comparing the engravings resulting from the rifling. While it is not clear from the paper whether the comparisons made are class characteristics (e.g. the spacing of land and groove engraved areas) or individual characteristics, this represents the first known foray into what might be termed "modern" examinations of fired bullets. In 1902, testimony concerning the markings on bullets was admitted into evidence (*Commonwealth v. Best*, 1902) in a criminal case for the first time, and in 1907, fired cartridge cases were also used as evidence (Frankford Arsenal Army Personnel, 1907). In 1912, photographic comparisons were used to examine individualized markings of land and groove engraved areas; enlarged photographs were also used to compare cartridge case marks. In the 1920s, the idea that individualizing characteristics are present on fired bullets and cartridges became more widely accepted across the country, accompanied by an increase in court cases involving evidence from bullets and cartridge cases. The uniqueness of striations or other marks is impossible to prove, but it seems clear that marks carry information about the firearm that fired the ammunition. A thorough historical review of the discipline of firearms examination can be found in Hamby and Thorpe (1999).

The process of comparing the striations of bullets recovered from a crime scene to test-fired bullets from recovered weapons became much more practical in 1925 with the invention of the comparison microscope, which allowed two bullets to be examined simultaneously and manually aligned (Conrad et al., 2008). Figure 16.4 shows an early model of a comparison microscope; portions of both bullets under examination are shown in a single unified view window. The same year, a series of two articles in the *Saturday Evening Post* described the potential uses of firearms examination in popular press (Stout, 1925a,b). Five years later, Goddard would testify about the firearms used in the Valentine's Day Massacre (Goddard, 1930); this testimony included analysis of the cartridge cases recovered from the scene as well as the markings left on the bullets.

Firearms examination became more of a discipline in the 1930s, with textbooks published on the subject in both the UK and the United States (Burrard, 1934; Hatcher, 1935; Gunther and Gunther, 1935). The 1930s also saw the foundation of a number of laboratories focused on the scientific examination of forensic evidence, a trend that continued into the 1940s. In the 1950s and 1960s, the field began to move toward greater quantification of firearms evidence, with the introduction of the striagraph (Davis, 1968), an early forerunner of laser and digital scanning of bullets. In 1959, Biasotti published a landmark paper discussing the use of visual features of pairwise bullet comparisons to produce a quantitative description of the strength of the match between the two bullets. Biasotti examined the striation marks on fired bullets under a virtual comparison microscope and determined that consecutive

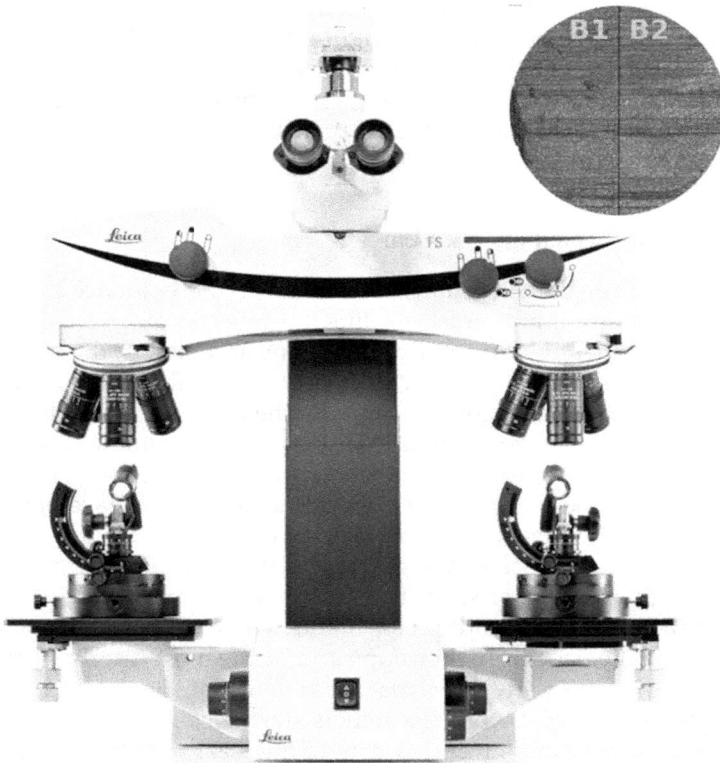

FIGURE 16.4
A comparison microscope, which consists of two identical microscopes connected to a single eyepiece. The viewer sees the images of both bullet surfaces simultaneously, facilitating alignment and comparison of the bullet striae, shown in the inset image.

matching striations could be used to determine the strength of a match between two bullets which were aligned based of visual assessment of striation patterns. Based on a sample of bullets fired from 24 Smith & Wesson revolvers, Biasotti determined that bullets fired from different guns were extremely unlikely to have more than 6 consecutively matching striae. This quantitative threshold for determining whether two pieces of evidence originate from the same firearm was the first attempt to derive an empirical threshold for the strength of a match in firearms examination.

Some methods have been developed to automatically match high-resolution photographs of bullet land engraved areas or cartridge cases (Gardner, 1978; Geradts et al., 2001), but in the past 20 years or so there has been a push to use 3D imaging technology to obtain actual measurements of the surface topology of land engraved areas on bullets and of cartridge case bases. Methods which depend on photographs rely on inferring the height of the surface from the color of the image pixels; topological measurements provide much greater precision in assessing the similarity of two samples. One of the first proponents of measuring features of the surface topology of bullets were Kinder and Bonfanti (1999). The authors used a laser profilometer to scan bullets and obtain measurements of the distance and depth of striations. They then computed a correlation between two aligned sets of features to quantify the similarity between two bullets. Since then, several new methods that

372*Handbook of Forensic Statistics*

rely on 2D and 3D imaging of bullets and cartridge cases have been proposed to quantify the similarity between two items (Tong et al., 2014; Fischer and Vielhauer, 2014; Chumbley et al., 2010). In 2017, Hare et al. demonstrated the use of supervised learning algorithms to construct a similarity score that can be used to assess the strength of a match between two bullets quantitatively, and (Tai and Eddy, 2018) constructed a similar scoring algorithm for cartridge case comparison.

16.3.1 Microscopic Imperfections

Forensic examiners will begin an examination of crime scene evidence by identifying class characteristics, such as ammunition type, rifling, number of grooves, shape of the firing pin and other features shared by many different guns and bullets; the second stage of the process is to establish whether a bullet was fired by a specific gun using individualizing characteristics. Most automatic algorithms bypass the comparison of class characteristics and only examine the microscopic individual striations that uniquely identify a specific barrel. The left panel of Figure 16.5 shows the engraved areas left behind by the lands and grooves of the barrel on the bullet. Small imperfections in the surface of the barrel create striations on the bullet as it leaves the gun. Marks are also imprinted to the base of the cartridge case by the breech face of the gun at the time it is fired; these marks can also be compared through a similar process. The right panel in Figure 16.5 shows a cartridge case with breech face, firing pin and ejector marks. Firearms examiners tend to focus on the breech face impressions because they appear to be the most discriminating.

It was only during the past few years that the role of statistics in firearm and toolmark examination began to surface. Before researchers started collecting 2D and 3D images of cartridge cases and bullets, there were no data to analyze, and the few experiments that were conducted did not include a rigorous experimantal design phase. Indeed, there was no discussion of the potential benefits of using principled quantitative analysis among firearm and toolmark examiners until the 2009 NRC report. The recommendations in the

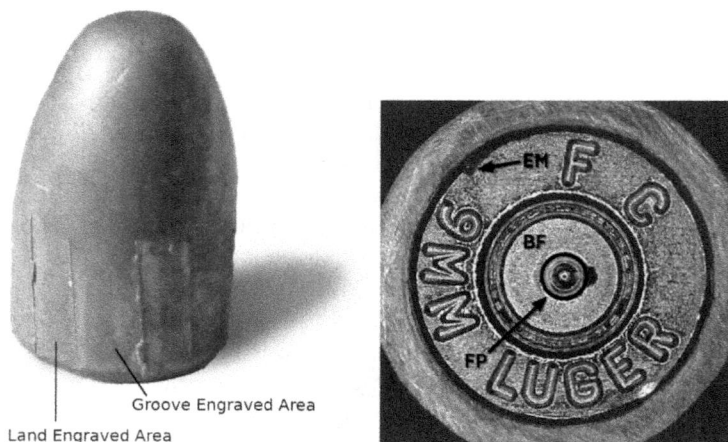

FIGURE 16.5
Left panel: Bullet showing engraved land areas and groove impressions. Right panel: Base of a cartridge case. EM denotes ejector marks, FP denotes firing pin impression and BF denotes breech face impressions. Cartridge case image downloaded from NIST, via Wikimedia Commons (https://commons.wikimedia.org/wiki/File:Cartridge_Case_(7788259910).jpg).

report, plus the gradual acceptance of a likelihood ratio framework for the interpretation of evidence in other forensic disciplines encouraged several research groups to begin designing experiments and collecting data, developing algorithms for classification and matching (Riva and Champod, 2014; Zheng et al., 2014; Vorburger et al., 2015), and proposing approaches to approximate likelihood ratios via empirical score-based ratios (Hare et al., 2017; Song et al., 2018; Tai and Eddy, 2018). We review some of this work in the next two sections.

16.4 Comparison of Cartridge Case Marks

With the advent of high resolution 2D and 3D microscopy, researchers in the last two decades have proposed several new methods to quantify the similarity between two bullets or cartridge cases. Much of the research has focused on cartridge cases rather than bullets, probably because bullets tend to get damaged on impact and are recovered with lower frequency (NRC, 2008). Research results presented in the last two decades include Bachrach (2002), Vorburger et al. (2011), Weller et al. (2012), Song (2013), Chu et al. (2013), Riva and Champod (2014), Zheng et al. (2014), Vorburger et al. (2015), and Song et al. (2018).

Song et al. (2018) review a method called Congruent Matching Cells (CMC) to compare the breech face impressions on two cartridge cases. The data with which the authors work are 3D images of breech face markings on the base of cartridge cases, obtained via confocal microscopy. The comparison method consists in first dividing the reference sample into a grid of cells and then implementing an automated search for the closest matching areas on the questioned sample. Figure 16.6 illustrates the method. In the figure, the left panels show the same cartridge case base A with a sequence of cells defined on a regular grid by the user. Note that jointly, the cells cover most of the breech face marked area. The two right panels show cartridge case B, to be compared to cartridge case A. On the top right panel, B appears to match A; after some rotation, it is possible to find congruent cells in B for almost all cells in A. In contrast, the image on the bottom right panel illustrates what happens when the two cartridge cases were fired from different guns. Even after systematically rotating image B by small increments of two degrees, it is not possible to fnd cells in B to correspond to the cells in A.

The number of CMC can be thought of as a score. When two cartridge cases are fired from the same gun, we expect to see a larger number of CMC than when the rounds were fired from different guns. To establish a threshold number of CMC to conclude that the samples were indeed fired from the same gun, the authors obtain the empirical distribution of the number among pairs of cases known to have been fired by the same gun and pairs of cases known to have been fired by different guns. Their method is promising in that the empirical distributions, at least for the cases they have considered, put mass on distinctly different ranges of number of CMC for known matching and known non-matching pairs of cases.

Instead of dividing the image into cells, a common alternative is to use a global measure of similarity between two images, for example the correlation between the corresponding pixels in the two images. For such methods to work, a series of processing and alignment steps is required. These methods are described by Tai and Eddy (2017) for the case of 2D cartridge case images.

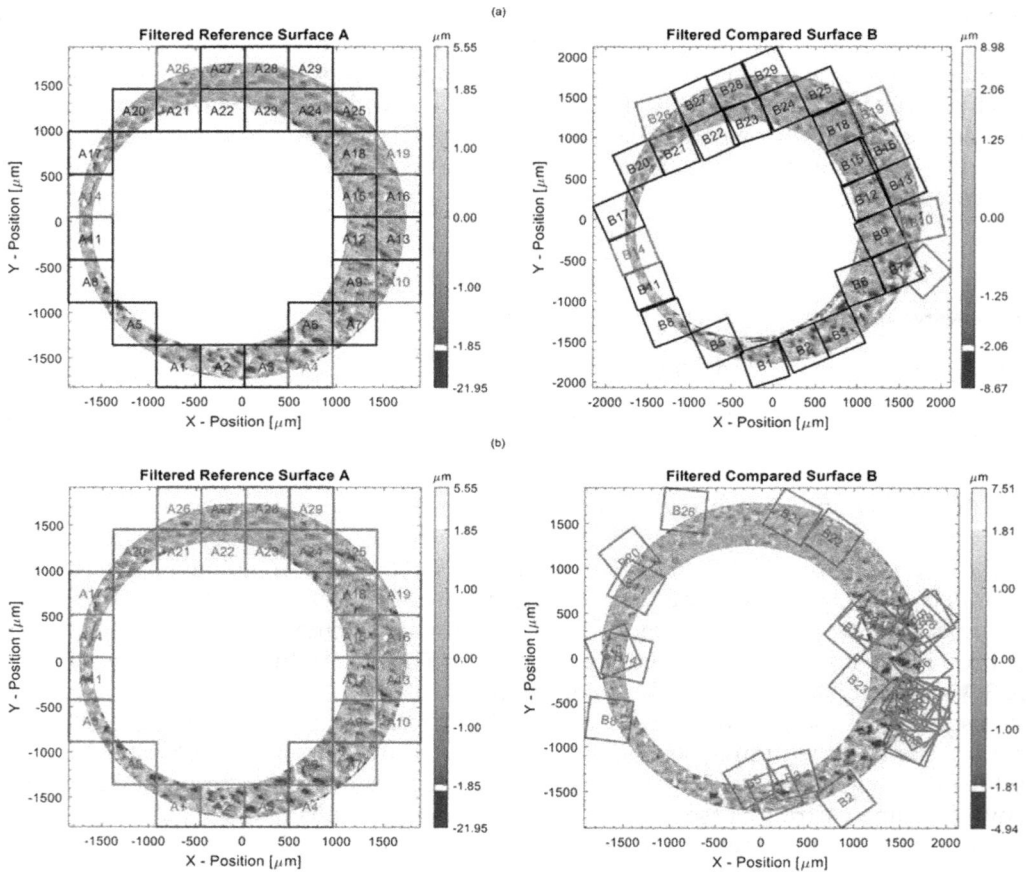

FIGURE 16.6
The method of Congruent Matching Cells. Top two panels illustrate the case where the two cartridge cases were fired from the same gun. The bottom two panels show an example of two cartridge cases fired from different guns. Image downloaded from NIST, (https://www.nist.gov/news-events/news/2018/02/how-good-match-it-putting-statistics-forensic-firearms-identification).

The first step in the pre-processing of the image is to select parts of the image that are of interest, in this case the breechface marks. This involves removing the firing pin impression, and any other extraneous details outside of the perimeter of the breechface marks. This can be done automatically either using image processing techniques, or in 3D by fitting a plane to the breechface area and setting a distance threshold beyond which information is discarded. Next, the image is further processed to highlight distinguishing features and remove sources of inaccuracies. For example, a Gaussian filter is commonly applied, which roughly corresponds to highlighting individual characteristics that examiners look for, while removing noise from the image acquisition process.

After pre-processing, a pair of images needs to be aligned to each other before a similarity score can be extracted. Assuming that images have been captured in a roughly similar manner, as is usually the case with equipment used in forensic laboratories, three parameters need to be estimated: the rotation angle, horizontal translation and vertical translation. This can be done using a grid search, looking for the set of parameters that maximizes the similarity measure. Other optimization techniques, such as the Lucas-Kanade algorithm (Lucas and Kanade, 1981), often used in object tracking in the computer vision literature, have also been tested. These have the advantage of much shorter run-times compared to a grid search, but often get stuck in local minima, thus producing poorer overall results.

Tai and Eddy (2017) test the above methods on various publicly available data sets maintained by NIST (https://tsapps.nist.gov/NRBTD/), and illustrate some of the results. Each data set contains a varying number of breech face images, collected for different purposes by groups in the firearms and toolmarks community. For example, a set called the NIST Ballistics Identification Database Evaluation (NBIDE) has images from 144 cartridge cases generated from three gun types (Sig Sauer, Ruger, and Smith & Wesson), four guns of each type, four ammunition types (Remington, Winchester, PMC, and Speer), and firing three repetitions for each ammunition type. Both 2D images (using reflectance microscopy) and 3D topographies (using confocal microscopy) were captured. Given these 144 cartridge cases, there are a total of 10,296 pairs of cartridge cases, meaning that the procedure for pairwise alignment needs to be done 10,296 times. Since these data were collected as part of an experiment, the ground truth for whether each pair of images comes from the same gun is known, and Tai and Eddy (2017) can generate plots of the empirical distribution of similarity scores for matching and non-matching pairs. These empirical distributions are shown in Figure 16.7. As we can see, non-matching pairs generally have low similarity scores, while matching pairs have either low or high scores.

If the problem of interest is to predict whether a specific pair of cartridge cases have the same source, we need to select a cutoff for the similarity score, above which we classify the pair to be a match. If distributions of similarity scores for matching and non-matching pairs were clearly separated for all data sets, such a cutoff would be straightforward to determine, but as we see in Figure 16.7, this is not the case. Tai and Eddy (2017) consider different cutoffs, and illustrate the results using precision and recall graphs in Figure 16.8.

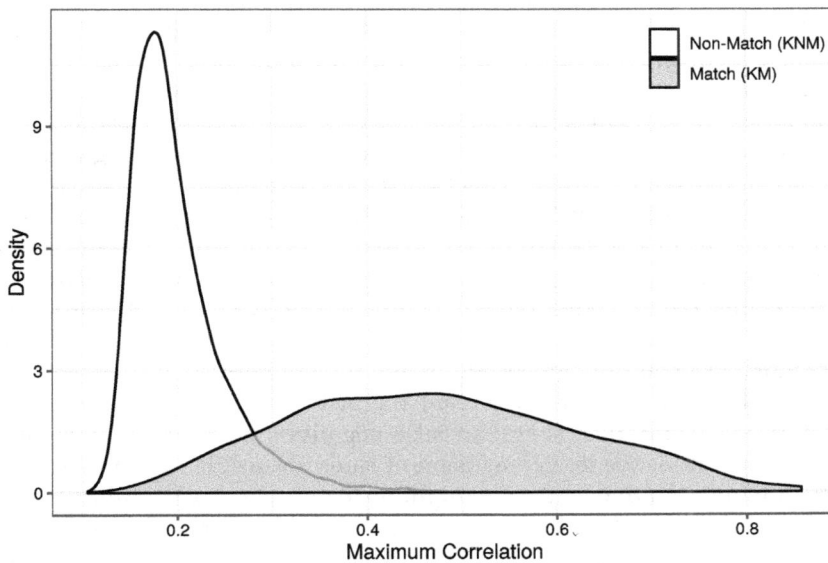

FIGURE 16.7
Distribution of similarity scores for matching and non-matching pairs for the NBIDE data set. Image provided by X-H Tai.

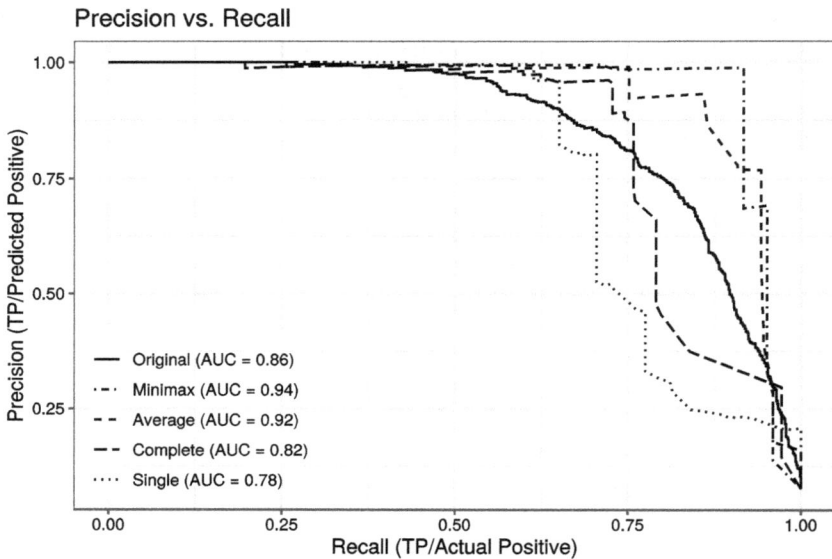

FIGURE 16.8
Precision and recall for various cutoffs on the similarity score. Additionally, precision and recall are also calculated for different cutoffs using different linkage methods in hierarchical clustering. Image provided by X-H Tai.

As a reminder, *precision* refers to the proportion of true matches among pairs that the algorithm predicted to have a common source, and *recall* is the proportion of same source pairs that were identified as such by the algorithm. A good reference for hierarchical clustering and the various linkage methods is Johnson and Wichern (2013).

In actual cases, forensic practitioners may have multiple cartridge cases to compare, so the issue of the transitivity of matches becomes important. Given that we are considering pairs of images, a classification method might produce the result that A matches B and B matches C, while A does not match C. To resolve this problem, Tai and Eddy (2017) propose using hierarchical clustering on the pairwise results, with various types of linkages (see Murtagh and Contreras, 2012). In the example above, using single linkage would simply add a link between A and C, concluding that A, B and C all come from the same gun. Results observed after applying various clustering techniques to the similarity scores obtained by Tai and Eddy (2017) are also plotted in Figure 16.8; average and minimax linkage produce the best results in terms of precision and recall on the NBIDE data set.

We conclude this section with several comments as well as suggestions for future research directions. In an actual implementation of the type of methodology described above, one might select cutoffs depending on the type of performance desired. For example, in a criminal case where the defendant might be implicated and prosecuted, false positives could be more undesirable than false negatives, in which case we might like a very high level of confidence in a conclusion of same source. If such a method is simply used to generate investigative leads, we might instead prefer high recall. Also, regardless of the type of algorithm used for comparing two breech face marks, different data sets produce varying quality of results (see, e.g., Figures 16.7 and 16.8). This is roughly consistent with observations by examiners that some gun brands or ammunition types produce varying quality of marks. Finally, in light of the criticism of subjectivity in pattern matching techniques, any algorithm that is fully automatic and produces reliable and reproducible results would be highly desirable. Before any one approach is selected for application in real case work, it will be critical to continue testing and fine-tuning the different approaches that look promising.

16.5 Comparison of Marks on Land Engraved Areas of Bullets

For most rifling types, fired bullets exhibit a sequence of land engraved areas separated by grooves. The striations that are used to compare bullets appear on the land engraved areas, so the measurements are obtained by scanning each area individually. Measurements on a land engraved area consist of heights on an xy grid in micron-level increments. The exact resolution at which images are taken depends on the microscope. In Hare et al. (2017), scans made available through the NIST Ballistics Research Database were used to develop their algorithm. These scans are taken at a resolution of 1.5625 μm × 1.5625 μm. The total area that is captured from each land engraved area is approximately 2.2 mm × 0.6 mm, and the data are the xyz coordinates of each point on the grid. For one land engraved area, the dimension of the xy matrix is 507 × 1001, which means that the number of z-values obtained for each land engraved area is slightly over half a million. For an entire bullet with six land engraved areas, the dataset has more than three million measurements. Data files in the format just described can be downloaded from the NIST ballistics database (Zheng, 2014). Figure 16.9 shows the 3D scan of a land engraved area of a bullet fired from a Smith & Wesson firearm. Well pronounced striations can usually be found close to the bottom of the bullet, in the area shaded in darker gray in the figure.

The algorithm proposed in Hare et al. (2017) uses the average height of striations on a set of consecutive cross-sections of the land engraved area at a specific value of y. This value is determined automatically, by computing the cross-correlations among heights at consecutive cross-sections and selecting the coordinate y at which the cross-correlations are less variable. This is the area on the land wher striations appear to be most stable. Figure 16.9 shows the selected value of y for a land engraved area scan. Figure 16.10 shows the same scanned area as in Figure 16.9, but viewed as a cross-section where y is fixed. The bottom panel is a single-pixel representation of the cross-section shown in the upper panel. In both panels, it is evident where the land engraved area begins and ends, but in some cases, finding the grooves in an automated way is challenging.

The dominant feature in the top panel of Figure 16.10 is the curvature of the surface of the land engraved area. This structure is so pronounced that the striations appear to be

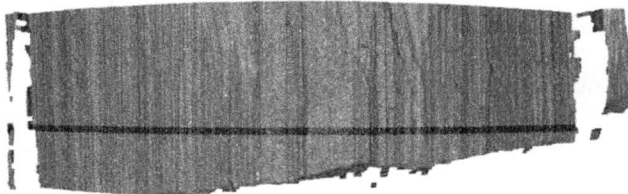

FIGURE 16.9
Scanned surface of a bullet land engraved area. Striations in the area shaded in dark gray are well pronounced and informative for the matching process.

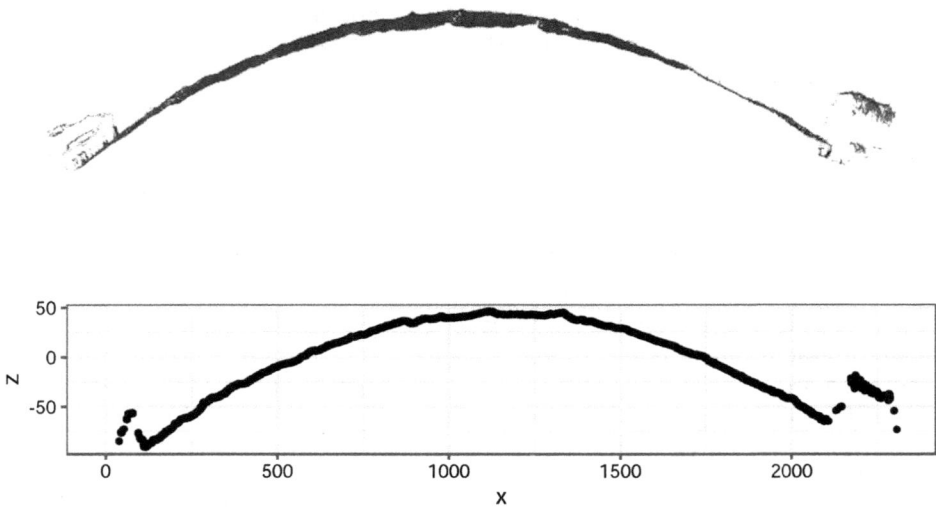

FIGURE 16.10
The top figure shows a sideways view of the scan from Figure 16.9. The figure in the bottom panel is a one-pixel representation of the sideways view of the cross-section shaded in dark gray in Figure 16.9, with measurements of the height of striations shown as the z value on the vertical axis. The horizontal axis represents the location in microns, relative to the leftmost edge of the image.

negligibly shallow. To bring the striations to the forefront, the first step in the algorithm proposed by Hare et al. (2017) is to subtract the curvature from the image by fitting a local polynomial regression model via Loess (Cleveland and Devlin, 1988). The residuals from the Loess function represent the striations and comprise the data that will be used to construct a signature. Figure 16.11 shows the fitted curve in gray in the top panel, and the Loess residuals in the bottom panel. Note the difference in scale of the two figures; the range of heights of the land engraved area on the top panel is about 120 μm, while the range of the residuals shown in the bottom panel is approximately 6 μm. For some combinations of ammunition and gun, the striations are shallower.

16.5.1 Pairwise Comparisons

The signature of a land engraved area is in the form of a text file with the ordered depth of the striae. The location at which each depth was measured can be inferred from the relative location of the measurement in the file and the size of the step between one measurement and the next. The question of interest is whether signatures extracted from a land engraved areas from two different bullets are similar enough to suggest that the bullets were fired from the same barrel. To quantify the degree of similarity between two signatures, Hare et al. (2017) propose overlaying the signatures and then computing several different features that represent the distance between the two signatures. For example, it is possible to compute a cross-correlation between two signatures, or to measure differences in the height of peaks or depth of valleys. Hare et al. (2017) define several different features that can be quantified and used to decide whether the signatures are similar enough to suggest that the bullets may have been fired from the same barrel. Figure 16.12 shows the overlain signatures for two bullets that were fired from the same barrel. Even though the signatures

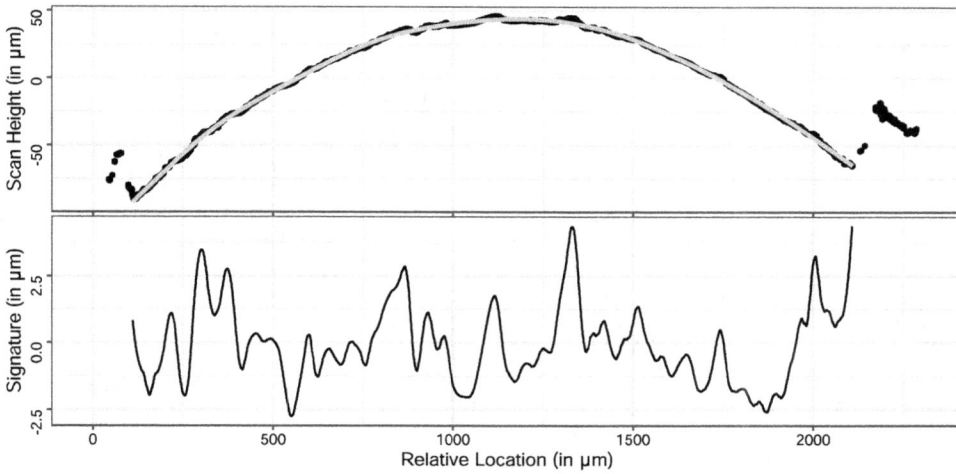

FIGURE 16.11
Loess fit to remove curvature shown in gray (top). Residuals from the Loess fit (bottom) show the relative heights of the engraved striae.

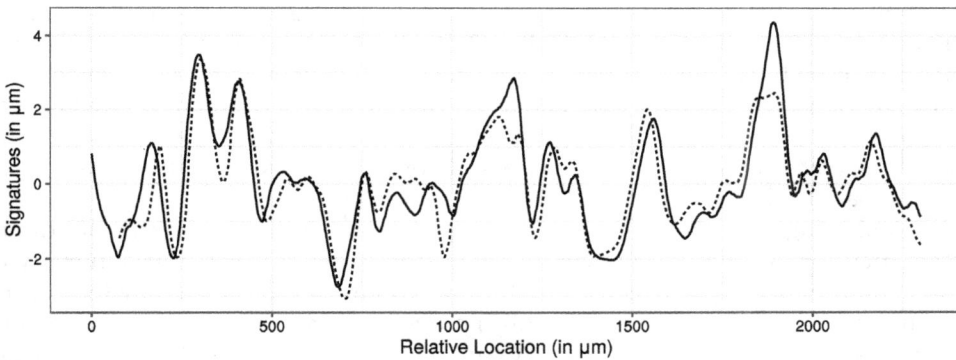

FIGURE 16.12
Overlaid signatures of two bullets fired from the same gun.

are not exactly the same, they are *similar enough* to suggest that they may have a common source. We revisit the definition of "similar enough" later in this chapter.

Hare et al. (2017) extracted seven features from the overlain signatures; to explore whether any of them would serve to accurately determine whether two bullets could have been fired from the same gun, they carried out the following experiment. First, they obtained 3D images of bullets that were fired by Hamby et al. using 10 consecutively rifled 9 mm Ruger barrels. The data consisted of two bullets known to have been fired by each of the 10 barrels, plus 15 "unknown" bullets, where we know that at least one and no more than three were fired by each of the test barrels. The design of this study has been criticized (President's Council of Advisors on Science and Technology, 2016) because it is *closed*, meaning that after the first few comparisons, the number of candidates becomes smaller and smaller. Next they constructed all possible pairs of images of land engraved areas, resulting in a total of 10,384 pairs of lands; of these, 172 pairs corresponded to the same

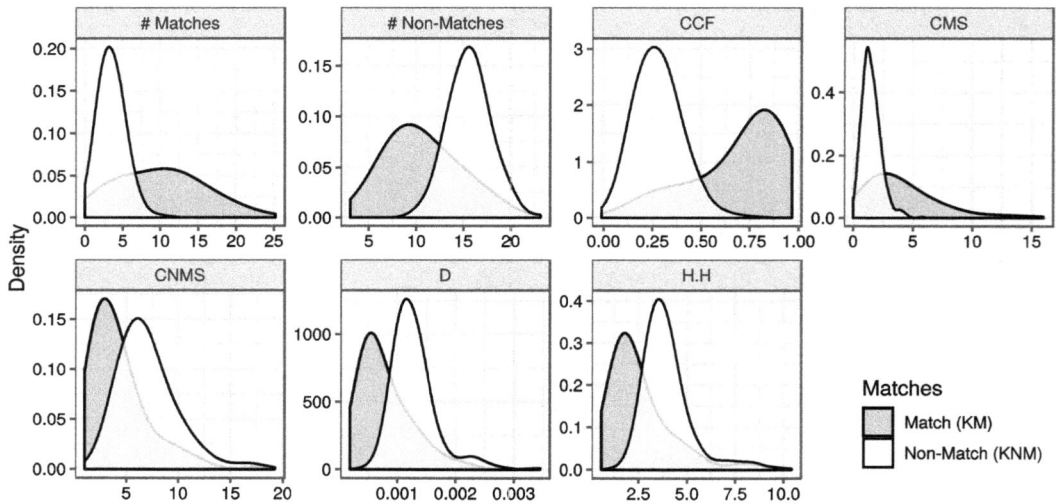

FIGURE 16.13
Empirical feature distributions.

land on bullets fired from the same barrel, and the rest (10,212) corresponded to either different pairs of land engraved areas on bullets fired from the same barrels or to pairs of land engraved areas on bullets fired from different barrels. Two damaged lands were excluded from the comparisons. Then, for each pair of images Hare et al. (2017) computed the value of each of seven features; the distribution of values of each of the features among pairs of known matching land engraved areas, and among pairs of known non-matching land engraved areas are shown in Figure 16.13.

Figure 16.13 suggests that none of the seven features individually can discriminate between known matching and known non-matching pairs. This is because in every case there is significant overlap between the distributions of feature values among the matching and the non-matching pairs. Hare et al. (2017) proposed instead that features be combined into a single score using a random forest. A random forest is an ensemble supervised learning method for classification or regression based on the idea of decision trees (Breiman, 2001). For classification, the random forest is composed of many decision trees (500 or more) that are grown using bootstrap samples of the original data, and selecting a random subset of the features for each binary split. The output the most likely class for each unit, and each unit is scored between 0 and 1, where larger scores denote a higher degree of similarity. Since the scores represent the average proportion of units of the predicted class in each terminal node, random forest scores can be thought of as the empirical probability of same class computed for each pair of images.

16.5.1.1 From Land-to-Land Comparisons to Bullet-to-Bullet Comparisons

Land-land comparisons lead to a set of scores for bullet-bullet comparisons. Figure 16.14 shows two matrices of scores for a pair of bullet-bullet matches. On the left, a matrix is shown that is typical for scores from two bullets from the same barrel. On the right, values for a pair of known non-matching bullets are shown.

When imaging bullets, operators scan one land at a time in a clockwise (left twisted rifling) or anti-clockwise (right twisted rifling) sequence. The order in which scans are

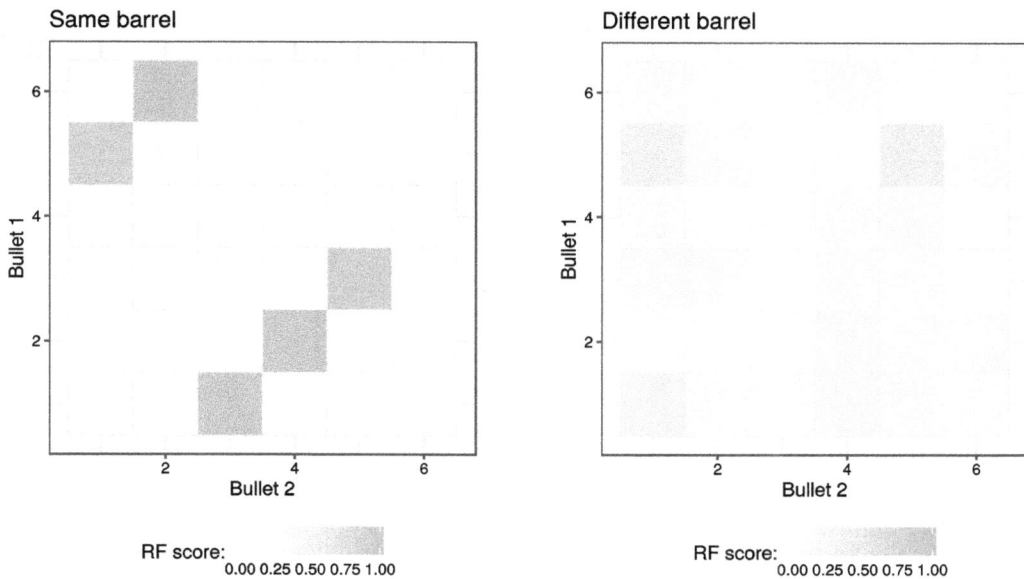

FIGURE 16.14
Overview of land-to-land matching scores for two pairs of bullet-to-bullet comparisons. On the left, the two bullets are known to come from the same source (barrel); on the right, the bullets are from two different sources (barrels).

acquired is kept as meta-information. Let us assume that lands on a bullet are labelled ℓ_i with $i = 1, ..., p$. A match between two bullets therefore results in an expected additional $p - 1$ matches between pairs of lands. These lands are also expected to be in a sequence, i.e. if there is a match between lands ℓ_i on bullet 1 and ℓ_j on bullet 2, we also expect lands $\ell_{i \oplus s}$ and $\ell_{j \oplus s}$ to match for all integers s, where $\oplus$ is defined as $a \oplus b \equiv ((a + b) \mod p) + 1$. Here, the symbol mod() refers to the *modulo* operator, which finds the remainder of the division of one number by another. This relationship gives rise to the sequence average maximum (Sensofar, 2017), also known as "max phase" in Chu et al. (2010) to quantify a bullet-to-bullet match. Figure 16.15 shows all possible in-phase combinations for bullets with 6 lands.

16.5.1.1.1 Definition (Sequence Average and its Maximum)
Let A be a square real-valued matrix of dimensions $p \times p$. The kth *sequence average* $SA(A, k)$ for $k = 0, ..., p - 1$ is defined as

$$SA(A, k) = \frac{1}{p} \sum_{i=1}^{p} a_{i, i \oplus k}, \text{ where } i \oplus k := ((i + k) \mod p) + 1.$$

The *Sequence Average Maximum* (Sensofar, 2017) of square matrix R is defined as

$$SAM(R) = \max_{k=1}^{p} SA(R, k).$$

Looking back at Figure 16.14, we see that for the two bullets from the same barrel, the sequence average for $k = 2$ is higher than the other sequence averages, and also higher than the sequence averages for the other pair of bullets shown on the right of the figure.

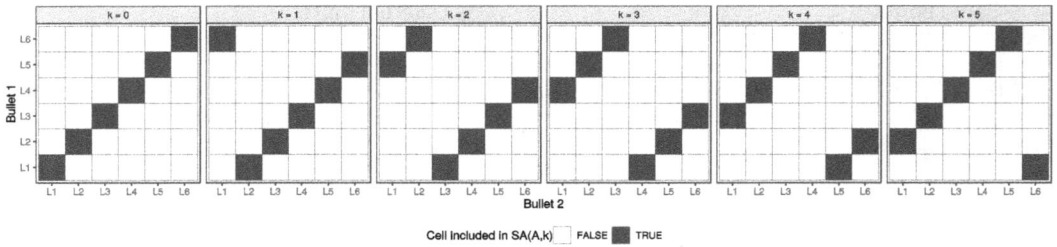

FIGURE 16.15
Sketch of all six land-to-land sequences between two bullets with six lands.

16.5.1.2 Results for the Hamby Set of Bullets

For the case of the Hamby bullets, the random forest produced a perfect classification of all pairs into the two classes. The separation between the values of the scores among known matching and known non-matching pairs was complete, as is shown in Figure 16.16.

A limitation of all learning algorithms is that they depend critically on the data used to train them and they tend to over-fit the training data. Consequently, the algorithm's performance when classifying a new set of units can be much worse and result in large misclassification errors. This occurs when the minimizing the bias in the training data is the dominating criterion. To guard against over-fitting, it is possible to set aside a portion of the training data that can then be used as a test dataset, but even then, the misclassification error tends to be underestimated, albeit to a lower extent.

To explore whether the random forest fitted to the Hamby et al. (2009) bullets has good performance when used to classify pairs of land engraved areas from bullets fired from other barrels or by guns of different make and model, and when the ammunition is also made by a different manufacturer, Hare et al. (2017) applied the model to thousands of pairs of known matching and known non-matching land engraved areas obtained from crime laboratories across the United States. Results have been promising; the random

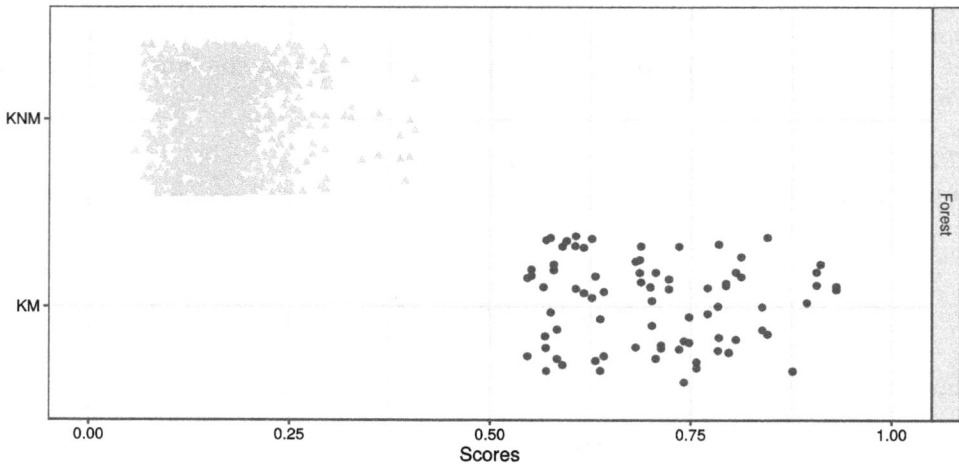

FIGURE 16.16
Random forest scores for the Hamby et al. (2009) known matching pairs of land engraved areas (dark gray) and known non-matching pairs (light gray).

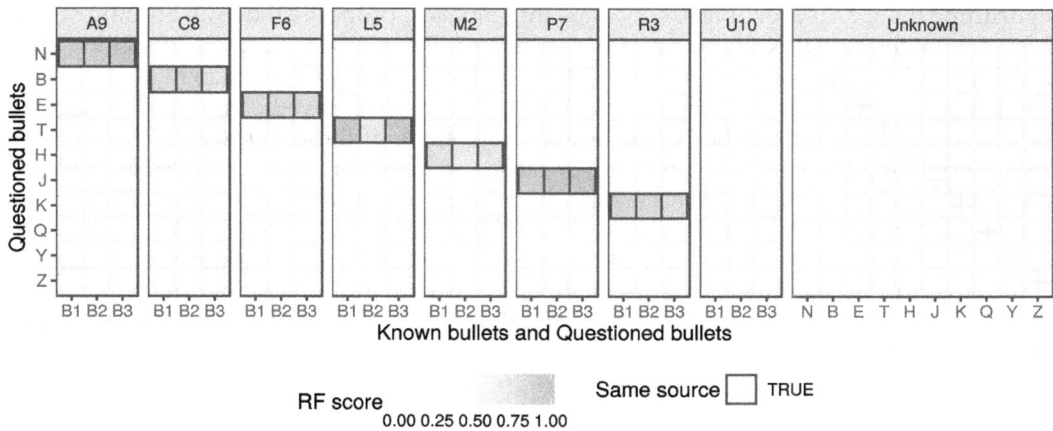

FIGURE 16.17
Random forest scores of the similarity between ten test shots and 24 reference shots from eight 9 mm barrels.

forest correctly determined every pair of bullets in every study when the gun barrel had conventional (rather than polygonal) rifling and striations were reasonably well marked, and when bullets were not coated with a polymer that flakes on contact with the barrel. Figure 16.17 shows the results observed in one such test set. Each pairwise comparison is represented as a square, with the center of the square shaded to represent the composite match score for the six lands on each bullet; squares are outlined with a dark rectangle if the bullets are from the same source. In the test set shown in Figure 16.17, there are three reference shots from each of eight barrels, and 10 fired rounds of unknown provenance. The test set was open, which means that it was possible that some of the test shots were not fired by any of the eight study barrels and that one or more of the barrels may have fired none of the 10 test shots; in this case, questioned bullets Q, Y, and Z did not originate from the 8 barrels in the study.

Hare et al. (2017) used a threshold of 0.5 for the SAM computed from the random forest score to categorize pairs of bullets into "same" and "different" gun. A row of three orange squares indicates that the corresponding questioned bullet matched all three of the reference shots fired by the barrel indicated in the column. In all cases, the algorithm classified the questioned bullets correctly. It correctly concluded, for example, that the last three questioned shots labeled Q, Y and Z were not similar enough to any of the reference shots to declare a match, and that the barrel labeled U10 had not fired any of the questioned rounds.

The comparison method proposed by Hare et al. (2017) is promising, but the method must undergo extensive additional testing and validation before it can be implemented in casework. Researchers working on this type of problem have found that the optimal score threshold for categorizing pairs of bullets as "same" or "different" varies from dataset to dataset. This may be a consequence of the fact that the random forest that Hare et al. (2017) have been applying to those sets of images was trained only once and only on a small training set. As more 3D images become available, it will be important to re-train the forest on a more diverse set that perhaps includes combinations of bullets and guns of the same caliber but manufactured by different companies. This issue brings up a more general statistical question: what are the factors that have an effect on the attributes of the striations that the gun imparts on a bullet and consequently, on the performance of the classification algorithm? It seems reasonable to speculate that physical properties such as the hardness of the inside of the barrel of the gun and of the bullet jacket might be important factors. Type of rifling and the degree of smoothness of the inside of the barrel are also likely to be factors that affect the characteristics of striations. To date, there have been no well-designed, appropriately powered studies to begin exploring some of these questions. We note that fully automating the process is difficult, because the detection of the grooves is challenging. At this moment, it is still necessary to carefully check the automatically identified groove locations, which not only slows down the process but also increases the chances for errors. Finally, adoption of any new technology in working crime labs will require significant changes in protocols and in the way in which firearm examiners interpret and present their evidence. Thus, it is not likely that we will see these methodologies used in practice for several more years.

16.6 Revisiting the Question of Source

The machine learning methods we have discussed to quantify the similarity between two items do not, by themselves, suffice to address the question of source. If a questioned bullet and a test shot from the suspect's gun are similar, all we can say is that the suspect's gun cannot be excluded as the source of the questioned bullet. To be able to conclude that the suspect's gun and no other fired the questioned round, we would need to also show that the degree of similarity we observed is probative: we observed the same degree of similarity only when two bullets were fired from the same gun. The same type of reasoning applies to all other forms of evidence.

In the case of DNA profiles, we know that barring laboratory error, a match between two samples is probative, because no two individuals (except identical twins) have the same set of alleles at every marker. For fingerprints, it is also assumed that each person is born with an individual pattern, so that perfect prints from two different individuals should be distinguishable. For all other types of evidence, we have no means to compute a probative value, and therefore, we do not know whether a high degree of similarity between two items is an indication of a common source. Until recently, forensic examiners in reports or testimony were able to focus on the similarity between two objects and conclude that high degree of similarity implied common source. But the blunders that led to questioning the validity of many forensic disciplines also led the public and the scientific community to revisit assumptions such as the uniqueness and repeatability of striations of rounds fired by a gun. As a result, juries and law professionals today are more likely to expect some discussion about the probative value of a "match" from forensic experts.

The probative value of evidence, or more precisely the *weight of evidence*, can be estimated using a likelihood ratio (Lindley, 1977; Grove, 1980). The likelihood ratio is a one-number summary of the probability of observing the evidence under the two competing hypotheses of same (H_{ss}) or different (H_{ds}) source. Let E denote "evidence". Computation of the numerator $\Pr(E|H_{ss})$ typically involves information collected and analyzed in the process of the investigation. The calculations required to quantify the denominator, on the other hand, are challenging, because they require information about the background population. This in turn raises other challenging questions including what is the relevant background population in each case, and whether available databases are representative of the specific background population.

In the absence of a statistical model that may permit estimating the likelihood of a pattern under different assumptions about provenance, it may be possible to compute an empirical estimate of the frequency with which a given degree of similarity between two items can be expected when the items have a common source and when they do not. Figure 16.18 illustrates this idea using firearms as an example.

The empirical distributions of similarity scores shown in light and dark gray in Figure 16.19 represent the values of the score that could have been obtained when comparing bullets fired by the same or by different guns, respectively. Suppose that these distributions of scores were obtained from a very large number of pairs of bullets representing the population of guns and ammunition of a certain caliber. A crime is committed and a bullet is recovered from the scene. The suspect's gun is test fired and the bullet in evidence is compared to the test shot using the algorithm that produced the background scores. Suppose that the resulting score is 0.8, as in Figure 16.19. Visual inspection suggests that a score of 0.8 appears to be likely if the two bullets were fired from the same gun and unlikely otherwise. A *score-based likelihood ratio* (SLR) is computed as the height of orange empirical density over the height of the gray distribution at 0.8, which in this example comes out to 371.4. The interpretation of this value is the following: a score of 0.8 is about 370 more likely when two bullets (or cartridge cases) were fired from the same gun than when fired from different guns, so the score represents strong support for the same source proposition. We could also calculate the probability of observing a score of 0.8 under each of the two empirical distributions. Regardless of the method, we would conclude that the two bullets were likely fired by the same gun. If instead the similarity score was 0.3, it would not be possible to reach a conclusion regarding source with any degree of confidence. The ratio of the heights of both distributions for a score of 0.3 is 0.3 indicating that it is about twice as likely to observe that score if bullets were fired by different guns, which is only weak evidence in favor of the hypothesis of different source. Lund and Iyer

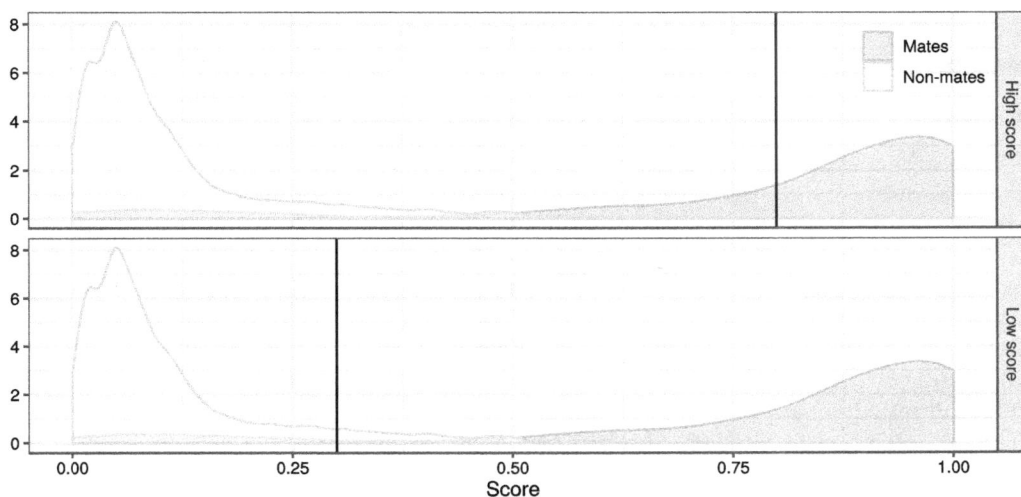

FIGURE 16.18
Empirical distributions of random forest scores for pairs of land engraved area scans known to have been produced from the same land (gray) and known to have originated from different lands (white). The vertical lines in the top and bottom panels represent the score that was obtained when comparing a questioned with a reference scan of a land engraved area.

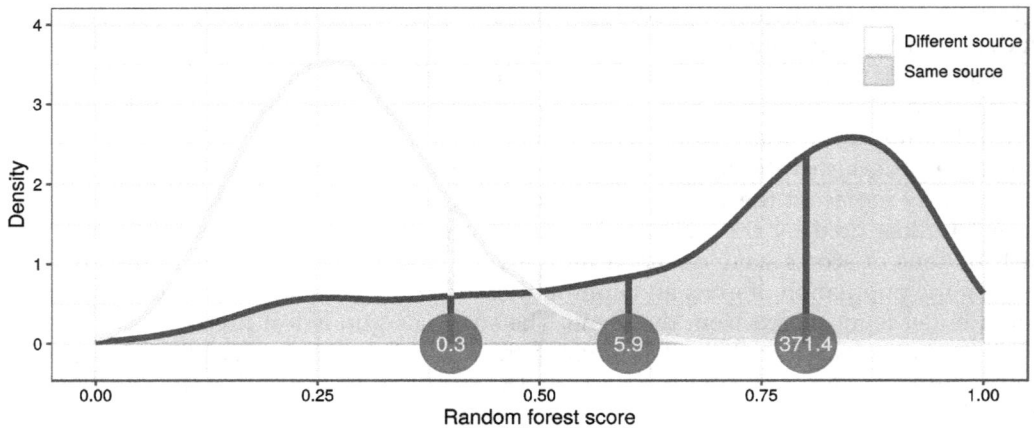

FIGURE 16.19
Densities of known matches (dark gray) and known non-matches (light gray) corresponding to all pair-wise land comparisons of the Hamby set. The large gray points along the x axis mark three hypothetical scores a comparison might result in. The number in white inside the points gives the corresponding score-based likelihood ratio, indicating the likelihood to observe this score from a known match rather than a known non-match.

(2017) and references therein discuss some potential limitations of score-based likelihood ratios to approximate model-based likelihood ratios.

The conclusion is that to assess the probative value of a match between two pieces of evidence, we need to have extensive background information about the population from which those two pieces of evidence might have originated. Thus, the construction of those reference databases must become a priority.

16.7 Some Final Thoughts

The role of statistics in the evaluation and interpretation of bullet and cartridge case evidence has only recently become apparent. While the idea of moving toward a likelihood ratio framework for estimating the weight of evidence has gradually been accepted in several forensic disciplines (e.g., latent prints, handwriting, trace), this is not the case among firearm and toolmark examiners. In part, this can be explained by the fact that we know very little about the factors that affect how striations are imparted on bullets and cartridge cases. Consequently, we do not really know how to construct the reference populations of bullets and cartridge cases that are needed to calculate the appropriate denominator of the likelihood ratio in a specific case.

This chapter has focused on the promising approaches that have been proposed to quantify similarity between two bullets or two cartridge cases. Much of the progress has relied on the development and implementation of data science tools such as supervised learning algorithms. What so far has received less attention is the critical role that more traditional statistical tools including experimental design, sampling, two- or higher-dimensional functional analysis, and non-parametric models to represent, for example, the breech face impression on a cartridge case, can play in the construction of the likelihood ratio infrastructure in firearms and toolmark examination.

Consider again the calculation of the "probability of a coincidental match", or the denominator of the likelihood ratio. How should we determine what is the relevant background population for computing $Pr(E|H_{ds})$? Suppose that the recovered gun in a crime scene is a 9 mm Sig-Sauer P320 and the cartridge case corresponds to ammunition manufactured by Winchester. One approach to computing $Pr(E|H_{ds})$ is to base calculations on similarity scores among known matches and known non-matches obtained using exclusively Sig Sauer P320 guns and Winchester ammunition. This approach necessitates the creation of a very large number of datasets, one for each possible combination of model of gun and make of ammunition. A principled, potentially more efficient approach is to understand which gun and ammunition attributes result in comparable striations. With this information, we can then group gun/ammunition combinations and greatly reduce the number of different reference populations that need to be built and maintained. Understanding the factors that are associated with variability in striations of ammunition fired by the same or by different guns will require carefully designed experiments, that are adequately powered.

Machine learning has the potential to greatly improve the way in which forensic scientists evaluate pattern and other types of evidence. Yet much needs to be done before any of these new methods are ready for implementation in real casework. Serious limitations for further development include the dearth of data available to researchers, and the fact that collecting additional data in the form of 3D images of bullets and cartridge cases is both costly and time consuming. Efforts including the NIBD database maintained by NIST, which is growing and continues to be publicly available, are beginning to pay off in terms of attracting more researchers to work in this area. Organizations such as the Center for Statistics and Applications in Forensic Evidence (CSAFE; www.forensicstats.org), that put data and algorithms in the public domain, are also contributing to transparency and provide resources that enable independent research.

No algorithm will completely replace humans in the analysis and interpretation of evidence, but with representative and large background databases, extensive testing, and validation, learning algorithms and other statistical tools can alleviate some of the subjectivity of forensic firearms examination as it is currently practiced and can serve to at least approximate the degree of uncertainty in forensic conclusions.

References

Bachrach, B. Development of a 3d-based automated firearms evidence comparison system. *Journal of Forensic Science*, **47**, 1–12, 2002.

Beattie, J.M. *The First English Detectives: The Bow Street Runners and the Policing of London, 1750–1840*. Oxford University Press, Oxford, England, 2012.

Biasotti, A. A statistical study of the individual characteristics of fired bullets. *Journal of Forensic Sciences*, **4**, 34, 1959.

Breiman, L. Random forests. *Machine Learning*, **45**(1), 5–32, October 2001. ISSN 1573.0565. https://doi.org/10.1023/A:1010933404324

Burrard, G. *The Identification of Firearms and Forensic Ballistics*. Herbert Jenkins Ltd., London, 1934.

Carriquiry, A., Hofmann, H., Tai, X.-H., and VanderPlas, S. Machine learning in forensic applications. *Significance*, **16**(2), 29–35, 2019.

Chu, W., Song, J., Vorburger, T., Yen, J., Ballou, S., and Bachrach, B. Pilot study of automated bullet signature identification based on topography measurements and correlations. *Journal of*

Forensic Sciences, **55**(2), 341–347, 2010. ISSN 1556.4029. http://dx.doi.org/10.1111/j.1556-4029. 2009.01276.x

Chu, W., Thompson, R.M., Song, J., and Vorburger, T.V. Automatic identification of bullet signatures based on consecutive matching striae (cms) criteria. *Forensic Science International*, **231**, 137–141, 2013.

Chumbley, L.S., Morris, M.D., Kreiser, M.J., Fisher, C., Craft, J., Genalo, L.J., Davis, S., Faden, D., and Kidd, J. Validation of tool mark comparisons obtained using a quantitative, comparative, statistical algorithm. *Journal of Forensic Sciences*, **55**(4), 953–961, July 2010. ISSN 1556.4029. http://onlinelibrary.wiley.com/doi/abs/10.1111/j.1556-4029.2010.01424.x.00040

Cleveland, W.S. and Devlin, S.J. Locally weighted regression: a approach to regression analysis by local fitting. *Journal of the American Statistical Association*, **83**(403), 596–610, 1988.

Commonwealth v. Best, 62 N.E. 748 (Mass. 1902).

Conrad, W., Dillon, Jr. J.H., Hamby, J., Hill, G., Hill, R., Jones, J.A., Savage, K., and Tilstone, W. *Firearm Examiner Training*. National Forensic Science Technology Center, 2008. https://projects.nfstc.org/firearms

Davis, J.E. *An Introduction to Tool Marks, Firearms and the Striagraph*. Charles C. Thomas Publisher, Springfield, Illinois, 1968. ISBN 978-0-398-08916-0.

Federal Bureau of Investigation. *Crime in the United States*. U.S. Department of Justice, 2017. https://ucr.fbi.gov/crime-in-the-u.s/2017/crime-in-the-u.s.-2017

Fischer, R. and Vielhauer, C. Digital crime scene analysis: automatic matching of firing pin impressions on cartridge bottoms using 2d and 3d spatial features. In *Proceedings of the 2nd ACM workshop on Information hiding and multimedia security – IH&MMSec '14*, pages 77–82, Salzburg, Austria, 2014. ACM Press. ISBN 978-1-4503-2647-6. http://dl.acm.org/citation.cfm?doid=2600918--2600930

Frankford Arsenal Army Personnel. Study of the fired bullets and shells in brownsville, texas, riot. In *Annual Report of the Chief of Ordnance, U.S. Army*. U.S. Government Printing Office, Washington, DC, 1907.

Gardner, G.Y. Computer identification of bullets. *IEEE Transactions on Systems, Man, and Cybernetics*, **8**(1), 69–76, January 1978. ISSN 0018.9472. doi:10.1109/TSMC.1978.4309834

Geradts, Z.J., Bijhold, J., Hermsen, R., and Murtagh, F. Image matching algorithms for breech face marks and firing pins in a database of spent cartridge cases of firearms. *Forensic Science International*, **119**(1), 97–106, June 2001. ISSN 0379.0738. https://doi.org/10.1016/S0379-0738(00)00420-5

Goddard, C. The valentine day massacre: a study in ammunition-tracing. *The American Journal of Police Science*, **1**(1), 60–78, 1930. ISSN 1547.6154. https://www.jstor.org/stable/1147257

Grove, D.M. The interpretation of forensic evidence using a likelihood ratio. *Biometrika*, **67**(1), 243–246, 1980. https://doi.org/10.1093/biomet/67.1.243

Gunther, J.D. and Gunther, C.O. *The Identification of Firearms from Ammunition Fired Therein: With an Analysis of Legal Authorities*. J. Wiley & Sons, New York, NY, 1935.

Hall, A.L. The missile and the weapon. *The Buffalo Medical Journal*, **39**, 727–736, 1900. ISSN 1547.6154. https://archive.org/details/buffalomedicaljo3918unse/page/n739

Hamby, J., Brundage, D.J., and Thorpe, J.W. The identification of bullets fired from 10 consecutively rified 9mm ruger pistol barrels: a research project involving 507 participants from 20 countries. *AFTE Journal*, **41**, 99–110, 2009.

Hamby, J.E. and Thorpe, J.W. The history of firearm and toolmark identification. *AFTE Journal*, **31**(3), 226. 284, 1999.

Hare, E., Hofmann, H., and Carriquiry, A. Algorithmic approaches to match degraded land impressions. *Law, Probability and Risk*, **16**(4), 203–221, 2017.

Hatcher, J.S. *Textbook of firearms investigation, identification and evidence*. Small-arms technical publishing company, 1935. https://books.google.com/books?id=pm3zAAAAMAAJ

Heard, B.J. *Handbook of Firearms and Ballistics: Examining and Interpreting Forensic Evidence*. John Wiley and Sons, Chichester, England, 1997.

Johnson, R. and Wichern, D. *Applied Multivariate Statistical Analysis*. New International Pearson Edition, London, England, 2013.

Kinder, J.D. and Bonfanti, M. Automated comparisons of bullet striations based on 3d topography. *Forensic Science International*, **101**(2), 85–93, April 1999. ISSN 03790738. https://doi.org/10.1016/S0379-0738(98)00212-6

Lindley, D.V. Problem in forensic science. *Biometrika*, **64**(2), 207–213, 1977.

Lucas, B.D. and Kanade, T. An iterative image registration technique with an application to stereo vision. *Proceedings of Imaging Understanding Workshop*, pages 121–130, 1981.

Lund, S.P. and Iyer, H. Likelihood ratio as weight of forensic evidence: a closer look. *Journal of Research of the National Institute of Standards and Technology*, **122**(27), 1–32, 2017. https://doi.org/10.6028/jres.122.027

Murtagh, F. and Contreras, P. Algorithms for hierarchical clustering: an overview. *Wiley Interdisciplinary Reviews: Data Mining and Knowledge Discovery*, **2**, 86–97, 2012. https//doi.org/10.1002/widm.53

National Research Council. *Strengthening forensic science in the United States: a path forward*. National Academies Press, Washington, DC, 2009. ISBN 978-0-309-13135-3. https://www.ncjrs.gov/pdffiles1/nij/grants/228091.pdf

President's Council of Advisors on Science and Technology. *Forensic Science in Criminal Courts: Ensuring Scientific Validity of Feature-Comparison Methods*. Executive Office of the President, 2016. https://obamawhitehouse.archives.gov/sites/default/files/microsites/ostp/PCAST/pcastfiforensicfisciencefireportfifinal.pdf

Riva, F. and Champod, C. Automatic comparison and evaluation of impressions left by a firearm on fired cartridge cases. *Journal of Forensic Science*, **59**, 637–647, 2014.

Sensofar Metrology. SensoMATCH bullet comparison software, 2017. https://www.sensofar.com/wp-content/uploads/2016/06/AppNote-SensoMATCH-Bullet-Comparison.pdf

Song, J. Proposed nist ballistics identification system (nbis) using 3d topography measurements on correlation cells. *AFTE Journal*, **45**, 184–194, 2013.

Song, J., Vorburger, T.V., Chu, W., Yen, J., Soons, J.A., Ott, D.B., and Zhang, N.F. Estimating error rates for firearm evidence identifications in forensic science. *Forensic Science International*, **284**, 15–322222, 2018. https://doi.org/10.1016/j.forsciint.2017.12.013

Stout, W.W. Fingerprinting Bullets: The Expert Witness. *The Saturday Evening Post*, pages 6–7, June 1925a.

Stout, W.W. Fingerprinting Bullets: The Silent Witness. *The Saturday Evening Post*, pages 18–19, June 1925b.

Tai, X.-H. and Eddy, W.F. A fully automatic method for comparing cartridge case images. *Journal of Forensic Science*, **63**(2), 440–448, 2017. https://doi.org/10.1111/1556-4029.13577

Tai, X.-H. and Eddy, W.F. A fully automatic method for comparing cartridge case images. *Journal of Forensic Sciences*, **63**(2), 440–448, March 2018. ISSN 00221198. https://doi.org/10.1111/1556-4029.13577

Tong, M., Song, J., Chu, W., and Thompson, R.M. Fired cartridge case identification using optical images and the congruent matching cells (CMC) method. *Journal of Research of the National Institute of Standards and Technology*, **119**, 575, November 2014. ISSN 2165.7254. https://doi.org/10.6028/jres.119.023

US Department of Justice and Bureau of Alcohol Tobacco Firearms and Explosives. National integrated ballistic information network (nibin). Technical report, ATF, 2018. https://www.atf.gov/firearms/national-integrated-ballistic-information-network-nibin

Vorburger, T.V., Song, J.F., Chu, W., Bui, S.H., Zheng, X.A., and Renegar, T.B. Applications of crosscorrelation functions. *Wear*, **271**, 529–533, 2011.

Vorburger, T.V., Song, J.F., and Petraco, N. Topography measurements and applications in ballistics and tool mark identifications. *Surface Topography: Metrology and Properties*, **4**, 013002, 2015. https://doi.org/10.1088/2051-672X/4/1/013002

Weller, T.J., Zheng, X.A., Thompson, R., and Tulleners, F.A. Confocal microscopy analysis of breech face marks on fired cartridge cases from 10 consecutively manufactured pistol slides. *Journal of Forensic Science*, **57**, 912–917, 2012.

Wilson, C. and Wilson, D. *Written in Blood: A History of Forensic Detection*. Carroll & Graf, 2003. ISBN 978-0-7867-1266-3.

Zheng, X.A. NIST Ballistics Toolmark Database, 2014. https://www.nist.gov/programs-projects/nist-ballistics-toolmark-database

Zheng, X.A., Soons, J., Vorburger, T.V., Song, J., Renegar, T., and Thompson, R. Applications of surface metrology in firearm identification. *Surface Topography: Metrology and Properties*, **2**, 014012, 2014.

17

Shoeprints: The Path from Practice to Science

Sarena Wiesner, Naomi Kaplan-Damary, Benjamin Eltzner, and Stephan Huckemann

CONTENTS

17.1 Introduction 391
17.2 The Process of Evaluating Shoeprints 393
17.3 The Challenges that Arise from Analyzing the Working Procedure 395
17.4 Requirements for Good Practice 396
17.5 How to Strengthen the Scientific Foundation for Footwear Analysis? 397
17.6 Wishlist for and Design of the Semi-Automated ElementAccidentalSensor 399
17.7 An Early Stage Implementation of the Semi-Automated
     ElementAccidentalSensor 400
    17.7.1 Preprocessing and Binarization 400
    17.7.2 Element Detection 401
    17.7.3 Detecting Accidentals and Wear 403
17.8 Discussion and Conclusions 406
Appendix 17A 406
References 407

17.1 Introduction

Up until several decades ago, forensic evidence was regarded as a set of well-established and trusted methods, and in many legal systems was even considered to be incontestable. The emergence of DNA analysis, together with changes in the law and advances in the broader scientific community, led to a serious questioning of the validity of many of the traditional forensic disciplines (Saks and Koehler, 2005), and revealed errors made in both forensics and the legal system.

The most significant legal watershed resulted from the United States Supreme Court case Daubert v. Merrell Dow Pharmaceuticals (Daubert v. Merrel Dow Pharmaceuticals, 1995). The Daubert ruling established four "illustrative factors" for use in determining whether evidence was based on a "scientific methodology":

1. The theory or technique has been empirically tested.

2. The theory or technique has been subjected to peer review and publication.

3. The known or potential rate of error and the maintenance of standards controlling the techniques operation should be considered.

4. The methods and techniques are generally accepted within the scientific community.

Articles have been published in most forensic fields and have been peer reviewed by forensic specialists, but review by the broader scientific community is a much rarer occurrence. In many of these studies, the sample sizes are relatively small since the collection of large samples and their analysis is extremely time consuming and is performed at the expense of routine case work. The major challenges remain in the testability of the theories and the calculation of their error rates, both of which require substantial research efforts. These include incorporating black-box studies of the performance of examiners under conditions as close as possible to actual casework and blind quality control cases as part of routine work.

The development of DNA analysis, today a well-established scientific field, triggered the demand that other forensic areas meet its standards. The 2009 National Academy of Sciences (NAS) report (National Research Council, 2009) which evaluated the state of the science in various forensic fields and proposed the development of rigorous standards. Its important recommendations include the standardization of the terminology used in reporting, lab accreditation and personal certification, including estimated probabilities and measures of uncertainty in reported results, and taking steps to minimize potential bias and human error. Some of these aspects' standards were addressed in the 2016 President's Council of Advisors on Science and Technology (PCAST) report (President's Council of Advisors on Science and Technology, PCAST) and identified as future thresholds for courts in the USA.

For most fields of forensic evidence, current methods fall short of a scientifically objective and quantitative analysis that can be presented in court when trying to determine a match between samples. DNA is utilized routinely to link suspects to crime scenes because of its perceived scientific objectivity, developed statistical models and accessible documentation (at least with analyses of DNA from single sources). However, the evaluation of other types of evidence such as shoeprints and the like have not reached this gold standard (Kafadar, 2015).

The identification of footwear impressions is based on the comparison of a print found at the crime scene with a test impression made from a suspect's shoe (Bodziak, 2000). This comparison is performed by a qualified examiner who was hands-on trained by a practicing examiner for over a year until he gained enough experience and passed a practical exam. At first the correlation of shoe-sole pattern, size and wear between the test impression and the crime scene print is examined. If they correspond, weight is given to the "accidental marks", also known as RACs (Randomly Acquired Characteristics) that result from random processes of wear and tear, such as holes, scratches, etc. One of the common methods currently used to present the level of confidence in the connection between the test impression and the crime scene print is an ordinal scale, for example, the SWGTREAD scale, used in the USA (SWGTREAD, 2013), which is a subjective approach to the estimation of confidence or significance, rather than one based on an explicit statistical model. The current method of evaluating class and accidental characteristics clearly reduces the potential suspects' shoe population but it lacks formal statistical analyses based on large databases, and fails to provide a scientific and quantitative scale for assessing the match between a crime scene print and a suspect's shoe. As a result, crime lab practitioners constantly face the challenge of quantifying the probability and explaining the evidentiary value of such matches in court (High Court of Justice Court of Appeal, 2010; Bodziak, 2012).

Majamaa and Ytti (1996), Ytti et al. (1997) and Shor and Wiesner (1999) found variation in the conclusions of shoeprint reports drawn from identical cases in different forensic crime laboratories. However, Hammer et al. (2012) have shown that in contrast to these studies, when experienced examiners used the same conclusion scale and compared the same features, there was little variability in their stated findings. This need for data-based statistics led to an attempt to study the distribution of RACs conducted by Yekutieli et al. (2016) and exposed the necessity for further research.

This chapter first surveys current challenges of evaluating shoeprint evidence from the collection stage to the submission of testimony in court from the perspective of one of the author's (S.W) shoeprint lab, then current working procedure is reviewed, and finally semi-automated pattern and RAC retrieval is addressed in detail as a suggested improvement step. On the ground of these three pillars, practical recommendations and initial scientific steps are proposed to carry the shoeprint comparison field along the path from practice to science.

17.2 The Process of Evaluating Shoeprints

In this section the process of documenting and evaluating shoeprints will be presented as a basis for understanding the pitfalls of the current practice.

1. CSIs (Crime Scene Investigators) who arrive at the crime scene, using oblique light search for locations where the perpetrators have likely left their shoeprints, preferably an area left untouched by other inhabitants. Visible shoeprints are identified and the area is then darkened while investigators using oblique light search for shoeprints (SWGTREAD, 2005a).

2. After locating a shoeprint, the CSI places an L-shaped photography scale next to it, illuminates it at an angle that reveals as much detail as possible and photographs it with the camera positioned on a tripod directly over the print (SWGTREAD, 2006a).

3. If possible, the surface on which the print appears is wrapped in paper and sent to the crime lab for further processing (SWGTREAD, 2005b).

4. If this is not possible, prints are lifted using one of the following methods:
   a. Two-dimensional print methods (SWGTREAD, 2007a; Manual for BVDA Gel-lifters – Product Information, 2020); white adhesive lifter (most common method used in this lab for 2D shoeprints), black gelatin lifter and electrostatic lifter.
   b. Three-dimensional prints are cast using dental stone (SWGTREAD, 2007b; Cohen et al., 2012).

5. Once a suspect is apprehended, the suspect's shoes are sent to the crime lab for comparison with the shoeprints. Usually the shoes will be removed from the suspect's feet, but occasionally, the suspect's house will be searched for shoes with soles that resemble the shoeprints from the crime scene.

6. All exhibits are registered by the investigating unit and are then sent to the crime lab through the "evidence office" which is responsible for assigning file numbers and passing on the exhibits to the relevant lab. The investigators add a letter that contains details about the crime and the collection of the relevant evidence.

7. All the exhibits and their packaging are documented and marked at the crime lab. The examiner documents all of the information concerning the exhibits as they

were received by the lab (date of reception, description of packaging and exhibit, etc.) (SWGTREAD, 2008). This is known as the "chain of custody". Keeping the chain of custody is important to ensure that the exhibits examined by the lab are the same exhibits confiscated in this case.

8. All the shoeprints (exhibits, white adhesive lifters, black gelatin lifters and casts) and the shoes are photographed at high resolution (1000 dpi). The photographs taken at the crime scene are calibrated to 1000 dpi as well. Minimal image processing is carried out if necessary (conversion to gray scale, contrast and brightness adjustment, etc.).

9. If applicable, the shoeprints collected at the crime scene go through a process of enhancement. White adhesive lifters are sprayed with Bromophenol Blue which has a yellow color that turns blue when reacting with the dust, thus creating the shoeprint (Glattstein et al., 1996; Shor et al., 1998). Shoeprints on items collected at the crime scene are lifted using the most appropriate method, black gelatin with a press (Shor et al., 2003) or white adhesive lifter.

10. The first step of evaluation is visual examination of the crime scene prints and the shoe soles. If the patterns differ, exclusion is declared. If the patterns are similar, test impressions are made from the suspect's shoes.

11. The shoe soles are dusted with fingerprint powder and then impressions are made, while wearing the shoe, by stepping onto a clear adhesive film. At least two test impressions are made from each shoe (Hilderbrand, 2007), first by walking and again by pressing the adhesive to the shoe sole while it is in the air. If necessary, for the comparison stage additional three-dimension test impressions are made using Biofoam ©(SWGTREAD, 2005c).

12. The two common comparison methods are:
    a. Overlay – a transparency of the test impression of the suspect's shoe is positioned over a photograph of the print from the crime scene.
    b. Side by Side – the test impression and the photograph of the print from the crime scene are laid out side by side in order to compare the similarities between them.

The method used in the lab directed by one of the authors (S.W.) is on screen overlay (using Lucia TrasoScan, by LIM ©) with 1000 ppi (pixel per inch) images.

The examination process includes comparison of class characteristics (pattern, size and wear) and the identification of accidentals (RACs and unique wear) (SWGTREAD, 2006b) as descried in Appendix 17A.1.

13. The examiner determines his or her level of certainty and writes a report. An ordinal scale is used to describe the degree of the match (ENFSI Expert Working Group Marks Conclusion Scale Committee, 2006). (Appendix 17A.2 provides the scale used in S.W.'s lab.)

Based on the conclusion distribution of the lab, in approximately 40% of all cases the suspects' shoes sent to the lab result in a non-match to the crime scene prints. A match of class characteristics ("Possible") is found in approximately a third of the cases and accidentals are found in nearly 25% of all cases. Only less than 2% of the crime scene prints lack sufficient details for comparison since the CSI are instructed to merely collect shoeprints with some visible detail, hence match of class characteristics with the suspects' shoe can be determined or eliminated.

17.3 The Challenges that Arise from Analyzing the Working Procedure

The working procedure described above does not specify the process of identifying the pattern and the RACs. Without semi-automated pattern and RAC retrieval systems, as proposed in Sections 17.6 and 17.7, this process is very difficult to document. Currently the process of differentiating the signal from the noise (i.e. the shoeprint from the background) is, to our understanding, a subconscious method that is a combination of born capabilities and training.

Crime scene shoeprints are very complex and many factors may degrade the information they contain. These factors include:

1. *Noisy background.* The substrate on which the shoeprint appears may vary dramatically from hard to soft, clean to messy (dusty, granular or oily), dry to wet, or uniform to diverse (contaminated by other materials or textures).

2. *The material creating the shoeprint*: dust, oil, blood, mud, etc.

3. *Shoeprint acquisition.* The methods to recover the shoeprints from different surfaces also vary, including direct photography, digital 3d reconstruction, chemical enhancements and developments, plaster casting, lifting, and others.

4. *Movement of the shoe* while creating the print, distorting the print pattern and the RACs.

5. *Partial shoeprints* (only a small section of the print is visible or recoverable) in which the location of the portion of the shoe leaving the print is not always known.

6. *Missing parts of the RACs* (even small RACs may be partially observable in the print).

7. *Multiple prints.* Several shoeprints one on top of the other, partially concealing each other.

This degraded information introduces additional challenges in determining whether the suspects' shoe is the source of the crime scene print and therefore increases the error rate.

In some cases, not all the RACs that are visible on the shoe appear on the crime scene print, and in others, suspected RACs may appear on the crime scene print but not on the suspect's shoe. This does not necessarily indicate the shoe didn't create the crime scene print. It may be a result of factors mentioned above.

Usually only partial prints are visible at the crime scene. In some cases, the shoeprint is so partial that there are several possible locations on the suspects' shoe from which the crime scene print could have originated. This complicates the analysis. In many cases, only partial RACs are visible on the crime scene print. As a result, the amount of information is reduced and so is the evidentiary value. In the time between the occurrence of the crime and confiscating the shoes, the shoe might acquire additional RACs or existing RACs may transform or wear, creating a difference between them (Wyatt et al., 2005; Sheets et al., 2013).

After identifying RACs during the comparison process, the amount of information in each single RAC and combination of RACs is evaluated. As of today, there is no quantified or statistical method for weighing and evaluating the RACs. The conclusion is drawn by the examiners based on their training and experience. Moreover, examiner personality has some influence as well and hence the conclusion is subjective to some degree. This may result in variability in the conclusions drawn by different examiners or even in those of the same examiner at different times. These variations are most likely to occur in borderline cases when the conclusions are usually relatively close (Hammer et al., 2012).

One of the main challenges that arises from examining the working procedure is potential bias of various kinds.

First, there is potential bias in choosing the shoeprints according to the suspect's shoes: in cases of arresting the suspect prior to searching the crime scene, the CSI occasionally observes the suspect's shoes and then focuses on the matching shoeprints at the crime scene. This issue could indeed improve the search but the probability calculations should be adjusted accordingly. Bias can arise also when choosing which of the suspect's shoes to search, in cases when a person becomes a suspect because his shoes are of the same pattern as the crime scene prints. In this case the probability calculations shouldn't take into account the pattern match, but merely the size, degree of wear and RACs.

Contextual bias may occur if the examiner is exposed to irrelevant information about the facts of the case (such as information about prior convictions) which could influence his/her judgment, especially in the determining the strength of the conclusion. As described in the working procedure above, identifying the exact pattern of the crime scene print and the RACs that exist on it, is based on comparison with an apparently matching exemplar. This leads to potential confirmation bias, a situation in which the search for information on the crime scene print is performed in a way that confirms the information on the suspects' shoe. The optimal case is that examiners complete marking the latent print before looking at the suspected exemplar. Alas, unlike the comparison of fingerprints (Langenburg et al., 2012) which have a standard appearance and typical minutiae, it is extraordinarily difficult to identify shoe patterns, and even more so the RACs, without comparison to the reference shoe because of the wide variety of the shoe sole patterns and the random nature of the RACs' shapes. Crime scene prints often appear on a noisy background and the dynamics of creating a shoeprint add to this noise.

17.4 Requirements for Good Practice

The complexity of the shoeprint comparison process as presented in Section 17.3 requires high practice standards both for the lab and for the practitioners.

A detailed Standard Operating Procedure (SOP) is needed to ensure that the work is performed professionally and based on a standard working procedure. This minimizes error, makes the procedure more objective, less operator dependent and increases the transparency of the process. Laboratory accreditation by an exterior institute is another safety net to mitigate the possibility of careless work. Forensic testing laboratories are assessed against the general criteria found in ISO/IEC 17025. The ISO/IEC 17025 standard comprises five elements: Scope, Normative References, Terms and Definitions, Management Requirements and Technical Requirements. Working by SOP in an accredited laboratory assures working standards. Proficiency tests ensure professional working quality. These tests are usually distributed by commercial companies. After performing the test and sending back the results, they are examined by the company and a summary of all results is included. Successfully performing proficiency test at least once a year by each practitioner in the lab is necessary to safeguard a basic level of professional competence. Blind proficiency tests can ensure an even higher level of competency.

Forensic laboratories should establish routine quality assurance and quality control procedures including blind testing and performance testing to guarantee the accuracy of forensic analyses and the work of forensic practitioners as well. Blind proficiency tests can be created internally or collaboratively among several cooperating laboratories. In such tests, the examiner is not aware that the test cases are not real ones. For performance testing, sample cases are re-examined by other experts or the same experts after a period of time and the results are compared.

Verification is a key step for preventing error during routine case work. After an examiner reaches a conclusion and writes the report, a second expert reviews the case. The optimal form of verification is to perform a blind comparison independently and then to compare the results with the original report (double blind). This method is very time consuming and would require doubling the manpower. The common method of verification is for the second expert to carefully review all the work performed by the first examiner and the conclusions drawn by him/her in all cases (not only positive conclusions). The flaw of this method is the potential influence of the first expert's conclusion which may bias the verification process. This type of bias is known as confirmation bias (Kassin et al., 2013).

As noted in Section 17.3, bias may influence decision making. Contextual bias during the working process should be minimized as much as possible. This issue has been studied and many solutions have been suggested (Dror, 2013).

Maintaining high lab standards cannot be complete without investing in the professional training of the examiners. A scientific background is recommended for all examiners. Academic scientific studies educate in logical and systematic thinking, and the limitations of human knowledge, which is a basic need of the forensic practitioners' work. Specific training in the profession of shoeprint comparison should be a major step before the examiner is allowed to write expert opinions. This should be limited to certified examiners. There are two kinds of certifications. The first is mandatory and is given by the laboratory once it is convinced the examiner is competent enough to perform the work. The second is international and can be achieved only after several years of experience and proven capabilities. Certification will ensure professional standards and will mitigate examiner error.

17.5 How to Strengthen the Scientific Foundation for Footwear Analysis?

A first step to become scientific is meeting good practice criteria as written above. In order to further address the challenges stated in Section 17.3, shoeprint comparisons must stand up to scientific standards in the following aspects.

A crucial issue, mentioned above, is that the current procedure involves almost no statistics that are based on large scale datasets related to either class characteristics or RACs, since these are not available.

The distribution of shoe-sole patterns in a specific geographic area constantly changes. The shoe industry changes the patterns sold periodically, and people regularly replace their old shoes with new ones. For this reason, keeping an up-to-date database of shoe-soles to assist in statistical calculations is nearly impossible and prohibitively expensive.

Nevertheless, several surveys have been conducted which indicate that the chance of finding more than one appearance of most shoe-sole patterns in a survey of less than 1000, is uncommon (Hancock et al., 2012; Johansson and Stattin, 2008).

Stone (2006) analyzes the individual characteristics (RACs) that appear on a shoeprint and quantifies them using their location, configuration and orientation, on the assumption that they are independent. The study is not intended to estimate the probability of these features in actual shoeprints, but rather to provide a hypothetical calculation about their degree of rarity. Kaplan Damary et al. (2018) found that the assumption of independence between the features of a RAC is unjustified and a more sophisticated statistical model for calculating the rarity degree of RACs is required. Evett et al. (1998) present a Bayesian analysis using likelihood ratios to compare the probability that the crime scene RAC would have the observed features, if it was left by the suspect's shoe, as opposed to if it was left by another shoe of the same sole pattern. This approach was later developed by the ENFSI guidelines (SWGTREAD, 2005c), and the probabilities used for the calculations are not based on datasets but rather determined subjectively by the examiner. Petraco et al. (2010) apply statistical techniques used in facial pattern recognition to RAC patterns using only the coordinate locations of the RACs. They quantify the similarity between two patterns without considering other available information such as the size and shape of RACs, and use principal component analysis in order to group prints from the same shoe (at different time points) and distinguish them from prints made by other shoes. An attempt to study the distribution of RACs has been conducted by Yekutieli et al. (2016) who also collected a large dataset of approximately 9000 RACs using a dedicated software application developed for identification of RACs. The assumption of independence among RACs and among the features of a RAC was used, as in Stone (2006), by Willis et al. (2015). An additional dataset was collected by Speir et al. (2016).

The aforementioned studies use very simple statistical tools. The work that has been done triggered the need for ongoing statistical research which includes developing a model that explains the creation of RACs, adjusting statistical calculations according to the relevant population in each casefile, estimating the covariance structure of RACs and analyzing the crime scene noise by the use of experiments. An example of the latter would be the collection of noisy shoeprints by conducting a controlled experiment and comparing the prints from the experiment to the test impressions made from the shoes under controlled conditions.

Much research has been done on automatic identification and classification of shoe sole patterns (Richetelli et al., 2017; Alizadeh and Kose, 2017; Kong et al., 2018; Wang et al., 2014; Huynh et al., 2003; Luostarinen and Lehmussola, 2014; Benedict et al., 2014) but these data sets should include information concerning the current distribution of shoe sole patterns and shoe sizes in a specific area. Sampling the shoe soles of every suspect arrested by the police should be part of the standard procedure, just as every suspect is fingerprinted and photographed. This will provide an up to date and relevant database of suspect shoe sole patterns for valid statistical calculations.

A semi-automated system could assist the examiners in reaching objective conclusions. For example, an automatic system that finds the shoe pattern on a crime scene print prior to observing the suspect shoes or an automatic system that will find the RACs and an examiner will approve them, would both reduce confirmation bias. Also, a semi-automated system can be used for automated database building, for offline statistical evaluation and online statistical learning. The next section will focus on the ongoing development of such a system.

17.6 Wishlist for and Design of the Semi-Automated ElementAccidentalSensor

Mitigating human bias, enhancing differentiation between signal and noise and allowing for building large databases, we propse a semi-automated ElementAccidentalSensor.

While the desired capabilities of the ElementAccidentalSensor are determined by the demands of forensic shoe print analysis, the technical framework is determined by the use case. It is instrumental that the software is easy to understand and to adjust and as robust as possible to a scenario where it is changed by a succession of programmers with different styles and levels of skill. From this, we derive the requirements that source code be freely accessible and reasonably documented. Furthermore, it should be structured in as modular a fashion as possible and written in a language which provides at least a basic level of fault tolerance.

The object centered language Java combines easy memory management with fairly good speed. The rather restricted syntax of the language makes it newcomer friendly and leads to a fairly uniform style, which facilitates reading other programmers' code. Also, a rich ecosystem of image processing libraries exist.

Shoeprint image processing decomposes into three major steps. The first step aims at enhancing the image, the following step identifies elements and the third step finds candidates for accidentals and wear. While the first and second step can run mostly automated, except for some parameter adjustment, the last step requires expert interaction.

For the first step, it is important to account for the variability in shoe print images. Lighting, background pattern, footprint type and applied pressure lead to a wide variety of image properties. Some of the resulting conditions require local image treatment. As the output of the first step, we provide a standardized model, where background is represented as dark and pattern is represented as bright.

In the second step, we would like to detect elements and attempt to classify them. We require that elements consist of extensive connected bright areas. The three basic element categories are polygons, ellipses and lines. Since line detection has to deal with lines of different width, it is mandatory to detect polygons and ellipses first, in order not to treat them as lines. At the same time we would like to record all contiguous bright regions, in order to be able to identify mismatches of fitted elements with the image.

Already the second step requires dealing with some intricacies since elements with simple shapes can have additional patterns, cf. Figures 17.4–17.6. Sometimes, this can even impede detection of true elements, either by leading to detection of very irregular non-convex polygons, or if an element appears as a cluster of small elements. Such layered structures demand a stratified model of shape element representation, with basic shape patterns at the bottom level and additionally layered patterns at higher levels.

The third step, identifying candidates for wear and accidentals, relies on, and is entangled with, element detection. Since wear is characterized by a loss of structure, among other things, wear will appear as large, irregularly shaped, contiguous flat regions on the shoe sole which will be detected as irregularly shaped elements. Thus, in the stratified framework outlined above, wear is expected at the bottom level, where elements with irregular shape and without substructure are detected.

Accidentals, on the other hand, are usually rather localized, since they represent damage or production errors in the sole. Therefore, accidentals are modeled at the higher levels of the stratified framework. As substructures of lower level elements, which usually follow regular patterns, accidentals are less regular. To reduce the number of false positive accidental candidates, it is therefore necessary to improve layered element detection.

Both wear as well as accidental candidates are prone to be false positives, either due to noise or due to non-regular shaped true elements. To reduce the number of false positives, it is important to build a library of known shape templates, which the program can use to identify structures in a shoe print. This requires a large database of high quality shoe print images and a protocol to build a shape template database. To avoid clutter, it appears reasonable to allow for manual addition, correction and removal of elements from the database. In conjunction with the element database, a shoe print database can be built, which stores sole templates of known shoe models.

The desired work flow of a fully developed program would roughly be as follows.

1. An image of an unknown shoeprint is given to the ElementAccidentalSensor.

2. The ElementAccidentalSensor identifies elements, making use of its template database as well as its algorithms.

3. Based on the found elements, the program proposes known shoe models from the database.

4. Finally, the program proposes candidates for wear and accidentals, ideally relying on a previously identified shoe model template and element templates. These are confirmed or rejected by the user.

The output of the fully developed program, ideally, consists of the identified shoe model, the shoe sole area covered by the print, and finally the identified wear and accidentals in this area.

17.7 An Early Stage Implementation of the Semi-Automated ElementAccidentalSensor

Following the above wishlist we now detail an early stage implementation of the ElementAccidentalSensor, written in Java, building on the FilamentSensor from Eltzner et al. (2015), also written in Java.

17.7.1 Preprocessing and Binarization

The original image is converted to grayscale, to simplify the treatment. If the background is brighter than the pattern, the image is inverted, such that a brighter pattern on a darker background is achieved.

Contrast adjustment: By default, contrast is enhanced locally by the algorithm outlined below, which one of the authors (B.E.) developed for this purpose. The aim of the procedure is to obtain two brightness maps I_- and I_+ so that we can transform the image I into $\widetilde{I}$, by calculating for every pixel position (x, y) the value

$$\widetilde{I}(x,y) = (I(x,y) - I_-(x,y))/(I_+(x,y) - I_-(x,y))$$

The brightness maps are obtained as follows. The image is binned into square blocks and for each of the block centers (x, y), the minimal, maximal and mean brightness $B_{\min}(x, y)$,

$B_{\max}(x,y)$, $B_{\text{mean}}(x,y)$ over the respective block are determined. To reduce the effect of bright outliers, determine $\widetilde{B}_+(x,y) = \min(B_{\max}(x,y), 2B_{\text{mean}}(x,y) - B_{\min}(x,y))$. To avoid amplifying noise in dark area without signal, we use a user-adjustable global minimal contrast value $d_{\min}$ and, whenever $\widetilde{B}_+(x,y) - B_{\min}(x,y) < d_{\min}$, we redefine

$$B_+(x,y) = (\widetilde{B}_+(x,y) + B_{\min}(x,y) + d_{\min})/2$$
$$B_-(x,y) = (\widetilde{B}_+(x,y) + B_{\min}(x,y) - d_{\min})/2.$$

If B_+ is higher than the maximum possible brightness or $B_- < 0$, both values are accordingly readjusted such that $B_+(x,y) - B_-(x,y) = d_{\min}$ is preserved.

As a final step, the maps I_- and I_+ are produced from B_- and B_+ as

$$I_\pm(a,b) = \frac{\sum_{(x,y)} \exp(-|(a,b) - (x,y)|^2/(2L^2))B_\pm(x,y)}{\sum_{(x,y)} \exp(-|(a,b) - (x,y)|^2/(2L^2))}.$$

Here, (a,b) denotes all pixel positions, while (x,y) in the sums runs only over block centers.

As a simpler alternative to this elaborate *local contrast adjustment*, it is possible to adjust contrast globally by determining constant values I_+ and I_- for the whole image. Indeed, if the background brightness varies strongly, then local contrast adjustment may lead to an undesirable enhancement of such variation, which usually means that the global contrast enhancement is preferable. If, however, the background is rather homogeneous but the brightness of the shoeprint is inhomogeneous, local contrast adjustment is often a highly important preprocessing step. Figure 17.1 illustrates the difference between local and global contrast adjustment.

Two other preprocessing steps use generic Gaussian and Laplacian filters with adjustable variance and magnitude (currently with empirically determined default values), respectively, the order of which can be interchanged.

The ElementAccidentalSensor will be extended to allow arbitrary ordering of preprocessing steps and available filters. Since the program has a generic interface for preprocessing filters, also new filters can be added.

Finally, the preprocessed image is binarized (every pixel is set to either white or black) using the cross-entropy threshold of Li and Lee (1993). The resulting binary image is then used for element detection.

17.7.2 Element Detection

To detect polygons and ellipses we use the BoofCV library (Abeles, 2016). In a first step, we use the function *boofcv.alg.shapes.ShapeFittingOps.fitPolygon* to identify polygons in the binary image. We expose the parameters of the function through the GUI, see Figure 17.2, so they can be adjusted by the user. For polygons with many edges, we then fit ellipses instead, as illustrated in Figure 17.3. Finally, we use the Line Sensor, (Gottschlich et al., 2009), to identify lines in the remaining structure.

In parallel to the above shape detection, the binary image is segmented into contiguous (arbitrarily shaped) white regions. These regions are compared to the identified shapes to identify misfits. If a particular element has a hole, e.g. due to an accidental, this will be

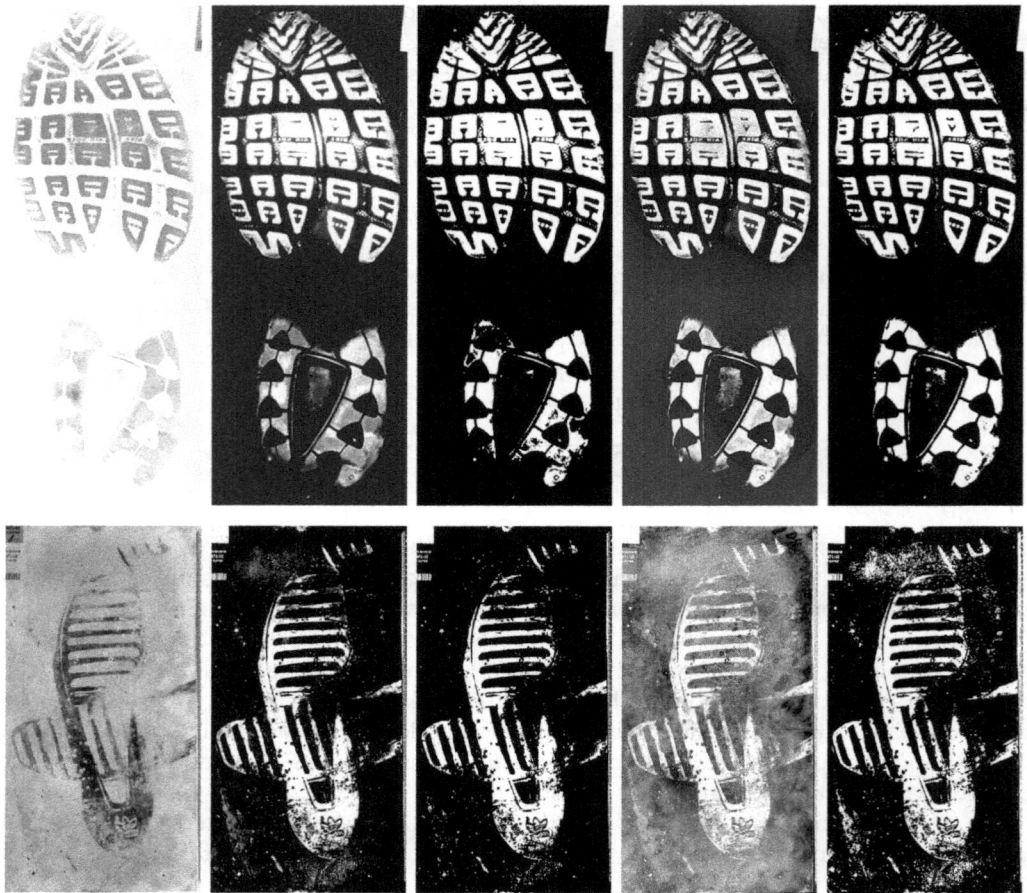

FIGURE 17.1

Image preprocessing for two shoeprints. Column 1: original, column 2: global contrast adjustment, column 3: binarization after global contrast adjustment, column 4: local contrast adjustment, column 5: binarization after local contrast adjustment. The original of the upper row has a smooth homogeneous background, so that local contrast adjustment recovers rather well elements and potential accidentals. In contrast, global contrast adjustment may light up specific areas at the cost of losing potential wear and accidentals (e.g., in the ball area right above the inscription "AIR"), while darkening other areas at the cost of introducing potential false accidentals (e.g., at the heel). Local contrast adjustment fares much better, giving a fairly even brightness of front and heel parts and leading to a better binary image. The original of the lower row has a much rougher background, so that global contrast adjustment fares much better (only losing some darker element parts) than local contrast adjustment, leading to a too bright background and thus to considerable noise and so to many potentially false accidentals in the binary image.

classified as missing polygon or ellipse pixels in the corresponding region, as illustrated in Figure 17.4.

An additional problem to consider concerns shapes with a pattern. An example of this would be a dotted square, thus consisting of a cluster of small elements, as illustrated in Figure 17.5. To identify such a pattern, we propose single linkage or k-nearest neighbor clustering of the regions identified above. These approaches lead to a cluster tree representation of the objects where branch length represents distance. If the cluster tree has a

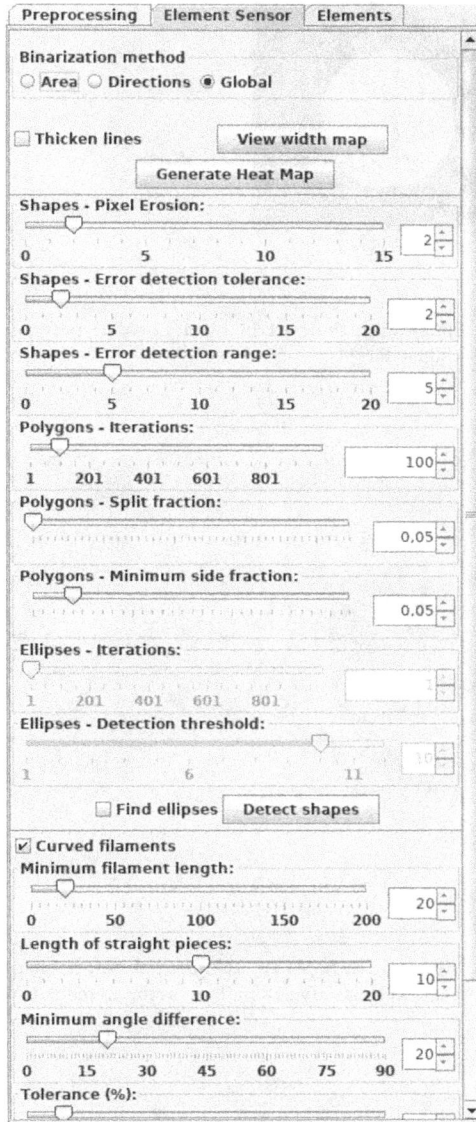

FIGURE 17.2
Screen shot of the ElementAccidentalSensor's menu for element classification.

"distance gap" with none or very few nodes, we propose truncating the tree at this gap and considering convex hulls of the clustered regions.

17.7.3 Detecting Accidentals and Wear

Every contiguous set of pixels that is part of a contiguous region but is not contained in any of the elements associated to that region, is considered a candidate for an accidental. So far, this approach is very coarse and tends produce too many tentative accidentals. Therefore,

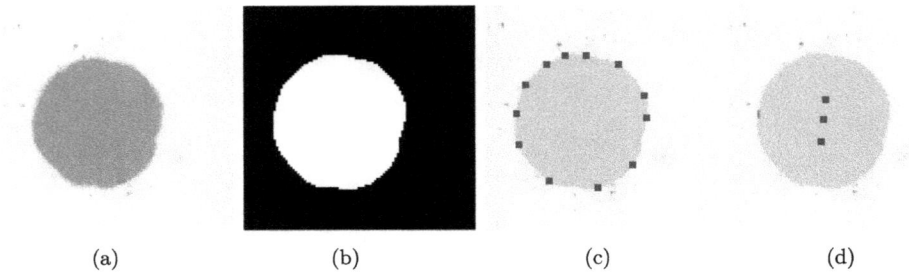

FIGURE 17.3
Image of an elliptical element, illustrating the best fit by a 12-sided polygon and an ellipse. (a) Original. (b) Binarized. (c) Polygon Fit. (d) Ellipse Fit.

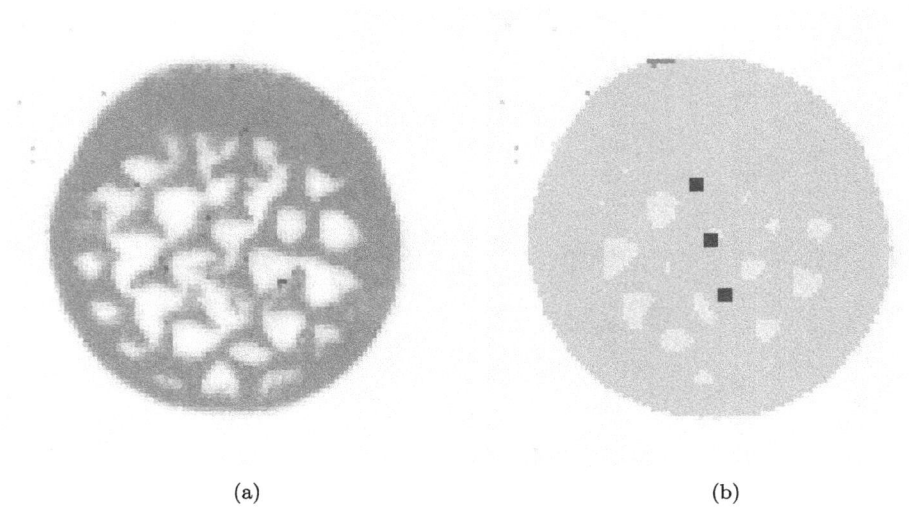

FIGURE 17.4
Image of an elliptical element with holes. The holes are identified as defects of an element and thus as tentative accidentals, which a human expert can well classify as a pattern on the element. (a) Original Element. (b) Proposed Accidentals in green.

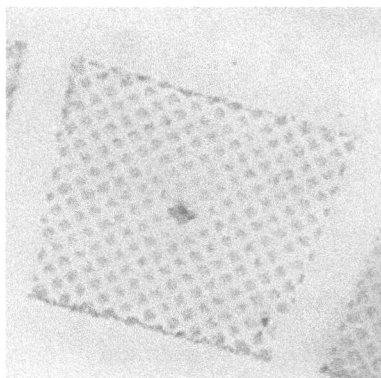

FIGURE 17.5
Image of a quadrilateral with a checkerboard pattern, which is prone to be identified as several small elements but should be found as one "cluster element".

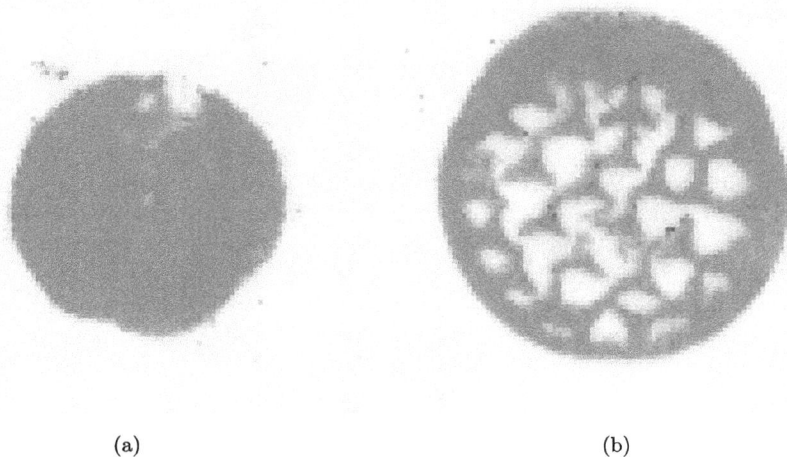

(a)   (b)

FIGURE 17.6
Detail images showing a true accidental as compared to a pattern on an element, which would not be considered an accidental. (a) True accidental. (b) False positives.

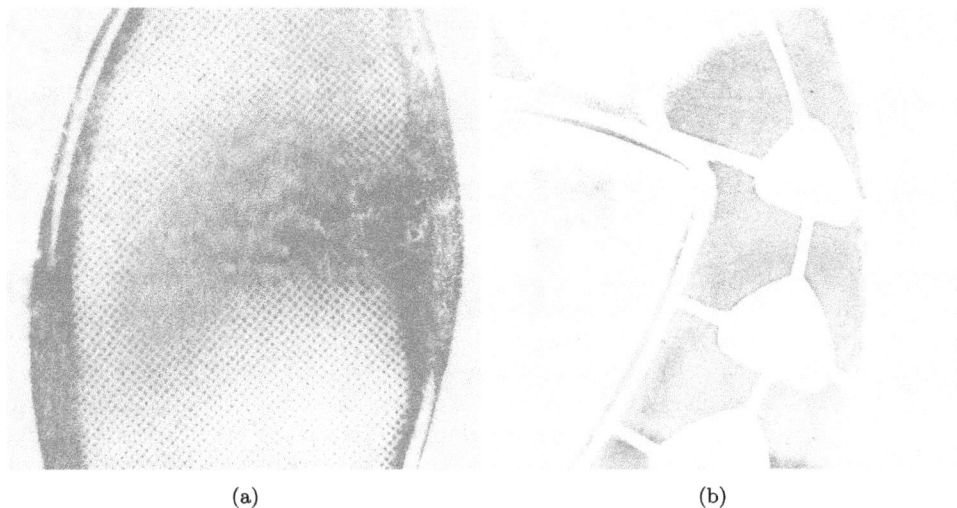

(a)   (b)

FIGURE 17.7
Image details showing true wear as compared to irregularly shaped elements, which are not due to wear. (a) True wear. (b) Irregularly shaped elements.

in this phase of database building, the detection of accidentals will require a fair amount of manual selection, as illustrated in Figure 17.6.

Irregularly shaped regions that cannot be classified by the means outlined above can be proposed as tentative wear patterns. Since some shoes do, however, feature very irregular patterns, this will also require manual selection, as illustrated by Figure 17.7.

17.8 Discussion and Conclusions

Like any science, forensics and shoeprint comparison in particular have a need to continuously upgrade and improve the performance of practitioners as well as the level of scientific understanding within the discipline as a whole. The conservative nature of the court systems, which value stability and consistency, conflicts with the constant improvements in methods seen in science, medicine and engineering.

A basic, yet not satisfactory, demand in order to meet the challenges presented in this chapter is that the criteria of good practice be fulfilled, but fundamental validity must be established first. The task requires that the personnel involved in the process be able to make the necessary changes in attitude towards a more scientific form of practice.

Within a framework of transparent and reproducible SOPs, we propose semi-automated systems that, while implementing expert knowledge, mitigate human bias.

A combined and long-term effort of scientists from multiple disciplines is obligatory for solving the underlying complex issues. The fields of physics, chemistry, computer science and image processing, on the one hand, as well as psychology, criminology, law and others, on the other hand, must be included in this process and linked via statistics at various levels.

Forensic scientists and practitioners are key figures in finding the solution. Their understanding of the importance of this change and their involvement in ongoing scientific research is essential for its implementation. Meeting the challenges mentioned above will not complete this scientific revolution. This will happen only when the results are presented to the court in a manner that is accessible by the judicial system and the weight of the evidence is fully understood. This leads to new challenges such as the court's understanding of the scientific terminology (Thompson and Schumann, 1987; Thompson et al., 2013).

It should be taken into consideration that the border between practice and science may remain in place. It seems that in the foreseeable future, humans will still be an essential part of the process since it cannot be fully automated.

The process of evaluating shoeprint evidence reduces drastically the potential shoe population, provides additional information which can be useful for the judicial system, and is very important as supportive evidence. Identifying the pattern, size and wear of the shoe which created the shoeprint omits all other shoes which have different patterns, sizes or wear. Furthermore, in many cases the RACs contain much information which considerably narrows down the potential shoe population. Identifying the suspects' population may serve both prosecution and defense. Thus, shoeprints have an important evidentiary value and should be presented to a court of law, even if the full scientific procedure, from practice to science, is not yet completed.

Appendix 17A

17A.1 Shoeprint Comparison Stages

1. Pattern – If the exact shoe pattern does not match, the result of the comparison is exclusion.

2. Size – If the shoe pattern matches, but the physical size of shoeprint differs from the corresponding area of the shoe-sole this might be explained by incorrect scaling, photography at an angle, movement of the shoe while producing the shoeprint, etc. If no explanation satisfies the examiner as a logical explanation for the dissimilarities, exclusion is determined.

3. Wear – If exclusion is not determined by now, the degree of general wear and local wear are compared. If the wear areas differ or crime scene print is worn more than the shoe, exclusion is determined. If the shoe is worn more than the crime scene print, the time elapsed between creation of the shoeprint and confiscation of the shoes is considered.

4. The crime scene print is searched for locations that the pattern is not complete and the reason might be the presence of RACs. If they indeed appear on both the test impression and the crime scene print, they are marked and their clarity, complexity and rarity are evaluated based on the examiner's experience.

17A.2 The Conclusion Scale

1. Negative – the shoe under investigation is significantly different from the crime scene print, and therefore the suspect's shoe couldn't have left the crime scene print.

2. Indication of non-association – Differences were found during the comparison but the quality of the print or the essences of the differences are not sufficient for total exclusion.

3. Lacks sufficient details – The crime scene shoeprint lacks information that would enable significant comparison.

4. Cannot be eliminated – Similarities in class characteristics were found, but the details on the crime scene print were limited and therefore, specific association is not possible.

5. Possible – there is a match regarding class characteristics between the crime scene print and the suspect shoe.

6. Probable – Matching class and identifying characteristics were both found, but the amount of information in the identifying characteristics is not sufficient to determine identification.

7. Highly probable – Similar to Probable, but a higher degree of certainty.

8. Identification – Besides match in class characteristics, the match in identifying characteristics is of sufficient quality and quantity.

References

Abeles, P. Boofcv v0–25, 2016. http://boofcv.org/

Alizadeh, S. and Kose, C. Automatic retrieval of shoeprint images using blocked sparse representation. *Forensic Science International*, **277**, 103–114, 2017.

Benedict, I., Corke, E., Morgan-Smith, R., Maynard, P., Curran, J.M., Buckleton, J., and Roux, C. Geographical variation of shoeprint comparison class correspondences. *Science & Justice*, **54**(5), 335–337, 2014.

Bodziak, W.J. *Footwear Impression Evidence: Detection, Recovery and Examination*. CRC Press, 2000.

Bodziak, W.J. Traditional conclusions in footwear examinations versus the use of the bayesian approach and likelihood ratio: A review of a recent uk appellate court decision. *Law, Probability and Risk*, **11**, 279, 2012.

Cohen, A., Wiesner, S., Grafit, A., and Shor, Y. A new method for casting three dimensional shoe prints and tire marks with dental stone. *Journal of Forensic Sciences*, **56**(S1), S210–S213, 2011.

Daubert v. Merrel Dow Pharmaceuticals, Inc. 43 F.3d 1311 (9th Cir. 1995).

Dror, I.E. Practical solutions to cognitive and human factor challenges in forensic science. *Forensic Science Policy & Management: An International Journal*, **4**, 105–113, 2013.

Eltzner, B., Wollnik, C., Gottschlich, C., Huckemann, S., and Rehfeldt, F. The filament sensor for near real-time detection of cytoskeletal fiber structures. *PloS one*, **10**(5), e0126346, 2015.

ENFSI Expert Working Group Marks Conclusion Scale Committee. Conclusion scale for shoeprint and toolmarks examinations. *Journal of Forensic Identification*, **56**(2), 255–280, 2006.

Evett, I.W., Lambert, J.A., and Buckleton, J.S. A bayesian approach to interpreting footwear marks in forensic casework. *Science and Justice*, **38**(4), 241–247, 1998.

Glattstein, B., Shor, Y., Levin, N., and Zeichner, A. pH indicators as chemical reagents for the enhancement of footwear marks. *Journal of Forensic Sciences*, **41**(1), 23–26, 1996.

Gottschlich, C., Mihailescu, P., and Munk, A. Robust orientation field estimation and extrapolation using line sensors. *IEEE Transactions on Information Forensics and Security*, **4**(4), 802–811, 2009.

Hammer, L., Duffy, K., Fraser, J., and Nic Daeid, N. A study of the variability in footwear impression comparison conclusions. *Journal of Forensic Identification*, **63**(2), 205–218, 2012.

Hancock, S., Morgan-Smith, R., and Buckleton, J. The interpretation of shoeprint comparison class correspondences. *Science and Justice*, **52**, 243–248, 2012.

Hilderbrand, D.S. *Footwear: The Missed Evidence*. (2nd ed.). Staggs Publishing, Wildomar, CA, USA, 2007.

Huynh, C., de Chazal, P., McErlean, D., Reilly, R. B., Hannigan, T. J., and Fleury, L. M. Automatic classification of shoeprints for use in forensic science based on the fourier transform. *Proceedings 2003 International Conference on Image Processing (Cat. No. 03CH37429)*, Vol. 3, pages III–569. IEEE, 2003.

Johansson, Å. and Stattin, T. *Footwear Impression as Forensic Evidence – Prevalence, Characteristics and Evidence Value*. Department of Mathematics, Linköping University, 2008.

Kafadar, K. Statistical issues in assessing forensic evidence. *International Statistical Review*, **83**, 111–134, 2015.

Kaplan Damary, N., Mandel, M., Wiesner, S., Yekutieli, Y., Shor, Y., and Spiegelman, C. Dependence among Randomly Acquired Characteristics on shoeprints and their Features. *Forensic Science International*, **283**, 173–179, 2000.

Kassin, S.M., Dror, I.E., and Kukucka, J. The forensic confirmation bias: Problems, perspectives, and proposed solutions. *Applied Research in Memory and Cognition*, **31**(1), 42–52, 2013.

Kong, B., Supancic, J., Ramanan, D., and Fowlkes, C.C. Cross-domain image matching with deep feature maps. *International Journal of Computer Vision*, **127**(11–13), 1738–1750, 2018.

Langenburg, G., Champod, C., Genessay, T., and Jones, J. Informing the judgment of fingerprint analysts using quality metric and statistical assessment tools. *Forensic Science International*, **219**(1), 183–198, 2012.

Li, C.H. and Lee, C.K. Minimum cross entropy thresholding. *Pattern Recognition*, **26**(4), 617–625, 1993.

Luostarinen, T. and Lehmussola, A. Measuring the accuracy of automatic shoeprint recognition methods. *Journal of Forensic Sciences*, **59**(6), 1627–1634, 2014.

Majamaa, H. and Ytti, A. Survey of the conclusions drawn of similar footwear cases in various crime laboratories. *Forensic Science International*, **82**, 109–120, 1996.

Manual for BVDA Gellifters – Product Information. BVDA International B.V., Holland. http://www.bvda.com/en/gellifters#lifting_sp (June 11, 2020).

National Research Council. *Strengthening Forensic Science in the United States: A Path Forward*. The National Academies Press, Washington, DC, 2009.

Petraco, N.D.K., Gambino, C., Kubic, T.A., Olivio, D., and Petraco, N. Statistical discrimination of footwear: A method for the comparison of accidentals on shoe outsoles inspired by facial recognition techniques. *Journal of Forensic Sciences*, **55**(1), 34–41, 2010.

President's Council of Advisors on Science and Technology (PCAST). Forensic Science in Criminal Courts: Ensuring Scientific Validity of Feature-Comparison Methods. Report to the president, Washington, DC, September 2016.

Regina vs. T, Case No: 2007/03644/D2, High Court of Justice Court of Appeal, Strand London, October 26, 2010, (EWCA [2010], Crim 2439).

Richetelli, N., Lee, M.C., Lasky, C.A., Gump, M.E., and Speir, J.A. Classification of footwear outsole patterns using fourier transform and local interest points. *Forensic Science International*, **275**, 102–109, 2017.

Saks, M.J. and Koehler, J.J. The coming paradigm shift in forensic identification science. *Science*, **309**(5736), 892–895, 2005.

Sheets, H.D., Gross, S., Langenburg, G., Bush, P.J., and Bush, M.A. Shape measurement tools in footwear analysis: A statistical investigation of accidental characteristics over time. *Forensic Science International*, **232**(1), 84–91, 2013.

Shor, Y., Tsach, T., Vinokurov, A., Glattstein, B., Landau, E., and Levin, N. Lifting shoeprints using gelatinlifters and a hydraulic press. *Journal of Forensic Sciences*, **48**(2), 368–372, 2003.

Shor, Y., Vinokurov, A., and Glattstein, B. The use of an adhesive lifter and pH indicator for the removal and enhancement of shoeprints in dust. *Journal of Forensic Sciences*, **43**(1), 182–184, 1998.

Shor, Y. and Wiesner, S. A survey on the conclusions drawn on the same footwear marks obtained in actual cases by several experts throughout the world. *Journal of Forensic Sciences*, **44**(2), 380–384, 1999.

Speir, J.A., Richetelli, N., Fagert, M., Hite, M., and Bodziak, W.J. Quantifying randomly acquired characteristics on outsoles in terms of shape and position. *Forensic Science International*, **266**, 399–411, 2016.

Stone, R.S. Footwear examinations: Mathematical probabilities of theoretical individual characteristics. *Journal of Forensic Identification*, **56**(4), 577, 2006.

SWGTREAD. Guide for the detection of footwear and tire impressions in the field. *Journal of Forensic Identification*, **55**(6), 766–769, 2005a.

SWGTREAD. Guide for the collection of footwear and tire impressions in the field. *Journal of Forensic Identification*, **55**(6), 770–773, 2005b.

SWGTREAD. Guide for the Preparation of Test Impressions from Footwear and Tires (03/2005c). http://www.swgtread.org/images/documents/standards/published/swgtreadfi 05fitestfiimpressionsfi200503.pdf

SWGTREAD. Guide for the Forensic Documentation and Photography of Footwear and Tire Impressions at the Crime Scene (03/2006a). http://www.swgtread.org/images/documents/ standards/published/swgtreadfi09fidocumentationfiandfiphotographyfi200603.pdf

SWGTREAD. SWGTREAD. Guide for the Examination of Footwear and Tire Impression Evidence (03/2006b). http://www.swgtread.org/images/documents/standards/published/swgtreadfi 08fitestfiexaminationfi200603.pdf

SWGTREAD. Guide for Lifting Footwear and Tire Impression Evidence (03/2007a). http:// www.swgtread.org/images/documents/standards/published/swgtreadfi12fiLiftingFootwear andTireImpressionEvidence200703.pdf

SWGTREAD. Guide for Casting Footwear and Tire Impression Evidence (03/2007b). http://www .swgtread.org/images/documents/standards/published/swgtreadfi11fiCastingFootwearand TireImpressionEvidence200703.pdf

SWGTREAD. Guide for Casework Documentation (03/2008). http://www.swgtread.org/images/ documents/standards/published/swgtreadfi13fiCaseworkDocumentation200803.pdf

SWGTREAD. Guide for the Range of Conclusions Standard for Footwear and Tire Impression Examinations (03/2013). http://www.swgtread.org/images/documents/standards/ published/swgtread

Thompson, W.C. and Cole, S.A. Psychological aspects of forensic identification evidence. *Expert Psycological Testimony for the Courts*, **31**, 34–36, 2007.

Thompson, W.C., Kaasa, S.O., and Peterson, T. Do jurors give appropriate weight to forensic identification evidence? *Journal Empirical Legal Studies*, **10**(2), 359–397, 2013.

Thompson, W.C. and Schumann, E.L. Interpretation of statistical evidence in criminal trials: The prosecutor's fallacy and the defense attorney's fallacy. *Law and Human Behavior*, **11**(3), 167, 1987.

Wang, X., Sun, H., Yu, Q., and Zhang, C. Automatic shoeprint retrieval algorithm for real crime scenes. *Asian Conference on Computer Vision*, pages 399–413. Springer, 2014.

Willis, S., McKkenna, L., McDermott, S., O'Donell, G., Barrett, A., Rasmusson, B., Nordgaard, A., Berger, C.E., Sjerps, M.J., Lucena-Molina, J., and Zadora, G. ENFSI guideline For Evaluatlive reporting in Forensic Science. European Network of Forensic Science Institutes, 2015.

Wyatt, J.M., Duncan, K., and Trimpe, M.A. Aging of shoes and its effect on shoeprint impressions. *Journal of Forensic Identification*, **55**(2), 181–188, 2005.

Yekutieli, Y., Shor, Y., Wiesner, S., and Tsach, T. Expert Assisting Computerized System for Evaluating the Degree of Certainty in 2D Shoeprints. Final report, NIJ Award Number: IAA-2009-DN-R-090, 2016. https://www.ncjrs.gov/pdffiles1/nij/grants/250336.pdf

Ytti, A., Majamaa, H., and Virtanen, J. Survey of the conclusions drawn of similar shoeprint cases, part II. *Proceedings of the European Meeting for shoeprint and Toolmarks Examiners*, The Hague, Netherlands, pages 157–169, 1997.

18

Forensic Glass Evidence

Karen Pan, Junqi Chen, and Karen Kafadar

CONTENTS

18.1 Introduction 411
    18.1.1 Early Cases of Forensic Glass Evidence 412
    18.1.2 Types of Glass 412
    18.1.3 Glass Breakage and Recovery 414
    18.1.4 Physical and Optical Properties 414
    18.1.5 Chemical Composition 416
18.2 Methods for Measuring Glass Elemental Concentrations 418
    18.2.1 ICP-MS and LA-ICP-MS 418
    18.2.2 Micro X-Ray Fluorescence (μ-XRF) 419
18.3 Multivariate Methods 421
    18.3.1 Glass Data 426
    18.3.2 Data Sets 427
    18.3.3 Variability 430
18.4 Validating Procedures 431
    18.4.1 Simulation Strategy 431
    18.4.2 Simulated Match Rates 432
    18.4.3 Another Interpretation of "Difference" 436
18.5 Conclusions 437
References 438

18.1 Introduction

20th century French forensic science pioneer Dr. Edmond Locard formulated the basis for all forensic science with his quote "Every contact leaves a trace." This has become known as Locard's exchange principle, which Kirk (1953) further detailed and expressed as:

> Wherever he steps, whatever he touches, whatever he leaves, even unconsciously, will serve as a silent witness against him. Not only his fingerprints or his footprints, but his hair, the fibers from his clothes, the glass he breaks, the tool mark he leaves, the paint he scratches, the blood or semen he deposits or collects. All of these and more, bear mute witness against him. This is evidence that does not forget. It is not confused by the excitement of the moment. It is not absent because human witnesses are. It is factual evidence. Physical evidence cannot be wrong, it cannot perjure itself, it cannot be wholly absent. Only human failure to find it, study and understand it, can diminish its value.

While physical evidence "does not forget," it can be degraded or contaminated by the environment or, more importantly, misinterpreted by humans. Moreover, the measurements on the evidence may not be sufficient to uniquely distinguish it from multiple sources. For example, glass falls under the category of trace evidence, which generally refers to small pieces of evidence left at a crime scene, along with bullets, paint, hair, fibers, blood, fire debris (arson), and soil. But measurements on these pieces may match multiple sources. Statistical methods are needed to characterize and quantify the effects of these sources of variation on the evidence.

18.1.1 Early Cases of Forensic Glass Evidence

Most early analyses of forensic glass evidence examined how glass physically broke. In his book *Criminal Investigation*, Hans Gross describes how to distinguish from which side of a glass pane a bullet was fired, by examining which side contains "shell-shaped fractures" that "result from the bullet taking away with it the last layer of glass as it passes through the pane" (Gross, 1907).

Joseph R. was murdered in his home in 1928. The evidence and environment, taken at face value, suggested a criminal broke and entered through a kitchen window. During the trial, the question arose as to whether or not someone had staged a broken window for deceptive purposes. The judge tasked Matwejeff with determining whether the window had broken from the inside or outside (Matwejeff, 1931). Matwejeff conducted a series of experiments using a fist or blunt object to strike a glass pane, and examined the cracks radiating from the point of force. He concluded that in some cases, the radial and concentric cracks that formed allowed determination of which side the glass was struck from (Matwejeff, 1931). This was later confirmed by the United States Federal Bureau of Investigation (U.S. FBI) (United States, 1936).

Despite research and analysis of glass chips and cracks, it was recognized very early on that being able to piece or fit together glass fragments was the only (or, at least very highly) conclusive proof of common origin (Tryhorn, 1939; Gamble et al., 1943; Nelson, 1959). This is often impossible as glass fragments may be missing or too small.

In 1933, the issue of identifying minute glass splinters arose in New Zealand in the form of a burglary case. While the glass fragments of interest were recovered from a broken window and the suspect's briefcase, other control fragments (other windows, bottle glass, etc.) were obtained for comparison. These were all examined with regard to refractive index, appearance under ultra-violet light, and specific gravity. Out of 65 pieces of glass, only glass splinters from the briefcase matched in all three properties to that of the broken window – "it was therefore considered highly probable that the splinters ... were obtained from the window concerned" (Marris and Hoff, 1934).

18.1.2 Types of Glass

ASTM C162-05 (2015), Standard Terminology of Glass and Glass Products, defines glass as "an inorganic product of fusion which has cooled to a rigid condition without crystallizing" (ASTM International, 2005). The first manmade glass likely originated from obsidian or faience (an older material of crushed quartz and alkali) about 3,500 years ago (Curran et al., 2000). Today soda-lime-silicate glass (soda-lime glass), one of the most prevalent types of glass used in windows and containers, is comprised of soda (Na_2O), lime (CaO), and silica (SiO_2) (ASTM International, 2005). A typical melting tank can produce several hundred tons of glass per day (Bottrell, 2009).

FIGURE 18.1
Pilkington Figure 5 (Pilkington, 1969), "Sheet glass has a perfect fire finish. Only the edges gripped by rollers are marked."

Flat/sheet glass. The modern sheet glass process, the Fourcault process, was developed in 1914 by Emile Fourcault and produced flat glass by pulling a glass sheet straight up many stories from a melting tank of molten glass (Tooley and Scholes, 1984). The glass sheet passed through cooled rollers to stiffen (Pilkington, 1969), and was cut into sheets at the top of the machine. Flat glass has a "fire-finish" achieved as the glass cools without touching anything solid, but is wavy due to minor variations in thickness and thus flatness. Because the thickness of the glass ribbon is controlled by the viscosity of the molten glass, any small inhomogeneities in the mixture can cause variations in the finished glass. While acceptable for window glasses, these distortions are unacceptable for windshield glass or mirrors. The plate process was developed to create distortion free glass (Pilkington, 1969), which involved casting molten metal on a table, rolling it into a plate, annealing, grinding, and polishing, and was much more expensive. See Figure 18.1.

Float glass. Sir Alastair Pilkington introduced the float glass process in 1959 (Pilkington, 1969, 2019). A continuous ribbon of glass moves through a melting furnace and along the surface of an enclosed bath of molten tin. It remains on the tin in high temperatures long enough for irregularities to melt and the surface to become flat. The ribbon continues passing across the molten tin until it hardens enough to be removed without marking the surface of the glass (Figure 18.2). The result is glass of nearly uniform thickness (i.e., the standard deviation of glass thickness is small but not zero) with a bright fire polished surface. Most flat glass is currently manufactured using this process. Float glass can be identified by luminescence on the surface originally in contact with tin (Curran et al., 2000).

Toughened/tempered glass. Toughened or tempered glass, also called fully tempered glass or heat-strengthened glass, is safety glass which shatters without sharp edges or points

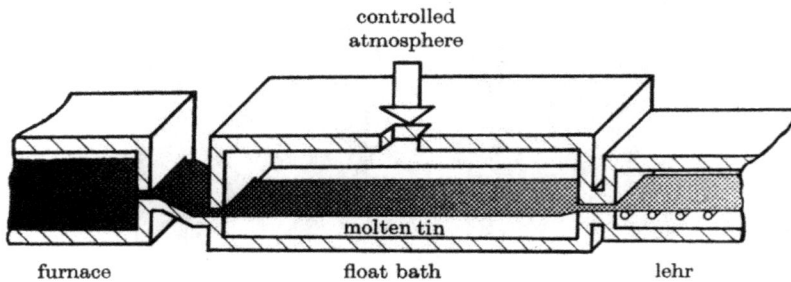

FIGURE 18.2
Pilkington Figure 9 (Pilkington, 1969), "Diagram of the float process."

(ASTM International, 2005). Heated glass is rapidly cooled with air jets, which compresses the surfaces and increases tension in the center. This process strengthens and toughens glass against breakage, which is thought to occur due to surface flaws (Curran et al., 2000). Generally, tempered glass is four to five times stronger (Varshneya and Tomozawa, 1994) and heat-strengthened glass (which is chilled to a lower temperature than tempered glass) is about twice as strong (Bottrell, 2009) as non-tempered glasses. Tempered glass is required for vehicle side and rear windows in the U.S.

Laminated glass. Laminated glass is made by heat-sealing a thin layer(s) of plastic between two panes of heat-strengthened glass and is known for its ability to resist penetration (Curran et al., 2000). The two sheets of glass may or may not originate from the same pallet, and in some cases are purposely chosen to be different colors. Laminated glass is required for vehicle windshields in the U.S. to help ensure passenger safety (Bottrell, 2009).

18.1.3 Glass Breakage and Recovery

Upon breakage, glass fragments may eject in all directions to land on clothing; yet the number of recoverable fragments will decrease as they fall off over time (Bottrell, 2009; Hicks et al., 1996; Curran et al., 1998). Nelson and Revell (1967) broke 19 glass panes and observed backward fragmentation in all, concluding that "[w]henever a window is broken it is to be expected that numerous fragments will strike a person within a few feet of the window"; the number of fragments retained depending in part on the type of fabric and time elapsed between glass breaking and glass collection (Nelson and Revell, 1967). In addition to the surface of fabrics, glass particles have been observed to embed themselves into the interior of clothing such as pockets, and may migrate from pockets to surfaces and vice versa (Brewster et al., 1985; O'Sullivan et al., 2011).

Fragments can be recovered by hanging a garment and scraping it, adhesive tape, tweezers, vacuuming, or shaking (Curran et al., 2000). After recovery, examiners must confirm that the fragment recovered is glass and not another material with similar appearance. For very large fragments, the thickness, color, and edges may be compared to see if they came from a specific source.

18.1.4 Physical and Optical Properties

Forensic glass examiners may use physical and optical properties to discriminate glass. Generally, these examinations are conducted before chemical analysis, as certain chemical procedures will destroy the sample or require larger sample sizes and expensive equipment. However, when none of the physical and optical properties can discriminate glass fragments, elemental analysis may be used.

Thickness. If the recovered glass fragments are large enough to view the full thickness of the source object, this dimension of the glass fragments can be measured (United States, 1936; Tryhorn, 1939; Bottrell, 2009). However, thickness should be considered as only a basic comparison, as thickness across an object may not be uniform (windshields, vases, etc.). Renshaw and Clarke (1974) measured seven vehicle windshields and observed standard deviations (corresponding residual standard deviations; RSDs) of 0.004–0.006 mm (0.08%–0.12%) for five float glass samples and 0.013–0.037 (0.26%–0.73%) for two non-float samples. Crockett and Taylor (1969) surveyed 100 samples of vehicle safety glass and noticed thickness was normally distributed with two modes of 0.197″ and 0.250″, one attributed to the requirements of British Standard 857 specifying nominal safety glass thickness (BS 857 specifies thickness of 0.197″ with limits 0.189″–0.205″, 0.236″ with limits 0.228″–0.244″, and 0.244″ with limits 0.236″–0.252″), and the other unknown. These two modes had standard deviation values of 0.0065 (3.30%) and 0.0098 (3.92%).

Color. Larger fragments may also be compared for color when placed on a neutral or complementary colored background and viewed side by side under various lighting conditions. While differences in glass color represent changes in glass chemistry, it can be hard to differentiate colors if the color density is low or the fragments are too small (Bottrell, 2009). Color comparison for small fragments can be uncertain at best; and it is recommended that examiners be blind tested for color comparisons (Curran et al., 2000). Colored glass may also be examined analytically by spectrography (determining the chemical elements by measuring wavelengths or spectral line intensity of a sample) to determine the coloring agent (Nickolls, 1966; Meggers et al., 1922). Glasses purposefully colored for decorative purposes may contain some common inorganic colorants: iron (green, brown, or blue), manganese (purple), cobalt (blue, pink, green), titanium (purple, brown), cerium (yellow), and gold (red) (Copley, 2002).

Curvature. The radius of curvature can be compared (Tryhorn, 1939), and may be determined visually with low power magnification for large fragments or quantitatively using interferometry for small particles (Bottrell, 2009). Locke and Elliott (1984) devised a numerical criterion for assessing the curvature of glass particles for various types of glass (Locke and Elliott, 1984). They also concluded that objects with curved surfaces produce distinctly curved particles in comparison to flat window glasses.

Edges and hatch marks. Larger fragments may be pieced together to see if the edges match and fit. Similarly, hackle or hatch marks that are created when glass is broken may be compared. Since these marks are created randomly in a multitude of ways, it is canonically asserted that any specific pattern has a very small chance of occurring and hence typically are viewed as highly distinctive and meaningful for comparison (Nickolls, 1966).

Surface features. Surface features can be divided into three types. Manufacturing features can be intentional (etching, texturing, frosting) or accidental (mold/polish marks, reams) (Bottrell, 2009; Kammrath et al., 2016). Post-manufacturing features are unintentional marks such as scratches and abrasions, and include any dirt or debris stuck to glass. Coatings, including thin films and mirrored backings, can have single or multiple layers, the precise number of which may be hard to detect.

Fluorescence. Fluorescence describes the radiation emitted by certain substances under incident radiation such as X-rays or ultraviolet (UV) light. The color that glass can fluoresce varies, and can be used for establishing dissimilarity (Tryhorn, 1939). It can also indicate the side of float glass in contact with a tin bath, which fluoresces under short UV (254 nm) light (Dabbs and Pearson, 1972; Bottrell, 2009). Some types of coating may also cause fluorescence.

Specific gravity. Otherwise known as relative density, this term was used interchangeably in early glass analysis with density, although the terms are not exactly the same (Kammrath et al., 2016). Tryhorn (Tryhorn, 1939) defined specific gravity (SG) as the "weight in grams of a cube of substance of 1 centimeter edge. Since the weight of such a cube of water at 20°C is 1 gram, the specific gravity measured at the same temperature gives the density of a substance relative to water." A fragment of a solid will remain suspended if immersed in a liquid with equal specific gravity.

Refractive index. Refraction refers to the slight change in direction of light as it passes through a surface at a non-perpendicular angle, and is measured by refractive index. The RI of common glasses is between 1.5 and 1.9 due to the composition of glass (Tryhorn, 1939). RI, which is non-destructive and requires only a small sample, was often used in analyses.

Dispersion. Dispersion is defined as a variation of refractive index (RI) with a change in wavelength of illumination (ASTM International, 2005), and may offer little improvement over RI (Slater and Fong, 1982; Locke et al., 1986).

Although RI and SG are correlated (Gamble et al., 1943; Nickolls, 1966; Dabbs and Pearson, 1972), some maintained that the discriminating power between the two differed and thus both should be measured (Cobb, 1968; Slater and Fong, 1982). However, RI alone may be sensitive enough to distinguish fragments (Gamble et al., 1943; Roche and Kirk, 1947). Crockett and Taylor (1969) also observed much variation in SG within one glass pane. With improvements in equipment and methodology, the discrimination advantage of RI over density increased, and many labs measured only RI (Curran et al., 2000). The first automated Glass Refractive Index Measurement (GRIM) system was released in 1982 by Locke and Underhill (1985).

With recent advances in the glass manufacturing process, variation in RI has been decreasing (Crockett and Taylor, 1969; Dabbs and Pearson, 1972; Lambert and Evett, 1984; Koons and Buscaglia, 2002). Davies et al. (1980) estimate the standard deviation of RI for sheet and float window glass as 0.00012, 0.00033 for pattered glasses, and 0.00008 for toughened float glasses (Davies et al., 1980). The consistency in RI across glass panes suggests that RI may no longer be as useful for discrimination nor for classifying glass types (Catterick and Hickman, 1981; Ryland, 1986); especially as RI may vary across a pane (Bennett et al., 2003). Furthermore, the RI of float glass surfaces in contact with molten tin during the glass production process will always be greater than the RI of the bulk (the rest of the glass never coming into contact with molten tin) (Davies et al., 1980; Underhill, 1980).

18.1.5 Chemical Composition

We assume that the examiner has conducted prior analyses using physical characteristics (such as color and thickness), and that the results of those physical tests led the examiner to pursue the chemical assessment of trace elements. Previously, measuring the exact chemical concentrations in glass required expensive machinery, larger glass fragments not often available, and much labor and time (Tryhorn, 1939). Physical and optical examinations, especially of refractive index, were deemed sufficient. However, for reasons mentioned above concerning RI, including manufacturing improvements and more widespread availability of chemistry equipment, a move has been made towards the use of elemental concentration analysis for forensic glass discrimination, which has been proposed for discrimination among glass samples from different sources when RI and other physical properties alone cannot (Blacklock et al., 1976; Hughes et al., 1976; Smale, 1973; Reeve et al., 1976; Haney, 1977; Calloway and Jones, 1978; Catterick and Hickman, 1981; Hickman, 1981; Hickman et al., 1983; Terry et al., 1984; Koons et al., 1988; Wolnik et al., 1989;

Zurhaar and Mullings, 1990; Koons et al., 1991; Buscaglia, 1994; Duckworth et al., 2000; Koons and Buscaglia, 1999, 2002; Duckworth et al., 2002). To date chemical analysis seems to be the most effective method for differentiating glass specimens (Koons and Buscaglia, 1999, 2002). Zurhaar and Mullings (1990) described the discriminatory power of inductively coupled plasma mass spectrometry (ICP-MS), a method for obtaining elemental compositions:

> Analysis by ICP-MS provides an elemental fingerprint of a glass sample. By using an initial discriminating group of 15 elements the uniqueness of a fragment can be established. By determining a further 25 elements, the results of the discrimination test can then be considered unquestionable.

Although the challenges of using chemical analyses remain today, Koons and Buscaglia (1999) state that:

> The forensic scientist should use the most discriminating technique available in the examination of glass or other form of trace evidence because it is the most effective means of both avoiding false associations and excluding two similar, but separate, sources. *It is in the best interest of the court for the scientist to use the most discriminating analytical technique even if this means that exact probability figures for a conclusion cannot be calculated.* In cases where the analytical discrimination is very good, as in compositional measurements of glass, factors such as manufacturer distribution of products and age and breakage of glass objects in the crime scene and suspect environments are more significant than the probability of two randomly selected sources from a large glass population having coincidentally indistinguishable characteristics. These factors can either be determined by standard investigative techniques or they involve everyday experiences of the nonscientist. As a result, their significance can be readily weighed by the trier of fact without resorting to statistical calculations. [emphasis added]

The job of a statistician is to understand the validity of the claims made in this paragraph; specifically: probabilities of an assessment are needed in order to assure that (1) the most discriminative analytical technique has been used; (2) experiments can and should be designed to quantify the magnitude of effects of factors such as "manufacturer distribution of products and age and breakage of glass objects in the crime scene and suspect environments" to determine whether these factors are indeed "more significant than the probability of two randomly selected sources from a large glass population having coincidentally indistinguishable characteristics"; and (3) the trier of fact cannot judge their significance without resorting to statistical calculations.

Many methods exist for obtaining elemental concentrations in glass, including spectrographic analysis (Gamble et al., 1943; Blacklock et al., 1976; Smale, 1973), neutron activated analysis (NAA) (Haney, 1977; Hughes et al., 1976; Calloway and Jones, 1978; Hickman et al., 1983), spark source mass spectrometry (Haney, 1977; Blacklock et al., 1976; Calloway and Jones, 1978), dilution spark source mass spectrometry (Haney, 1977), flame atomic emission spectrometry (Hickman et al., 1983), energy dispersive X-ray (EDX), inductively coupled plasma atomic emission spectrometry (ICP-AES) (Catterick and Hickman, 1981; Hickman et al., 1983; Koons et al., 1988, 1991; Buscaglia, 1994; Koons and Buscaglia, 1999, 2002; Duckworth et al., 2002), inductively coupled plasma mass spectrometry (ICP-MS) (Zurhaar and Mullings, 1990; Duckworth et al., 2000, 2002), laser ablation ICP-MS (LA-ICP-MS) (Trejos et al., 2003; Bajic et al., 2005; Latkoczy et al., 2005; Trejos and Almirall, 2005; Berends-Montero et al., 2006), and micro X-ray fluorescence (μ-XRF) (ASTM International, 2013; Trejos et al., 2013b).

Three of these methods, ICP-MS, LA-ICP-MS, and μ-XRF will be described in the following section. These are of interest because three ASTM standards have been proposed for comparing forensic glass evidence fragments using these techniques. The strengths and limitations of the methods need to be well understood by the statisticians analyzing the data, and different glass samples will dictate the choice of measurement method.

18.2 Methods for Measuring Glass Elemental Concentrations

The introduction and utilization of elemental analysis techniques helped remedy the lack of discriminatory power from using refractive index measurements alone.

18.2.1 ICP-MS and LA-ICP-MS

Inductively coupled plasma mass spectrometry (ICP-MS) was first introduced for trace element determination in 1980 by Professor Harry J. Svec, and remains a powerful method for quantitative elemental analysis today (Houk et al., 1980). The ICP-MS machine is comprised of two parts: ICP, which can efficiently ionize elements in samples; and MS, a detector that can distinguish different elements from their positive ions.

The ICP is a type of plasma source in which the energy is supplied by electric currents that are produced by electromagnetic induction. The high frequency current flows through a coil-shaped electrode at the top of the machine, creating high temperatures and a strong alternating magnetic field. Cooling gas is passed through the outside of the electrode to prevent the system from melting. The working gas (usually argon) is first discharged and then sent through the inside of the electrode. Due to the high temperature and interaction with the strong alternating magnetic field, the electrons, ionized argon atoms, and ground state argon, are all moving very quickly. They collide with each other, which induces a plasma torch. The temperature of the plasma can reach 6000–10000°K (Shunko et al., 2014), similar to the surface of the sun. The sample solution (in this case, a prepared glass sample) will be introduced to the center of the plasma and will be ionized with it. The degree of ionization (proportion of neutral particles ionized to charged particles) for about 60 elements exceeds 90%, which allows for observation with suitable detectors (Dunnivant and Ginsbach, 2017), such as OES (optical emission spectrometer) and MS (mass spectrometry).

MS is an analytical technique that ionizes chemical samples and sorts ions based on their mass-to-charge ratio. ICP-MS uses ICP to ionize samples and different types of mass analyzers (such as time-of-flight, quadrupole mass filter, and ions trap) to distinguish elements according to their mass-to-charge ratio. For example, in the time-of-flight mass analyzer, all ions will be accelerated with the electric field through the same potential, and the time needed to reach the detector is measured. If the ions all have the same charge, their kinetic energies will be identical and their velocities will depend only on their masses. Ions with lower mass will reach the detector first (Wollnik, 1993). By this method, different elements in the sample can be distinguished.

The ICP has a very high energy which makes it possible to ionize all the elements of interest in a sample. Since every element and its isotopes have unique masses, MS can be used to separate and quantitatively analyze them simultaneously. ICP-MS has the advantages of nearly simultaneous multi-elemental capability, reduced matrix interference effects, wide linear dynamic ranges, and excellent precision and sensitivity. These attributes result in superior discriminatory power compared to other methods of glass analysis. The repeatability of ICP-MS is often below 5% RSD, reproducibility between laboratories lower than 10%, bias lower than 10%, and limits of detection between 0.03 and 9 $\mu g/g$ for the majority of the elements monitored (Trejos et al., 2013a).

However, the protocol for using ICP-MS for glass sample analysis requires a time-consuming and hazardous sample preparation process. The glass sample must first be washed with concentrated nitric acid then dissolved in an acid mixture containing concentrated hydrofluoric acid, hydrochloric acid, and concentrated nitric acid with a 2:1:1 ratio. The samples are thoroughly dried to eliminate most of the silicate matrix and excess acids. The residue will be dissolved again with nitric acid and mixed with the internal standard Rh. This digestion step is also destructive, and around 6 mg of the sample is consumed during the analysis (Trejos and Almirall, 2005). To simplify this process, laser ablation inductively coupled plasma mass spectrometry (LA-ICP-MS) was introduced to this field.

As the name suggests, LA-ICP-MS uses a laser for sample volatilization and to introduce the sample into the plasma. The setup involves the ICP-MS machine as described above with an additional laser and ablation cell. After a solid sample is placed in the ablation cell, the sample surface is struck with a laser beam, producing particles that are then transported to the ICP (usually by argon or helium). LA-ICP-MS has several advantages over ICP-MS, mainly minimal sample preparation, eliminating the need for hazardous materials such as acids, reduced sample consumption, and faster digestion methods (up to eight times faster) (Trejos and Almirall, 2005).

18.2.2 Micro X-Ray Fluorescence (μ-XRF)

X-ray fluorescence (XRF) is the emission of characteristic "secondary" fluorescence or X-rays from an excited atom. The underlying principle of XRF relies on the structure of an atom. Each atom contains a nucleus (with neutrally-charged neutrons and positively-charged protons) and one or more electrons (negatively charged) that orbit the nucleus. Since the nucleus and electrons have different charges, an electromagnetic force exists between them which helps stabilize the electrons near the nucleus. Electrons are localized on the different orbitals of the atom which contain differing amounts of energy. The electrons located on inner orbitals will have less energy than those located on outer orbitals. Lower energy electrons bond more tightly to the nucleus.

When materials are exposed to short-wavelength X-rays or gamma rays, the ionization of atoms may take place. If the radiation is energetic enough, it could expel electrons from the inner orbitals of the atom. The removal of electrons in this manner makes the electronic structure of the atom unstable, and electrons in outer orbitals have a strong potential to fill the hole left behind. Since electrons in outer orbitals have a higher potential energy than electrons in the inner electron shell, the filling process will release energy in the form of a photon or X-ray whose energy is equal to the energy difference between the two orbitals involved. Each chemical element has electronic orbitals of characteristic energy, so the energy of the emitted radiation has characteristics of the atoms present. The emitted radiation is called "fluorescence" due to the formation phenomena in which the absorption or radiation (X-rays or gamma rays here) of a specific energy results in the re-emission of radiation of a different energy. The re-emitted "fluorescence" is detected in the X-ray fluorescence spectrometer with an energy dispersive X-ray detector (EDS).

The electron orbitals can be grouped into electron shells, which are named K, L, M, N, and O from the inner shell to the outer shell, respectively. There are a number of rules which limit the filling transitions. The name of the transitions depends on which electron shell the electron is removed from by radiation. For example, if the removal of the electron happens in the K layer, the re-emitted fluorescence is called K-series. An L→K transition is traditionally called K_α, an M→K transition is called K_β, and so on. If the removal of the electron happens in the L layer, then the re-emitted fluorescence is called L-series. Each of these transitions yields a fluorescent photon or X-ray with a characteristic energy equal to the difference in energy of the initial and final orbitals. The wavelength of this fluorescent radiation can be calculated using Planck's Law.

Conventional XRF has a typical spatial resolution ranging in diameter from several hundred micrometers up to several millimeters, which is unsuitable for forensic analysis. The newly developed technology Micro X-ray Fluorescence (μ-XRF) overcomes this disadvantage by restricting the excitation beam size or focusing the excitation beam to a sub-μm spot (Behrends and Kleingeld, 2009). Conventional μ-XRF equipment uses a pinhole aperture to restrict the excitation beam, but this method blocks the majority of the X-ray flux which severely affects the sensitivity of trace elemental analysis (Bichlmeier et al., 2001). All commercial μ-XRF instruments utilize the effective diffraction of X-rays at the surface of glass capillaries to ensure a small spot size. Two different designs are used for μ-XRF applications: mono-capillary and poly-capillary. Mono-capillary design has the advantages of less crucial working distance, better sample volume definition, and smaller spot size. However, it has less photon flux than mono-capillary design (Behrends and Kleingeld, 2009).

In μ-XRF analysis, the qualitative analysis is based on elements' characteristic X-ray energies in the specimen. Semi-quantitative analysis is accomplished by comparing the relative area under the characteristic X-ray's peak. Typical limit of detections (LODs)* range from parts per million (μgg-1) to percent (%) (ASTM International, 2013). Compared to other methods such as inductively coupled plasma optical emission spectrometry (ICP-OES) and inductively coupled plasma mass spectrometry (ICP-MS), μ-XRF has higher LODs, but has the advantages that its non-destructive capability permits the re-analysis of test samples as well as preservation of valuable materials. For forensic glass analysis applications, μ-XRF requires smaller sample sizes and much shorter sample preparation time. LA-ICP-MS offers similar specimen size and lower LODs, but the instrument is much more expensive and complicated to operate (ASTM International, 2013).

In general, μ-XRF instruments offer a nondestructive, simultaneous qualitative analysis for multiple elements in small specimens, but is not suitable for analysis of elements lighter than sodium (Na). For transition elements, detection limits between 10 and 100 ppm seem to be achievable, and can be considered the lowest possible detection limits of current μ-XRF instruments. Analysis of the sample under vacuum or under an He flux can increase the sensitivity of the instrument for lighter elements such as Al and Si.

If concentrations of elements of interest are near the LOD of the EDS system, the forensic examiner must decide whether or not to utilize this chemical peak information in analysis. Signal-to-noise ratios (SNRs) of elemental data from glass samples may provide additional objective information on (1) peak identification/labeling decisions (i.e., whether a peak is present); (2) selection of elements for semiquantitative ratio comparisons; (3) comparisons of μ-XRF instruments/configurations; (4) calculation of LODs; and (5) quality control (i.e., whether an instrument is sufficiently sensitive to a group of relevant elements) (Ernst et al., 2014).

---

* The limit of detection (LOD), or detection limit, is used to describe the smallest concentration of a substance that can be reliably measured and distinguished from the absence of the substance (sometimes called a blank or blank value) by an analytical procedure (Armbruster and Pry, 2008).

18.3 Multivariate Methods

Trace element analysis has been used to evaluate the source of bullets, glass, paint, copper wire, and other types of physical evidence. The "working hypothesis" is that concentrations of certain elements (presumably highly specific to different pieces of evidence) may provide a distinctive "signature" that allows for comparison of evidence found at a crime scene with that found in the possession of a suspect. Typically, the forensic glass examiner will chemically analyze fragments recovered from the crime scene and fragments found on or in the possession of a suspect, and will try to decide between two hypotheses (or, which of the two hypotheses is likely to hold): either the two sets of fragments came from the same source, or they came from different sources. Here, "source" refers to the specific piece or pane of glass, not a piece or pane with a chemically "similar" signature that came from a different location or is associated with a different crime scene than the one under investigation. Statistical issues surround this approach. If the batch of material from which the evidence is manufactured is extremely homogeneous, then the measurements on many pieces that were manufactured from the same batch may be deemed "not distinguishable," even though they may have come from different pieces (ASTM International, 2012, 2013, 2016). Depending on the level of error in the measurements themselves, this may lead one to erroneously interpret "not distinguishable" as "came from the same source" and hence to potential false positives. Conversely, if the specific piece of evidence is itself rather inhomogeneous, then concentrations in pieces from two different parts of the same evidence may be different, leading to false negatives. For example, companies constructing large neighborhood tracts may purchase architectural float glass in bulk from a single manufacturer, resulting in many houses containing glass from possibly the same batch or batches produced on the same day. Using error rates determined from a diverse population could be extremely misleading given the recovered glass fragment could have come from one of many window panes in the neighborhood. Trace element analysis of forensic evidence may be unsatisfactory for both inclusion and exclusion purposes (the NAS report for comparing bullet lead evidence describes such situations (National Research Council, 2004)).

Over the years, various researchers have proposed procedures for assessing the similarity in the compositions of glass from two locations or sources. Some of these procedures (e.g., Parker, 1966, 1967; Lindley, 1977; Evett and Lambert, 1977; Evett and Buckleton, 1990; Curran, 2003; Aitken and Lucy, 2004) suggest a multi-step approach; all of them however involve some statistical inference procedure. It is this inference procedure that is in need of evaluation and validation. Because two of the ASTM standards were recently approved for the OSAC Registry,* we focus here on only the evaluation of the approved standard ASTM E2927-16e1; but an evaluation and validation procedure would apply to any of these methods, including those with multi-step processes.

Although the ASTM standards do not explicitly mention "inclusion/exclusion," jurors in the United States may well understand "analytically indistinguishable" to mean "from the same source" (Gabel-Cino, 2017). Wording in the standards, "[i]f the samples are distinguishable... in any of these observed and measured properties, [for example, color, refractive index, density, elemental composition], it may be concluded that they did not

---

* At the time of this writing, the third ASTM standard, "Standard Test Method for Determination of Concentrations of Elements in Glass Samples using Inductively Coupled Plasma Mass Spectrometry (ICP-MS) for Forensic Comparisons" (E2330-19), remains up for vote on whether to be posted on the OSAC Registry.

originate from the same source of broken glass," also provides conditions under which "distinguishable" samples can be used for exclusion (ASTM International, 2012, Section 1.1; ASTM International, 2013, Introduction; ASTM International, 2016, Introduction). ASTM E2927-16e1 (ASTM International, 2016) further asserts that following the technique as described "yields high discrimination among sources of glass" and "provides high discriminating value in the forensic comparison of glass fragments." In addition to potential misunderstandings by jurors, phrasing in the standards themselves suggest that these procedures may have probative value.[*]

The three glass standards proposed by ASTM International describe a procedure for comparing forensic glass evidence found at a crime scene (sometimes called the "Known" (K) fragment) with glass found on or in connection with a potential suspect (sometimes called the "Recovered" (R) or "Questioned" (Q) fragment). Using measured trace elemental concentrations from these glass fragments, the standards provide an inferential method for determining if two fragments are "analytically distinguishable" and "it may be concluded that they did not originate from the same source of broken glass" (ASTM International, 2012, 2013, 2016). These standards all outline the same general process but differ in the instruments used to measure and process glass evidence and the number of elements that should be analyzed (8–17).[†]

- ASTM E2330-19, *Standard Test Method for Determination of Concentrations of Elements in Glass Samples Using Inductively Coupled Plasma Mass Spectrometry (ICP-MS) for Forensic Comparisons.*
- ASTM E2926-17, *Standard Test Method for Forensic Comparison of Glass Using Micro X-ray Fluorescence (μ-XRF) Spectrometry.*
- ASTM E2927-16e1, *Standard Test Method for Determination of Trace Elements in Soda-Lime Glass Samples Using Laser Ablation Inductively Coupled Plasma Mass Spectrometry for Forensic Comparisons.*

Each standard includes a "Calculation and Interpretation of Results" section describing the creation of a "mean ± n·SD" "match interval." The steps are similar for each standard; below are those for ASTM E2927-16e1 (LA-ICP-MS) (ASTM International, 2016).

11.1   *The procedure below shall be followed to conduct a forensic glass comparison using the recommended match criteria is as follows (Weis et al., 2011; Trejos et al., 2013a,b; Dorn et al., 2015).*

11.1.1 *For the Known source fragments, using a minimum of 9 measurements (from at least 3 fragments, if possible), calculate the mean for each element.*

11.1.2 *Calculate the standard deviation for each element. This is the Measured SD.*

---

[*] We thank Professor William Thompson (Professor Emeritus of Criminology, Law, and Society; Psychology and Social Behavior; and Law; at the University of California Irvine School of Social Ecology) (see https://protect-us.mimecast.com/s/-isSC9rm1mS0Q9VSocPzV?domain=faculty.sites.uci.edu) for these remarks.

[†] The elements in the "signature" differ for each standard. Standard ASTM E2330-19 for ICP-MS recommends 14 elements: magnesium (Mg), aluminum (Al), iron (Fe), titanium (Ti), manganese (Mn), rubidium (Rb), strontium (Sr), zirconium (Zr), barium (Ba), lanthanum (La), cerium (Ce), neodymium (Nd), samarium (Sm), and lead (Pb). ASTM E2927-16e1 for LA-ICP-MS recommends all of the same except Sm, plus lithium (Li), potassium (K), calcium (Ca), and cerium (Ce) (17 elements). The standard for XRF is less specific; see ASTM E2926-17 Section 10.6.2.1.

11.1.3 *Calculate a value equal to at least 3% of the mean for each element. This is the Minimum SD.*

11.1.4 *Calculate a match interval for each element with a lower limit equal to the mean minus 4 times the SD (Measured or Minimum, whichever is greater) and an upper limit equal to the mean plus 4 times the SD (Measured or Minimum, whichever is greater).*

11.1.5 *For each Recovered fragment, using as many measurements as practical, calculate the mean concentration for each element.*

11.1.6 *For each element, compare the mean concentration in the Recovered fragment to the match interval for the corresponding element from the Known fragments.*

11.1.7 *If the mean concentration of one (or more) element(s) in the Recovered fragment falls outside the match interval for the corresponding element in the Known fragments, the element(s) does not match and the glass samples are considered distinguishable.*

For ASTM E2330-19 (ICP-MS), "Calculation and Interpretation of Results" appears as Section 10, also with a "4-SD match interval" but requiring a minimum of three measurements instead of nine; for E2926-17 (μ-XRF), Section 10.7.3.2 uses a "3-SD match interval":

> For each elemental ratio, compare the average ratio for the questioned specimen to the average ratio for the known specimens $\pm 3s$. This range corresponds to 99.7% of a normally distributed population. If, for one or more elements, the average ratio in the questioned specimen does not fall within the average ratio for the known specimens $\pm 3s$, it may be concluded that the samples are not from the same source.

Note that "99.7%" coverage applies only if the standard deviations are known, not estimated – as it is here, from possibly as few as three measurements (ASTM International, 2013) – and only if the measurements come from a Gaussian distribution. Furthermore, this procedure accounts for variability from only the K fragments, and not additional variability in the standard deviation. Correlations between the elements, which are not insignificant, are also not taken into consideration. It is risky to ignore these correlations entirely, and exploratory data analysis plots have indeed shown that certain corresponding pairs of elements are highly correlated. Individual "match intervals" cannot be treated as independent.

With any measurement procedure – including those for glass – a standard should characterize the following quantities, which are related to false positive and false negative error rates:

- Sensitivity (true positive rate): given that the true concentrations differ by less than a prescribed "difference threshold," the probability the procedure correctly concludes "same source"
- Specificity (true negative rate): given that the true concentrations differ by more than a prescribed "difference threshold," the probability the procedure correctly concludes "difference source"
- False positive rate (FPR): given that two samples come from different sources, the probability the procedure incorrectly concludes "same source," calculated as $1 -$ *specificity*

TABLE 18.1

An Illustration of ASTM E2330 with a Hypothetical Data Set

	Mg	Al	Fe	Ti	Mn	La	Ce	Nd	Sm	Pb
K fragment 1	30500	2217	4169	206	112	2.994	5.728	2.54	0.542	1.086
K fragment 2	30110	2150	4213	194	111	3.034	5.648	2.81	0.68	1.056
K fragment 3	30580	2226	4155	208	115	2.954	5.69	2.58	0.89	1.13
$mean_k(x)$	30396.67	2197.67	4179	202.67	112.67	2.99	5.69	2.64	0.7	1.09
sd_k	251.46	41.53	30.27	7.57	2.08	0.04	0.04	0.15	0.18	0.04
$0.03x$	911.9	65.93	125.37	6.08	3.38	0.09	0.17	0.08	0.02	0.03
$\max\{0.03x, SD\}$	911.9	65.93	125.37	7.57	3.38	0.09	0.17	0.15	0.18	0.04
lower bound	26749.07	1933.95	3677.52	172.39	99.15	2.63	5.01	2.04	−0.02	0.93
upper bound	34044.27	2461.39	4680.48	232.95	126.19	3.35	6.37	3.24	1.42	1.25
R fragment mean	30400	2321	4590	**240**	112	2.4	5.8	3	0.7	**1.3**

Note: The measured concentrations in two elements from the recovered glass fragment, Ti and Pb, fall outside the upper bound of the $4 \times \max\{0.03x, SD\}$ interval. Thus, these glass samples "are considered distinguishable."

- False negative rate (FNR): given that two samples come from the same source, the probability the procedure incorrectly concludes "different source," calculated as $1 - sensitivity$

A hypothetical data set is shown in Table 18.1 to illustrate the creation of such a match interval. The first three rows indicate three measurements taken on a known glass sample. The next three sets of rows are calculations of, as the standard instructs, the means, standard deviations, and 3% of the mean to give a lower bound for the SD. The lower and upper bounds of the created interval can then be calculated using "mean $\pm 4 \cdot \max\{0.03 \cdot \text{mean}, SD\}$" to create the match interval. The last row indicates elemental concentration means from a recovered (R) glass fragment. As concentrations for two of the elements fall outside the interval (in this case larger than the upper bound), according to step 11.1.7 (10.1.7 in ASTM International, 2012), these glass samples "are considered distinguishable."

Justification for these procedures (Dorn et al., 2015; Trejos et al., 2013a; Weis et al., 2011; Koons and Buscaglia, 2002) appears to be based on empirically observed "error rates" calculated from all possible pairs of glass samples from different sources. For example, Weis et al. (2011) measured 62 different glass samples, mostly from different manufacturers, but some from the same manufacturer produced from different batches at different time periods. The "error rate" was then calculated as the proportion of all pairs that satisfied the "match" criterion, even though the two came from different sources. Comparing each one of the 62 samples as the K with any one of the other 61 samples as the R, they found two of the 1,891 pairs satisfied their "modified n-sigma criterion with fixed relative standard deviations (FRSDs)," giving a Type II error rate of 0.11%, where the FRSDs varied between 3.0% and 8.9%. Dorn et al. (2015) used a similar "4-SD match criterion," but with an RSD_{min} set to 3% for the concentrations of the 10 elements in their study. They found similarly small error rates: 0.27% (6/2256, 48 same-source samples)* for their Type I error rate (two samples

---

* Note that Dorn et al. (2015) actually measured 24 fragments 9 times each, and a 25th fragment 24 times, which is quite different from 48 fragments. See page 87, "*Group 1*", for details.

from same source failed to satisfy the "match" criterion), and 0.11% (7/6642, 82 different-source samples) for their Type II error rate (two different samples satisfied the "match" criterion).

Reported error rates from four commonly referenced papers are very low, typically less than 1% (Dorn et al., 2015; Trejos et al., 2013a; Weis et al., 2011; Koons and Buscaglia, 2002). These error rates are calculated, as above, by comparing two different-source samples from among all possible pairs in the data set, and marking a comparison as a "false positive" if sample means for all 17 elements from the R fragment fall within the "mean $\pm 4 \cdot$ SD" interval created from the K fragment. In this process, the same sample is used for multiple different-source comparisons, and the match or no match conclusion may differ if samples i and j are the R or K fragments, respectively. False positive rates (FPR) calculated in this manner also depend heavily on the specific data set used, some of which are purposefully created to be diverse. If samples in the data set are all highly similar samples (e.g., Toyota windshields) manufactured at similar times, the expected mean difference between these samples will be lower than if samples were very different (e.g., car windshields and baby food jars). The estimated FPR for the first data set may well be higher than that of the second data set.

To better understand the true probability of a false positive, the dependence on data set must be eliminated and the true difference in concentrations, δ_0, between two samples must be known (without measurement variation). If δ_0 is large (as may be for some different-source samples), the procedure should differentiate well between samples and the false positive probability (FPP) is expected to be low. As δ_0 decreases and the samples become more similar, the FPP may increase. A more formal statistical modeling approach, such as that of Pan and Kafadar (2018), is required to provide more accurate estimates of error rates for assessing "distinguishability" between two glass samples.

Statistical modeling is especially critical here because (i) glass databases are small; typically a lab has the facilities to measure and store at most only a few hundred samples; and (ii) the procedure uses a "minimum SD" method – the maximum of the calculated SD from all K fragments ("at least three" replicates) and 3% of the mean (from ideally three fragments, but not required). Such a procedure is not easily analyzed theoretically[*], but error rates can be calculated through statistical modeling of the available data, validation of this model, and then simulation of samples according to this model. Two critical advantages of this approach are that (a) "measurements" can be simulated from a Gaussian distribution as well as from more realistic distributions (e.g., distributions with heavier tails or outliers more often than the presumed idealistic Gaussian); and (b) the error rate when the true difference in elemental concentrations is a known stated percentage (e.g., if the true difference is 3%) can be estimated more precisely, because samples are simulated to have concentrations at specific levels. Thus, the level of the difference in concentrations at which the error rate of a "4-SD match interval" procedure will fall below a specific target can be more precisely quantified.

The best that we can ever do with forensic glass trace element concentration examinations is determine whether the mean of the multivariate distribution of the measurements of the trace element concentrations in fragments recovered from a crime scene are the same as the mean of the multivariate distribution of the measurements of the trace element concentrations in fragments associated with a suspect. Formally, if the multivariate

---

[*] If the data are guaranteed to come from a lognormal distribution (which is rarely true due to outliers and other causes for unusual departures), theory would require the distribution of $max\{s, 3\%\}$, which is the square root of a truncated chi-squared distribution.

measurements of the n recovered fragments are denoted by $X_1, \ldots, X_n \sim G$ with mean μ_G and the multivariate measurements of the m fragments associated with a suspect are denoted by $Y_1, \ldots, Y_m \sim H$ with mean μ_H, then the best we can ever test is whether $\mu_G = \mu_H$. Even if the statistical test that we use concludes that the hypothesis $\mu_G = \mu_H$ is more likely than the hypothesis $\mu_G \neq \mu_H$, we cannot conclude that the recovered fragments and the questioned fragments came from the same source: they may have come from two panes of glass for which the true difference in the mean concentrations, $\mu_G - \mu_H$, was not assessed to be different by the statistical test. Ideally, if $\mu_G - \mu_H$ indeed is not the zero vector, but is small, then we prefer the test that recognizes the difference over one that fails to assert the difference.

18.3.1 Glass Data

For at least two important reasons, we are interested in logarithms of the glass data. First, chemists tend to refer to the "relative standard deviation" (RSD) rather than raw SD, as the SD of elemental concentrations tends to be related to the mean. For example, six measurements of ^{7}Li might be very different from six measurements of ^{90}Zr, which have means and SDs around 11 times larger, but whose RSDs are very similar (see Table 18.2). By transforming the measurements via logarithms, the estimated SDs are approximately equivalent to the RSDs. Second, elemental concentration measurements may have slightly skewed distributions, while the distributions of the logarithms tends to be more symmetric.

Ideally, a multivariate version of Student's t such as Hotelling's T^2 is performed to account for correlated elements and variability in both fragments. Unfortunately, sample sizes are usually not large enough to allow for estimating the correlation matrix. ASTM E2330 requires a minimum of three measurements per sample, while ASTM E2927-16e1 requires "a minimum of 9 measurements (from at least 3 fragments, if possible)" (ASTM International, 2016). None of the data sets available to us for analysis has more than nine replicates, yet the number of elements varies from 8–17 depending on the standard. Weis et al. (2011) dismiss the use of Hotelling's T^2 statistic for assessing the significance of the measurement difference in elemental concentrations in two samples:

> Hotelling's T^2-test, a multivariate equivalent of Student's t-test, has the disadvantage that for mathematical reasons the number of factors (replicate measurements on the control sample plus replicate measurements on the recovered sample) must be at least larger by two than the number of dimensions (i.e. the number of element concentrations, in our case 18). Therefore, at least 10 replicate measurements of both samples to be compared must be conducted for the Hotelling's T^2-test to be applicable. If only six replicate

TABLE 18.2

Means, SDs, and RSDs for Six Measurements of ^{7}Li and ^{90}Zr

Raw Data	1	2	3	4	5	6	SD/Mean	RSD
^{7}Li	4.56	4.68	4.79	4.25	4.33	4.49	0.205/4.517	**4.54%**
^{90}Zr	54.16	55.25	51.93	50.13	49.97	49.44	2.416/51.813	**4.66%**

Log Data	1	2	3	4	5	6	Mean	SD
$\log(^7\text{Li})$	1.517	1.543	1.567	1.447	1.466	1.502	1.507	**4.54**
$\log(^{90}\text{Zr})$	3.992	4.012	3.950	3.915	3.911	3.901	3.947	**4.62**

measurements are carried out for each of the two samples to be compared, the number of elements used for the comparisons has to be reduced to 10, which leads to a loss of evidential value. Hence, Hotelling's T^2-test calculations will not be addressed in this paper.

However, lack of data does not eliminate the need to consider correlations and additional variation in calculating error rates. Rather, as correlations between certain elements are known to be high, we need a procedure that takes these correlations into account.

18.3.2 Data Sets

Four glass data sets are described below. The first two were used in the statistical modeling approach of Pan and Kafadar (2018), the third allows for estimating realistic δ values of variability and differences in elemental concentrations, and the last confirms the high correlations between elements.

LA-ICP-MS Data Set 1. Dr. David Ruddell (Centre of Forensic Sciences, Toronto, Canada) kindly shared the data from Dorn et al. (2015). Data from the "pane study" consisted of 48 "samples" taken from a single 4′ × 6′ pane of glass. 24 fragments were cut and measured 9 times each, and a 25th fragment was measured 24 times (see page 87, "*Group 1,*" for details). The 23 elements measured included all 17 cited in ASTM E2927-16e1 plus silicon (^{29}Si), cobalt (^{59}Co), tin (^{118}Sn), thorium (^{232}Th), and uranium (^{238}U). These data can be used to corroborate the estimates of measurement variability found from Weis' 33 same-source samples as well as within-fragment variability (from the 25th fragment measured 24 times). Table 18.3 shows very good agreement (except for ^{7}Li) in the estimated standard deviations from these two data sets.

LA-ICP-MS Data Set 2. Dr. Peter Weis (Bundeskriminalamt/Federal Criminal Police Office, Forensic Science Institute, KT 42–Inorganic Materials and Microtraces, Coatings, Wiesbaden, Germany) kindly shared the data that were published in Weis et al. (2011). The elements include all 17 of the elements cited in ASTM E2927-16e1, plus sodium (^{23}Na), tin (^{118}Sn), and silicon (^{29}Si, the constant standard). Each sample had six measurements. Data set (A) "Same" consisted of 33 fragments from the same pane of glass, plus a 34th fragment that was measured six times on each of 11 consecutive days, permitting rough estimates of between-fragment variability (among the 33 fragments) and between-day variability

TABLE 18.3

Within-Fragment Measurement Variability in LA-ICP-MS Measurements on 17 Elements Using Data from Canada (CAN) and Germany (GER)

	CAN	GER		CAN	GER		CAN	GER
^{7}Li	10.48	2.41	^{55}Mn	2.22	1.90	^{139}La	2.01	2.57
^{25}Mg	1.59	0.88	^{57}Fe	2.56	0.96	^{140}Ce	2.09	1.66
^{27}Al	1.60	2.54	^{85}Rb	2.61	2.10	^{146}Nd	2.65	3.39
^{39}K	1.57	1.32	^{88}Sr	1.76	1.65	^{178}Hf	3.27	4.06
^{42}Ca	1.37	1.35	^{90}Zr	2.24	2.84	^{208}Pb	4.57	2.55
^{49}Ti	1.77	1.74	^{137}Ba	2.77	2.39			

Note: Standard deviations on log measurements (approximately relative standard deviations on raw scale).

(among the 11 days). This also allows for verification that within-fragment variability (among the six replicates) is consistent with within-day variability (six replicates on 11 days). Data set (B) "Different" consisted of 62 samples mostly from different sources, but some from the same manufacturer but different year.

LA-ICP-MS Data Set 3. Dr. Soyoung Park and Dr. Alicia Carriquiry of Iowa State University (ISU) kindly shared data on float glass samples from two manufacturers (Park and Carriquiry, 2018, 2019). The 18 elements measured included all 17 cited in ASTM E2927-16e1 as well as sodium (^{23}Na). Data from 48 glass panes were collected, 31 from Company A produced over a three-week period, and 17 from Company B produced over a two-week period. Twenty-four fragments were randomly sampled from each pane with 21 fragments having five replicate measurements and three fragments having 20 measurements. These data provide valuable information on within-manufacturer variability. Exploratory analyses suggest that for within-manufacturer panes, at least half the elements vary by at least $\delta = 0.1$, or roughly 10% on the log scale. Between-manufacturer panes differ in certain elemental concentrations for just over half of the elements (see Figure 18.3).

ICP-MS Data Set. Data were obtained from Florida International University (FIU) (Almirall et al., 2003) in which concentrations of 16 elements were measured on multiple glass samples via ICP-MS. 13 of the 14 elements in E2330-19 are included, minus neodymium (Nd) plus antimony (^{121}Sb and ^{123}Sb), gallium (^{71}Ga), and hafnium (^{178}Hf). Each sample had three replicate measurements, and the collection of 590 samples included seven types of glass: 160 Container, 189 Float Architecture, 46 Float Autowindow (CFS), 97 Float Autowindow (non-CFS), 45 Headlamp, 10 Laboratory, and 43 "Rare." Not all types of glass had elemental concentration measurements for all 16 elements. Because the seven types of glass are so different, covariances and correlations were estimated separately for each type. Table 18.4 shows several pairs of elements with consistently high correlations across all types. Because these data originated from such variable sources, these were not used in the statistical modeling approach, but they do provide information about correlations among trace elements in glass.

Gaussian quantile-quantile plots for all data sets suggest most means are Gaussian, but high outliers are quite common. Ideally, we estimate the covariance (or correlation) matrix among the p elements measured in the data set. Because few data sets have at least (and preferably much more than) p replicate measurements (where p is the number of trace elements being measured), and because outliers are typical, we cannot rely

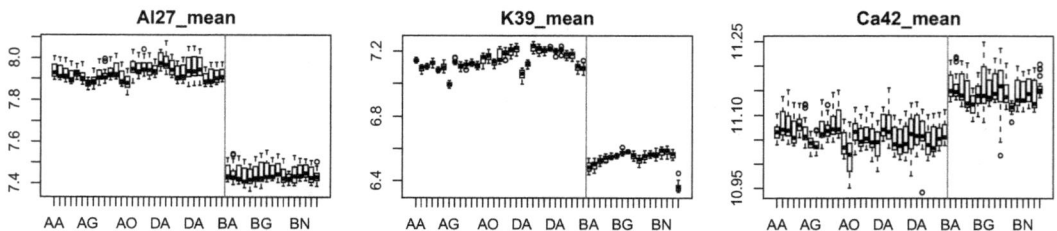

FIGURE 18.3
Log concentrations for three elements from Company A (left of vertical line) and Company B (right of vertical line). Each vertical boxplot represents concentrations from each pane. Relatively clear splits between manufacturers similar to those shown here can be observed in just over half of the 18 elements.

TABLE 18.4

Robust Estimates of Pairwise Correlations for FIU Data (*Italic:* $0.7 \leq |x| < 0.8$, **Bold:** $0.8 \leq |x| \leq 1$)

	Ce-La	Ce-Sm	La-Sm	Mn-Sm	Ba-Mn	Ba-Sm	Mn-Ti	La-Mn	Sm-Ti
Container	**0.98**	**0.92**	**0.94**	**0.83**	0.47	0.65	0.63	**0.83**	*0.74*
Float Arch[a]	**0.96**	**0.92**	**0.95**	*0.76*	*0.70*	**0.82**	*0.77*	*0.70*	0.43
Float Auto (CFS)[a]	—	0.37	—	**0.87**	**−0.83**	*−0.75*	*−0.77*	—	*−0.73*
Float Auto (non CFS)[a]	**0.89**	**0.92**	**0.95**	**0.83**	**0.83**	**0.92**	*0.79*	**0.81**	**0.87**
Headlamp	**0.98**	**0.96**	**0.92**	−0.32	0.17	0.37	−0.23	−0.29	0.48
Lab[b]	**0.98**	**1.00**	**0.98**	*0.71*	**0.97**	**0.86**	**0.88**	**0.82**	**0.96**
Rare	**0.99**	**0.92**	**0.95**	0.42	**0.90**	*0.72*	0.54	0.41	*0.80*

Note: Robust estimates were used to downweight the effects of outliers on the estimates, see (Pan and Kafadar, 2018).

[a] Float Auto (CFS) does not have measurements for La; all three Float glass types do not have measurements for Sb.

[b] Lab has only 10 samples (and some missing values), not enough to calculate robust correlations. Classical correlation values are shown.

on calculating pairwise Pearson correlations to provide a reliable estimate of the true correlation matrix. This issue is addressed by using the two LA-ICP-MS data sets on which multiple measurements were taken on the same pane of glass on multiple occasions:

1. Weis et al. (2011): Sample 104G was measured six times on 11 separate days to assess day-to-day variability. If this day effect is absent, one would have 66 measurements to estimate a 17×17 covariance matrix.

2. Dorn et al. (2015): A 3 cm × 3 cm fragment of glass was measured nine times on 24 separate occasions with generally three sets of nine measurements performed per day. If this occasion effect is absent, 210 measurements are available to estimate pairwise covariances and correlations (one data run only had three measurements and was removed to keep the data nicely balanced, for a total of 207 measurements).

3. Park and Carriquiry (2018, 2019): Three of the 24 fragments sampled from 48 glass panes had 20 replicate measurements. As elemental concentrations do appear to vary by Company, assuming no day or week effect, one would have 93 measurements from Company A and 51 from Company B to estimate covariances and correlations.

These data sets are large enough to estimate correlations if day and occasion effects are absent. Since outliers are likely to occur, a robust pooled estimate of the covariance matrix that downweights obviously discrepant observations is computed. Fast MCD (minimum covariance determinant Rousseeuw and Driessen, 1999) was computed for each data set using the package MASS in statistical software R (Ripley, 2015). Robust correlation matrices were also computed, but are very similar to the classical correlations shown in Figure 18.4. Correlation matrices vary largely between different data sets. Note that the German data set measured only the upper triangular half of a pane of glass, while the Canadian and ISU data sets measured an entire pane. This may account for some of the differences in correlation matrices.

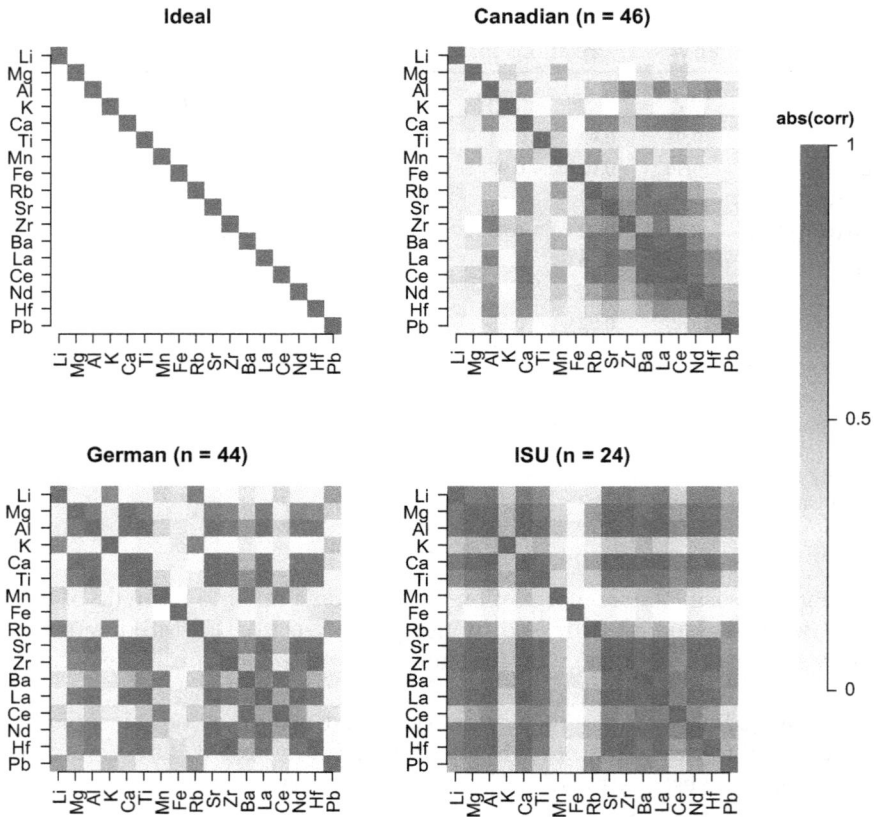

FIGURE 18.4
Classical correlation heatmaps for four datasets. Top row: data from Dorn et al. (2015) (left) and data from Weis et al. (2011) (right). Bottom row: data from Iowa State University (ISU) Manufacturer A (left) and Manufacturer B (right) (Park and Carriquiry, 2018, 2019).

18.3.3 Variability

Measurements of trace elements in materials such as glass have four main sources of variability. Listed in generally increasing order of magnitude, these are:

1. Measurement variation (σ_e): variability in measurements on a single fragment taken at nearly the same time (i.e. minutes apart).

2. Time variation (σ_t): variability in measurements taken on a single fragment at different times (i.e., on different days, or even weeks apart).

3. Fragment variability (σ_f): variability in measurements on different fragments from the same pane of glass.

4. Source variability (σ_B): variability in measurements of samples from different panes of glass.

Data from Weis et al. (2011) suggests σ_e is quite small, around 1–4% (RSD) for most elements. Time variation on fragments measured a few days apart suggests $\sigma_t \approx 3$–8% for all elements except ^{90}Zr (12%) and ^{178}Hf (13.6%). Because a measurement on a single fragment involves both sources of variation (to properly characterize the range of variability if it had been measured again and/or on another day), the root mean square of these two sources (i.e. $\sqrt{\sigma_e^2 + \sigma_t^2}$) is approximately 3–9%.

The underlying justification for using trace element concentrations in glass as forensic evidence rests with the idea that σ_B far exceeds σ_e, σ_t, and σ_f combined; that is, the procedure can correctly distinguish fragments that originated from different sources from those that came from the same source. Error rates depend crucially on good estimates of the magnitudes of these sources of variation, along with the correlations on the measurements in pairs of elements.

We estimate variances and covariances among all pairs of elements in a given sample and pool these estimates across all samples for both measurement variation (1) and time variation (2). These pooled covariance matrices are denoted as V_e and V_t, respectively. Some of our data sets allow for the estimation of fragment variability (3), V_f. Hence, the difference in the concentrations between two fragments from the same glass pane can be expected to vary due to V_e, V_t, and V_f.

18.4 Validating Procedures

18.4.1 Simulation Strategy

Let p denote the number of elements in each standard.* The logarithms of the p measured concentrations are modeled initially as Gaussian with mean μ and covariance matrix Σ, where Σ is the sum of three covariance matrices: V_e (representing variances due to measurements), V_t (representing variances due to different days on which the fragment might have been measured), and V_f (variances between fragments). Thus, a vector X of p concentrations has a distribution that we will denote as $N_p(\mu, V)$, where $V = V_e + V_t + V_f$. The limited data available suggest $\sigma_t \approx \sigma_f$ for many elements. Because most trace element concentrations in glass evidence are measured on the same day, only the effects of variation due to measurements and fragments (source (1) and (3), respectively) are considered in the modeling approach.

All three standards begin with two samples. For the R (or Q) sample, three measurements (the minimum number of replicates required by the ASTM standards) are simulated from $N_p(\mu, V^* = V_f^* + V_e^*)$, where μ is a vector of length p of all zeroes, and V_f^* and V_e^* are estimates of the between-fragment and within-fragment covariance matrices, respectively. For the K sample, another three measurements are simulated, this time from $N_p(\mu + \delta, V^*)$, where δ is a vector of length p of differences in the means between the logarithms of the measurements.[†] Values of δ, the absolute difference in the two means on the log scale or the relative difference on the original scale, will be set to specific values so the true difference in means is known. The proportion of simulated samples that meet the "match" criterion is tabulated and the "match rate" is calculated for a theoretical set of thousands of samples, not dependent on a particular set of collected samples.

---

* In our analyses, p varies from 8–17.

[†] A change of δ on the log scale corresponds to a relative difference in the means between the two fragments on the original scale of $\exp(\delta) - 1$. For example, if $\delta = 1.5$, then the means of an element of the K and R samples may be $\exp(3) = 20.1$ and $\exp(4.5) = 90.0$, for a relative difference of $(90.0 - 20.1)/20.1 = \exp(1.5) - 1 = 3.48$.

Formally, the simulation proceeds as follows, for p elements ($10 \leq p \leq 17$, depending on the data set):

1. Set `matchcount` to 0.

2. Simulate two covariance matrices $\hat{V}_1$, $\hat{V}_2$ from a Wishart distribution, assuming V^* is the "true" covariance matrix, to account for variability in estimating V^* from data.*

3. Generate a sample of 3 (or 6, or 9) measurements from $N_p(0, \hat{V}_1)$, representing 3 (or 6, or 9) measurements of concentrations on p elements for the K fragment. Let $\bar{X} = (\bar{x}_1, \ldots, \bar{x}_p)$ and $S_x = (s_1, \ldots, s_p)$ represent the vector of means and standard deviations, respectively, for each of these p elements, and let $S_x^* = (s_1^*, \ldots, s_p^*)$ where each $s_i^* = max(0.03, s_i)$.

4. Calculate the "match interval" for the ith element as $(\bar{x}_i - 4s_i^*, \bar{x} + 4s_i^*)$.

5. Generate another sample of 3 (or 6, or 9) measurements from $N_p(\delta, \hat{V}_2)$, representing 3 (or 6, or 9) measurements of concentrations on p elements for the R fragment. Let $\bar{Y} = (\bar{y}_1, \ldots, \bar{y}_p)$ represent the vector of means for each of these p elements.

6. If $\bar{y}_i \geq \bar{x}_i - 4s_i^*$ and $\bar{y}_i \leq \bar{x}_i + 4s_i^*$ for each element $i = 1, \ldots, p$, then increase `matchcount` by 1.

Steps 1–5 are repeated 100,000 times for various values of δ between 0.00 (true matches) and 1.00 (relative change in means on raw scale is $\exp(1) - 1 = 1.71828$). Note that $\delta = 0.3$ corresponds to a relative change in raw means of 35%, and $\delta = 0.5$ is a relative change in raw means of 65%. Note also that the "match rate" at $\delta = 0$ provides the probability of false exclusions; in fact, if measurements typically vary 10–15% anyway, one may wish to consider samples whose means differ by no more than $\delta = 0.15$ (16% relative difference) as "indistinguishable," and consider the "match rate" at $\delta \leq 0.15$ the "false exclusion rate." The probability of a "match" should fall to zero as δ increases, because large differences in means should be increasingly easy to detect.

The steps above are repeated, but where the log(concentrations) come from a heavier-tailed distribution than the Gaussian. The family of t-distributions satisfies this purpose: t_{30} (30 degrees of freedom) is rather close to Gaussian, while t_3 (3 degrees of freedom) is considerably heavier-tailed (see Figure 18.5). While these distributions look quite similar except for the tails, the effect of heavy-tailed distributions on error rates is quite substantial.

18.4.2 Simulated Match Rates

The following figures show some simulated match rates using the procedure described above. Because our data are known to be roughly normal but contain outliers, simulations were performed assuming both Gaussian distributed data and data from several t distributions. In each simulation run, we simulated not only the two sets of r (number of replicates) p-dimensional vectors (representing the lognormally distributed concentrations from the

---

* When the data arise from a truly multivariate Gaussian distribution having covariance matrix Σ, the distribution of the sample covariance matrix V can be shown to have a Wishart distribution; see Wishart (1928), Wishart and Bartlett (1932), Wishart and Bartlett (1933). Many statistical software packages, including R (that we used in Pan and Kafadar, 2018) have built-in routines for simulating from Wishart distributions, thus making it a convenient model for our purposes.

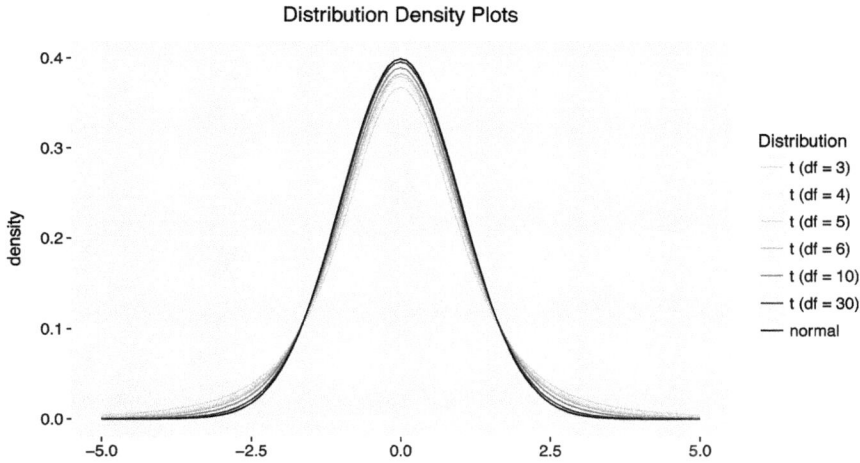

Distribution Density Plots

FIGURE 18.5
Family of *t* distributions with varying degrees of freedom. A t_{30} distribution with 30 degrees of freedom is very similar to the Gaussian distribution. Decreasing the degrees of freedom results in distributions with lower peaks and heavier tails.

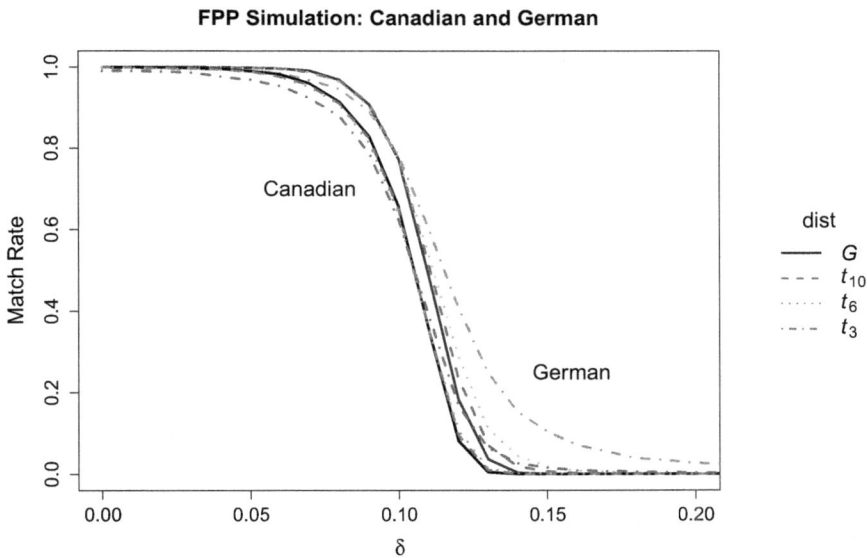

FPP Simulation: Canadian and German

FIGURE 18.6
Figure 2 from Pan and Kafadar (2018). Match rates from German (Data Set 1) and Canadian (Data Set 2) simulations for data from four different distributions. δ gives the approximate relative change in means.

K and R fragments), but also generated a covariance matrix from a Wishart distribution with mean $(df) \cdot V$, where df = degrees of freedom on which V is based.* See Figure 18.6.

---

* The simulated Wishart covariance had $df = 100$ and 40 (degrees of freedom) for the covariance matrices estimated from Data Sets 1 and 2, respectively, less than the nominal $207 - 17 = 190$ and $66 - 17 = 49$ degrees of freedom, to account for other (unknown) sources of variability. Similarly, simulated Data Set 3A and 3B covariances used 700 and 350 degrees of freedom instead of the nominal 707 and 388, respectively.

FPP Simulation (G)

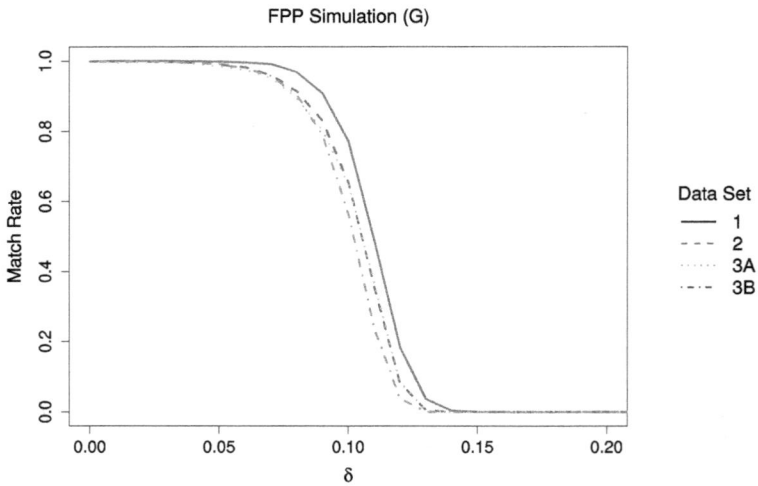

(a) Match rates with log(concentrations) simulated from a Gaussian distribution.

FPP Simulation (t_3)

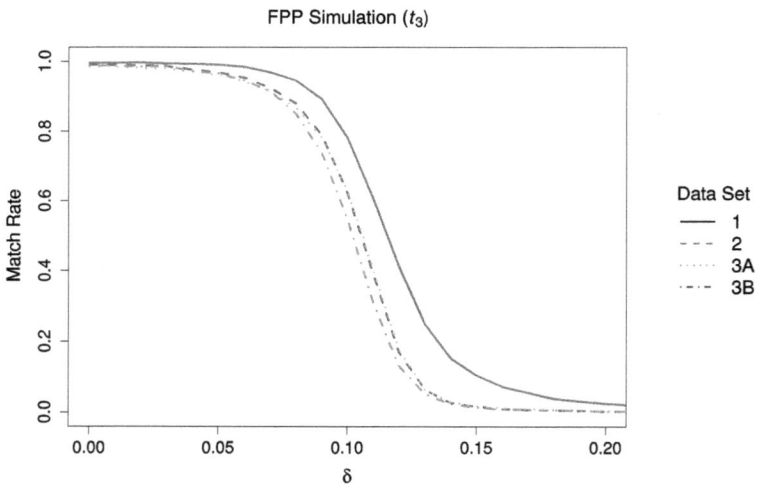

(b) Match rates with log(concentrations) simulated from a t_3 distribution.

FIGURE 18.7
Match rates from Data Sets 1, 2, and 3. Match rates for Data Set 3 are estimated separately for Companies A and B due to differences in covariance matrices. Data Set 2 and 3A have very similar error rates.

Figure 18.7 shows match rates for simulations using covariance matrix information from LA-ICP-MS Data Sets 1, 2, and 3 (estimated separately for the two manufacturers due to observed differences in covariance matrices). Match rates from Data Sets 2 and 3A are very similar, and Data Set 3B has overall lower match rates. For consistency with the data sets, nine measurements were simulated from each distribution for Data Set 1 and six for Data Sets 2 and 3. The abscissa indicates δ, the "known" set difference in elemental concentrations, and the ordinate shows the match rates. Ideally, we see high match rates at low values of δ (around 0), which decrease as δ increases. Table 18.5 provides numerical match rates corresponding to Figure 18.7 for δ values from 0.05 to 0.3.

TABLE 18.5

An Extension of Table 3 from Pan and Kafadar (2018) Detailing Simulated Match Rates at Various δ for Data Sets 1, 2, and 3

(δ)	0.05	0.1	0.15	0.2	0.25	0.3
		(a) Data Set 1 Match Rates				
t_3	0.991	0.781	0.103	0.024	0.006	0.003
t_6	0.997	0.773	0.018	0.000	0.000	0.000
t_{10}	0.998	0.775	0.006	0.000	0.000	0.000
G	0.999	0.772	0.000	0.000	0.000	0.000
		(b) Data Set 2 Match Rates				
t_3	0.968	0.622	0.016	0.003	0.001	0.001
t_6	0.988	0.638	0.001	0.000	0.000	0.000
t_{10}	0.989	0.648	0.000	0.000	0.000	0.000
G	0.990	0.654	0.000	0.000	0.000	0.000
		(c) Data Set 3A Match Rates				
t_3	0.968	0.622	0.016	0.003	0.001	0.001
t_6	0.988	0.638	0.001	0.000	0.000	0.000
t_{10}	0.989	0.648	0.000	0.000	0.000	0.000
G	0.990	0.654	0.000	0.000	0.000	0.000
		(d) Data Set 3B Match Rates				
t_3	0.961	0.627	0.017	0.004	0.001	0.000
t_6	0.981	0.640	0.001	0.000	0.000	0.000
t_{10}	0.985	0.648	0.000	0.000	0.000	0.000
G	0.985	0.648	0.000	0.000	0.000	0.000

The next series of figures and tables show simulated match rates from Data Set 1, which had more observations for estimating the covariance matrix, seemed more stable, and sampled glass fragments from an entire glass pane instead of just the upper triangular half.

Figure 18.8 shows a comparison of the match rates from Hotelling's T^2 test and the ASTM 4-SD approach using generated data of sample sizes 12 and 20. The leftmost set of lines corresponding to match rates from Hotelling's T^2 test indicate much lower match rates at larger values of δ; i.e., Hotelling's T^2 test has a higher likelihood of claiming "distinguishable" those samples which truly differ in their mean concentrations.

Table 18.6 provides simulated match rates at various δ using $n = 4$, which is the current choice for ASTM standard E2927-16e1 (ASTM International, 2016). Table 18.7 shows detailed match rates for some alternative values of n at $\delta = 0.10$ and 0.20 for both Gaussian distributed and t distributed data with three degrees of freedom.

Figure 18.9 overlays 95%, 99.8%, and 99.9% t confidence intervals onto Figure 18.8. Match rates for the latter two intervals overlap slightly with those from Hotelling's T^2 with 20 replicates, but are overall more conservative. The 4-SD approach, as the name suggests, uses the standard deviation in place of the standard error. This lack of normalization by $\sqrt{n}$ results in higher match rates as n increases because the confidence level is also increasing.

Figure 18.10 shows simulated match rates from the ICP-MS Data Set. The data and match rates indicate that the Container and Float Architecture categories are comprised

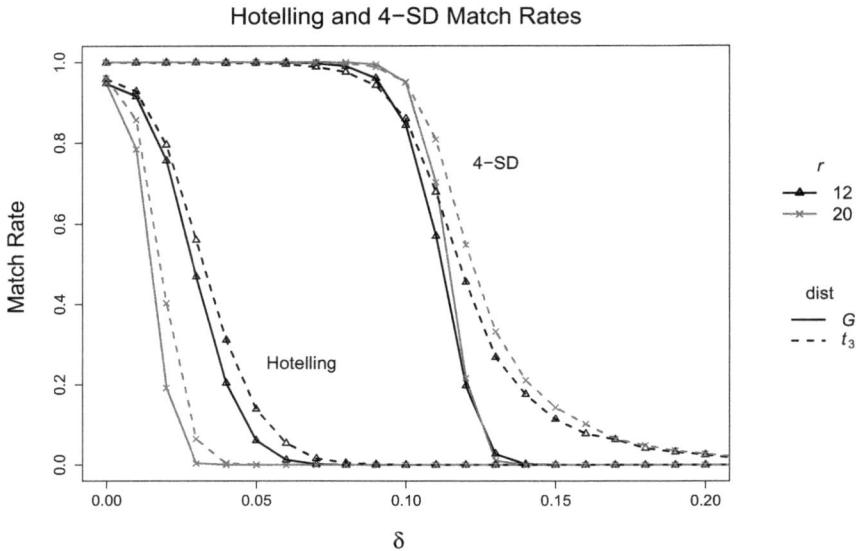

FIGURE 18.8
Figure 4 from Pan and Kafadar (2018). Match rates for Hotelling's T^2 vs. 4-SD approach for G and t_3 distributed data. Distribution has a much larger effect on the 4-SD approach.

TABLE 18.6

Table 4 from Pan and Kafadar (2018)

(δ)	0.05	0.1	0.15	0.2	0.25	0.3
3 G	0.829	0.354	0.006	0.000	0.000	0.000
3 t_3	0.754	0.370	0.051	0.013	0.005	0.003
9 G	0.999	0.768	0.000	0.000	0.000	0.000
9 t_3	0.991	0.779	0.105	0.023	0.007	0.003
12 G	1.000	0.852	0.000	0.000	0.000	0.000
12 t_3	0.997	0.856	0.115	0.025	0.007	0.003

Note: Match rates by sample size at $n = 4$ (G = Gaussian, $t_3 = t$ with df = 3) for various δ.

of a diverse population of glass samples. Indeed, samples labeled "Container" include multiples types of bottles and jars, generally separable into beverage, alcoholic beverages, and other. Float Architecture samples come mainly from four manufacturers: Cardinal, Guardian, PPG, and TempGlass. Correlation matrices of these subpopulations indicate significant differences, and may explain the larger match rates in comparison to Float Autowindow glasses.

18.4.3 Another Interpretation of "Difference"

The previous simulation study assumed the "difference" between two fragments of glass varied by a constant δ_0 in all elements. However, it may be more realistic to consider higher variability in certain as opposed to all elements – Dorn et al. (2015) observed higher variability in ^{39}K and ^{57}Fe than other elements. Park and Carriquiry (2018) observed similar

TABLE 18.7

Table 5 from Pan and Kafadar (2018)

(n)	1.00	1.50	2.00	2.50	3.00	3.50	4.00
			(a) $r = 3$				
$\delta = 0.1\ G$	0.000	0.000	0.045	0.638	4.437	15.883	35.335
$\delta = 0.1\ t_3$	0.091	0.426	1.456	4.002	9.847	21.156	36.788
$\delta = 0.2\ G$	0.000	0.000	0.000	0.000	0.000	0.001	0.007
$\delta = 0.2\ t_3$	0.010	0.033	0.080	0.197	0.408	0.765	1.281
			(b) $r = 6$				
$\delta = 0.1\ G$	0.000	0.000	0.007	0.211	4.227	26.491	62.176
$\delta = 0.1\ t_3$	0.100	0.611	2.320	6.766	17.616	38.684	64.383
$\delta = 0.2\ G$	0.000	0.000	0.000	0.000	0.000	0.000	0.001
$\delta = 0.2\ t_3$	0.008	0.037	0.095	0.237	0.527	1.073	2.024

Note: n-SD approach match rates where $\delta = 0.1, 0.2$ for certain values of n. Values are multiplied by 100.

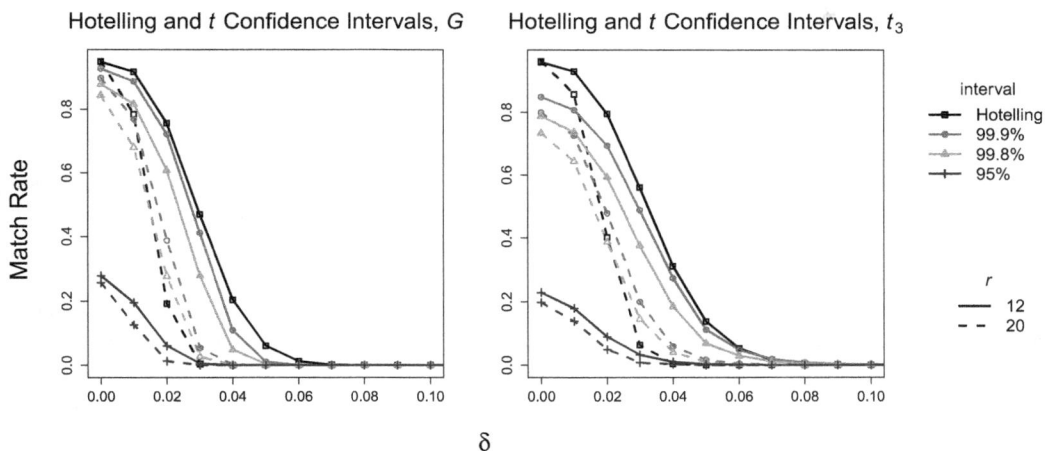

FIGURE 18.9

Figure 5 from Pan and Kafadar (2018). Match rates for Hotelling T^2 and t (95%, 99.8%, 99.9%) intervals.

high variability in ^{39}K and ^{57}Fe, but comparing data from Company A and B suggests larger variability in almost half of the 18 elements measured.

18.5 Conclusions

The procedures outlined by three ASTM standards for the analysis of forensic glass evidence reference four papers which quote error rates of less than 1%. However, these error rates depend heavily on the database used for calculations. Error rates are more accurately estimated by statistical modeling, which eliminates any dependency on database

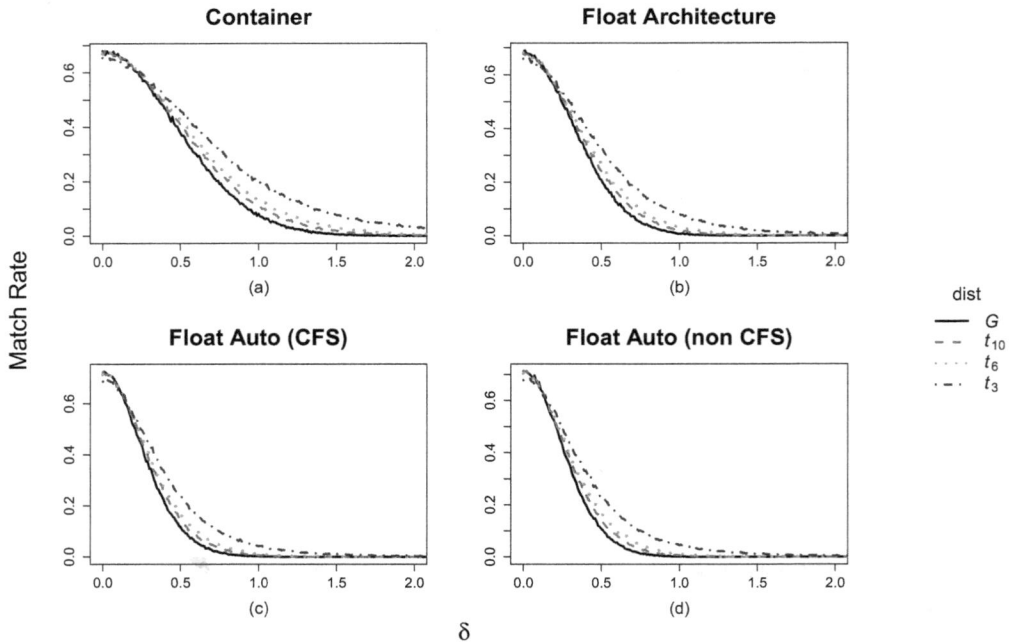

FIGURE 18.10
Figure 8 from Pan and Kafadar (2018). Match rates for FIU (ICP-MS) data for four different glass categories. The number of Wishart degrees of freedom by category are: Container (75), Float Architecture (75), Float Autowindow non CFS (75), Float Autowindow CFS (25), Headlamp (25). The match rates for Headlamp are not shown, but are similar to those in Figure 8(c).

and suggest expected error rates may be higher than 1%. Use of any of these standards without a proper statistical evaluation such as the one described in this chapter can result in underestimation of the probability of a false match; and the possibility of an exaggeratedly low error rate increases significantly when the databases used are not representative of the entire population of specimens that may be manufactured (which may likely be the case if databases are obtained by convenience).

In this chapter, we have offered an evaluation strategy for estimating error rates that are not biased by the characteristics of a specific dataset. Unless a procedure has undergone a thorough validation process like the one proposed here, the researcher will not have an honest assessment of the consequences of the techniques in use. As statisticians, it is important to assess and characterize as accurately as possible true error rates.

References

Aitken, C.G.G. and Lucy, D. Evaluation of trace evidence in the form of multivariate data. *Journal of the Royal Statistical Society*, 53(1), 109–122, 2004.

Almirall, J., Duckworth, D., Bayne, C., and Furton, K. Discrimination of forensic glasses via trace element analysis by inductively coupled plasma mass spectrometry and statistical treatment. Technical Report DAAD05-99-T-0734, International Forensic Research Institute, Florida

International University; Department of Chemistry, University Partk, Miami, FL; Oak Ridge National Laboratory, Oak Ridge, TN. Mission Area: (R-555) Improved Forensic Glass Analysis and Database Development, 2003.

Armbruster, D.A. and Pry, T. Limit of blank, limit of detection and limit of quantitation. *The Clinical Biochemist Reviews*, **29**(Suppl 1), S49, 2008.

ASTM International. *ASTM C162-25(2015) Standard Terminology of Glass and Glass Products*. ASTM International, 2015. DOI: 10.1520/C0162-05R15

ASTM International. *ASTM E2330-19 Standard Test Method for Determination of Concentrations of Elements in Glass Samples Using Inductively Coupled Plasma Mass Spectrometry (ICP-MS) for Forensic Comparisons*. ASTM International, 2019. DOI: 10.1520/E2330-19

ASTM International. *ASTM E2926-17 Standard Test Method for Forensic Comparison of Glass Using Micro X-ray Fluorescence (fi-XRF) Spectrometry*. ASTM International, 2017. DOI: 10.1520/E2926-17

ASTM International. *ASTM E2927-16e1 Standard Test Method for Determination of Trace Elements in Soda-Lime Glass Samples Using Laser Ablation Inductively Coupled Plasma Mass Spectrometry for Forensic Comparisons*. ASTM International, 2016. https://doi.org/10.1520/E2927-16E01

Bajic, S.J., Aeschliman, D.B., Saetveit, N.J., Baldwin, D.P., and Houk, R.S. Analysis of glass fragments by laser ablation-inductively coupled plasma-mass spectrometry and principal component analysis. *Journal of Forensic Science*, **50**(5), JFS2005088–5, 2005.

Behrends, T. and Kleingeld, P. Bench-top micro-xrf – a useful apparatus for geochemists. *Geochemical News*, **138**, 1–5, 2009.

Bennett, R.L., Kim, N.D., Curran, J.M., Coulson, S.A., and Newton, A.W. Spatial variation of refractive index in a pane of float glass. *Science and Justice*, **43**(2), 71–76, 2003.

Berends-Montero, S., Wiarda, W., de Joode, P., and van der Peijl, G. Forensic analysis of float glass using laser ablation inductively coupled plasma mass spectrometry (la-icp-ms): validation of a method. *Journal of Analytical Atomic Spectrometry*, **21**(11), 1185–1193, 2006.

Bichlmeier, S., Janssens, K., Heckel, J., Gibson, D., Hoffmann, P., and Ortner, H.M. Component selection for a compact micro xrf spectrometer. *X Ray Spectrometry: An International Journal*, **30**(1), 8–14, 2001.

Blacklock, E.C., Rogers, A., Wall, C., and Wheals, B.B. The quantitative analysis of glass by emission spectrography: a six element survey. *Forensic Science*, **7**(2), 121–130, 1976.

Bottrell, M.C. Forensic glass comparison: background information used in data interpretation. *Forensic Science Communications*, **11**(2), 2009.

Brewster, F., Thorpe, J.W., Gettinby, G., and Caddy, B. The retention of glass particles on woven fabrics. *Journal of Forensic Science*, **30**(3), 798–805, 1985.

Buscaglia, J.A. Elemental analysis of small glass fragments in forensic science. *Analytica Chimica Acta*, **288**(1-2), 17–24, 1994.

Calloway, A.R. and Jones, P.F. Enhanced discrimination of glass samples by phosphorescence analysis. *Journal of Forensic Science*, **23**(2), 263–273, 1978.

Catterick, T. and Hickman, D.A. The quantitative analysis of glass by inductively coupled plasma-atomicemission spectrometry: a five-element survey. *Forensic Science International*, **17**(3), 253–263, 1981.

Cobb, P.G.W. A survey of the variations in the physical properties of glass. *Journal of the Forensic Science Society*, **8**(1), 29–31, 1968.

Copley, G.J. The composition and manufacture of glass and its domestic and industrial applications. In *Forensic Examination of Glass and Paint*, pages 39–58. CRC Press, 2002.

Crockett, J.S. and Taylor, M.E. Physical properties of safety glass. *Journal of the Forensic Science Society*, **9**(3-4), 119–122, 1969.

Curran, J.M. The statistical interpretation of forensic glass evidence. *International Statistical Review*, **71**(3), 497–520, 2003.

Curran, J.M., Champod, T.N.H., and Buckleton, J.S.. *Forensic Interpretation of Glass Evidence*. CRC Press, Boca Raton, Florida, 2000.

Curran, J.M., Triggs, C.M., Buckleton, J.S., Walsh, K., and Hicks, T. Assessing transfer probabilities in a Bayesian interpretation of forensic glass evidence. *Science and Justice*, **38**(1), 15–21, 1998.

Dabbs, M.D.G. and Pearson, E.F. Some physical properties of a large number of window glass specimens. *Journal of Forensic Science*, **17**(1), 70–78, 1972.

Davies, M.M., Dudley, R.J., and Smalldon, K.W. An investigation of bulk and surface refractive indices for flat window glasses, patterned window glasses and windscreen glasses. *Forensic Science International*, **16**(2), 125–137, 1980.

Dorn, H., Ruddell, D.E., Heydon, A., and Burton, B.D. Discrimination of float glass by la-icp-ms: assessment of exclusion criteria using casework samples. *Canadian Society of Forensic Science Journal*, **48**(2), 85–96, 2015.

Duckworth, D.C., Bayne, C.K., Morton, S.J., and Almirall, J. Analysis of variance in forensic glass analysis by icp-ms: variance within the method. *Journal of Analytical Atomic Spectrometry*, **15**(7), 821–828, 2000.

Duckworth, D.C., Morton, S.J., Bayne, C.K., Koons, R.D., Montero, S., and Almirall, J.R. Forensic glass analysis by icp-ms: a multi-element assessment of discriminating power via analysis of variance and pairwise comparisons. *Journal of Analytical Atomic Spectrometry*, **17**(7), 662–668, 2002.

Dunnivant, F.M. and Ginsbach, J.W. *Flame Atomic Absorbance and Emission Spectrometry and Inductively Coupled Plasma – Mass Spectrometry*. Whitman College, 2017. https://protect-us.mimecast.com/s/NKX7CjRPZPir8P1f7QlQj?domain=people.whitman.edu

Ernst, T., Berman, T., Buscaglia, J., Eckert Lumsdon, T., Hanlon, C., Olsson, K., Palenik, C., Ryland, S., Trejos, T., Valadezm, M., and Almirall, J.R. Signal to noise ratios in forensic glass analysis by micro x ray fluorescence spectrometry. *X Ray Spectrometry*, **43**(1), 13–21, 2014.

Evett, I.W. and Buckleton, J. The interpretation of glass evidence. a practical approach. *Journal of the Forensic Science Society*, **30**(4), 215–223, 1990.

Evett, I.W. and Lambert, J.A. The interpretation of refractive index measurements. *Forensic Science*, **9**, 209–217, 1977.

Gabel-Cino, J. Expert witnesses and lawyers: can we all get along? presentation to the second annual conference of the national center for forensic science. Orlando, FL, October 2017.

Gamble, L., Burd, D.Q., and Kirk, P.L. Glass fragments as evidence. a comparative study of physical properties. *Journal of Criminal Law and Criminology (1931-1951)*, **33**(5), 416–421, 1943.

Gross, H. *Criminal investigation*: lawyers' co-operative publishing Company [Madras printed], 1907.

Haney, M.A. Comparison of window glasses by isotope dilution spark source mass spectrometry. *Journal of Forensic Science*, **22**(3), 534–544, 1977.

Hickman, D.A. A classification scheme for glass. *Forensic Science International*, **17**(3), 265–281, 1981.

Hickman, D.A., Harbottle, G., and Sayre, E.V. The selection of the best elemental variables for the classification of glass samples. *Forensic Science International*, **22**(2-3), 189–212, 1983.

Hicks, T., Vanina, R., and Margot, P. Transfer and persistence of glass fragments on garments. *Science and Justice*, **36**(2), 101–107, 1996.

Houk, R.S., Fassel, V.A., Flesch, G.D., Svec, H.J., Gray, A.L., and Taylor, C.E. Inductively coupled argon plasma as an ion source for mass spectrometric determination of trace elements. *Analytical Chemistry*, **52**(14), 2283–2289, 1980.

Hughes, J.C., Catterick, T., and Southeard, G. The quantitative analysis of glass by atomic absorption spectroscopy. *Forensic science*, **8**, 217–227, 1976.

Kammrath, B.W., Koutrakos, A.C., McMahon, M.E., and Reffner, J.A. The forensic analysis of glass evidence: past, present, and future. In *Forensic Science: A Multidisciplinary Approach*, 2016.

Kirk, P.L. Crime investigation; physical evidence and the police laboratory, 1953.

Koons, R.D. and Buscaglia, J. The forensic significance of glass composition and refractive index measurements. *Journal of Forensic Science*, **44**(3), 496–503, 1999.

Koons, R.D. and Buscaglia, J.A. Interpretation of glass composition measurements: the effects of match criteria on discrimination capability. *Journal of Forensic Science*, **47**(3), 505–512, 2002.

Koons, R.D., Fiedler, C., and Rawalt, R.C. Classification and discrimination of sheet and container glasses by inductively coupled plasma-atomic emission spectrometry and pattern recognition. *Journal of Forensic Science*, **33**(1), 49–67, 1988.

Koons, R.D., Peters, C.A., and Rebbert, P.S. Comparison of refractive index, energy dispersive x-ray fluorescence and inductively coupled plasma atomic emission spectrometry for forensic characterization of sheet glass fragments. *Journal of Analytical Atomic Spectrometry*, **6**(6), 451–456, 1991.

Lambert, J.A. and Evett, I.W. The refractive index distribution of control glass samples examined by the forensic science laboratories in the united kingdom. *Forensic Science International*, **26**(1), 1–23, 1984.

Latkoczy, C., Becker, S., Dücking, M., Günther, D., Hoogewerff, J.A., Almirall, J.R., Buscaglia, J., Dobney, A., Koons, R.D., Montero, S., and van der Peijl, G.J. Development and evaluation of a standard method for the quantitative determination of elements in float glass samples by la-icp-ms. *Journal of Forensic Science*, **50**(6), JFS2005091–15, 2005.

Lindley, D.V. A problem in forensic science. *Biometrika*, **64**(2), 207–213, 1977.

Locke, J. and Elliott, B.R. The examination of glass particles using the interference objective. part 2: a survey of flat and curved surfaces. *Forensic Science International*, **26**(1), 53–66, 1984.

Locke, J. and Underhill, M. Automatic refractive index measurement of glass particles. *Forensic Science International*, **27**(4), 247–260, 1985.

Locke, J., Underhill, M., Russell, P., Cox, P., and Perryman, A.C. The evidential value of dispersion in the examination of glass. *Forensic Science International*, **32**(4), 219–227, 1986.

Marris, N.A. and Hoff, R.W. Identification of glass splinters. *Analyst*, **59**(703), 686–689, 1934.

Matwejeff, S.N. Criminal investigation of broken window panes. *American Journal of Police Science*, **2**, 148–0, 1931.

Meggers, W.F., Kiess, C.C., and Stimson, F.J. Practical spectrographic analysis. *Journal of the Franklin Institute*, **194**(3), 382–383, 1922.

National Research Council. *Forensic Analysis: Weighing Bullet Lead Evidence (K. O. MacFadden, Chair)*. The National Academies Press, Washington, DC, 2004. ISBN 978-0-309-09079-7. doi:10.17226/10924.

Nelson, D.F. Illustrating the fit of glass fragments. *The Journal of Criminal Law, Criminology, and Police Science*, **50**(3), 312–314, 1959.

Nelson, D.F. and Revell, B.C. Backward fragmentation from breaking glass. *Journal of the Forensic Science Society*, **7**(2), 58–61, 1967.

Nickolls, L.C. Glass as evidence. *Journal of the Forensic Science Society*, **6**(4), 180–182, 1966.

O'Sullivan, S., Geddes, T., and Lovelock, T.J. The migration of fragments of glass from the pockets to the surfaces of clothing. *Forensic Science International*, **208**(1-3), 149–155, 2011.

Pan, K. and Kafadar, K. Statistical modeling and analysis of trace element concentrations in forensic glass evidence. *The Annals of Applied Statistics*, **12**(2), 788–814, 2018.

Park, S. and Carriquiry, A. Glass data description, 2018. Retrieved from https://github.com/CSAFEISU/AOAS-2018-glass-manuscript

Park, S. and Carriquiry, A. Learning algorithms to evaluate forensic glass evidence. *The Annals of Applied Statistics*, In press 2019.

Parker, J.B. A statistical treatment of identification problems. *Journal of the Forensic Science Society*, **6**(1), 33–39, 1966.

Parker, J.B. The mathematical evaluation of numerical evidence. *Journal of the Forensic Science Society*, **7**(3), 134–144, 1967.

Pilkington, L.A.B. Review lecture. the float glass process. In *Proceedings of the Royal Society of London. Series A, Mathematical and Physical Sciences*, volume 314 of *1516*, pages 1–25, 1969.

Pilkington, L.A.B. Sir alastair pilkington, 2019. https://www.pilkington.com/en/global/about/education/siralastair-pilkington

Reeve, V., Mathiesen, J., and Fong, W. Elemental analysis by energy dispersive x-ray: a significant factor in the forensic analysis of glass. *Journal of Forensic Science*, **21**(2), 291–306, 1976.

Renshaw, G.D. and Clarke, P.D.B. The variation in thickness of toughened glass from car windows. *Journal of the Forensic Science Society*, **14**(4), 311–317, 1974.

Ripley, B. Mass: support functions and datasets for venables and Ripleys mass. R package version 7.3-45, 2015.

Roche, G.W. and Kirk, P.L. Applications of microchemical techniques: differentiation of similar glass fragments by physical properties. *Journal of Criminal Law and Criminology*, **38**(2), 168–171, 1947.

Rousseeuw, P.J. and Driessen, K.V. A fast algorithm for the minimum covariance determinant estimator. *Technometrics*, **41**(3), 212–223, 1999.

Ryland, S.G. Sheet or container?–forensic glass comparisons with an emphasis on source classification. *Journal of Forensic Science*, **31**(4), 1314–1329, 1986.

Shunko, E.V., Stevenson, D.E., and Belkin, V.S. Inductively coupling plasma reactor with plasma electron energy controllable in the range from ∼6 to ∼100ev. *IEEE Transactions on Plasma Science*, **42**(3), 774–785, 2014.

Slater, D.P. and Fong, W. Density, refractive index, and dispersion in the examination of glass: their relative worth as proof. *Journal of Forensic Science*, **27**(3), 474–483, 1982.

Smale, D. The examination of paint flakes, glass and soils for forensic purposes, with special reference to electron probe microanalysis. *Journal of the Forensic Science Society*, **13**(1), 5–15, 1973.

Terry, K.W., Van Riessen, A., Lynch, B.F., and Vowles, D.J. Quantitative analysis of glasses used within Australia. *Forensic Science International*, **25**(1), 19–34, 1984.

Tooley, F.V. and Scholes, S.R. *The Handbook of Glass Manufacture (Vol. 1)*. Ashlee Publishing, 1984.

Trejos, T. and Almirall, J.R. Sampling strategies for the analysis of glass fragments by la-icp-ms: part i. microhomogeneity study of glass and its application to the interpretation of forensic evidence. *Talanta*, **67**(2), 388–395, 2005.

Trejos, T., Koons, R., Becker, S., Berman, T., Buscaglia, J., Duecking, M., Eckert-Lumsdon, T., Ernst, T., Hanlon, C., Heydon, A., and Mooney, K. Cross-validation and evaluation of the performance of methods for the elemental analysis of forensic glass by μ-xrf, icp-ms, and la-icp-ms. *Analytical and bioanalytical chemistry*, **405**(16), 5393–5409, 2013a.

Trejos, T., Koons, R.,Weis, P., Becker, S., Berman, T., Dalpe, C., Duecking, M., Buscaglia, J., Eckert-Lumsdon, T., Ernst, T., and Hanlon, C. Forensic analysis of glass by fl-xrf, sn-icp-ms, la-icp-ms and la-icp-oes: evaluation of the performance of different criteria for comparing elemental composition. *Journal of Analytical Atomic Spectrometry*, **28**(8), 1270–1282, 2013b.

Trejos, T., Montero, S., and Almirall, J.R. Analysis and comparison of glass fragments by laser ablation inductively coupled plasma mass spectrometry (la-icp-ms) and icp-ms. *Analytical and bioanalytical chemistry*, **276**(8), 1255–1264, 2003.

Tryhorn, F.G. The examination of glass. *The Police Journal*, **12**(3), 301–318, 1939.

Underhill, M. Multiple refractive index in float glass. *Journal of the Forensic Science Society*, **20**(3), 169–176, 1980.

United States. FBI law enforcement bulletin – October 1936. Washington, DC: U.S. Dept. of Justice, Federal Bureau of Investigation, October 1936.

Varshneya, A.K. and Tomozawa, M. Fundamentals of inorganic glasses. *Journal of Non Crystalline Solids*, **170**(1), 112, 1994.

Weis, P., Dücking, M., Watzke, P., Menges, S., and Becker, S. Establishing a match criterion in forensic comparison analysis of float glass using laser ablation inductively coupled plasma mass spectrometry. *Journal of Analytical Atomic Spectrometry*, **26**(6), 1273–1284, 2011.

Wishart, J. The generalised product moment distribution in samples from a normal multivariate population. *Biometrika*, **20A**(1/2), 32–52, Jul. 1928.

Wishart, J. and Bartlett, M.S. The distribution of second order moment statistics in a normal system. *Mathematical Proceedings of the Cambridge Philosophical Society*, **28**(4), 455–459, 1932. Cambridge University Press.

Wishart, J. and Bartlett, M.S. The generalised product moment distribution in a normal system. *Mathematical Proceedings of the Cambridge Philosophical Society*, **29**(2), 260–270, 1933. Cambridge University Press.

Wollnik, H. Time of flight mass analyzers. *Mass spectrometry reviews*, **12**(2), 89–114, 1993.

Wolnik, K.L., Gaston, C.M., and Fricke, F.L. Analysis of glass in product tampering investigations by inductively coupled plasma atomic emission spectrometry with a hydrofluoric acid resistant torch. *Journal of Analytical Atomic Spectrometry*, **4**(1), 27–31, 1989.

Zurhaar, A. and Mullings, L. Characterisation of forensic glass samples using inductively coupled plasma mass spectrometry. *Journal of Analytical Atomic Spectrometry*, **5**(7), 611–617, 1990.

19

Estimation of Insect Age for Assessing Minimum Post-Mortem Interval in Forensic Entomology Casework

Davide Pigoli, Martin J.R. Hall, and John A.D. Aston

CONTENTS

19.1 Background 443
19.2 Isomorphen and Isomegalen Diagrams 444
19.3 Thermal Summation Models 444
19.4 Curvilinear Models 444
19.5 Spectral Measurements 445
19.6 Growth Curve Reconstruction 445
19.7 Conclusion 448
References 449

19.1 Background

It has been apparent for many years that there are problems in estimating the post-mortem interval (PMI) in death investigations based on insect evidence, for example, insufficient information on fly activity and weather conditions (Catts, 1992). Since that time many efforts have been made to reduce the problems but they still exist. A good review of entomology-based methods for estimation of post-mortem interval is a paper of the same name (Harvey et al., 2016).

The major two ways that insect evidence is used to estimate PMI are through application of knowledge of rates of insect development or insect succession. The former is more accurate and is the focus of this review. The data that are collected can be categorical, i.e., the stage of the insect, egg, larval instar and so forth, or continuous, e.g., larval weight or length. Continuous data can give greater temporal resolution but is also open to error introduced by the response of specimens to different rearing and preservation techniques (e.g., Adams and Hall, 2003).

In this chapter, we will give a brief overview and comparison of existing methods to estimate the PMI, namely isomorphen and isomegalan diagrams, thermal summation models, curvilinear models and spectral measurement. Then, we will present some more recent development on the reconstruction of the specimen growth process which, in addition to providing a lower bound for the PMI, offer a diagnostic tool to check that the conditions at which the specimens developed at the crime scene is consistent with the experimental ones.

19.2 Isomorphen and Isomegalen Diagrams

The isomorphen (same shape) and isomegalen (same size) diagrams were developed to illustrate the relationship between the stage or size, respectively, of immature blow flies at a range of constant temperatures (Reiter, 1984; Grassberger and Reiter, 2001, 2002). They are based on laboratory studies at several constant temperatures, with the values for sizes between these temperatures being filled in by interpolation. Examples of these diagrams for *Lucilia sericata* can be found in Grassberger and Reiter (2001, Figures 2 and 3). Their weakness is that temperatures at most crime scenes are not constant, but vary in a diurnal pattern. Therefore, while one can estimate an average daily temperature, it can be difficult to apply these models with confidence. On the other hand, the main strength of this kind of approach is its relative simplicity with respect to the other existing methods, thus making the results of the analysis easier to present.

19.3 Thermal Summation Models

Thermal summation models were developed as tools in ecology, mainly to study and predict the development of populations of agricultural pests across a season (Wagner et al., 1984; Petitt et al., 1991; Saulich, 1999). They have become among the most used tools to estimate insect age in forensic entomology because they can be applied to fluctuating temperatures and work reasonably well between experimentally determined or estimated lower and upper developmental thresholds (LDT and UDT, respectively) of temperature, when the rate of development is linear in relation to temperature. Age can be calculated as a product of temperature (in degrees Celsius above the LDT) and time (in hours or days), termed accumulated degree hours (ADH) or days (ADD).

In theory, due to the linear relationship between rate of development and temperature between the two thresholds, the number of ADH or ADD required for an insect to develop to any particular stage is constant. However, this is not always the case. For example, the blow fly Calliphora vicina (Diptera: Calliphoridae) has a LDT of about 1.0°C and yet the ADH requirement rises at temperatures below about 15°C (Amendt et al., 2007, Figure 2). Care should thus be taken in applying thermal constants to a range of temperatures that have not been experimentally tested (e.g., as in Marchenko, 2001).

The manner in which insect sampling is performed to gather developmental data can affect the result (Wells et al., 2015; Wells and LaMotte, 2017) and so is an important factor in designing studies.

ADD models have also been applied to estimate PMI based solely on body decomposition, without the inclusion of entomology (Megyesi et al., 2005; Marhoff et al., 2016).

19.4 Curvilinear Models

Sharpe and DeMichele (1977) developed a complex biophysical model that describes the non-linear response in developments at both high and low temperatures and the linear response at intermediate temperatures. A model for estimating larval age from weight using logistic equations has been proposed (Williams, 1984). Tarone and Foran (2008) also developed non-linear models based on Generalized Additive Model (GAM) techniques.

A new model based on the non-linear development of the blow fly, *Lucilia sericata*, is described in Reibe et al. (2010). In an example of application of the model and existing techniques to casework, they estimated egg-laying with very similar values, i.e., of 107 hours (isomegalen diagram and average temperature for period), 101 hours (ADH and hourly temperatures) and 99 hours (from new model). The known PMI from suspect confession was approximately 96 hours. It is possible that the longer durations estimated from the three techniques were due to inaccuracies in scene temperature estimation as temperatures were based on the nearest weather station, which was 10 km from the scene.

19.5 Spectral Measurements

A different approach based on spectral measurements from immature stages of *Protophormia terraenovae* is proposed by Warren et al. (2017). In experimental data, they linked the observed spectrum with the developmental time via a functional linear model. This can then allow one to predict the developmental age of immature larvae observed at the crime scene. They show this method to accurately predict the day of development for specimens that have been exposed to a constant laboratory temperature. One of the benefits of this approach is it being a non-destructive and non-invasive method which allows other ageing methods to be used on the same specimens for corroboration.

19.6 Growth Curve Reconstruction

A complementary approach has been proposed by Pigoli et al. (2017), which focuses on the reconstruction of the average growth curve for the blow flies at the scene. This may be helpful when larvae are observed in the middle of the development process and it also provides a visual diagnostic tool to evaluate how well the developmental model fits the observed data. The main challenge is that laboratory experiments are bound to measure blow fly development only for a relatively small number of simple (usually constant) temperature profiles and therefore Pigoli et al. (2017) use techniques from functional data analysis and nonparametric regression to estimate the expected growth curve corresponding to more realistic temperature profiles. Once the growth lengths are available for any temperature profile, one is able to estimate the hatching time and hence the interval between hatching and body discovery from data available at the scene. In a typical larval developmental dataset, we have K experimental temperatures $T_1^* < T_2^* < \cdots < T_K^*$ and, for each species of interest, one observes the larval lengths Y_{kjl} measured at time t_j after hatching, $t_1, \ldots, t_{n_k}$, for $l = 1, \ldots, N_{kj}$ individual larvae which have been exposed to a constant experimental temperature T_k^*, $k = 1, \ldots, K$. The observation times $t_1, \ldots, t_{n_k}$ may also differ across experimental temperature.

We can then assume that the observed lengths satisfied a nonparametric regression model

$$Y_{kjl} = L_{T_k^*}(t_j) + \epsilon_{kjl},$$

where ϵ_{kjl} are independent, zero mean random variables with the same variance and the mean larval length curve $L_{T_k^*}(t)$ depends on the experimental temperature $T_k^*, k = 1, \ldots, K$. It is then possible to estimate $L_{T_k^*}(t)$ by means of the preferred non-parametric smoothing estimator. In this work, we use a local linear polynomial estimator with a Gaussian kernel. The same equivalent degrees of freedom are used for all the temperatures and the number is chosen by visual inspection. This provides estimates $\widehat{L}_{T_k^*}(t)$ for all the constant temperature growth curves.

The question is now how to produce an estimate for the growth curve associated to a generic temperature profile. We claim that, if we consider a small enough time interval, the temperature can be considered roughly constant and therefore the growth process would be bound to follow the local dynamics of the correspondent constant temperature growth curve *at the same length stage*. This suggests the following model for the local growth process:

$$\left.\frac{d\,L(t)}{dt}\right|_{t=t_k} = \left.\frac{d\,L_{T(t_k)}(u)}{du}\right|_{u=(L_{T(t_k)})^{-1}(L(t_k))}, \tag{19.1}$$

where $L_{T(t_k)}(\cdot)$ is the growth profile at constant temperature $T(t_k)$ and $L(t)$ is the growth profile with varying temperature. This means that the (expected) local increment in length at time t_k is the one that would occur in the growth profile at constant temperature $T(t_k)$ when the length is equal to $L(t_k)$. This allows us to reconstruct the varying temperature growth profile iteratively as

$$\begin{cases} L(t_1) = L_{T(t_1)}(t_1) \\ L(t_{k+1}) = L(t_k) + \left(\left.\frac{d\,L_{T(t_k)}(u)}{du}\right|_{u=(L_{T(t_k)})^{-1}(L(t_k))}\right)(t_{k+1} - t_k), \end{cases}$$

for $k = 2, \ldots, n$ and with $T(t_1), \ldots T(t_n)$ being the varying temperature profile.

This would solve the problem if we knew the expected growth curve L_T for any temperature T we can observe, but in practice we can have experimental data for only a relatively small set of temperatures. We therefore need to estimate first the growth profile L_T for a generic temperature T from a set of estimated growth curves $\widehat{L}_{T_1^*}, \ldots, \widehat{L}_{T_K^*}$. The main difficulty here is that the temperature influences the speed of the growth process. Using the language of functional data analysis, these curves present both amplitude and phase variation (Ramsay and Silverman, 2005; Marron et al., 2014).

At this stage, one postulates that the experimental growth curves $L_{T_1^*}(t), \ldots, L_{T_K^*}(t)$ have corresponding growth shapes $S_{T_1^*}, \ldots, S_{T_K^*}$ that can be obtained by computing warping transformations $\tau_{T_1^*}, \ldots, \tau_{T_K^*}$ acting over the time in such a way that, for any $k = 1, \ldots, K$, $S_{T_k^*}(\tau_{T_k^*}(t)) = L_{T_k^*}(t)$. This (potentially non-linear) transformation of time scale is called a warping function (see Ramsay and Silverman, 2005, Chapter 7, for more details) and it can be estimated via a landmark registration procedure that aligns hatching time, peak time and pupation time of the growth processes. It is of course possible to use more advanced registration methods but the relatively simple shape of the growth curves usually allows the landmark registration to perform well.

From the registration procedure, one has at hand K estimated growth shapes $\widehat{S}_{T_1^*}, \ldots, \widehat{S}_{T_K^*}$ and K estimated warping functions $\widehat{\tau}_{T_1^*}, \ldots, \widehat{\tau}_{T_K^*}$. These are now defined on the same domain and this allows us to interpolate them linearly across temperatures to approximate the growth process for any constant temperature T^*.

For any $T^* \in [T_k^*; T_{k+1}^*]$,

$$\widehat{S}_{T^*}(u) = \frac{T_{k+1}^* - T^*}{T_{k+1}^* - T_k^*} \widehat{S}_{T_k^*}(u) + \frac{T^* - T_k^*}{T_{k+1}^* - T_k^*} \widehat{S}_{T_{k+1}^*}(u),$$

and

$$\widehat{\tau}_{T^*}(t) = \frac{T_{k+1}^* - T^*}{T_{k+1}^* - T_k^*} \widehat{\tau}_{T_k^*}(t) + \frac{T^* - T_k^*}{T_{k+1}^* - T_k^*} \widehat{\tau}_{T_{k+1}^*}(t).$$

Finally, the estimated growth process at constant temperature T^* will be $\widehat{L}_{T^*}(t) = \widehat{S}_{T^*}(\widehat{\tau}_{T^*}(t))$.

To select the most likely hatching date given a set of length measurements taken at a time t^* (where the reference time is the local one), we are going to compare the growth profiles that would be expected if the hatching time t_h was at any time between the last time the victim has been seen alive t_a and t^*. Let $L(t - t_h)$, $t_h \le t \le t^*$, be the growth curve for hatching time equal to t_h and temperature profile $\{T(t^* - t); t_h \le t \le t^*\}$. Let then $Y_i^*, = 1, \ldots, n_{\text{obs}}$ be the measured larval lengths

$$Y_i^* = L(t^* - t_h) + \epsilon_i, \tag{19.2}$$

with ϵ_i, $i = 1, \ldots, n_{\text{obs}}$ independent random errors with zero mean and unknown variance σ^2. Then, we can estimate t_h as

$$\widehat{t}_h = \arg\min_{t_a \le t \le t^*} \sum_{i=1}^{n_{\text{obs}}} \left(\widehat{L}(t^* - t) - Y_i \right)^2 = \arg\min_{t_a \le t \le t^*} \left(\widehat{L}(t^* - t) - \overline{Y} \right)^2, \tag{19.3}$$

i.e., we choose the hatching time whose expected length at time t^* best fits the observed values. However, we may also want to include some expert knowledge in the estimation procedure. First, the forensic entomologist collecting the sample may recognise that the larvae have reached (or not) the post-feeding phase, i.e., the region of the growth curve after the peak where larvae stop feeding in preparation for pupariation and therefore they decrease in length. We can easily integrate this piece of information in the estimation procedure by restricting the admissible region for the minimisation problem (19.3) to the hatching times whose associated growth process at time t^* has already reached (or not) the post-feeding region, i.e. the estimated derivative is negative at some time $t \le t^*$. Using this framework, it is also possible to assess the asymptotic uncertainty about the hatching time as well as include additional prior information in the estimation procedure (see Pigoli et al., 2017 for more details).

We demonstrate here the procedure by considering an anonymized investigative case in the UK, where $n_{\text{obs}} = 70$ *Calliphora vicina* post-feeding larvae were collected from the body. The temperature time series for the 371 hours before the time the larvae were killed, prior to subsequent measurement, together with the estimated constant temperature growth profiles corresponding to each observed temperature in the series, can be seen in Figure 19.1.

Temperature time series at the crime scene

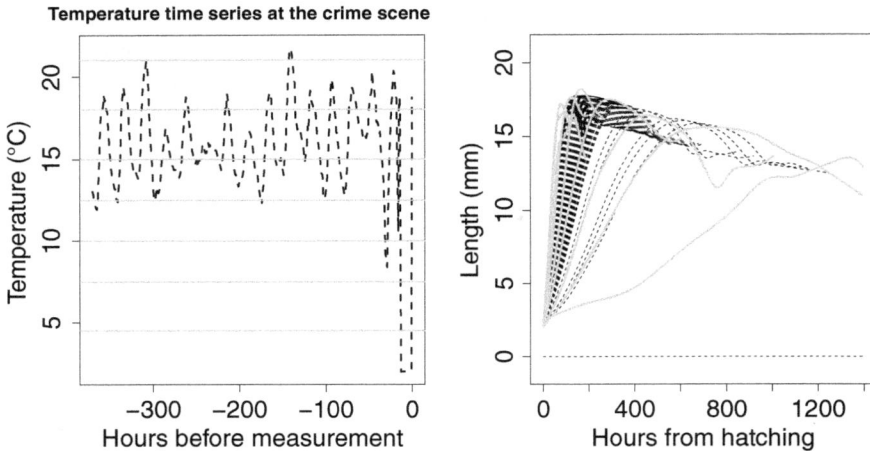

FIGURE 19.1
Left: Temperature profile to which the body was subjected in the hours before larval lengths were measured. Note that between the body discovery and the collection of the larvae the body was stored in a refrigerator for a few hours. Horizontal gray lines denote the temperatures for which experimental developmental data are available. Right: Estimated constant-temperature growth curves for each temperature in the observed interval (dashed black lines) and constant-temperature growth curves for the temperatures for which experimental developmental data are available (gray lines). Note the constant (zero) growth corresponding to the temperature of the refrigerator in which the body was kept between the time of discovery and the time the measurements were taken.

These were obtained from developmental data for *Calliphora vicina* collected at the Natural History Museum, London. The interval of 371 hours was the largest one that was considered possible by forensic scientists at the scene.

For this case, the application of the ADH method suggested as plausible interval for eggs' hatching was the one between 240 and 192 hours before the measurements were taken (Donovan et al., 2006).

Figure 19.2 shows the growth curves for the hatching time estimated with (19.3) and the profile of the objective function in the minimisation problem (19.3). The estimated hatching time is −256 hours (before the larval measurement), which is a bit earlier than what it was obtained from the ADH method. However, the flat plateau in the criterion suggests little stability for the estimate. Indeed, if we assume a Gaussian distribution for the measurement errors, the 95% approximated confidence interval procedure (Pigoli et al., 2017) gives us an interval of [−260, −190], which includes the range suggested as plausible by the ADH method. Note that here the computation of the criterion is restricted to the admissible region of hatching times for which the expected growth curve would have reach the post-feeding stage by the time the larvae were collected.

19.7 Conclusion

In this chapter, we discussed the most relevant methods to assess post-mortem interval based on insects evidence, and we highlighted their strengths and limitations. We also presented some more recent developments that show some promise in improving the accuracy

Estimated larval growth curve

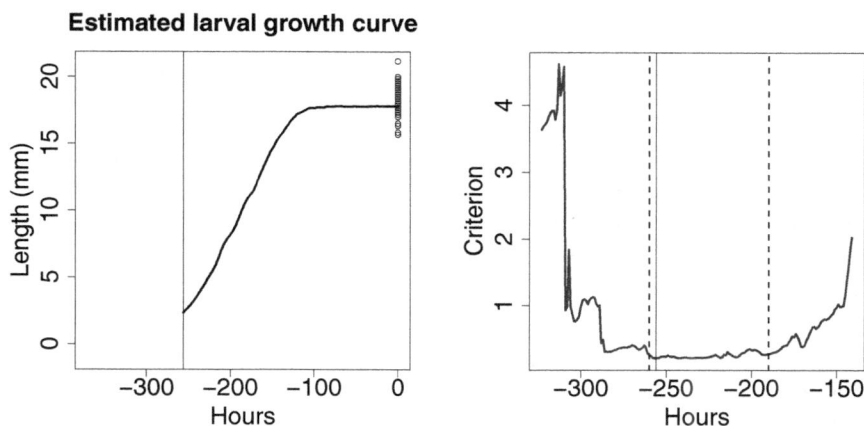

FIGURE 19.2

Left: Estimated average growth curve for the larvae of *Calliphora vicina* collected at the crime scene (dots indicate the observed lengths). The most likely hatching time is 256 hours before the measurements were taken (vertical solid line). Right: Profile of the criterion to be minimized as a function of hatching time. Vertical dashed lines indicate the boundaries of the uncertainty region obtained in Pigoli et al. (2017). The asymmetry of the region with respect to the estimated hatching time depends on the constraint for the larvae to be in the post-feeding stage at time of collection.

of the estimation, as well as providing a more rigorous way to evaluate the uncertainty and some model diagnostics. In addition to the sources we cited for each class of methods, more details on quantitative methods in forensic entomology can be found in Tomberlin and Benbow (2015).

References

Adams, Z.J.O. and Hall, M.J.R. Methods used for the killing and preservation of blowfiy larvae, and their effect on post-mortem larval length. *Forensic Science International*, **138**(1), 50–61, 2003.

Amendt, J., Campobasso, C.P., Gaudry, E., Reiter, C., LeBlanc, H.N., and Hall, M.J.R. Best practice in forensic entomology.standards and guidelines. *International Journal of Legal Medicine*, **121**(2), 90–104, 2007.

Catts, E.P. Problems in estimating the postmortem interval in death investigations. *Journal of Agricultural Entomology*, **9**(4), 245–255, 1992.

Donovan, S.E., Hall, M.J.R., Turner, B.D., and Moncrieff, C.B. Larval growth rates of the blowfiy, calliphora vicina, over a range of temperatures. *Medical and Veterinary Entomology*, **20**(1), 106–114, 2006.

Grassberger, M. and Reiter, C. Effect of temperature on lucilia sericata (diptera: Calliphoridae) development with special reference to the isomegalen-and isomorphen-diagram. *Forensic Science International*, **120**(1), 32–36, 2001.

Grassberger, M. and Reiter, C. Effect of temperature on development of the forensically important Holarctic blow fiy protophormia terraenovae (robineau-desvoidy)(diptera: Calliphoridae). *Forensic Science International*, **128**(3), 177–182, 2002.

Harvey, M., Gasz, N., and Voss, S. Entomology-based methods for estimation of postmortem interval. *Research and Reports Forensic Medical Science*, **6**, 2016.

Marchenko, M.I. Medicolegal relevance of cadaver entomofauna for the determination of the time of death. *Forensic Science International,* **120**(1), 89–109, 2001.

Marhoff, S.J., Fahey, P., Forbes, S.L., and Green, H. Estimating post-mortem interval using accumulated degree-days and a degree of decomposition index in Australia: a validation study. *Australian Journal of Forensic Sciences,* **48**(1), 24–36, 2016.

Marron, J.S., Ramsay, J.O., Sangalli, L.M., and Srivastava, A. Statistics of time warpings and phase variations. *Electronic Journal of Statistics,* **8**(2), 1697–1702, 2014.

Megyesi, M.S., Nawrocki, S.P., and Haskell, N.H. Using accumulated degree-days to estimate the post-mortem interval from decomposed human remains. *Journal of Forensic Science,* **50**(3), 1–9, 2005.

Petitt, F.L., Allen, J.C., and Barfield, C.S. Degree-day model for vegetable leafminer (diptera: Agromyzidae) phenology. *Environmental Entomology,* **20**(4), 1134–1140, 1991.

Pigoli, D., Aston, J.A.D., Ferraty, F., Mazumder, A., Richards, C., and Hall, M.J.R. Estimation of temperaturedependent growth profiles for the assessment of time of hatching in forensic entomology. arXiv preprint arXiv:1709–00623, 2017.

Ramsay, J.O. and Silverman, B.W. *Functional Data Analysis.* Springer, 2005.

Reibe, S., Doetinchem, P.V., and Madea, B. A new simulation-based model for calculating post-mortem intervals using developmental data for lucilia sericata (dipt.: Calliphoridae). *Parasitology Research,* **107**(1), 9–16, 2010.

Reiter, C. Zum wachstumsverhalten der maden der blauen schmeißfiiege calliphora vicina. *International Journal of Legal Medicine,* **91**(4), 295–308, 1984.

Saulich, A.K. The law of effective temperature sum: Limitations and scope for application. *Entomological Review,* **79**(1), 81–94, 1999.

Sharpe, P.J.H. and DeMichele, D.W. Reaction kinetics of poikilotherm development. *Journal of Theoretical Biology,* **64**(4), 649–670, 1977.

Tarone, A.M. and Foran, D.R. Generalized additive models and lucilia sericata growth: assessing confidence intervals and error rates in forensic entomology. *Journal of Forensic Sciences,* **53**(4), 942–948, 2008.

Tomberlin, J.K. and Benbow, M.E. *Forensic Entomology: International Dimensions and Frontiers.* CRC Press, 2015.

Wagner, T.L., Wu, H.-I., Sharpe, P.J.H., Schoolfield, R.M., and Coulson, R.N. Modeling insect development rates: a literature review and application of a biophysical model. *Annals of the Entomological Society of America,* **77**(2), 208–220, 1984.

Warren, J.-A., Ratnasekera, T.D.P., Campbell, D.A., and Anderson, G.S. Initial investigations of spectral measurements to estimate the time within stages of protophormia terraenovae (robineaudesvoidy) (diptera: Calliphoridae). *Forensic Science International,* **278**, 205–216, 2017.

Wells, J. and LaMotte, L. The role of a pmi-prediction model in evaluating forensic entomology experimental design, the importance of covariates, and the utility of response variables for estimating time since death. *Insects,* **8**(2), 47, 2017.

Wells, J.D., Lecheta, M.C., Moura, M.O., and LaMotte, L.R. An evaluation of sampling methods used to produce insect growth models for postmortem interval estimation. *International Journal of Legal Medicine,* **129**(2), 405–410, 2015.

Williams, H. A model for the aging of _ y larvae in forensic entomology. *Forensic Science International,* **25**(3), 191–199, 1984.

20

<hr style="border-top: 4px solid black;">

Statistical Models in Forensic Voice Comparison

Geoffrey Stewart Morrison, Ewald Enzinger, Daniel Ramos,
Joaquín González-Rodríguez, and Alicia Lozano-Díez

CONTENTS

20.1 Introduction 452
20.2 Feature Extraction 455
    20.2.1 Mel-Frequency Cepstral Coefficients (MFCCs) 455
    20.2.2 Deltas and Double Deltas 457
    20.2.3 Voice-Activity Detection (VAD) and Diarization 458
20.3 Mismatch Compensation in the Feature Domain 458
    20.3.1 Cepstral-Mean Subtraction (CMS) and Cepstral-Mean-and-Variance
            Normalization (CMVN) 459
    20.3.2 Feature Warping 460
20.4 GMM-UBM 461
    20.4.1 Training the Relevant-Population Model (the UBM): Expectation
            Maximization (EM) Algorithm 462
    20.4.2 Training the Known-Speaker Model: Maximum a Posteriori (MAP)
            Adaptation 465
    20.4.3 Calculating a Score 465
    20.4.4 Remarks Regarding UBM Training Data 467
20.5 i-Vector PLDA 467
    20.5.1 i-Vectors 467
    20.5.2 i-Vector Domain Mismatch Compensation (LDA) 470
    20.5.3 PLDA 470
20.6 DNN-Based Systems 474
    20.6.1 DNN Senone Posterior i-Vector Systems 477
    20.6.2 Bottleneck-Feature Based Systems 477
    20.6.3 DNN Speaker Embedding Systems (x-Vector Systems) 478
20.7 Score-to-Likelihood-Ratio Conversion (Calibration) 479
20.8 Validation 483
    20.8.1 List of Published Validation Studies 487
20.9 Conclusion 488
Acknowledgments 488
Appendix 20A: Mathematical Details of T Matrix Training and i-Vector Extraction .. 489
References 491
Legal References 497

20.1 Introduction

The purpose of *forensic voice comparison* (aka forensic speaker comparison, forensic speaker recognition, and forensic speaker identification) is to assist a court of law in deciding whether the voices on two (or more) recordings were produced by the same speaker or by different speakers. For simplicity, in the present chapter we will assume that there is one recording of a speaker of known identity (the *known-speaker recording*) and one recording of a speaker whose identity is in question (the *questioned-speaker recording*). Common scenarios include that the questioned-speaker recording is of a telephone call made to a call center, is of an intercepted telephone call, or is made using a covert recording device, and the known-speaker recording is of a police interview with a suspect or is of a telephone call made by a person who is in custody. The known-speaker recording is often an existing recording in which the identity of the speaker is not disputed, but sometimes a recording is made specifically for the purpose of conducting a forensic-voice-comparison analysis (practice varies depending on jurisdiction, laboratory policy, and the circumstances of the particular case).

There is usually a mismatch between the questioned- and known-speaker recordings in terms of speaker-intrinsic conditions, or speaker-extrinsic conditions, or both. Speaker-intrinsic variability can be due to multiple factors including differences in speaking style (e.g., casual versus formal and quiet versus loud), emotion (e.g., calm, angry, happy, sad), tiredness or physical stress (e.g., being out of breath), and elapsed time (the way a speaker speaks varies from minute to minute, hour to hour, day to day, etc. with larger differences occurring over longer time periods, Kelly and Hansen, 2016). In addition, the words and phrases that a speaker says vary from occasion to occasion. Speaker-extrinsic variability can be due to multiple factors including background noise that can vary in loudness and type (e.g., office noise, ventilation system noise, traffic noise, crowd noise – as well as noise being captured on the recording, speaking in a noisy environment also causes speakers to change the way they speak), reverberation (e.g., echoes in rooms with hard walls and floors), distance of the speaker from the microphone, the quality of the microphone and other components of the recording equipment, transmission of the recording through different communication channels that distort and degrade the signal in different ways (e.g., landline telephone, mobile telephone, Voice over Internet Protocol VoIP), and the format in which the recording is saved (to reduce the file size formats such as MP3 distort and degrade the signal). Intrinsic and extrinsic variability leads to mismatches between questioned- and known-speaker recordings within cases, and leads to different conditions and different mismatches from case to case.

Historically and in present practice, several approaches have been used to extract information from voice recordings and different frameworks have been used to draw inferences from that information. In *auditory* and *spectrographic* approaches information is extracted using subjective judgment, by listening to the recordings and by looking at graphical representations of parts of the recordings respectively (spectrograms are time by frequency by intensity plots of the acoustic signal). In conjunction with auditory and spectrographic approaches, inferences have almost invariably been drawn on the basis of subjective judgment. In *acoustic-phonetic* and *human-supervised automatic* approaches information is extracted in the form of quantitative measurements of the acoustic properties of the audio recordings. For the acoustic-phonetic approach, inferences can be drawn via statistical models, although in practice it is more common for practitioners of this approach to draw inferences on the basis of subjective judgment, e.g., by making a plot of the measured values from different recordings and then looking at the plot. For the automatic approach, inferences are invariably drawn via the use of statistical models. Even if inferences are drawn via statistical models, rather than directly reporting the output of the statistical model many practitioners use the output of the statistical model as input to a subjective judgment that also includes consideration of other information such as their subjective judgment based on auditory and acoustic-phonetic approaches (for arguments against this practice see Morrison and Stoel, 2014). It should be noted that even if the output of a quantitative-measurement and statistical-model approach is directly reported, such an approach still requires subjective judgments in decisions such as the choice of data used to train the statistical models. These subjective judgments are, however, as far removed as possible from the forensic practitioner's final conclusion, so of all the approaches this one is most resistant to cognitive bias (for recent reviews of cognitive bias in the context of forensic science see Found, 2015; National Commission on Forensic Science, 2015; Stoel et al., 2015; Edmond et al., 2017).

As of June 1, 2020, there are no published national or international standards specific to forensic voice comparison. The England and Wales Forensic Science Regulator's Codes of Practice and Conduct (Forensic Science Regulator, 2020) and their appendices relating to specific branches of forensic science are effectively national standards in that forensic laboratories in the UK can seek accreditation to the Codes, usually in combination with accreditation to ISO 17025:2015 General Requirements for The Competence of Testing and Calibration Laboratories. There is an appendix to the codes for Speech and Audio Forensic Services (Forensic Science Regulator, 2016). The European Network of Forensic Science Institutes (ENFSI) has published Methodological Guidelines for Best Practice in Forensic Semiautomatic and Automatic Speaker Recognition (Drygajlo et al., 2015). The Speaker Recognition Subcommittee of the Organization of Scientific Area Committees for Forensic Science (OSAC SR) is in the process of developing standards.

The ENFSI guidelines include use of the likelihood-ratio framework for drawing inferences, and empirical validation of system performance under conditions reflecting those of the cases to which they are applied. The Forensic Science Regulator's codes also require methods to be validated. In forensic voice comparison these are not new ideas: quantitative-measurement and statistical-model based implementations of the likelihood-ratio framework date back to the 1990s, and calls for forensic-voice-comparison systems to be empirically validated under conditions reflecting casework conditions date back to the 1960s (for reviews see Morrison, 2009, and Morrison, 2014, respectively). The ENFSI guidelines also include the use of transparent and reproducible methods and procedures, and the use of procedures that reduce the potential for cognitive bias. Morrison and Thompson (2017) and Morrison (2018a) have argued that transparent quantitative-measurement and statistical-model based implementation of the likelihood-ratio framework with empirical validation under casework conditions would be the only practical way to comply with the admissibility criteria set out in United States Federal Rules of Evidence 702 and the *Daubert* trilogy of Supreme Court rulings (Daubert v. Merrell Dow Pharmaceuticals, 1993; General Electric v. Joiner, 1997; Kumho Tire v. Carmichael, 1999), and those set out in England and Wales Criminal Practice Directions (2015) section 19A.

To reduce the potential for cognitive bias, increase transparency and reproducibility, facilitate validation under conditions reflecting casework conditions, and to draw inferences that are logically correct, we believe that the most practical approach is the *human-supervised automatic approach* used in conjunction with statistical-model based implementation of the *likelihood-ratio framework* with direct reporting of the output of the statistical model. The performance of acoustic-phonetic approaches have been found to be much poorer than the performance of automatic approaches (see Enzinger et al., 2012, 2014; Zhang et al., 2013; Enzinger and Kasess, 2014; Jessen et al., 2014; Enzinger and Morrison, 2017). Acoustic-phonetic approaches are also much more time consuming and costly in skilled human labor, which makes empirical validation practically difficult. In the present chapter, we therefore discuss only the human-supervised automatic approach and statistical-model based implementation of the likelihood-ratio framework. We assume the reader is familiar with the likelihood-ratio framework, which has been described elsewhere in the present volume. Our aim is to provide an overview of signal-processing and statistical-modeling techniques that are commonly used to calculate likelihood ratios in human-supervised automatic approaches to forensic voice comparison. We aim to bridge the gap between general introductions to forensic voice comparison and the highly technical (and often fragmented) automatic-speaker-recognition literature from which the signal-processing and statistical-modeling techniques are mostly drawn. The automatic-speaker-recognition literature is often fragmented because many influential papers are short conference-proceedings papers that do not provide fully detailed descriptions of the techniques they apply.

For readers unfamiliar with forensic voice comparison, we recommend general introductions such as Morrison and Thompson (2017), Morrison et al. (2018), and Morrison and Enzinger (2019). These include discussions of the likelihood-ratio framework, empirical validation, and legal admissibility. In the present chapter we go into greater technical detail regarding the calculation of likelihood ratios than is provided in such general introductions, but still attempt to make the material relatively accessible to an audience with a limited background in signal processing and statistical modeling. We have in mind researchers from other branches of forensic science and students of forensic voice comparison. There are alternatives to and multiple variants of many of the feature-extraction and statistical-modeling techniques we describe below. We do not attempt to be comprehensive, and describe only some of the variants that have commonly been used in forensic voice comparison. For overviews of automatic speaker recognition in general see Kinnunen and Li (2010), Hansen and Hasan (2015), Fernández Gallardo (2016) §2.4, Ajili (2017) ch. 4, Matějka et al. (2020) (and see Hansen and Bořil, 2018, for a review of intrinsic and extrinsic variability in the context of automatic speaker recognition). These publications cover some of the same topics as the present chapter but are aimed at an audience with a background in signal processing and machine learning.

The chapter is structured as follows:

- Section 20.2 describes extraction of features from voice recordings, in particular extraction of mel-frequency cepstral coefficients (MFCCs).
- Section 20.3 describes mismatch compensation in the feature domain, in particular cepstral-mean subtraction (CMS), cepstral-mean-and-variance normalization (CMVN), and feature warping.
- Section 20.4 describes the Gaussian mixture model - universal background model (GMM-UBM) approach.
- Section 20.5 describes the identity-vector - probabilistic linear discriminant analysis (i-vector PLDA) approach, including mismatch compensation in the i-vector

domain using canonical linear discriminant functions (CLDF). In the automatic-speaker-recognition literature, the latter is known as linear discriminant analysis (LDA).

- Section 20.6 describes deep neural network (DNN) based approaches, in particular those using senone posterior i-vectors, bottleneck features, and speaker embeddings (aka x-vectors).
- Section 20.7 describes score-to-likelihood-ratio conversion (aka calibration) using logistic regression. This can be applied to the output of GMM-UBM, i-vector PLDA, or DNN-based systems.
- Section 20.8 discusses empirical validation of forensic-voice-comparison systems.

GMM-UBM is an older approach, in use from about 2000. It has mostly been replaced by the i-vector PLDA approach, in use from about 2010. We describe GMM-UBM because it is still used by some forensic practitioners, it is somewhat more straightforward to describe than i-vector PLDA, and part of the discussion of the GMM-UBM approach provides a foundation for understanding the i-vector PLDA approach. Since about 2015 state-of-the-art systems in automatic-speaker-recognition research have been based on DNNs, and commercial DNN-based forensic-voice-comparison systems were first released around 2018. DNNs are generally used to create vectors (alternatives to i-vectors) that are then fed into PLDA models.

Color versions of the figures in this chapter and other material related to the chapter are available at http://handbook-of-forensic-statistics.forensic-voice-comparison.net/.

20.2 Feature Extraction

In the context of forensic voice comparison, *features* are the acoustic measurements made on voice recordings. We describe the most commonly used features in automatic approaches to forensic voice comparison, mel-frequency cepstral coefficients (MFCCs), and their derivatives, deltas and double deltas.

For reviews of feature extraction methods in automatic speaker recognition, see Chaudhary et al. (2017), Dişken et al. (2017), and Tirumala et al. (2017).

20.2.1 Mel-Frequency Cepstral Coefficients (MFCCs)

Mel-frequency cepstral coefficients (MFCCs; see Davis and Mermelstein, 1980) are spectral measurements made at regular intervals, i.e., once every few milliseconds, during the sections of the recording corresponding to the speech of the speaker of interest. The noun "cepstrum" and adjective "cepstral" were coined by Bogert et al. (1963) via rearrangements of the letters in the words "spectrum" and "spectral". "Mel" refers to a frequency scaling that, unlike hertz, reflects human perception of frequency (Stevens et al., 1937). MFCCs are standard in speech processing in general, not just in automatic speaker recognition. In contrast to its use in automatic speech recognition, there is no principled reason for using mel scaling for automatic speaker recognition. Other variants of cepstral coefficients could be used, but MFCCs work well and are the most popular (Tirumala et al., 2017).

The steps for extracting MFCC measurements from voice recordings are described below, see also Figure 20.1 in which the numbers within black circles correspond to the numbered steps below.

1. The speech signal is multiplied by a bell-shaped window (e.g., a hamming window) with a duration typically on the order of 20 ms. The shape of the window is designed to reduce the impact of the window itself on the measurement of the spectrum (see Harris, 1978).

2. The power spectrum of the windowed signal is calculated using a discrete Fourier transform (DFT, or for computational efficiency a fast Fourier transform, FFT). A complex waveform, such as a windowed speech signal, can be constructed by

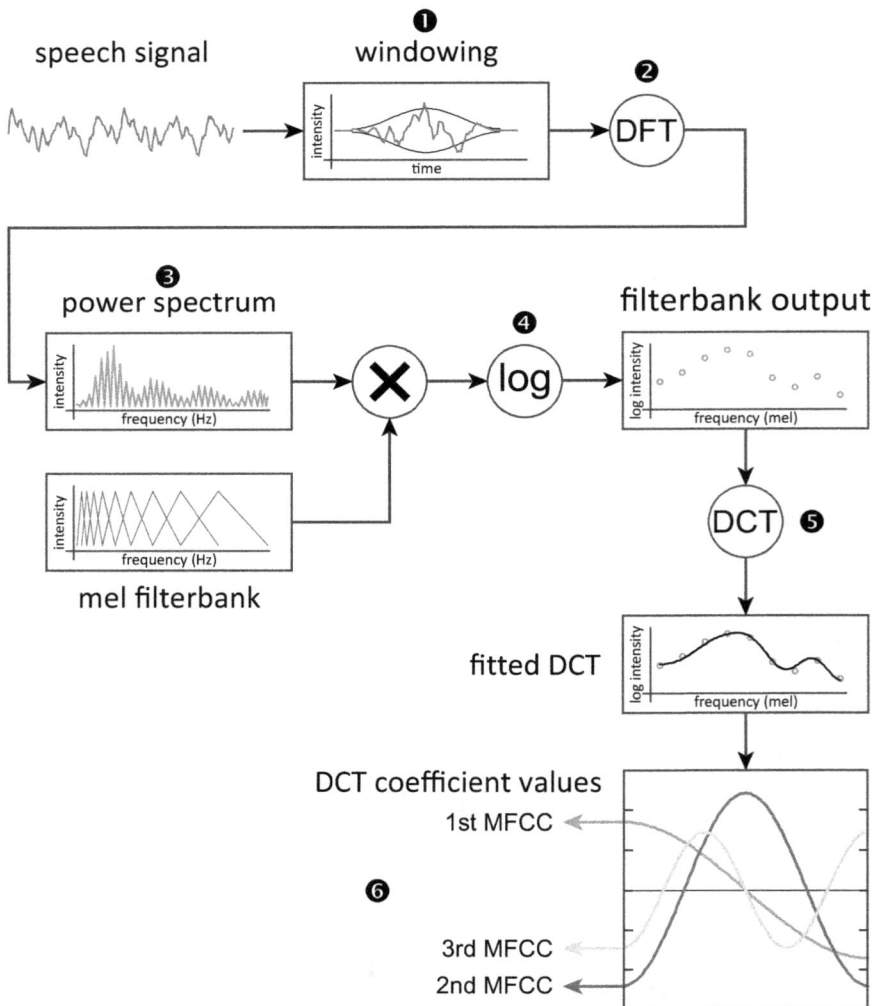

FIGURE 20.1
Procedure for the calculation of MFCCs. The numbers in black circles correspond to the numbered steps in the main text. DFT = discrete Fourier transform. DCT = discrete cosine transform.

adding together at each point in time the instantaneous intensities of a series of simple sine waves (the idea that any arbitrary waveform can be constructed using such a series is credited to Fourier, 1808). A Fourier analysis determines the frequencies, intensities, and phases of the sine waves that would have to be added together to make the observed complex waveform. The power spectrum consists of the frequencies and intensities of the sine wave components. For automatic speaker recognition, information about phase is usually discarded.

3. The power spectrum is multiplied by a filterbank. This is a series of triangular shaped filters, e.g., 26 filters, that are equally spaced on the mel-frequency scale. Each filter has a 50% overlap with each of its neighbors. Since forensic voice comparison often involves telephone recordings, the frequency range covered by the filters may be restricted to the traditional landline telephone bandpass (300 Hz – 3.4 kHz).

4. The intensities of the filterbank outputs are scaled logarithmically. The fact that this is log intensity will be relevant for feature domain mismatch compensation as described in Section 20.3 below.

5. A discrete cosine transform (DCT) is fitted to the output of the filterbank. A DCT is similar to a Fourier transform but all components are cosines of the same phase (or the opposite phase if the coefficient value is negative). The components are orthogonal: a constant, half a period of a cosine, one period of a cosine, one and a half periods of a cosine, etc. The process of fitting a DCT involves selecting values for the weights, or coefficients, on each component which result in the best fit to the data. Using all the DCT coefficients, the values of the original data can be recovered. Using only the first few DCT coefficients results in a smoothed version of the original data, i.e., a smoothed spectrum (aka a cepstrum). The higher order coefficients tend to capture statistical noise.

6. The first few DCT coefficients (e.g., the 1st through 14th DCT coefficients) are used as a vector of MFCCs. The 0th coefficient encodes the mean intensity of the signal and is usually discarded since this can be affected by factors not related to who is speaking (e.g., the distance of the speaker to the microphone or automatic gain control on the recording system).

The window is advanced in time, e.g., a 20-ms long window is advanced by 10 ms. Each range of time covered by a window is called a *frame*, and there is usually a 50% overlap between adjacent frames. Steps 1 through 6 are then repeated to produce another vector of MFCC values. The window is repeatedly advanced until MFCC vectors have been extracted from all sections of the recording corresponding to the speech of the speaker of interest. Since one MFCC vector is extracted every few milliseconds, a relatively large number of feature vectors is extracted, e.g., 100 feature vectors per second if the frame advance is 10 ms.

20.2.2 Deltas and Double Deltas

Deltas are derivatives of MFCCs and encode the local rate of change of MFCC values over time (see Furui, 1986). *Double deltas* are the second derivatives of MFCCs and encode the local rate of change of delta values over time. Vectors of deltas and double deltas are concatenated with the MFCC vectors to produce longer *feature vectors*, e.g., if the original

MFCC vector has 14 values, the MFCC + delta vector will have 28, and the MFCC + delta + double delta vector will have 42.

Delta values are calculated separately for each MFCC dimension. In each dimension, a linear regression is fitted to a contiguous set of MFCC values, e.g., the MFCC value at the frame in time where the delta is being measured plus the MFCC values from the same dimension in the two preceding and the two following frames. Measurements are usually made over the ± 2 or ± 3 adjacent frames. The value of the slope of the fitted linear regression is used as the delta value.

Double deltas are calculated in the same way as deltas, but based on the delta values rather than on the MFCC values.

The statistical modeling steps of automatic-speaker-recognition systems usually discard information regarding the original time sequence of the feature vectors; hence the deltas and double deltas encode the only time-sequence information that is exploited. DNN embedding systems are an exception to this.

20.2.3 Voice-Activity Detection (VAD) and Diarization

Either prior to or after extracting MFCCs, the speech of the speaker of interest should be separated from periods of silence, transient noises, and the speech of other speakers. Either only the speech of the speaker of interest should be measured, or only measurements corresponding to the speech of the speaker of interest should be kept. This is done by either manually or automatically marking the beginning and end of each utterance (stretch of speech) of interest. An automatic procedure may be followed by manual checking and correction as needed.

The process of finding utterances is called *voice-activity detection* (VAD, aka speech-activity detection, SAD). A simple automatic *voice-activity detector* (also abbreviated as VAD) may be based solely on root-mean-square (RMS) amplitude, and thus simply find louder parts of the recording. Such simple VADs may not, however, perform well under conditions that include background noise, as is common in forensic voice comparison (see Mandasari et al., 2012). More sophisticated VADs employ algorithms to distinguish speech from other sounds and noises (e.g., Sohn et al., 1999; Beritelli and Spadaccini, 2011; Sadjadi and Hansen, 2013).

The process of attributing different utterances in the recording to different speakers is called *diarization*. Automatic diarization is itself a form of automatic speaker recognition and may itself make use of MFCCs.

20.3 Mismatch Compensation in the Feature Domain

The acoustic properties of voice recordings can vary because they are produced by different speakers, but also because of other factors such as differences in speaking styles (e.g., casual, formal, whispering, shouting), differences in background noise and/or reverberation, differences in the distance of the speaker to the microphone, different types of microphones, transmission through different communication channels (e.g., landline telephone, mobile telephone, VoIP), and being saved in different formats (to reduce the size of the files stored, formats such as MP3 use lossy compression which discards some acoustic information and distorts remaining acoustic information). Mismatch compensation techniques seek to maximize between-speaker differences and minimize differences due to other factors. Feature domain mismatch compensation techniques generally attempt to reduce differences due to recording and transmission channels and differences due to acoustic noise.

We describe three commonly used feature domain mismatch compensation techniques: cepstral-mean subtraction (CMS), cepstral-mean-and-variance normalization (CMVN), and feature warping.

20.3.1 Cepstral-Mean Subtraction (CMS) and Cepstral-Mean-and-Variance Normalization (CMVN)

Cepstral-mean subtraction (CMS; Furui, 1981) as a mismatch compensation technique is based on the premise that the speech signal, which is changing rapidly over time, is convolved with a channel that is invariant over time (it is linear time invariant, LTI). This is a good model for a traditional landline telephone system, the effect of which is essentially to pass the speech signal through a bandpass filter. Different microphones have different frequency responses. Some may be more sensitive to lower frequency sounds, some to higher frequency sounds, etc. Microphones and other basic components of recording systems can be treated as linear time invariant. Differences in the signal due to differences in the distance from the speaker to the microphone can also be treated as linear time invariant effects, as long as the distance does not change during the recording.

Convolution in the time domain is equivalent to multiplication in the frequency domain. Since MFCCs are frequency domain representations they can be considered the result of multiplying the dynamic speech signal and the invariant channel. But note that in generating MFCCs, intensity was logarithmically scaled (Step 4 in Section 20.2.1 above). Multiplication on a linear scale is equivalent to addition on a logarithmic scale, hence we should consider MFCCs the result of adding the speech signal and the channel in the log frequency domain.

Assuming the channel is invariant over time, it can be estimated by taking the mean of the cepstral coefficients over time. For the value in each dimension in each frame, the mean for that dimension is subtracted. What remains are the dynamic aspects of the original MFCC features. Deltas and double deltas are calculated on the raw MFCCs before CMS is applied to the MFCCs, and (although not obviously theoretically motivated) CMS is then usually applied to the deltas and double deltas.

As well as removing the effect of the invariant channel, CMS also removes the speech signal's mean. If there is a substantial channel mismatch between the questioned- and known-speaker recordings, removing the effect of the channel will tend to lead to better performance despite the loss of the speech-signal mean information (but CMS will tend to lead to worse performance if there is in fact no channel mismatch, Reynolds, 1994). For each recording, the mean of the feature values in each dimension will be 0, hence statistical models applied to the post-CMS features are designed to exploit what remains: non-normalities (in terms of skewness, kurtosis, and multimodalities), variance, and multidimensional correlations.

Although the theoretical motivation is less immediately obvious, *cepstral-mean-and-variance normalization* (CMVN; Vikki and Laurila, 1998) is a statistically obvious extension to CMS in which, as well as subtracting the mean, the variance of each MFCC dimension is scaled to 1.

CMS and CMVN can be applied globally, i.e., using the mean and variance of MFCC values from the whole of the speech of the speaker of interest in a particular recording, or can be applied locally, i.e., using the mean and variance of the MFCC values from the speech of the speaker of interest within a window that is a few seconds long. Such local application is discussed below in the context of feature warping.

20.3.2 Feature Warping

In contrast to assuming that a channel effect is invariant throughout the recording, *feature warping* (Pelecanos and Sridharan, 2001) is based on the assumption that there may be a slowly changing channel effect convolved with a rapidly changing speech signal. It also assumes that there may be slowly changing additive noise. Since these are assumed to be slowly changing, their effect on the MFCC values is assumed to be stable locally, i.e., over a period of a few seconds.

The implementation of feature warping is described below, see also Figure 20.2. Feature warping is applied separately to each feature vector dimension. It is based on finding where the current feature value lies relative to the empirical distribution of the values in, for example, the preceding 1.5 s and the following 1.5 s (±150 frames), and then warping that value to the corresponding value in a target distribution. The target distribution is usually a Gaussian distribution with a mean of 0 and variance of 1.

1. Obtain the feature values from the 150 frames preceding the current frame, and the 150 frames following the current frame.

2. Sort the 301 values (including the current value) in ascending order, and calculate each value's rank proportional to the total number of values. This can be represented graphically as a plot of the empirical cumulative distribution of the feature values.

3. Find the value of the empirical cumulative distribution, $y_{empirical}$, which corresponds to the current feature value, $x_{original}$.

4. On the target cumulative probability distribution, locate the point $y_{target} = y_{empirical}$.

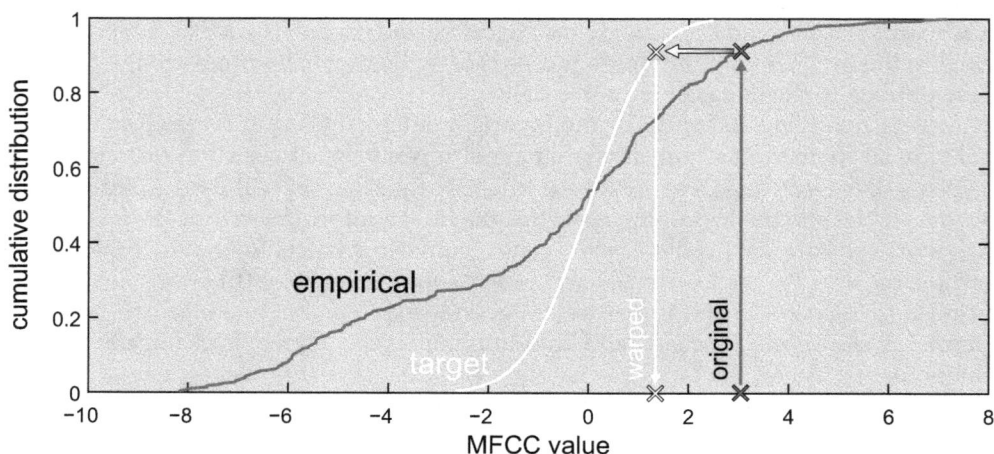

FIGURE 20.2

Feature warping. The original feature value is replaced by the warped feature value. The warping is achieved by mapping from the empirical cumulative probability distribution of the original feature values to a parametric target cumulative probability distribution; in this case, the standard cumulative Gaussian distribution.

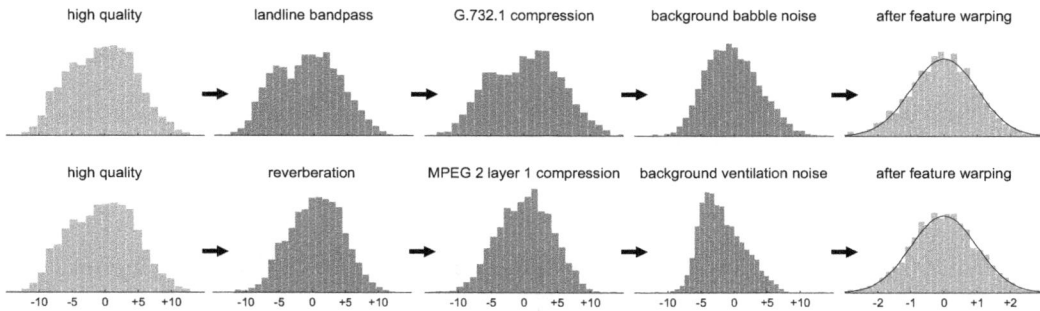

FIGURE 20.3
Examples of the effects of channel and feature warping on the distribution of MFCCs. The first column represents MFCC values extracted from a high-quality audio signal (the same in both rows). The next three columns
represent the results of sequentially applying various signal processing techniques to simulate casework conditions. The top row represents a simulated questioned-speaker recording condition and the bottom row represents
a simulated known-speaker recording condition. The final column represents the results of applying feature
warping to the values represented in the immediately preceding column, applied separately to the simulated
questioned-speaker recording condition and to the simulated known-speaker recording condition.

    5. Read off the corresponding x_{warped} value from the target cumulative probability
       distribution.

    Deltas and double deltas are calculated on the raw MFCCs before feature warping is
applied to the MFCCs, and feature warping is then applied to the deltas and double deltas.
    Figure 20.3 shows an example of the effect of applying feature warping to mismatched
recording conditions. The histograms in the leftmost column represent the distribution
of the 1st order MFCC values extracted from a studio-quality speech recording. The
histograms in the next three columns show the distributions of the 1st order MFCC values extracted after having applied various signal processing techniques to simulate the
different conditions of a questioned-speaker recording and a known-speaker recording
from a case (top and bottom rows respectively). Notice how the distributions change
after each step and how the distributions under the simulated questioned- and known-
speaker conditions diverge from one another. The histograms in the rightmost column
show the distributions after feature warping is applied. Notice how the distributions are
now warped to approximate the same target distribution.
    Making all univariate distributions the same may seem extreme, but the ordering within
vectors is not altered and so correlations across dimensions are not lost. The correlations
of interest are ultimately due to time and frequency patterns in the acoustic signal resulting from the articulation of speech sounds. Particular time and frequency patterns result
from the articulation of particular speech sounds. Statistical models applied to the warped
features are designed to exploit these multidimensional correlations. Empirically, feature
warping can outperform CMS and CMVN (see for example: Pelecanos and Sridharan, 2001;
Silva and Medina, 2017).

20.4 GMM-UBM

This section describes the *Gaussian mixture model - universal background model* approach
(GMM-UBM; see Reynolds et al., 2000).

The GMM-UBM model is a *specific-source model*,[*] i.e., it builds a model for the specific known-speaker recording in the case, and thus answers the following two-part question:

> What would be the probability of obtaining the feature vectors of the questioned-speaker recording if the questioned speaker were the known speaker?

> versus

> What would be the probability of obtaining the feature vectors of the questioned-speaker recording if the questioned speaker were <u>not</u> the known speaker, but some other speaker selected at random from the relevant population?

Conceptually, the answer to the first part of the question provides the numerator for the likelihood ratio and the answer to the second part provides the denominator.

A *Gaussian mixture model* (GMM), is a probability density function constructed from a weighted linear combination of Gaussian probability density functions. Assuming a multivariate distribution, each of the G component Gaussian distributions, $f(x \mid \boldsymbol{\mu}_g, \boldsymbol{\Sigma}_g)$, in the mixture has a mean vector $\boldsymbol{\mu}_g$, a covariance matrix $\boldsymbol{\Sigma}_g$, and a weight w_g, see Equation 20.1. Usually, diagonal-only covariance matrices are used – a larger number of Gaussian components with diagonal-only covariance matrices can achieve the same degree of fit as a smaller number of components with full covariance matrices, and the former approach is more computationally efficient. In the present context, x is a feature vector after application of feature domain mismatch compensation, and y is the probability density (likelihood) of the model evaluated at x.

$$y = \sum_{g=1}^{G} w_g f(x \mid \boldsymbol{\mu}_g, \boldsymbol{\Sigma}_g), \qquad \sum_{g=1}^{G} w_g = 1, \quad w_g \geq 0 \qquad (20.1)$$

20.4.1 Training the Relevant-Population Model (the UBM): Expectation Maximization (EM) Algorithm

The mean vectors, covariance matrices, and weights for a GMM are trained using the *expectation maximization* (EM) algorithm (Dempster et al., 1977; Hastie et al., 2009, §8.5). Although the data in forensic voice comparison are multivariate, for simplicity of exposition we describe the EM algorithm below assuming univariate data. Also see Figure 20.4 which shows an example of a two-component GMM being trained on artificial data that were created for illustrative purposes. In the one dimensional case, assume that we have scalar training data, $x_i, i \in \{1, \ldots, N\}$.

1. The number of Gaussian components, G, to use must be specified. A commonly used number of Gaussian components is 1024 (the optimal value will depend on the particular application and amount of training data available).

2. A starting point is needed. A common approach is to use a k-means clustering algorithm (see Arthur and Vassilvitskii, 2007; Hastie et al., 2009, §14.3) to cluster

---

[*] For a discussion of the distinction between *specific-source* and *common-source* models, see Ommen and Saunders (2018).

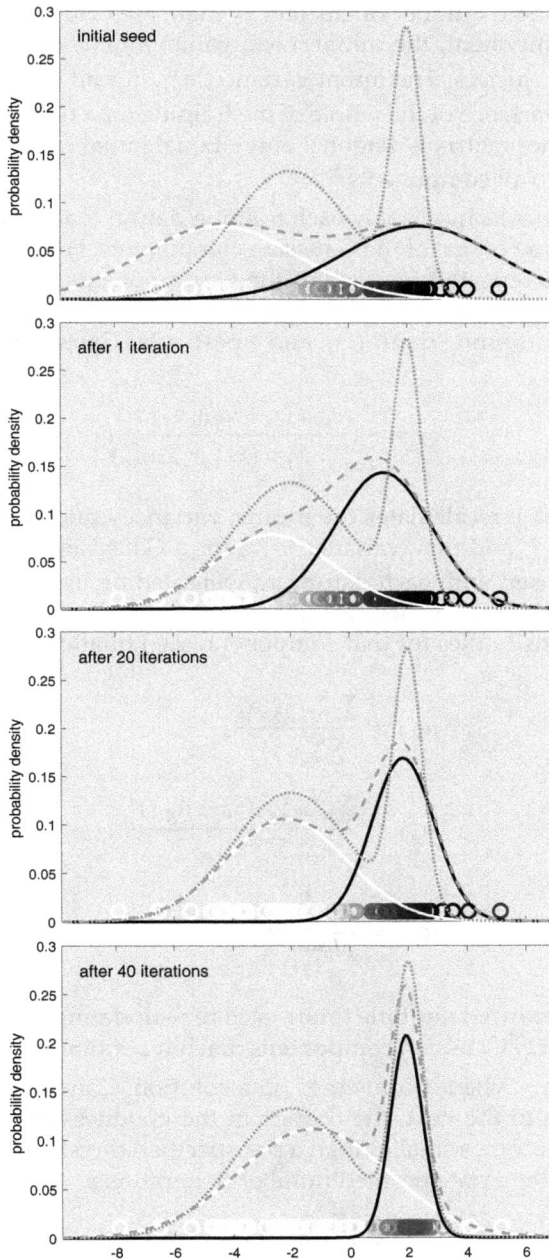

FIGURE 20.4
Example of using the EM algorithm to train a two-component GMM. The example is based on artificial data that were created for illustrative purposes. The dotted curve represents the distribution that was used to generate the data. The dashed curve represents the fitted GMM, and the black curve and the white curve represent the two Gaussian components of the fitted GMM. The top panel shows the initial GMM distribution based on a random seed. The data points (the circles on the *x* axis) are shaded according to their responsibilities with respect to the two components. The second panel shows the fitted GMM distribution after 1 iteration of maximization (along with the responsibilities after that iteration). The third panel shows the results after 20 iterations, and the bottom panel the results after 40 iterations.

the data into the same number of clusters as there are Gaussian components. For each Gaussian component, the initial mean value, $\mu_{g,0}$, is trained using the data from one of these clusters. The initial variance, $\sigma_{g,0}^2$, for all components is usually set to the sample variance of the whole of the training data (in the multivariate case the initial covariance matrix is diagonal only). Equal initial weights, $w_{g,0} = 1/G$, are usually assigned to all components.[†]

3. The *expectation step* asks how likely each training datum is assuming it had it come from the distribution of one of the Gaussian components. For each training datum, x_i, this question is repeated for each of the G components. The answers, $\gamma_{g,i}$, are called *responsibilities*. Each is calculated as the relative likelihood for the combination of a particular training datum, x_i, and a particular Gaussian component, g, see Equation 20.2.

$$\gamma_{g,i} = \frac{w_{g,0}f(x_i \mid \mu_{g,0}, \sigma_{g,0})}{\sum_{j=1}^{G} w_{j,0}f(x_i \mid \mu_{j,0}, \sigma_{j,0})} \qquad (20.2)$$

4. The *maximization step* recalculates the means, variances, and weights. To calculate the new mean, $\mu_{g,1}$, and new variance, $\sigma_{g,1}^2$, for a Gaussian component, g, all the training data are used with each datum, x_i, weighted by its responsibility, $\gamma_{g,i}$, see Equations 20.3 and 20.4. The new weight for the Gaussian component, $w_{g,1}$, is the mean of the responsibilities for that component, see Equation 20.5.

$$\mu_{g,1} = \frac{\sum_{i=1}^{N} \gamma_{g,i} x_i}{\sum_{i=1}^{N} \gamma_{g,i}} \qquad (20.3)$$

$$\sigma_{g,1}^2 = \frac{\sum_{i=1}^{N} \gamma_{g,i}(x_i - \mu_{g,1})^2}{\sum_{i=1}^{N} \gamma_{g,i}} \qquad (20.4)$$

$$w_{g,1} = \frac{1}{N} \sum_{i=1}^{N} \gamma_{g,i} \qquad (20.5)$$

5. Steps 3 and 4 are repeated multiple times, each time updating the means, variances, and weights for all G Gaussian components. Each repetition is called an *iteration*.

6. The algorithm stops when it converges on a solution. Convergence occurs when, from one iteration to the next, the change in the goodness of fit of the model to the training data becomes smaller than a pre-specified threshold. Alternatively, the algorithm stops after a pre-specified number of iterations.

The first step in using a GMM-UBM is to train a GMM using feature vectors extracted from recordings of a sample of speakers representative of the relevant population for the case. Feature vectors from recordings of all the speakers in the sample are pooled and used to train a GMM which is called a *universal background model* (UBM). This is the model which will be used to calculate the denominator of the likelihood ratio.

[†] Other approaches include binary splitting (see Ueda et al., 2000), and random seed. For the latter, the initial mean value, $\mu_{g,0}$, of each component is set to a randomly selected x_i value from the training data. If a random seed is used, each random seed will lead to a different solution. Training may be repeated multiple times using multiple random seeds and the result with the best fit to the data used.

20.4.2 Training the Known-Speaker Model: Maximum a Posteriori (MAP) Adaptation

The model for calculating the numerator of the likelihood ratio, a *speaker model*, is also a GMM, but rather than training the model from scratch, the known-speaker GMM is adapted from the UBM (Reynolds et al., 2000). One reason for this is that successful training of a GMM with a large number of Gaussian components in a high dimensional space requires a large amount of training data. The UBM is trained using data from a large number of speakers, but the known-speaker GMM has to be trained using data from one speaker, and often only one relatively short recording of that speaker is available. The procedure for adapting a single-speaker GMM from the UBM is a form of *maximum a posteriori adaptation* (MAP). It is similar to the EM algorithm, but usually only one iteration of MAP adaptation is applied, usually only the means are adapted, and the means are only partially adapted. A new mean is the result of a weighted mixture of the original mean from the UBM, $\mu_{g,\text{UBM}}$, and what would be the new mean if the standard EM algorithm were applied, $\mu_{g,\text{EM}}$ (the latter is $\mu_{g,1}$ in Equation 20.3 above). The new mean is calculated using Equation 20.6, in which α_g, the weight for mixing the EM and UBM means, is known as the *adaptation coefficient*, and τ is known as the *relevance factor*.

$$\mu_{g,1} = \alpha_g \mu_{g,\text{EM}} + (1 - \alpha_g)\mu_{g,\text{UBM}} = \alpha_g \frac{\sum_{i=1}^{N} \gamma_{g,i} x_i}{\sum_{i=1}^{N} \gamma_{g,i}} + (1 - \alpha_g)\mu_{g,0} \qquad (20.6)$$

$$\alpha_g = \frac{N w_{g,1}}{N w_{g,1} + \tau}$$

If the new weight for a Gaussian component, $w_{g,1}$, is large, i.e., lots of adaptation training data are associated with that Gaussian component, then α_g is larger and the new mean depends more on the EM mean, whereas if $w_{g,1}$ is small, α_g is smaller and the new mean depends more on the UBM mean.* This can therefore be thought of as a form of Bayesian adaptation with the UBM mean as the prior mean. The more sample data associated with a Gaussian component the closer its posterior mean will be to the sample mean. Increasing τ gives globally greater weight to the UBM means (the value for τ used in Reynolds et al., 2000, was 16).

Figure 20.5 shows an example of a two-dimensional UBM and a MAP adapted known-speaker GMM.

20.4.3 Calculating a Score

Assume that a multivariate UBM population model was trained on a sample of data from the relevant population, and has mean vectors μ_{r_j}, covariance matrices Σ_{r_j}, and weights w_{r_j}, $j \in \{1\ldots G\}$. Assume that a multivariate GMM known-speaker model was trained on data from the known speaker, and has mean vectors μ_{k_j}, covariance matrices Σ_{k_j}, and weights w_{k_j}, $j \in \{1\ldots G\}$. Also, assume that the data from the questioned-speaker recording consists of N_q feature vectors: x_{q_i}, $i \in \{1\ldots N_q\}$.

To calculate a likelihood ratio, $\Lambda_{q_i,k}$, for a single feature vector from the questioned-speaker recording, x_{q_i}, the likelihood of the known-speaker model is evaluated given that feature vector, the likelihood of the population model is evaluated given that feature

---

* If only the means are adapted, the "new" weights are only used for this calculation, and it is actually the old UBM weights that are used for the speaker model.

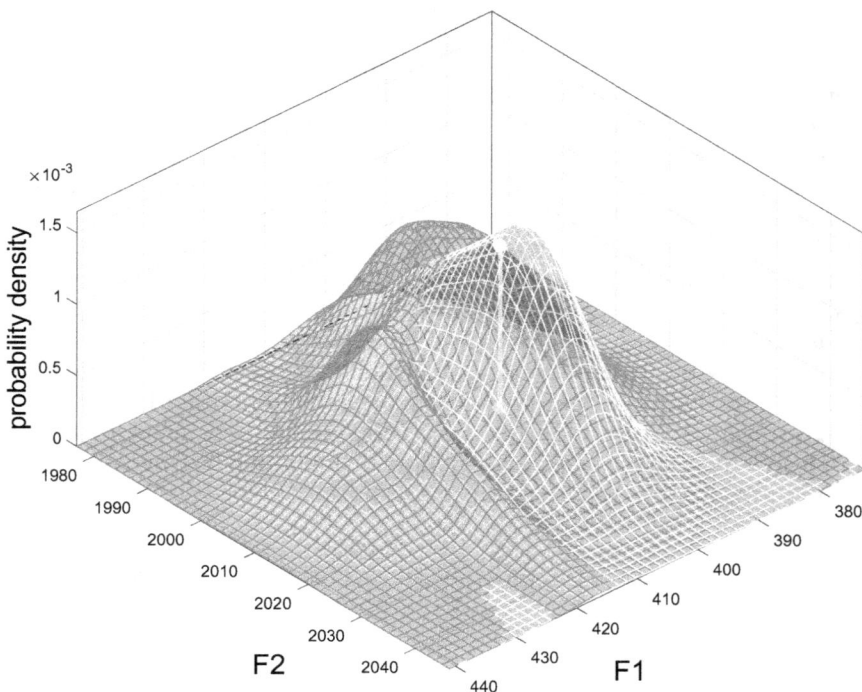

FIGURE 20.5
Example of using a GMM-UBM model to calculate a likelihood ratio for a single questioned-speaker-recording feature vector. This example is based on artificial two-dimensional data generated for illustrative purposes. The UBM is represented by the surface drawn with the darker mesh and the MAP adapted known-speaker GMM is represented by the surface drawn with the lighter mesh. The $x - y$ location of the vertical line indicates a questioned-speaker-recording feature vector value at which the likelihoods of the UBM and the known-speaker GMM are being evaluated (the likelihoods are given by the z values of the intersections of the vertical line with each of the surfaces).

vector, and the former is divided by the latter, see Equation 20.7 and a graphical example in Figure 20.5.

$$\Lambda_{q_i,k} = \frac{\sum_{j=1}^{G} w_{k_j} f\left(x_{q_i} \mid \boldsymbol{\mu}_{k_j}, \boldsymbol{\Sigma}_{k_j}\right)}{\sum_{j=1}^{G} w_{r_j} f\left(x_{q_i} \mid \boldsymbol{\mu}_{r_j}, \boldsymbol{\Sigma}_{r_j}\right)} \tag{20.7}$$

Note, however, that if a feature vector is extracted every 10 ms, then there will be 100 feature vectors for every second of speech in the questioned-speaker recording. The total number of feature vectors, N_q, from the questioned speaker recording may be in the thousands or tens of thousands. We do not want to report a likelihood ratio at the first 10-ms mark, a likelihood ratio at the second 10-ms mark, etc. We want to report a single value quantifying the strength of evidence associated with the whole of the questioned-speaker speech. Our next step toward this is to calculate the mean of the per-feature-vector log likelihood ratios, as in Equation 20.8.

$$S_{q,k} = \frac{1}{N_q} \sum_{i=1}^{N_q} \log\left(\Lambda_{q_i,k}\right) \tag{20.8}$$

We will call the mean of the per-feature-vector log likelihood ratios, $S_{q,k}$, a *score*. We will not call it a likelihood ratio. Multiplying all the per-feature-vector likelihood ratios together, or adding all the per-feature-vector log likelihood ratios together, is known as naïve Bayes fusion. It is naïve because it assumes there is no correlation between the likelihood-ratio values being combined. So that score values are not heavily dependent on the duration of the questioned-speaker recording, a score is calculated as the mean of the per-feature-vector log likelihood-ratio values rather than as their sum, but this still ignores correlation. In fact, the likelihood-ratio values come from a series of feature vectors taken from adjacent frames of speech recording with 50% overlap between adjacent frames (see Section 20.2.1). Substantial correlation between the frame-by-frame likelihood-ratio values is therefore expected. In addition, in training the UBM and the GMM speaker models we have estimated a large number of parameter values, e.g., 42 dimensions $\times$ (1024 means + 1024 variances) + 1024 $-$ 1 weights = 87,039 parameter values. Unless we had an extremely large amount of data, those estimates may be poor. Thus, we are not safe to treat the value of $S_{q,k}$ as an appropriate answer to the question posed by the same-speaker and different-speaker hypotheses in the case. To fix this problem, we will implement an additional step: score-to-likelihood-ratio conversion (aka calibration). This is the topic of Section 20.7 below – some readers may wish to skip directly to Section 20.7 and return to Sections 20.5 and 20.6 later.

20.4.4 Remarks Regarding UBM Training Data

The UBM is the model which represents the relevant population. The data used to train the UBM should therefore be a sample that is representative of the relevant population in the case. In addition, the training data should reflect the speaking style and recording conditions of the known-speaker recording. Any mismatch between the questioned-speaker data and the data used to train the known-speaker model, the model in the numerator of the likelihood ratio, will then be the same as the mismatch between the questioned-speaker data and the data used to train the population model, the model in the denominator of the likelihood ratio. Feature-level mismatch compensation techniques cannot be assumed to be 100% effective, and a difference in the conditions of the data used to train the model in the numerator and the model in the denominator would be expected to bias the calculated value of the score (see Morrison, 2018b).

20.5 i-Vector PLDA

20.5.1 i-Vectors

An *i-vector* is a single vector representing the speaker information extracted from a single recording. The "i" stands for "identity." The lengths of i-vectors extracted from different recordings are the same irrespective of the lengths of the recordings. i-Vectors are described in: Kenny et al. (2005), Matrouf et al. (2007), Dehak et al. (2011), Matějka et al. (2011), Bousquet et al. (2013).

Below we describe a set of procedures for calculating i-vectors that is one of the sets of procedures tested in Bousquet et al. (2013). These are not the most commonly used procedures, but are approximately equivalent to the more commonly used procedures and easier to explain and understand. Further below, we give a brief overview of the more commonly used procedures, with additional details provided in Appendix A.

To generate an i-vector from a speech recording, we begin by training a UBM on feature vectors extracted from recordings of a large number of speakers under a variety of recording conditions. The standard approach is to use a very large diverse set of speakers in a diverse range of recording conditions. Ideally, the training data should include multiple recordings from each speaker, the multiple recordings including different recording conditions. In the standard approach the training data for the UBM do not represent the case-specific relevant population or the case-specific conditions.*

After training the UBM, the next step is to train a GMM for each recording in a set of recordings that are representative of the relevant population for the case and that reflect the recording conditions of the questioned- and known-speaker recordings in the case (when the amount of case-relevant data is small, domain adaptation may be used, see García-Romero and McCree, 2014). The GMM for each recording is trained using mean-only MAP adaptation from the UBM. Since each GMM is adapted from the same UBM, the mean vectors in different GMMs have a parallel structure (they lie in the same vector space), e.g., the first mean of the mean vector of the first component in the GMM for one recording is parallel to the first mean of the mean vector of the first component in the GMM for another recording because they were both adapted from the first mean of the mean vector of the first component in the UBM, *mutatis mutandis* for every mean in the mean vector of every component. We take all the mean vectors from all the components in the GMM, and concatenate them to form a *supervector*. For example, if we have 42 dimensional features, hence 42 dimensional mean vectors, we take the mean vector from the first component, the mean vector from the second component, and concatenate them to form a vector that has $42 + 42 = 84$ dimensions. We then concatenate the latter vector with the mean vector from the third component to form a vector that has $84 + 42 = 126$ dimensions. We continue until we have concatenated the mean vectors from all the components. With 1024 components, the supervector has 43,008 dimensions. We generate one supervector for each recording of each speaker.

Next, we reduce the number of dimensions using *principal component analysis* (PCA). PCA finds new dimensions which are linear combinations of the original dimensions such that the first PCA dimension accounts for the largest amount of variance in the training data, the second PCA dimension accounts for the largest remaining amount of variance in the training data after the first dimension is removed, the third PCA dimension accounts for the largest remaining amount of variance in the training data after the first and second dimensions are removed, etc. The number of dimensions is reduced by only using the first few PCA dimensions. These dimensions capture most of the variance in the data. Figure 20.6 shows an example of a reduction from two original dimensions to one PCA dimension. The PCA dimension is in the direction of maximum variance in the original two-dimensional space. (The initial development of PCA is credited to Pearson, 1901.)

The supervectors are reduced from tens of thousands of dimensions to a much smaller number of dimensions, e.g., 400 dimensions. The reduced-dimension vectors are the i-vectors. The PCA dimensions are in the directions of maximum variance in the original space, irrespective of whether the variance is primarily due to speaker or condition (or other) differences, hence the result is sometimes called the *total variability space*.

---

* Enzinger (2016) ch. 4, obtained favorable results when a relatively small amount of case-specific data were used for training the UBM, i.e., data that represented the relevant population for the case and reflected the conditions of the questioned- and known-speaker recordings in the case. One should be cautious, however, because training with small amounts of data may give unstable results.

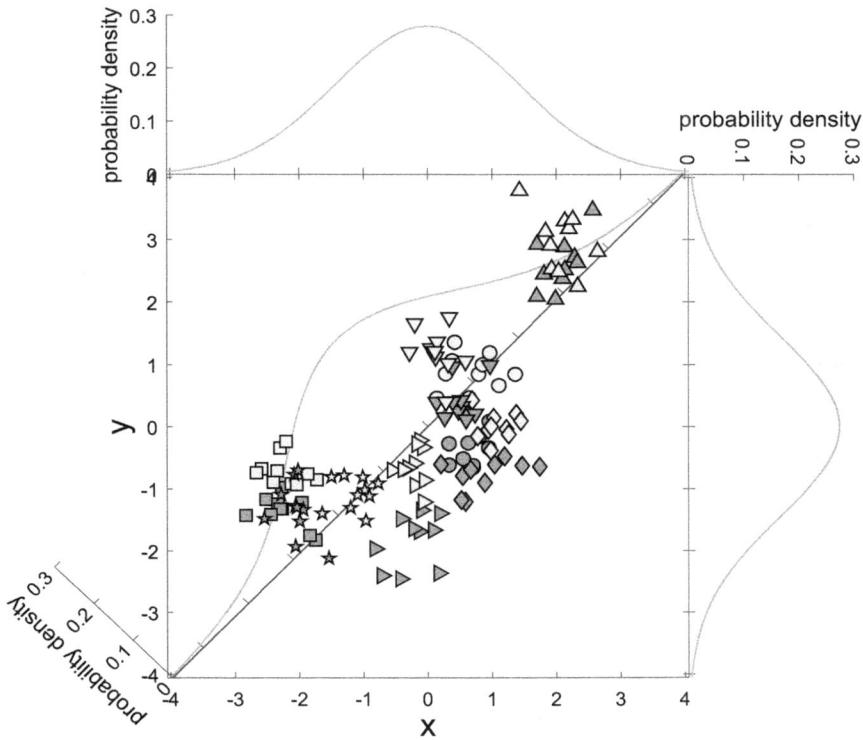

FIGURE 20.6
Example of using PCA to reduce the data from two dimensions to one dimension. The example uses artificial data created for illustrative purposes. Different shaped symbols represent different speakers and the two different intensities of shading represent two different conditions, e.g., a questioned-speaker-recording condition and a known-speaker-recording condition. The PCA dimension (represented by the oblique axis) is in the direction of maximum variance in the original $x - y$ space.

Conceptually, i-vectors result from mapping supervectors onto a linear subspace with a lower number of dimensions. This is represented in Equation 20.9, in which s is a recording-specific supervector, m is the supervector for the UBM, and v is the i-vector for the specific recording. $\mathbf{T}$ is a low-rank matrix representing the linear subspace of the supervector space in which the i-vectors reside. The $\mathbf{T}$ matrix is designed such that the i-vectors have a few hundred dimensions rather than the tens of thousands of dimensions of the supervectors.

$$s = m + Tv \tag{20.9}$$

The more commonly used procedures for extracting i-vectors do not use PCA for dimension reduction, but instead a form of factor analysis. The $\mathbf{T}$ matrix is trained using an iterative maximum likelihood technique (the EM algorithm). Rather than first adapting a GMM from the UBM, then concatenating a supervector, and then reducing its dimensions, a more direct procedure is used to map from feature vectors to i-vectors. The first step is to calculate the *Baum-Welch statistics* for each recording. The 0th order Baum-Welch statistics are based on the probability that a feature vector of a recording j would belong to component g of the UBM (like the responsibilities in the EM algorithm). For each feature vector, one value is calculated for each of the G components, thus we have a matrix with N_j columns (as many columns as there are feature vectors in the recording) and G rows of responsibilities. We then sum over the columns to create a set of G 0th order statistics for recording j. The 1st order Baum-Welch statistics are based on the deviations of the feature vectors of a recording from the mean vector of each component g of the UBM, weighted by the probabilities that the feature vectors of the recording would belong to component g (the responsibilities of component g for the feature vectors). Deviations are calculated on a per-feature-dimension basis, hence for each of the N_j feature vectors, M values are calculated for each of the G components (where M is the number of feature dimensions). We thus have an N_j by M by G matrix. We sum over the first of these dimensions to arrive at a set of 1st order statistics for recording j, consisting of G vectors each of length M. The **T** matrix is trained using the 0th and 1st order Baum-Welch statistics from a set of training recordings, and an i-vector for a given recording can then be directly calculated using the **T** matrix and the 0th and 1st order Baum-Welch statistics from that recording. We provide mathematical details in Appendix A. For greater detail, see the references cited at the beginning of this section.

i-vectors are usually "whitened," i.e., subjected to radial Gaussianization and length normalization, so that they conform better to the assumptions of subsequent statistical modeling procedures (see García-Romero and Espy-Wilson, 2009). After whitening only the direction of an i-vector away from the origin of the i-vector space is relevant (whitened i-vectors define points that lie on the hypersurface of a hypersphere).

20.5.2 i-Vector Domain Mismatch Compensation (LDA)

A common i-vector domain mismatch compensation technique is known in the automatic-speaker-recognition literature as *linear discriminant analysis* (LDA). In practice, linear discriminant analysis is not actually performed (i.e., neither likelihoods, nor posterior probabilities, not classification results are output), but *canonical linear discriminant functions* (CLDFs) are calculated and used to transform and reduce the dimensionality of the i-vectors (see Klecka, 1980; Fisher, 1936, is credited with the initial development of discriminant analysis and discriminant functions). In contrast to PCA which finds new dimensions that maximize the total variance, LDA finds dimensions that maximize the ratio of between-category variance versus within-category variance. CLDFs are trained, using i-vectors from training data that include multiple recordings of each speaker, with different recordings in different conditions. Each speaker is treated as a category, thus the CLDFs maximize the ratio of between-speaker variance to within-speaker variance, much of the within-speaker variance being due to variability in speaking styles and recording conditions. Only the first few CLDFs are used so as to primarily capture between-speaker variance. The CLDFs are used to transform the i-vectors into a new smaller set of dimensions, e.g., 50 dimensions rather than 400.

Figure 20.7 shows an example of CLDF reduction from two dimensions to one dimension. The same data are used for Figures 20.6 and 20.7 in order to illustrate the differences between the PCA and CLDF procedures. CLDFs could be calculated for supervectors rather than i-vectors, but there can be problems with attempting to train CLDFs in such a high-dimensional space.

20.5.3 PLDA

An i-vector approach produces a single i-vector for each recording. Hence, unlike in the case of GMM-UBM, there is a single vector for the known-speaker recording rather

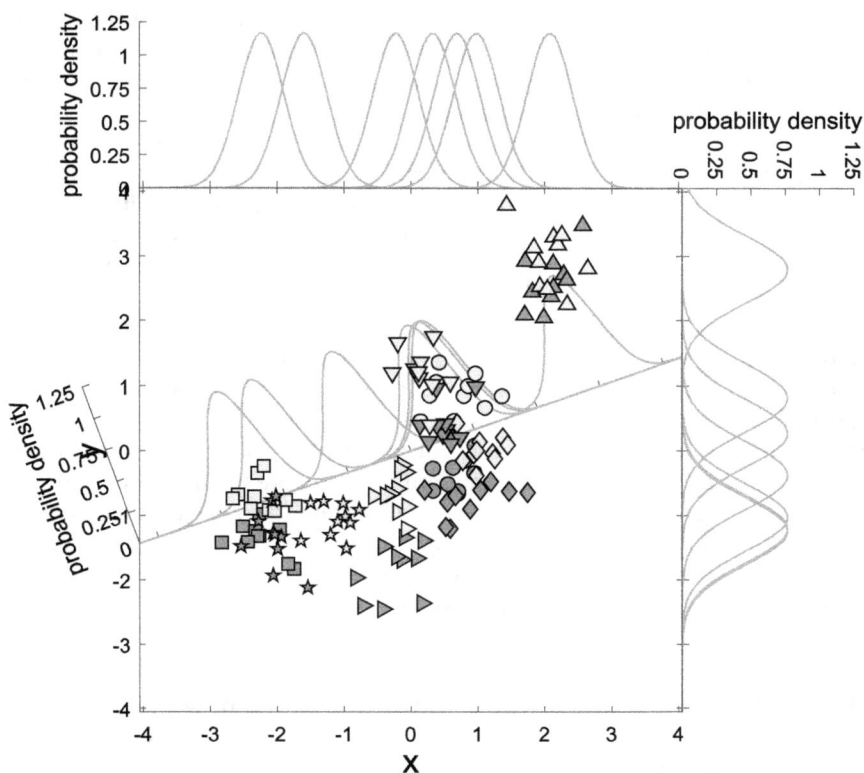

FIGURE 20.7
Example of using CLDF for mismatch compensation and to reduce the data from two dimensions to one dimension. The example uses artificial data created for illustrative purposes (the same data as used for Figure 20.6). Different shaped symbols represent different speakers and the two different intensities of shading represent two different conditions, e.g., a questioned-speaker-recording condition and a known-speaker-recording condition. The CLDF dimension (represented by the oblique axis) is in the direction in the original $x - y$ space that has the maximum ratio of between- versus within-speaker variance.

than a set of vectors that could be used to train a known-speaker model. A different modeling approach is therefore used to calculate likelihood ratios from i-vectors. In the automatic-speaker-recognition literature, this is known as *probabilistic linear discriminant analysis* (PLDA). PLDA is described in: Prince and Elder (2007), Kenny (2010), Brümmer and de Villiers (2010), Sizov et al. (2014).

PLDA is a *common-source model*,* i.e., it does not build a model for the specific known-speaker recording in the case, but instead answers the following two-part question:

> What would be the probability of obtaining the feature vectors of the questioned-speaker recording and the feature vectors of the known-speaker recording if the questioned speaker and the known speaker were the same speaker, a speaker selected at random from the relevant population?

> versus

---

* For a discussion of the distinction between *specific-source* and *common-source* models, see Ommen and Saunders (2018).

What would be the probability of obtaining the feature vectors of the questioned-speaker recording and the feature vectors of the known-speaker recording if the questioned speaker and the known speaker were different speakers, each selected at random from the relevant population?*

Conceptually, the answer to the first part of the question provides the numerator for the likelihood ratio and the answer to the second part provides the denominator.

Assume that we have a sample of N speakers from the relevant population with multiple recordings of each speaker, and that some recordings of each sample speaker are in the questioned-speaker condition and others are in the known-speaker condition. We calculate an i-vector for each recording and apply i-vector domain mismatch compensation. Imagine that we also have a known-speaker recording and a questioned-speaker recording, and we calculate an i-vector for each of the two recordings and apply i-vector domain mismatch compensation, resulting in i-vectors v_k and v_q respectively.

In practice, we fit a multivariate model, but to simplify the explanation below we will describe a univariate model. We assume that both the within-speaker and between-speaker distributions are Gaussian. We also assume equal within-speaker variance for all speakers.

If v_k and v_q came from the same speaker ($q = k$), they would both come from a within-speaker distribution with a mean μ_k and variance σ_k^2. If they came from different speakers ($q \neq k$), one would come from a within-speaker distribution with a mean μ_k and variance σ_k^2, and the other from a within-speaker distribution with a mean μ_q and variance σ_q^2. We want to calculate a likelihood ratio as shown in Equations 20.10–20.12, i.e., the joint likelihood for v_k and v_q if they both came from the same within-speaker distribution versus the joint likelihood for v_k and v_q if each came from a different within-speaker distribution. If we assume equal variance for all speakers, including the questioned- and known-speaker, we can pool the i-vectors from all speakers in the sample of the population, calculate the pooled within-speaker variance, $\hat{\sigma}_w^2$, and then substitute that for σ_k^2 and σ_q^2, as in Equation 20.12.

$$\Lambda_{q,k} = \frac{f(v_q, v_k \mid H_{q=k})}{f(v_q, v_k \mid H_{q\neq k})} \tag{20.10}$$

$$\Lambda_{q,k} = \frac{f\left(v_q \mid \mu_k, \sigma_k^2\right) f\left(v_k \mid \mu_k, \sigma_k^2\right)}{f\left(v_q \mid \mu_q, \sigma_q^2\right) f\left(v_k \mid \mu_k, \sigma_k^2\right)} \tag{20.11}$$

$$\Lambda_{q,k} = \frac{f\left(v_q \mid \mu_k, \hat{\sigma}_w^2\right) f\left(v_k \mid \mu_k, \hat{\sigma}_w^2\right)}{f\left(v_q \mid \mu_q, \hat{\sigma}_w^2\right) f\left(v_k \mid \mu_k, \hat{\sigma}_w^2\right)} \tag{20.12}$$

We do not know the values for μ_k and μ_q. For each speaker, r, in the sample of the relevant population, let us calculate the mean of the speaker's i-vectors: $\hat{\mu}_r$, $r \in \{1...N\}$. Since we do not know the value for μ_k, let us randomly pick one of the speakers from the sample of the population and use their mean $\hat{\mu}_i$, and since we do not know the value for μ_q, let us randomly pick another of the speakers from the population and use their mean $\hat{\mu}_j$. We

---

* An objection may be raised that the known speaker was not selected at random, but on the basis of other evidence. The forensic practitioner's task, however, is to assess the strength of the particular evidence they have been asked to analyze – they should not consider the other evidence in the case, considering all the evidence is the role of the trier of fact. For the forensic practitioner's task, it is therefore acceptable to make the statistical assumption that the known speaker was selected at random from the relevant population.

impose the condition that $i \neq j$. We can now calculate a likelihood ratio as in Equation 20.13.

$$\Lambda_{q,k,i,j} = \frac{f\left(v_q \mid \hat{\mu}_i, \hat{\sigma}_w^2\right) f\left(v_k \mid \hat{\mu}_i, \hat{\sigma}_w^2\right)}{f\left(v_q \mid \hat{\mu}_j, \hat{\sigma}_w^2\right) f\left(v_k \mid \hat{\mu}_i, \hat{\sigma}_w^2\right)} \tag{20.13}$$

If we pick two more speakers at random, we will calculate a different value for the likelihood ratio. Let us imagine that we pick lots of random pairs of speakers and consider the average likelihood-ratio value.

If v_k and v_q are far apart, then the numerator of the likelihood ratio will always be relatively small because μ_i cannot be close to both v_k and v_q. The denominator of the likelihood ratio will sometimes be relatively high because it is possible for μ_i to be close to v_k and μ_j to be close to v_q. On average, the denominator will be larger than the numerator and the average value of the likelihood ratio will therefore be low.

If v_k and v_q are close to each other and atypical with respect to the population, i.e., both out on the same tail of the distribution, then the value of the numerator of the likelihood ratio will usually be small because the probability of μ_i being close to v_k and v_q is small, but when μ_i is close to v_k and v_q the value of the numerator will be large. On average the value of the denominator will be even lower because the probability of both μ_i and μ_j being out on the same tail of the distribution and therefore one being close to v_k and the other being close to v_q will be lower than the probability of only μ_i being out on that tail of the distribution. On average, the numerator will be larger than the denominator and the average value of the likelihood ratio will therefore be high.

If v_k and v_q are close to each other but typical with respect to the population, i.e., both in the middle of the distribution, then the value of the numerator of the likelihood ratio will usually be large because the probability of μ_i being in the middle of the distribution and therefore being close to v_k and v_q will be large, but the value of the denominator will also usually be large because the probability of both μ_i and μ_j being in the middle of the distribution and therefore one being close to v_k and the other being close to v_q will also be large. On average, the values of numerator and denominator will be about the same and the average value of the likelihood ratio will therefore be close to 1.

In general, the average calculated value of the likelihood ratio will reflect how similar v_k and v_q are to each other, and how typical they are with respect to the sample of the population.

Rather than selecting pairs of speakers at random, we could systematically go through all possible combinations of speakers in our sample of the population and calculate the mean likelihood ratio, as in Equation 20.14 (the sum in parenthesis in the denominator is from 1 to $N - 1$ since $j = i$ is skipped).

$$\Lambda_{q,k} = \frac{\frac{1}{N} \sum_{i=1}^{N} \left(f\left(v_q \mid \hat{\mu}_i, \hat{\sigma}_w^2\right) f\left(v_k \mid \hat{\mu}_i, \hat{\sigma}_w^2\right) \right)}{\frac{1}{N} \sum_{i=1}^{N} \left(f\left(v_k \mid \hat{\mu}_i, \hat{\sigma}_w^2\right) \frac{1}{N-1} \sum_{j=1}^{N-1} f\left(v_q \mid \hat{\mu}_j, \hat{\sigma}_w^2\right), j \neq i \right)} \tag{20.14}$$

Rather than a discrete summation using the mean of each speaker in the sample of the relevant population, we can model the distribution of the speaker means and use integration. The speaker means are treated as *nuisance parameters* and "integrated out." We calculate the mean and the variance of the speaker means: $\hat{\mu}_b$ and $\hat{\sigma}_b^2$ – this gives us the between-speaker distribution. Converting from discrete summation to integration, we arrive at Equation 20.15. For Gaussian distributions, there is a closed-form solution for the integrals, as given in Equation 20.16. Equation 20.16 uses bivariate Gaussian distributions. The covariance matrix in the numerator has positive off-diagonal elements, hence if the values of v_k and v_q are similar, e.g., both high or both low, the likelihood is greater. In contrast, the covariance matrix in the denominator is diagonal only, hence the likelihood is not greater if similarity is greater, e.g., the likelihood would be the same if both v_k and v_q are high or if one is high and the other is low. The smaller the within-speaker variance, $\hat{\sigma}_w^2$, the larger the off-diagonal elements, $\hat{\sigma}_b^2$, will be relative to the on-diagonal elements, $\hat{\sigma}_w^2 + \hat{\sigma}_b^2$, and the greater the effect of similarity on the value of the likelihood ratio.

$$\Lambda_{q,k} = \frac{\int f\left(v_q \mid \mu_r, \hat{\sigma}_w^2\right) f\left(v_k \mid \mu_r, \hat{\sigma}_w^2\right) f\left(\mu_r \mid \hat{\mu}_b, \hat{\sigma}_b^2\right) d\mu_r}{\int f\left(v_q \mid \mu_r, \hat{\sigma}_w^2\right) f\left(\mu_r \mid \hat{\mu}_b, \hat{\sigma}_b^2\right) d\mu_r \int f\left(v_k \mid \mu_r, \hat{\sigma}_w^2\right) f\left(\mu_r \mid \hat{\mu}_b, \hat{\sigma}_b^2\right) d\mu_r} \tag{20.15}$$

$$\Lambda_{q,k} = \frac{f\left(\begin{bmatrix} v_q \\ v_k \end{bmatrix} \mid \begin{bmatrix} \hat{\mu}_b \\ \hat{\mu}_b \end{bmatrix}, \begin{bmatrix} \hat{\sigma}_w^2 + \hat{\sigma}_b^2 & \hat{\sigma}_b^2 \\ \hat{\sigma}_b^2 & \hat{\sigma}_w^2 + \hat{\sigma}_b^2 \end{bmatrix}\right)}{f\left(v_q \mid \hat{\mu}_b, \hat{\sigma}_w^2 + \hat{\sigma}_b^2\right) f\left(v_k \mid \hat{\mu}_b, \hat{\sigma}_w^2 + \hat{\sigma}_b^2\right)}$$

$$= \frac{f\left(\begin{bmatrix} v_q \\ v_k \end{bmatrix} \mid \begin{bmatrix} \hat{\mu}_b \\ \hat{\mu}_b \end{bmatrix}, \begin{bmatrix} \hat{\sigma}_w^2 + \hat{\sigma}_b^2 & \hat{\sigma}_b^2 \\ \hat{\sigma}_b^2 & \hat{\sigma}_w^2 + \hat{\sigma}_b^2 \end{bmatrix}\right)}{f\left(\begin{bmatrix} v_q \\ v_k \end{bmatrix} \mid \begin{bmatrix} \hat{\mu}_b \\ \hat{\mu}_b \end{bmatrix}, \begin{bmatrix} \hat{\sigma}_w^2 + \hat{\sigma}_b^2 & 0 \\ 0 & \hat{\sigma}_w^2 + \hat{\sigma}_b^2 \end{bmatrix}\right)} \tag{20.16}$$

Example PLDA numerator and denominator distributions are shown in Figure 20.8: values of $\hat{\mu}_b = 0$, $\hat{\sigma}_w^2 = 0.25$, $\hat{\sigma}_b^2 = 1$, $v_k = -1.5$, and $v_q = -1$ were selected for illustrative purposes only; the calculated likelihood-ratio value is 2.4.

To calculate the likelihood ratio for actual i-vectors, which are multivariate vectors, we use Equation 20.17, which is the multivariate equivalent of Equation 20.16 ($\hat{\Sigma}_w$ and $\hat{\Sigma}_b$ are within- and between-speaker covariance matrices respectively).

$$\Lambda_{q,k} = \frac{f\left(\begin{bmatrix} v_q \\ v_k \end{bmatrix} \mid \begin{bmatrix} \hat{\mu}_b \\ \hat{\mu}_b \end{bmatrix}, \begin{bmatrix} \hat{\Sigma}_w + \hat{\Sigma}_b & \hat{\Sigma}_b \\ \hat{\Sigma}_b & \hat{\Sigma}_w + \hat{\Sigma}_b \end{bmatrix}\right)}{f\left(v_q \mid \hat{\mu}_b, \hat{\Sigma}_w + \hat{\Sigma}_b\right) f\left(v_k \mid \hat{\mu}_b, \hat{\Sigma}_w + \hat{\Sigma}_b\right)} \tag{20.17}$$

In practice, the output of PLDA is treated as a score and subjected to score-to-likelihood-ratio conversion, see Section 20.7.

What we have described above is known as the *two-covariance* version of PLDA. There are at least two other versions of PLDA (that Sizov et al., 2014, label *standard* and *simplified*). The latter are more complex than the two-covariance version in that they include dimension reduction to work in speaker-variability and session-variability subspaces. Since the latter include subspace modeling, they are not usually preceded by CLDF transformation, whereas since the two-covariance version does not include subspace modeling it should be preceded by CLDF transformation. For further details of the two-covariance version of PLDA and for details of the standard and simplified versions of PLDA, see the references cited at the beginning of this section. For a comparison of the three, see Sizov et al. (2014).

20.6 DNN-Based Systems

Artificial neural networks are machine learning models that consist of *nodes* (neurons) that are organized in *layers*, and *connections* between the nodes (synapses). An example of the

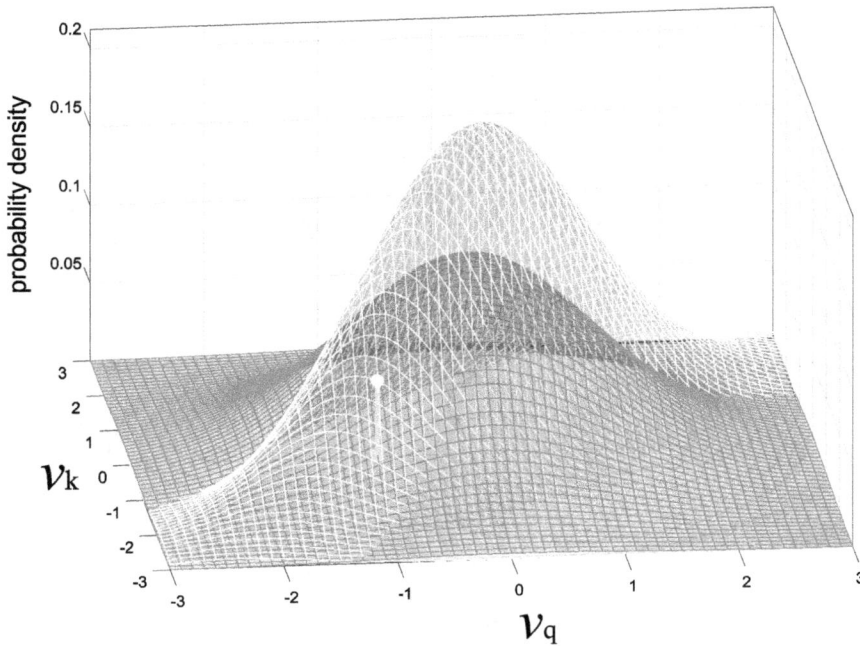

FIGURE 20.8
Example of using a PLDA model to calculate a score for a pair of i-vectors. The numerator of the PLDA model is represented by the surface drawn with the lighter mesh and the denominator is represented by the surface drawn with the darker mesh. The $x - y$ location of the vertical line indicates the values of the i-vectors v_q and v_k (extracted from the questioned- and known -speaker recordings respectively) at which the likelihoods of numerator and denominator of the PLDA model are being evaluated (the likelihoods are given by the z values of the intersections of the vertical line with each of the surfaces).

architecture of an artificial neural network is shown in Figure 20.9. This is an example of a standard feedforward fully connected architecture, i.e., all units in a given layer are connected to all the units in the preceding layer and all the units in the following layer. The example has an input layer, two hidden layers, and an output layer. A hidden layer is any layer other than the input or output layer. The number of hidden layers is a design choice (this example happens to have two hidden layers). An artificial neural network with more than one hidden layer is known as a *deep neural network* (DNN). A value is presented at each input node, e.g., the set of input nodes accepts a feature vector, x. Each input node then has an activation level, h_{0,n_a}, corresponding to an x value. Each input node is connected to each node in the first hidden layer via connections each of which is given a weight, w_{l_{j-1},n_a,l_j,n_b}, where l_{j-1} and l_j index the layers (in this case the input layer, $l_{j-1} = 0$, and the first hidden layer, $l_j = 1$), and n_a and n_b index the particular nodes that are connected. Each node in the first hidden layer then has an activation level, h_{l_j,n_b}, that is a function, φ, of the weighted sum of the values presented to the input layer, see Equation 20.18 (b_{l_j,n_b} is a node-specific bias term). A non-linear function, e.g., a sigmoidal function, is usually used.

$$h_{l_j,n_b} = \varphi \left(b_{l_j,n_b} + \sum_{n_a=1}^{N_a} w_{l_{j-1},n_a,l_j,n_b} h_{l_{j-1},n_a} \right) \tag{20.18}$$

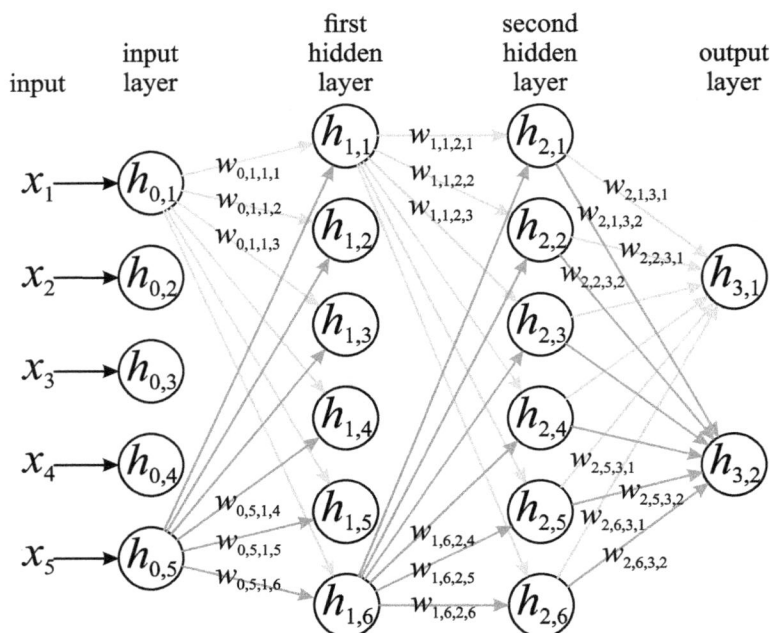

FIGURE 20.9

Simplified example of the architecture of a feed-forward DNN consisting of an input layer, two hidden layers, and an output layer. Not all connection between nodes have been drawn, and weights have been indicated for only a few connections.

The activation levels of nodes in higher layers are in turn based on weighted sums of the activation levels of the nodes in the immediately preceding layer.

Usually, the task of the network is to predict the category that the input belongs to, and each node in the output layer represents a category. For example, for an optical character recognition system, the input could be an image of a character (i.e., a letter of the alphabet, a number, or a punctuation mark), each output node would represent a particular character, and the relative level of activation of an output node would represent the strength of the model's prediction that the input image is of the character corresponding to that output node. All else being equal, to classify the image, one would select the character corresponding to the output node with the highest level of activation. The activations of the output nodes can be scaled to represent posterior probabilities.

Training an artificial neural network is the process of setting the connection weights (and node-specific biases) so as to optimize the network's performance on the classification task. A standard supervised training algorithm is *backpropagation*. This is an iterative process similar to the EM algorithm. Starting from an initial state (e.g., a random seeding of the weights), it involves presenting inputs for which the categories are known, comparing the activation levels of the output nodes with the desired activation levels given the known category of the input, and adjusting the weights so as to reduce the error, i.e., to lead to higher relative activation of the output node corresponding to the known category of the input. For more detailed introductions to artificial neural networks, including network training, see Duda et al. (2000) ch. 6, and Hastie et al. (2009) ch. 11.

Below, we briefly describe three DNN approaches: (1) *DNN senone posterior i-vector systems*, (2) *Bottleneck-feature based systems*, and (3) *DNN speaker embedding (x-vector) systems*.

All three were common in state-of-the-art automatic-speaker-recognition systems when we began writing the present chapter in 2017. By 2019 the x-vector approach had emerged as the clear winner.

20.6.1 DNN Senone Posterior i-Vector Systems

A senone posterior i-vector system can be considered a variant of an i-vector PLDA system, but using a DNN to train the UBM rather than using the EM algorithm (see García-Romero et al., 2014; Kenny et al., 2014; Lei et al., 2014). In contrast to standard i-vectors, senone posterior i-vectors are designed to explicitly capture information about how speakers pronounce speech sounds.

The input to the DNN consists of a series of feature vectors (e.g., MFCC + delta + double delta). The DNN is trained to classify the input into sequences of speech sounds, e.g., triphones. For simplicity, we present an example based on orthography rather than speech sounds: The word "forensic" contains the trigraphs "for," "ore," "ren," "ens," "nsi," and "sic." In general, the sequences of speech sounds are called *senones*. (Although there has been a shift and broadening of meaning to its current use in automatic speaker recognition, the term "senone" was originally coined in Hwang and Huang, 1992, to mean a subphonetic unit.)

Standard supervised training techniques are used to train the DNN to classify a large number of different senones (e.g., between 5000 and 10,000). A "true" category label for the senone corresponding to each feature vector is typically computed using an automatic-speech-recognition system. The DNN is trained using a large number of recordings for a large number of speakers in diverse recording conditions. The activation of each output node of the DNN is scaled to represent the posterior probability for the senone it was trained on.

The UBM consist of a GMM in which each Gaussian component corresponds to a DNN output node. Feature vectors from a new set of multiple recordings of multiple speakers are presented to the DNN. The activation of each DNN output node, g, is obtained for each feature vector, x_i, and is used as the responsibility, $\gamma_{g,i}$, in Equations 20.3–20.5 (the version of Equations 20.3–20.5 given in Section 20.4.1 was for univariate data, the corresponding multivariate version is actually used). The x_i and $\gamma_{g,i}$ provide all the information necessary to train the UBM in a single iteration.

20.6.2 Bottleneck-Feature Based Systems

In a bottleneck-feature based system (see Yaman et al., 2012; García-Romero and McCree, 2015; Lozano-Díez et al., 2016; Matějka et al., 2016) a DNN is trained in the same way as in the senone posterior i-vector system but one layer in the DNN, the *bottleneck layer*, has a substantially smaller number of nodes than the other hidden layers, e.g., 60–80 nodes in the bottleneck layer compared to 1000–1500 nodes in other layers. The bottleneck layer is designed to capture information about the phonetic content of voice recordings. The small number of nodes provides a compact representation of this information. The bottleneck layer may be between other hidden layers, or may be immediately before the output layer.

For each input feature vector (e.g., MFCC + delta + double delta), the activations of the nodes in the bottleneck layer are used as a new feature vector, a *bottleneck-feature vector*. The bottleneck-feature vectors are usually concatenated with the original MFCC + delta + double delta feature vectors and the concatenated feature vectors then used as input to a standard UBM-based i-vector PLDA system (note that the activations of the DNN's output layer are not used).

20.6.3 DNN Speaker Embedding Systems (x-Vector Systems)

A DNN speaker embedding system is described in Snyder et al. (2017), and specific values given below are those of that system (other variants are described in: Peddinti et al., 2015a,b; Snyder et al., 2018; Lee et al., 2020; Matějka et al., 2020; Villalba et al., 2020). In this context, *embeddings* are fixed-length vectors that are used instead of i-vectors as input to PLDA. These embeddings are also known as *x-vectors*. In contrast to senone posterior i-vector systems and bottleneck-feature systems that are trained to classify senones, a DNN embedding system is trained to classify speakers.

Rather than the input layer of a DNN speaker embedding system accepting a single feature vector, it is organized to accept a two-dimensional matrix in which one dimension is the elements of the MFCC vectors (that represent the frequency components of the signal at the time corresponding to each vector), and the other dimension is time. Time is discretized according to the frame shifts used when extracting the series of MFCC vectors from the voice recording. Deltas and double deltas are not included in the input feature vectors, instead each node in the input layer is connected to the MFCC vectors from 5 contiguous frames, i.e., the MFCC vectors corresponding to the frames at times $t-2$, $t-1$, t, $t+1$, $t+2$. Nodes in the first hidden layer are connected to nodes at times $t-2$, t, $t+2$ of the input layer, and nodes in the second hidden layer are connected to nodes at times $t-3$, t, $t+3$ of the first hidden layer, see Figure 20.10. The input layer ultimately spans frames $t-7$ to $t+7$.* The second hidden layer is followed by two more hidden layers that are unidimensional – their nodes are only connected to nodes at time t of the preceding layer, hence the time dimension has been collapsed. All the layers described so far are called *frame-level layers*. Layers that receive inputs from multiple times are said to *splice* those inputs together.

The next layer in the DNN is called a *statistics-pooling layer*. The whole sequence of MFCC vectors from the speech of the speaker of interest in a recording is presented to the system, the input layer sees 15 frames at a time (frames $t-7$ through $t+7$) and t is advanced one frame at a time through the entire sequence of MFCC vectors. For each node in the immediately preceding layer, the statistics-pooling layer has two nodes, one calculates the mean of the activation of the preceding layer's node over all t, and the other calculates the standard deviation of the activation of the preceding layer's node over all t. The statistics-pooling layer provides a fixed-length vector irrespective of the length of the recording.

The values of the statistics-pooling layer are used as an input vector to a set of three more layers, including the output layer. These are called *segment-level layers*.† The second of these layers has fewer nodes than the first, e.g., 300 versus 512 nodes, hence the second is a bottleneck. The activations of either the first layer or the second layer (or both) can be used as an embedding (an x-vector). As with i-vectors, CLDF for mismatch compensation and dimension reduction can be applied before PLDA.

Training the DNN makes use of recordings of several thousand speakers. Multiple recordings of each speaker are used, with different recordings reflecting different

---

* The weights for the connections from the input layer to the nodes in the first hidden layer are not independent of each other. The weights on different connections entering a first-layer node generally differ from each other, but the same set of weights are used for the set of connections entering each and every first-layer node.
† In the automatic-speaker-recognition literature, the term "segment" refers to all the speech of the speaker of interest in the whole recording. It is the latter meaning which is intended, as opposed to the meaning of "segment" in phonetics, which is an individual speech sound, a realization of a phoneme.

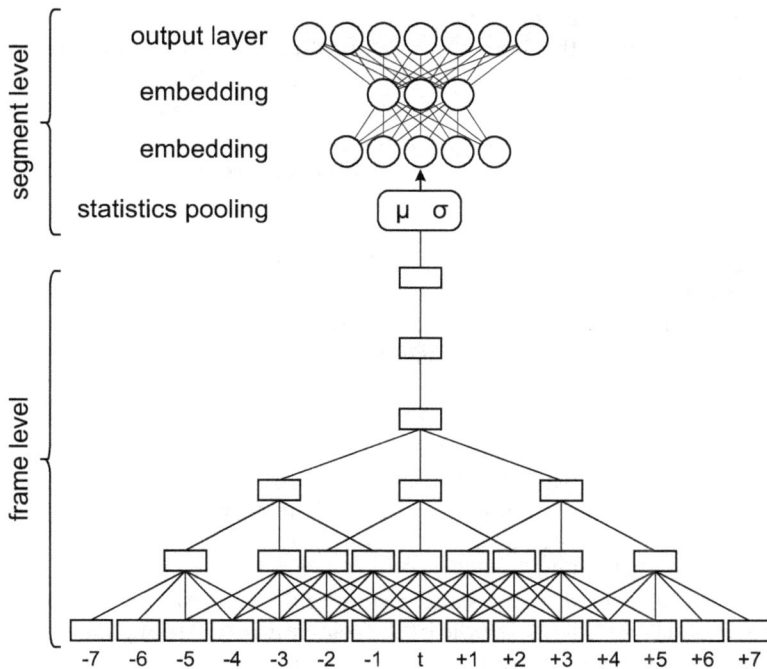

FIGURE 20.10

Illustration of the architecture of a DNN embedding system. Only the time-dimension is shown for the frame level (the frequency dimension is not shown). The input layer (at the bottom) spans $t - 7$ through $t + 7$. The time dimension is collapsed by the third hidden layer of the frame level. The statistics-pooling layer calculates the mean and standard deviation of each node in the preceding layer when the whole of the speech of interest in a recording is passed by the input layer. The frequency dimension is shown for the segment level.

conditions. In addition to recordings genuinely recorded under different conditions, the range of variability is often increased by entering the same recordings multiple times after they have undergone processing to simulate multiple other conditions, e.g., addition of different types and levels of noise, lossy compression, and reverberation (see McLaren et al., 2018). The aim is for the DNN to learn the speaker-dependent properties of the input and to learn to ignore the condition-dependent properties. The activations of the nodes in the output layer indicate posterior probabilities for the speakers in the training set. The ultimate aim, however, is not to build a system that classifies those particular speakers, hence the activations of the DNN's output layer are not used as input to the PLDA. Instead, the aim is to generate fixed-length vectors that characterize the different speech properties of different speakers in general. Those are the embeddings (x-vectors), and it is the embeddings (x-vectors) that are used as input to the PLDA.

20.7 Score-to-Likelihood-Ratio Conversion (Calibration)

Score-to-likelihood-ratio conversion (aka *calibration*), particularly using logistic regression, is described in: Pigeon et al. (2000), Ramos Castro (2007), González-Rodríguez et al. (2007), Morrison (2013), Morrison and Poh (2018), Ferrer et al. (2019).

Scores are similar to likelihood ratios in that, for a specific-source model, they take account of the similarity of the voice on the questioned-speaker recording with respect to the voice of the known speaker and the typicality of the voice on the questioned-speaker recording with respect to the voices of speakers in the relevant population, or, for a common-source model, they take account of the similarity of the voices on the questioned- and known-speaker recordings and their typicality with respect to the voices of speakers in the relevant population.*

In forensic voice comparison, scores that take account of both similarity and typicality are generated using models such as GMM-UBM and PLDA. Because of violations of modeling assumptions or lack of sufficient training data, however, scores are not interpretable as meaningful likelihood-ratio values answering the question posed by the same- and different-speaker hypotheses in the case. An additional step is needed to convert scores to interpretable likelihood ratios. Another way of looking at the problem is that scores are likelihood ratios, but their values are not interpretable because they are not calibrated. The score-to-likelihood-ratio-conversion process is therefore also known as calibration, and for brevity we will henceforth tend to use the latter term.

To obtain data for training a calibration model, feature vectors from a large number of same-speaker pairs of recordings and a large number of different-speaker pairs of recordings are input to the GMM-UBM system, the i-vector PLDA system, or the DNN-based system, and a set of *same-speaker scores* and a set of *different-speaker scores* are obtained. The training data should come from a sample of the relevant population, and should be distinct from the data that were used to train earlier models in the system. Also, one member of each pair should have conditions reflecting those of the known-speaker recording and the other should have conditions reflecting those of the questioned-speaker recording. Training using data that do not reflect the relevant population and the conditions of the case will lead to miscalibrated results and poorer performance. Mandasari et al. (2013) and Mandasari et al. (2015) illustrated that training using recordings that matched the durations of the test recordings resulted in better performance than using longer-duration training recordings, and that training using recordings whose signal to noise ratios matched those of the test recordings resulted in better performance than using training recordings whose signal to noise ratios did not match those of the test recordings.[†]

A simple score-to-likelihood-ratio conversion model makes use of two Gaussian distributions with equal variance. We calculate the mean for the *same-speaker model*, $\hat{\mu}_s$, using the same-speaker training scores, and the mean for the *different-speaker model*, $\hat{\mu}_d$, using the different-speaker training scores. We calculate a single variance, $\hat{\sigma}^2$, using data pooled from both categories, i.e., a *pooled variance*. To calculate a likelihood ratio, we first calculate a score, $S_{q,k}$, for the comparison of the voices on the questioned-speaker and known-speaker recordings. We then calculate the likelihood of the same-speaker model at the value of $S_{q,k}$, the likelihood of the different-speaker model at the value of $S_{q,k}$, and divide the former by the latter, see Equation 20.19.

$$\Lambda_{q,k} = \frac{f\left(S_{q,k} \mid \hat{\mu}_s, \hat{\sigma}^2\right)}{f\left(S_{q,k} \mid \hat{\mu}_d, \hat{\sigma}^2\right)} \tag{20.19}$$

* Scores that are based only on similarity do not account for typicality with respect to the relevant population and this cannot be remedied at the score-to-likelihood-ratio-conversion stage (see Morrison and Enzinger, 2018).
† The aim of Mandasari et al. (2013) and Mandasari et al. (2015) was to find automatic procedures to address this problem rather than relying on a human expert to select data on a case-by-case basis; see also McLaren et al. (2014).

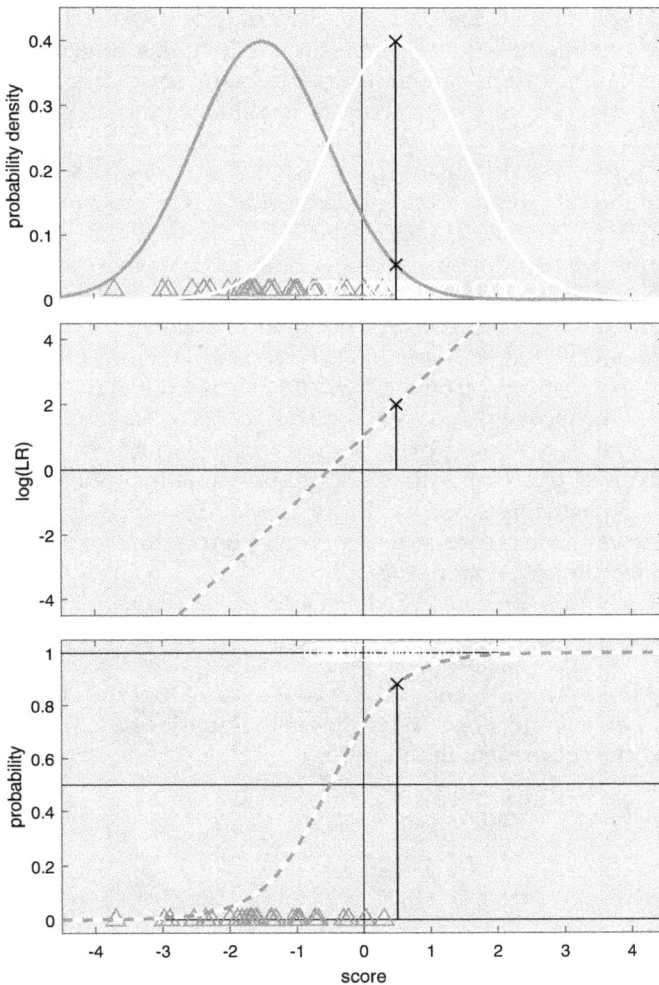

FIGURE 20.11
Example of using linear discriminant analysis or logistic regression to convert a score to a likelihood ratio. The example uses artificial data created for illustrative purposes. Different-speaker training scores are shown as grey triangles and same-speaker training scores are shown as white circles. The top panel shows a linear discriminant analysis model fitted to the data. The middle and bottom panels can be derived from the top panel. The middle and bottom panels can also be derived by fitting a logistic regression model to the same data. The vertical line represents a score value that is being converted to a likelihood-ratio value. In the top panel the output is clearly the ratio of two likelihoods. The middle panel shows that the conversion from scores to log-likelihood-ratio values is a linear function. (In reality the plotted training data are illustrative only and the plotted functions show ideal values based on specified parametric distributions.)

The calculation of a likelihood ratio using this pooled-variance two-Gaussian model is illustrated in the top panel of Figure 20.11 in which, for illustrative purposes, $\hat{\mu}_s = +0.5$, $\hat{\mu}_d = -1.5$, and $\hat{\sigma}^2 = 1$. The likelihood-ratio value calculated for an illustrative test score of $+0.5$ is $0.399/0.054 = 7.39$.

Note that whereas in the feature domain the data were multivariate and had complex distributions and we fitted models requiring a large number of parameter values to be estimated, in the score domain the data are univariate and we fit models requiring only a small number of parameter values to be estimated (e.g., two means and one variance). We can therefore obtain better parameter estimates in the score domain than in the feature domain, and hence the output of the score-to-likelihood-ratio-conversion model is well calibrated.

The value of a score has no meaning by itself, but if one score has a higher value than another score then the higher-valued score indicates greater relative support for the same-speaker hypothesis over the different-speaker hypothesis than does the lower-valued score. Thus, although we do not know whether both scores correspond to likelihood ratios less than 1, or both to likelihood ratios greater than 1, or one to a likelihood ratio less than 1 and the other to a likelihood ratio greater than 1, and we do not know whether they correspond to likelihood ratios that are close to each other or far from each other, we do know that the higher-valued score corresponds to a higher-valued likelihood ratio than does the lower-valued score. The model for converting from scores to likelihood ratios should therefore be *monotonic*, that is, it should preserve the ranking of scores, a lower-ranked score should not be converted to a higher likelihood-ratio value than the likelihood-ratio value corresponding to a higher-ranked score. In the model described above, using two Gaussians with the same variance results in a monotonic conversion. Using a different variance for each Gaussian would not be monotonic.

The pooled-variance two-Gaussian model is a *linear discriminant analysis* model, and can be shown to reduce to a linear model in a logged-odds space or in a log-likelihood-ratio space. A linear model has the form $y = a + bx$, and requires the estimation of only two parameter values, an intercept, a, and a slope, b. If natural logarithms are used in calculating the score, the intercept and slope are as shown in Equations 20.20–20.22, and the output is the natural logarithm of the likelihood ratio.

$$\log(\Lambda_{q,k}) = a + bS_{q,k} \tag{20.20}$$

$$a = -\frac{\hat{\mu}_s^2 - \hat{\mu}_d^2}{2\hat{\sigma}^2} = -b\left(\hat{\mu}_s + \hat{\mu}_d\right)/2 \tag{20.21}$$

$$b = \frac{\hat{\mu}_s - \hat{\mu}_d}{\hat{\sigma}^2} \tag{20.22}$$

This version of the model is illustrated in the middle panel of Figure 20.11, in which $a = +1$ and $b = +2$. The illustrative test score of $+0.5$ converts to a natural log likelihood ratio of $1 + 2 \times 0.5 = 2$, which as before is a likelihood ratio of 7.39.

In practice, it is more common to use *logistic regression* rather than linear discriminant analysis. If the assumptions of Gaussian distributions with equal variances hold, then logistic regression and linear discriminant analysis will give the same results (Hastie et al., 2009, §4.4.5). Logistic regression, however, is more robust to violations of those assumptions, thus in general it results in better performance. The same-speaker and the different-speaker scores are used to train the logistic-regression model (see Figure 20.11 bottom panel), which is linear in the log-likelihood-ratio space (see Figure 20.11 middle panel), hence training results in calculation of the values for the intercept, a, and slope, b. These values are then used in Equation 20.20 to convert the score value to a log likelihood-ratio value.

Logistic regression is trained using an iterative procedure that minimizes the deviance statistic (see Menard, 2010; Hosmer et al., 2013). It is usually used to calculate a posterior probability (see Figure 20.11 bottom panel), but by converting from posterior probability to posterior odds and training the model using equal priors for the same-speaker and the different-speaker categories, the output is interpretable as a log likelihood ratio (see Figure 20.11 middle panel). Logistic regression is a discriminative procedure rather than a generative procedure and hence does not provide separate values that could be interpreted as the numerator and denominator of the likelihood ratio. Its output is not literally the ratio of two likelihoods, but the interpretation of its output as a likelihood ratio is justified by its analogy with linear discriminant analysis.

20.8 Validation

Empirical validation of a forensic analysis system under conditions reflecting those of the case to which it is to be applied is required to inform admissibility decisions under United States Federal Rules of Evidence 702 and the *Daubert* trilogy of Supreme Court rulings (Daubert v. Merrell Dow Pharmaceuticals, 1993; General Electric v. Joiner, 1997; and Kumho Tire v. Carmichael, 1999), and under England and Wales Criminal Practice Directions (2015) section 19A. This has been emphasized by President Obama's Council of Advisors on Science and Technology (2016) and by the England and Wales Forensic Science Regulator (2014). As mentioned in the introduction, there have been calls from the 1960s onward for the performance of forensic-voice-comparison systems to be empirically validated under casework conditions (for a review see Morrison, 2014), and such validation is advocated by the ENFSI guidelines (Drygajlo et al., 2015).

The results of a validation study depend on both the system that is tested and the conditions under which it is tested. To provide information that will be helpful in understanding how well a forensic-voice-comparison system will perform when applied in a particular case, what must be tested is the system that will actually be used in the case, and it must be tested using test recordings that are sufficiently representative of the relevant population and sufficiently reflective of the conditions of the questioned- and known-speaker recordings in the case. Results of tests of the system with other populations or under other conditions (or tests of other systems) will in general not be informative as to the performance of the system when applied in the particular case, and could be highly misleading. Since there can be substantial variation in relevant population and recording conditions from case to case, case-specific validation may be required.

In black-box validation studies, pairs of recordings are entered into the forensic-voice-comparison system, and in response to each input pair the system outputs a likelihood ratio. In each pair, one member of the pair must have conditions that reflect those of the questioned-speaker recording and the other must have conditions that reflect those of the known-speaker recording. Some pairs must be same-speaker pairs, and others must be different-speaker pairs. For each pair, the goodness (or badness) of the likelihood-ratio output is assessed with respect to the tester's knowledge of whether the input was a same-speaker or a different-speaker pair. Traditional performance metrics such as false-alarm rate and miss rate depend on making binary decision by applying a threshold to a posterior probability, and are therefore inconsistent with the likelihood-ratio framework. An appropriate metric would be based directly on the likelihood-ratio value of the output and would take account of the magnitude of the likelihood-ratio value. Given a same-speaker input, a good output would be a likelihood-ratio value that is much larger than 1, a less good output would be a value that is only a little larger than 1, a bad output would be a

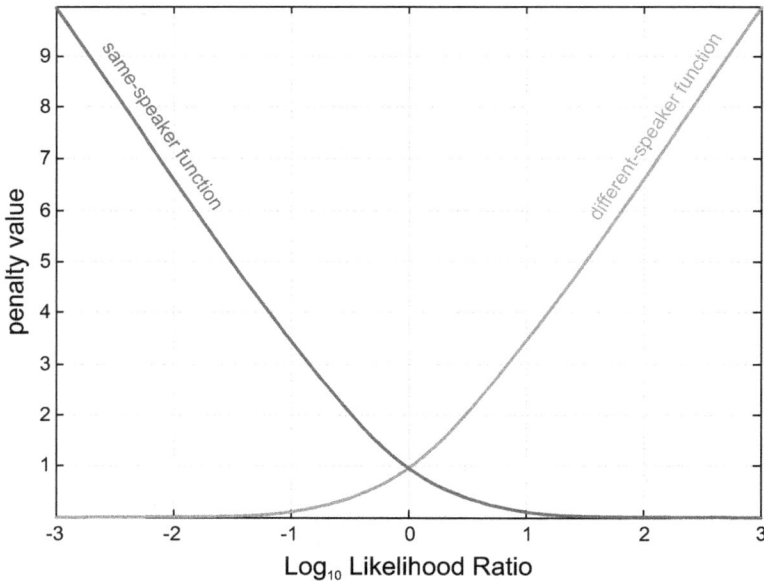

FIGURE 20.12

Penalty functions for calculating C_{llr}. The same-speaker and different-speaker curves correspond to the functions within Equation 20.23's left and right summations respectively.

value less than 1, and a worse output would be a value much less than 1. *Mutatis mutandis* for a different-speaker input for which a good output would be a value much less than 1.

In the forensic-voice-comparison literature, the log-likelihood-ratio cost (C_{llr}) is a popular metric for quantifying system performance (Brümmer and du Preez, 2006; González-Rodríguez et al., 2007; Morrison, 2011; Drygajlo et al., 2015; Morrison and Enzinger, 2016; Meuwly et al., 2017). The formula for calculating C_{llr} is given in Equation 20.23, in which Λ_s and Λ_d are likelihood-ratio outputs corresponding to same- and different-speaker input pairs respectively, and N_s and N_d are the number of same- and different-speaker input pairs respectively.

$$C_{llr} = \frac{1}{2} \left(\frac{1}{N_s} \sum_{i=1}^{N_s} \log_2 \left(1 + \frac{1}{\Lambda_{s_i}} \right) + \frac{1}{N_d} \sum_{j=1}^{N_d} \log_2 \left(1 + \Lambda_{d_j} \right) \right) \qquad (20.23)$$

Figure 20.12 plots the penalty functions for log likelihood-ratio outputs corresponding to same- and different-speaker input pairs. These are the functions within Equation 20.23's left and right summations respectively. If the input is a log likelihood-ratio from a same-speaker pair and its value is much greater than 0 it receives a small penalty value, but if its value is lower it receives a higher penalty value. If the input is a log likelihood-ratio from a different-speaker pair and its value is much less than 0 it receives a small penalty value, but if its value is higher it receives a higher penalty value. C_{llr} is calculated as the mean of the penalty values with equal weight given to the set of same-speaker penalty values as to the set of different-speaker penalty values.

Smaller C_{llr} values indicate better performance. C_{llr} values generally lie in the range 0 to 1. C_{llr} values cannot be less than or equal to 0. A system that gave no useful information and always responded with a likelihood ratio of 1, irrespective of the input, would have a C_{llr} value of 1. C_{llr} values substantially greater than 1 can be produced by uncalibrated or miscalibrated systems.

In the forensic-voice-comparison literature, a popular graphical representation of system performance is a Tippett plot (Meuwly, 2001; González-Rodríguez et al., 2007; Drygajlo et al., 2015; Morrison and Enzinger, 2016; Meuwly et al., 2017; Morrison and Enzinger, 2019). Tippett plots consist of plots of the empirical cumulative probability distribution of the likelihood-ratio values resulting from same-speaker inputs and of the empirical cumulative probability distribution of the likelihood-ratio values resulting from different-speaker inputs. The tradition is to plot lines joining the data points rather than to plot the data points themselves. Examples are shown in Figure 20.13. The y-axis values corresponding to the curves rising to the right give the proportion of same-speaker test results with log likelihood-ratio values less than or equal to the corresponding value on the x-axis. The y-axis values corresponding to the curves rising to the left give the proportion of different-speaker test results with log likelihood-ratio values greater than or equal to the corresponding value on the x-axis. In general, shallower curves with greater separation between the two curves indicates better performance. Tippett plots can also reveal problems such as bias in the output.

In the top panel of Figure 20.13 the separation between the same-speaker and different-speaker curves is small and the system is clearly biased – both the same-speaker and different-speaker curves are too far to the right. The middle panel of Figure 20.13 shows the results of a validation of a better performing system. This Tippett plot has somewhat greater separation between the same-speaker and different-speaker curves and the results are not obviously biased. The second system was actually the same as the first system except that it included a logistic-regression calibration stage whereas the first did not. The bottom panel of Figure 20.13 shows the results of a validation of a system with substantially better performance – the same-speaker and different-speaker curves have greater separation and are shallower. The C_{llr} values calculated on the same validation results as shown in the top, middle, and bottom panels of Figure 20.13 are 1.068, 0.698, and 0.307 respectively.

Given the range of possible forensic-voice-comparison systems and the case-to-case variability in relevant population and in recording conditions, there is no validation study that is representative across systems and across casework conditions. Below we list some published studies that report on empirical validation of several different systems under several different sets of conditions (including a series of papers in a journal virtual special issue in which multiple different systems were tested under the same conditions). Full references are provided in the list. These references are not repeated in the main reference list at the end of the chapter unless the publications are referenced elsewhere in the chapter. The references are listed by year of publication, and those published in the same year are listed in alphabetical order. Publications are included in the list only if they report on empirical validation of human-supervised automatic forensic-voice-comparison systems under conditions that reasonably closely reflect forensic casework conditions, the systems output likelihood-ratio values, and the validation results are reported using metrics and graphics that are consistent with the likelihood-ratio framework. We apologize if we have inadvertently omitted any studies that meet these criteria. The list was current as of June 1, 2020. We expect more validation studies to be published in the future.

Unless the relevant population and conditions of a published study reflect those of a particular case and the system validated in the study is the one that will be used in that particular case, the published study should not be used in a debate as to whether testimony derived from that forensic-voice-comparison system should be admitted in that particular

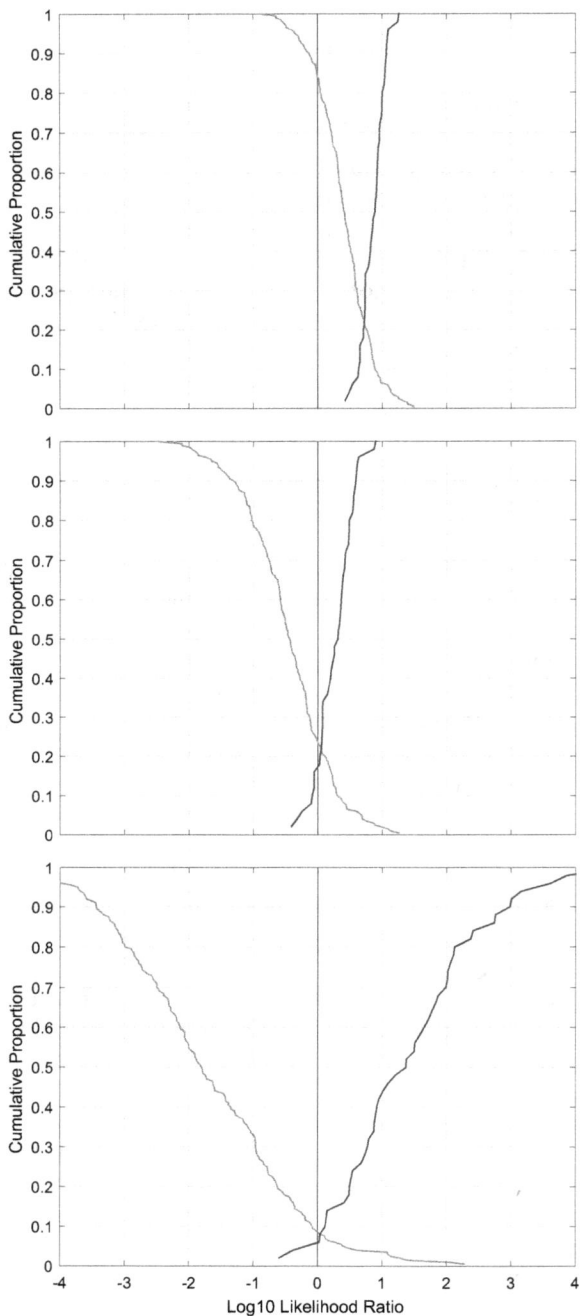

FIGURE 20.13
Examples of Tippett plots. The three plots represent three different systems tested on the same test data. The examples are based on artificial data created for illustrative purposes. The data represent 50 same-speaker test pairs and 200 different-speaker test pairs – imbalance in the number of same-speaker and different-speaker test pairs is usual since it is easier to construct different-speaker pairs than same-speaker pairs. The C_{llr} values corresponding to the results shown in the top, middle, and bottom panels are 1.068, 0.698, and 0.307 respectively.

case. If no validation study exists that sufficiently closely reflects the relevant population and conditions for the case, a case-specific validation study will have to be conducted. We would also suggest that whether the level of performance of a system is sufficient should be decided on a case by case basis.* Poorer recording conditions will lead to poorer performance, but a system with relatively poor performance may still be capable of providing useful information to the court in the form of a moderately sized likelihood-ratio value – Tippett plots will give an indication of the range of possible likelihood-ratio values that the system could generate under the test conditions.

20.8.1 List of Published Validation Studies

Solewicz, Y.A., Becker, T., Jardine, G., and Gfroerer, S. 2012. Comparison of speaker recognition systems on a real forensic benchmark. In *Proceedings of Odyssey 2012, The Speaker and Language Recognition Workshop*, pp. 85–91. International Speech Communication Association. http://isca-speech.org/archive/odyssey_2012/od12_086.html

Solewicz, Y.A., Jardine, G., Becker, T., and Gfroerer, S. 2013. Estimated intra-speaker variability boundaries in forensic speaker recognition casework. In *Proceedings of Biometric Technologies in Forensic Science (BTFS)*, pp. 30–33. http://www.ru.nl/clst/btfs/proceedings/

van der Vloed, D., Bouten, J., and van Leeuwen, D. 2014. NFI-FRITS: A forensic speaker recognition database and some first experiments. In *Proceedings of Odyssey 2014, The Speaker and Language Recognition Workshop*, pp. 6–13. International Speech Communication Association. http://cs.uef.fi/odyssey2014/program/pdfs/21.pdf

Enzinger E. 2016. Comparison of GMM-UBM and i-vector models under casework conditions: Case 1 revisited. In *Implementation of forensic voice comparison within the new paradigm for the evaluation of forensic evidence*, Ch. 4. Doctoral dissertation, University of New South Wales. http://handle.unsw.edu.au/1959.4/55772

Enzinger, E., Morrison, G.S., and Ochoa, F. 2016. A demonstration of the application of the new paradigm for the evaluation of forensic evidence under conditions reflecting those of a real forensic-voice-comparison case. *Science & Justice*, **56**: 42–57. http://dx.doi.org/10.1016/j.scijus.2015.06.005

van der Vloed, D. 2016. Evaluation of Batvox 4.1 under conditions reflecting those of a real forensic voice comparison case (forensic_eval_01). *Speech Communication*, **85**: 127–130. http://dx.doi.org/10.1016/j.specom.2016.10.001 [errata published in (2017) **92**: 23. http://dx.doi.org/10.1016/j.specom.2017.04.005]

Enzinger E. and Morrison G.S. 2017. Empirical test of the performance of an acoustic-phonetic approach to forensic voice comparison under conditions similar to those of a real case. *Forensic Science International*, **277**: 30–40. http://dx.doi.org/10.1016/j.forsciint.2017.05.007

Silva, G.D. da and Medina, C.A. 2017. Evaluation of MSR Identity Toolbox under conditions reflecting those of a real forensic case (forensic_eval_01). *Speech Communication*, **94**: 42–49. http://dx.doi.org/10.1016/j.specom.2017.09.001

van der Vloed, D., Jessen, M., and Gfroerer, S. 2017. Experiments with two forensic automatic speaker comparison systems using reference populations that (mis)match the

---

* At a consensus-development meeting held in September 2019, a number of experts in validation of forensic-voice-comparison systems concluded that the only logically justified validation threshold for C_{llr} was 1. As of June 1, 2020, a written statement of the consensus is still under development.

test language. *Proceedings of the Audio Engineering Society Conference on Forensic Audio*, paper 2-1.

Morrison, G.S. 2018. The impact in forensic voice comparison of lack of calibration and of mismatched conditions between the known-speaker recording and the relevant-population sample recordings. *Forensic Science International*, **283**: e1–e7. http://dx.doi.org/10.1016/j.forsciint.2017.12.024

Zhang, C. and Tang, C. 2018. Evaluation of Batvox 3.1 under conditions reflecting those of a real forensic voice comparison case (forensic_eval_01). *Speech Communication*, **100**: 13–17. https://doi.org/10.1016/j.specom.2018.04.008

Kelly, F., Fröhlich, A., Dellwo, V., Forth, O., Kent, S., and Alexander, A. 2019. Evaluation of VOCALISE under conditions reflecting those of a real forensic voice comparison case (forensic_eval_01). *Speech Communication*, **112**: 30–36. https://doi.org/10.1016/j.specom.2019.06.005

Jessen, M., Bortlík, J., Schwarz, P., and Solewicz, Y.A. 2019. Evaluation of Phonexia automatic speaker recognition software under conditions reflecting those of a real forensic voice comparison case (forensic_eval_01). *Speech Communication*, **111**: 22–28. https://doi.org/10.1016/j.specom.2019.05.002

Jessen, M., Meir, G., and Solewicz, Y.A. 2019. Evaluation of Nuance Forensics 9.2 and 11.1 under conditions reflecting those of a real forensic voice comparison case (forensic_eval_01). *Speech Communication*, **110**: 101–107. https://doi.org/10.1016/j.specom.2019.04.006

van der Vloed, D., Kelly, F., and Alexander, A. 2020. Exploring the effects of device variability on forensic speaker comparison using VOCALISE and NFI-FRIDA, a forensically realistic database. In *Proceedings of Odyssey, The Speaker and Language Recognition Workshop*, pp. 402–407. International Speech Communication Association. http://dx.doi.org/10.21437/Odyssey.2020-57

20.9 Conclusion

We have described signal-processing and statistical-modeling techniques that are commonly used to calculate likelihood ratios in human-supervised automatic approaches to forensic voice comparison. We hope that we have fulfilled our aim of bridging the gap between general introductions to forensic voice comparison and the highly technical and often fragmented automatic-speaker-recognition literature from which the signal-processing and statistical-modeling techniques are mostly drawn. We hope that this has been of value to students of forensic voice comparison and to researchers interested in understanding statistical modeling techniques that could potentially also be applied to data from other branches of forensic science.

Acknowledgments

We dedicate this chapter in memory of David Lucy, who died on June 20, 2018. David was perhaps best known in the forensic science community for his book *Introduction to Statistics for Forensic Scientists* (Lucy, 2005) and for his work on statistical models for the calculation of likelihood ratios, including his work on multivariate kernel density models, e.g., Aitken and Lucy (2004). These models were used extensively by the first two authors of the current chapter in their earlier work on acoustic-phonetic approaches to forensic voice comparison. David will be greatly missed. We extend our condolences to his family and friends.

Earlier collaboration between the first four authors that ultimately led to the writing of this chapter benefited from support provided by the Australian Research Council, Australian Federal Police, New South Wales Police, Queensland Police, National Institute of Forensic Science, Australasian Speech Science and Technology Association, and the Guardia Civil through Linkage Project LP100200142. The contributions of the last three authors were supported by the Spanish Ministry of Economy and Competitiveness through project TEC2015-68172-C2-1-P, and by collaboration over the last 20 years between AUDIAS and the Guardia Civil. Contributions by the first two authors to late stages of the preparation of the chapter were supported by Research England's Expanding Excellence in England Fund as part of funding for the Aston Institute for Forensic Linguistics 2019–2022.

Appendix 20A: Mathematical Details of T Matrix Training and i-Vector Extraction

This appendix provides details of the procedures usually used for **T** matrix training and extraction of i-vectors. It is based on the factor analysis procedure described in Dehak et al. (2011).

1. Feature vectors from all the speakers in the training set are pooled and then used to train a UBM with the parameter set $\Omega = \{w_g, \mu_g, \Sigma_g\}_{g=1}^{G}$, i.e., G Gaussians with mean vectors μ_g, diagonal-only covariance matrices Σ_g, and weights w_g.

2. For each recording j in the training and test sets, centralized 0th and 1st order Baum-Welch statistics, $n_{g,j}$ and $f_{g,j}$, are computed with respect to the UBM, see Equations 20A.1–20A.3 in which: x_{j_i} is the ith feature vector of recording j; $\gamma_{g,i}$ is a responsibility (Equation 20A.3 is the multivariate equivalent of Equation 20.2 in Section 20.4.1); N_j is the number of feature vectors extracted from recording j; $n_{g,j}$ is the probability that the feature vectors of recording j would belong to Gaussian g of the UBM; and $f_{g,j}$ is the sum, over all vectors in the recording, of the deviation of each vector from the UBM mean vector, weighted by the probability that it belongs to Gaussian g.

$$n_{g,j} = \sum_{i=1}^{N_j} \gamma_{g,i} \tag{20A.1}$$

$$f_{g,j} = \sum_{i=1}^{N_j} \gamma_{g,i} \left(x_{j_i} - \mu_g\right) \tag{20A.2}$$

$$\gamma_{g,i} = \frac{w_g f \left(x_{j_i} \mid \mu_g, \Sigma_g\right)}{\sum_{l=1}^{G} w_l f \left(x_{j_i} \mid \mu_l, \Sigma_l\right)} \tag{20A.3}$$

3. A maximum-likelihood estimate of the **T** matrix is trained using the training data and an EM algorithm (Kenny et al., 2005; Matrouf et al., 2007). First, each element of the **T** matrix is initialized by random draws from the standard Gaussian distribution. In the expectation step, for each recording $j \in \{1 \ldots J\}$ in the training set, we evaluate the posterior distribution of ϕ_j given the Baum-Welch statistics $\{n_{g,j}, f_{g,j}\}_{g=1}^{G}$ from that recording. Assuming a standard Gaussian prior for ϕ_j and using the current estimate of the **T** matrix (denoted as $\mathbf{T}_{\text{old}}$, the first and second moments, (ϕ_j) and $(\phi_j \phi_j^t)$, of the posterior are calculated as in Equations 20A.4 and 20A.5. $\mathbf{L}_j^{-1}$ is the posterior covariance matrix. $\boldsymbol{\Sigma}_g$ is the covariance matrix of the gth Gaussian of the UBM. In the maximization step, the **T** matrix is updated as in Equation 20A.7 (see Proposition 3 in Kenny et al., 2005, §III). Often, a minimum divergence step is added after the maximization step, in which the **T** matrix is updated so that the empirical i-vector distribution better conforms to a standard Gaussian prior (Kenny, 2008, §II-B; Glembek, 2012, §3.6.4). This is achieved by computing the dot-product of the matrix $\mathbf{T}_{g,\text{ML}}$ after the maximization step and the matrix **Q** obtained using a symmetrical decomposition (e.g., Cholesky decomposition) of $\mathbf{P}^{-1}$, the covariance matrix of the prior, Equations 20A.8–20A.10. Empirically, adding the minimum divergence step results in faster convergence of the training algorithm. The expectation-maximization and minimum divergence steps are repeated for multiple iterations.

$$\mathbf{L}_j = \mathbf{I} + \sum_{g=1}^{G} n_{g,j} \mathbf{T}_{g,\text{old}}^{t} \boldsymbol{\Sigma}_g^{-1} \mathbf{T}_{g,\text{old}} \tag{20A.4}$$

$$(\phi_j) = \mathbf{L}_j^{-1} \sum_{g=1}^{G} \mathbf{T}_{g,\text{old}}^{t} \boldsymbol{\Sigma}_g^{-1} \mathbf{f}_{g,j} \tag{20A.5}$$

$$(\phi_j \phi_j^t) = \mathbf{L}_j^{-1} + (\phi_j)(\phi_j)^t \tag{20A.6}$$

$$\mathbf{T}_{g,\text{ML}} = \left(\sum_{j=1}^{J} n_{g,j} (\phi_j \phi_j^t) \right)^{-1} \left(\sum_{j=1}^{J} \mathbf{f}_{g,j} (\phi_j)^t \right) \tag{20A.7}$$

$$\mathbf{P}^{-1} = \frac{1}{J} \sum_{j=1}^{J} \mathbf{L}_j^{-1} + (\phi_j)(\phi_j)^t \tag{20A.8}$$

$$\mathbf{Q}\mathbf{Q}^t = \mathbf{P}^{-1} \tag{20A.9}$$

$$\mathbf{T}_{g,\text{MD}} = \mathbf{T}_{g,\text{ML}} \mathbf{Q} \tag{20A.10}$$

$$\mathbf{T}_{\text{new}} = \begin{bmatrix} \mathbf{T}_{1,\text{MD}} \\ \vdots \\ \mathbf{T}_{G,\text{MD}} \end{bmatrix} \tag{20A.11}$$

4. For each recording, j, in the training and test sets, an i-vector is obtained as the posterior mean (ϕ_j) using the Baum-Welch statistics $\{n_{g,j}, \mathbf{f}_{g,j}\}_{g=1}^{G}$ computed from recording j. The calculations are the same as in Equations 20A.4 and 20A.5, but using the final estimate of the $\mathbf{T}$ matrix.

References

Aitken, C.G.G. and Lucy, D. Evaluation of trace evidence in the form of multivariate data. *Applied Statistics*, **53**, 109–122, 2004. http://dox.doi.org/10.1046/j.0035-9254.2003.05271.x [Corrigendum: 2004. 53: 665–666. http://dox.doi.org/10.1111/j.1467-9876.2004.02031.x]

Ajili, M. Reliability of voice comparison for forensic applications. Doctoral Dissertation, University of Avignon, 2017.

Arthur, D. and Vassilvitskii, S. k-means++: The advantages of careful seeding. In *Proceedings of the Eighteenth Annual ACM-SIAM Symposium on Discrete Algorithms*, pp. 1027–1035, 2007.

Beritelli, F. and Spadaccini, A. The role of voice activity detection in forensic speaker verification. In *Proceedings of the 17th International Conference on Digital Signal Processing*, pp. 1–6, 2011. https://doi.org/10.1109/ICDSP.2011.6004980

Bogert, B.P., Healy, M.J.R., and Tukey, J.W. The quefrency alanysis of time series for echoes: Cepstrum, pseudo autocovariance, cross-cepstrum and saphe cracking. In M. Rosenblatt, editor, *Proceedings of the Symposium on Time Series Analysis*, pp. 209–243. Wiley, New York, 1963.

Bousquet, P.-M., Bonastre, J.-F., and Matrouf, D. Identify the benefits of the different steps in an i-vector based speaker verification system. In *Progress in Pattern Recognition, Image Analysis, Computer Vision, and Applications*, pp. 278–285. Springer, Berlin, 2013. http://dox.doi.org/10.1007/978-3-642-41827-3_35

Brümmer, N. and de Villiers, E. The speaker partitioning problem. In *Proceedings of Odyssey 2010: The speaker and language recognition workshop*, pp. 194–201. International Speech Communication Association, 2010. https://www.isca-speech.org/archive_open/odyssey_2010/od10_034.html

Brümmer, N. and du Preez, J. Application independent evaluation of speaker detection. *Computer Speech and Language*, **20**, 230–275, 2006. https://doi.org/10.1016/j.csl.2005.08.001

Chaudhary, G., Srivastava, S., and Bhardwaj, S. Feature extraction methods for speaker recognition: A review. *International Journal of Pattern Recognition and Artificial Intelligence*, **31**(12), 1750041-1–1750041-39, 2017. http://dx.doi.org/10.1142/S0218001417500410

Davis, S. and Mermelstein, P. Comparison of parametric representations for monosyllabic word recognition in continuously spoken sentences. *IEEE Transactions on Acoustics, Speech, and Signal Processing*, **28**(4), 357–366, 1980. https://doi.org/10.1109/TASSP.1980.1163420

Dehak, N., Kenny, P.J., Dehak, R., Dumouchel, P., and Ouellet, P. Front-end factor analysis for speaker verification. *IEEE Transactions on Audio, Speech, and Language Processing*, **19**(4), 788–798, 2011. https://doi.org/10.1109/TASL.2010.2064307

Dempster, A.P., Laird, N.M., and Rubin D.B. Maximum likelihood from incomplete data via the EM algorithm. *Journal of the Royal Statistical Society, Series B*, **39**(1), 1–38, 1977. http://www.jstor.org/stable/2984875

Dişken, G., Tüfekçi, Z., Saribulut, L., and Çevik, U. A review on feature extraction for speaker recognition under degraded conditions. *IETE Technical Review*, **34**(3), 321–332, 2017. https://doi.org/10.1080/02564602.2016.1185976

Drygajlo, A., Jessen, M., Gfroerer, S., Wagner, I., Vermeulen, J., and Niemi, T. *Methodological guidelines for best practice in forensic semiautomatic and automatic speaker recognition, including guidance on the conduct of proficiency testing and collaborative exercises.* European Network of Forensic Science

Institutes, 2015. http://enfsi.eu/wp-content/uploads/2016/09/guidelines_fasr_and_fsasr_0. pdf

Duda, R.O., Hart, P.E., and Stork, D.G. *Pattern Classification* (2nd ed.). Wiley, New York, 2000.

Edmond, G., Towler, A., Growns, B., Ribeiro, G., Found, B., White, D., Ballantyne, K. et al. Thinking forensics: Cognitive science for forensic practitioners. *Science & Justice*, **57**, 144–154, 2017. http://dx.doi.org/10.1016/j.scijus.2016.11.005

Enzinger, E. A first attempt at compensating for effects due to recording-condition mismatch in formant-trajectory-based forensic voice comparison. In *Proceedings of the 15th Australasian International Conference on Speech Science and Technology*, pp. 133–136. Australasian Speech Science and Technology Association, 2014. http://www.assta.org/sst/SST-14/6.A. %20FORENSICS%202/1.%20ENZINGER.pdf

Enzinger, E. Implementation of forensic voice comparison within the new paradigm for the evaluation of forensic evidence. Doctoral dissertation, University of New South Wales, 2016. http:// handle.unsw.edu.au/1959.4/55772

Enzinger, E. and Kasess, C.H. Bayesian vocal tract model estimates of nasal stops for speaker verification. In *Proceedings of the International Conference on Acoustics, Speech, and Signal Processing (ICASSP 2014)*, pp. 1685–1689, 2014. http://dx.doi.org/10.1109/ICASSP.2014.6853885

Enzinger, E. and Morrison, G.S. Empirical test of the performance of an acoustic-phonetic approach to forensic voice comparison under conditions similar to those of a real case. *Forensic Science International*, **277**, 30–40, 2017. http://dx.doi.org/10.1016/j.forsciint.2017.05.007

Enzinger, E., Zhang, C., and Morrison G.S. Voice source features for forensic voice comparison— An evaluation of the Glottex® software package. In *Proceedings of Odyssey 2012, The Language and Speaker Recognition Workshop*, pp. 78–85. International Speech Communication Association, 2012. http://isca-speech.org/archive/odyssey_2012/od12_078.html [Errata and addenda available at: https://box.entn.at/pdfs/enzinger2012_odyssey_vsferradd.pdf]

Fernández Gallardo, L. *Human and Automatic Speaker Recognition over Telecommunication Channels.* Springer, Singapore, 2016. https://doi.org/10.1007/978-981-287-727-7

Ferrer, L., Nandwana, M.K., McLaren, M., Castan, D., and Lawson A. Toward fail-safe speaker recognition: Trial-based calibration with a reject option. *IEEE/ACM Transactions on Audio, Speech, and Language Processing*, **27**, 140–153, 2019. http://dx.doi.org/10.1109/TASLP.2018.2875794

Fisher, R.A. The use of multiple measurements in taxonomic problems. *Annals of Eugenics*, **7**, 179–188, 1936. http://dx.doi.org/10.1111/j.1469-1809.1936.tb02137.x

Forensic Science Regulator. *Guidance on validation (FSR-G-201 Issue 1).* Forensic Science Regulator, Birmingham, UK, 2014. https://www.gov.uk/government/publications/forensic-science-providers-validation

Forensic Science Regulator. *Codes of Practice and Conduct—Appendix: Speech and Audio Forensic Services (FSR-C-134 Issue 1).* Forensic Science Regulator, Birmingham, UK, 2016. https://www.gov.uk/government/publications/speech-and-audio-forensic-services

Forensic Science Regulator. *Codes of Practice and Conduct for Forensic Science Providers and Practitioners in the Criminal Justice System (Issue 5).* Forensic Science Regulator, Birmingham, UK, 2020. https://www.gov.uk/government/collections/forensic-science-providers-codes-of-practice-and-conduct

Found, B. Deciphering the human condition: The rise of cognitive forensics. *Australian Journal of Forensic Sciences*, **47**, 386–401, 2015. http://dx.doi.org/10.1080/00450618.2014.965204

Fourier, J. Mémoire sur la propagation de la chaleur dans les corps solides, presented 21 December 1807 at the Institut national. *Nouveau Bulletin des sciences par la Société philomatique de Paris*, **1**(6), 112–116, 1808. https://www.biodiversitylibrary.org/item/24789#page/120/mode/ 1up [Although the presentation was given by Fourier, the published report of the presentation was written by Poisson.]

Furui, S. Cepstral analysis technique for automatic speaker verification. *IEEE Transactions on Acoustics, Speech, and Signal Processing*, **29**(2), 254–272, 1981. https://doi.org/10.1109/TASSP.1981. 1163530

Furui, S. Speaker-independent isolated word recognition using dynamic features of speech spectrum. *IEEE Transactions on Acoustics, Speech, and Signal Processing*, **34**, 52–59, 1986. https://doi.org/10.1109/TASSP.1986.1164788

García-Romero, D. and Espy-Wilson, C.Y. Analysis of i-vector length normalization in speaker recognition systems. In *Proceedings of Interspeech 2011*, pp. 249–252. International Speech Communication Association, 2011. https://www.isca-speech.org/archive/interspeech_2011/i11_0249.html

García-Romero, D. and McCree, A. Supervised domain adaptation for i-vector based speaker recognition. In *Proceeding of the IEEE International Conference on Acoustics, Speech, and Signal Processing*, pp. 4828–4831, 2014. http://dx.doi.org/10.1109/ICASSP.2014.6854362

García-Romero, D. and McCree, A. Insights into deep neural networks for speaker recognition. In *Proceedings of Interspeech 2015*, pp. 1141–1145. International Speech Communication Association, 2015. https://www.isca-speech.org/archive/interspeech_2015/i15_1141.html

Garcia-Romero, D., Zhang, X., McCree, A., and Povey, D. Improving speaker recognition performance in the domain adaptation challenge using deep neural networks. In *Proceedings of the 2014 IEEE Spoken Language Technology Workshop (SLT)*, pp. 378–383, 2014.

Glembek, O. Optimization of Gaussian mixture subspace models and related scoring algorithms in speaker verification. Doctoral dissertation, Brno University of Technology, 2012.

González-Rodríguez, J., Rose, P., Ramos, D., Toledano, D.T., and Ortega-García, J. Emulating DNA: Rigorous quantification of evidential weight in transparent and testable forensic speaker recognition. *IEEE Transactions on Audio, Speech, and Language Processing*, **15**, 2104–2115, 2007. https://doi.org/10.1109/TASL.2007.902747

Hansen, J.H.L. and Bořil, H. On the issues of intra-speaker variability and realism in speech, speaker, and language recognition tasks. *Speech Communication*, **101**, 94–108, 2018. https://doi.org/10.1016/j.specom.2018.05.004

Hansen, J.H.L. and Hasan, T. Speaker recognition by machines and humans: A tutorial review. *IEEE Signal Processing Magazine*, November, 74–99, 2015. http://dx.doi.org/10.1109/MSP.2015.2462851

Harris, F.J. On the use of windows for harmonic analysis with the discrete Fourier transform. *Proceedings of the IEEE*, **66**(1), 51–83, 1978. http://dx.doi.org/10.1109/PROC.1978.10837

Hastie, T., Tibshirani, R., and Freidman, J. *The Elements of Statistical Learning Data Mining, Inference, and Prediction*, 2nd edition. Springer, New York, 2009.

Hosmer, D.W. Jr., Lemeshow, S., and Sturdivant, R.X. *Applied Logistic Regression* (3rd ed.). Wiley, Hoboken NJ, 2013. http://dx.doi.org/10.1002/9781118548387

Hwang, M.Y. and Huang, X. Subphonetic modeling with Markov states—Senone. In *Proceedings of the IEEE International Conference on Acoustics, Speech, and Signal Processing (ICASSP)*, pp. I-33–I-36, 1992. https://doi.org/10.1109/ICASSP.1992.225979

Jessen, M., Alexander, A., and Forth, O. Forensic voice comparisons in German with phonetic and automatic features using Vocalise software. In *Proceedings of the 54th Audio Engineering Society (AES) Forensics Conference*, pp. 28–35. Audio Engineering Society, 2014.

Kelly, F. and Hansen, J.H.L. Score-aging calibration for speaker verification. *IEEE/ACM Transactions on Audio, Speech, and Language Processing*, **24**, 2414–2424, 2016. http://dx.doi.org/10.1109/TASLP.2016.2602542

Kenny, P. Joint factor analysis of speaker and session variability: Theory and algorithms. Technical Report CRIM-06/08-13. Centre de Recherche Informatique de Montréal, 2005.

Kenny, P. Bayesian speaker verification with heavy tailed priors. In *Proceedings of Odyssey 2010: The Speaker and Language Recognition Workshop, paper 014*. International Speech Communication Association, 2010. https://www.isca-speech.org/archive_open/odyssey_2010/od10_014.html

Kenny, P., Boulianne, G., and Dumouchel, P. Eigenvoice modeling with sparse training data. *IEEE Transaction on Speech and Audio Processing*, **13**, 345–354, 2005. https://doi.org/10.1109/TSA.2004.840940

Kenny, P., Gupta, V., Stafylakis, T., Ouellet, P., and Alam, J. Deep neural networks for extracting Baum-Welch statistics for speaker recognition. In *Proceedings of Odyssey 2014*, pp. 293–298. International Speech Communication Association, 2014. https://www.isca-speech.org/archive/odyssey_2014/abstracts.html#abs28

Kinnunen, T. and Li, H. An overview of text-independent speaker recognition: From features to supervectors. *Speech Communication*, **52**, 12–40, 2010. http://dx.doi.org/10.1016/j.specom.2009.08.009

Klecka, W.R. *Discriminant Analysis*. Sage, Beverly Hills, CA, 1980.

Lee, K.A., Yamamoto, H., Okabe, K., Wang, Q., Guo, L., Koshinaka, T., Zhang, J., and Shinoda, K. NEC-TT System for mixed-bandwidth and multi-domain speaker recognition. *Computer Speech & Language*, **61**, article 101033, 2020. https://doi.org/10.1016/j.csl.2019.101033

Lei, Y., Scheffer, N., Ferrer, L., and McLaren, M. A novel scheme for speaker recognition using a phonetically-aware deep neural network. In *Proceedings of the 2014 IEEE International Conference on Acoustics, Speech and Signal Processing (ICASSP)*, pp. 1695–1699, 2014. https://doi.org/10.1109/ICASSP.2014.6853887

Lozano-Díez, A., Silnova, A., Matějka, P., Glembek, O., Plchot, O., Pesan, J., Burget, L., and González-Rodríguez, J. Analysis and Optimization of Bottleneck Features for Speaker Recognition. In *Proceedings of Odyssey 2016*, pp. 352–357. International Speech Communication Association, 2016. http://dx.doi.org/10.21437/Odyssey.2016-51

Lucy, D. *Introduction to Statistics for Forensic Scientists*. Wiley, Chichester, UK, 2005.

Mandasari, M.I., McLaren, M., and van Leeuwen, D.A. The effect of noise on modern automatic speaker recognition systems. In *Proceedings of the 2012 IEEE International Conference on Acoustics, Speech and Signal Processing (ICASSP)*, pp. 4249–4252, 2012. https://doi.org/10.1109/ICASSP.2012.6288857

Mandasari, M.I., Saeidi, R., McLaren, M., and van Leeuwen, D.A. Quality measure functions for calibration of speaker recognition systems in various duration conditions. *IEEE Transactions on Audio, Speech, and Language Processing*, **21**, 2425–2438, 2013. http://dx.doi.org/10.1109/TASL.2013.2279332

Mandasari, M.I., Saeidi, R., and van Leeuwen, D.A. Quality measures based calibration with duration and noise dependency for speaker recognition. *Speech Communication*, **72**, 126–137, 2015. http://dx.doi.org/10.1016/j.specom.2015.05.009

Matějka P., Glembek O., Castaldo F., Alam M., Plchot O., Kenny P., Burget L., and Černocký, J. Full-covariance UBM and heavy-tailed PLDA in i-vector speaker verification. In *Proceeding of the 2011 IEEE International Conference on Acoustics, Speech, and Signal Processing (ICASSP)*, pp. 4828–4831, 2011. http://dx.doi.org/10.1109/ICASSP.2011.5947436

Matějka, P., Glembek, O., Nomotiv, O., Plchot, O., Grézl, F., Burget, L., and Černocký, J. Analysis of DNN approaches to speaker identification. In *Proceeding of the 2016 IEEE International Conference on Acoustics, Speech, and Signal Processing (ICASSP)*, pp. 5100–5104, 2016. http://dx.doi.org/10.1109/ICASSP.2016.7472649

Matějka, P., Plchot, O., Glembek, O., Burget, L., Rohdin, J., Zeinali, H., Mošner, L., Silnova, A., Novotný, O., Diez, M., and Černocký, J.H. 13 years of speaker recognition research at BUT, with longitudinal analysis of NIST SRE. *Computer Speech & Language*, **63**, article 101035, 2020. https://doi.org/10.1016/j.csl.2019.101035

Matrouf, D., Scheffer, N., Fauve, B., and Bonastre, J. A straightforward and efficient implementation of the factor analysis model for speaker verification. In *Proceedings of Interspeech*, pp. 1242–1245. International Speech Communication Association, 2007. https://www.isca-speech.org/archive/interspeech_2007/i07_1242.html

McLaren, M., Castan, D., Nandwana, M.K., Ferrer, L., and Yılmaz, E. How to train your speaker embeddings extractor. In *Proceedings of Odyssey, The Speaker and Language Recognition Workshop*, pp. 327–334. International Speech Communication Association, 2018. https://doi.org/10.21437/Odyssey.2018-46

McLaren, M., Lawson, A., Ferrer, L., Scheffer, N., and Lei, Y. Trial-based calibration for speaker recognition in unseen conditions. In *Proceedings of Odyssey 2014: The Speaker and Language Recognition*

Workshop, pp. 19–25. International Speech Communication Association, 2014. https://www.isca-speech.org/archive/odyssey_2014/abstracts.html#abs43

Menard, S. *Logistic Regression: From Introductory to Advanced Concepts and Applications*. Sage, Thousand Oaks, CA, 2010. http://dx.doi.org/10.4135/9781483348964

Meuwly, D. Reconnaissance de locuteurs en sciences forensiques: l'apport d'une approche automatique [Speaker recognition in forensic science: The contribution of an automatic approach]. Doctoral dissertation, University of Lausanne, 2001.

Meuwly, D., Ramos, D., and Haraksim R. A guideline for the validation of likelihood ratio methods used for forensic evidence evaluation. *Forensic Science International*, **276**, 142–153, 2017. http://dx.doi.org/10.1016/j.forsciint.2016.03.048

Morrison, G.S. Measuring the validity and reliability of forensic likelihood-ratio systems. *Science & Justice*, **51**, 91–98, 2011. http://dx.doi.org/10.1016/j.scijus.2011.03.002

Morrison, G.S. Tutorial on logistic-regression calibration and fusion: Converting a score to a likelihood ratio. *Australian Journal of Forensic Sciences*, **45**, 173–197, 2013. http://dx.doi.org/10.1080/00450618.2012.733025

Morrison, G.S. Distinguishing between forensic science and forensic pseudoscience: Testing of validity and reliability, and approaches to forensic voice comparison. *Science & Justice*, **54**, 245–256, 2014. http://dx.doi.org/10.1016/j.scijus.2013.07.004

Morrison, G.S. 2018a. Admissibility of forensic voice comparison testimony in England and Wales. *Criminal Law Review*, **2018**(1): 20–33.

Morrison, G.S. 2018b. The impact in forensic voice comparison of lack of calibration and of mismatched conditions between the known-speaker recording and the relevant-population sample recordings. *Forensic Science International*, **283**: e1–e7. http://dx.doi.org/10.1016/j.forsciint.2017.12.024

Morrison, G.S. and Enzinger, E. Multi-laboratory evaluation of forensic voice comparison systems under conditions reflecting those of a real forensic case (forensic_eval_01)—Introduction. *Speech Communication*, **85**, 119–126, 2016. http://dx.doi.org/10.1016/j.specom.2016.07.006

Morrison, G.S. and Enzinger, E. Score based procedures for the calculation of forensic likelihood ratios—Scores should take account of both similarity and typicality. *Science & Justice*, **58**, 47–58, 2018. http://dx.doi.org/10.1016/j.scijus.2017.06.005

Morrison, G.S. and Enzinger, E. Forensic voice comparison. In W.F. Katz and P.F. Assmann, editors, *The Routledge Handbook of Phonetics*, pp. 599–634. Routledge, Abingdon, UK, 2019. https://doi.org/10.4324/9780429056253

Morrison, G.S., Enzinger, E., and Zhang, C. Forensic speech science. In I. Freckelton and H. Selby, editors, *Expert Evidence*, ch. 99. Thomson Reuters, Sydney, Australia, 2018.

Morrison, G.S. and Poh, N. Avoiding overstating the strength of forensic evidence: Shrunk likelihood ratios/Bayes factors. *Science & Justice*, **58**, 200–218, 2018. http://dx.doi.org/10.1016/j.scijus.12.005

Morrison, G.S. and Stoel, R.D. Forensic strength of evidence statements should preferably be likelihood ratios calculated using relevant data, quantitative measurements, and statistical models—A response to Lennard (2013) Fingerprint identification: How far have we come? *Australian Journal of Forensic Sciences*, **46**, 282–292, 2014. http://dx.doi.org/10.1080/00450618.2013.833648

Morrison, G.S. and Thompson, W.C. Assessing the admissibility of a new generation of forensic voice comparison testimony. *Columbia Science and Technology Law Review*, **18**, 326–434, 2017.

National Commission on Forensic Science. *Ensuring That Forensic Analysis is based upon Task-Relevant Information*, 2015. https://www.justice.gov/ncfs/file/818196/download

Ommen, D.M. and Saunders, C.P. Building a unified statistical framework for the forensic identification of source problems. *Law, Probability and Risk*. **17**, 179–197, 2018. http://dx.doi.org/10.1093/lpr/mgy008

Pearson, K. On lines and planes of closest fit to systems of points in space. *The London, Edinburgh, and Dublin Philosophical Magazine and Journal of Science*, Series 6, **2**(11), 559–572, 1901. http://dx.doi.org/10.1080/14786440109462720

Peddinti, V., Chen, G., Povey, D., and Khudanpur, S. Reverberation robust acoustic modeling using i-vectors with time delay neural networks. In *Proceedings of Interspeech 2015*, pp. 2440–2444. International Speech Communication Association, 2015a. https://www.isca-speech.org/archive/interspeech_2015/i15_2440.html

Peddinti, V., Povey, D., and Khudanpur, S. A time delay neural network architecture for efficient modeling of long temporal contexts. In *Proceedings of Interspeech 2015*, pp. 3214–3218. International Speech Communication Association, 2015b. https://www.isca-speech.org/archive/interspeech_2015/i15_3214.html

Pelecanos, J. and Sridharan, S. Feature warping for robust speaker verification. In *Proceedings of Odyssey 2001: The Speaker Recognition Workshop*, pp. 213–218. International Speech Communication Association, 2001. https://www.isca-speech.org/archive_open/odyssey/odys_213.html

Pigeon, S., Druyts, P., and Verlinde, P. Applying logistic regression to the fusion of the NIST'99 1-speaker submissions. *Digital Signal Processing*, **10**, 237–248, 2000. http://dx.doi.org/10.1006/dspr.1999.0358

President's Council of Advisors on Science and Technology. *Forensic Science in Criminal Courts: Ensuring Scientific Validity of Feature-Comparison Methods*, 2016. https://obamawhitehouse.archives.gov/administration/eop/ostp/pcast/docsreports/

Prince, S.J.D. and Elder, J.H. Probabilistic linear discriminant analysis for inferences about identity. In *Proceedings of the IEEE 11th International Conference on Computer Vision*, pp. 1–8, 2007. https://doi.org/10.1109/ICCV.2007.4409052

Ramos Castro, D. Forensic evaluation of the evidence using automatic speaker recognition systems. Doctoral dissertation, Autonomous University of Madrid, 2007.

Reynolds, D.A. Speaker identification and verification using Gaussian mixture speaker models. In *Proceedings of the ESCA Workshop on Automatic Speaker Recognition, Identification, and Verification*, pp. 27–30, 1994.

Reynolds, D.A., Quatieri, T.F., and Dunn, R.B. Speaker verification using adapted Gaussian mixture models. *Digital Signal Processing*, **10**, 19–41, 2000. https://doi.org/10.1006/dspr.1999.0361

Sadjadi, S.O. and Hansen, J.H.L. Unsupervised speech activity detection using voicing measures and perceptual spectral flux. *IEEE Signal Processing Letters*, **20**(3), 197–200, 2013. https://doi.org/10.1109/LSP.2013.2237903

Silva, D.G. and Medina, C.A. Evaluation of MSR Identity Toolbox under conditions reflecting those of a real forensic case (forensic_eval_01). *Speech Communication*, **94**, 42–49, 2017. http://dx.doi.org/10.1016/j.specom.2017.09.001

Sizov, A., Lee, K.A., and Kinnunen, T. Unifying probabilistic linear discriminant analysis variants in biometric authentication. In P. Fränti, G. Brown, M. Loog, F. Escolano, and M. Pelillo, editors, *Structural, Syntactic, and Statistical Pattern Recognition*, pp. 464–475. Springer, Berlin, 2014. https://doi.org/10.1007/978-3-662-44415-3_47

Snyder, D., Garcia-Romero, D., Povey, D., and Khudanpur, S. Deep neural network embeddings for text-independent speaker verification. In *Proceedings of Interspeech 2017*, pp. 999–1003. International Speech Communication Association, 2017. http://dx.doi.org/10.21437/Interspeech.2017-620

Snyder, D., Garcia-Romero, D., Sell, G., Povey, D., and Khudanpur, S. X-vectors: Robust DNN embeddings for speaker recognition. In *Proceedings of IEEE International Conference on Acoustics, Speech and Signal Processing (ICASSP) 2018*, pp. 5329–5333, 2018. http://dx.doi.org/10.1109/ICASSP.2018.8461375

Sohn, J., Kim, N.S., and Sung, W. A statistical model-based voice activity detection. *IEEE Signal Processing Letters*, **6**(1), 1–3, 1999. https://doi.org/10.1109/97.736233

Stevens, S.S., Volkmann, J., and Newman, E.B. A scale for the measurement of the psychological magnitude pitch. *Journal of the Acoustical Society of America*, **8**, 185–190, 1937. https://doi.org/10.1121/1.1915893

Stoel, R.D., Berger, C.E.H., Kerkhoff, W., Mattijssen, E.J.A.T., and Dror E.I. Minimizing contextual bias in forensic casework. In K.J. Strom and M.J. Hickman, editors, *Forensic Science and the Administration of Justice: Critical Issues and Directions*, pp. 67–86. Sage, Thousand Oaks, CA, 2015. http://dx.doi.org/10.4135/9781483368740.n5

Tirumala, S.S., Shahamiri, S.R., Garhwal, A.S., and Wang, R. Speaker identification features extraction methods: A systematic review. *Expert Systems with Applications*, **90**, 250–271, 2017. https://doi.org/10.1016/j.eswa.2017.08.015

Ueda, N., Nakano, R., Ghahramani, Z., and Hinton, G.E. Split and merge EM algorithm for improving Gaussian mixture density estimates. *The Journal of VLSI Signal Processing-Systems for Signal, Image, and Video Technology*, **26**, 133–140, 2000. http://dx.doi.org/10.1023/A:1008155703044

Viikki, O. and Laurila, K. Cepstral domain segmental feature vector normalization for noise robust speech recognition. *Speech Communication*, **25**, 133–147, 1998. https://doi.org/10.1016/S0167-6393(98)00033-8

Villabla, J., Chen, N., Snyder, D., García-Romero, D., McCree, A., Sell, G., Borgstrom, J., García-Perera, L.P., Richardson, F., Dehak, R., Torres-Carrasquillo, P.A., and Dehak, N. State-of-the-art speaker recognition with neural network embeddings in NIST SRE18 and Speakers in the Wild evaluations. *Computer Speech & Language*, **60**, article 101026, 2020. https://doi.org/10.1016/j.csl.2019.101026

Yaman, S., Pelecanos, J., and Sarikaya, R. Bottleneck features for speaker recognition. In *Proceedings of Odyssey 2012*, pp. 105–108. International Speech Communication Association, 2012. https://www.isca-speech.org/archive/odyssey_2012/od12_105.html

Zhang, C., Morrison, G.S., Enzinger, E., and Ochoa F. Effects of telephone transmission on the performance of formant-trajectory-based forensic voice comparison—Female voices. *Speech Communication*, **55**, 796–813, 2013. http://dx.doi.org/10.1016/j.specom.2013.01.011

Legal References

Criminal Practice Directions [2015] EWCA Crim 1567.

Daubert v. Merrell Dow Pharms., 509 U.S. 579 (1993).

Federal Rules of Evidence as amended Apr. 17, 2000, eff. Dec. 1, 2000; Apr. 26, 2011, eff. Dec. 1, 2011.

General Electric Co. v. Joiner, 522 U.S. 136 (1997).

Kumho Tire Co. v. Carmichael, 526 U.S. 137 (1999).

21

Bringing New Statistical Approaches to Eyewitness Evidence

Alice J. Liu, Karen Kafadar, Brandon L. Garrett, and Joanne Yaffe

CONTENTS

21.1 Introduction 499
21.2 The Eyewitness Task 501
    21.2.1 System and Estimator Variables 504
21.3 Current Statistical Methodologies 507
    21.3.1 Diagnosticity Ratio, Discriminability Index, and ROC Curves 507
    21.3.2 Calibration Curve 508
    21.3.3 ROC Curves 509
    21.3.4 Logistic Regression 512
    21.3.5 Expected Utility 514
21.4 Statistical Models From Diagnostic Medicine 514
    21.4.1 Logitnormal Bivariate Random-Effects Model 515
    21.4.2 Nonparametric Meta-Analysis for Diagnostic Accuracy Studies 517
21.5 Supervised Learning Classification Methods 519
    21.5.1 Machine Learning Classification Models 519
    21.5.2 Graphical Models 521
21.6 Tools Based on ROC Methods 522
    21.6.1 Methodology Development 522
    21.6.2 Predictive Receiver Operating Characteristic Curve 524
    21.6.3 Multivariate ROC Curves 525
    21.6.4 AUC Estimation 526
    21.6.5 Confidence Intervals for ROC 527
21.7 Estimating Probability of Eyewitness Accuracy 528
    21.7.1 Probability of Accuracy 528
21.8 Example 530
21.9 Discussion 533
Acknowledgements 534
References 534

21.1 Introduction

In July 1984, Jennifer Thompson was sexually assaulted by an assailant, who, later that night, sexually assaulted a second woman. Thompson helped create the composite sketch that led to the assembly of a live line-up in which she positively identified Ronald Cotton as the perpetrator. "Yeah. This is the one… I think this is the guy," said Thompson at the live line-up (Garrett, 2012). A second line-up was assembled, with Cotton as the only repeated person. "This looks the most like him," Thompson confirms, stating that she was "absolutely sure" Cotton was the culprit. Cotton was convicted of sexual assault and burglary based on circumstantial evidence and Thompson's identification. He was sentenced to life in prison plus 54 years. In 1995, after 10 years in prison, Cotton was exonerated through DNA testing with help from the Innocence Project.* This is a particularly well-known example of a common problem. In 360+ post-conviction DNA exonerations documented by the Innocence Project since 1989, approximately 71% of these exonerations involved one or more mistaken eyewitness identifications.[†]

Eyewitness identification (EWID) plays a critical role in criminal cases, from the investigation to the prosecution of the crime. The core element of EWID is memory – remembering the suspect, the proceedings of the crime, and the emotions associated. Memory is first encoded, then consolidated with existing information in the brain, and then retrieved (i.e., reconstructed) at a later time (Howe and Knott, 2015). Each stage can cause memories to degrade or mutate over time, depending on the purpose for retrieving the information, to whom, and how it is recalled, can alter one's recall of the actual events. In addition to internal factors, such as the person's own memory processes, external factors can distort one's information retrieval, such as length of time between the event and need for retrieval of the memory, intermediate events during that time, and identification procedures. This fallibility has detrimental consequences for those who fall victim to EWID mistakes. We need experiments that faithfully represent EWID processes to assess which factors can be varied and set at levels that minimize the probabilities of grievous EWID errors.

Statistical methods, used to analyze datasets[‡] concerning eyewitness choices in experiments or in the field, can allow one to better understand what factors affect the likelihood that an eyewitness will choose correctly. Statisticians can work in conjunction with psychologists to conduct tests with high ecological validity[§] to identify factors that improve the reliability of EWID evidence. Statistically designed experiments[¶] help identify factors that are more likely to lead to errors as well as those that are less likely to result in mistakes, by encouraging efficient experimental practices, integration of variability measures, and application of existing statistical models from other fields to EWID data. The National Academy of Sciences (NAS) emphasized both needs in its important report on the subject issued in 2014 (National Research Council, 2014): "The committee recommends a broad exploration of the merits of different statistical tools for use in the evaluation of eyewitness performance" (p. 108).

We review potential statistical models to quantify the effects of factors influencing the accuracy of eyewitness identification in controlled experiments as well as explore methods for analyzing the results from these experiments, using statistical models and intuitive displays of the effects of these factors. For example, while the receiver operating characteristic (ROC) curve has been used for decades in statistical quality control, diagnostic medicine, and many other fields where methods or techniques are being compared, the ROC curves using data from eyewitness identification experiments are constructed using

---

* https://www.innocenceproject.org/cases/ronald-cotton/, Accessed 02-Jan-2020.
† https://www.innocenceproject.org/eyewitness-identification-reform/, Accessed 18-Dec-2019.
‡ EWID datasets are further addressed in Section 21.2.
§ We define "ecological validity" to mean that the study (including methods, materials, setting, etc.) approximates the real-world, so that study findings can be generalized to real-world settings.
¶ More generally referred to as the design of experiments or experimental design. It is a procedure used to plan experiments so that the resulting data can be analyzed to produced valid and objective conclusions.

the experimental participant's expressed confidence level (ECL) in the identification, which can be affected by error and variation. We present alternative statistical approaches, some of which have been used in similar scenarios (e.g., comparing medical diagnostic imaging modalities) with the aim of developing more powerful analyses to better quantify the effects of variables (including or modifying the ECL) influencing the accuracy of EWID procedures. These statistical tools may offer powerful ways of identifying factors that affect EWID accuracy, beyond the conventional tools of diagnosticity ratios and ROC.

This chapter serves to provide information on existing and viable statistical methods for analyzing EWID experiments. Whichever technique is used, proper characterization of the uncertainties associated with inferences must be calculated. Background information of the eyewitness task and EWID data is provided in Section 21.2. In Sections 21.3 and 21.4, current statistical methods in EWID research and potential methods from the field of diagnostic medicine are reviewed. Alternative statistical methods to the conventional ROC curve are provided in Section 21.6. A new approach to examining EWID data, along with a modeling procedure, is proposed in Section 21.7. This new approach re-examines the perceived structure of EWID data, and results in a tool that could potentially be used as an in-field assessment of eyewitness reliability. Section 21.8 presents an example of analysis of variance to compare EWID procedures. Finally, Section 21.9 provides a discussion of the methods in terms of their adaptability for EWID experiments as well as to suggest improved models. We recommend statistical approaches in the final section, depending on the data, experimental conditions, and concomitant information available.

21.2 The Eyewitness Task

The task of the eyewitness is to try to identify the perpetrator of a crime that (s)he witnessed. With a single suspect, the identification decision is binary: either the presented suspect is, or is not, the person whom (s)he saw commit the crime. The binary choice results in a binary outcome: either the suspect was, or was not, the true perpetrator, and either the eyewitness does, or does not, implicate that suspect.

In the standard paradigm of EWID, the two correct outcomes are the conviction of the truly guilty (true positive) and the exoneration of the truly innocent (true negative). The two incorrect outcomes are the conviction of the truly innocent (false positive) and the exoneration of the truly guilty (false negative). Table 21.1 shows these outcomes from the eyewitness, who serves as the "binary classifier" for this task.[*] For many people, minimizing false positives is the key priority, as the consequences to the wrongfully convicted are profound. Law enforcement personnel also want to minimize false negatives, to prevent perpetrators from committing further crimes.

Of course, the perpetrator may not be in the lineup at all. Thus, the target may be present (TP) or absent (TA) in the lineup.[†] Figure 21.1 shows an example of a simultaneous lineup with photos of six possible suspects that might be shown to an eyewitness; "Not Present" is also offered as an option.[‡] For more examples of simultaneous lineups used in such

---

[*] The 2015 NAS report (National Research Council, 2014) called attention to the consideration of the eyewitness as a binary classifier for analysis purposes: "It is important that practitioners in this field broadly explore the large and rich field of statistical tools for evaluation of binary classifiers" (p. 91).

[†] Law enforcement hopes that the target is present in the lineup, but no data exist on the proportion of lineups that are "TP." Likely it varies substantially by jurisdiction.

[‡] This lineup was provided by Chad Dodson.

TABLE 21.1

The Eyewitness Task Shown Visually as a Two-by-Two Table,
Assuming the Eyewitness Serves as the "Binary Classifier"

		Witness' Decision	
		"Guilty"	**"Innocent"**
Suspect's True Status	Guilty	True Positive (TP)	False Negative (FN)
	Innocent	False Positive (FP)	True Negative (TN)

Not Present

FIGURE 21.1
Example of a fair, target present simultaneous lineup in an experimental setting, target suspect (shown as the perpetrator in a video of the "crime") is in the top-left.

laboratory experiments, see Wells et al. (2011). If the target is present in the lineup, the eyewitness can make three possible decisions:

(P1) Make the right decision and choose the guilty suspect;

(P2) Make a wrong decision and choose an innocent foil*;

(P3) Make a wrong decision and state that the guilty suspect (i.e., target) is not present.

If the target is absent in the lineup, the eyewitness can make two possible decisions:

(A1) Make the right decision and state that the guilty suspect is not present;

(A2) Make a wrong decision and choose the innocent suspect or a foil.

---

* A "foil" is an innocent person in a police lineup. It is also sometimes referred to as "filler" in the literature.

	Accurate	Not Accurate	
Choose Target (P1)	Choose Foil (P2)	Don't Choose (P3)	**Target Present**
Don't Choose (A1)	Choose Foil (A2)		**Target Absent**

FIGURE 21.2
A display of the eyewitness decision outcome space, which takes into account the underlying status of the lineup.

Thus, five possible decision outcomes can occur, only two of which (P1 and A1) are correct; see Figure 21.2.

Researchers often include a designated "innocent suspect" to serve as the "target" in TA lineups. Based on this set-up, there are four categories of classification:

- Correct suspect identification;
- Innocent suspect identification;
- Foil identification;
- Lineup rejection (suspect not present).

Table 21.2 shows three different approaches to EWID data structure.

Given an identification, memory theory from the paradigm of signal detection theory (SDT) indicates that the eyewitness applies "a simple rule to make an identification

TABLE 21.2

This Table Provides the Three Possible Structures Assumed for EWID Data, from the Two Previously Addressed Structures in Table 21.1 and Figure 21.2 to the Inclusion of an Innocent Suspect

			Suspect's True Status	
		Scenario	*"Guilty"*	*"Innocent"*
Eyewitness's Decision	Errors treated equally	*"Guilty" Suspect*	TP	FP
		Not the "Guilty" Suspect	FN	TN
	No designated innocent suspect	*"Guilty" Suspect*	TP	Forced 0
		Foil (Known Innocent)	Incorrect	FP
		Not Present	FN	TN
	Designated innocent suspect	*"Guilty" Suspect*	TP	Forced 0
		"Innocent" Suspect	Forced 0	FP
		Foil (Known Innocent)	Incorrect	Incorrect
		Not Present	FN	TN

The table provides the possible EWID outcomes based on the eyewitness's decision versus the true underlying status of the lineup, which could affect the analysis approaches used by researchers.

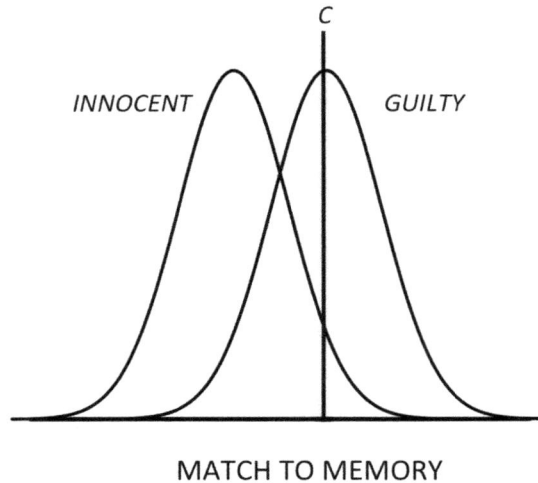

FIGURE 21.3
Distribution of "memory strength" for identification of guilty and innocent suspects (Clark et al., 2015). This illustration conforms with memory theory in assuming a Gaussian distribution for "memory strength" and the individual's memory threshold c as the mean of the right-hand curve. Other models for the distribution of "memory strength" can be proposed, and individual thresholds may fall at different quantiles of the distribution other than 50%.

decision" (Clark et al., 2015). If the association between the suspect and the eyewitness's reconstruction (via memory) of the perpetrator exceeds a "threshold" of memory strength, say c, then the witness will identify that suspect as the perpetrator. If it falls below that individual's threshold, then the eyewitness will exclude the suspect as a perpetrator. This paradigm assumes that the decision is based on the individual's threshold for a single variable, "memory strength": a false identification occurs if the suspect is innocent but the individual's "memory strength" falls above c, and false exclusions occur if the "memory strength" falls below c for an innocent suspect. According to Gronlund and Benjamin (2018), SDT provides a cohesion for decision-making with ambiguous evidence, with a link to metacognition.[*] Figure 21.3 displays this memory theory paradigm: the eyewitness's decision comes from one of these two distributions, often conveniently assumed to be Gaussian, and the memory "threshold" is the mean (median) of the "Guilty" distribution. The larger the separation between these two distributions, and the higher the quantile of the distribution of "Guilty" for the individual's threshold c, the lower the error rates (false negatives, false positives).

21.2.1 System and Estimator Variables

The accuracy of this eyewitness task depends on many factors. Some factors are under the control of law enforcement (e.g., type of lineup), while others arise by the circumstances (e.g., lighting). A summary of these factors is shown in Figure 21.4. Factors that can affect accuracy of eyewitness identification and are under the control of law enforcement

---

[*] Awareness, understanding, analysis, and control of one's own cognitive (learning, thinking, reasoning, etc.) processes.

Environmental
Variables

Culprit
Variables

Estimator
Variables

Victim
Variables

Eyewitness
Identification
Accuracy

System
Variables

Instruction
Variables

Lineup
Variables

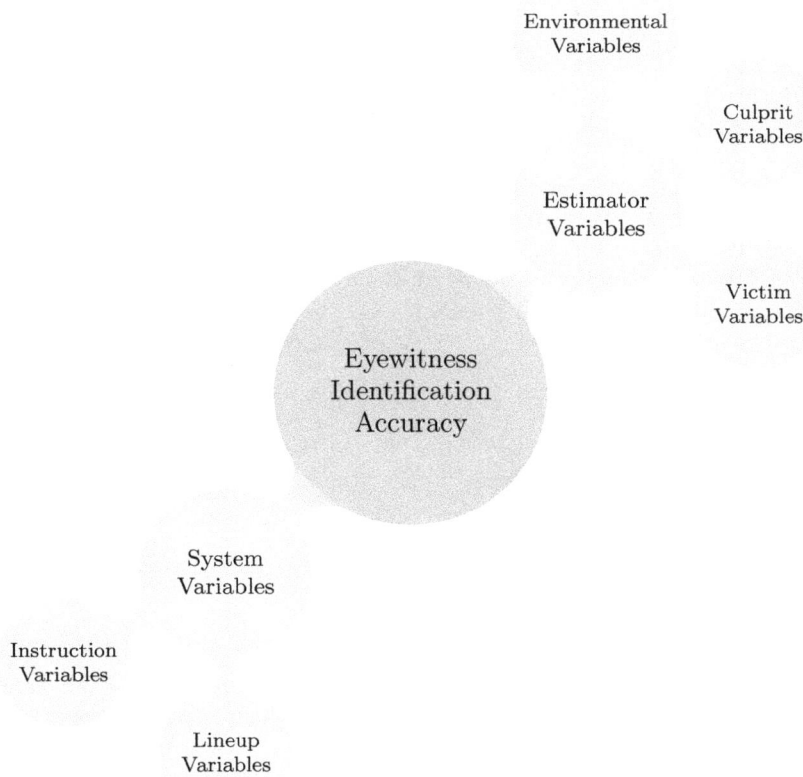

FIGURE 21.4
Examples of variables that could affect eyewitness identification accuracy.

have been called "system variables" in the eyewitness identification literature ("control" variables in the experimental design literature); they include:

System Variables: Controllable by Law Enforcement

- Type of lineup (or photo array, if it is a photos are used), typically "sequential" (suspects or photos show sequentially, one at at time), or "simultaneous" (shown together);
- Size of lineup (e.g., 5 or 6 or more suspects shown), ideally chosen so that the probability of identifying an innocent suspect by chance is low (Brigham et al., 1999);
- Fairness (i.e., subjective similarity of appearance of people) of lineup: In a truly fair lineup, the probability that any one of the suspects is selected is equal; increasingly biased lineups are those for which the probabilities are not equal;
- Delay (or retention interval): Time between incident and eyewitness' identification task;

- Lineup instructions (degree of detail in guidance to the eyewitness in the identification procedure);
- "Unblinded" or "blind" lineups: Law enforcement officer conducting the lineup either is, or is not, within view of the eyewitness and the photos (s)he is viewing. (The concern is that the "unblinded" law enforcement officer may unconsciously deliver subtle cues that affect the eyewitness's selection of a suspect.)

Many other "environmental" factors affect eyewitness accuracy; they arise from the circumstances and are not under the control of law enforcement officers conducting the lineup. Factors that can affect accuracy of eyewitness identification and are *not* under the control of law enforcement have been called "estimator variables" in the eyewitness identification literature ("noise" variables in the experimental design literature); they include:

- Presence or absence of weapon at time of incident (gun, knife, etc.);
- Presence or absence of a distinctive feature of the perpetrator;
- Lighting (which can affect visibility);
- Distance between eyewitness and perpetrator at time of incident;
- Length of exposure (seconds, minutes, etc.);
- Same- or cross-race (perpetrator and eyewitness are same or different races; studies suggest higher accuracy for the former.[*]

Datasets commonly used in EWID research proceed from designed experiments from the field of psychology, where some of the aforementioned variables are purposefully manipulated. Much EWID research has focused on comparing sequential versus simultaneous lineups (Amendola and Wixted, 2014; Lindsay and Wells, 1985; Carlson et al., 2008; Rotello et al., 2015). Relatively few studies considered the effects of several variables simultaneously in one experiment. Multi-factor experiments can be very informative in this context: if the effects of weapon presence, or cross-race, or delay (to identification) hugely dominate the effect of lineup type (sequential or simultaneous), then law enforcement officers will know to focus their energy on, say, minimizing the delay between incident and lineup, and less attention to the type of lineup. Law enforcement officers can also assess potential accuracy of an eyewitness in view of the conditions, such as presence or absence of a weapon or lighting that can affect visibility. Multi-factor experiments allow the estimation of jointly varying effects from different sources.[†] The National Academy of Sciences report (2015) recommended the conduct of more multi-factor experiments, to better characterize the effect of presence (or absence) of weapon, relative to the choice of lineup (sequential or simultaneous) (National Research Council, 2014).

Limitations of designed experimental data include lack of ecological validity, where some aspects of the reality of the EWID may not be reflected in the experimental situation. In order to address this issue, some researchers coordinated with law enforcement agencies to provide field datasets (Wixted et al., 2016). However, field data lacks the underlying truth of suspect guilt. While the conviction and/or conclusions from evidence provide

---

[*] For references to these cross-race studies, please refer to Sporer (2001), Meissner and Brigham (2001), Wilson et al. (2013).

[†] Studies that considered additional variables jointly include Dodson and Dobolyi (2016), Wixted et al. (2016), Mickes et al. (2017), Sauerland et al. (2018), Clark (2005), Sauer et al. (2010), Palmer et al. (2013), Humphries and Flowe (2015), Carlson et al. (2016a,2016b), Colloff et al. (2016), Colloff et al. (2017), Steblay (1997).

an estimated classification of guilt, using field data to train models to assess accuracy could lead to biased models. In this situation, the models would address the relationships of a particular law enforcement agency in identifying what they believe is a guilty suspect. Trade-offs exist for either forms of data, and the form used should be justified and consistent with the goal or research question.

Much literature exists on EWID from many perspectives (experimental, theory of memory, etc.); we give only a very brief background on that literature. Our main focus is on the myriad of statistical tools that may offer powerful ways of identifying factors that affect EWID accuracy, which follow in the subsequent sections.

21.3 Current Statistical Methodologies

Surprisingly few statistical approaches have been used in analyzing data from EWID experiments. Some psychologists have historically used the diagnosticity ratio and the discriminability index (d') as measures of comparison across different eyewitness procedures (Wixted and Mickes, 2012; Mickes et al., 2014, 2017). Other psychologists have been proponents of the point-biserial correlation coefficient (r_{pb}) and Goodman and Kruskal's gamma (G), which tend to result in misleading conclusions. Additional methods include: calibration curves (Juslin et al., 1996; Krug, 2007; Gronlund et al., 2015; Clark et al., 2015; Brewer and Wells, 2006; Sporer et al., 1995); receiver operating characteristic (ROC) curve analysis based on the signal detection theory (SDT) paradigm using expressed confidence levels (ECLs) as the cutpoints (Clark et al., 2015; Wixted and Mickes, 2010, 2012; Pepe, 2000; Wixted and Mickes, 2015); partial area under the curve (pAUC) as an extension of ROC analysis (Walter, 2005; Mickes et al., 2014; Wixted et al., 2017; Lampinen et al., 2019); estimation of posterior probability of guilt based on Bayes' Theorem (Wells and Lindsay, 1980; Wells et al., 2015a,b); expected utility (Lampinen et al., 2019; Smith et al., 2019); logistic regression (Wetmore et al., 2015; Andersen et al., 2014); and log-linear models (Luby, 2016, 2017). Overall, psychologists are exploring these methods to further the theory of eyewitness cognition, which consists of memory judgments (making a selection in a lineup) and accompanying metacognitive context (the associated confidence statement) (Gronlund and Benjamin, 2018). We describe these approaches in this section.

21.3.1 Diagnosticity Ratio, Discriminability Index, and ROC Curves

The diagnosticity ratio (DR), equivalently positive likelihood ratio LR_+, is the ratio of the odds of the suspect being guilty relative to the odds of the suspect being innocent. It measures the probative value, which is how much information is available in the evidence, of a lineup procedure. The DR provides the posterior odds of guilt or the likelihood that a guilty suspect is identified in a lineup (Wixted and Mickes, 2012; Mickes et al., 2014).

$$\text{DR} = \frac{\text{Correct ID Rate}}{\text{False ID Rate}} \tag{21.1}$$

$$= \frac{\text{HR}}{\text{FAR}}$$

$$= \frac{P(\text{suspect identified} \mid \text{suspect guilty})}{P(\text{suspect identified} \mid \text{suspect innocent})}$$

The discriminability* index d' (also known as the sensitivity index), which originates from signal detection theory, is a popular estimate signal strength (Swets, 2014). A higher d' indicates a larger partial area under the curve (pAUC),[†] leading pAUC proponents to prefer it over DR. Equation 21.2 shows the relationship of d' to AUC ($z(\cdot)$ represents the normal score[‡] associated with the function inputs):

$$d' = z(\text{Correct ID Rate}) - z(\text{False ID Rate}) = \sqrt{2} \cdot z(\text{AUC}) \qquad (21.2)$$

Both the DR and d' are summaries that characterize EWID performance across all "levels" of system and estimator variables. However, any such measure oversimplifies performance: a single index cannot capture all the information in a comparison between two procedures. In experimental settings, maximizing the DR may lead to more conservative responding (i.e., more likely to choose a "not present" response) (Wixted and Mickes, 2012). For example, more extreme instructions designed to protect the innocent induced a higher DR, but did not necessarily lead to a better accuracy result. The DR has a tendency to naturally increase even if discriminability is constant (Mickes et al., 2017). This could result in misleading conclusions, since a different lineup instruction would not (and should not) change the witness's memory, which should be constant across conditions. The DR was a popular performance metric for comparing procedures (e.g., simultaneous versus sequential), until some researchers (e.g., Wixted and Mickes, 2015) observed that a third variable, "expressed confidence level" (ECL), can affect this ratio, and that DR could confound changes in accuracy with changes in "response bias."

The ROC curve is a plot of the numerator of the DR (hit rate, HR) versus its denominator (false alarm rate, FAR) for various levels of ECL (e.g., "at least 10% confident," ..., "at least 40% confident," ...); the slope of the ROC curve at one of these points corresponds to HR/FAR, i.e., the DR, at that ECL. (Hence, a straight line indicates the same DR for all ECLs; i.e., DR does not depend on ECL in this case.) Because ROC curves incorporate additional information (e.g., ECL), they are viewed as more useful for comparing methods than the simple DR collapsed over all ECL categories. The DR allows the researcher to disregard suspect identifications that are categorized as "untrustworthy" (i.e., identifications made with low confidence) (Wixted and Mickes, 2015). Both DR and d' should be accompanied by measures of variability (but often are not). The NRC report acknowledged advantages of ROC over DR in some circumstances, but emphasized that other statistical analyses of EWID experimental data are more powerful (e.g., logistic regression, binary classifiers); see below.

21.3.2 Calibration Curve

The calibration curve is a graph that plots accuracy on the x-axis and confidence on the y-axis. Calibration is the agreement between objective (accuracy) and subjective (ECL) variables (Juslin et al., 1996). All participants with $c\%$ confidence should have $c\%$ accuracy, which indicates well-calibrated participants. In the graph, well-calibrated participants

---

* Discriminability is defined as the ability to perceive and respond to differences among stimuli.
† More information on pAUC in Section 21.3.3 in part 21.2.1.
‡ The normal score for some value x is found by normalizing the value such that $z = \frac{x-\mu}{\sigma/\sqrt{n}}$, where μ is the population mean, σ is the population standard deviation, and n is the number of observations available.

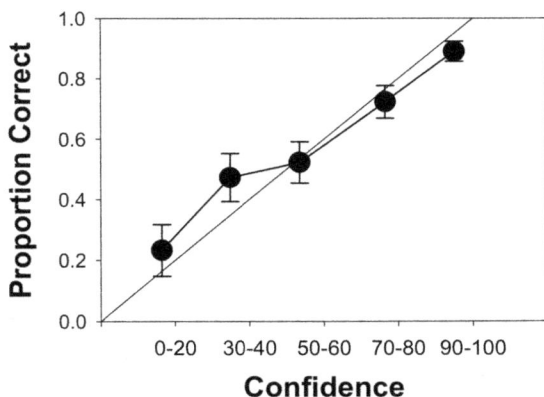

FIGURE 21.5
This plot shows the observed relationship between proportion of correct decisions and expressed confidence levels using data from Juslin et al. (1996), Wixted et al. (2015).

would fall on a diagonal line where accuracy is equal to confidence (i.e., slope $b_1 = 1$ and intercept $b_0 = 0$). Overconfident participants would fall below this line, and underconfident participants would fall above this line. The over/under-confidence statistic ω is a supplementary statistic from the calibration curve, with a $\omega \in [-1, 1]$. Well-calibrated participants receive a score of 0 (i.e., perfect calibration). Underconfidence is indicated with a negative score, and overconfidence is indicated with a positive score (Figure 21.5).

These curves provide information on only the average captured confidence-accuracy (CA) relationship. While calibration curves may indicate "fair" and well-calibrated data, they may not clarify the impact of the various system and estimator variables on eyewitness choice and accuracy. Much calibration research suggests that participants are usually overconfident in their assessments of their memory accuracy (Krug, 2007). Some psychologists view the calibration curve as a measure of the CA relationship, with good calibration as indicating a strong relationship (Gronlund et al., 2015). The calibration curve is a primary reason for strong belief in the CA relationship.

21.3.3 ROC Curves

The ROC curve was originally developed in the 1950s and used with electronic signal detection theory (SDT), with first applications in radar (Hajian-Tilaki, 2013). Since then, researchers in many other fields, including psychology, diagnostic radiology, medical diagnostics, and machine learning, use it to compare different techniques, often by its area under the ROC curve (AUC) or partial area under the curve (pAUC). An ROC curve plots the hit rate (sensitivity) against the false alarm rate (1 - specificity). The curve is based on some decision variable, and the counts of good and bad results will vary based on the chosen threshold of that decision variable. It is a descriptive device that demonstrates the range of trade-offs between true positive rates (TPRs) and false positive rates (FPRs) within a particular test (Pepe, 2000). An ROC curve with better discriminant capacity will appear as a curve closer to the upper left-hand corner in the ROC space. A curve lying on a straight diagonal line with a slope $b_1 = 1$ indicates the test has a performance similar to that of chance. The slope of the tangent line each point of the ROC curve is equal

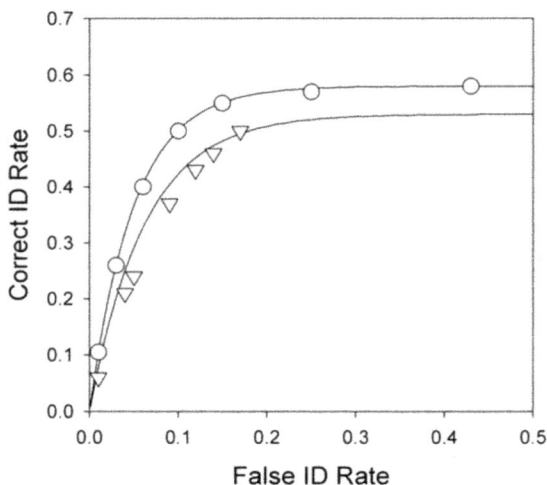

FIGURE 21.6
Hypothetical ROC curves for simultaneous (circles) and sequential (triangles) procedures. In this case, the plot concludes that simultaneous procedures are diagnostically superior (Gronlund et al., 2014).

to the likelihood ratio, which is the ratio of the two density functions describing the two distributions* of the decision variable in population one and population two.

$$\text{Sensitivity} = \frac{\text{\# true positives}}{\text{\# true positives} + \text{\# false negatives}} \quad (21.3)$$

$$\text{Specificity} = \frac{\text{\# true negatives}}{\text{\# true negatives} + \text{\# false positives}} \quad (21.4)$$

Use of ROC analysis in EWID research was first proposed in 2012 by Wixted and Mickes because a lineup procedure is characterized by a range of DRs, rather than a single DR (Wixted and Mickes, 2012). Wixted and Mickes state that the ROC can show which of two procedures is diagnostically preferable (Figure 21.6). Researchers disagree if ROC analysis is the best method to measure underlying discriminability (Wells et al., 2015a,b).

The points on ROC curves (HR versus FAR) constructed from data in EWID experiments can be based on many third variables. A common variable is the eyewitness's *expressed confidence level (ECL)*[†] at the time of the lineup. In most lab experiments, the "mock eyewitness" reports an ECL often as numerical response along a scale (0 to 1) to the question, "How confident are you in your identification?" with discrete choices; e.g., "0.0," "0.1," "0.2", ..., "1.0" (11 categories), or, more coarsely, "0.0," "0.2," "0.4", ..., "1.0" (6 categories); as with any scale, the difference in a respondent's reactions of, say, "0.0" versus "0.2" may be more clear to a respondent than the difference in the respondent's reactions of "0.4" to "0.6" (which the respondent may possibly view as less distinguishable). In real life, police officers recognize that typical eyewitnesses are not comfortable with numerical scales, so they solicit their responses as verbal descriptors. The officer's translation of those descriptors as a numeric value may depend on the officer.

---

* The distributions are usually assumed to be normal.
[†] Researchers have stated that only confidence recorded immediately after the identification should be used (Sauer et al., 2019).

Sauer et al. (2019) state, "The extent of variation in the confidence-accuracy relation precludes us from making strong, generalized claims about the accuracy of high confidence identification decisions, even under pristine conditions*, when evaluating individual identifications." They note that an individual identification differs from "aggregate level" confidence-accuracy relationship, which is equivalent to an ensemble of eyewitnesses. Either way, the ECL is likely subject to uncertainty, depending on many factors (such as high levels of stress) whose effects on ECL remain largely unstudied. These effects deserve further study so the uncertainty in ECL can be incorporated in the analysis of data from EWID experiments.

Confidence-based ROC analysis has some connection to ROC analysis in diagnostic medicine, used to compare the diagnostic superiority of different systems (e.g., MRI vs. mammography). Target present [absent] lineups may be viewed as "condition present [absent]" (e.g., presence or absence of tumor) (Wixted and Mickes, 2015). The analog of the ROC points in EWID (ECL) are ranges of assessment of condition (e.g., "definitely not malignant" to "definitely malignant"); cf. (Park et al., 2004; Mickes et al., 2012). Note that radiologists are trained professionals, with their training based on medical standards, whereas eyewitnesses are rarely "trained" in face recognition and likely have no prior practice nor experience. Kantner and Dobbins (2019) suggest that a given confidence report is largely (if not completely) determined by individual differences.[†] Nonetheless, ROC analysis may have value in the analysis of EWID experiments, if sources of uncertainty are properly taken into account. In the subsequent sections, we discuss statistical methodology alternatives to ROC analysis.

Construction. A confidence-based ROC curve is constructed by plotting the number of correct identifications versus the number of false identifications, with each point of this curve within an ordinal category of expressed confidence level (ECL) from 0% to 100% (Gronlund et al., 2015). The number of categories of confidence varies among researchers. The correct identification rate[‡] at a given ECL, say c%, is estimated as the proportion of people who correctly chose the perpetrator in the "target present" condition and expressed a confidence level of at least c%. The false identification rate[§] at a given ECL, say c%, is estimated as the proportion of people who chose the "innocent" suspect incorrectly within the target absent population and expressed confidence of at least c%. This is done for each ECL in $0\% \leq c \leq 100\%$. The slope (i.e., tangent at each plotted point) of the ROC curve is equal to the DR for that ECL.

Area Under the Curve. The area under the curve (AUC) is a standard summary of an ROC curves for purposes of comparing procedures, with preference for the procedure with the larger AUC. Some authors (e.g., Mickes et al., 2014; Wixted et al., 2017) prefer to summarize the method's performance via a partial AUC (pAUC). The AUC represents the average value of sensitivity over all possible FARs $\in [0, 1]$ (Walter, 2005), and is related to the Mann-Whitney U-statistic, which evaluates the significance of the difference between the sample distribution of positive and negative decisions (Pepe, 2000). Some authors (e.g., Mickes et al., 2014; Wixted et al., 2017) prefer to summarize the method's performance via a partial AUC (pAUC), particularly in situations where the maximum value on the x-axis (here,

---

* viz., only one suspect in the lineup, the suspect did not stand out, the witness was cautioned that the culprit may not be present, double-blind testing was used, and the confidence statement was obtained at the time of testing.

† These differences are broadly defined as self-efficacy, use of the confidence scale, and/or other factors.

‡ Hit rate (HR) or true positive rate (TPR).

§ False alarm rate (FAR) or false positive rate (FPR).

(false ID rate, or FAR) is guaranteed to be less than 1. In a target present lineup, five possible false identifications and one correct identification exist, so the maximum possible false ID rate is $(n-1)/n$, where n is the number of people or photos in the lineup. The pAUC has limitations also,* and a comparison of procedures based on either AUC or pAUC may not be straightforward if one curve is not consistently higher than the other across the entire range of HR and FAR (Streiner and Cairney, 2007).

Variability. Researchers realize that the decision criteria (in this case, ECL) may vary among participants, and use the term *criterial variance* to represent the variance in decision criteria (i.e., the differences among eyewitnesses in their criteria for making identification decisions). This is also known as criterial noise or criterion variance. Decision criteria refers to the cutoff that is used for making an identification or responding "not present." Since people use different criteria for their individual cutoffs, there is variability across people. Researchers also assume variance in the underlying distributions for target and fillers in the SDT model. The variances assumed are the equal variance versus unequal variance model for the underlying normal distributions for memory strength; see Figure 21.7. These distributions are estimated for the latent variable of memory strength. For each ECL c, the DR d_c is theoretically calculated based on these assumed normal distributions for target and foil decisions; see Equation 21.5.

$$d_c = \frac{\mu_{\text{Target}} - \mu_{\text{Foil}}}{\sqrt{(\sigma^2_{\text{Target}} + \sigma^2_{\text{Foil}})/2}}. \tag{21.5}$$

This variability aims to represent the between-participant, versus the within-participant, variability. Within-participant variability is a measure of a single participant's ECL across many lineups, of many or the same stimuli. The within-participant variability may be considered as a type of "measurement error." In this case, the measurer is not necessarily the experiment conductor or the LEO, but rather the eyewitness. Russ et al. (2018) examined the phenomenon, and reached the conclusion that a more realistic "field encounter" does not necessarily engender robust eyewitness identifications due to development of "limited cognitive representations of a target." A more controlled setting results in more consistent and correct identifications. They suggest that the degree of familiarity a participant has with a target could be a potential index for EWID accuracy. Kantner and Dobbins (2019) reiterate the point that ROC curves should be fitted to individual participants rather than in aggregate form across a large group. They found large inter-subject differences, and expect group ROC curves to be variable (i.e., noisy).

Both measures of variability differ from methods that provide intervals for point estimates. Each DR serves as a point estimate, which should have some measure of variability that captures underlying true DR. This point is explored further in Section 21.6.5 of Section 21.6.

21.3.4 Logistic Regression

The logistic regression model has a binary dependent variable and one or multiple continuous and/or categorical independent variables. In the EWID paradigm, the dependent variable is accuracy (correct or false identification) and the independent variables are relevant system and estimator variables. An estimated coefficient in logistic regression provides the change in the odds ratio for a one unit increase (for a continuous variable) or the

* See S.D. Walter, "The partial area under the summary ROC curve," *Statistics in Medicine* 2005:**24**(13), 2025–2040.

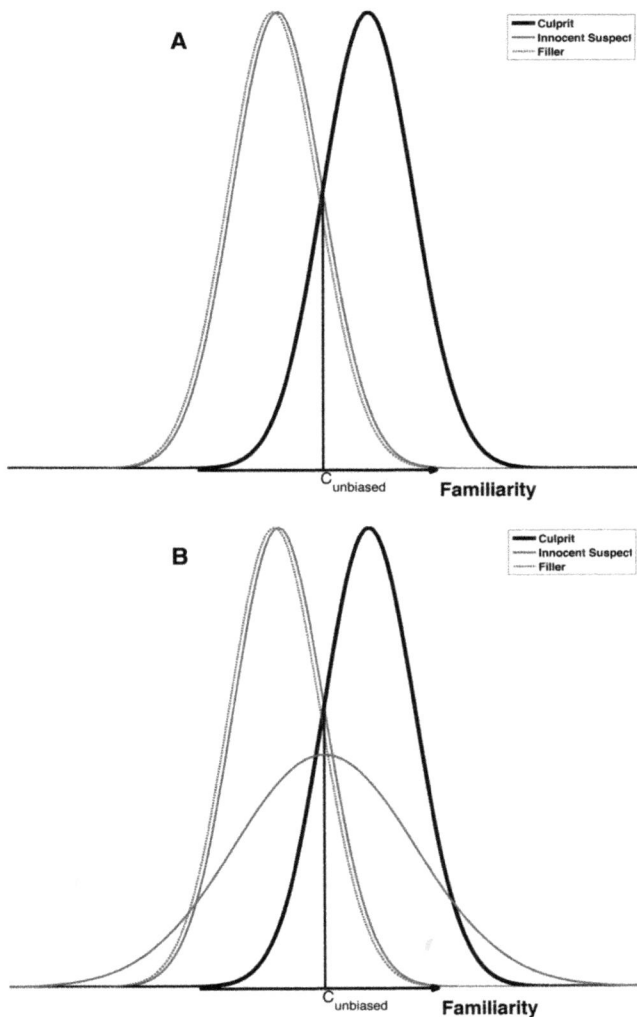

FIGURE 21.7
Plot A shows a lineup procedure under the assumption of equal variance for the culprit and innocent suspect Gaussian distributions and plot B shows a lineup procedure under the assumption of unequal variance (i.e., assuming criterial variance). C represents the ECL (Smith et al., 2017).

change in odds for one category versus the reference category (for a categorical variable). Variables selected in the model are deemed the informative variables of discriminability. The predicted accuracy from the fitted logistic regression model can be viewed in a contingency table with the observed accuracy, providing model performance. Cross-validation can be used as well.

As noted earlier, using a binary response variable may not be the most realistic choice. Researchers are interested in understanding what influences a witness to choose the true suspect, to choose an "innocent" suspect, to choose a foil, or to not choose at all. Multinomial logistic regression or some other multiple classification method may work better in this respect.

21.3.5 Expected Utility

Some researchers adopted the idea of expected utility and decision theory in the analysis of ROC curves (Lampinen et al., 2019). Since the dominating ROC curve is not always clear, some researchers have suggested an approach based on decision theory and estimation of base rates (BRs, or prior probabilities). The possible EWID task decisions result in some subjective benefit or cost, leading to the notion of a procedure's "utility," defined as the product of the probability of the outcome, BR, and cost or benefit. Other proposed measures of utility include terminal point utility (utility calculated at the rightmost point on the ROC curve), high-confidence utility (based on only those participants expressing high utility), average utility (averaged over all confidence-level utilities), and maximum utility.

Smith et al. discuss a metric for ROC curve analysis based on expected utility, which distinguishes diagnostic utility and ECL (Smith et al., 2019). This metric, known as the deviation from perfect performance (DPP), claims to consistently indicate which of two lineup procedures has higher expected utility. DPP is based on the global measure of predictive performance r discussed by Shiu and Gatsonis (2008). The modified measure for ROC curve use is defined as

$$DPP(c) = [1 - \text{Suspect}(c)] + [\text{Innocent}(c)] \tag{21.6}$$

$$\text{where } 0 \leq \text{Suspect}(c), \text{Innocent}(c) \leq 1$$

Suspect(c) is the suspect ID rate at a given point and Innocent(c) is the "innocent" suspect ID rate at the same point. Perfect performance is achieved when Suspect(c) = 1 and Innocent(c) = 0 or $DPP(c) = 0$. The index is computed as the average $DPP(c)$ of the entire ROC curve, providing a value of how much a lineup procedure deviates from "perfect performance." The $DPP(c)$ is not tied to a specific region of the ROC space, which means it does not force researchers to make comparisons that are confounded by an EWI's ECL. In both of these expected utility approaches, the assumed cost and benefit of decisions is subjective and may not accurately portray the information. Both methods rely on a single value to summarize the entirety of a lineup, similar to DR and d'.

21.4 Statistical Models From Diagnostic Medicine

The tasks in diagnostic medicine, to identify abnormalities in an image, bear resemblance to the EWID task, to identify a perpetrator from a lime-up. Accordingly, we discuss approaches that have been developed for comparing detection modalities and conducting meta-analyses in diagnostic medicine that may be suitable for comparing procedures (e.g., lineup format) in EWID. A successful model in diagnostic medicine is a bivariate random-effects statistical model for sensitivity and specificity ("hit rate" and "1 − false alarm rate" in EWID, respectively), which can lead to models for positive predictive value (PPV) and negative predictive value (NPV). These methods apply to meta-analysis for combining data from similar studies.

Meta-analysis is used to provide synthesized statistics across similar studies, including multiple tests of diagnostic accuracy. Research synthesis, when done well, also provides a determination of study validity based on study design and execution of included studies, and can also be used to test effects of patient and test characteristics and to identify areas for further research (Irwig et al., 1994). Likewise in EWID, several meta-analyses and other forms of research synthesis have been conducted, and a database identifying studies related to both single and combinations of variables (e.g., presence/absence of weapon, or retention interval) is under development. This new database, which will be publically accessible, is anticipated to identify gaps in the existing knowledge base and facilitate new research syntheses (see https://protect-us.mimecast.com/s/OL7gCJ6RyRfJEXLHGizEj? domain=osf.io).

A bivariate random-effects model (in the form of a hierarchical Bayesian model) was originally suggested by DuMouchel in a 1994 technical report as a compromise between those who used the traditional fixed-effect meta-analytic methods and those who argued against meta analysis (i.e., that data from across studies should never be combined) (DuMouchel, 1994; Junaidi et al., 2012). Other researchers support Bayesian methods in meta-analysis for the study of fixed and random effects (Sutton and Abrams, 2001; Rutter and Gatsonis, 2001). Most of these methods compare test results to a "gold" reference standard, which does not necessarily exist in the EWID paradigm. Certain methods that can overcome the lack of the reference standard could be adapted for the use in EWID data analysis, perhaps by simply comparing two experimental methods.

The framework of meta-analysis is natural for the EWID paradigm. Meta-analysis requires the combination of data from various sources (i.e., studies and experiments) that may have been performed using the same or similar settings with a common result, but were performed at different times. In the EWID field, each individual court case or eyewitness could be viewed as an individual "study." We are interested in combining the information obtained across many of these "studies" (i.e., persons) or court cases or for different experiments from various researchers in the EWID field. Should these models be adapted to EWID research, diagnostic test accuracy literature could provide a solid foundation for the work. For example, the Cochrane Collaboration, a non-profit organization formed to organize medical research findings, may provide guidelines to application through their Cochrane Handbook (Higgins et al., 2019).

The sections below review statistical models for meta-analysis from the field of diagnostic medicine, which can be adapted to be used with EWID data.

21.4.1 Logitnormal Bivariate Random-Effects Model

A popular approach to assessing the impact of several variables on accuracy in diagnostic medicine, and hence also EWID experiments, is a bivariate model for the logit transformation* sensitivity and specificity proposed by Reitsma et al. (2005), and generalized by Chu and Cole (2006). For example, in comparing diagnostic technologies in a meta-analysis, the Reitsma model is a linear mixed effects model and assumes the logit-transformed sensitivity and specificity marginally follow a normal distribution, then combines the pair into a bivariate normal distribution.

The proposed bivariate model is a logitnormal bivariate random-effects model that relies on a two-level structure, which estimates the between-study variation as well as the correlation between sensitivity and specificity. The correlation provides information on the heterogeneity of the studies. Let $\theta_{A,i}$ be the true logit sensitivity of individual study i, with common mean value θ_A and within-study variance σ_A^2. Similar notation is used for the true

---

* Logit transformation is defined $\text{logit}(p) = \log(p/(1-p))$.

logit specificity using $\theta_{B,i}$. Let σ_{AB} represent the covariance between logit sensitivity and logit specificity. Then the model is:

$$\begin{pmatrix} \theta_{A,i} \\ \theta_{B,i} \end{pmatrix} \sim N\left[\begin{pmatrix} \theta_A \\ \theta_B \end{pmatrix}, \Sigma\right], \quad \text{where } \Sigma = \begin{pmatrix} \sigma_A^2 & \sigma_{AB} \\ \sigma_{AB} & \sigma_B^2 \end{pmatrix} \tag{21.7}$$

The Chu and Cole extension reduces some Reitsma model assumptions. First, they assume the number of true negatives n_{00} and the number of true positives n_{11} follow binomial distributions,

$$n_{11,i} \mid \theta_{A,i} \sim \text{Binomial}(N_{A,i}, p_{A,i})$$

$$n_{00,i} \mid \theta_{B,i} \sim \text{Binomial}(N_{B,i}, p_{B,i})$$

Let $p_{A,i}$ and $p_{B,i}$ be the observed proportions for sensitivity and specificity, respectively. The logit-transformation can be specified as

$$\text{logit}(p_{A,i}) = X_i\alpha + \theta_{A,i}$$

$$\text{logit}(p_{B,i}) = Z_i\beta + \theta_{B,i}$$

Here, X_i and Z_i are vectors of covariates that are related to sensitivity and specificity, which are possibly overlapping. The Chu and Cole extension assumes the following structure

$$\begin{pmatrix} \theta_{A,i} \\ \theta_{B,i} \end{pmatrix} \sim N\left[\begin{pmatrix} 0 \\ 0 \end{pmatrix}, \Sigma\right], \quad \text{where } \Sigma = \begin{pmatrix} \sigma_A^2 & \rho\sigma_A\sigma_B \\ \rho\sigma_A\sigma_B & \sigma_B^2 \end{pmatrix} \tag{21.8}$$

This model for sensitivity and specificity was adapted in 2012 by Leeflang et al. for PPV and NPV (Leeflang et al., 2012); it is identtical to that in Chu and Cole except for replacing PPV and NPV for sensitivity and specificity, respectively. PPV and NPV take account of prevalence, so Leeflang et al. (2012) chose to incorporate prevalence in their model by allowing it to vary, thereby avoiding its estimation.

Chu et al. (2009) proposed a trivariate model that jointly models PPV, NPV, and prevalence (Chu et al., 2009). Ma et al. (2014) modified the trivariate model to handle a missing reference test outcome (i.e., missing disease status) (Ma et al., 2014). The model extends the Reitsma model and Chu and Cole (2006) by adding prevalence as an additional random variable, assuming a trivariate normal distribution. The latent class bivariate model is another way to evaluate the accuracy of diagnostic tests in the absence of a "gold standard" (Eusebi et al., 2014); this approach models the between-study heterogeneity by assuming each study in the meta-analysis belongs to one of K latent classes.

The logitnormal bivariate random-effects model performs well in characterizing the performance of different diagnositic modalities (EWID procedures), in part because it models the logits of the probabilities; models for dependent outcomes restricted to a range (like [0,1]) must incorporate constraints in the parameter estimation. The model does involve only one correlation parameter, although extensions to incorporate additional correlation structures are straightforward. Finally, parameter estimation via maximum likelihood (ML) estimation may require computational methods, such as numerical integration or Markov chain Monte Carlo (MCMC) techniques. All of these issues can be readily addressed.

Due to the occasional non-convergence with the standard likelihood method, Chen et al. proposed a composite likelihood (CL) function that uses an independent working assumption between sensitivity and specificity (Chen et al., 2017). The method specifies a pseudo-likelihood for sensitivity and specificity based on the marginal distributions. Equation 21.9 defines the pseudo-likelihood, and $\log L_B(\theta_B)$ is defined similarly, shown below

$$\log L_p(\theta_A, \theta_B) = \log L_A(\theta_A) + \log L_B(\theta_B), \quad \text{where} \tag{21.9}$$

$$\log L_A(\theta_A) = \sum_{i=1}^{m} \log P(n_{i,11} \mid n_{i,1}; \theta_A)$$

$$= \sum_{i=1}^{m} \left\{ \log \int \text{Binomial}(n_{i,11} \mid n_{i,1}, \text{Se}_i) \phi(\text{Se}_i; \theta_A) d\,\text{Se}_i \right\} \tag{21.10}$$

The authors note that approximation errors decrease in this method as only one-dimensional integrals are involved in the calculation. This method also relies on the marginal normality of the logit sensitivity and logit specificity, allowing the estimation to be more robust to the misspecification of the joint distribution assumption. Nikoloulopoulos compared CL versus ML estimation methods, and found that the CL method is nearly as efficient as the ML method (Nikoloulopoulos, 2018). Neither estimation method is robust to marginal distribution misspecification. The CL method proposed by Chen will always converge because the proposed pseudolikelihood has a closed form.

21.4.2 Nonparametric Meta-Analysis for Diagnostic Accuracy Studies

Zapf et al. proposed a non-parametric method for meta-analysis (Zapf et al., 2015). The authors assume fixed effects only, using a vector of individual test results that is a multivariate Bernoulli distribution

$$(\mathbf{X}'_{i0}, \mathbf{X}'_{i1}) = (X_{i01}, \dots, X_{i0n_{i0}}, X_{i11}, \dots, X_{i1n_{i1}}) \tag{21.11}$$

This format is based on Lange and Brunner's unified, nonparametric approach for sensitivity, specificity, and ROC curves (Lange and Brunner, 2012). Overall sensitivity and specificity are given as

$$\hat{\text{Se}} = \frac{1}{n_1} \sum_{i=1}^{I} \sum_{s=1}^{n_{i1}} X_{i1s}$$

$$\hat{\text{Sp}} = \frac{1}{n_0} \sum_{i=1}^{I} \sum_{s=1}^{n_{i0}} (1 - X_{i0s}) \tag{21.12}$$

where 1 indicates "diseased" and 0 indicates "non-diseased." Then, a multivariate normal distribution is defined from the overall sensitivity and specificity, using asymptotic theory, as shown below

$$\sqrt{I}\left[\begin{pmatrix} \hat{\text{Se}} \\ \hat{\text{Sp}} \end{pmatrix} - \begin{pmatrix} \text{Se} \\ \text{Sp} \end{pmatrix}\right] \sim MVN(\mathbf{0}, \mathbf{V}), \quad \text{where } \mathbf{V} = \text{Cov}\left(\sqrt{I}\left[\begin{pmatrix} \hat{\text{Se}} \\ \hat{\text{Sp}} \end{pmatrix} - \begin{pmatrix} \text{Se} \\ \text{Sp} \end{pmatrix}\right]\right) \tag{21.13}$$

The covariance matrix can be estimated by the following unbiased estimator

$$\hat{\mathbf{V}} = \frac{I^2}{I-1}\sum_{i=1}^{I}(\mathbf{Y}_i - \mathbf{S}_i)\cdot(\mathbf{Y}_i - \mathbf{S}_i)' \tag{21.14}$$

where

$$\mathbf{Y}_i = \left(\frac{\mathrm{TP}_i}{n_1}, \frac{\mathrm{TN}_i}{n_0}\right)$$

$$\mathbf{S}_i = \left(\frac{n_{i1}}{n_1^2}\cdot\mathrm{TP}, \frac{n_{i0}}{n_0^2}\cdot\mathrm{TN}\right) \tag{21.15}$$

TP and TN are the total counts across individual tests of TPs and TNs. No assumptions are made regarding the distribution of the data or the correlation structure. But the model assumes homogeneity of sensitivities and specificities across studies, and the method does not yet have a way to include covariates.

Quadrivariate Logistic Regression Model. Hoyer and Kuss proposed the quadrivariate logistic regression model to compare different diagnostic tests via meta-analysis (Hoyer and Kuss, 2018). For EWID data, researchers seek to compare different lineup procedures to determine the diagnostically superior one. This methodology could work well in the EWID paradigm. Each study reports two four-fold tables with TP_{ij}, TN_{ij}, FP_{ij}, and FN_{ij} for the ith study and jth diagnostic test, $j = 1, 2$. The TPs and TNs are still assumed to binomially distributed,

$$\mathrm{TP}_{ij} \mid \mathrm{Se}_{ij} \sim \mathrm{Binomial}(\mathrm{TP}_{ij} + \mathrm{FN}_{ij}, \mathrm{Se}_{ij})$$

$$\mathrm{TN}_{ij} \mid \mathrm{Sp}_{ij} \sim \mathrm{Binomial}(\mathrm{TN}_{ij} + \mathrm{FP}_{ij}, \mathrm{Sp}_{ij}) \tag{21.16}$$

The models for the logit transformations of sensitivity and specificity are additive in two effects: an effect, μ_j and ν_j, respectively, for the method j, and a random effect, ϕ_{ij} and ψ_{ij}, respectively; viz., $\mathrm{logit}(\mathrm{Se}_{ij}) = \mu_j + \phi_{ij}$ and $\mathrm{logit}(\mathrm{Sp}_{ij}) = \nu_j + \psi_{ij}$. Four random effects ϕ_{i1}, ψ_{i1}, ϕ_{i2}, and ψ_{i2} are assumed to follow a quadrivariate normal distribution, such as

$$\begin{pmatrix}\phi_{i1}\\\psi_{i1}\\\phi_{i2}\\\psi_{i2}\end{pmatrix} \sim N\left[\begin{pmatrix}0\\0\\0\\0\end{pmatrix}, \begin{pmatrix}\sigma_{\phi_{i1}}^2 & \rho_{\phi_1\psi_1}\sigma_{\phi_1}\sigma_{\psi_1} & \rho_{\phi_1\phi_2}\sigma_{\phi_1}\sigma_{\phi_2} & \rho_{\phi_1\psi_2}\sigma_{\phi_1}\sigma_{\psi_2}\\ & \sigma_{\psi_1}^2 & \rho_{\psi_1\phi_2}\sigma_{\psi_1}\sigma_{\phi_2} & \rho_{\psi_1\psi_2}\sigma_{\psi_1}\sigma_{\psi_2}\\ & & \sigma_{\phi_2}^2 & \rho_{\phi_2\psi_2}\sigma_{\phi_2}\sigma_{\psi_2}\\ & & & \sigma_{\psi_2}^2\end{pmatrix}\right] \tag{21.17}$$

The model captures the potential between-study heterogeneity of sensitivities and specificities, as well as the corresponding correlation among the random effects. The main parameters of interest are the differences of sensitivities and specificities between the meta-analyses. The difference in the logistic transformations (i.e., inverse logit) of sensitivity and specificity between the two studies provide the following formula for the parameter of interest,

$$\Delta\mathrm{Se} = \frac{\exp(\hat{\mu}_1)}{1+\exp(\hat{\mu}_1)} - \frac{\exp(\hat{\mu}_2)}{1+\exp(\hat{\mu}_2)}$$

$$\Delta\mathrm{Sp} = \frac{\exp(\hat{\nu}_1)}{1+\exp(\hat{\nu}_1)} - \frac{\exp(\hat{\nu}_2)}{1+\exp(\hat{\nu}_2)} \tag{21.18}$$

Similarly, Dimou et al. proposed a multivariate method for the meta-analytic comparison of diagnostic tests (Dimou et al., 2016). It is an extension of the bivariate model for the comparison of two or more tests.

21.5 Supervised Learning Classification Methods

The classification problem has a long history in the statistical literature; it has reappeared in the machine learning field as "supervised learning" but the goal is the same: create "rules" by which to categorize new observations into groups. We provide a brief overview of some common classification methods as potential models for eyewitness identification accuracy. The algorithms result in predicted decisions that can be compared to the underlying truth, as well as the influential predictors for the decisions. The goal is to minimize all types of errors. The resulting model can be adjusted by changing the thresholds of errors, depending on whiche error is considered more grievous. Once the model has been trained properly under the supervised learning framework, and validated with representative test data, it can be applied to real world data.

The difference between the methods mentioned in this section and the methods mentioned in the previous sections is the lack of a meta-analytic framework. At this time, the described methods cannot accomodate the meta-analytic framework. Some researchers are exploring methods of integrating machine learning algorithms to aid in study selection and data extraction for systematic reviews and meta-analysis. Methods have not yet been developed for computational purposes.

In classification methods, point estimates, which can be characterized by finding variance estimates using simulation and/or repetition, are obtained per data set. The true value in classification methods is how easily they are applied, which could be helpful for law enforcement agents, lawyers, and jurors. How well these models work in practice is yet unknown, but can be determined through simulation or application to real data sets.

21.5.1 Machine Learning Classification Models

Common classification methods include linear discriminant analysis (LDA), quadratic discriminant analysis (QDA), boosted logistic regression (in addition the standard logistic regression), decision trees, random forests, graphical models via Bayesian networks, support vector machines (SVMs), and neural networks. Brief descriptions of these methods, as well as graphical approaches, are provided in the following sections; see also *The Elements of Statistical Machine Learning, 2nd Edition* by Hastie et al. (2016) for in-depth discussions on all methods. Some of these methods (SVMs, random forests, and neural networks) suffer from "black box" syndrome, where the the results are not necessarily interpretable due to injected randomness, etc. Machine learning researchers have developed methodologies to mitigate this issue, which is beyond the scope of this chapter. These methodologies include the partial dependence plot (PDP) from Friedman (2001), local interpretable model-agnostic explanations (LIME) from Ribeiro et al. (2016), and Shapley additive explanations (SHAP) from Lundberg and Lee (2017).

Discriminant Analysis. LDA and QDA are conventional classification methods proposed by Fisher (1936) that use linear and quadratic decision boundaries, respectively, in the space spanned by the covariates that influence the outcome. In the framework of EWID, the outcome is "accuracy" (hit rate or 1 – false alarm rate), using vectors of covariates to predict eyewitness' decisions. The choice between LDA and QDA depends heavily on the structure and amount of data, and the assumption of normally-distributed covariates; QDA for an underlying linear model results in highly biased predictions.

(Boosted) Logistic Regression. Logistic regression (Section 21.3.4 in 21.2.1) can be made more powerful by "boosting," which was originally proposed Schapire (1990). The idea was further adapted to gradient boosting machines by Friedman et al. (2000). Boosting combines the performance of many "weak" classifiers to produce a more powerful "committee." For EWID analysis, covariates are added to account for differences in probability for a correct or incorrect identification. For more on boosting, see Hastie et al. (2016).

Decision Trees and Random Forests. Decision or classification trees provide the foundation for random forests. The goal of decision trees is to create a model that predicts a value of a target variable based on several covariates. Nodes on the tree are the decision points that provide the path for the particular datum considered. Decision trees are simple to understand and easy to interpret. Classification trees are the individual units of random forests. Given the data, the covariates will be used as splitting variables to branch the data into sorted clusters. The splits are determined based on the homogeneity of observations in the resulting child nodes from the parent node. The resulting terminal nodes will be the decision determined by classification and regression tree (CART) algorithm (Breiman et al., 1984).

Random forests are ensemble classifiers based on decision trees. Votes arise from groups of decision trees. Tree bagging (bootstrap aggregating) draws repeated samples from the original data. Each sample is drawn randomly with replacement, and creates a classification tree. One generates B such trees. When one wants to classify a new observation, one uses each of the B trees in the "forest" (collection of de-correlated trees) and uses majority (or plurality) rule to assign the classification. This decreases the variance in the model. Random forests are also generated using feature bagging, where random samples of covariates are used for each tree rather than the entire set of covariates. For each candidate (observation), a random subset of features is obtained. An observation is classified by majority vote from all the trees. Explaining the concept of a random forest can be done using visualizations.

Support Vector Machines (SVMs). Similar to other supervised learning algorithms, SVMs take as input the covariates for EWID to build the model based on training data. SVMs construct a hyperplane that is used to separate the data. A high-dimensional divider classifies the data into groups based on the interaction of several covariates. SVMs rely to classifying using hyperplanes (i.e., some sort of separator) in high dimensions, depending on the number of included covariates. Conveying this concept of high-dimensionality to laypeople may be difficult, which may affect its use in EWID and law enforcement settings. While SVMs can be effective and accurate in prediction in some circumstances, both the SVM algorithm and the output are difficult to interpret, making SVMs possibly problematic for a court setting.

Neural Networks. Neural networks is a black box method that uses layers or neurons $p_j(t)$, which receive input. These neurons then change their internal state (activation) $a_j(t)$ based on that input, and produces output. Some threshold θ_j determines activation, which is an input to some activation function $a_j(t+1) = f(a_j(t), p_j(t), \theta_j)$. The output function is expressed as $o_j(t) = f_{out}(a_j(t))$. The network is formed by the connection of several of these neurons. Neural networks are flexible and can model a variety of functional forms, making it useful for complex and/or abstract problems. Like other machine learning algorithms, neural networks require training and computational resources. The covariates in an EWID experiment are used to determine the hidden units of the neural network, which are processed by the output function, resulting in a decision for each person. The decision from the algorithm for each person can then be compared to the person's actual outcome.

21.5.2 Graphical Models

Graphical models, used in other forensic analysis, are also useful for the EWID paradigm (Dawid and Mortera, 2017). Luby explored this approach with log-linear analysis (Luby, 2016). In this model, the data are in the form of a multi-way table with Target Absence/ Presence (2 levels) × Eyewitness Decision (2 levels) × ECL (11 levels) × Witness instructions (2+ levels); additional variables can be included without changing the theoretical foundation for the analysis. The model is fit iteratively to find the expected counts for each cell using a training set of data. Based on the experiment and corresponding data, we generate different graphical models as follows. Let α represent the main effects, β represent the two-way interactions, subscript wc represent witness choice, subscript t represent target absence or presence, i represent witness instructions, and c represent ECL. Equation 21.19 shows an example of a fitted model (Figure 21.8). The model can include system and estimator variables, previously discussed in Section 21.2 of part 21.1.

$$\log m_{wc,t,i,c} = \alpha_{wc} + \alpha_t + \alpha_i + \alpha_c + \alpha_e + \beta_{wc,t} + \beta_{wc,i} + \beta_{wc,c} + \beta_{e,i} + \beta_{c,i} \tag{21.19}$$

Garbolino discusses the use of Bayesian networks for evaluating testimony (Garbolino, 2016); Garbolino's model is actually very general, and applies to testimony of any kind, not just from an eyewitness. The proposed model assumes that the witness is: (1) accurate, (2) objective, and (3) truthful. Each of these characteristics corresponds to an inference about the witness' personality:

1. Senses give evidence of what is seen;
2. Belief in the evidence from the senses;
3. Belief in what is said.

In the end, Garbolino proposes an object-oriented Bayesian network class for the analysis of the reliability of human witnesses. D'Agostini notes that Bayesian networks are a technical tool, but their true value is as a very powerful conceptual tool that can handle

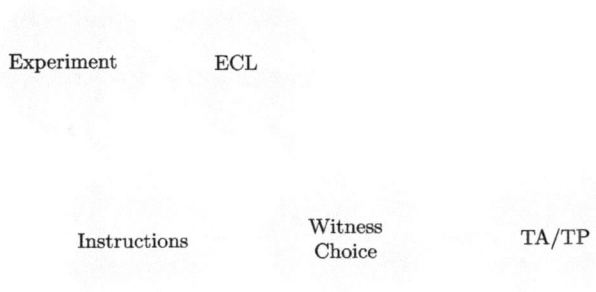

FIGURE 21.8
This is the graphical model corresponding to log-linear model in Equation 21.19.

complex problems with variables related by both probabilistic and causal links (D'Agostini, 2016). Even with subjective probability (i.e., eyewitness testimony), the intuitive idea of probability can be recovered.

21.6 Tools Based on ROC Methods

The popularity of the ECL-based ROC curve to compare lineup procedures, together with its limitations (see Section 1.3.1), leads us to consider other methods that augment and improve upon ROC curves for a more complete comparison between methods. We discuss the PROC curve (which utilizes PPV and NPV in a similar way that HR and FAR are used in ROC curves), multivariate ROC curves, and AUC estimation for these curves. We also discuss the inclusion of variability measures for ROC curves that could also be adapted for the PROC curve and multivariate ROC curves.

21.6.1 Methodology Development

The NRC report called for a broader "exploration of the merits of different statistical tools for use in the evaluation of eyewitness performance" as an important area of research (National Research Council, 2014).

The analysis of EWID experimental data needs to consider three aspects:

1. Sensitivity: the probability that an eyewitness correctly identifies the true perpetrator;
2. Specificity: the probability that an eyewitness correctly dismisses an innocent person;
3. Prevalence: the proportion of individuals who might be the culprit.

Positive predictive value (PPV) and negative predictive value (NPV) are functions of all three aspects, so accurate estimation of all three quantities is essential (Kafadar, 2015).

PPV is the probability that, when the eyewitness makes an identification, the identified person is truly the perpetrator. Similarly, NPV is the probability that, when the eyewitness fails to identify a person as a perpetrator, that person was truly not the perpetrator. In general, and in real life, we don't know if the eyewitness's decision is correct, but we can estimate the probability (PPV, NPV). PPV can be rewritten in terms of the odds ratio (OR) and positive likelihood ratio (LR$_+$) as follows. Let Se and Sp denote sensitivity and specificity, respectively, and p the probability that the suspect is truly the perpetrator. For example, in a lineup with six photos, p could be equal to $1/6$. We define PPV as the conditional probability that the identification was correct, given that the eyewitness selected a person from the lineup:

$$\text{PPV} = \frac{\#\text{ true positives}}{\#\text{ true positives} + \#\text{ false positives}}$$

$$= \frac{\text{Se} \cdot p}{\text{Se} \cdot p + (1 - \text{Sp}) \cdot (1 - p)}$$

$$= \frac{1}{1 + 1/(\text{OR} \cdot \text{LR}_+)} \tag{21.20}$$

where $\text{OR} = p/(1-p)$ denotes the ratio of probabilities that a suspect P is guilty versus is innocent, and LR_+ is the likelihood ratio of a positive call:

$$\text{LR}_+ = \frac{P\{\text{ eyewitness selects } P \mid P \text{ is perpetrator}\}}{P\{\text{eyewitness selects } P \mid P \text{ is innocent}\}}$$

$$= \frac{\text{Se}}{1-\text{Sp}} = \frac{\text{HR}}{\text{FAR}} \tag{21.21}$$

Because LR_+ can be written as $\text{Se}/(1-\text{Sp}) = $ (hit rate/false alarm rate), it is equivalent to the diagnosticity ratio (DR); as discussed in Section 1.3.1. Note also that the slope of the ECL-based ROC curve at ECL level C is the diagnosticity ratio (LR_+) for those persons who expressed confidence of at least C. A higher LR_+ leads to a higher PPV for the same prevalence. Thus, if the probability that the guilty suspect is in the lineup (i.e., population under consideration), then the lineup procedure with the higher DR yields a higher PPV. We can also see that PPV is more affected by specificity.

NPV is the probability that the excluded person is truly not the perpetrator. NPV can also be written in terms of OR and the negative likelihood ratio (LR_-). Similar to how a higher LR_+ results in a higher PPV, a lower LR_- would result in a higher NPV. We define NPV as

$$\text{NPV} = \frac{\text{\# true negatives}}{\text{\# true negatives} + \text{\# false negatives}}$$

$$= \frac{\text{Sp} \cdot (1-p)}{(1-\text{Se}) \cdot p + \text{Sp} \cdot (1-p)}$$

$$= \frac{1}{1 + (\text{OR} \cdot \text{LR}_-)} \tag{21.22}$$

LR_- is the likelihood ratio of a negative call. A smaller LR_- leads to a higher NPV for the same prevalence. NPV is normally not considered, as most researchers in EWID, as well as legal and law practitioners, consider a foil choice as not dangerous, because foils are known to be innocent (Amendola and Wixted, 2014) practitioners and policymakers are less concerned with the probabilities associated with choosing innocent foils (Wixted et al., 2015). We define LR_- as

$$\text{LR}_- = \frac{1-\text{Se}}{\text{Sp}} = \frac{1-\text{HR}}{1-\text{FAR}} \tag{21.23}$$

Provided the conditions for comparing two lineup procedures are the same (e.g., OR is the same), then procedure one is preferred over procedure two if PPV_1 (PPV for procedure one) is greater than PPV_2 (PPV for procedure two). This is true if and only if

$$\frac{\text{OR}_1}{\text{DR}_1} < \frac{\text{OR}_2}{\text{DR}_2} \equiv \frac{\text{DR}_1}{\text{OR}_1} > \frac{\text{DR}_2}{\text{OR}_2} \tag{21.24}$$

Using PPV (i.e., LR_+) as the criterion, procedure one is preferred over procedure two if

$$\text{DR}_1 > \text{DR}_2 \equiv (\text{LR}_+)_1 > (\text{LR}_+)_2 \tag{21.25}$$

Similarly, for NPV, method one is preferred if

$$\frac{1}{(\text{LR}_-)_1} > \frac{1}{(\text{LR}_-)_2} \tag{21.26}$$

Thus, both LR$_+$ and LR$_-$ need to be be considered when choosing "optimal" procedures. In the EWID paradigm, a vector of match-to-the-witness's-memory values (i.e., memory strength) for $n-1$ alternatives with a lineup size of n between the eyewitness's memory of the perpetrator and the lineup member could be used in conjunction with the NPV (Clark, 2005). This emulates the framework for ROC curves in the SDT model. Since the PROC curve is an extension of the ROC curve, we can use some of the same ideas.

Some EWID researchers state that as responding becomes more conservative, both LR$_+$ and LR$_-$ increase, suggesting these values depict the tradeoff related to liberal versus conservative responding, not discriminability (Mickes et al., 2017). LR$_+$ and LR$_-$ are functions of only sensitivity and specificity. The targeted values in EWID accuracy experiments are PPV and NPV; hence they need to be jointly considered also in the analysis of identification accuracy.

21.6.2 Predictive Receiver Operating Characteristic Curve

Shiu and Gatsonis (2008) proposed a display PPV and NPV jointly via a *predictive receiver operating characteristic* (PROC) curve (Shiu and Gatsonis, 2008). The PROC curve is defined as $\{(1-\text{NPV}(c), \text{PPV}(c))\}$ for $c \in R$, where R is the set of all possible thresholds for test positivity. This curve is affected by prevalence p. Specifically speaking, PPV increases and NPV decreases when prevalence increases. Thus, with increasing prevalence, a point on the PROC curve will move towards the upper-right direction.

The PROC curve lacks monotonicity, which occurs if a one-to-one correspondence between PPV and NPV exists. The criteria for monotonicity is established using hazard rate order, reverse hazard rate order, and likelihood ratio order. The likelihood ratio order says the ratio $f(c)/g(c)$ is a monotonic function of c; this is a sufficient condition for monotonicity of the PROC curve. But the monotonicity properties are complex in certain cases. It seems that in the binormal case, if the scaling parameter $b=1$ for the binormal model, then we can see an obvious trade-off between PPV and NPV, and the PROC curve is monotone. For $b \neq 1$, monotonicity is guaranteed for only certain segments along the curve.

Figure 21.9 demonstrates this complicated pattern of monotonicity. The middle plot shows the clear monotonicity – an increase in PPV has a corresponding decrease in NPV. For the other two plots, overlap is visible depending on the location along the

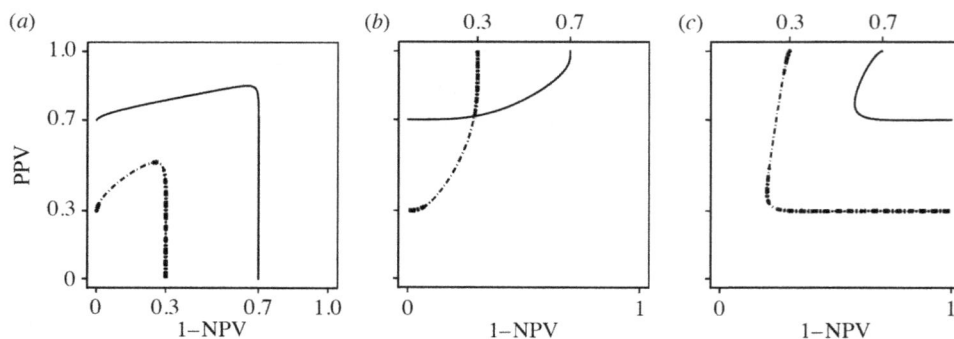

FIGURE 21.9
This plot shows the predictive curves with $a = 0.8$: (a) $b = 0.7$, (b) $b = 1$, (c) $b = 1.5$. The solid line represents high prevalence ($p = 0.7$) and the dot-dashed line represents low prevalence ($p = 0.3$) (Shiu and Gatsonis, 2008).

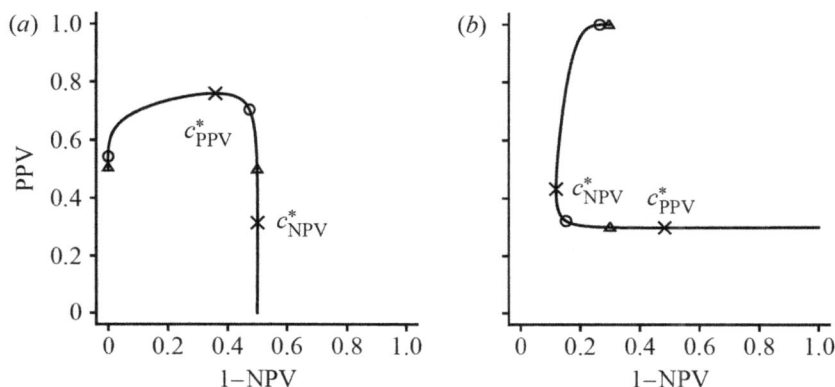

FIGURE 21.10

Predictive curves with (a) $a = 1$, $b = 0.5$, $p = 0.5$; (b) $a = 2$, $b = 2$, $p = 0.3$. Circles denote points corresponding to operating thresholds at -1 and $a + b$, triangles denote points corresponding to operating thresholds at -2 and $a + 2b$, and crosses denote c_{PPV}^* and c_{NPV}^* (Shiu and Gatsonis, 2008).

curve. Figure 21.10 shows this phenomenon, where monotonicity is defined on $(-\infty, c_{PPV}^*)$, $[c_{PPV}^*, c_{NPV}^*]$, and (c_{NPV}^*, ∞). The visually vertical and horizontal lines in this figure result from either PPV or NPV converging faster than the other.

21.6.3 Multivariate ROC Curves

The standard ROC curve cannot account for more than two covariates. In 2009, Jin and Lu (2009) proposed the ROC region, which plots the TPR over the FPR for all possible choices of the decision thresholds for two continuous covariates. The thresholds arise from a tree-based nonlinear combination of multiple predictors. In 2012, Wang and Li (2012) proposed a bivariate ROC and a bivariate weighted ROC (WROC) for biomarker measurements. The authors defined a bivariate ROC as a conditional expectation of TP as a function of the two continuous biomarkers given the FP as a function of the two biomarkers. Let $0 \leq q \leq 1$, (U_0, V_0) be a pair of bivariate markers from a non-diseased group,

$$\text{ROC}(q) = E\left[\text{TP}(U_0, V_0) \mid \text{FP}(U_0, V_0) = q\right] \qquad (21.27)$$

The authors further defined WROC as the unconditional expectation of TP as a function of the two continuous biomarkers given the FP as a function of the two biomarkers. This idea was extended in 2013 to multivariate biomarkers (Wang and Li, 2013). In the multivariate markers extension, the decision thresholds for the continuous biomarkers were decided by classification tree-based methods. Another similar method, proposed by Pundir and Amala (2015), is the bivariate lognormal ROC curve for detecting the accuracy of two biomarkers. The WROC is a plot of the TPRs and FPRs as functions of the two thresholds from the biomarkers. Possibly we can adapt these multivariate ROC methods to both continuous and categorical predictors, and apply them to EWID contexts.

Hong (2012) proposed a bivariate ROC model, which assumes a sliced bivariate normal distribution function for two predictors, X_1 and X_2. In the method, $X_2 = h(X_1)$ by using some linear function that passes through the mean vector of the X_2 pseudo-random variable. The points for the ROC come from the cumulative distribution function defined for the ROC curve. Hong and Jeong (2012) proposed An optimal classification function for this bivariate ROC curve (Hong and Jeong, 2012).[*]

21.6.4 AUC Estimation

The AUC for any one of the bivariate ROC curves can be modeled as a function of eyewitness, and lineup procedure, and any other variables at play, using a hierarchical model, similar to the model proposed by Wang and Gatsonis (2008). In that article, the authors propose a hierarchical model for multi-reader, multi-modality[†] studies in diagnostic medicine. Heterogeneity is introduced at the first level of the hierarchy. Effects for some covariates, such as reader variability, are treated as random (not fixed), and Markov chain Monte Carlo (MCMC) can be used to estimate model parameters. In the model, three levels (i.e., types) of correlation exist:

- Within-reader variability;
- Between-reader variability;
- Variation of hyperparameters.

The within-reader variability represents the correlation due to readers between AUC estimates for a reader in two modalities. Let Z_j represent the reader level covariates and β represent the reader random effects. The authors assume correlation r_{1j} is common across all readers. Letting γ be a vector of regression coefficients, following independent normal prior distributions with mean zero and large variances, the authors model the within-reader variability as

$$\begin{pmatrix} y_{1,j} \\ y_{2,j} \end{pmatrix} \Bigg| \boldsymbol{\beta}_j \sim N\left[\begin{pmatrix} \mu_{1,j} \\ \mu_{2,j} \end{pmatrix}, \Sigma_j \right], \tag{21.28}$$

where

$$\Sigma_j = \begin{bmatrix} f(\mu_{1j}) & r_{1j}\sqrt{f(\mu_{1j})f(\mu_{2j})} \\ r_{1j}\sqrt{f(\mu_{1j})f(\mu_{2j})} & f(\mu_{2j}) \end{bmatrix}$$

$$\text{and} \quad \boldsymbol{\mu}_j = \begin{pmatrix} \mu_{1j} \\ \mu_{2j} \end{pmatrix} = g(\mathbf{Z}_j^T \boldsymbol{\gamma} + \boldsymbol{\beta}_j)$$

The between-reader variability represents the correlation from two different readers in the same modality. This model assumes that the random effects follow normal distributions, where $\beta_{1j} \mid \sigma_1^2 \sim N(0, \sigma_1^2)$, $\beta_{2j} \mid \sigma_2^2 \sim N(0, \sigma_2^2)$, and $\boldsymbol{\beta}_j = (\beta_{1j}, \beta_{2j})^T$. The variation of hyperparameters represents the correlation due to cases between any two AUC estimates since the estimates arise from the same set of subjects. We also assume proper prior distributions on the hyperparameters, such that σ_1^2 has an inverse gamma distribution with parameters a_1 and b_1 and, similarly, σ_2^2 has an inverse gamma distribution with parameters a_2 and b_2. Additionally, r_{1j} has a beta distribution with parameters c_j and d_j.

---

[*] Both papers, Hong (2012) and Hong and Jeong (2012), are in Korean; their abstracts are in English.
[†] A modality is a particular diagnostic procedure.

We can easily extend the model to include additional covariates. The approach may be a more efficient alternative to stratified analyses, and it also is sufficiently flexible to accommodate complex correlation structures. Finally, the model fitting can be done on publicly available software.

In 2012, Lange and Brunner proposed a unified, nonparametric approach to multi-reader diagnostic trials based on ranks, which allows them to estimate the AUC as a vector for different modalities (Lange and Brunner, 2012). The authors suggest the approach is equivalent to a factorial chi-squared test on repeated measures. In the factorial design, the reader (i.e., eyewitness) and modality (i.e., lineup procedure) are the two factors. They use a nonparametric approach to show that sensitivity and specificity are areas under particular ROC curves. Additional nonparametric estimations for AUC could also be considered.

21.6.5 Confidence Intervals for ROC

Based on the considerations of previous work done with ROC curves in the EWID paradigm, variability is an increasingly important consideration for more robust conclusions. We continue the discussion of variability in EWID data and for confidence-based ROC curves. These methods acknowledge that the calculated DR values for ROC curves are point estimates at the center of an interval that captures the true population DR value. The methods presented below are possible ways to incorporate measures of variability.

Point-wise confidence intervals for ROC curves are the intervals of sensitivity at a given value of specificity. We can construct confidence bands for a range of specificity or for the entire ROC curve (Yin, 2014). Some EWID researchers have used bootstrap resampling to estimate standard errors for their ROC curves (Mickes et al., 2017). Luby used confidence boxes for the HR and FAR (Luby, 2017). Confidence bands are routinely calculated for medical applications, and should be used for EWID applications also.*

Macskassy and Provost (2008) provide an empirical study of methods for the estimation of confidence bands (Macskassy and Provost, 2008). These methods include vertical averaging (VA), threshold averaging (THA), simultaneous joint confidence regions (SJR), Working-Hotelling based bands (WHB), and fixed-width confidence bands (FWB). These rely on sweep methodology, which samples the observed ROC point and the confidence boundary around it to generate upper and lower confidence bands.

The VA, THA, and point-wise WHB did not translate well to confidence bands, and failed to perform robustly for varied parameters used for the generation of ROC curves. The authors attribute the failure of VA and THA to the naive methodology. Both the SJR and FWB worked well, and quite robustly, given the data. The SJR does not require any samples to generate the confidence bands, but has a higher variance. The FWB uses the bootstrap to empirically determine the proper displacement for the confidence band generation, but was found to be stable and consistent. In 2012, Demidenko introduced parametric confidence bands for the binormal ROC curve, and the ellipse-envelope (EE) confidence band construction based on the Working-Hotelling approach, with variation calculated via the delta method (Demidenko, 2012). The EE confidence band has a shorter width than the WHB, under the assumption of a binormal curve. More details of these methods are described in Table 21.3.

---

* See examples in Appendix D of the NRC 2015 report.

TABLE 21.3

This Table Provides a Summary of ROC Curve Confidence Interval/Band Estimation Methods

Method	Description
VA	This is a sweep method that looks at successive FP rates and averages TPs for multiple bootstrapped ROC curves at a specific FP rate.
THA	This is a sweep method that freezes the threshold of the test rather than the FP rate, by identifying the set of ROC points that would be generated using a particular threshold on each of multiple ROC curves.
SJR	This method utilizes the Kolmogorov-Smirnov (KS) one-sample test statistic to find the global confidence interval (i.e., simultaneous confidence rectangles) for TP and FP, which are generated by freezing FP to identify the respective TP.
WHB	This method fits a a regression line $y = a - b \cdot x$, of the form $\ell(x, \pm k) = a - b \cdot x \pm k \cdot \sigma(x)$ for $k \geq 0$ and $\sigma(x) = \sqrt{\sigma_a^2 - 2\rho\sigma_a\sigma_b \cdot x + \sigma_b^2 \cdot x^2}$. The line is estimated using maximum likelihood estimation (MLE). Other estimation methods include coaxial ellipses based on an envelop of a system of ellipses.
FWB	This method identifies a slope $b = -\sqrt{m/n} < 0$, where m and n are sample sizes, along which to displace the original ROC curve to generate confidence bands, and sweeps along the FP axis to identify the TP value at that FP.

21.7 Estimating Probability of Eyewitness Accuracy

So far, we have discussed methods to model data to assess influences of variables on EWID accuracy. Importantly, law enforcement officers seek a tangible, usable tool that can immediately assess the inherent reliability of an eyewitness. What statistical framework can we use to provide an objective substantiation to the reliability of eyewitnesses? Can we reliably estimate an eyewitness's accuracy given the conditions of the identification process?

We use the system and estimator variables to estimate an individual's *probability of choosing* dependent upon target presence, which form the components for estimating the probability of eyewitness accuracy.

21.7.1 Probability of Accuracy

We decompose the probability of accuracy $P(\text{Accurate})$ into two components, which depend on the rate of target presence and the probability of making certain decisions. Let

$$p_1 = P(\text{Choose Target} \mid \text{TP})$$
$$p_2 = P(\text{Choose Foil} \mid \text{TP}) \tag{21.29}$$
$$p_3 = P(\text{Don't Choose} \mid \text{TP}) \cdot P(\text{TP})$$

where $p_1 + p_2 + p_3 = 1$. Further, let

$$a_1 = P(\text{Don't Choose} \mid \text{TA})$$
$$a_2 = P(\text{Choose Foil} \mid \text{TA}) \tag{21.30}$$

where $a_1 + a_2 = 1$. Also, let

$$\theta = P(\text{TP}) = P(T = 1)$$
$$1 - \theta = P(\text{TA}) = P(T = 0)$$

(21.31)

The decomposition of the $P(\text{Accurate}) = P(Y_i = 1)$ is given below.

$$P(\text{Accurate}) = P(Y_i = 1)$$
$$= P(Y_i = 1 \cap T = 1) + P(Y_i = 1 \cap T = 0)$$
$$= p_{1i} \cdot \theta + a_{1i} \cdot (1 - \theta)$$

(21.32)

From these equations, $P(Y_i = 1)$ depends on p_1, a_1, and θ. We estimate p_1 and a_1 by fitting classification models to the subsets of target present and target absent data, respectively, from an well-designed experiment. Given the target is present, we have a multiple classification problem: (1) choose target; (2) choose foil; and (3) don't choose. Given the target is absent, we have a binary classification problem: (1) choose foil; and (2) don't choose.

We can estimate the rate of target presence θ using a decomposition of a subset of the decision outcome space. In this case, we decompose a quantity that will always be observed, the probability of choosing $P(\text{Choose})$. Let the probability of choosing X_i follow a Bernoulli distribution with parameter $p_i(\theta)$ for each person $i = 1, \ldots, n$ where $p_i(\theta)$ is the probability of "success" (i.e., choosing). Let $a_i = (1 - a_{1i})$ and $b_i = p_{1i} + p_{2i} + a_{1i} - 1$.

$$P(\text{Choose}) = P(X_i = 1)$$
$$= P(X_i = 1 \mid T = 1) \cdot P(T = 1) + P(X_i = 1 \mid T = 0) \cdot P(T = 0)$$
$$= (1 - a_{1i}) + (p_{1i} + p_{2i} + a_{1i} - 1) \cdot \theta$$
$$= a_i + b_i \cdot \theta$$
$$= p_i(\theta)$$

(21.33)

We estimate θ using maximum likelihood estimation (MLE). The joint likelihood of X_i for n observations is,

$$L[p_i(\theta)] = \prod_{i=1}^{n} [p_i(\theta)]^{x_i} \cdot [1 - p_i(\theta)]^{1-x_i}$$

(21.34)

The joint log-likelihood,

$$\ell[p_i(\theta)] = \sum_{i=1}^{n} \left[x_i \log(a_i + b_i \cdot \theta) + (1 - x_i) \log(1 - a_i - b_i \cdot \theta) \right]$$

(21.35)

Taking the first partial derivative of the joint log-likelihood with respect to θ and setting equal to zero to solve for the maximum,

$$\ell[p_i(\theta)] = \sum_{i=1}^{n} \left[\frac{x_i \cdot b_i}{a_i + b_i \cdot \theta} - \frac{(1 - x_i) \cdot b_i}{1 - a_i - b_i \cdot \theta} \right] \triangleq 0$$

(21.36)

Solving Equation 21.36 explicitly for $\hat{\theta}_{MLE}$ would be difficult. Using a numerical method can provide a point estimate such that $\hat{\theta}_{MLE} \in [0,1]$; e.g., the routine `uniroot()` in R.

With an estimate of the probability of target presence $\hat{\theta}_{MLE}$, we have all of the components to estimate the underlying probability of accuracy from Equation 21.32.

This proposed procedure has two immediate benefits:

- It provides a way to estimate the underlying base rate of target presence in a lineup;
- It provides an estimate for an eyewitness's accuracy based on given information.

From discussion with law enforcement officers, judges, and lawyers in the EWID community, both of the above quantities are of extreme interest for matters of policy and understanding factors affecting EWID accuracy.

21.8 Example

As noted in Section 21.3.1, the diagnosticity ratio [HR/FAR = sensitivity / (1 – specificity)] can depend not only on an eyewitness's tendency towards "conservative" or "liberal" identification (as measured by "expressed confidence level"), but also on numerous other factors, including:

(A) Lineup procedure (e.g., 2 levels: simultaneous versus sequential);

(B) Weapon presence/absence (2 levels; more levels could be considered, such as *gun, knife, towel, none*);

(C) Stress (e.g., 3 levels such as high, medium, low);

(D) Elapsed time between incident and exam (e.g., 3 levels: 30 min, 2 hours, 1 day);

(E) Race difference (e.g., 2 levels: same or different race; or 4 levels: eyewitness/ culprit = white/white; white/non-white; non-white/white; non-white/non-white; non-white/white);

(F) Subject (e.g., N levels, corresponding to N subjects).

If a study is sufficiently large, one could construct an ROC for each participant corresponding to each of these conditions; i.e., plot *HR* versus *FAR* at different expressed confidence levels for each participant of the N participants. (To avoid running such an enormous experiment, one would sensibly consider running a fraction of all possible combinations.[*]) One can then summarize the information in the ROC via different measures, such as the logarithm of the AUC, or log(AUC). Consider the following approach:

Let $y_{ijklmnr}$ denote the log(AUC) for the rth trial using participant n ($n = 1,\ldots,N$) for procedure i, weapon level j, stress level k, time condition ℓ, and cross-race effect m. Then we could write:

$$y_{ijklmnr} = \mu + \alpha_i + \beta_j + \gamma_k + \delta_l + \phi_m + (\alpha\beta)_{ij} + \ldots \text{(interactions)} \ldots + \epsilon_{ijklmnr}$$

---

[*] See, for example, Box et al. (2005) on constructing *fractional factorial designs*.

where μ represents the overall average log(AUC) across all conditions, the next six terms reflect the main effects of A (lineup procedure: $i = 1$ for sequential and $i = 2$ for simultaneous); B (weapon: $j = 1$ for presence and $j = 2$ for absence of weapon); C (stress level: $k = 1$ for low, $k = 2$ for medium, $k = 3$ for high); D (elapsed time between incident and report: $\ell = 1$ for 30 minutes, $\ell = 2$ for 2 hours, $\ell = 3$ for 1 day); E (cross-race effect: $m = 1$ for same race, $m = 2$ for different races); F (participant effect: $n = 1, 2, \ldots, N$ participants); "(interactions)" reflects the joint effect of two or more factors together, and the last term, $\epsilon_{ijklmnr}$ represents any random error in the rth trial that is not specified from the previous terms (e.g., measurement, ECL, multiple trials, etc.). This approach would allow separation of the effects of the different factors, enable one to assess which factors have the greatest influence on the outcome (here, logarithm of the area under the ROC curve: bigger is better), and especially to evaluate the importance of these factors relative to variation among "eyewitnesses." It may be that eyewitnesses are the greatest source of variability, dominating the effects of all other factors. Or it may be that, in spite of person-to-person variability, one or more factors still stand out as having strong influence on the outcome. Note that (i) other covariates could be included, such as age and gender of participant; and (ii) the ROC curve need not be defined in terms of "expressed confidence level" thresholds, if a more sensitive measure of "response bias" (tendency towards "liberal" versus "conservative" identifications) can be developed.

An Example Carlson and Carlson (2014) use pAUC, as a summary measure of the information in an ROC curve (bigger is better), for each of 12 different conditions defined by three factors:

(A) *Procedure*, 3 levels: simultaneous (SIM: suspect in position 4), sequential (SEQ2: suspect in position 2), sequential (SEQ5: suspect in position 5);

(B) *Weapon focus*, 2 levels: present versus absent;

(C) *Distinctive feature*, 2 levels: present versus absent.

The data are provided in their Table 3, along with 95% confidence intervals. Because the length of a confidence interval is proportional to the standard error, pAUC values with shorter confidence intervals correspond to having smaller standard errors and hence should have higher weights. The logarithms of the reported pAUC values and weights (reciprocals of the lengths of the reported confidence intervals) are given Table 21.4.

For this study, the data on all $N = 2675$ participants (720 undergraduates and 1955 "Survey Monkey" respondents) were combined, and ECLs were solicited on a 7-point scale. Variations in the 12 log(pAUC) values can be decomposed into three main effects (one each for procedure, weapon, and feature), and their two-way interactions. (The raw data may permit a more detailed analysis.) The data can be analyzed using a less complex model than that stated above (because we have fewer terms):

$$y_{ijk} = \mu + \alpha_i + \beta_j + \gamma_k + (\alpha\beta)_{ij} + (\alpha\gamma)_{ik} + (\beta\gamma)_{jk} + \epsilon_{ijk}$$

where y_{ijk} denotes $(5 + \log(\text{pAUC}))$ for procedure i ($i = 1, 2, 3$), weapon condition j ($j = 1, 2$), and feature k ($k = 1, 2$); μ represents the overall average log(pAUC) across all conditions; α_i represents the effect of procedure i; β_j represents the effect of weapon condition j; γ_k represents the effect of feature condition k; and the next three terms reflect the three two-factor interactions between the main factors. The analysis of variance, where log(pAUC) values are weighted according to the values in the last column

TABLE 21.4

Logarithms of the Reported pAUC Values and Weights (Reciprocals of the Lengths of the Reported Confidence Intervals)

Condition	Procedure	Weapon	Feature	5 + log(pAUC)	Weight
1	SIM	Yes	Yes	1.31112	47.6
2	SIM	Yes	No	1.72983	33.3
3	SIM	No	Yes	0.92546	55.6
4	SIM	No	No	1.87643	45.5
5	Seq2	Yes	Yes	1.49344	47.6
6	Seq2	Yes	No	1.22774	47.6
7	Seq2	No	Yes	1.08798	52.6
8	Seq2	No	No	1.58875	41.7
9	Seq5	Yes	Yes	1.70316	38.5
10	Seq5	Yes	No	0.98262	58.8
11	Seq5	No	Yes	0.65719	66.7
12	Seq5	No	No	1.49344	55.6

TABLE 21.5

Analysis of Variance Table for log(pAUC)[a]

Source of Variation	Degrees of Freedom	Sum of Squares	Mean Square	F-Statistic	p-Value
Procedure	2	8.04	4.02	1.129	0.470
Weapon	1	2.94	2.94	0.826	0.460
Feature	1	14.72	14.72	4.138	0.179
Procedure × Weapon	2	0.59	0.30	0.083	0.923
Procedure × Feature	2	10.41	5.21	1.463	0.406
Weapon × Feature	1	34.80	34.80	9.780	0.089
Residuals	2	7.12	3.56		

[a] Data on pAUC from Table 3 in Carlson and Carlson (2014). This table appeared in Appendix C of the NAS report (pp. 150–154) (National Research Council, 2014).

of Table 21.5. None of the factors are significant.* The result is surprising, because all three factors in Table 21.5 (lineup type, presence/absence of weapon, presence/absence of distinctive feature) appear in the literature as having consequential effects on accuracy. So the lack of significance could be due to low power in detecting small effect sizes, the use of expressed confidence level (ECL) in the ROC, or the insensitivity of pAUC in characterizing a condition.* A complete set of raw data may yield a more powerful analysis with different results, as might a different summary measure of the ROC curve, such as the AUC.

---

* We can decompose the two degrees of freedom in the sum of squares for *Procedure* (3 levels), 8.04, into two single degree of freedom contrasts, *seq2* versus *seq5* (4.14) and *sim* versus the average of *seq2* and *seq5* (3.90), and consider all pairwise interaction terms among the four "main effects." All single-degree-of-freedom effects remain non-significant, in either this weighted analysis or in an unweighted analysis.
* For a discussion of the advantages and disadvantages of using AUC versus pAUC as a summary measure, see Walter (2005).

21.9 Discussion

Long-standing conventional statistical methodologies, including logistic regression and, more generally, generalized linear models, particularly for bivariate outcomes (sensitivity and specificity), remain valuable and appropriate tools for analyzing EWID experiments, especially when the experiment includes concomitant information, such as environmental variables of the experiment and demographic characteristics of the "eyewitness." In the absence of such information, ROC curves remain a useful comparison of two methods in diagnostic medicine, statistical process control, and eyewitness experiments. Newer approaches from statistical machine learning may be useful with very large experiments, though the impact of specific variables on the outcome may not always be as interpretable as with conventional linear models. Whichever technique is used, proper characterization of the uncertainties associated with the inferences must be calculated.

As recommended by the NRC report, a broader "exploration of the merits of different statistical tools for use in the evaluation of eyewitness performance" is an important area of research (National Research Council, 2014). The goal is to encourage the use of more discerning statistical models and analytical methods for assessing EWID procedures used to increase the accuracy of IDs, decrease the number of false convictions, and ensure guilty perpetrators are properly convicted.

The methodologies that we discussed in this chapter provide a foundation for future work, and raise several issue that remain for future research. We have indicated statistical approaches to evaluating current methods that law enforcement officers use in the field (e.g., sequential versus simultaneous lineups, presence or absence of an officer during the eyewitness's deliberations, etc.). We note that procedures from other fields, such as diagnostic medicine, can apply to data from EWID experiments, with some modifications as needed. adapted from other fields, such as diagnostic medicine.

Due to the popularity of the ECL-based ROC curve to compare accuracies of EWID procedures, we offered several alternatives related to the quantities used in ROC curves, namely sensitivity and specificity, with a focus on the real targets of interest, positive predictive value (PPV) and negative predictive value (NPV). This exploration led to the following questions.

First, what kind of curve can we use to describe PPV and NPV in an intuitive manner that will also hold up theoretically? Are there multivariate ROC curves that will display comparisons among several procedures simultaneously? How informative is a plot of LR_+ versus LR_- as "proxies" for PPV and NPV. For the classification methods, the determination of effectiveness and accuracy depend on useable and good data that replicate real world scenarios so that our proper assessment of the method's efficacy is valid. Confidence intervals or confidence bands should also always be included with any point estimates.

Second, what role can supervised learning classification methods play in predicting the accuracy of an eyewitness's decision? In a meta-analytic framework, how can we adapt the established bivariate and/or hierarchical modeling methods to the EWID framework? The answers to these questions require simulations and experimental data with the underlying truth known.

Alternative ways of examining the data could also lead to new modeling procedures or algorithms that would be useful in practice. We proposed a method for estimating the probability of accuracy for eyewitnesses that takes proper account of individuals' probabilities of choosing or not choosing a suspect from a lineup. This method is a potential tool that could provide an in-field assessment of eyewitness reliability, which can be explained to and understood by juries, judges, lawyers, law enforcement officers, and any other non-statisticians working in EWID. Further methods depend on the available types of EWID data, which could include recordings of eyewitness proceedings by working in conjunction with police departments.

Larger, more "ecologically valid" studies, may more properly reflect real-world scenarios by encompassing the realism of the stress, timelines, etc., than lab-based experiments where subjects know they are part of studies. Some ideas include staging a minor crime such as a robbery in a convenience store, and then asking participants if they recall the target. Another idea might be to have participants walk around with a camera on their heads, and ask them to recall targets or faces they may have seen. The camera would provide an objective source of comparison to eyewitness reports.

Researchers who conduct more varied and complex types of experiments will produce sets of observational data (National Research Council, 2014), leading to the development of novel modeling procedures and statistical methods.

Eyewitness identification plays a crucial role in the prosecution of crimes. The tools and procedures for analyzing the data in meaningful and utilitarian ways can provide thoughtful and valid conclusions. Such methodologies require ease of use and interpretation, as well flexibility and efficient implementation. The methods proposed in this chapter will help lead to better EWID procedures, hopefully resulting in fewer errors, both in false convictions and false acquittals.

Acknowledgements

This chapter was written with partial support from a grant from Arnold Ventures awarded to the University of Virginia. The views expressed are those of the authors and do not necessarily represent those of the funder.

References

Amendola, K.L. and Wixted, J.T. Comparing the diagnostic accuracy of suspect identifications made by actual eyewitnesses from simultaneous and sequential lineups in a randomized field trial. *Journal of Experimental Criminology*, **11**, 263–284, 2014.

Andersen, S.M., Carlson, C.A., Carlson, M.A., and Gronlund, S.D. Individual difference predict eyewitness identification performance. *Personality and Individual Differences*, **60**, 36–40, 2014.

Box, G.E.P., Hunter, W.G., and Stuart Hunter, J. *Statistics for Experimenters: Design, Innovation, and Discover* (2nd ed.). Wiley, Hoboken, NJ, 2005.

Breiman, L., Friedman, J., Olshen, R., and Stone, C. *Classification and Regression Trees*. Chapman & Hall, Boca Raton, FL, 1984.

Brewer, N. and Wells, G.L. The confidence-accuracy relationship in eyewitness identification: effects of lineup instructions, foil similarity, and target-absent base rates. *Journal of Experimental Psychology*, **12**(1), 11–30, 2006.

Brigham, J.C., Meissner, C.A., and Wasserman, A.W. Applied issues in the construction and expert assessment of photo lineups. *Applied Cognitive Psychology*, **13**, S73–S92, 1999.

Carlson, C.A. and Carlson, M.A. An evaluation of lineup presentation, weapon presence, and a distinctive feature using ROC analysis. *Journal of Applied Research in Memory and Cognition*, **3**(2), 45–53, 2014.

Carlson, C.A., Dias, J.L., Weatherford, D., and Carlson, M.A. An investigation on the weapon focus effect and the confidence-accuracy relationship for eyewitness identification. *Journal of Applied Research in Memory and Cognition*, **6**(1), 82–92, 2016a.

Carlson, C.A., Gronlund, S.D., and Clark, S.E. Lineup composition, suspect position and the sequential lineup advantage. *Journal of Experimental Psychology: Applied*, **14**(2), 118–128, 2008.

Carlson, C.A., Young, D.F., Weatherford, D., Carlson, M.A., Bednarz, J.E., and Jones, A.R. The influence of perpetrator exposure time and weapon presence/timing on eyewitness confidence and accuracy. *Applied Cognitive Psychology*, **30**, 898–910, 2016b.

Chen, Y., Liu, Y., Ning, J., Nie, L., Zhu, H., and Chu, H. A composite likelihood method for bivariate metaanalysis in diagnostic systematic reviews. *Statistical Methods in Medical Research*, **26**(2), 914–930, April 2017.

Chu, H. and Cole, S.R. Bivariate meta-analysis of sensitivity and specificity with sparse data: a generalized linear mixed model approach. *Journal of Clinical Epidemiology*, **59**, 1331–1333, 2006.

Chu, H., Nie, L., Cole, S.R., and Poole, C. Meta-analysis of diagnostic accuracy studies accounting for disease prevalence: alternative parameterizations and model selection. *Statistics in Medicine*, **28**, 2384–2399, 2009.

Clark, S.E. A re-examination of the effects of biased lineup instructions in eyewitness identification. *Law and Human Behavior*, **29**(4), 395–424, August 2005.

Clark, S.E., Benjamin, A.S., Wixted, J.T., Mickes, L., and Gronlund, S.D. Eyewitness identification and the accuracy of the criminal justice system. *Policy Insights from the Behavioral and Brain Sciences*, **2**(1), 175–186, 2015.

Colloff, M.F., Wade, K.A., and Strange, D. Unfair lineups make witnesses more likely to confuse innocent and guilty suspects. *Psychological Science*, **27**(9), 1227–1239, 2016.

Colloff, M.F., Wade, K.A., Wixted, J.T., and Maylor, E.A. A signal-detection analysis of eyewitness identification across the adult lifespan. *Psychology and Aging*, **32**(3), 243–258, 2017.

D'Agostini, G. Basic probabilistic issues in sciences and in forensics (hopefully) clarified by a toy experiment modeled by a BN. Presented at Isaac Newton Institute for Mathematical Sciences at the conference on Bayesian networks and argumentation in evidence analysis, September 2016.

Dawid, A.P. and Mortera, J. *Graphical Models for Forensic Analysis*. Chapman & Hall/CRC Handbooks of Modern Statistical Methods Series, 2017.

Demidenko, E. Confidence intervals and bands for the binormal ROC curve revisited. *Journal of Applied Statistics*, **39**(1), 67–79, January 2012.

Dimou, N.L., Adam, M., and Bagos, P.G. A multivariate method for meta-analysis and comparison of diagnostic tests. *Statistics in Medicine*, **35**(20), 3509–3523, 2016.

Dodson, C.S. and Dobolyi, D.G. Confidence and eyewitness identifications: the cross-race effect, decision time, and accuracy. *Applied Cognitive Psychology*, **30**, 113–125, 2016.

DuMouchel, W. Hierarchical Bayes linear models for meta-analysis. Technical Report 27, National Institute of Statistical Sciences, September 1994.

Eusebi, P., Reitsma, J.B., and Vermunt, J.K. Latent class bivariate model for the meta-analysis of diagnostic test accuracy studies. *BioMed Central Medical Research Methodology*, **14**(88), 2014. https://protect-us.mimecast.com/s/-QzfCKrR1RSo5Jxh3Rj2D?domain=doi.org

Fisher, R.A. The use of multiple measurements in taxonomic problems. *Annals of Eugenics*, **7**, 179–188, 1936.

Friedman, J. Greedy function approximation: a gradient boosting machine. *The Annals of Statistics*, **29**(5), 1189–1232, 2001.

Friedman, J., Hastie, T., and Tibshirani, R. Additive logistic regression: a statistical view of boosting. *The Annals of Statistics*, **28**(2), 337–407, 2000.

Garbolino, P. Bayesian networks for the evaluation of testimony. Presented at Isaac Newton Institute for Mathematical Sciences at the Conference on Bayesian Networks and Argumentation in Evidence Analysis, September 2016.

Garrett, B.L. *Convicting the Innocent: Where Criminal Prosecutions Go Wrong*. Harvard University Press, Cambridge, MA, 2012.

Gronlund, S.D. and Benjamin, A.S. The new science of eyewitness memory. *The Psychology of Learning and Motivation*, **69**, 241–284, 2018.

Gronlund, S.D., Mickes, L., Wixted, J.T., and Clark, S.E. Conducting an eyewitness lineup: how the research got it wrong. *Journal is Psychology of Learning and Motivation*, **63**, 1–43, 2015.

Gronlund, S.D., Wixted, J.T., and Mickes, L. Evaluating eyewitness identification procedures using receiver operating characteristic analysis. *Current Directions in Psychological Science*, **23**(1), 3–10, 2014.

Hajian-Tilaki, K. Receiver operating characteristic (ROC) curve analysis for medical diagnostic test evaluation. *Caspian Journal of Internal Medicine*, **4**(2), 627–635, 2013.

Hastie, T., Tibshirani, R., and Friedman, J. *The Elements of Statistical Machine Learning* (2nd ed.). Springer, New York, NY, 2016.

Higgins, J., Thomas, J., Chandler, J., Cumpston, M., Li, T., Page, M., and Welch, V., editors. *Cochrane Handbook for Systematic Reviews of Interventions*. The Cochrane Collaboration, Hoboken, NJ, 2019.

Hong, C.S. Bivariate ROC curve. *The Korean Journal of Applied Statistics*, **19**, 277–286, 2012.

Hong, C.S. and Jeong, J.A. Bivariate ROC curve and optimal classification function. *Communications for Statistical Applications and Methods*, **19**, 629–638, 2012.

Howe, M.L. and Knott, L.M. The fallibility of memory in judicial processes: lessons from the past and their modern consequences. *Memory*, **23**(5), 633–656, July 2015.

Hoyer, A. and Kuss, O. Meta-analysis for the comparison of two diagnostic tests to a common gold standard: a generalized linear mixed model approach. *Statistical Methods in Medical Research*, **27**(5), 1410–1421, 2018 (e-pub in 2016).

Humphries, J.E. and Flowe, H.D. Receiver operating characteristic analysis of age-related changes in lineup performance. *Journal of Experimental Child Psychology*, **132**, 189–204, 2015.

Irwig, L., Tosteson, A.N.A., Gatsonis, C., Lau, J., Colditz, G., Chalmers, T.C., and Mosteller, F. Guidelines for meta-analyses evaluating diagnostic tests. *Annals of Internal Medicine*, **120**(8), 667–676, April 1994.

Jin, H. and Lu, Y. The ROC region of a regression tree. *Statistics and Probability Letters*, **79**, 936–942, 2009.

Junaidi, J., Nur, D., and Stojanovski, E. Bayesian estimation of a meta-analysis model using Gibbs sampler. In *Proceedings of the Fifth Annual ASEARC Conference - Looking to the future - Programme and Proceedings*, 2012.

Juslin, P., Olsson, N., and Winman, A. Calibration and diagnosticity of confidence in eyewitness identification: comments on what can be inferred from low confidence-accuracy correlation. *Journal of Experimental Psychology*, **22**(5), 1304–1316, 1996.

Kafadar, K. Statistical issues and reliability of eyewitness identification as a forensic tool: presentation, September 2015.

Kantner, J. and Dobbins, I.G. Partitioning the sources of recognition confidence: the role of individual differences. *Psychonomic Bulletin & Review*, **26**, 1317–1324, 2019.

Krug, K. The relationship between confidence and accuracy: current thoughts of the literature and a new area of research. *Applied Psychology in Criminal Justice*, **3**(1), 7–41, 2007.

Lampinen, J.M., Smith, A.M., and Wells, G.L. Four utilities in eyewitness identification practice: dissociations between receiver operating characteristic (ROC) analysis and expected utility analysis. *Law and Human Behavior*, **43**(1), 26–44, 2019.

Lange, K. and Brunner, E. Sensitivity, specificity, and ROC curves in multiple reader diagnostic trials – a unified nonparametric approach. *Statistical Methodology*, **9**, 490–500, 2012.

Leeflang, M.M.G., Deeks, J.J., Rutjes, A.W.S., Reitsma, J.B., and Bossuyt, P.M. Bivariate meta-analysis of predictive values of diagnostic tests can be an alternative to bivariate meta-analysis of sensitivity and specificity. *Journal of Clinical Epidemiology*, **65**, 1088–1097, 2012.

Lindsay, R.C.L. and Wells, G.L. Improving eyewitness identifications from lineups: simultaneous versus sequential lineup presentation. *Journal of Applied Psychology*, **70**(3), 556–564, 1985.

Luby, A.S. A graphical model approach to eyewitness identification. Presented at Isaac Newton Institute for Mathematical Sciences at the Conference on Bayesian Networks and Argumentation in Evidence Analysis, September 2016.

Luby, A.S. Strengthening analyses of line-up procedures: a log-linear model framework. *Law, Probability and Risk*, **16**, 241–257, 2017.

Lundberg, S.M. and Lee, S.-I. A unified approach to interpreting model predictions. In *31st Conference on Neural Information Processing System (NIPS 2017)*, 2017.

Ma, X., Suri, M.F.K., and Chu, H. A trivariate meta-analysis of diagnostic studies accounting for prevalence and non-evaluable subjects: re-evaluation of the meta-analysis of coronary CT angiography studies. *BioMed Central Medical Research Methodology*, **14**(128), 2014. https://protect-us.mimecast.com/s/72WUCM86w6Hr4lVHQmfsA?domain=doi.org

Macskassy, S.A. and Provost, F. Confidence bands for ROC curves: methods and an empirical study. In *Conference: ROC Analysis in Artificial Intelligence, 1st International Workshop*, 2008.

Meissner, C.A. and Brigham, J.C. Thirty years of investigating the own-race bias in memory for faces: a meta-analytic review. *Psychology, Public Policy, and Law*, **7**, 3–35, 2001.

Mickes, L., Flowe, H.D., and Wixted, J.T. Receiver operating characteristic analysis of eyewitness memory: comparing the diagnostic accuracy of simultaneous vs. sequential lineups. *Journal of Experimental Psychology: Applied*, **18**(4), 361–376, 2012.

Mickes, L., Moreland, M.B., Clark, S.E., and Wixted, J.T. Missing the information needed to perform ROC analysis? then compute $d0$, not the diagnosticity ratio. *Journal of Applied Research in Memory Cognition*, **3**, 58–62, 2014.

Mickes, L., Seale-Carlisle, T.M., Wetmore, S.A., Gronlund, S.D., Clark, S.E., Carlson, C.A., Goodsell, C.A., Weatherford, D., and Wixted, J.T. ROCs in eyewitness identification: instructions versus confidence ratings. *Applied Cognitive Psychology*, **31**(5), 467–477, September/October 2017.

National Research Council, editor. *Identifying the Culprit: Assessing Eyewitness Identification*. The National Academies Press, Washington, DC, 2014.

Nikoloulopoulos, A.K. On composite likelihood in bivariate meta-analysis of diagnostic test accuracy studies. *Advances in Statistical Analysis*, **102**(2), 211–227, April 2018.

Palmer, M.A., Brewer, N., Weber, N., and Nagesh, A. The confidence-accuracy relationship for eyewitness identification decisions: effects of exposure duration, retention interval, and divided attention. *Journal of Experimental Psychology: Applied*, **19**(1), 55–71, 2013.

Park, S.H., Goo, J.M., and Jo, C.-H. Receiver operating characteristic (ROC) curve: practical review for radiologists. *Korean Journal of Radiology*, **5**, 11–18, 2004.

Pepe, M.S. Receiver operating characteristic methodology. *Journal of the American Statistical Association*, **95**(449), 308–311, 2000.

Pundir, S. and Amala, R. Detecting diagnostic accuracy of two biomarkers through a bivariate lognormal ROC curve. *Journal of Applied Statistics*, **42**(12), 2671–2685, 2015.

Reitsma, J.B., Glas, A.S., Rutjes, A.W.S., Scholten, R.J.P.M., Bossuyt, P.M., and Zwinderman, A.H. Bivariate analysis of sensitivity and specificity produces informative summary measures in diagnostic reviews. *Journal of Clinical Epidemiology*, **58**, 982–990, 2005.

Ribeiro, M.T., Singh, S., and Guestrin, C. – Why Should I Trust You? Explaining the Predictions of Any Classifier. *arXiv*, August 2016.

Rotello, C.M., Heit, E., and Dube, C. When more data steer us wrong: replications with the wrong dependent measure perpetuate erroneous conclusions. *Psychonomic Bulletin & Review*, **22**(4), 944–954, 2015.

Russ, A.J., Sauerland, M., Lee, C.E., and Bindermann, M. Individual differences in eyewitness accuracy across multiple lineups of faces. *Cognitive Research: Principles and Implications*, **3**(30), 2018. https://protect-us.mimecast.com/s/U8l8CNkRLRcWmR5h0QCMx?domain=doi.org

Rutter, C.M. and Gatsonis, C.A. A hierarchical regression approach to meta-analysis of diagnostic test accuracy evaluations. *Statistics in Medicine*, **20**, 2865–2884, 2001.

Sauer, J., Brewer, N., Zweck, T., and Weber, N. The effect of retention interval on the confidence-accuracy relationship for eyewitness identification. *Law and Human Behavior*, **34**, 337–347, 2010.

Sauer, J.D., Palmer, M.A., and Brewer, N. Pitfalls in using eyewitness confidence to diagnose the accuracy of an individual identification decision. *Psychology, Public Policy, and Law*, **25**(3), 147–165, 2019.

Sauerland, M., Sagana, A., Sporer, S.L., and Wixted, J.T. Decision time and confidence predict choosers' identification performance in photographic showups. *PLoS One*, **13**(1), 2018. https://protect-us.mimecast.com/s/iWsMCOYRMRU8mGrcArPRt?domain=journals.plos.org

Schapire, R.E. The strength of weak learnability. *Machine Learning*, **5**(2), 197–227, 1990.

Shiu, S.-Y. and Gatsonis, C.A. The predictive receiver operating characteristic curve for the joint assessment of the positive and negative predictive values. *Philosophical Transactions of the Royal Society A*, **366**, 2313–2333, 2008.

Smith, A.M., Lampinen, J.M., Wells, G.L., Smalarz, L., and Mackovichova, S. Deviation from perfect performance measures the diagnostic utility of eyewitness lineups but partial area under the ROC curve does not. *Journal of Applied Research in Memory and Cognition*, **8**(1), 50–59, 2019.

Smith, A.M., Wells, G.L., Lindsay, R.C.L., and Penrod, S. Fair lineups are better than biased lineups and showups, but not because they increase underlying discriminability. *Law and Human Behavior*, **41**(2), 127–145, 2017.

Sporer, S.L. The cross-race effect: beyond recognition of faces in the laboratory. *Psychology, Public Policy, and Law*, **7**(1), 170–200, 2001.

Sporer, S.L., Penrod, S., Read, D., and Cutler, B. Choosing, confidence, and accuracy: a meta-analysis of the confidence-accuracy relation in eyewitness identification studies. *Psychological Bulletin*, **118**(3), 315–327, 1995.

Steblay, N.M. Social infiuence in eyewitness recall: a meta-analytic review of lineup instruction effects. *Law and Human Behavior*, **21**(3), 283–297, 1997.

Streiner, D.L. and Cairney, J. What's under the ROC? An introduction to receiver operating characteristics curves. *The Canadian Journal of Psychiatry*, **52**(2), 121–126, February 2007.

Sutton, A.J. and Abrams, K.R. Bayesian methods in meta-analysis and evidence synthesis. *Statistical Methods in Medical Research*, **10**, 277–303, 2001.

Swets, J.A. *Signal Detection Theory and ROC Analysis in Psychology and Diagnostics: Collected Papers.* Psychology Press, New York, NY, 2014.

Walter, S.D. The partial area under the summary ROC curve. *Statistics in Medicine*, **24**, 2025–2040, 2005.

Wang, F. and Gatsonis, C.A. Hierarchical models for ROC curve summary measures: design and analysis of multi-reader, multi-modality studies of medical tests. *Statistics in Medicine*, **27**, 243–256, 2008.

Wang, M.-C. and Li, S. Bivariate marker measurements and ROC analysis. *Biometrics*, **68**, 1207–1281, December 2012.

Wang, M.-C. and Li, S. ROC analysis for multiple markers with tree-based classification. *Lifetime Data Analysis*, **19**, 257–277, 2013.

Wells, G.L. and Lindsay, R.C.L. On estimating the diagnosticity of eyewitness nonidentifications. *Psychological Bulletin*, **88**(3), 776–784, 1980.

Wells, G.L., Smalarz, L., and Smith, A.M. ROC analysis of lineups does not measure underlying discriminability and has limited value. *Journal of Applied Research in Memory and Cognition*, **5**, 313–317, 2015a.

Wells, G.L., Smith, A.M., and Smalarz, L. ROC analysis of lineups obscures information that is critical for both theoretical understanding and applied purposes. *Journal of Applied Research in Memory and Cognition*, **4**, 324–328, 2015b.

Wells, G.L., Steblay, N.K., and Dysart, J.E. A test of the simultaneous vs. sequential lineup methods. Technical report, American Judicature Society, 2011.

Wetmore, S.A., Neuschatz, J.S., Gronlund, S.D., Wooten, A., Goodsell, C.A., and Carlson, C.A. Effect of retention interval on showup and lineup performance. *Journal of Applied Research in Memory and Cognition*, **4**, 8–14, 2015.

Wilson, J.P., Hugenbert, K., and Bernstein, M.J. The cross-race effect and eyewitness identification: how to improve recognition and reduce decision errors in eyewitness situations. *Social Issues and Policy Review*, **7**(1), 83–113, 2013.

Wixted, J.T. and Mickes, L. A continuous dual-process model of remember/know judgments. *Psychological Review*, **117**(4), 1025–1054, 2010.

Wixted, J.T. and Mickes, L. The field of eyewitness memory should abandon probative value and embrace receiver operating characteristic analysis. *Perspectives on Psychological Sciences*, **7**(3), 275–278, 2012.

Wixted, J.T. and Mickes, L. Evaluating eyewitness identification procedures: ROC analysis and its misconceptions. *Journal of Applied Research in Memory Cognition*, **4**, 318–323, 2015.

Wixted, J.T. and Mickes, L. ROC analysis measures objective discriminability for any eyewitness identification procedure. *Journal of Applied Research in Memory and Cognition*, **4**(4), 329–334, 2015.

Wixted, J.T., Mickes, L., Clark, S.E., Gronlund, S.D., and Roediger III, H.L. Initial eyewitness confidence reliably predicts eyewitness identification accuracy. *American Psychologist*, **70**(6), 515–526, September 2015.

Wixted, J.T., Mickes, L., Dunn, J.C., Clark, S.E., and Wells, W. Estimating the reliability of eyewitness identifications from police lineups. *PNAS*, **113**(2), 304–309, January 2016.

Wixted, J.T., Mickes, L., Wetmore, S.A., Gronlund, S.D., and Neuschatz, J.S. ROC analysis in theory and practice. *Journal of Applied Research in Memory and Cognition*, **6**(3), 343–351, September 2017.

Wixted, J.T., Read, J.D., and Stephen Lindsay, D. The effect of retention interval on the eyewitness identification confidence-accuracy relationship. *Journal of Applied Research in Memory and Cognition*, **5**, 192–203, 2016.

Yin, J. Overview of inference about ROC curve in medical diagnosis. *Biometrics and Biostatistics International Journal*, **1**(3), 2014. DOI: 10.15406/bbij.2014.01.00013.

Zapf, A., Hoyer, A., Kramer, K., and Kuss, O. Nonparametric meta-analysis for diagnostic accuracy studies. *Statistics in Medicine*, **34**, 3831–3841, 2015.

Index

A

Aadhaar, 278
ABC model, *see* Approximate Bayesian
    Computation model
Abduction, 25
ABFDE, *see* American Board of Forensic
    Document Examiners
Abusive head trauma (AHT), 138–140
  triad, 138
Accumulated degree days (ADD), 444
Accumulated degree hours (ADH), 444
ADD, *see* Accumulated degree days
ADH, *see* Accumulated degree hours
Adventitious matches, 331–332; *see also* DNA
    database
Advisory Committee on Rules of Evidence,
    233, 350
Aggregation errors, 211–212
American Board of Forensic Document
    Examiners (ABFDE), 350
American Society for Testing and Materials,
    see ASTM International
American Statistical Association, 53, 243
Analysis, Comparison, Evaluation, and
    Verification (ACE-V), xi, 280, 287, 353;
    see also Fingerprint examinations
  analysis stage, 280–282
  comparison stage, 282
  confirmation bias, 282
  decision making in, 283, 285, 286, 304–308
  evaluation stage, 282
  exclusion, 283
  first level of details, 281
  for handwriting, 353–354
  identification, 283
  minutiae, 281
  point-standard method, 284
  posterior beliefs, 285
  score-based likelihood ratios, 286
  subjectivity, 284
  verification stage, 282–283
Anthropology, 103
Approximate Bayesian Computation model
    (ABC model), 302
  Bayes factor, 302, 303, 305
  ROC-ABC algorithm, 304
  selection algorithm, 302
  Area under the curve (AUC), 154, 509,
    511
  Area under the Receiver Operating
    Characteristic curve, 154, 509, 511
Arson investigation, 137
Artificial neural networks, *see* Neural networks

ASTM International
  Glass C162–05, 412
  Glass E2926–17, 234, 236, 422
  Glass E2927–16e1, 421–422, 426–428, 435
  Glass E2330–19, 421–422, 422, 434–436
  Handwriting, 350, 354
Attained significance probability, 59–60; *see also*
    Hypothesis tests
AUC, *see* Area under the curve

B

Ballistics, 370; *see also* Firearms examination
Base rates (BRs), 12, 514
Baum-Welch statistics, 469; *see also*
    i-vector—PLDA
Bayes factor, 65, 84–85, 98–99, 243–244, 284–287;
    290, 293–295, 301-305, 313–314, 317;
    see also Weight of evidence
Bayesian decision networks, 28–30
Bayesian, Fiducial, and Frequentist Conferences
    (BFF Conferences), 94
Bayesian inference, 6–8, 11, 28, 61, 89, 91, 92–93,
    336
Bayesian methods, 73–90, 92–93; *see also* Bayes'
    Theorem
  beta-binomial, 76–79
  beta distribution densities, 77
  binomial distribution, 76
  credible interval, 73
  fixed effects model, 87
  gamma-Poisson, 79–85
  hierarchical Bayesian models, 88
  Markov chain Monte Carlo, 85–86
  mixed effects models, 87
  posterior density, 78
  posterior variance, 79
  random effects models, 86
Bayesian Networks (BNs), 26, 165, 193
  in case modeling, 27–28
  in criminal identification, 175–176, 178–180
  DAG, 193, 194
  hit-and-run accident, 168–175
  independence properties, 194
  inference patterns, 28
  kinship, 182–185
  mixed DNA profiles, 185–189
  object-oriented, 176–177
  in paternity cases, 180–184
Bayes' Theorem, 6, 74–76, 254; *see also* Bayesian
    methods
  DNA evidence, 254
  in Dreyfus case, 6–7

Bayes' Theorem (*Continued*)
  integral form of Bayes' Rule, 75–76
  odds form of, 7, 166–168, 254
Bernoulli process, 41–43, 63, 236, 517
Bertillon, Alphonse, 5–7
Best evidence rule, 226
Best Linear Unbiased Predictor (BLUP), 87
Beta-binomial, 76–79; *see also* Bayesian methods
BFF Conferences, *see* Bayesian, Fiducial, and
        Frequentist Conferences
Bias
  confirmation, 282, 307, 396–397
  contextual, 19, 283, 396–397
  in estimation, 18, 44, 87, 92, 95–96, 261, 314,
        518
  in human decision making, 9, 204, 282–283,
        308, 392, 396–398, 453
  measurement, 139, 307, 419, 505, 508, 531
  selection, 336, 396
Binomial distribution, 43, 52, 76, 236, 331,
        355–356, 516–518
  normal and other approximations, 45
  probability model, 43, 236
  proportion, 47–49
Bivariate random-effects model, 515; *see also*
        Statistical models
Bite-marks, 227
Blue bus case, 172–174, 204
BLUP, *see* Best Linear Unbiased Predictor
Bootstrap estimates, 66–68; *see also* Resampling
        methods
Bonferroni, 62
Bottleneck-feature based systems, 477–478;
        see also DNN-based systems
Burden of persuasion, 61, 112, 201, 217, 240, 243
Brier's rule, *see* Quadratic scoring rule

C
CABL, *see* Compositional analysis of bullet lead
CAI, *see* Case Assessment and Interpretation
Calibration curve, 508–509; *see also* Eyewitness
        identification
Canonical linear discriminant functions (CLDF),
        455, 470
CART, *see* Classification and regression tree
Case Assessment and Interpretation (CAI), 19
Center for Statistics and Applications in
        Forensic Evidence (CSAFE), 387
Centimorgan (cM), 272
Cepstral-mean-and-variance normalization
        (CMVN), 459–460; *see also* Forensic
        voice comparison
Cepstral-mean subtraction (CMS), 459–460;
        see also Forensic voice comparison
Chain of custody, 394
Class characteristics of trace evidence, 133
Classical hypothesis tests, 54; *see also* Hypothesis
        tests

type I errors and test size, 54–56
type II errors and test power, 56–59
Classification and regression tree (CART), 520
Classifier algorithm, 368; *see also* Firearms
        examination
CLDF, *see* Canonical linear discriminant
        functions
Clique of variables, 195
cM, *see* Centimorgan
CMC, *see* Congruent Matching Cells
CMS, *see* Cepstral-mean subtraction;
        Consecutively matching striae
CMVN, *see* Cepstral-mean-and-variance
        normalization
Cochrane Collaboration, 515
CODIS, *see* Combined DNA Index System
CODIS expanded set, 255; *see also* DNA
        frequencies and probabilities
Coincidence probability, 14
Combined DNA Index System (CODIS), 255
Common source scenario, 288
Composite hypothesis, 57, 243
Composite likelihood, 516
Compositional analysis of bullet lead (CABL),
        54, 86, 421
Conditional
  error rates, 244
  probability, 15–18, 26, 60, 74–76, 98, 108, 111,
        121, 132, 166, 169, 171–172, 189,
        193–195, 205, 209, 213, 238, 243, 253,
        356, 522
Confidence-accuracy relationship, 509, 511
Confidence distributions, 94
Confidence interval, 40; *see also* Frequentist
        estimation
  coefficient, 45–46
  hypothesis testing with, 59
  interpreting, 49–50, 60–65
  Wilson, 46
Confirmation bias, *see* Bias
Congruent matching cells (CMC), 373–374;
        see also Firearms examination
Consecutively matching striae (CMS), 366–367;
        see also Firearms examination
Contextual bias, *see* Bias
Correlation and causation, 132
Criminalistics, 5
Criterial noise, 512
Criterion variance, *see* Criterial noise
Critical region, 54–56
CSAFE, *see* Center for Statistics and
        Applications in Forensic Evidence
Cumulative distribution, 68

D
DAG, *see* Directed acyclic graph
Data, 39–40
Databases for trace evidence, 135–136

Database match, 332; *see also* DNA database
    matches
Daubert, 231, 235–238, 349, 350, 391
DCT, *see* Discrete cosine transform
Decision theory, 29–30, 74, 79, 103–105, 135
    decision space, 105
    decision-theoretic criterion, 112
    in forensic science, 111–125
    in law, 111–113
    loss function, 108–109
    monetary interpretation of losses, 122–123
    normative standards, 104
    quadratic scoring rule, 121–122
    utility function, 106–108
Deep neural network (DNN), 455, 475
Defense attorney's fallacy, 12, 210–211
Defense hypothesis, 287; *see also* Fingerprint
    examinations
DET, *see* Detection Error Tradeoff
Detection Error Tradeoff (DET), 154
Detection limit, *see* Limit of detection
Deviation from perfect performance (DPP), 514
DFT, *see* Discrete Fourier transform
Diagnosticity ratio, 507–508, 523
Directed acyclic graph (DAG), 193
Directional errors, 211
Disaster victim identification, 271, 335
Discrete cosine transform (DCT), 457
Discrete Fourier transform (DFT), 456
Discriminability, 507–508
Discriminant analysis, 519–520
Discriminating power, 13
Discrimination, 13, 132, 133
Distribution, *see* Statistical models
DNA database matches, 16, 325
    base-rate neglect, 334
    controversy, 16, 63, 279, 326, 328–331
    data-dependence of hypotheses, 16, 329–330
    dealing with selection effects, 16, 63, 335–336
    interpretation, 63, 333–335
    probable cause match, 332–333
    reporting, 332
DNA profiling, 251, 265, 325–326; *see also*
    Kinship, Mixtures, Population genetics,
    Probabilistic genotyping
    Bayes' Theorem, 254
    likelihood ratios, 252–254
    lineage markers, 257–258
    mixed profiles, 185–189, 258
    profile probability, 16–17
    sequence data and SNPs, 260, 272
    touch DNA, 258
DNN, *see* Deep neural network
DNN-based systems, 474–479; *see also* Forensic
    voice comparison
DPP, *see* Deviation from perfect performance
Dreyfus, Alfred, 6–7, 175
Dynamic model for handwriting, 357–358

E
ECE, *see* Empirical cross-entropy
ECEmin, *see* Minimum Empirical Cross-Entropy
ECL, *see* Expressed confidence level
Ecological validity, 203, 500, 506, 534
EDS, *see* Energy dispersive X-ray detector
EDX, *see* Energy dispersive X-ray
EE, *see* Ellipse-envelope
ELISA, *see* Enzyme-linked immunosorbent
    assay
Ellipse-envelope (EE), 527
EM algorithm, *see* Expectation maximization
    algorithm
Empirical cross-entropy (ECE), 24, 153
Empirical distribution of sample, 68
Energy dispersive X-ray (EDX), 417
Energy dispersive X-ray detector (EDS), 419
ENFSI, *see* European Network of Forensic
    Science Institutes
ENVSI, *see* Expected net value of sample
    information
Enzyme-linked immunosorbent assay (ELISA),
    75
Error rates, xii, 238–244
    confidence intervals for, 49
    in admissibility determination, 238–241
    methods for estimating, 99–100, 238–241,
        308–310
    misleading evidence, 23
    studies of, xii, 48–49, 240–241, 306–310, 483,
        487–488
Expert testimony, *see* Testimony
European Network of Forensic Science
    Institutes (ENFSI), 21, 144, 150, 152,
    394, 453, 483
Evidence, 40; *see also* Testimony
    Advisory Committee on Rules of Evidence,
        350
    Anglo-American legal rules, 225
    -centric testimony, 242
    evaluation, 5–6, 201
Evidentiary reliability, 231
EVOI, *see* Expected value of information
EVPI, *see* Expected value of perfect information
EVSI, *see* Expected value of sample information
Exact hypothesis, *see* Simple hypothesis
Expectation maximization algorithm, 462–464
Expected; *see also* Decision theory
    loss, 121
    utility, 514
    utility maximization principle, 108, 109–110
    net value of sample information (ENVSI),
        125
    value of information (EVOI), 127
    value of perfect information (EVPI), 124
    value of sample information (EVSI), 125
Expert witnesses, *see* Testimony

Expressed confidence level (ECL), 500, 507, 508, 511

Eyewitness identification, 500–501
  probability of accuracy, 528–530
  system and estimator variables, 504–507

F

False alarm rate (FAR), 55, 483, 508, 511, 514, 520, 523

False discovery rate (FDR), 63, 83

False negative probability, 47

False negative rate (FNR), 23, 47, 48, 55, 58, 134, 238–240, 273, 376, 421, 423–424, 501–502

False-positive probability (FPP), 17–18, 40, 46–49, 58, 59, 233, 238, 240, 242, 425

False positive rate (FPR), 23, 55, 134, 136, 208, 238, 239, 241, 254, 273, 304, 309, 311, 376, 399, 400, 405, 421, 423–425, 501–502, 509, 511

False report probability, 201–202

Family-wise error rate, 62

FAR, *see* False alarm rate

Fast Fourier transform (FFT), 456

FBI, *see* Federal Bureau of Investigation

FDR, *see* False discovery rate

Feature warping, 460–461; *see also* Forensic voice comparison

Federal Bureau of Investigation (FBI), xi, 368, 412

Federal Rules of Evidence (FRE), 225–226, 349
  Rule 401, 9, 225
  Rule 402, 225
  Rule 403, 225–226
  Rule 702, 226, 233
  Rules 801–807, 225
  Rules 1001–1003, 226

FFT, *see* Fast Fourier transform

Fiducial inference, 93–94

Filler, *see* Foil

Finger impressions, 278

Fingermarks, 279

Fingerprint evidence, 277, 293, 304; *see also* Fingerprint examinations
  Bayes factor for, 284–287, 301–304, 305
  Madrid train bombing, xi, 241
  Mayfield, Brandon, xi
  score-based likelihood ratios, 294–301
  similarity metrics and kernel functions, 293–294
  weight of, 293–304

Fingerprint examinations, 277, 317–318; *see also* ACE-V, Error rates
  ABC paradigm, 317
  common source scenario, 287, 288, 291–293
  control impressions, 278
  decision making during ACE, 304–308
  defense hypothesis, 287

factors affecting decision making and error rates, 304, 312
  FRStat, 312–317
  mean of population of sources, 289
  PCAST controversy, 308–310
  prosecution hypothesis, 287
  simulations, 289–291
  specific source scenario, 287, 288–289, 291–293

Fingerprint variability analysis, 277

Firearms examination, 365, 386–387
  bullet-to-bullet comparisons, 380–382
  classifier algorithm, 368
  comparison microscope, 370, 371
  comparison of cartridge case marks, 373–377
  comparison of marks on land engraved areas of bullets, 377–378
  congruent matching cells, 373–374
  consecutively matching striae, 366–367
  empirical feature distributions, 380
  history of firearms examination, 370–372
  machine learning, 367, 384, 387
  microscopic imperfections, 372–373
  optimization techniques, 375
  overlaid signatures of bullets, 379
  pairwise comparisons, 378–384
  pattern evidence evaluation, 365–366
  precision and recall for cutoffs on similarity score, 376
  similarity scores for NBIDE data set, 375
  variance-bias tradeoff, 367–368

FIU, *see* Florida International University

Fixed-width confidence bands, 527

Flat glass, 413; *see also* Glass types

Float glass, 413; *see also* Glass types

Florida International University (FIU), 428

FNR, *see* False negative rate

Foil, 502; *see also* Eyewitness identification

Forensic automatic likelihood ratio methods, 143, 161
  calibration, 155
  data-processing inequality, 157
  discrimination and calibration, 154–155
  empirical validation, 147–148, 483
  generalization, 161
  likelihood ratio, 155–158
  limiting factors, 157
  logarithmic scoring rule, 152
  monotonicity, 161
  performance characteristics, 147, 150, 158–159
  probability performance, 151–154
  reliability plots, 155
  robustness, 159–161
  scoring rule, 150, 151, 152
  standardization, 145–146
  validation, 145–150

Forensic document examination, *see* Handwriting analysis

Forensic entomology, 443, 448–449
  curvilinear models, 444–445
  growth curve reconstruction, 445–448
  insect evidence to estimate PMI, 443
  isomorphen and isomegalen diagrams, 444
  post-feeding, 447
  spectral measurements, 445
  temperature profile, 448
  thermal summation models, 444
Glass evidence , 411, 437–438
  breakage and recovery, 414
  chemical composition, 416–418
  data sets, 427–430
  early cases of, 412
  elemental concentrations, 418–419
  ICP-MS and LA-ICP-MS, 418–419
  micro X-Ray fluorescence, 419
  physical and optical properties, 414–416
  refractive index, 416
  simulated match rates, 431–436
  standards, 422–423
  types of glass, 412–414
Forensic Science Regulator's Code (England), 143, 453
Forensic voice comparison, 452, 488
  cepstral-mean-and-variance normalization, 459
  deltas and double deltas, 457–458
  diarization, 458
  DNN-based systems, 474–479
  ENFSI guidelines, 453
  feature extraction, 455
  feature warping, 460–461
  Forensic Science Regulator's codes, 453
  Gaussian mixture model-universal background model, 461–467
  i-vector PLDA, 467–474
  known-speaker recording, 452
  legal references, 497
  likelihood-ratio framework, 454, 479
  logistic regression, 482–483
  log-likelihood-ratio cost, 484
  mel-frequency cepstral coefficients, 455–457
  mismatch compensation in feature domain, 458
  pooled-variance two-Gaussian model, 482
  questioned-speaker recording, 452
  reverberation, 452
  speaker-extrinsic variability, 452
  T-matrix training and i-vector extraction, 489–491
  validation, 483, 487–488
  voice-activity detection, 458
  X-vector system, 478–479
Forgery, 342; *see also* Howland will
Foucault process, 413
FPP, *see* False-positive probability
FPR, *see* False positive rate

FRE, *see* Federal Rules of Evidence
Frequentist estimation, 44; *see also* Frequentist methods
  bootstrap confidence interval, 66–68
  confidence interval, 45–46, 49–50, 66-68
  design of experiments, 46–48
  estimating allele proportions, 44
  estimating false positive probability, 46
  point estimate, 44–45
  Wilson confidence interval, 46
Frequentist methods, 40; *see also* Statistical inference
  arbitrary lines, 63–64
  bootstrap estimates, 66–68
  confidence distributions, 94
  hypothesis tests, 53–60
  multiple tests, 61–63
  permutation tests, 68–70
  probability distributions, 41
  probability models, 42
  p-values, 50–53
  random variables, 40
  resampling methods, 65
  sampling from distribution, 42–44
  standard deviation, 42
  standard error, 67
  transposition, 60–61
  variance, 42
FRStat, 280, 287, 316–317; *see also* Fingerprint examinations
  Anderson-Darling test statistic, 316
  comparisons of FRStat-like values with LR, 315
  Kolmogorov-Smirnov test, 316
  limitations, 313–314
  technical issues, 316
Frye, 230, 231, 319, 349; *see also* General acceptance standard

G
GAM, *see* Generalized Additive Model Gamma family, 80
Gamma-Poisson, 79–85; *see also* Bayesian methods
Gaussian mixture model-universal background model (GMM-UBM), 461; *see also* Forensic voice comparison
  adaptation coefficient, 465
  calculating score, 465–467
  relevance factor, 465
  UBM training data, 467
  training known-speaker model, 465
  training relevant-population model, 462–464
  UBM training data, 467
Generalized Additive Model (GAM), 445
General-acceptance standard, 230–231, 319, 341
Genetic genealogy, 272–273

GMM-UBM, *see* Gaussian mixture
    model-universal background model
Graphical models, 521–522; *see also* Supervised
    learning classification methods

H
Handwriting
    ACE-V process, 353–354
    analysis principles, 352, 360
    admissibility of identifications, 342, 346,
        348–350, 363
    dynamic model for, 357–358
    differences in, 346
    statistical analysis of, 355
    variability determination, 357
    Bayesian approach to comparing signatures,
        356–357
    forgery, 342
    Howland will signatures, 5, 176, 355–356
    principal differential analysis, 358
    probability scale for expressing conclusions,
        354–355
    proficiency studies and error rates, 359
Hanratty, James, 184–185
Hardy-Weinberg Equilibrium (HWE), 255
Hearsay, 225
Hit rate, 508–509, 511, 514, 520, 523; *see also*
    Sensitivity
Hotelling's T2-test, 426
Howland will, 5, 176, 355–356
HWE, *see* Hardy-Weinberg Equilibrium
Hypothesis tests, 40, 53; *see also* Frequentist
    methods
    attained significance probability, 59–60
    classical hypothesis tests, 54–59
    with confidence intervals, 59
    interpreting results of, 60–65
    with p-values, 59
    statistical decision rule, 54
    Type I error, 54–56, 62–63, 92, 95–96, 99,
        313–314, 424–425
    Type II error, 55–56, 64, 92, 99, 313–314,
        424–425

I
IAI, *see* International Association for
    Identification
IBD, *see* Identical by descent
ICP-AES, *see* Inductively coupled plasma atomic
    emission spectrometry
ICP-MS, *see* Inductively coupled plasma mass
    spectrometry
ICP-MS and LA-ICP-MS, 418–419
ICP-OES, *see* Inductively coupled plasma optical
    emission spectrometry
Identical by descent (IBD), 272; *see also* Kinship
Identity matrix, 88
iid, *see* Independent identically distributed

Incidence rate, 12
Independent identically distributed (iid), 93
Inductively coupled plasma atomic emission
    spectrometry (ICP-AES), 417
Inductively coupled plasma mass spectrometry
    (ICP-MS), 417, 418, 420
Inductively coupled plasma optical emission
    spectrometry (ICP-OES), 420
Inference, *see* Statistical inference
Influence diagrams, *see* Bayesian decision
    networks
Innocence Project, 500
International Association for Identification
    (IAI), 48
International Organization for Standardization
    (ISO), 144
Inverse probability, 61
Inversion fallacy, *see* Transposed conditional
    fallacy
Iowa State University, 428
Island problem, 327, 329
ISO, *see* International Organization for
    Standardization
i-vector, 467–470; *see also* Forensic voice
    comparison
    Baum-Welch statistics, 469
    domain mismatch compensation, 470
    extraction, 489–491
    PLDA, 467, 470–474
    PLDA model to calculate score for pair of,
        475
    total variability space, 468

K
Kinship, 265, 273–274; *see also* Paternity
    Bayesian networks, 182–185
    genetic genealogy, 272–273
    identifying remains, 271
    joint genotypic probabilities for pairs of
        relatives, 269
    likelihood ratio, 266, 270
    parentage calculations for structured
        populations, 270–271
    paternity indices for structured populations,
        271
    relatedness, 268–271
    relationship probabilities for common
        relatives, 269
    SNP data, 272
Kolmogorov-Smirnov test, 316

L
Laboratory error (LE), 211
LA-ICP-MS, *see* Laser ablation inductively
    coupled plasma mass spectrometry
Laser ablation inductively coupled plasma
    mass spectrometry (LA-ICP-MS),
    417, 419

Latent prints, 279
LDA, *see* Linear discriminant analysis
LDT, *see* Lower developmental thresholds
Likelihood inference, 94
Likelihood ratio (LR), 4, 6–7, 13, 15, 65, 84,
        98–99, 111, 120, 135, 144, 155, 201, 205,
        327, 384, 386; *see also* Probative value,
        Score-based likelihood ratio, Weight of
        evidence
    in Bayes net, 171
    calibration, 157–158
    in decision framework, 111
    DNA frequencies and probabilities, 252–254
    investigative role of, 25–26
    performance, 155–157
    in speaker identification, 454
    uncertainty about, 130, 254
    verbal scale, 10, 20–22, 84, 394, 398, 453
LIME, *see* Local interpretable model-agnostic
        explanations
Limit of detection (LOD), 420
Lineage markers, 257–258; *see also* Kinship
Linear discriminant analysis (LDA), 455, 470,
        519
Linear time invariant (LTI), 459
Local interpretable model-agnostic explanations
        (LIME), 519
Locard's exchange principle, 411
LOD, *see* Limit of detection
Logistic regression, 482–483, 512–513, 520
Logitnormal bivariate random-effects model,
        515–517; *see also* Statistical models
Loss function, 108–109; *see also* Decision theory
Lower developmental thresholds (LDT), 444
LTI, *see* Linear time invariant

M
Machine learning, 367, 384, 387; *see also*
        Supervised learning classification
        methods
    classification models, 519
    decision trees and random forests, 520
    discriminant analysis, 519–520
    logistic regression, 520
    neural networks, 474–476, 520–521
    support vector machines, 520
MAP, *see* Maximum a posteriori
Markov chain Monte Carlo (MCMC), 85–86,
        516, 526
Massively parallel sequencing (MPS), 260;
        see also Next Generation Sequencing
Mass spectrometry (MS), 418
Maximizes the expected utility (MEU), 108
Maximum a posteriori (MAP), 465
Maximum likelihood (ML), 516
Maximum likelihood estimation (MLE), 529
MCD, *see* Minimum covariance determinant
McKreith, 236

MCMC, *see* Markov chain Monte Carlo
Mel-frequency cepstral coefficients (MFCCs),
        455–457; *see also* Forensic voice
MEU, *see* Maximizes the expected utility
Micro X-ray fluorescence (μ-XRF), 417, 419
Minimum covariance determinant (MCD), 429
Minimum Empirical Cross-Entropy (ECEmin),
        154
Mitochondria, 257; *see also* lineage markers
Mixed DNA profiles, 185–189, 258
ML, *see* Maximum likelihood
MLE, *see* Maximum likelihood estimation
MPS, *see* Massively parallel sequencing
MS, *see* Mass spectrometry
Mutation, 185, 237, 256, 258
μ-XRF, *see* Micro X-ray fluorescence

N
NAA, *see* Neutron activated analysis
NAS, *see* National Academy of Sciences
NASEM, *see* National Academies of Science,
        Engineering and Medicine
National Academies of Science, Engineering
        and Medicine (NASEM), 98
National Academy of Sciences (NAS), xii, 392,
        421, 500
National Bureau of Standards (NBS), 137
National Institute of Standards and Technology
        (NIST), 137, 266, 306
National Research Council (NRC), 353
NBIDE, *see* NIST Ballistics Identification
        Database Evaluation
NBS, *see* National Bureau of Standards
Negative predictive value (NPV), 514, 522, 533
Netherlands Forensic Institute (NFI), 334
Neural networks, 474–476, 520–521; *see also*
        Machine learning
Neutron activated analysis (NAA), 417
Next Generation Identification (NGI), 278–279
Next Generation Sequencing (NGS), 260
Neyman-Pearson hypothesis tests, 95
NFI, *see* Netherlands Forensic Institute
NGI, *see* Next Generation Identification
NGS, *see* Next generation sequencing
NIST, *see* National Institute of Standards and
        Technology
NIST Ballistics Identification Database
        Evaluation (NBIDE), 375; *see also*
        Firearms examination
Nonparametric meta-analysis, 517–519
NPV, *see* Negative predictive value
NRC, *see* National Research Council

O
Object-oriented Bayesian networks (OOBNs),
        166, 176; *see also* Bayesian networks
Odds ratio (OR), 133, 522; *see also* Bayes factor
OES, *see* Optical emission spectrometer

Offense-level propositions, 12
Omic data, 261
OOBNs, *see* Object-oriented Bayesian networks
Optical emission spectrometer (OES), 418
Organization of Scientific Area Committees for Forensic Science (OSAC), 309, 453
OSAC, *see* Organization of Scientific Area Committees for Forensic Science

P
Parameters, 44
Partial area under the curve (pAUC), 507, 508, 509
Partial dependence plot (PDP), 519
Paternity; *see also* Kinship
   Bayes net for, 180–182
   Index (PI), 270–271
pAUC, *see* Partial area under the curve
PAV, *see* Pool Adjacent Violators
PCA, *see* Principal component analysis
PCAST, *see* President's Council of Advisors on Science and Technology
PDP, *see* Partial dependence plot
Performance assessment, 22–34
Performance characteristics, 147
   data-processing inequality, 157
   generalization, 161
   limiting factors, 157
   reliability plots, 155
   robustness, 159–161
Permutation tests, 68–70; *see also* Resampling methods
Photo Response Non-Uniformity (PRNU), 335
PI, *see* Paternity Index
Pixels per inch (PPI), 394
PLDA, *see* Probabilistic linear discriminant analysis
PMI, *see* Post-mortem interval
Poincaré, Henri, 6–7
Point estimate, 44–45, 93
Pool Adjacent Violators (PAV), 154
Pooled-variance two-Gaussian model, 482
Population genetics, 254–258; *see also* Kinship
   loci, 255–256
Population structure, 17, 256–257, 266–268
Positive predictive value (PPV), 240, 514, 522, 533
Posterior
   beliefs, 285
   density, 78
   probability, *see* Bayes Theorem
Post-mortem interval (PMI), 443; *see also* Forensic entomology casework
Power, 56–59
PPI, *see* Pixels per inch
PPV, *see* Positive predictive value
Predictive receiver operating characteristic curve (PROC curve), 524

President's Council of Advisors on Science and Technology (PCAST), 5, 17, 143, 239–240, 306, 308–310, 392
Principal component analysis (PCA), 468
Principal differential analysis, 358; *see also* Handwriting and document analysis
Prior distribution, 97; *see also* Bayes Theorem
PRNU, *see* Photo Response Non-Uniformity
Probabilistic
   analysis, 326–328
   Expert Systems, 165
   genotyping, 241, 259–260, 262
   reasoning, 9–10
Probabilistic graphical models, 26; *see also* Bayesian networks
Probabilistic linear discriminant analysis (PLDA), 454, 470–474; *see also* Forensic voice comparison; i-vector
Probability, 97
   calibration, 155
   coincidence, 14
   density function, 41
   discrimination, 154
   distributions, 41
   models, 42
   personal probability, 120
   of proposition, 11–12
Probable cause cases, 332; *see also* DNA database
Probative value, 241–244
PROC curve, *see* Predictive receiver operating characteristic curve
Proper scoring rule, 151
Prosecution hypothesis, 287
Prosecutor's fallacy, *see* Transposed conditional fallacy
p-value, 40, 50, 64, 67–70, 92, 235, 254, 331; *see also* Frequentist methods
   in comparison of glass fragments, 50–52
   hypothesis testing with, 59
   interpreting, 52–53
   interpreting results of, 60–65
   multiple tests, 61–63
   tail-end probability, 51

Q
QDA, *see* Quadratic discriminant analysis
Quadratic discriminant analysis (QDA), 519
Quadratic scoring rule, 121–122
Quadrivariate logistic regression model, 518; *see also* Statistical models
Qualifications of expert witnesses, *see* Testimony

R
Random match probability, 16, 44, 63, 201–202, 204, 211–212, 214–215, 219–220, 228, 235, 251, 253, 326, 327, 329, 336, 337
   seed, 464
   variable, 40, 44
Randomization test, 68

Randomly Acquired Characteristics, 392
Reasonable scientific certainty, 229
Reasonable statistical certainty, 229
Receiver Operating Characteristic curves (ROC),
    154, 500, 507, 509–512; *see also*
    Eyewitness identification
    AUC estimation, 526–527
    confidence-based analysis, 511
    confidence intervals for, 527–528
    methodology development, 522–524
    modified measure for, 514
    multivariate, 525–526
    predictive receiver operating characteristic
      curve, 524–525
    summary of, 528
    tools based on ROC methods, 522
Refractive index, 54, 416
Rejection region, 55–56
Relatedness, 268–271; *see also* Bayes nets,
    Kinship, Paternity
Relative density, *see* Specific gravity
Relative risk, 133
Relative standard deviation (RSD), 426
Relevant evidence, 9, 229
Relevant population, 7, 12, 191
Reliability, 60, 154–155, 219–221
Repeatability, 61, 147, 287, 366, 419
Reproducibility, 61, 147, 238, 287, 317, 366,
    419, 453
Resampling methods, 65; *see also* Frequentist
    methods
    bootstrap estimates, 66–68
    permutation tests, 68–70
Residual standard deviations (RSDs), 415
Reverberation, 452; *see also* Forensic voice
    comparison
RMS, *see* Root-mean-square
ROC, *see* Receiver Operating Characteristic
    curves
Root-mean-square (RMS), 458
Route diagrams, 27
RSD, *see* Relative standard deviation

S
SAD, *see* Speech activity detection
Sample, 44
Sampling, 42–44; *see also* Frequentist methods
Scientific Working Group on Friction Ridge
    Analysis, Study and Technology
    (SWGFAST), 280
Score, 121, 150–152
Score-based likelihood ratio, 294, 385
    common source, 295–297
    suspect-centered, 301
    suspect-centered score-based models,
      297–300
    trace-centered, 301
    trace-centered score-based models, 301

Score-to-likelihood-ratio conversion, 479; *see also*
    Forensic voice comparison
Screening, 336
SDT, *see* Signal detection theory
SE, *see* Standard error
Semi-automated element accidental sensor,
    399–400; *see also* Shoeprints
    accidental sensor implementation, 400
    contrast adjustment, 400–401
    detecting accidentals and wear, 403–405
    ElementAccidentalSensor, 403
    element detection, 401–403
    elliptical element, 404
    image processing, 399
    preprocessing and binarization, 400–401, 402
    quadrilateral with checkerboard pattern, 404
    true accidental, 405
    true wear, 405
Sensitivity
    for classifications, 47, 48, 56, 75, 134, 206–209,
      423–424, 509–511, 514–518, 522, 530,
      532–533
    for inferences, 190–191
    to quantity of chemical, 260, 418, 420
    to weight of evidence, 203, 206–208, 216
SHAP, *see* Shapley additive explanations
Shaken baby syndrome, *see* Abusive Head
    Trauma
Shapley additive explanations (SHAP), 519
Sheet glass, 413; *see also* Glass types
Shoeprints 391–410
Short Tandem Repeat (STR), 18, 255, 260, 265
Signal detection theory (SDT), 503, 507, 509;
    see also ROC
Signal-to-noise ratio, 420
Simple hypothesis, 57
Simultaneous joint confidence regions, 527
Single nucleotide polymorphism (SNP), 260,
    265; *see also* Kinship
Single nucleotide variation (SNV), 260
Speaker identification, *see* Forensic voice
    comparison
Specific source scenario, 288–289
Specificity, 47–48, 56, 75, 134, 282, 423, 509–510,
    514–518, 523, 524, 527, 530, 533
Speech activity detection (SAD), 458
Standard deviation, 42
Standard error (SE), 67
Standards
    ASTM C162–05, 412
    ASTM E2926–17, 234, 236, 422
    ASTM E2927–16e1, 421–422, 426–428, 435
    ASTM E2330–19, 421–422, 422, 434–436
    ASTM Handwriting, 350, 354
    ISO/EIC 17025, 145
Statistic, 44
Statistical genetics, 261
Statistical experts, 226–228; *see also* Testimony

Statistical inference; *see also* Bayesian Methods, Bayes Theorem, Frequentist methods
    Bayesian, 92–93
    comparing approaches, 94–97
    comparing philosophies, 91–100
    fiducial, 93–94
    frequentist, 92
    confidence distributions, 94
    likelihood, 94
Statistical significance, 14, 55; *see also* Frequentist methods
Statistical models, 42, 44, 514; *see also* Bernoulli process
    bivariate random-effects model, 515
    logitnormal bivariate random-effects model, 515–517
    nonparametric meta-analysis, 517–519
    Normal, 45, 51–52, 55, 57, 67, 76, 80, 88, 93–94, 96–97, 234, 236, 255, 290, 415, 423, 425, 432, 435, 508, 510, 512, 515–518, 520, 526
    quadrivariate logistic regression model, 518
    uniform, 123, 228
Statistical statements, 201–202, 221
Support, *see* likelihood
Subjective feature-based comparison methods, 306
Supervised learning classification methods, 519; *see also* Eyewitness identification
    graphical models, 521–522
    machine learning classification models, 519–521
Support vector machines (SVMs), 519, 520; *see also* Machine learning
Suspect-centered score-based likelihood ratios, 301
SVMs, *see* Support vector machines
SWGDOC, *see* Technical Working Group for Document Examination
SWGFAST, *see* Scientific Working Group on Friction Ridge Analysis, Study, and Technology
SWGTREAD scale, 392; *see also* Shoeprints

T
Technical Working Group for Document Examination (TWGDOC), 350
Testimony
    Bayes net for, 176–177
    Bayesian view of expert's role, 244
    conclusion-centric, 242, 243
    error rates and admissibility, 238–241
    distinguishing data from evaluations and conclusions, 242
    evidence-centric, 242
    evidentiary-value approach, 243
    forms of expert testimony, 228–229
    lay and expert testimony, 225

    qualifications of experts, 226–227
    reasonable scientific or statistical certainty, 229
    statistical analysis as evidence, 234–235
Threshold averaging, 527
Tippett plot, 23, 150, 158, 485–487
T matrix training, 489–491; *see also* Forensic voice comparison
Touch DNA, 258
Toughened glass, 413–414; *see also* Glass types
TPR, *see* True positive rate
Trace-centered score-based likelihood ratios, 301
Training known-speaker model, 465; *see also* Gaussian mixture model-universal background model
Training relevant-population model, 462–464; *see also* Gaussian mixture model-universal background model
Transposed conditional fallacy, 7, 11, 132–133, 136, 209–210
Trees and random forests, 520
True positive rates (TPRs), 509; *see also* Sensitivity
TWGDOC, *see* Technical Working Group for Document Examination
Two-stage procedure, 13, 99–100, 241
Type I error, 54–56, 62–63, 92, 95–96, 99, 313–314, 424–425; *see also* Hypothesis tests
Type II error, 55–56, 64, 92, 99, 313–314, 424–425; *see also* Hypothesis tests

U
UDT, *see* Upper developmental thresholds
Uniform Language for Testimony and Reporting, 84, 352, 353
Upper developmental thresholds, 444
Defense Forensic Science Center laboratory (DFSC), 280, 287
    FRStat, 312–317
Utility; *see also* Decision theory
    function, 106
    theory, 106–108

V
VA, *see* Vertical averaging
VAD, *see* Voice-activity detection
Validity, xi–xii, 22–23, 40, 46, 100, 135, 137, 143–145, 219, 260, 304, 317, 383, 406, 431–437, 483–488; *see also* Evidentiary reliability, Error rates, Reliability
    for admissibility in court, 230–233, 235–241, 349
    ecological, 203, 500, 506, 534
    of likelihood ratios, 146–161
Value of sample information (VSI), 125
Variance, 42
    -bias tradeoff, 367–368

Verbal scale, *see* likelihood ratio
Vertical averaging, 527
Voice-activity detection (VAD), 458; *see also* Forensic voice comparison
VSI, *see* Value of sample information

W
Weak evidence effect, 211
Weight of evidence, xii, 65, 85, 111, 135, 207, 317, 384; *see also* Bayes factor, Likelihood ratio, Probative value
   Good, I. J., 12, 65, 85, 111
Weighted ROC (WROC), 525
WHB, *see* Working-Hotelling based bands

Wilson confidence interval, 46
Working-Hotelling based bands (WHB), 527
WROC, *see* Weighted ROC

X
X-ray fluorescence (XRF), 419
X-vector system, 478–479; *see also* DNN-based systems, Forensic voice comparison

Y
Y chromosome, 257; *see also* lineage markers
Y-STR haplotype, 258; *see also* lineage markers

For Product Safety Concerns and Information please contact our EU
representative GPSR@taylorandfrancis.com
Taylor & Francis Verlag GmbH, Kaufingerstraße 24, 80331 München, Germany

* 9 7 8 0 3 6 7 5 2 7 7 2 3 *